Handbook of Engaged Sustainability

Satinder Dhiman
Editor-in-Chief

Joan Marques
Editor

Handbook of Engaged Sustainability

Volume 1

With 99 Figures and 54 Tables

Editor-in-Chief
Satinder Dhiman
School of Business
Woodbury University
Burbank, CA, USA

Editor
Joan Marques
School of Business
Woodbury University
Burbank, CA, USA

ISBN 978-3-319-71311-3 ISBN 978-3-319-71312-0 (eBook)
ISBN 978-3-319-71313-7 (print and electronic bundle)
https://doi.org/10.1007/978-3-319-71312-0

Library of Congress Control Number: 2018941852

Printed on acid-free paper

This Springer imprint is published by the registered company Springer International Publishing AG part of Springer Nature.
The registered company address is: Gewerbestrasse 11, 6330 Cham, Switzerland

Humbly Dedicated to the
Well-Being of All Beings,
Our Shared Destiny, and
the Well-Being of Our Mother Nature!

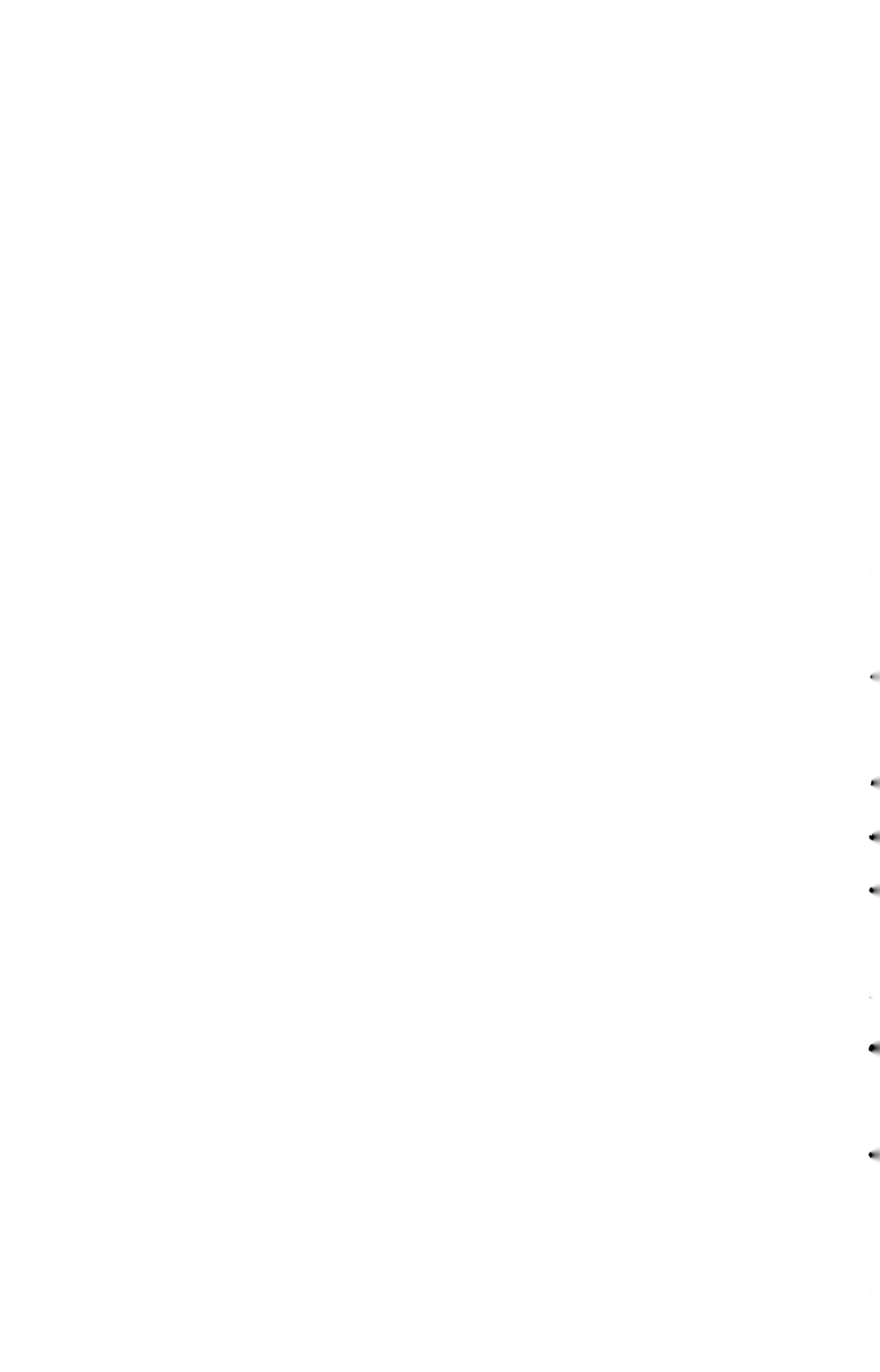

Preface

Sustainability *Matters*: Treading *Lightly* on Our Planet

> *The work an unknown good person has done is like a vein of water flowing hidden underground, secretly making the ground green* – Thomas Carlyle

As we sail through the precarious first decades of the twenty-first century, a new vision is emerging to end poverty, protect the planet, and ensure well-being for all as part of our shared destiny. Achieving these goals will require the concerted efforts of governments, society, educational institutions, the business sector, and informed citizens. To help achieve these goals, we propose that organizations and B-schools highlight *sustainability* so that they can become willing contributors to the vision of cultivating a harmonious society and a sustainable planet.

The uniqueness of this *Handbook* lies in its emphasis on the "engaged" dimension of sustainability.

The questions that we explore in this august volume are:

1. What can *I* do to change the future of the planet?
2. What can *organizations* do to improve the well-being of our society/planet?
3. What can we *all* do that is good for us, good for society, and good for the planet?

Within this *rethink* of sustainability, the ethico-spiritual basis of sustainability is never out of the frame. By definition, the topic of sustainability requires a broad interdisciplinary approach – as in "our total footprint on the planet," not just "our carbon footprint." Our work proposes to bring together the two allied areas of sustainability and spirituality in a dialectical manner, with ethics acting as a balancing force and spirituality playing the role of the proverbial invisible hand guiding our quest for sustainability.

Why Sustainability Matters?

The Exorbitant Economic Costs of Unsustainability

Hurricanes Harvey, Irma, and Maria are not the only disasters to ravage the world, says *US News and World Report*: natural disasters have hit nearly every continent in 2017, claiming the lives and livelihoods of hundreds of thousands of people in South Asia, North America, Central America, and Africa.[1] Commenting on the state of affairs of recent devastating disasters in North America, meteorologists and environmental risk experts predict that our planet faces a continuing grave risk from natural disasters.[2]

AccuWeather predicts that Hurricane Harvey paired with Hurricane Irma collectively cost the United States nearly $300 billion.[3] This is just one example of the exorbitant economic costs of our unsustainable footprint. The foregoing context underscores even more that *sustainability matters.*

Experts agree that the issues of climate change, dwindling biodiversity, and land degradation need to be urgently addressed. Responsible public policy changes at the macro level and mindful consumption at the individual level can go a long way to lessen the damage done to our planet. Business also needs to share the responsibility of protecting the environment in a major way. These observations have far-reaching implications for developing a sustainability mindset and also indicate the great potential for sustainability studies in management education.

United Nations Environment Programme (UNEP) offers the following overall recommendations:

1. Enhance sustainable consumption and production to reduce environmental pressures by addressing drivers associated with manufacturing processes and consumer demand.
2. Implement measures to reduce pollution and other environmental pressures.
3. Reduce dependence on fossil fuels and diversify energy sources.
4. Enhance international cooperation on climate, air quality, and other environmental issues.

---

[1]See: 10 of the Deadliest Natural Disasters of 2017. US News and World Report, Sept 20, 2017. Retrieved Sept 27, 2017: https://www.usnews.com/news/best-countries/slideshows/10-of-the-deadliest-natural-disasters-of-2017

[2]Annie Sneed, Hurricane Irma: Florida's Overdevelopment Has Created a Ticking Time Bomb, *Scientific American*, September 12, 2017. Retrieved September 27, 2017: https://www.scientificamerican.com/article/hurricane-irma-floridas-overdevelopment-has-created-a-ticking-time-bomb/

[3]Andrew Soergel, Forecast Predicts Heavy Economic Damage from Harvey, Irma. US News and World Report, Sept 11, 2017. Retrieved Sept 27, 2017: https://www.usnews.com/news/articles/2017-09-11/irma-harvey-forecast-to-do-290-billion-in-economic-damage

5. Low-carbon, climate-resilient choices in infrastructure, energy and food production coupled with effective and sustainable natural resource governance are key to protecting the ecological assets that underpin a healthy society.[4]

Cultivating a Sustainability Mind-Set

The contributors in this volume examine myriad sustainability issues objectively and offer diverse perspectives for reflection. A vibrant vision is presented to the readers in the form of a series of engaging *affirmations*:

1. We believe that it is impossible to separate economic developmental issues from environmental issues. In its most practical aspect, sustainability is about understanding close interconnections among environment, society, and economy.
2. We believe that the way to achieve sustainable, harmonious living in all spheres is through lived morality and spirituality at the personal level, team level, and societal level. We call it *engaged sustainability.*
3. We believe that excessive desire, anger, and greed are subtle forms of violence against oneself, others, and the planet.
4. We are not only unaware of these mental pollutants; we are often unaware how unaware we are.
5. We believe that a focus on engaged sustainability will help us harness what is good for us, good for society, and good for the planet.
6. We believe that achieving this goal will require a shift from being a consumer to being a contributor.
7. Only an individual life rooted in the continuous harmony with nature – a life based on moral and spiritual awareness – can be sustainable for the entire creation.

In this *Handbook*, our overarching goal has been to explore the application of sustainability to a wide variety of contemporary contexts – from economics of consumption and growth to government policy to achieving a sustainable planet. These three perspectives – ecology, equity, and economics – can serve as guiding principles. *This framework has far-reaching implications for developing a sustainability mindset and also indicate the great potential for garnering a higher awareness about sustainability.* As a point of departure, we can start with and build upon the World Commission on Environment and Development's 1987 definition of sustainability as economic development activity that "meets the needs of the present without compromising the ability of future generations to meet their own needs."

[4]Rate of Environmental Damage Increasing Across Planet but Still Time to Reverse Worst Impacts. Retrieved September 27, 2017: http://www.un.org/sustainabledevelopment/blog/2016/05/rate-of-environmental-damage-increasing-across-planet-but-still-time-to-reverse-worst-impacts/

In the web of life, everything is linked with everything else: you cannot pluck a flower without disturbing a star, as the poet Francis Thompson observed. It is, therefore, incumbent on tomorrow's leaders to examine these issues holistically and objectively and to seek out diverse perspectives for reflection.

Our economic system is committed to maximizing productivity and profits. This new credo for sustainability asks for an additional commitment: examine existing belief systems in light of the evidence presented, rather than scrutinizing the evidence in the light of preexisting notions. Believe nothing; research everything. This expectation is at the heart of every scientific endeavor. It requires us to be aware of confirmation-bias and premature cognitive commitment. There is a difference between being on the side of the evidence and insisting that the evidence be on your side.

This is the most important key to understanding all profound questions of life. Aristotle is reported to have said the following of his teacher, Plato: "Plato is dear; still dearer is the truth."

Human Undertakings: Prime Driver of Climate Change

When the real enemy is within, why fight an external war? – Gandhi

Calling the post 1950s the *Anthroposene* (literally, the "era of humans"), Ricard Matthieu tells us that this is the first era in the history of the world when human activities are profoundly modifying and degrading the entire system that maintains life on earth. He states that the wealthy nations are the greatest culprits: An Afghan produces 2500 times less CO_2 than a Qatari and a thousand times less than an American.[5] With a note of urgency, Ricard rightly observes:

> If we continue to be obsessed with achieving growth, with consumption of natural resources increasing at its current exponential rate, we will need three planets by 2050. We do not have them. In order to remain within the environmental safety zone in which humanity can continue to prosper, we need to curb our endless desire for "more."[6]

Likewise, observations throughout the world make it clear that climate change is occurring, and rigorous scientific research demonstrates that the greenhouse gases emitted by human activities are the primary driver.[7]

These conclusions are based on multiple independent lines of evidence, and contrary assertions are inconsistent with an objective assessment of the vast body of peer-reviewed science. Moreover, there is strong evidence that ongoing climate

[5]Matthieu Ricard, *Altruism: The Power of Compassion to Change Yourself and the World*, trans. by Charlotte Mandell and Sam Gordon (New York: Little, Brown and Company, 2015), 8.

[6]Ibid.

[7]See: Climate Change 2007: Synthesis Report. Retrieved August 19, 2015: http://www.ipcc.ch/publications_and_data/ar4/syr/en/spm.html

change will have broad impacts on society, including the global economy and on the environment. For the United States, climate change impacts include sea level rise for coastal states, greater threats of extreme weather events, and increased risk of regional water scarcity, urban heat waves, western wildfires, and the disturbance of biological systems throughout the country. The severity of climate change impacts is expected to increase substantially in the coming decades.[8]

If we are to avoid the most severe impacts of climate change, emissions of greenhouse gases must be dramatically reduced. In addition, adaptation will be necessary to address those impacts that are already unavoidable. Adaptation efforts include improved infrastructure design, more sustainable management of water and other natural resources, modified agricultural practices, and improved emergency responses to storms, floods, fires, and heat waves.[9]

Some Positive Initiatives About Climate Change Adaptability

While climate change is a global issue, it is felt on a local scale. In the absence of national or international climate policy direction, cities and local communities around the world have been focusing on solving their own climate problems. They are working to build flood defenses, plan for heat waves and higher temperatures, install water-permeable pavements to better deal with floods and storm water, and improve water storage and use.

According to the 2014 report on Climate Change Impacts, Adaptation and Vulnerability from the United Nations Intergovernmental Panel on Climate Change, governments at various levels are also getting better at adaptation. Climate change is starting to be factored into a variety of development plans: how to manage the increasingly extreme disasters we are seeing and their associated risks, how to protect coastlines and deal with sea-level encroachment, how to best manage land and forests, how to deal with and plan for reduced water availability, how to develop resilient crop varieties, and how to protect energy and public infrastructure.[10]

On June 18, 2015, Pope Francis officially issued 184-page encyclical *Laudato si',* Italian for *"Praise Be to You."* Subtitled as *On Care for Our Common Home*, it is a new appeal from Pope Francis addressed to "every person living on this planet" for

[8]The conclusions in this paragraph reflect the scientific consensus represented by, for example, the Intergovernmental Panel on Climate Change and US Global Change Research Program. Many scientific societies have endorsed these findings in their own statements, including the American Association for the Advancement of Science, American Chemical Society, American Geophysical Union, American Meteorological Society, and American Statistical Association. See: Statement on climate change from 18 scientific associations (2009).

[9]Statement on climate change from 18 scientific associations (2009). Retrieved February 25, 2018: http://www.aaas.org/sites/default/files/migrate/uploads/1021climate_letter1.pdf

[10]NASA: *Global Climate Change: Vital Signs of the Planet*. Retrieved August 19, 2015: http://climate.nasa.gov/solutions/adaptation-mitigation/

an inclusive dialogue about how we are shaping the future of our planet. As Yardley and Goldstein review in the *New York Times*, the encyclical boldly calls for:

> ...a radical transformation of politics, economics and individual lifestyles to confront environmental degradation and climate change, blending a biting critique of consumerism and irresponsible development with a plea for swift and unified global action.[11]

Laudato si' has a wider appeal; it has found resonance with Buddhists, Hindus, with Jews, Muslims, Protestant and Orthodox Christians, as well as with atheists and agnostics.

Regarding the Entire World as One Family

When we see unity in diversity, it helps us develop universal outlook in life which is so essential sustaining the sanctity of our war ravaged planet. By developing universal pity and contentious compassion toward all and everything, one is then able to make peace with the world and feel at home in the universe. Let us seek and share the underlying truth of mutuality that does not lead to unnatural differences and disharmony. That is the truth of our identity behind diversity – the essential oneness of all that exists. By seeking the truth that is equally good to all existence, we will be able to revere all life and truly redeem our human existence. Only then can we ensure equally the happiness and welfare of all beings. That will be our true gift of sustainability to the universe.

Concluding Reflections

Our organizations and societies are human nature writ large; therefore, we believe that the solution to society's current chaos lies in the spiritual transformation of each one of us. For material development to be sustainable, spiritual advancement must be seen as an integral part of the human development algorithm. The choice we face is between conscious change and chaotic annihilation. The last century has highlighted both the creative and the destructive power of human ingenuity. Whereas humanity's greatest gains in this century came in the areas of science and technology, we also witnessed the horror of two world wars, the rise of international terrorism, and economic and financial meltdowns. Many believe that the greatest harm occurred in the erosion of moral and spiritual values.

In his splendid little book, *The Compassionate Universe*, Eknath Easwaran notes the *urgency* of the responsibility of humans to *heal* the environment – "the only creatures on Earth who have the power – and, it sometimes seems, the inclination –

[11]Jim Yardley and Laurie Goodstein "Pope Francis, in Sweeping Encyclical, Calls for Swift Action on Climate Change." *The New York Times*. 18 June 2015.

to bring life on this planet to an end."[12] As human beings, we are given the power to think and power to do. We are also given the power to choose, the free will. We can choose to live differently and create our own reality. This is perhaps the most unique gift we have that needs to be harnessed and realized if humanity has to have a shared future.

We have one planet to live. Let's cultivate it together.

Satinder Dhiman
Joan Marques

[12]Eknath Easwaran, *The Compassionate Universe: The Power of Individual to Heal the Environment* (Petaluma, California: Nilgiri Press, 1989), 7.

Contents

Volume 1

Volume 2

About the Editor-in-Chief

Satinder Dhiman, Ph.D., Ed.D., M.B.A, M.Com., serves as Associate Dean, Chair, and Director of the MBA Program and Professor of Management at Woodbury University's School of Business. He holds a Ph.D. in Social Sciences from Tilburg University, Netherlands; a Doctorate in Organizational Leadership from Pepperdine University, Los Angeles; an M.B.A. from West Coast University, Los Angeles; and an M.Com. (with gold medal) from the Panjab University, India. *He has also completed advanced Executive Leadership programs at Harvard, Stanford, and Wharton*. In 2013, Dr. Dhiman was invited to be the opening speaker at the prestigious TEDx Conference @ College of the Canyons in Santa Clarita, California. He serves as the President of International Chamber of Service Industry (ICSI).

Professor Dhiman teaches courses pertaining to ethical leadership, sustainability, organizational behavior and strategy, and spirituality in the workplace in the MBA program. He has authored, co-authored, and co-edited over 17 management, leadership, and accounting related books and research monographs, including most recently authoring: *Holistic Leadership* (Palgrave 2017); *Gandhi and Leadership* (Palgrave 2015), and *Seven Habits of Highly Fulfilled People* (Personhood 2012); and co-editing and co-authoring, with Dr. Marques, *Spirituality and Sustainability* (Springer 2016) and *Leadership Today* (Springer 2016).

He is the *Editor-in-Chief* of two multi-author *M*ajor *R*eference *W*orks: *Springer Handbook of Engaged Sustainability* and *Palgrave Handbook of Workplace Spirituality and Fulfillment* and *Editor-in-Chief* of *Palgrave Studies in Workplace Spirituality and Fulfillment* and editor of *Springer Series in Management, Change, Strategy* and *Positive Leadership*. Some of his forthcoming titles include: *Leading without Power: A New Model of Highly Fulfilled Leaders*; *Bhagavad Gītā: A Catalyst for Organizational Transformation* (both by Palgrave MacMillan); and *Conscious Consumption: Diet, Sustainability and Wellbeing* (Routledge, 2019).

Recipient of several national and international professional honors, Professor Dhiman is also the winner of Steve Allen Excellence in Education Award and the prestigious ACBSP International Teacher of the Year Award. He has presented at major international conferences and published research with his colleagues in *Journal of Values-Based Leadership*, *Organization Development Journal*, *Journal of Management Development*, *Journal of Social Change*, *Journal of Applied Business and Economics*, and *Performance Improvement*. He also serves as Accreditation Mentor and Site Visit Team Leader for the Accreditation Council for Business Schools and Programs (ACBSP) for various universities in America, Canada, Europe, and India.

Professor Dhiman is the Founder-Director of Forever Fulfilled, a Los Angeles-based Wellbeing Consultancy, that focuses on workplace wellness, sustainability, and self-leadership.

About the Editor

Joan Marques has reinvented herself from a successful media entrepreneur in Suriname, South America, to a groundbreaking "edupreneur" (educational entrepreneur) in California, USA. She currently serves as Dean at Woodbury University's School of Business, in Burbank, California, where she works on infusing and nurturing the concept of "Business with a Conscience" into internal and external stakeholders, using every reputable resource possible. She is also a Full Professor of Management and teaches business courses related to Leadership, Ethics, Creativity, and Organizational Behavior in graduate and undergraduate programs.

Joan holds a Ph.D. in Social Sciences from Tilburg University's Oldendorff Graduate School (2011); and an Ed.D. in Organizational Leadership from Pepperdine University's Graduate School of Education and Psychology (2004). She also holds an M.B.A. from Woodbury University (2000) and a B.Sc. equivalent degree (HEAO) in Economics from MOC, Suriname (1987). Additionally, she has completed post-doctoral work at Tulane University's Freeman School of Business (2010).

Dr. Marques is a frequent speaker and presenter at academic and professional venues. In 2016, she gave a TEDx-Talk at College of the Canyons in California, titled "An Ancient Path Towards a Better Future," in which she analyzed the Noble Eightfold Path, one of the foundational Buddhist practices, within the realm of contemporary business performance.

Joan's research interests pertain to Awakened Leadership, Buddhist Psychology in Management, and Workplace Spirituality. Her works have been widely published and cited in both academic and popular venues. She has written more than 150 scholarly articles and has (co)authored more than 20 books, among which are *Ethical Leadership, Progress with a Moral Compass* (Routledge, 2017); *Leadership, Finding Balance Between Acceptance and Ambition* (Routledge, 2016); *Leadership Today: Practices for Personal and Professional Performance* (with Satinder Dhiman - Springer, 2016); *Spirituality and Sustainability: New Horizons and Exemplary Approaches* (with Satinder Dhiman – Springer, 2016); *Business and Buddhism* (Routledge, 2015); and *Leadership and Mindful Behavior: Action, Wakefulness, and Business* (Palgrave MacMillan, 2014).

Contributors

Inés Alegre Managerial Decisions Sciences Department, IESE Business School – University of Navarra, Barcelona, Spain

Eugene Allevato Woodbury University, Burbank, CA, USA

Sachi Arakawa Portland State University, Portland, OR, USA

Poonam Arora Department of Management, School of Business, Manhattan College, Riverdale, NY, USA

Paul A. Becks Welty Building Company, Akron, OH, USA

Christopher G. Beehner Center for Business, Legal and Entrepreneurship, Seminole State College of Florida, Heathrow, FL, USA

Jasmina Berbegal-Mirabent Department of Economy and Business Organization, Universitat Internacional de Catalunya, Barcelona, Spain

Apryl Bergstrom Department of Resource Economics and Environmental Sociology, University of Alberta, Edmonton, AB, Canada

Ritamoni Boro Justice K. S. Hegde Institute of Management, Nitte University, Nitte, India

Rajesh Buch Walton Sustainability Solutions Initiatives, Arizona State University, Tempe, AZ, USA

John E. Carroll Department of Natural Resources and Environment, University of New Hampshire, Durham, NH, USA

Judith Cavazos-Arroyo Business, Universidad Popular Autónoma del Estado de Puebla, Puebla, Puebla, Mexico

Sooyeon Choi Department of Consumer Science, Purdue University, West Lafayette, IN, USA

Valeria B. Cuevas-Albarrán Centro Intercultural de Proyectos y Negocios, Universidad Intercultural Maya de Quintana Roo, José Ma. Morelos, Quintana Roo, Mexico

Michael Dalrymple Walton Sustainability Solutions Initiatives, Arizona State University, Tempe, AZ, USA

George L. De Feis Department of Management, Healthcare Management, and Business Administration, Iona College, School of Business, New Rochelle, NY, USA

Annick De Witt Copernicus Institute of Sustainable Development, Utrecht University, Utrecht, The Netherlands

Mara DeFilippis Walton Sustainability Solutions Initiatives, Arizona State University, Tempe, AZ, USA

Satinder Dhiman School of Business, Woodbury University, Burbank, CA, USA

Elif Üstündağlı Erten Faculty of Economics and Administrative Sciences, Department of Business Administration, Ege University, İzmir, Turkey

André M. Everett Department of Management, University of Otago, Dunedin, New Zealand

Richard A. Feinberg Department of Consumer Science, Purdue University, West Lafayette, IN, USA

Mateus Possati Figueira Arceburgo, MG, Brazil

Dolors Gil-Doménech Department of Economy and Business Organization, Universitat Internacional de Catalunya, Barcelona, Spain

Elizabeth Gingerich Valparaiso University, Valparaiso, IN, USA

Ebru Belkıs Güzeloğlu Faculty of Communication, Department of Public Relations and Publicity, Ege University, İzmir, Turkey

Patricia Hammer IPS Department International Planning Systems, Faculty of Spatial and Environmental Planning, Technische Universität Kaiserslautern, Kaiserslautern, Germany

Howard Harris School of Management, University of South Australia, Adelaide, SA, Australia

Dennis Heaton Maharishi University of Management, Fairfield, IA, USA

Jorge Huchin Chan Desarrollo Empresarial, Universidad Intercultural Maya de Quintana Roo, José Ma. Morelos, Quintana Roo, Mexico

Hans-Henrik Hvolby Department of Materials and Production, Centre for Logistics, Aalborg University, Aalborg, Denmark

Katariina Koistinen School of Energy Systems, Lappeenranta University of Technology, Lappeenranta, Finland

Helen Kopnina Faculty Social and Behavioural Sciences, Institute Cultural Anthropology and Development Sociology, Leiden University, Leiden, The Netherlands

Leiden University and The Hague University of Applied Science (HHS), Leiden, The Netherlands

Naomi T. Krogman Department of Resource Economics and Environmental Sociology, Faculty of Graduate Studies and Research, University of Alberta, Edmonton, AB, Canada

Lassi Linnanen School of Energy Systems, Lappeenranta University of Technology, Lappeenranta, Finland

Cassandra Lubenow Walton Sustainability Solutions Initiatives, Arizona State University, Tempe, AZ, USA

Anil K. Maheshwari Maharishi University of Management, Fairfield, IA, USA

Joan Marques School of Business, Woodbury University, Burbank, CA, USA

Will McConnell Woodbury University, Burbank, CA, USA

Mirja Mikkilä School of Energy Systems, Lappeenranta University of Technology, Lappeenranta, Finland

Francisco J. Moo-Xix Centro Intercultural de Proyectos y Negocios, Universidad Intercultural Maya de Quintana Roo, José Ma. Morelos, Quintana Roo, Mexico

Peter Newman Curtin University Sustainability Policy (CUSP) Institute, School of Design and Built Environment, Curtin University, Perth, Australia

Dan O'Neill Walton Sustainability Solutions Initiatives, Arizona State University, Tempe, AZ, USA

Branka V. Olson School of Architecture, Woodbury University, Burbank, CA, USA

Sindik Olson Associates, Los Angeles, CA, USA

Karina Pallagst IPS Department International Planning Systems, Faculty of Spatial and Environmental Planning, Technische Universität Kaiserslautern, Kaiserslautern, Germany

William Paolillo Welty Building Company, Akron, OH, USA

Josi Paz Consultant - Communication and International Cooperation, University of Brasília, Brasilia, DF, Brazil

Carol Pomare Ron Joyce Center for Business Studies, Mount Allison University, Sackville, NB, Canada

Julia M. Puaschunder The New School, Department of Economics, Schwartz Center for Economic Policy Analysis, New York, NY, USA

Graduate School of Arts and Sciences, Columbia University, New York, NY, USA

Princeton University, Princeton, NJ, USA

Schwartz Center for Economics Policy Analysis, New York, NY, USA

Raghavan 'Ram' Ramanan Desert Research Institute, Dallas, TX, USA

Parag Rastogi Management Development Institute, Gurgaon, India

Francisco J. Rosado-May Universidad Intercultural Maya de Quintana Roo, José Ma. Morelos, Quintana Roo, Mexico

Janet L. Rovenpor Department of Management, School of Business, Manhattan College, Riverdale, NY, USA

Sonya Sachdeva Northern Research Station, USDA Forest Service, Evanston, IL, USA

Manjula S. Salimath Department of Management, College of Business, University of North Texas, Denton, TX, USA

Harpinder Sandhu College of Science and Engineering, Flinders University, Adelaide, SA, Australia

Sukhbir Sandhu School of Management, University of South Australia, Adelaide, SA, Australia

K. Sankaran Justice K. S. Hegde Institute of Management, Nitte University, Nitte, India

Kristen Schiele California State Polytechnic University, Pomona, CA, USA

Vivek Shandas Portland State University, Portland, OR, USA

Radha Sharma Management Development Institute, Gurgaon, India

Claudine Soosay School of Management, University of South Australia, Adelaide, SA, Australia

Edward R. Straub Velouria Systems LLC, Brighton, MI, USA

Satu Teerikangas Turku School of Economics, Turku University, Turku, Finland

School of Construction and Project Management, Bartlett Faculty of the Built Environment, University College London, London, UK

Jason Thistlethwaite School of Environment, Enterprise and Development (SEED), University of Waterloo, Waterloo, ON, Canada

Joachim Timlon Ronneby, Sweden

Mine Üçok Hughes Department of Marketing, California State University, Los Angeles, Los Angeles, CA, USA

Rohana Ulluwishewa (Former) Massey University, Palmerston North, New Zealand

José G. Vargas-Hernández University Center for Economic and Managerial Sciences, University of Guadalajara, Guadalajara, Jalisco, Mexico

Núcleo Universitario Los Belenes, Zapopan, Jalisco, Mexico

Stephanie Watts Susilo Institute for Ethics in a Global Economy, Associate Professor of Information Systems, Questrom School of Business at Boston University, Boston, MA, USA

Michael O. Wood School of Environment, Enterprise and Development (SEED), University of Waterloo, Waterloo, ON, Canada

Babak Zahraie Research Management Office, Lincoln University, Lincoln, New Zealand

Justyna Zdunek-Wielgołaska Faculty of Architecture, University of Technology, Otwock, Poland

Ragna Zeiss Faculty of Arts and Social Sciences, Department of Technology and Society Studies, Maastricht University, Maastricht, The Netherlands

Part I

Mapping the Engaged Sustainability Terrain

Selfishness, Greed, and Apathy

Spiritual Impoverishment at the Root of Ecological Sustainability

Satinder Dhiman

The greatest threat to our planet is the belief that someone else will save it.
Robert Swan
(Robert Swan is a polar explorer, environmentalist, and the first man in history to walk unsupported to both the North and South Poles)

Contents

Partially based on author's works: *Holistic Leadership* (Palgrave MacMillan, 2017); *Spirituality and Sustainability* (Springer Nature, 2016); *Gandhi and Leadership* (Palgrave MacMillan, 2015); and "Ethics and Spirituality of Sustainability: What Can We All Do?" *The Journal of Values-Based Leadership*: Vol. 9: Iss.1, Article 11.

S. Dhiman (✉)
School of Business, Woodbury University, Burbank, CA, USA
e-mail: Satinder.Dhiman@woodbury.edu

S. Dhiman, J. Marques (eds.), *Handbook of Engaged Sustainability*,
https://doi.org/10.1007/978-3-319-71312-0_1

Abstract

This introductory chapter explores the broad moral and spiritual basis of sustainability. It garners the view that the starting point for safeguarding the health of our planet is assuming full responsibility of our total footprint on the planet and purging our mind of the toxic emotions. It proposes that we need to recognize and address the spiritual decadence at the root of environmental degradation. An impartial survey of the contemporary ecological landscape reveals that the real global environmental problems are not biodiversity loss, ecosystem collapse, or climate change. The real environmental problems are selfishness, greed, violence, apathy, and lack of awareness which are playing havoc with the integrity of our planet. Mental pollution is increasingly becoming the greatest threat to our planet. And to address these problems root branch and all, we need a systemic spiritual and cultural transformation. This chapter suggests that sustainability is not possible without a deep change of values and commitment to a sustainable lifestyle at the "being" level. It cannot be achieved simply as an expression of economic functionality or legislative contrivance.

Drawing upon the findings of two recent graduate sustainability seminars by way of *cases in point*, the chapter also presents strategies geared toward cultivating essential sustainability awareness. These findings have far-reaching implications for developing a sustainability mindset and also indicate the great potential for sustainability studies in management education. The chapter specially focuses on the virtue of nonviolence as framework to achieve sustainable living. Excessive desire, anger, and greed are subtle forms of violence against oneself, others, and the planet. This chapter presents a philosophy of oneness of all life (regarding the world as an extended family) – as enunciated in the Hindu Vedas and the Bhagavad Gītā – as a holistic solution to address excessive desire, anger, and greed that rob our peace of mind and in turn disturb the peace of the planet.

Keywords

Corporate greed · Environmental degradation · Overconsumption · Ecological sustainability · Moral and spiritual impoverishment · Compassion · Nonviolence

Introduction

Western consumer culture is creating a psycho-spiritual crisis that leaves us disoriented and bereft of purpose. How can we treat our sick culture and make ourselves well?[1]

This chapter proposes that we need to recognize and address the moral and spiritual decadence at the root of environmental degradation. An impartial survey of the contemporary ecological landscape reveals that the real global environmental problems are not biodiversity loss, ecosystem collapse, or climate change. The real environmental problems are selfishness, greed, violence, apathy, and lack of awareness that are playing havoc with the sanctity of our planet. When the spiritual dimension of our being is underdeveloped, we turn into pleasure-seeking automatons, plundering the planet in a mindless race called progress. This makes us self-centered and greedy for material wealth which leads to social disharmony and overexploitation of natural resources, ignoring a vital fact that unlimited growth on a finite planet cannot be possible. It also creates disempowering social disparities and disenfranchises the majority, further compromising the peace and sanctity of the planet.

How have we come to be in an adversarial relationship with Mother Nature? How can we address the disempowering stance of selfishness, greed, violence, and apathy? What good is monetary wealth when the health of the planet is compromised? How can we overcome this syndrome of "more-ism" – wanting to have more and more of more and more in more and more ways?

We believe that to address these questions root branch and all, we need a complete spiritual and cultural transformation. We believe that it is more gratifying to enjoy the satisfaction of a contented mind than to constantly want more – a more expensive car, name-brand clothes, or a luxurious house. We believe that unless people's moral and spiritual qualities are nurtured and developed, the best of sustainability efforts will not work. Our political and economic thinking needs to be attuned to spirituality rather than materialism, for no economics is any good that does not make sense in terms of morality. We need to refuse to treat economics and politics as if people do not matter. After all, we are "Homo moralis" and not "Homo economicus." We believe that the way to achieve sustainable, harmonious living in all spheres is through *lived* morality and spirituality at the personal level.

When we live a life of greater self-awareness, we tend to consume less and, more so, less mindlessly. Accordingly, the chapter relies on the spiritual power of individuals to heal themselves and the environment. Its central thesis is that in order for sustainability to be sustainable, it must help us transition from being a consumer to

---

[1]John F Schumaker, The demoralized mind, *New Internationalist*, April 01, 2016. Retrieved on July 30, 2017: https://newint.org/columns/essays/2016/04/01/psycho-spiritual-crisis/

Schumaker concludes his insightful essay in a circumspect manner, thusly: "With its infrastructure firmly entrenched, and minimal signs of collective resistance, all signs suggest that our obsolete system – what some call 'disaster capitalism' – will prevail until global catastrophe dictates for us new cultural directions."

becoming a contributor. "I" is the beginning of "illness," while "we" is the beginning of "wellness." When we change our orientation from "I" to "we," we transition from illness to wellness – *individually* and *collectively.*

This chapter makes a case for transforming what the Buddhist call the three poisons of mind – delusion, greed, and hatred. Greed is born of selfishness and misplaced desire and results in hankering for happiness outside of ourselves. Hatred is born of unaddressed anger and our aversion toward unpleasant people and situations. Delusion refers to metaphysical ignorance, our mistaken notions of reality as discreet separate entities. We can counterbalance delusion by wisdom, greed by generosity, hatred by forgiveness, and violence by compassion. Above all, we can resolve these negative emotions through an understanding of oneness of all life.

The journey of world transformation starts at the individual level. It begins with the understanding that all life is essentially one. This is where *spiritual outlook* becomes paramount. It confers upon us the wisdom to see all existence as the expression of our very own self and help us to spontaneously act for the well-being of all beings. Selfless love and compassion naturally flow out of the understanding of unity and oneness of all life. The teachings of Vedānta and Buddhist psychology can go a long way to help us understand the essential oneness and interconnectedness of all life. When we truly realize the same essential Truth in everyone and everything, we begin to have correct valuation of things. This leads to a profound change not just in our thinking but in our being and behavior as well. We start seeing the terror of the situation more vividly – at our cellular level.

With this understanding comes the liberating realization that "there is no sustainability without . . . spirituality."[2] We need to place the responsibility of developing a high moral sense on the individual and on the power of individuals to heal the society. Only "an individual life rooted in the continuous harmony with life as a whole"[3] – a life based on wisdom, selfless service, and contribution – is a life worth living. The chapter builds on the premise that the best service we can offer to the universe in the realm of *engaged sustainability* is to purge our own mind of the toxic emotions of greed, anger, and self-centeredness. It specially focuses on the virtue of nonviolence as framework to achieve sustainable living. Excessive desire, anger, and greed are subtle forms of violence against oneself, others, and the planet.

As a case in point, the chapter also draws upon the findings distilled from teaching two recent graduate-level seminars on sustainable living which underscored the message that true sustainability is always "engaged" since it brings about a transformation about how we live our life, moment to moment. These findings have far-reaching implications for developing a sustainability mindset and also indicate the great potential for sustainability studies in business education.

---

[2]John E. Carroll, *Sustainability and Spirituality* (New York: State University of New York Press, 2004), 6. Also see Dhiman and Marques (Eds.), *Spirituality and Sustainability.*

[3]Eknath Easwaran, *The Compassionate Universe: The Power of the Individual to Heal the Environment* (California: Nilgiri Press, 1989), 10.

Defining Sustainability

Definitions of sustainability abound. The most frequently quoted definition is from *Our Common Future*, also known as the Brundtland Report, published in 1987 by the United Nations' World Commission on Environment and Development: "Sustainable development is development that meets the needs of the present without compromising the ability of future generations to meet their own needs."[4] There has been a growing realization in national governments and multifaceted institutions that it is impossible to separate economic development issues from environment issues. For example, poverty is the major cause as well as effect of global environmental problems. It is therefore futile to attempt to deal with environmental problems without a broader perspective that encompasses the factors underlying world poverty and international inequality.[5]

In simple terms, sustainability means utilizing natural resources in a manner that we do not end up, during the process, destroying the set up. In its most practical aspect, sustainability is about understanding the interconnections among environment, society, and economy. According to the US Environmental Protection Agency (EPA), "Sustainability is based on a simple principle: Everything that we need for our survival and well-being depends, either directly or indirectly, on our natural environment. Sustainability creates and maintains the conditions under which humans and nature can exist in productive harmony, that permit fulfilling the social, economic and other requirements of present and future generations."[6]

Harmlessness: Help Ever, Hurt Never

The chapter specially focuses on the virtue of nonviolence as framework to achieve sustainable living. If there is one inhumane cause that cuts across most human issues in depth and scale, it is perhaps violence. If there is one value that holds the keys to most problems that currently plague humanity, it is perhaps compassion. The virtue of compassion is the direct expression of the value of nonviolence.

Nonviolence is the most universally cherished value. It has been called the highest religion and the supreme good (अहिंसा परमो धर्मः अहिंसा परमा गतिः). It is the absolute form of religious austerity and love toward all human beings. *Noninjury is the therefore first and last universal value that can serve as sure-footed guide to our conduct.* All other values depend upon it. Non-stealing is a value because stealing *causes* a mental *hurt* to the person whose things are stolen. Similarly, speaking truth

[4]For further details, see the Report of the World Commission on Environment and Development: *Our Common Future*. Retrieved on August 2, 2017: http://www.un-documents.net/our-common-future.pdf

[5]Ibid., 12.

[6]What is Sustainability? United States Environmental Protection Agency. Retrieved on June 2, 2017: http://www.epa.gov/sustainability/basicinfo.htm

and keeping a promise are values because lying and breaking a promise cause hurt to the person who is lied to. This is the reason nonviolence is called the cardinal value on which all other values depend. It fosters kindness, love, compassion, and all other values. This is also the first step on the spiritual ladder – making sure that our actions (mental, verbal, and physical) do not cause any harm to any being in any shape or form. This principle of non-harming extends to all forms of life – animate and inanimate. It will not be an exaggeration to say that nonviolence, *ahiṁsā*, is the single most value to ensure the overall well-being of our planet at all levels.

What is the moral and spiritual basis of sustainability? For a most profound answer, let's turn to the great Indian epic, Mahābhārata, which contains 100,000 verses and is considered to be the world's longest epic poem (it is more than seven times the combined size of Homer's *Iliad* and *Odyssey*).[7] Its claim to greatness, however, does not rest on its length but in the breadth of its message. It has been rightly said about the Mahābhārata, "Whatever is here, on Law, on Profit, on Pleasure, and on Salvation, is found elsewhere. But what is not here is nowhere else."[8]

A. R. Orage, the British philosopher whom George Bernard Shaw once called the "the most brilliant English editor and critic of last 100 years,"[9] who studied Mahābhārata concertedly for 15 years, understood that it contained absolute truths. Orage extolls its glory as follows:

> The Mahabharata is the greatest single effort of literary creation of any culture in human history. The *Iliad* and the *Odyssey* are episodes in it; and the celebrated Bhagavad Gita is simply the record of a single conversation on the eve of one of its many battles....(It is) the most colossal work of literary art ever created. It contains every literary form and device known to all the literary schools, every story ever enacted or narrated, every human type and circumstance ever created or encountered.
>
> Unlike the reading of derivative works of art, the reading of the Mahabharata is a first-hand experience. One ends it differently, just as one emerges differently from everything real. The Mahabharata towers over all subsequent literature as the pyramids look over the Memphian sands. . ..More real Mysticism can be gathered from the Mahabharata than from the whole of modern mystical writings.[10]

[7]See Krishna Maheshwari, Mahabharata, *Hindupedia*, retrieved on January 22, 2016, http://www.hindupedia.com/en/Mahabharata#cite_note-0

[8]J. A. B. van Buitenen, trans., *The Mahabharata, Volume 1: Book 1: The Book of the Beginning* (Chicago, IL.: University of Chicago Press, 1980), 130.

[9]Philip Mairet, *A.R Orgage: A Memoir* (New Hyde Park, NY: University Books, 1966), 121. Also see Wallace Martin, *The New Age under Orage: Chapters in English Cultural History* (New York, Manchester University Press, 1967).

[10]See Avin Deen's response: Mahabharata (Hindu epic): Why do some Indians think Mahabharata is superior to all other epics ever written? Retrieved on January 28, 2017: https://www.quora.com/Mahabharata-Hindu-epic/Why-do-some-Indians-think-Mahabharat-is-superior-to-all-other-epics-ever-written

The legendary author of this epic, Veda Vyāsa, when asked about the most important single verse that represented the essence of the work, replied:

श्लोकार्धेन प्रवक्ष्यामि यदुक्तं ग्रन्थकोटिभिः ।

परोपकारः पुण्याय, पापाय परपीडनम् ॥

ślokārdhena pravakṣyāmi yaduktaṁ granthakoṭibhiḥ | paropakāraḥ puṇyāya, pāpāya parapīḍanam ||

Veda Vyāsa, the great composer of Vedas, says: I will present the gist of a million treatises in half a verse:

The greatest virtuous act is doing good to others; and the greatest evil is causing pain to others.

It is important to reflect on this verse singled out by the learned author of this great work. In order to understand the spiritual depth of this observation that "hurting others is the greatest evil," we have to dig further into the truth of our existence. Let's consider the following illustration: At its most basic level, all existence – from a piece of rock to the most developed specimen of living beings – is composed of five fundamental elements: earth, water, fire, air, and space. Each of these elements within our body, for example, exists with reference to the totality of its corresponding element outside our body. Take, for example, the element of air. The air that we breathe in exists by virtue of its relationship and interaction with the totality of air that exists outside in the environment. Likewise, the water that exists in our body in the form of various liquids cannot exist without the totality of water that exists outside. The same is true of the remaining elements: they exist in the microcosm of our body by virtue of their relationship with the totality of these elements in the macrocosm. If this is true regarding these physical elements, how much more so must it be of the consciousness which is the fundamental building block of all existence?

This short excursion into the interconnected nature of reality perhaps provides the simplest compelling reason yet to understand the oneness of the whole of existence. In our human terms, it means that we are inseparably one with the rest of existence. So, in effect, *to hurt others is to hurt ourselves*. That is why sages of humanity have always advocated helping others; they understood that, essentially, there are no others, and all life is inseparably interlinked and interconnected. This is perhaps the highest moral and spiritual foundation of sustainability.

Why is this simple existential fact not obvious to everyone? Why are we not naturally able to sense it and warm up to this view of reality? Perhaps in the perennial fight for self-preservation, this type of thinking does not further any evolutionary agenda. Perhaps by some sort of optical illusion, we are not able to see beyond the façade of self-centered, separately existing objects vying for their survival and flourishing. This leads to self-defeating strategies that disempower at best and seriously impinge upon the mutual preservation of everyone's interest, the mutual maintenance of the universe. Albert Einstein captures the issue succinctly and suggests a solution to come out of this prison of separateness, as follows:

> A human being is part of a whole called by us 'Universe,' a part limited in time and space. He experiences himself, his thoughts and feelings, as something separated from the rest...a kind of optical delusion of consciousness. This delusion is a kind of prison for us, restricting us to our personal desires and to affection for a few persons nearest to us. Our task must be to free ourselves from this prison by widening our circle of compassion to embrace all living creatures and the whole of nature in its beauty.[11]

David Bohm, Einstein's colleague and successor at Princeton, believed that the quantum theory reveals the "unbroken wholeness of the universe."[12] According to Bohm, this is the natural state of the human world – separation *without* separateness. This understanding garners reverence for life and helps us to realize our kinship with the entire universe. Harry Palmer highlights the fact that we are inseparably one and that there are no "individual" gains or losses:

> When you adopt the viewpoint that there is nothing that exists that is not part of you, that there is no one who exists who is not part of you, that any judgment you make is self-judgment, that any criticism you level is self-criticism, you will wisely extend to yourself an unconditional love that will be the light of the world. As your own true nature is realized – undefined and ever present – all will recognize that there is no gain that we do not all participate in and no loss for which we do not all share the sacrifice.[13]

However, we continue to delight in differences and fail to see what is essentially the same in all of us. It is abundantly evident that the divisions of race, religion, color, creed, and culture have contributed to the most heinous horrors of humankind. This will continue unabated, as history testifies, until we see the tyranny of our disempowering stance. Let's seek and share the underlying truth of mutuality that does not lead to unnatural differences and disharmony. That is the truth of our identity behind diversity – the essential oneness of all that exists. By seeking the truth that is equally good to all, we will be able to revere all life and truly redeem our human existence. Only then can we ensure equally the happiness and welfare of all beings. That will be our true gift of harmlessness to the sustainability of our planet. This gift will be rooted in the concept of mutual interdependence. The Buddhist concept of *pratītyasamutpāda* splendidly captures the essence of reality as the mutually interdependent web of cause and effect.

---

[11]From a compilation of Einstein quotes published from multiple online sources and credited to Kevin Harris (1995).

[12]See David Bohm, *Wholeness and the Implicate Order* (London: Routledge Classics, 2002). For general background, see also Ken Wilber, ed., *Quantum Questions: Mystical Writings of World's Great Physicists* (Boston: Shambhala, 1984).

[13]Harry Palmer, What is Enlightenment? Retrieved on July 29, 2017: https://aboutharrypalmer.com/faqs.html

Thich Nhat Hanh, a Vietnamese Buddhist whom Martin Luther King, Jr. nominated for Nobel Peace Prize, has coined the word "interbeing" to represent the interdependence of all existence. In his talks and seminars, Thich Nhat Hanh usually underscores the principle of "interbeing" by inviting his audience to look deeply at a piece of paper:

> If you are a poet, you will see clearly that there is a cloud floating in this sheet of paper. Without a cloud, there will be no rain; without rain, the trees cannot grow; without trees we cannot make paper. . ..If we look into this sheet of paper even more deeply, we can see sunshine in it. If the sunshine is not there, the forest cannot grow. . ..And if we continue to look, we can see the logger who cut the trees and brought it to the mill to transform into paper. . ..When we look in this way, we see that without all of these things, the sheet of paper cannot exist. . ..This sheet of paper is, because everything is. . ..As thin as this sheet of paper is, it contains everything in the universe in it.[14]

This understanding of the *interbeing* nature of all things lies at the heart of eco-spirituality and sustainability. This understanding predicates on intuiting and realizing the unity and oneness of all life. The next sections explore the metaphysical basis and the practical application of this thought position.

Understanding Dharma, *ṚTAṂ*, the Cosmic Order

We are quintessentially integral with the universe.[15]

Dharma is the most important and pivotal concept in the spiritual tradition of India. Etymologically, the word *dharma* comes from the root *dhṛ* which means "to bear, to support, to uphold," – *dhārayate uddhāryateva iti dharma* – that which "supports, sustains, and uplifts" is *dharma*. We have a Vedic injunction: धर्मो रक्षति रक्षिता: *Dharmo rakṣati rakṣitāḥa*: Dharma protects those who uphold dharma. There is another Vedic concept which is closely related to *dharma*, called *ṛta*. *Ṛta* is the *order* behind the manifest world, the harmony among all aspects of manifestation, each of which obeys its own truth. There is physical order, biological order, and psychological order.

Everything in the universe follows its own inner order, *ṛtaṃ*. Actually, *dharma* is conceived as an aspect of *Ṛta*. As John Warne explains in his editorial preface to Taittirīya Upaniṣad, "*Ṛta* is the universal norm identified with truth which, when brought to the level of humanity, become known as *dharma*, the righteous order here on earth."[16] Indian seers and sages prescribed that one should fulfill one's desires

[14]Thich Nhat Hanh, *The Heart of Understanding: Commentaries on the Prajnaparamita Heart Sutra* (CA, Berkeley: Parallax Press), 3–5.

[15]Thomas Berry, *The Great Work: Our Way Into the Future* (New York: Harmony/Bell Tower, 1999), 32.

[16]See *Taittiriya Upanisad* by Swami Dayananda Saraswati (Saylorsburg, Pennsylvania: Arsha Vidya Gurukulum, 2005), transcribed and edited by John Warne, iv.

(*kāma*) or pursue wealth and security (*artha*) within the framework of *dharma*, which ensures the good of everyone.[17]

In Indian philosophy and religion, *dharma*[18] has multiple meanings such as religion, duty, virtue, moral order, righteousness, law, intrinsic nature, cosmic order, and nonviolence (*ahiṁsā paramo dharmaḥ*[19]). *Dharma* also means the invariable, intrinsic nature of a thing (*svadharma*) from which it cannot deviate, like there cannot be a cold fire. In the realm of ethics and spirituality, *dharma* denotes conduct that is in accord with the cosmic order, the order that makes life and creation possible. When our actions are in harmony with the cosmic order, *Ṛta*, and in accord with the dictates of inner law of our being, *dharma*, they are naturally and spontaneously good and sustaining. Alexander Pope was right: "He can't be wrong whose life is in the right."

The Bhagavad Gītā, the loftiest philosophical poem that forms a part of the epic, Mahābhārata, is a well-known Indian spiritual and philosophical text, and its message is universal, nonsectarian. Both the Gītā and the Vedas base their philosophy on the understanding that the whole existence essentially forms one single unitary movement despite the apparent variegated diversity. What universal vision of ethical conduct is presented by the Gītā and the Upaniṣads which fosters sustainable lifestyle and growth? In the next section, we present some spiritual values and virtues based on the teachings of the Gītā and Upaniṣads that can contribute significantly to sustainable existence.

Key Ethical and Spiritual Virtues

Ethics deals with choosing actions that are right and proper and just. Ethics is vital in commerce and in all aspects of living. Society is built on the foundation of ethics. Without adherence to ethical principles, businesses are bound to be unsuccessful in

[17]Bangalore Kuppuswamy, *Dharma and Society: A Study in Social Values* (Columbia, Mo: South Asia Books, 1977).

[18]There is no single word in any Western language that can capture the multiple shades and subtle nuances of the word *dharma*. Like the words *karma* and *yoga*, it has been left untranslated in this chapter for the most part, with their contextual meaning presented in the parentheses where necessary. These words have found wide currency and familiarity in the Western culture. Similar confusion also exists regarding the meaning of the word *yoga*, as used in the Bhagavad Gītā. According to the preeminent Sanskrit scholar, J. A. B. van Buitenen, "The word *yoga* and cognates of it occur close to 150 times in the Gītā, and it needs attention." See J. A. B. van Buitenen, ed. and trans., *The Bhagavad Gītā in the Mahābhārata: A Bilingual Edition* (Chicago: University of Chicago Press, 1981), p. 17.

Etymologically, the word *yoga* comes from the Sanskrit root "*yuj*," which is cognate with the word "yoke." The *yoga*, "yoking," that is intended in the Gītā is the union of individual self, *jivātmā*, with the Supreme Self, *Paramātmā*.

[19]*ahiṁsā paramo dharmaḥ, ahiṁsā paramo tapaḥ | ahiṁsā paramo satyaṁ yato dharmaḥ pravartate | | ahiṁsā paramo dharmaḥ, ahiṁsā paramo damaḥ| ahiṁsā parama dānaṁ, ahiṁsā parama tapaḥ|| ahiṁsā parama yajñaḥ ahiṁsā paramo phalam| ahiṁsā paramaṁ mitraḥ ahiṁsā paramaṁ sukham||*

~*Mahābhārata/Anuśāsana Parva* (115-23/116-28-29).

the long run. As has become abundantly evident from the recent events, without ethics, a business degenerates into a mere profit-churning machine, inimical to both the individual and the society.

Virtues are *lived* values. In this context, Aristotle, the great Greek philosopher and author of *Nicomachean Ethics*, employs the word "hexis"[20] (from Latin "habitus") in a very special sense, denoting "moral habituation" or a dynamically "active state of moral virtue." For Aristotle, happiness is the "virtuous activity of the soul in accordance with reason." Urmson clarifies that, in Aristotle's view, "the wise man who wishes for the best life will accept the requirements of morality."[21] Aristotle further clarifies that, to be happy, we should seek what is good for us in the long run for we cannot become happy by living for the pleasures of the moment. Aristotle includes among the main constituents of happiness such things as health and wealth, knowledge and friendship, good fortune, and good moral character. For him, a life lived in accordance with excellence in moral and intellectual virtue constitutes the essence of a happy life: "He is happy who lives in accordance with complete virtue and is sufficiently equipped with external goods, not for some chance period but throughout a complete life. . . .*A good life is one that has been lived by making morally virtuous choices or decisions*."[22]

In the same manner, Buddha uses the word "compassion" in the sense of "wisdom in action" – right understanding flowering into right action. Chinese use a word called "te" or "teh" which means virtue in the same sense. *Tao Te Ching* is a great classic book of wisdom by Lao Tzu. It means the Way, the way of virtue and power – denoting that one has to walk upon it.

Harmlessness: The Highest Virtue

The highest goodness is nonviolence. This principle of non-harming extends to all forms of life – animate and inanimate. *Ahiṁsā* is the basis for vegetarianism within Hinduism and Buddhism though it goes well beyond just being vegetarian. This recommendation is also repeated to the seeker after truth in the Upaniṣads. According to the Chandōgya Upaniṣad (7.26.2), "If the food is pure, the mind becomes pure. If the mind is pure, memory becomes firm. When memory becomes strong, one is released from all knots of the heart and liberation is attained."[23] Since plant-based diet involves minimum or no amount of violence in its wake, its purity brings peace of mind and ensures our psychological well-being.

---

[20]W.D. Ross rendered "hexis" as a *state of character*. See David Ross, translation of Aristotle's *Nicomachean Ethics* (Oxford: Oxford University Press, 1980).

[21]J.O. Urmson, *Aristotle's Ethics*, 2.

[22]Mortimer Adler, *Arsitotle for Everybody: Difficult Thought Made Easy* (New York: Bantam Books, 1980). Emphasis added.

[23]*Ahara-suddhau sattva-suddhih, Sattva-suddhau dhruva-smritih, Smritilabhe sarvagranthinam vipramokshah.*

Vasudhaiva Kutumbakam: The Entire World as One Family

In the Vedic vision, for the magnanimous, the entire world constitutes but a single family (*udāracaritānām tu vasudhaiva kuṭumbakam*).[24] This vision is in stark contrast to viewing the world as a giant marketplace. Selfless love is the glue that holds members a family together, while self-centered profit is the basis of market-place. This vision calls for a certain awareness and a worldview that is predicated on our understanding of the universe as a divine manifestation. The Bhagavad Gītā and the Upaniṣads, the oldest wisdom texts of the world, regard this entire world, *jagat*, as the manifestation of the Lord, *Īśvara*. This thinking invests all existence with a deeper moral basis and a higher spiritual purpose.

We also find similar understanding in various other philosophical traditions of the world. The great Stoic philosopher and Roman emperor, Marcus Aurelius, has this to say about the unity of all existence:

> Constantly regard the universe as one living being, having one substance and one soul; and observe how all things have reference to one perception, the perception of this one living being; and how all things act with one movement; and how all things are the cooperating causes of all things which exist; observe too the continuous spinning of the thread and the contexture of the web.[25]

When this vision dawns, we understand the true meaning of such terms as compassion, contribution, and harmony. When we see unity in diversity, it helps us develop universal outlook in life which is so essential sustaining the sanctity of our war-ravaged planet. By developing universal pity and contentious compassion toward all and everything, one is then able to make peace with the world and feel at home in the universe. For this vision to become a reality, we have to understand that runaway economic growth is no longer an option. We either secure or discard our place in the biosphere. And this isn't some idealistic, romantic notion. It is preparation for a profounder life which is dramatically different from the one we are living now.

If we are "busy putting on our oxygen mask first," let's not forget that a larger system provided us with that oxygen mask to begin with. Let's seek and share the underlying truth of mutuality that does not lead to self-centeredness and unnatural differences and disharmony. That is the truth of our unity behind diversity – the essential oneness of all that exists. By seeking the truth that is equally good to all existence, we will be able to revere all life and truly redeem our human existence. Only then can we ensure equally the happiness and welfare of all beings – a

[24]*Mahōpaniṣad* – VI.73 (a). Alternative rendering: "For those who live magnanimously, the entire world constitutes but a family." See Dr. A. G. Krishna Warrier, trans., *Maha Upanishad* (Chennai: The Theosophical Publishing House, n.d.). Accessed: July, 31, 2015: http://advaitam.net/upanishads/sama_veda/maha.html

[25]Marcus Aurelius, in *Meditations* (c. 161–180 CE), Book IV, 40.

necessary precondition to a state of "happy individuals and harmonious society."[26] That will be our true gift of sustainability to the universe.

Fivefold Offerings to the Universe: Pancha Mahā Yajñās

Ethical conduct in the Upaniṣads revolves around the five *Yajñās* or offerings/sacrifices. These sacrifices are described as a person's duty toward gods, seers, ancestors, fellow humans, and animals. The *Pancha Mahā Yajñas* are extremely versatile set of religious-cum-spiritual disciplines. They have a religious (ritualistic) dimension as well as a spiritual (non-ritualistic) dimension. We provide the spiritual version of these "offerings" as follows:

These *Pancha Mahā Yajñās* – five areas of contribution – are[27]:

1. *Pitru Yajña* (offering/service to parents/elderly and ancestors)
2. *Manushya Yajña* (offering/service to all human beings)
3. *Bhūta Yajña* (offering/service to the animals and plants)
4. *Deva Yajña* (offering/service to the Lord)
5. *Brahma Yajña* (offering/service to all seers/saints and scriptures)

Our first contribution is in the form of the *Pitru Yajña* – whatever we do for the preservation of the family and for the protection and honoring of our ancestors and our senior citizens in general. The maintenance of family structure with love and care is *Pitru Yajña.* A society is considered mature only when it takes care of its elderly people properly with respect and reverence. The next contribution is the *Manushya Yajña* which is in the form of all kinds of social service that we do through varieties of organizations, charities, and associations. In this offering, we help the fellow human beings. The *Bhūta Yajña* represents our reverential attitude toward all the plants and animals and our contribution for the protection of nature, protection of environment, and protection of ecological balance. The *Deva Yajña* represents our reverential attitude toward the five basic elements, *Pancha Mahā Bhūtāni* – space, fire, air, water, and earth. These elements are looked upon as the divine expression of the Lord. The worship is offered to the Lord conceived in the form of a Universal Being, *Vishva rūpa Īśvara.* Finally, the *Brahma Yajña* represents our reverential contribution to the preservation and propagation of scriptural learning by supporting the teachers, *ācāryas*, and the spiritual institutions which support and propagate such activities.

[26]The expression "a happy individual and a harmonious society" is coined by Dr. Vemuri Ramesam, author of *Religion Mystified* and *Yogavasistha*. Dr. Vemuri runs a remarkable blog called Beyond Adviata: http://beyond-advaita.blogspot.com/

[27]The section on *Pancha Mahā Yajñās* draws upon Swami Paramarthananda ji's discourse, *The Spiritual Journey*. Retrieved on July 20, 2015: http://talksofswamiparamarthananda.blogspot.com/ Also see https://www.youtube.com/watch?v=a4jfDxMaXWQ

When we follow the *Pancha Mahā Yajñā*, it brings about an all-round harmony through spiritual, *dhārmik*, activities. We conclude this section with a Peace Invocation:

ॐ सर्वे भवन्तु सुखिनः सर्वे सन्तु निरामयाः ।

सर्वे भद्राणि पश्यन्तु मा कश्चिद्दुःखभाग्भवेत् ।

ॐ शान्तिः शान्तिः शान्तिः ॥

oṁ sarve bhavantu sukhinaḥ sarve santu nirāmayāḥ|sarve bhadrāṇi paśyantu mā kaś cidduḥkhabhāgbhavet | oṁ śāntiḥ śāntiḥ śāntiḥ ||

May all be happy, May all be free from misery.
May all realize goodness, May none suffer pain.
Oṁ! Peace. Peace. Peace.

Turning the Wheel of Cosmic *Co*-creation: Our Life as an Offering!

In order to grow spiritually, enjoin the Vedas, one has to convert one's whole life into an offering to the Divine, as a sort of cosmic sacrifice (*yajñārthāt karmaṇo*: BG 3.9). According to the Gītā (3.10–3.13), all beings are a part of the cosmic wheel of creation, sustained by the principle of mutual contribution and mutual maintenance. Therefore, every action should be performed in the spirit of sacrifice, *Yajña*, which sustains all beings, as an offering to the Universal Lord. They are great thieves, according to the Gītā, who do not help in the turning of this cosmic wheel of sacrifice (3.12). Thus, the Gītā does not stop at concern for humans alone; it is cosmic in its scope and universal in its view.

The Gītā (18.5) mandates threefold acts of sacrifice (*Yajña*), charity (*dānaṃ*), and austerity (*tapas*) and considers these as the "purifiers of the wise" (*pāvanāni manīṣiṇām*). "*Yajña*" literally means a sacrifice or an offering. The highest form of offering is living a life of sincerity – a life marked by being good and doing good. A sincere life is characterized by doing what we love and loving what we have to do. "*Dānaṃ*" means charity and denotes much more than writing a check to a favorite cause or organization. At the deepest level, it means the gift of "expressed love."

The Vedic philosophy of India has always emphasized the human connection with nature. The sacred literature of India – the Vedas, Upaniṣads, Purāṇas, Mahābhārata, Rāmāyaṇa, and Bhagavad Gītā – contain some of the earliest teachings on ecological balance and harmony and the need for humanity's ethical treatment of Mother Nature. The Vedic seers recognized that the universe is intelligently put together which presupposes knowledge and intelligence. They underscored interdependence and harmony with nature and recognized that all natural elements hold divinity. They posit the Lord as the *maker* as well as the *material* of the world, thus investing all creation with divine significance. Vedas do not view creation as an *act* of "creation" per se as many theologies postulate, but an *expression* or manifestation (*abhivyakti*) of what was unmanifest before.

The following excerpt from Chāndogya Upaniṣad, one of the most important Upaniṣads, explains the process of creation in amazingly simple and scientific terms and puts the irreducible minimum of spirituality based on this understanding within the compass of one short paragraph. By way of universal spirituality, it also represents its pinnacle:

> In the beginning, there was Existence alone –
> *One only, without a second.*
> It, the One, thought to Itself:
> "Let Me be many, let Me grow forth."
> *Thus, out of Itself, it projected the universe,*
> *and having projected the universe out of Itself,*
> *It entered into every being.*
> All that is has its self in It alone.
> Of all things It is the subtle essence.
> It is the truth. It is the Self.
> And you *are* That!
> ~Chāndogya Upaniṣad[28]

How can God be both the material (*upādāna*) and efficient (*nimita*) cause of the universe? Are there any parallels of this phenomenon in the familiar world? The Vedas provide two examples to show how the maker and the material can be one. The first example is of a spider and the spider web. Spiders produce silk from their spinneret glands located at the tip of their abdomen.[29] There is a subtle difference though between how the maker and the material of the creation are one and the spider and spider web: the spider web can exist independently of the spider, whereas the creator and the creation are inseparably one. The second example is dream objects and their creation by the dreamer. During dream, the "dreamer" is the single material and efficient cause (*abhina nimita upādāna kāranaṃ*) of dream creations. When a dreamer dreams about being afraid seeing a lion, the outside world, lion, jungle, and so forth are the creations of dreamer's mind. The emotion of fear is also within dreamer's mind.

The great practical advantage of viewing the Lord as both the material and the maker of the universe is the attainment of spiritual outlook regarding the entire creation. When everything becomes divine in our eyes, we develop a reverence for all life. Equipped with this understanding of One Self in All and All in One Self

---

[28]Adapted from Eknath Easwaran, trans., *The Upanishads, Translated for the Modern Reader* (Berkeley, CA: Nilgiri Press, 1987) and Swami Nikhalananda, trans. and ed., *The Upanishads: A One Volume Abridgement* (New York: Harper & Row Publishers, 1964).

[29]Spider Web. Retrieved on July 31, 2017: https://en.wikipedia.org/wiki/Spider_web

(*sarvātmabhāva*), we can live a life of harmony, benevolence, and compassion toward all existence.

Our dignity as humans should lie in protecting those who are weaker than us. Those who have more power ought to be more kind to those who are weak. All spiritual traditions teach us not to do to others what we don't want to be done to us. No living being wants to be hurt, to die. Moreover, this cruelty to animals is not environmentally sustainable. That time does not seem to be too far when we will have to stop this, if only as an environmental necessity.

Creative Altruism vs. Destructive Egocentrism

Matthieu Ricard in his recent book titled *Altruism: The Power of Compassion to Change Yourself and the World* (2015) presents a vision revealing how altruism can answer the key challenges of our times: economic inequality, life satisfaction, and environmental sustainability. With a rare combination of the mind of a scientist and the heart of a sage, he makes a robust case for cultivating altruism – a caring concern for the well-being of others – as the best means for simultaneously benefiting ourselves and our global society.

Ricard notes that Daniel Batson was the first psychologist to investigate rigorously whether real altruism existed and was not limited to disguised selfishness.[30] He notes the emphasis placed by Darwin on the importance of cooperation in nature and emphasizes the fact, as evidenced by coming together of human spirit during times of catastrophe, that human beings are essential "super-cooperators." He summarizes his view stating that:

> Altruism seems to be a determining factor of the quality of our existence, now and to come, and should not be relegated to the realm of noble utopian thinking maintained by a few big-hearted, naïve people. We must have the perspicacity to acknowledge this and the audacity to say it.[31]

Ricard cites decades of research conducted by the American psychologist, Tim Kasser, highlighting the high price of materialist values.[32] Representative studies spread over 20 years have shown that individuals who concentrated their existence on wealth, image, social status, and other materialistic values promoted by the consumer society are less satisfied with their existence. They are in less good health than the rest of the society. . ..Even in sleep their dreams seem to be infected with anxiety and distress. Thus, in so far as people seem to have adopted the "American dream" of stuffing their pockets, they seem to that extent to be emptier of soul and

[30]Ibid., 6.

[31]Ibid., 11–12.

[32]Cited in Ricard, *Altruism*, p. 9. See Tim Kasser, *The High Price of Materialism* (Cambridge, MA.: MIT Press, 2003).

self.[33] Tim Kasser goes on to show how desires or needs to have more or consume more are deeply and dynamically connected with feelings of personal insecurity.

In a foreword to Tim Kasser's book, *The High Price of Materialism*, Richard M. Ryan points out that the cultural climate of consumerism makes everyone vulnerable to what he calls "affluenza," an infectious disease in which everyone gets addicted to *having more*. He calls it "the tragic tale of modernity – we are the snakes eating our own tales."[34] Noting the widely held view by humanistic and existential thinkers such as Abraham Maslow, Erich Fromm, and Carl Rogers – that *focus on materialistic values detracts from well-being and happiness* – Ryan recounts that Kasser highlights how materialism actually contributes to unhappiness:

> Desires to have more and more material goods drive us into an ever more frantic pace of life. Not only must we work harder, but, once possessing the goods, we have to maintain, upgrade, replace, insure and constantly manage them. Thus in the journey of life, materialists end up carrying an ever-heavier load, one that expends the energy necessary for living, loving and learning – the really satisfying aspects of that journey. Thus materialism, although promising happiness, actually creates stress and strain.[35]

Nevertheless, recent advances in neuroscience confirm the experience of thousands of years of contemplative practice that individual transformation is possible through training and practice. Any form of training induces a restructuring in human brain at both the functional and structural levels. This is also, Ricard contends, what happens when one trains in developing altruistic love and compassion.[36]

Human Activities: Prime Driver of Climate Change

> Human activities are changing the climate in dangerous ways. Levels of carbon dioxide which heat up our atmosphere are higher than that they have been in 800,000 years. 2014 was planet's warmest year on record. And we have been setting several records in terms of warmest years over the last decade. One year does not make a trend but 14 out of 15 warmest years on record have fallen within the first 15 years of this century.
>
> Climate change is no longer just about the future we are predicting for our children or grandchildren. It is about the reality we are living with every day, right now. While we cannot say that any single weather event is entirely caused by climate change, we have seen stronger storms, deeper droughts, and longer wild fire seasons. Shrinking ice caps forced National Geographic to make the biggest change in its atlas since the Soviet Union broke apart.[37]

[33]See Tim Kasser, *The High Price of Materialism* (Cambridge, MA: MIT Press, 2003), xi.

[34]Ibid., xii.

[35]Ibid., xi.

[36]Ibid., 10.

[37]Climate change and President Obama's Action Plan. Video retrieved on August 3, 2016: https://www.youtube.com/watch?t=182&v=r4lTx56WBv0

Environmentalists continue to point out that the current state of our planet is alarming – from the standpoint of economic development, social justice, or the global environment – and that sustainable development has hardly moved beyond rhetoric since it was first used in the 1980s. It is fairly evident to anyone who has a nodding acquaintance with the world affairs that humanity is hardly closer to eradicating extreme poverty, respecting the dignity and rights of all peoples, or resolving environmental challenges, climate change, or the extinction of plants and animals.[38] And to add insult to the injury, strangely, we find ourselves in an era of "sustainababble" marked by wildly proliferating claims of sustainability. Even as adjectives like "low-carbon," "climate-neutral," "environment-friendly," and "green" abound, there is a remarkable absence of meaningful tests for whether particular governmental and corporate actions actually merit such description.[39]

For many experts, the increasing level of carbon dioxide in the environment is the most worrisome. The Emissions Database for Global Atmospheric Research, EDGAR, a database created by European Commission and Netherlands Environmental Assessment Agency, released its recent estimates, providing global past and present-day anthropogenic emissions of greenhouse gases and air pollutants by country. According to these estimates, the United States has the second highest CO2 emissions, trailing behind China, and one of the highest CO2 emissions per capita.[40]

According to a recent report by NASA, "Despite increasing awareness of climate change, our emissions of greenhouse gases continue on a relentless rise. In 2013, the daily level of carbon dioxide in the atmosphere surpassed 400 parts per million for the first time in human history. The last time levels were that high was about 3–5 million years ago, during the Pliocene era."[41]

This situation calls for creative solutions both at the collective and individual level. At the same time, we cannot wait for and rely on legislative measures alone; something fundamental needs to change in terms of how we live and view the world. According to the NASA report, responding to climate change involves a two-pronged approach involving mitigation and adaptation:

1. Reducing emissions of and stabilizing the levels of heat-trapping greenhouse gases in the atmosphere ("mitigation")
2. Adapting to the climate change ("adaptation")[42]

[38]David Biello, State of the Earth: Still Seeking Plan A for Sustainability?

[39]Robert Engelman cited in Michael Renner, "The Seeds of Modern Threats," in *World Watch Institute State of the World 2015: Confronting Hidden Threats to Sustainability* (Washington, DC: Island Press, 2013), 2.

[40]EDGAR: Trends in global CO2 emissions: 2014 report. Retrieved on August 1, 2017: http://edgar.jrc.ec.europa.eu/news_docs/jrc-2014-trends-in-global-co2-emissions-2014-report-93171.pdf

[41]NASA: *Global Climate change: Vital Signs of the Planet*. Retrieved on August 1, 2017: http://climate.NASA.gov/solutions/adaptation-mitigation/

[42]Ibid.

The Theravada Buddhist monk, Ven. Bhikkhu Bodhi, underscores the environmental urgency and suggests an effective solution:

> Today we face not merely a climate emergency but a single multidimensional crisis whose diverse facets – environmental, social, political, and economic – intersect and reinforce each other with dizzying complexity. ...The realization that human activity is altering the earth's climate assigns to human beings the gravest moral responsibility we have ever faced. *It puts the destiny of the planet squarely in our own hands* just at a time when we are inflicting near-lethal wounds on its surface and seas and instigating what has been called 'the sixth great extinction.'[43]

The Earth has entered a new period of extinction, concludes a study by the universities of Stanford, Princeton, and Berkeley, and humans could be among the first casualties.

According to *BBC News*, one of the study's authors said: "We are now entering the sixth great mass extinction event." The last such event was 65 million years ago, when dinosaurs were wiped out, in all likelihood by a large meteor hitting the Earth. The lead author of the study, Gerardo Ceballos, stated: "If it is allowed to continue, life would take many millions of years to recover and our species itself would likely disappear early on."[44]

It is believed that the five mass extinctions recorded in the last 600 million years were precipitated by natural causes. According to some scientists, we may have just one more generation before everything collapses. In fact, in a recently published research article titled *Accelerated modern human-induced species losses: Entering the sixth mass extinction*,[45] Ceballos et al. state, unequivocally, that the planet has officially entered its sixth mass extinction event. The study shows that species are already being killed off at rates much faster than they were during the other five extinction events and warns ominously that humans could very likely be among the first wave of species to go extinct.[46]

Calling the post-1950s the *Anthropocene* (literally, the "era of humans"), Ricard Matthieu notes that this is the first era in the history of the world when human

---

[43]Ven. Bhikkhu Bodhi, Climate change as a Moral Call to Social Transformation. The Buddhist Global Relief. Retrieved on January 28, 2016: http://buddhistglobalrelief.me/2015/12/02/climate-change-as-a-moral-call-to-social-transformation/ [emphasis added].

[44]Source: *BBC News*: Earth "entering new extinction phase" – *US study*, June 20, 2015. From the section Science and Environment. Retrieved on August 5, 2017: http://www.bbc.com/news/science-environment-33209548

[45]Gerardo Ceballos, Paul R. Ehrlich, Anthony D. Barnosky, Andrés García, Robert M. Pringle, Todd M. Palmer, "Accelerated modern human–induced species losses: Entering the sixth mass extinction," *Environmental Sciences* June, 19, 2015, 1–5. Retrieved on July 10, 2017: http://advances.sciencemag.org/content/advances/1/5/e1400253.full.pdf

[46]For further details, also see Dahr Jamail, Mass Extinction: It's the End of the World as We Know It. Retrieved on July 15, 2017: http://www.truth-out.org/news/item/31661-mass-extinction-it-s-the-end-of-the-world-as-we-know-it

activities are profoundly modifying and degrading the entire system that maintains life of earth. He states that the wealthy nations are the greatest culprits: An Afghan produces 2,500 times less CO2 than a Qatari and a thousand times less than an American.[47] With a note of urgency, Ricard rightly observes:

> If we continue to be obsessed with achieving growth, with consumption of natural resources increasing at its current exponential rate, we will need three planets by 2050. We do not have them. In order to remain within the environmental safety zone in which humanity can continue to prosper, we need to curb our endless desire for 'more.'[48]

Likewise, observations throughout the world make it clear that climate change is occurring, and rigorous scientific research demonstrates that the greenhouse gases emitted by human activities are the primary driver.[49]

These conclusions are based on multiple independent lines of evidence, and contrary assertions are inconsistent with an objective assessment of the vast body of peer-reviewed science. Moreover, there is strong evidence that ongoing climate change will have broad impacts on society, including the global economy, and on the environment. For the United States, climate change impacts include sea level rise for coastal states, greater threats of extreme weather events, and increased risk of regional water scarcity, urban heat waves, Western wildfires, and the disturbance of biological systems throughout the country. The severity of climate change impacts is expected to increase substantially in the coming decades.[50]

If we are to avoid the most severe impacts of climate change, emissions of greenhouse gases must be dramatically reduced. In addition, adaptation will be necessary to address those impacts that are already unavoidable. Adaptation efforts include improved infrastructure design, more sustainable management of water and other natural resources, modified agricultural practices, and improved emergency responses to storms, floods, fires and heat waves.[51]

---

[47]Matthieu Ricard, *Altruism: The Power of Compassion to Change Yourself and the World*, trans. by Charlotte Mandell and Sam Gordon (New York: Little, Brown and Company, 2015), 8.

[48]Ibid.

[49]See Climate change 2007: Synthesis Report. Retrieved on August 3, 20157: http://www.ipcc.ch/publications_and_data/ar4/syr/en/spm.html

[50]The conclusions in this paragraph reflect the scientific consensus represented by, for example, the Intergovernmental Panel on Climate change and US Global Change Research Program. Many scientific societies have endorsed these findings in their own statements, including the American Association for the Advancement of Science, American Chemical Society, American Geophysical Union, American Meteorological Society, and American Statistical Association. See statement on climate change from 18 scientific associations (2009).

[51]Statement on climate change from 18 scientific associations (2009). Retrieved on August 4, 2017: http://www.aaas.org/sites/default/files/migrate/uploads/1021climate_letter1.pdf

Conscientious Compassion

Compassion is the positive expression of the universal value of nonviolence. In Buddhist psychology, compassion is the flowering of wisdom. Compassion born of wisdom can be called conscious compassion. The American scholar and Theravada monk Venerable Bhikkhu Bodhi combines the two concepts of justice and compassion to form a distinct ethical ideal called "conscientious compassion." According to Bhikkhu Bodhi, "When compassion and justice are unified, we arrive at what I call conscientious compassion. This is compassion, not merely as a beautiful inward feeling of empathy with those suffering, but a compassion that gives birth to a fierce determination to uplift others, to tackle the causes of their suffering, and to establish the social, economic, and political conditions that will enable everyone to flourish and live in harmony."[52] He warns us about the dangers of taking an instrumental view of people, products, and planet:

> The major threat that I see today lies in the ascendency of a purely utilitarian worldview driven by a ruthless economic system that rates everything in terms of its monetary value and sees everything as nothing more than a source of financial profit. Thus, under this mode of thinking, the environment turns into a pool of 'natural resources' to be extracted and turned into profit-generating goods, and people are exploited for their labor and then disposed of when they are no longer of use.[53]

If humanity is to avoid a horrific fate, Bhikkhu Bodhi concludes, a double transformation is necessary. First, we must undergo an "inner conversion" away from the quest to satisfy proliferating desires and the constant stimulation of greed or craving. But change is also needed in our institutions and social systems. Finally, Bhikkhu Bodhi suggests that people turn away from an economic order based on incessant production and consumption and move toward a steady-state economy managed by people themselves for the benefit of their communities, rather than by corporate executives bent on market dominance and expanding profits.[54] Essentially, the bird of change needs two wings to rise up and fly: one wing is *moral vision* and the other wing is a *commitment to action*. We need greater moral awareness of empowering values such as justice, equality, loving-kindness, compassion, and self-restraint and the necessity for constant struggle against injustice, violence, hatred, cruelty, self-centeredness, and narcissist self-indulgence. All great spiritual traditions remind us that the responsibility for creating such a world rests with us and not with others.[55]

---

[52]Raymond Lam, *Conscientious Compassion* – Bhikkhu Bodhi on Climate change, Social Justice, and Saving the World. An e-Interview published in *Buddhistdoor Global*. Posted on August 14, 2015. Retrieved on August 1, 2017: http://buddhistglobalrelief.me/2015/08/14/conscientious-compassion-bhikkhu-bodhi-on-climate-change-social-justice-and-saving-the-world/

[53]Ibid.

[54]Ibid.

[55]Ven. Bhikkhu Bodhi, "On Hope and Hype: Reflections on a New Year's Tradition," *Buddhist Global Relief*. January 11, 2016. Retrieved on January 20, 2016: http://buddhistglobalrelief.me/2016/01/11/on-hope-and-hype-reflections-on-a-new-years-tradition/

In the following sections, we share some findings distilled from teaching two recent graduate-level seminars on sustainable living which underscored the message that true sustainability is always "engaged" since it brings about a transformation about how we live our life, moment to moment.

Sustainability Matters!

> ***WHAT CAN WE ALL DO TO MAKE THE PLANET MORE SUSTAINABLE!***
> *The work an unknown good person has done is like a vein of water flowing hidden underground, secretly making the ground green.* ~Thomas Carlyle

This author taught two graduate (MBA) Summer Seminars (2016 and 2017) on Sustainable Business and Sustainable Living using the tripartite framework of economy, equity, and ecology.[56] There were a total of 33 dynamic working professional enrolled in these seminars who hailed from four different continents. We explored a wide range of topics such as green economics, clean technology, toxic emotions, urban ecology, green luxury, deep ecology, smart cities, transforming waste into renewal energy, meat- vs. plant-based diet, GMOs, etc. About 30% of the participants switched to the plant-based diet and several students resolved to opt for a lesser intake of meat and to choose smaller hybrid or electric vehicle to be their next car, to say nothing of the shorter showers, zero plastic water bottles, living more mindfully and compassionately.

In the following sections, we will share the *modus operandi* of the course and some insights gleaned from the course geared toward cultivating sustainability mindset.

Engaged Sustainability à la an MBA Course!

As we sail through the precarious decades of the twenty-first century, a new vision is emerging to end poverty, protect the planet, and ensure well-being for all as part of our shared destiny. Achieving these goals will require the concerted efforts of governments, society, education institutions, the business sector, and informed citizens. To help the business sector contribute to achieving these goals, an MBA program at a small, innovative private university uniquely underscored *sustainability matters* dimension of business education through two graduate-level seminars offered during summer 2016 and 2017.

A vibrant vision was presented to the seminar participants in the form of a series of engaging affirmations through syllabus, class discussions, and presentations:

[56]This author is indebted to all 33 participants in these two graduate seminars who enriched our understanding of sustainability through their insightful discussion and comments. Parts of this section have benefitted from the suggestions of Mallory Quiroa, who was one of the participants in the course.

1. We believe that it is impossible to separate economic development issues from environment issues. In its most practical aspect, sustainability is about understanding close interconnections among environment, society, and economy.
2. We believe that the way to achieve sustainable, harmonious living in all spheres is through *lived* morality and spirituality at the personal level. We call it *engaged sustainability.*
3. We want our MBA grads to be *willing contributors* to the vision of cultivating happy individuals and harmonious society.
4. We believe that excessive desire, anger, and greed are subtle forms of violence against oneself, others, and the planet.
5. We are not only unaware of these mental pollutants; we are often unaware *how* unaware we are.
6. We believe that a focus on engaged sustainability will help us harness what is good for us, good for the society, and good for the planet.
7. We believe that achieving this goal will require a *shift* from being a consumer to becoming a contributor.
8. Only an individual life rooted in the continuous harmony with nature – a life based on moral and spiritual awareness – can be sustainable for the entire creation.

Got Sustainability?

This course has been designed to operationalize our current mission: *Cultivating Transformational Leaders for Sustainable Business*. We take sustainability seriously. We do not just track *carbon footprint* of business, but its *total* footprint. We view business *holistically* through the triple lens: economy, equity, and ecology. For us, there is no sustainability bereft of *ethics* and *spirituality.*

In the web of life, everything is linked with everything else, and everything affects everything: *You cannot pluck a flower without disturbing a star, as the poet Francis Thompson observed.* There are no "weeds" in the Garden of Nature. Weed is a *distinctly* human invention. In Japanese gardening, for example, when a weed that was near a plant is removed, it is not thrown away or destroyed. It is replanted elsewhere in the garden, realizing its importance in the overall scheme of things. After all, a weed is a plant whose medicinal power we have not discovered yet! The business today needs such holistic thinking and vision.

To operationalize this vision, our MBA students tackled projects from plant-based diet to smart cities to clean technology to eco-friendly community gardens and everything in between.

Our Modus Operandi

We focused on the practice of environmental sustainability – making responsible decisions that will reduce business' negative impact on the environment.

Throughout, our emphasis was on "engaged" sustainability: that is, what can we all do to "tread lightly on the planet." We used a case studies-based approach that utilized real-life business examples to illustrate the need and importance of sustainability.

Our overarching goal was to explore the application of sustainability in a wide variety of contemporary contexts – from economics of consumption and growth to government policy and sustainable planet. We examined sustainability from three perspectives: ecology, equity, and economics. As a point of departure, we began with and built upon the 1987 definition by the World Commission on Environment and Development of sustainability as economic development activity that "meets the needs of the present without compromising the ability of future generations to meet their own needs."

A punctilious participant in this course will notice that throughout, the ethico-spiritual basis of sustainably was never lost sight of. This course approached the topic of sustainability in a broad interdisciplinary fashion – in the possible manner of our *total footprint* on the planet, not just our *carbon footprint*. It proposed to bring together the two allied areas of sustainability and spirituality in a dialectical manner, with ethics acting as a balancing force and spirituality playing the role of the proverbial invisible hand guiding our quest for sustainability. It took the view that, in essence, spirituality and sustainability are vitally interlinked and that *there is no sustainability without spirituality*.

We tackled following questions:

1. What can *I do* to change the future of the planet?
2. What can *businesses* do to change the fate of our planet?
3. What can *we all do* that is good for us, good for the society, and good for the planet?

Every issue was examined objectively and diverse perspectives were presented for reflection. The course offered food for thought *without* interfering with participants' intellectual appetite. We believe that at best, teachers can only open the door; students have to enter of their own volition.

This course asked for only one commitment from its participants: examine your belief system in the light of the evidence presented, rather than scrutinizing the evidence in the light of your pre-existing notions. *Believe nothing; research everything*. This expectation is at the heart of every scientific endeavor. In fine, be aware of your confirmation bias and premature cognitive commitment. There is a difference between being on the side of the evidence and insisting that the evidence be on your side. This is the most important key to understanding all profound questions of life and leadership. Aristotle is reported to have said the following of his teacher, Plato: "Plato is dear; still *dearer is the truth*."

Interested in becoming a contributor…
Come and see (*ehipassiko*)!

Some Findings: Building the Plane While Flying It!

We used Ray Anderson's famous *Ted Talk* "The business logic of sustainability" as a launching pad to show the power of a leader's vision to trigger profound change. Participants found Ray's message refreshing and encouraging – a harbinger of hope and possibility. Ray founded a carpet company, *Interface*, which grew to be the first company to achieve 100% sustainability. By his abiding commitment to sustainable business, Ray Anderson increased sales and doubled profits while turning the traditional "take/make/waste" industrial system on its head.[57] It is a testimony to Ray's resonant vision that *Interface* personifies a "clear, compelling, and irrefutable case – business case – for sustainability." In an interview with the editorial staff of *The Journal of Values-Based Leadership,* Ray was succinct in his commitment to sustainable commerce: "If that product cannot be made sustainable, we have no business making that product. For that matter, neither does anyone else."[58] His *Keynote to Second International Conference on Gross Happiness* had a gentle ring of urgency:

> There is no more strategic issue for a company, or any organization, than its ultimate purpose. For those who think business exists to make a profit, I suggest they think again. Business makes a profit to exist. Surely it must exist for some higher, nobler purpose than that.[59]

Leyla Acaroglu's TEDxMelbourne talk *Why We Need to Think Differently About Sustainability* was another video that sparked some good discussion about the concept of systemic life cycle-based sustainability. Her mantra of "doing more with less," reminiscent of TQM, also resonated well with the participants.[60]

GMOs was another topic that we explored at some length. Dr. Vandana Shiva challenges the dominant paradigm of non-sustainable, industrial agriculture and explains why we need an "organic" future.[61] That both corporations and governments are accomplices in this current attack on human and plant genetics came loud and clear from the film[62] as well as Dr. Vandana's interview with the BBC reporter.[63]

---

[57]Ray Anderson, The business logic of sustainability. A Ted Talk. Retrieved on July 1, 2017: https://www.ted.com/talks/ray_anderson_on_the_business_logic_of_sustainability

[58](2009) "Progress Toward Zero: The Climb to Sustainability – Interview with Ray Anderson," *The Journal of Values-Based Leadership*: Vol. 2: Iss. 1, Article 3. Available at: http://scholar.valpo.edu/jvbl/vol2/iss1/3

[59]Cited by Samuel Mann, Some of my favorite Ray Anderson quotes. In Computing for Sustainability Saving the earth one byte at a time. Retrieved on July 30, 2017: https://computingforsustainability.com/2011/08/12/some-of-my-favourite-ray-anderson-quotes/ This view is in stark contrast with.

[60]See https://www.youtube.com/watch?v=5lOSIHWOp2I&t=42s

[61]Dr. Vandana Shiva: "Why We Need an Organic Future" (NOFA-VT 2017 Keynote Address). Retrieved on August 1, 2017: https://www.youtube.com/watch?v=gof7vdQI6OM

[62]Seeds of Death: Unveiling The Lies of GMO's – Full Movie https://www.youtube.com/watch?v=a6OxbpLwEjQ

[63]A Billion Go Hungry Because of GMO Farming: Vandana Shiva https://www.youtube.com/watch?v=vbIQF72IDuw

It also became clear that the statement that we need GMOs to feed the world is the biggest fallacy.[64] GMOs are promoted for their ability to help alleviate world hunger. While we do not deny the possibility of technology improving the crops, "much of the inability of GM technology to provide relief for the poorest nations seems to have less to do with the technology and more with social and political issues," according to Paul Diehl.[65] Acknowledging that the genetic modification is already part of the crop improvement tool kit, Diehl offers a balanced view: "The real question is if, *in addition to helping make many wealthier in the industrialized world*, this advanced technology provides part of the solution to help improve a lot of the poorest regions of the world."[66]

Another video that generated much heat and light was Zeitgeist: Moving Forward.[67] Amazingly, this video had over 24 million views on the YouTube as of this writing. The first half of this 2-h and 42-min video focuses on delineating the social, economic, and ecological issues that plague our planet. It becomes evident early on that the movie is aimed at exposing the dark side of capitalism in its myriad forms. All participants felt it to be a great eye-opener on many fronts and wondered that why such information is not more widely disseminated. One recurring theme in the participants' comments was the need to change our ways of thinking about consumerism, national debt, poverty, and health-care issues and explore new sustainable and equitable economic and ecological models. After all, excessive desire, anger, and greed are subtle forms of violence against oneself, others, and the planet. Once one sees through it without the scaffold of rationalization, one realizes the terror of the situation. One can then move on living a life marked by simplicity of heart, purity of mind, and clarity of spirit. This is the herald of coming good we long for, as promised by the last half of the movie. By the time all participants had critiqued the movie, there emerged good consensus that it is possible to live with minimum negative impact on the environment or ourselves by thinking about sustainability in every step we take and having natural alternatives in addition to minimizing the waste.

Plant-Based vs. Meat-Based Diet

Although plant- vs. meat-based diet was just one of the many topics that we explored, somehow, the participants took a special interest in exploring this topic

---

[64]See Maggie Hennessy, The biggest fallacy about GMOs is that we need them to feed the world: WFP May 7, 2014. Retrieved on August 1, 2017: http://www.foodnavigator-usa.com/Suppliers2/The-biggest-fallacy-about-GMOs-is-that-we-need-them-to-feed-the-world-WFP

[65]Paul Diehl, Can Genetically Modified Food Feed the World? What You Need To Know About Genetically Modified Crops. *The Balance*: Biotech Industry, June 03, 2017. Retrieved on August 1, 2017: https://www.thebalance.com/can-genetically-modified-food-feed-the-world-375634

[66]Ibid. Emphasis added.

[67]Zeitgeist: Moving Forward, Official Release, 2011 Retrieved on August 1, 2017: https://www.youtube.com/watch?v=4Z9WVZddH9w

at a deeper personal level. Through videos, readings, and class discussions,[68] three broad reasons for transitioning to a plant-based diet were identified: health, sustainability, and compassion. And surprisingly enough, it was compassion for the living beings that resonated the most with the participants and not medical and sustainability reasons. The plant-based diet was not presented as all-or-nothing option. It was clarified to the students at the very outset that the purpose of the class is "to provide some food for thought and not to interfere with students' intellectual appetite."

We explored the moral basis of several values, including non-harming and compassion. As a prelude to various class projects, students were asked to do a simple survey to find out whether "non-harming" and "compassion" would qualify as universal values. A universal value is one which is valued universally, regardless of one's religion, race, caste, creed, or gender. It is something that is cherished by all living forms, not just human beings. This exercise involved asking oneself and others, "Would you like to be hurt?" At first, it may seem to be a strange question to ask. Reflecting on this exercise, it soon dawned upon the participants that no living being *ever* wants to be hurt. Not only that, every living being longs to be treated kindly and compassionately. Thus, non-harming and compassion were established to be universally cherished values. This understanding served as a guiding star for all of our class discussions, projects, and activities.

This exercise proved to be so effective that during the first 2–3 weeks of class attendance, students voluntarily pledged to reduce the meat intake, and by the fifth week, 6 of the 16 students had switched to a full plant-based diet. Not only that, some of them became bit of advocates, enthusiastically championing the cause of compassion toward animals. This author regularly sees some of these students on campus and asks them how their pledge is faring along. It is comforting to learn that they are

---

[68]Paul McCartney – *If Slaughterhouses had Glass Walls*. Available at https://www.youtube.com/watch?v=p_UpyY2MIOc

Why Vegan? – Amazing Presentation by Gary Yourofsky. Available at https://www.youtube.com/watch?v=UROxRLbVils

The food we were born to eat: John McDougall at TEDxFremont. Retrieved on July 18, 2017: https://www.youtube.com/watch?v=d5wfMNNr3ak

Tackling diabetes with a bold new dietary approach: Neal Barnard at TEDxFremont. Retrieved on August 1, 2017: https://www.youtube.com/watch?v=ktQzM2IA-qU

Jennifer Dillard, A Slaughterhouse Nightmare: Psychological Harm Suffered by Slaughterhouse Employees and the Possibility of Redress through Legal Reform. *Georgetown Journal on Poverty Law & Policy*. Available at SSRN: https://ssrn.com/abstract=1016401

Michael Greger, *How Not to Die: Discover the Foods Scientifically Proven to Prevent and Reverse Disease* (New York: Flatiron Books, 2015).

In this informative book, Dr. Greger describes which foods to eat to prevent the leading causes of disease-related death and shows how a diet based on fruits, vegetables, tubers, whole grains, and legumes might even save your life. Dr. Greger runs the popular website NutritionFacts.org and serves as the Director of Public Health and Animal Agriculture at the Humane Society of the United States.

still steadfast about it. What is truly remarkable is that some of these students happen to be from cultures where meat is a major part of their diet and social life.

During our research on GMOs, we also discovered what is called best and worst fruits and vegetables.

EWG Shopper's Guide to Pesticides in Produce

Although, buying organic is the safest way to ensure no-GMO and food safety, many of us may not be able to afford everything organic – even if we know it is better to pay the farmer than the doctor.[69] The Environmental Working Group (EWG) ranks pesticide contamination of popular fruits and vegetables based on more than 36,000 samples of produce tested by the US Department of Agriculture and the Food and Drug Administration. EWG Shopper's Guide to Pesticides in Produce was released in March 2017, starting with the highest amounts of pesticide residue:

> The 2017 list of 'dirty dozen' featured strawberries, spinach, nectarines, apples, peaches, celery, grapes, pears, cherries, tomatoes, sweet bell peppers and potatoes. The 2017 'clean fifteen' included produce that had relatively fewer pesticides and lower total concentrations of pesticide residues, in order, sweet corn, avocados, pineapples, cabbage, onions, frozen sweet peas, papaya, asparagus, mangoes, eggplant, honeydew melon, kiwis, cantaloupe, cauliflower and grapefruit. Only 1% of samples showed any detectable pesticides in avocados and sweet corn, which were deemed the cleanest produce. More than 80% of pineapples, papaya, asparagus, onions and cabbage that were sampled showed no pesticide residue.

The Anatomy of Change: The Key Driver

This author wondered about what could have been the key trigger for these students to switch to the plant-based diet. Most students already knew that the processed meat is injurious to health. Sustainability argument, though rationally compelling, is not always the strongest. Perhaps it was a combined effect of all the three reasons, compassion possibly weighing more than the other two. It seems that when it comes to changing human behavior, emotions play a far greater role than logic. And compassion is the highest emotion that we as humans can experience, for it touches our being at the deepest level. There might be another factor that could have played its due part in the behavior change equation. At the very outset of the discussion, it was mentioned that what we eat is largely a matter of habit. It was clarified that habits can be acquired and can be given up as well if one decides to do so. Participants were

[69]Retrieved on July 30, 2017: https://www.ewg.org/foodnews/dirty_dozen_list.php#.WYeHm4TytQI

Also see Johanzynn Gatewood, Strawberries remain at top of pesticide list, report says. CNN. March 10, 2017.

Retrieved on July 19, 2017: http://www.cnn.com/2017/03/08/health/dirty-dozen-2017/index.html

told that they are *not* their habits. We also discussed the tyranny of *confirmation bias* and learned how to watch for it. This created an open environment of exploration without any judgment attached to what one decides to eat or not eat. Although shift to plant based is a personal choice, it was very gratifying to see how this 7-week elective course in sustainability made such a profound impact on students – from their eating habits to sustainable cars and homes. This reinforced author's conviction that journey of world transformation starts at the individual level. It begins with the understanding that all life is essentially and fundamentally one.

These observations have far-reaching implications for developing a sustainability mindset. They also indicate the great potential of sustainability field in management education. There are plenty of cases, personal stories, and corporate examples to choose from for a management educator.[70]

Concluding Thoughts

> *We therefore have a historic responsibility as we are the first generation to really become aware of the problem and yet the last generation that can deal with it.*[71]

This chapter approached the topic of sustainability in the broad manner of our *total footprint* on the planet, not just our *carbon footprint*. We believe that this approach is critical in addressing the profound issues of environmental sustainability and in mapping our plenary future. For sustainability to be "engaged," we need to focus on the spiritual power of individuals to heal themselves and the environment. Sensitive minds have always recognized that the most important issues confronting organizations and society at large are so profound and pervasive that they can only be resolved at the fundamental level of the human spirit – *at the level of one's authentic self.*[72] Politics without morality, even so as economics without ethics, is a dead-end

[70]Some more examples of the resources used for this seminar: Ray Anderson, sustainable-business pioneer, provides a compelling case for business rationale of sustainability: Ray Anderson, *The business logic of sustainability*. Available at: https://www.ted.com/talks/ray_anderson_on_the_business_logic_of_sustainability

The Girl Who Silenced the World for 5 Minutes! https://www.youtube.com/watch?v=XdK0uYjy85o

Vice President of Citi Bank: https://www.youtube.com/watch?v=U0VnpFJmkL0

Ethics in a meat-free world – Philip Wollen at TEDxMelbourne: https://www.youtube.com/watch?v=ApeIUzKLkuo

Sustainable development: what, where and by whom?: Kitty van der Heijden at TEDxHaarlem: https://www.youtube.com/watch?v=4sJ-uixn7Jg

What's wrong with our food system | Birke Baehr | TEDxNextGenerationAsheville https://www.youtube.com/watch?v=F7Id9caYw-Y

[71]Laurent Fabius, COP21 President and Minister of Foreign Affairs and International Development.

[72]See Satinder Dhiman, *Gandhi and Leadership: New Horizons in Exemplary Leadership* (New York: Palgrave MacMillan, 2015); Satinder Dhiman, *Holistic Leadership: A New Paradigm for Today's Leaders* (New York: Palgrave MacMillan, 2017).

road for it ignores the humanity of who we are. Accordingly, sustainability is no longer seen just as a scientific or political problem; it becomes a matter of individual moral choice, with profound spiritual significance. Hence, this chapter proposed to bring together the two allied areas of *sustainability* and *spirituality* in a *dialectical* manner, with ethics as a *balancing force*.

There is African saying that a person is a person because of other persons. The phrase "by benefiting others, we benefit ourselves," represents the idea of interconnectedness. This idea is also well described by Jose Ortega y Gasset: "I am myself plus my circumstance, and if I do not save it, I cannot save myself."[73] Thus, if we are to secure our survival as a species on this planet, there is a need to move from a mentality of competition to one of cooperation, from a lifestyle of *being a consumer to becoming a contributor*, based on the interconnectedness of all life. Therefore, to the question, "How to improve the state of the Planet?" we reply: "Everybody can do something!"[74]

At the managerial level, we need to start viewing our organizations as "living systems" rather than as "machines for producing money."[75] Long-lived companies were also found to be supremely sensitive to their environment. Thus true sustainability is not possible without a deep change of *values* and commitment to a lifestyle at the individual and organizational level. It cannot be achieved simply as an expression of economic functionality or legislative contrivance. Therefore, to the question, "How to improve the state of the Planet?" we reply: "Everybody can do something!"[76]

We believe that *engaged sustainability* is not possible without a deep change of values and commitment to a lifestyle at the "being" level. It cannot be achieved simply as an expression of economic functionality or legislative contrivance. The journey of world transformation starts at the individual level, with the understanding that all life is essentially and fundamentally one. With this understanding comes the liberating realization that "there is no sustainability without...spirituality."[77] We need to place the responsibility of developing a high moral sense on the individual and on

[73]José Ortega y Gasset, *Meditations on Quixote*, trans. Evelyn Rugg and Diego Marín (New York: W.W. Norton and Company, 1961), 45.

[74]David Biello, State of the Earth: Still Seeking Plan A for Sustainability? How to improve the state of the planet: "everybody can do something," *Scientific American*, Oct. 12, 2012. Retrieved on August 1, 2017: http://www.scientificamerican.com/book/planet-seeks-plan-for-sustainability/

[75]Arie de Gues, *The Living Company: Habits for Survival in a Turbulent Business Environment* (Boston, MA: Harvard Business Review Press, 2002), 91, 176.

[76]David Biello, State of the Earth: Still Seeking Plan A for Sustainability? How to improve the state of the planet: "everybody can do something." *Scientific American*, Oct. 12, 2012. Retrieved on July 31, 2017: http://www.scientificamerican.com/book/planet-seeks-plan-for-sustainability/

[77]John E. Carroll, *Sustainability and Spirituality* (New York: State University of New York Press, 2004), 6. Also see Dhiman and Marques (Eds.), *Spirituality and Sustainability.*

the power of individuals to heal the society. When everyone contributes their respective share in the cosmic scheme of things, it unexpectedly brings about the intended change in the entire world. Only "an individual life rooted in the continuous harmony with life as a whole"[78] – a life based on wisdom, selfless service, and contribution is a life worth living.

Observation and reflection dictate that the universe was not created for humans alone. If our universe is approximately 13.8 billion years old, as science informs us, and earth is about 4.5 billion years old, while Homo sapiens appeared 200,000 years ago in Africa having evolved leisurely from 3.2 million-year-old hominin (human-like primate), then obviously we cannot assume that the universe was created for humans alone![79] In the grand scheme of things, all forms of life are equally precious and so are their needs. It is a matter of great concern that as humans, we are the least sustainable of all species. Jonas Salk is reported to have said, "If all the insects were to disappear from the Earth, within fifty years all life on Earth would end. If all human beings disappeared from the Earth, within fifty years all forms of life would flourish."[80]

It is our bounden duty to act as caretakers of the planet's precious resources. If not we, who? If not now, when? *We only have one planet to live. Let's cultivate it together.*

We conclude by a quote that presents a clarion call to nourish our planet[81]:

> Even
> After
> All this time
> the sun never says to the earth:
> "You owe
> Me."
> Look
> What happens
> with a love like that,
> It lights the
> Whole
> Sky.

[78]Eknath Easwaran, *The Compassionate Universe: The Power of the Individual to Heal the Environment* (California: Nilgiri Press, 1989), 10.[79] See: Pallab Ghosh, 'First human' discovered in Ethiopia, BBC News, 4 March 2015. Retrieved November 26, 2017: http://www.bbc.com/news/science-environment-31718336

[79]See: Pallab Ghosh, 'First human' discovered in Ethiopia, BBC News, 4 March 2015. Retrieved November 26, 2017: http://www.bbc.com/news/science-environment-31718336

[80]"If all the insects were to disappear from the earth." Quoted during a Ted Talk by Sir Ken Robinson. Retrieved on August 2, 2017: https://www.youtube.com/watch?v=0JYW2JFkXsg

[81]Daniel Ladinsky, *The Gift: Poems by Hāfiz the Great Sufi Master* (New York: Penguin Compass, 1999), 34.

Cross-References

- ▶ Education in Human Values
- ▶ Expanding Sustainable Business Education Beyond Business Schools
- ▶ Just Conservation
- ▶ Responsible Investing and Corporate Social Responsibility for Engaged Sustainability
- ▶ The Spirit of Sustainability
- ▶ The Sustainability Summit
- ▶ To Eat or Not to Eat Meat

References

Adler, M. (1980). *Aristotle for everybody: Difficult thought made easy*. New York: Bantam Books.

Bangalore Kuppuswamy. (1977). *Dharma and society: A study in social values*. Columbia: South Asia Books.

Berry, T. (1999). *The great work: Our way into the future*. New York: Harmony/Bell Tower.

Bohm, D. (2002). *Wholeness and the implicate order*. London: Routledge Classics.

Carroll, J. E. (2004). *Sustainability and spirituality*. New York: State University of New York Press.

Ceballos, G., Ehrlich, P. R., Barnosky, A. D., García, A., Pringle, R. M., & Palmer, T. M. (2015). Accelerated modern human–induced species losses: Entering the sixth mass extinction. *Environmental Sciences*, 19 June 2015, pp. 1–5. http://advances.sciencemag.org/content/advances/1/5/e1400253.full.pdf. Retrieved 10 Jul 2017.

de Gues, A. (2002). *The living company: Habits for survival in a turbulent business environment*. Boston: Harvard Business Review Press.

Dhiman, S. (2015). *Gandhi and leadership: New horizons in exemplary leadership*. New York: Palgrave Macmillan.

Dhiman, S. (2017). *Holistic leadership: A new paradigm for today's leaders*. New York: Palgrave Macmillan.

Easwaran, E. (Trans.). (1987). *The Upanishads* (Translated for the modern reader). Berkeley: Nilgiri Press.

Easwaran, E. (1989). *The compassionate universe: The power of the individual to heal the environment*. California: Nilgiri Press.

Greger, M. (2015). *How not to die: Discover the foods scientifically proven to prevent and reverse disease*. New York: Flatiron Books.

José Ortega y Gasset. (1961). *Meditations on Quixote* (trans: Rugg, E. & Marín, D.). New York: W. W. Norton.

Kasser, T. (2003). *The high price of materialism*. Cambridge, MA: MIT Press.

Krishna Warrier, A. G. (Trans.). (n.d.) *Maha Upanishad*. Chennai: The Theosophical Publishing House. http://advaitam.net/upanishads/sama_veda/maha.html. Accessed 31 Jul 2015.

Ladinsky, D. (1999). *The gift: Poems by Hāfiz the great Sufi master*. New York: Penguin Compass.

Mairet, P. (1966). *A.R Orgage: A memoir*. New Hyde Park: University Books.

Martin, W. (1967). *The new age under Orage: Chapters in English cultural history*. New York: Manchester University Press.

Nhat Hanh, T. (1998). *The heart of understanding: Commentaries on the Prajnaparamita Heart Sutra*. Berkeley: Parallax Press.

Progress toward zero: The climb to sustainability – Interview with Ray Anderson. (2009). *The Journal of Values-Based Leadership, 2*(1), Article 3. http://scholar.valpo.edu/jvbl/vol2/iss1/3

Renner, M. (2013). The seeds of modern threats. In *World Watch Institute state of the world 2015: confronting hidden threats to sustainability*. Washington, DC: Island Press.

Ricard, M. (2015). *Altruism: The power of compassion to change yourself and the world* (trans: Mandell, C. & Gordon, S.). New York: Little, Brown and Company.
Ross, D. (1980). Translation of Aristotle's *Nicomachean ethics*. Oxford: Oxford University Press.
Schumaker, J. F. (2016). The demoralized mind. *New Internationalist*, 1 April 2016. https://newint.org/columns/essays/2016/04/01/psycho-spiritual-crisis/. Retrieved 30 Jul 2017.
Swami Dayananda Saraswati. (2005). *Taittiriya Upanishad* (trans and ed: Warne, J.). Saylorsburg: Arsha Vidya Gurukulum.
Swami Nikhalananda (Trans. and Ed.). (1964). *The Upanishads: A one volume abridgement*. New York: Harper & Row Publishers.
van Buitenen, J. A. B. (Trans.). (1980). *The Mahabharata, Vol. 1. Book 1: The book of the beginning*. Chicago: University of Chicago Press.
van Buitenen, J. A. B. (Ed. and Trans.). (1981). *The Bhagavad Gītā in the Mahābhārata: A bilingual edition*. Chicago: University of Chicago Press.
Wilber, K. (Ed.). (1984). *Quantum questions: Mystical writings of world's great physicists*. Boston: Shambhala.

To Eat or Not to Eat Meat

Striking at the Root of Global Warming!

Satinder Dhiman

We are quintessentially integral with the universe. (Berry 1999, p. 32)

Contents

Partially based on author's works: *Spirituality and Sustainability* (Springer Nature, 2016); and "Ethics and Spirituality of Sustainability: What Can We All Do?" *The Journal of Values-Based Leadership*: Vol. 9: Iss. 1, Article 11.

S. Dhiman (✉)
School of Business, Woodbury University, Burbank, CA, USA
e-mail: satinder.dhiman@woodbury.edu

S. Dhiman, J. Marques (eds.), *Handbook of Engaged Sustainability*,
https://doi.org/10.1007/978-3-319-71312-0_2

Abstract

This chapter explores the vital role each individual can play to improve the state of the planet. It focuses on understanding the economics, ethics, and spirituality of a meat-based vs. plant-based diet – something that concerns everyone and something over which everyone has complete choice and control. It offers a unique perspective that all *food* is essentially vegetarian (see Taittirīya Upaniṣad 2.1.2: ओषधीभ्यो अन्नम् Food comes from vegetation), although one can have a *meal* that is nonvegetarian. This chapter explores three main reasons to turn to a plant-based diet: health, sustainability, and compassion. It offers a perspective that switching to a plant-based diet or reducing the meat and dairy intake represents one of the most effective solutions to global warming. The uniqueness of this approach lies in its humanity and its locus of control: It depends upon each one of us.

Many spiritual traditions recommend a plant-based diet based on moral and compassionate grounds. Nonviolence, *ahiṁsā*, is the basis for the vegetarianism within Jainism, Hinduism, and Buddhism though it goes well beyond just being vegetarian. The universal value of harmlessness is a core virtue derived from the Vedic injunction "*mā hiṁsyāt sarvabhūtāni*" मा हिंस्यात् सर्वभूतानि – do no harm to living creatures.

Keywords

Conscious Consumption · Ethics of plant-based diet · Sustainable food systems · Ecological cost of food · Global warming · Health hazards of meat-based diet

Introduction

When nourishment is pure, reflection and higher understanding becomes pure. When reflection and higher understanding are pure, memory becomes steady. When memory becomes steady, there is release from all the knots of the heart.[1]

This chapter explores the vital role each individual can play to improve the state of the planet. It focuses on understanding the economics, ethics, and spirituality of a meat-based vs. plant-based diet – something that concerns everyone and something over which everyone has complete choice and control. It offers a unique perspective that all *food* is essentially vegetarian,[2] although one can have a *meal* that is nonvegetarian. This chapter explores three main reasons to turn to a plant-based diet: health, sustainability, and compassion. It offers a perspective that switching to a plant-based diet or reducing the meat and dairy intake represents one of the most

[1]Adi Śaṅkara, the pre-eminent commentator of Indian wisdom texts, Upaniṣads, in his commentary to Chāndogya Upaniṣad 7.26.2 states: आहारशुद्धौ सत्त्वशुद्धिः सत्त्वशुद्धौ ध्रुवा स्मृतिः स्मृतिलम्भे सर्वग्रन्थीनां विप्रमोक्षस्तस्मै
See Swāmī Swāhānanda (1996), pp. 546–547.

[2]See Taittirīya Upaniṣad 2.1.2 ओषधीभ्यो अन्नम्: Food comes from vegetation.

effective solutions to global warming. The uniqueness of this approach lies in its humanity and its locus of control: It depends upon each one of us.

The environmental and health risks of industrially produced red meat are well-documented. Research shows that eating red and processed meat increase the risk of cancer and heart disease and that a single individual by simply not consuming meat prevents the equivalent of 1.5 tons CO2 emissions in a year. There is also some evidence that an increase in industrially produced beef correlated with a decline in shareholder returns (Stashwick 2016). Since no living being ever wants to get hurt, the value of harmlessness or its positive form, compassion, is naturally recognized. Many spiritual traditions recommend a plant-based diet based on moral and compassionate grounds. Nonviolence, *ahiṁsā*, is the basis for the vegetarianism within Jainism, Hinduism, and Buddhism though it goes well beyond just being vegetarian. The universal value of harmlessness is a core virtue derived from the Vedic injunction "*mā hiṁsyāt sarvabhūtāni*" मा हिंस्यात् सर्वभूतानि – do no harm to living creatures.

This chapter answers the vital question: What can we all do to make our planet more sustainable?

As a point of departure, let's consider the following alarming statistics:

- Every year in the USA, more than 27 billion animals are slaughtered for food. Raising animals on factory farms is cruel and ecologically devastating.[3]
- By switching to a vegetarian diet, you can save more than 100 animals a year from this misery.[4]
- According to the United Nations, a global shift toward a vegan diet is necessary to combat the worst effects of climate change.[5]
- Producing 1 cal from animal protein requires 11 times as much fossil fuel input – releasing 11 times as much carbon dioxide – as does producing 1 cal from plant protein.[6]
- Of all the agricultural land in the USA, 80% is used to raise animals for food and grow grain to feed them – that's almost half the total land mass of the lower 48 states! On top of that, nearly half of all the water used in the USA goes to raising animals for food (Ibid.).
- Using land to grow crops for animals is vastly inefficient. It takes almost 20 times less land to feed someone on a plant-based (vegan) diet than it does to feed a meat-eater since the crops are consumed directly instead of being used to feed animals. According to the U.N. Convention to Combat Desertification, it takes up

---

[3]For further details, see http://www.humanesociety.org/news/resources/research/stats_slaughter_totals.html?referrer=https://www.google.com/

[4]See https://www.peta.org/living/food/vegetarian-101/

[5]See Lesser consumption of animal products is necessary to save the world from the worst impacts of climate change. Retrieved June 3, 2017, http://www.unep.org/climatechange/

According to UN Environment report, "World must urgently up action to cut a further 25% from predicted 2030 emissions." http://www.unep.org/emissionsgap/

[6]See Climate Change and Animal Agriculture, Explained. Retrieved June 5, 2017, https://www.peta.org/features/climate-change-animal-agriculture-explained/

to 10 lb of grain to produce just 1 lb of meat, and in the United States alone, 56 million acres of land are used to grow feed for animals, while only 4 million acres are producing plants for humans to eat.[7]

- More than 90% of all Amazon rainforest land cleared since 1970 is used for grazing livestock. In addition, one of the main crops grown in the rainforest is soybeans used for animal feed. (The soybeans used in most veggie burger, tofu, and soy milk products sold in the United States are grown right here in the U.S.) (Ibid.)
- Farming and raising livestock take up 40% of the world's land, use 70% of freshwater, and are responsible for 30% of the greenhouse gases pumping into the atmosphere. If every American skipped meat and cheese just 1 day every week, they would cause carbon emissions to decrease by 40 million metric tons, the equivalent of taking 7.6 million cars off the road for a year (Cassidy and Van Hoesen 2015).
- The *U.S. Environmental Protection Agency* shows that animal agriculture is the single largest source of methane emissions in the USA. A staggering 51% *or more* of global greenhouse gas emissions are caused by animal agriculture, according to a report published by the Worldwatch Institute (Goodland and Anhang 2009).
- According to *the U.S. Environmental Protection Agency*, Animals raised for food produce approximately 130 times as much excrement as the entire human population, and animal farms pollute our waterways more than all other industrial sources combined. Runoffs of animal waste, pesticides, chemicals, fertilizers, hormones, and antibiotics are contributing to dead zones in coastal areas, degradation of coral reef, and health problems (also see Foer 2009, p. 174).
- The US food production system uses about 50% of the total US land area, 80% of the freshwater, and 17% of the fossil energy used in the country. The production of one calorie of animal protein requires more than ten times the fossil fuel input as a calorie of plant protein (Pimentel and Pimentel 2003).
- NRDC (Natural Resources Defense Council) estimates that if Americans ate just one less quarter-pound of beef a week, it would be like taking ten million cars off the road for 1 year! (Stashwick 2016).

In the following sections, we will examine the case for plant vs. meat-based diet as one of the most effective resolutions to global warming. The uniqueness of this approach lies in its humanity and its locus of control: It depends upon each one of us. Whenever there is choice, a free will, it *becomes* a *human* problem. Hunger is not a human problem. Thirst is not a human problem. Humans have no choice over these natural desires. But we have full choice over *what* we eat. We can exercise our *free will*, *if* we *chose* to and *want* to.

[7]See Meat and the Environment. Retrieved June 6, 2017, https://www.peta.org/issues/animals-used-for-food/meat-environment/

From observing the animal behavior, we infer that they have no choice over what they eat. Their behavior is instinctually programmed and directed. A cow has no choice over being a vegetarian just as a lion has no choice over being a meat-eater. No one blames a lion for being violent and glorifies a cow for being kind and compassionate. They are just following their instinctual nature. And these animals face no moral dilemma since they have no freedom of choice. One cannot appeal to lions and tigers to consider plant-based diet. As humans, we have a choice – to eat or not to eat meat. How we exercise our choice is up to us.

To Eat or Not to Eat Meat: Addressing Global Warming!

> *It just takes just one second to decide to stop. The main reason not to eat meat and fish is to spare others' life. This is not an extreme perspective. This is a most reasonable and compassionate point of view.* (Ricard 2016a, b)

Matthieu Ricard, French writer and Buddhist monk, in his latest book, *A Plea for the Animals*, provides compelling scientific and moral reasoning for treating all of the animals with whom we share this planet with respect and compassion. He avers that compassion toward all beings, including our fellow animals, is a moral obligation and the direction toward which any enlightened society must aspire. Eating meat reveals another selfishness in terms of other fellow human beings. Rich countries consume the most meat: about 200 kl per year per inhabitant in the USA, compared to about 3 kl in India (Ibid.). The more the GDP of a country increases, usually so does the amount of meat consumption. While the health evidence and sustainability logic is strongly in favor of meatless diet, however, "the main reason to stop eating animals is to spare others' life." Today, 150 billion land animals and 1.5 trillion sea animals are killed for our consumption. We treat them like rats and vermin and cockroaches to be eliminated. This would be called genocide or dehumanization if they were human beings (Ibid.). In a recent post on his blog, Ricard shares even more disturbing statistics:

> Humans kill six million land animals and 120 million sea animals *every hour* for their so-called "needs." That is a lot of animals and a lot killing. In fact, in one week, that makes more killing than all the deaths in all the wars in human history. (Ricard 2017a)

Ricard further notes that "it is estimated that annually 2.5 million dogs and thousands of cats are brutally slaughtered and eaten in South Korea and, throughout Asia, this figure increased to over 30 million." Then he makes an important point that we should treat all animals alike, *i.e.*, with utmost compassion:

> Of course, it is not dogs alone that suffer from our cruelty. Compassion should know no boundaries. Calling for an end to barbarian treatment of dogs, baby seals, and whales does not mean that it is fine to tolerate the mass killing of pigs, cows, and chicken. (Ibid.)

At its barest minimum, what role can each individual play to improve the state of the planet? First and foremost, one can resolve to be well-informed on this issue. To be aware of the extent and veracity of the problem. It is essential to be accurately informed so as to avoid inveterate cluelessness that is rampant in some business circles. Ricard cites a strange statement about the rising level of the oceans by the American magnate Stephen Forbes, who declared on *Fox News*: "To change what we do because something is going to happen in one hundred years is, I would say, profoundly weird. . . .What matters is we sell our meat" (Ricard 2017b). One wonders what could be more bizarre than a statement like this from the head of the largest meat company in the United States?

Ricard rightly diagnoses the malady by stating that short-termism and self-centeredness lie at the root of the problem: "Selfishness is at the heart of most of the problems we face today: the growing gap between rich and poor, the attitude of "everybody for himself," which is only increasing, and indifference about the generations to come." What panacea does he offer? It is just this: "Altruism is this thread that will allow us naturally to connect the three scales of time – short, middle and long term – by reconciling their demands. We must have the perspicacity to acknowledge this and the audacity to say it" (Ibid.). If there is a single inhumane cause that cuts across most human issues in depth and scale, it is perhaps violence – in its overt and covert forms. If there is one value that holds the keys to most problems that humanity is heir to, it is perhaps compassion. The Buddha was right. And so was Gandhi.

A single individual by simply not consuming meat prevents the equivalent of 1.5 tons CO_2 emissions in a year (Ricard 2016a). In a 2006 UN Report entitled, *Livestock's Long Shadow: environmental issues and options*, Steinfeld H et al. note, "Livestock production is responsible for 18% of global greenhouse gas (GHG) emissions from all human activities, measured in CO_2 equivalent" (Steinfeld et al. 2006). According to this report, raising animals for food generates more greenhouse gases than all the cars and trucks in the world combined. Nitrous oxide and methane emissions from animal manure, methane emissions from the animals' digestion, and nitrous oxide emissions from mineral fertilizer used to grow feed crops for farmed animals make up the majority of this 18%. The livestock sector is responsible for the following proportions of global anthropogenic emissions of the main greenhouse gases:

- Thirty-seven percent of total methane (CH_4)
- Sixty-five percent nitrous oxide (N_2O) emissions
- Nine percent of methane (CO_2) emissions (Ibid.)

"The released methane," the UN Report notes, "has 23 times the global warming potential of CO_2" (Ibid.). How can we resolve this alarming situation? What role can each individual play in ameliorating this problem? All that we have to do is avoid eating meat. In the absence of demand for meat, there is no more need for breeding millions of animals for daily slaughter. The reversal of global warming becomes a certainty.

Herein lies the power of this strategy. It depends entirely upon us. By changing our eating habits, we can make this planet more sustainable.

Sustainable Diet: Animal vs. Plant Based?

In their groundbreaking book *Population, Resources, Environment*, Stanford Professors Paul R. and Anne H. Ehrlich state that the amount of water used to produce 1 lb of meat ranges from 2,500 to as much as 6,000 gallons (also see Robbins 2012). Let's say you take a shower every day. . .and your showers average 7 min. . .and the flow rate through your shower head is 2 gal per minute. . .. You would use, at that rate, [5,110] gallons of water to shower every day for a year. When you compare that figure, [5,110] gallons of water, to the amount the Water Education Foundation calculates is used in the production of every pound of California beef (2,464 gal), you realize something extraordinary.

In California today, you may save more water by not eating a pound of beef than you would by not showering for 6 entire months (Robbins 2010, p. 231). According to the U.S. Department of Commerce, Census of Agriculture, "While 56 million acres of U.S. land are producing hay for livestock, only 4 million acres are producing vegetables for human consumption."

One of the cardinal principles of sustainability is that, in the name of progress, we should not *upset* the *setup* carelessly. At its most fundamental level, that entails paying attention to what we eat since our bodies are "food bodies" and we are what we eat. Sri Ramana Maharshi, the great Indian sage of twentieth century, used to say that of all the yogic rules and regulations, the best one is taking of *sāttvic* foods in moderate quantities. This view is consistent with that expressed in the Bhagavad Gītā, and indeed most of the sacred literature of India. According to the Bhagavad Gītā (17.8), *sāttvic* foods are those foods which nourish the body and purify the mind:

> Foods that contribute to longevity, purify one's mind, and provide strength, health, happiness, and satisfaction. Such foods are sweet, juicy, fatty, and palatable. On the other hand, the Gītā (17.9–10) continues, foods which are too bitter, sour, salty, pungent, dry, and hot can lead to pain, distress, and disease of the body.

What is the moral basis of a vegetarian diet? It is the understanding that no living being wants to get hurt or to die. Our self is the dearest of all to us. Love of self comes as a natural endowment that has its roots perhaps in the instinct of self-preservation. An important verse in Bṛhadāraṇyaka Upaniṣad states that we do not love our husband, wife, son, or any other being for their sake, but for our own sake: "It is not for the sake of all, my dear, that all is loved, but for one's own sake that it is loved."[8] However, in our bid to push our self-interest, we often tend to forget the simple fact that, likewise, everyone's self is also most dear to them.

Is there also a metaphysical basis of a vegetarian diet? Metaphysically speaking, all life is one. There is single essential reality that pervades the entire universe and enlivens all beings. According to the Hindu Vedic tradition, all creatures form limbs

---

[8] *na vā are sarvasya kāmāya sarvam priyam bhavati, ātmanastu kāmāya sarvam priyam bhavati*: Swāmī Mādhavānanda (2008, pp. 246–247).

of a single, all-pervading divine being. To benefit any one limb is to benefit the divine being, and to harm any is to harm the integrity of the divine being. Therefore, every one of our actions should be performed for the welfare of all beings. All the great spiritual traditions of India, drawing upon this root idea, dictate that a spiritual aspirant must abstain as much as possible from causing any harm to any living being. However, at the same time, it is recognized that life inherently involves harm of some form or another.

High Cost of Raising Livestock

> *The current industrialized and corporate-led system is doomed to fail. We need a radical overhaul of food and farming if we want to feed a growing world population without destroying the planet.*[9]
> ~Magda Stoczkiewicz Director, Friends of the Earth Europe

Raising livestock for meat comes at a very high cost to the environment. Climate-impacting emissions are produced not just by the animals' digestive systems but also by the fertilizers and manure used to produce feed and the deforestation taking place to provide grazing lands. To add insult to injury, livestock animals consume large amounts of water, agricultural, and land resources that could be deployed to support a higher quality of life for humans.

It is now a well-known fact that compared with vegetables and grains, animal farming requires much more land, water, and energy. From the water and grain needed to feed livestock to the emissions created by huge herds of cattle, farming animals has an enormous negative impact on the environment. Desertification, soil erosion, contaminated groundwater, and greenhouse gas emissions are just a few of the effects caused by raising animals for food (INFOGRAPHIC 2016).

The True Environmental Cost of Eating Meat

> *The current industrialized and corporate-led system is doomed to fail. We need a radical overhaul of food and farming if we want to feed a growing world population without destroying the planet.* (Chemnitz and Becheva 2014)

Research shows that meat and dairy are hiking our carbon footprint (Thean 2011). According to energy expert Jamais Cascio, you can indirectly put up to 1,340 g of greenhouse gas when you order that burger for lunch (Ibid.). And "if a four-person family skips steak 1 day a week [for a year], it's like taking a car off the road for almost 3 months."[10] Of all the agricultural land in the USA, 80% is used to raise

[9]Cited in MEAT ATLAS (2014, p. 7).

[10]See Meat Eater's Guide to Climate Change +Health: Lifecycle Assessments: Methodology & Results. Environmental Working Group. Retrieved July 29, 2017, http://static.ewg.org/reports/2011/meateaters/pdf/methodology_ewg_meat_eaters_guide_to_health_and_climate_2011.pdf?_ga=2.195427749.867980501.1500842024-30149929.1499039227

animals for food and grow grain to feed them – that's almost half the total land mass of the lower 48 states! On top of that, nearly half of all the water used in the USA goes to raising animals for food.[11] According to a report published by the United Nations Food and Agriculture Organization:

- Globally, we consume 308.2 million ton of meat a year.
- Thirty percent of the planet's ice-free land is used for livestock production.
- Twenty-six percent of land is used for animal grazing.
- Thirty-three percent arable land is used for feed crop production.[12]

Researchers have found that the total supply of crop being fed to animals could feed at least four billion people instead. Meat and dairy production are extremely water intensive. A single cow used for milk can drink up to 50 gal of water per day – or twice that amount in hot weather – and it takes 683 gal of water to produce just 1 gal of milk. It takes more than 2,400 gal of water to produce 1 lb of beef, while producing 1 lb of tofu only requires 244 gal of water. By going vegan, one person can save approximately 219,000 gal of water a year.[13]

Partnering with CleanMetrics, the Environmental Working Group (EWG) has conducted a meat lifecycle assessment of 20 popular types of meat (including fish), dairy, and vegetable proteins. Capturing the environmental impacts of meat production at each stage of the supply chain, this assessment provides the full "cradle-to-grave" carbon footprint of each food item based on the greenhouse gas (GHG) emissions generated before and after the food leaves the farm, from the pesticides and fertilizer used to grow animal feed through the grazing, animal raising, processing, transportation, cooking and, finally, disposal of unused food.[14] Many of the EWG's findings are quite eye-opening – like some revealing facts about beef, which produces twice the emissions of pork, four times as much as chicken, and 13 times that of vegetable protein such as beans, lentils, and tofu. That's especially alarming since we waste so much meat – ultimately throwing away about 20% of what we produce – meaning that all that carbon was generated for nothing (see Thean 2011).

Examining about 50 years of data for 100 of the world's more populous nations to analyze global dietary trends and their drivers, University of Minnesota Professor of Ecology G. David Tilman and graduate student Michael Clark illustrate how

---

[11]Climate Change and Animal Agriculture, Explained. PETA/Features. Retrieved July 2, 2017, https://www.peta.org/features/climate-change-animal-agriculture-explained/

[12]According to a new report published by the United Nations Food and Agriculture Organization.

[13]Meat and the Environment. PETA/ISSUES/ANIMALS USED FOR FOOD. Retrieved July 2, 2017, https://www.peta.org/issues/animals-used-for-food/meat-environment/

[14]For details, see http://www.ewg.org/meateatersguide/interactive-graphic/

Also see A comparative Life Cycle Assessment of plant-based foods and meat foods: Assessing the environmental benefits of plant-based dietary choices through: a comparison of meal choices, and a comparison of meat products and MorningStar Farms® veggie products. Prepared for Morning Star Farms by Quantis. Retrieved July 23, 2017, https://www.morningstarfarms.com/content/dam/morningstarfarms/pdf/MSFPlantBasedLCAReport_2016-04-10_Final.pdf "ruminant meats (beef and lamb) have emissions per gram of protein that are about 250 times those of legumes" (David Tilman and Clark 2014). Professor Tilman, the lead author of the study, clarifies: "This is the first time this data has been put together to show these links are real and strong and not just the *mutterings of food lovers and environmental advocates*."

In the study, published in the online edition of *Nature*, these researchers write:

> Rising incomes and urbanization are driving a global dietary transition in which traditional diets are replaced by diets higher in refined sugars, refined fats, oils and meats. By 2050 these dietary trends, if unchecked, would be a major contributor to an estimated 80% increase in global agricultural greenhouse gas emissions from food production and to global land clearing.
>
> Moreover, these dietary shifts are greatly increasing the incidence of type II diabetes, coronary heart disease, and other chronic noncommunicable diseases that lower global life expectancies. (Tilman and Clark 2014)

The researchers indicate that the solution to what the ecologists call the "diet-environment-health trilemma" will require choosing plant-based, whole foods diets: "there would be no net increase in food production emissions if by 2050 the global diet had become the average of the Mediterranean, pescetarian and vegetarian diets" (Ibid.). Additionally, this research revealed that the "same dietary changes can add about a decade to our lives can also prevent massive environmental damage" (Smith 2014).

In a 2014 research report by Chatham House, the Royal Institute of International Affairs, an independent policy institute based in London, Rob Bailey, Antony Froggatt, and Laura Wellesley provide a comprehensive overview of high environmental cost of raising livestock. They also review the findings of the United Nations Food and Agriculture Organization (FAO) over last 10 years. Their research indicates that livestock industry produces more greenhouse gas emissions than all cars, planes, trains, and ships combined:

> Livestock production is the largest global source of methane (CH4) and nitrous oxide (N2O) – two particularly potent GHGs.…The global livestock industry produces more greenhouse gas emissions than all cars, planes, trains and ships combined.…Emissions from livestock, largely from burping cows and sheep and their manure, currently make up almost 15% of global emissions. Beef and dairy alone make up 65% of all livestock emissions. Average global estimates suggest that, per unit of protein, GHG emissions from beef production are around 150 times those of soy products, by volume. (Bailey et al. 2014)

According to a 2006 report by the United Nations Food and Agriculture Organization (FAO), our diets and, specifically, the meat in them cause more greenhouse gases methane (CO2), methane, nitrous oxide, and the like to spew into the atmosphere than either transportation or industry. The FAO report found that current production levels of meat contribute between 14% and 22% of the 36 billion tons of "CO2-equivalent" greenhouse gases the world produces every year (Fiala 2009).

Although experts have known the heavy impact on the environment of meat production but recent research shows a new scale and scope of impact, particularly for beef. The popular red meat requires 28 times more land to produce than pork or chicken, 11 times more water, and results in five times more climate-warming emissions. When compared to staples like potatoes, wheat, and rice, the impact of beef per calorie is even more extreme, requiring 160 times more land and producing 11 times more greenhouse gases, with one expert saying that eating less red meat would be a better way for people to cut carbon emissions than giving up their cars (Carrington 2014). According to Professor Tim Benton, at the University of Leeds, "The biggest intervention people could make towards reducing their carbon footprints would not be to abandon cars, but to eat significantly less red meat" (Ibid.).

In this regard, Environmental Working Group (EWG) recommends the following to reduce climatic and environmental impacts: Make meatless and cheese-less Mondays part of your life[15]; on at least two other days, make meat a side dish, not a main course. Eat "greener" meat when you do eat it. ... When you buy less meat overall, you can afford healthier, greener meat. Eat more plants: Good, low-impact protein foods include grains, legumes, nuts, and tofu. Choose organic when possible. Waste less meat: Buy right-size portions and eat what you buy. *On average, uneaten meat accounts for more than 20% of meat's greenhouse gas emissions*! Eat lower-fat dairy products. Choose organic when possible. Finally, EWG recommends to speak up – ask your representatives to change policies, such as:

- Strengthening regulation of concentrated animal feeding operation (CAFOs) to prevent pollution and unnecessary use of antibiotics and hormones
- Cutting taxpayer subsidies for animal feed and funding programs that support pasture-raised livestock and diversified, organic crop production
- Strengthening conservation requirements on farms that collect subsidies
- Serving less meat and more fresh fruits and vegetable in school lunch programs
- Enacting comprehensive energy and climate policies[16]

Which Diet Is Best Suited for Humans?

Which diet is most suitable for humans – meat-based diet or plant-based diet? While most humans are clearly "behavioral" omnivores, the question still remains as to whether humans are anatomically suited for a diet that includes animal as well as plant foods. One important argument in favor of a vegetarian diet is based on the idea that human anatomy and physiology is best suited to a plant-based diet (see Messina and Messina 1996, p. 16). If you just look at human physiology and compare to other herbivores, we are undoubtedly herbivorous, and we are best suited to eat a whole-

[15]The Meat Free Mondays movement has gained momentum and has now been established in 29 countries around the world. See: MEAT ATLAS (2014, p. 58).

[16]For more details, see http://www.ewg.org/meateatersguide/helpful-tips-for-meat-eaters/ [emphasis added].

food plant-based diet. Our intestine to trunk length (shoulder to butt) ratio matches to other herbivores at about ten to one.[17]

A.D. Andrews in his book, *Fit Food for Men*, presents the following information comparing meat-eaters, herbivores, and humans[18]:

Meat-eaters	Herbivores	Humans
Have claws	Have no claws	Have no claws
Have no skin pores and perspire through the tongue	Perspire through skin pores	Perspire through skin pores
Have sharp front teeth for tearing, with no flat molar teeth for grinding	No sharp front teeth, but flat rear molars for grinding	No sharp front teeth, but flat rear molars for grinding
Have intestinal tract that is only 3 times their body length so that rapidly decaying meat can pass through quickly	Have intestinal tract 10–12 times their body length	Have intestinal tract 10–12 times their body length
Have strong hydrochloric acid in stomach to digest meat	Have stomach acid that is 20 times weaker than that of a meat-eater	Have stomach acid that is 20 times weaker than that of a meat-eater
Salivary glands in mouth not needed to predigest grains and fruits	Well-developed salivary glands which are necessary to predigest grains and fruits	Well-developed salivary glands which are necessary to predigest grains and fruits
Have acid saliva with no enzyme ptyalin to predigest grains	Have alkaline saliva with ptyalin to predigest grains	Have alkaline saliva with ptyalin to predigest grains

In his perceptive essay,[19] "The Comparative Anatomy of Eating," Milton Mills notes that "observation" is not the best technique to use when trying to identify the most "natural" diet for humans. Mills suggests that a better and more objective technique is to look at human anatomy and physiology. Humans are vegetarian by design. Our flat teeth are perfect for grinding grains and vegetables, not for tearing apart animal flesh. Similarly, our hands are designed for gathering, not for flesh-tearing. Our saliva contains the enzyme alpha-amylase, the sole purpose of which is to digest the complex carbohydrates in plant foods. (This enzyme is not found in the saliva of carnivores.) Basically, we have all the right apparatus to consume vegetarian products, and none of the right apparatus for flesh foods. After a detailed comparative analysis of the oral cavity, stomach, small intestines, and colon structure of carnivores, herbivores, and omnivores, Mills, on balance, states that:

> In conclusion, we see that human beings have the gastrointestinal tract structure of a "committed" herbivore. Humankind does not show the mixed structural features one expects and finds in anatomical omnivores such as bears and raccoons. Thus, from comparing the

[17]See Mic, the Vegan, My 'Humans are Herbivores' Video Was Debunked. https://www.youtube.com/watch?v=vQyQS3d86BA

[18]Based on a chart by A.D. Andrews (1970).

[19]See Milton R. Mills, The Comparative Anatomy of Eating. Retrieved October 21, 2015, http://www.adaptt.org/Mills%20The%20Comparative%20Anatomy%20of%20Eating1.pdf

> gastrointestinal tract of humans to that of carnivores, herbivores and omnivores we must conclude that humankind's GI tract is designed for a purely plant-food diet. (Ibid.)

One of the arguments frequently advanced by meat-eaters to explain their food choices is that meat gives the body strength, builds muscle, and so on. However, the evidence proves otherwise. For example, Dave Scott, a U.S. triathlete and the first six-time Ironman Triathlon Hawaii Champion, followed a strict vegetarian diet during his entire training period.[20] Another great example of the power of a vegetarian diet is Hawaii legend Ruth E. Heidrich. Ruth not only overcame the cancer, she went on to become an award-winning, record-breaking triathlete (see Heidrich 2000). Ruth has run six Ironman triathlons, over 100 triathlons, and 66 marathons and won more than 900 trophies and medals since her diagnosis of breast cancer in 1982 at the age of 47!

Likewise, many gorgeous creatures of the animal kingdom explode this myth that meat begets strength, muscle, or size.[21] Some of the big, beautiful, strong, and powerful animals are herbivorous such as elephants, rhinos, hippos, horses, and yaks. They do not seem to have any protein deficiency either.

Plant-Based Diet: A Healthier, Sustainable Course for Future

> For millennia, large segments of the world's population thrived on diets with little or no meat. In the past century, however, the concept of eating meat as the paramount source of protein has become deeply engrained in the psyche and culture of Western countries and now pervades many other cultures and nations. ...This nutritional paradigm has changed in the past few decades as data now support that most plant-based diets are healthier than meat-based diets and yield greater longevity and lower chronic diseases among those who consume vegetarian diets. Furthermore, there is growing evidence linking meat consumption, in particular red meat and processed meat, with detrimental health outcomes. *From a strict health perspective, there is no need to consume meat.* (Sabaté and Soret 2004)

In the past, meatless diets have been advocated on the basis of religious, ethical, or philosophical values, not science. It is only in the past 150 years that empirical evidence has yielded dietary recommendations. The world's demographic explosion and the increase in the appetite for animal foods render the food system unsustainable. Plant-based diets in comparison to meat-based diets are more sustainable because they use substantially less natural resources and are less taxing on the environment. Changing course will require extreme downward shifts in meat and dairy consumption by large segments of the world population.

According to Joan Sabaté and Sam Soret, a sustainable diet has been defined as "protective and respectful of biodiversity and ecosystems, culturally acceptable,

[20]See Dave Scott (triathlete) entry in *Wikipedia*. Retrieved January 24, 2017, https://en.wikipedia.org/wiki/Dave_Scott_(triathlete)

[21]Top 10 Vegan Animals. Retrieved on June 20, 2017, http://www.vegansouls.com/top-vegan-animals

accessible, economically fair and affordable; nutritionally adequate, safe and healthy; while optimizing natural and human resources" (Ibid.). After combing through a large body of evidence on meat vs. plant-based diet and the changing world demographics, the authors conclude that reverting to plant-based diets worldwide seems to be a reasonable alternative for a sustainable future:

> "Going back" to plant-based diets worldwide seems to be a reasonable alternative for a sustainable future. Policies in favor of the global adoption of plant-based diets will simultaneously optimize the food supply, health, environmental, and social justice outcomes for the world's population. Implementing such nutrition policy is perhaps one of the most rational and moral paths for a sustainable future of the human race and other living creatures of the biosphere that we share. (Ibid.)

How a Plant-Based Diet Could Save the Planet

There is a mounting body of research that shows that a plant-based diet could save the planet, besides saving millions of lives and dollars. Recently, a group of researchers at Oxford University published their findings in the Proceedings of the National Academy of Sciences, PNAS, comparing the future effects of three different dietary scenarios out to the year 2050. This study provides a comparative analysis of the health and climate change benefits of global dietary changes for all major world regions. According to the lead author of the study, "There is huge potential from a *health* perspective, an *environmental* perspective and an *economic* perspective, really" (Worland 2016).

Transitioning toward more plant-based diets that are in line with standard dietary guidelines could reduce global mortality by 6–10% and food-related greenhouse gas emissions by 29–70% compared with a reference scenario in 2050. We find that the monetized value of the improvements in health would be comparable with, or exceed, the value of the environmental benefits although the exact valuation method used considerably affects the estimated amounts. Overall, we estimate the economic benefits of improving diets to be 1–31 trillion US dollars, which is equivalent to 0.4–13% of global gross domestic product (GDP) in 2050. Changing diets may be more effective than technological mitigation options for avoiding climate change and may be essential to avoid negative environmental impacts such as major agricultural expansion and global warming of more than 2 °C while ensuring access to safe and affordable food for an increasing global population (Springmanna et al. 2016).

Leo Tolstoy: Vegetarianism and Deep Compassion

> *As long as there are slaughter houses there will always be battlefields.* ~Leo Tolstoy

Leo Tolstoy (1828–1910), perhaps the greatest of Russian novelists, championed vegetarianism and animal rights. He advocated vegetarianism as one of the *first steps* toward a good life and self-restraint. After his conversion to vegetarianism, Tolstoy lived simply on bread, porridge, fruit, and vegetables. Tolstoy's main reason for becoming vegetarian was his conviction that eating flesh is "simply immoral as it involves the performance of an act which is contrary to moral feeling – killing; and is called forth only by greediness and the desire for tasty food" (Tolstoy 1909). Tolstoy believed that "deeply seated in the human heart is the injunction not to take life" and that "Vegetarianism is the taproot of humanitarianism." He averred that eating flesh is not *necessary*, but is only a *luxury*.

It is believed that one of the experiences that made Tolstoy such a confirmed vegetarian and supporter of animal rights was a visit he paid to a slaughterhouse in the early 1890s. He records his experience in *The First Step*, which he wrote as a preface to Howard Williams' *The Ethics of Diet*.[22] Tolstoy regarded the choice of vegetarianism as the first step toward virtue: If a man's aspirations towards a righteous life are serious…if he earnestly and sincerely seeks a righteous life, his first act of abstinence is from animal food, because, not to mention the excitement of the passions produced by such food, it is plainly immoral, as it requires an act contrary to moral feeling, i. e., killing – and is called forth only by greed. (Ibid.)

Leo Tolstoy is one of the first widely known vegetarians. Though a meat-eater in his early life, by the time he turned 50, he decided that it was immoral for someone to kill on his behalf just so he could enjoy a piece of meat for lunch. Although his lovely wife, Sofia, herself never became a vegetarian, she considerately used to cook delicious and nutritious vegetarian dishes for her husband.[23] We are told that Leo Tolstoy's wife Sofia once invited one of her relatives to dinner who was a nonvegetarian. Tolstoy wanted to have only vegetarian dishes for the dinner. At his wife's insistence, and in consideration of their guest's preference, he relented and offered a creative solution. A live rooster was brought and tied to the chair of the guest, with sharp knifes placed next to guest's plate. The idea was that if the guest really wanted to eat meat, she can cut the rooster and they will then cook it for her. When the guest arrived and saw this, she was shocked and could not even conceive of doing something so cruel. And the family did not have to go through the ordeal of cooking meat! One of the strange realities of the industrial farming is that the meat-eaters do not get to see the stark and ruthless process that the animals have to go through. The modern consumer remains blissfully ignorant of what goes behind the closed walls of slaughter houses.

There are at least three main reasons for anyone to turn to a plant-based diet: health, sustainability, and compassion. Research shows that eating red and processed meat increase the risk of cancer and heart disease (Rocheleau 2015). Says Dr. Neal D. Barnard: "The beef industry has contributed to more American deaths than all the wars of this century, all natural disasters, and all automobile accidents combined. If beef is your idea of 'real food for real people' you'd better live real close to a real good hospital."[24] Dr. Greger, who runs the popular website NutritionFacts.org and

[22]It was originally written, in Russian, as the Preface to the Russian translation of *The Ethics of Diet* by Howard Williams, first published 1883, Russian version from 1892.

[23]This recipes are now available in a book form: S. Pavlenko (2016).

[24]Cited in Tallman (2015, p. 19).

serves as the Director of Public Health and Animal Agriculture at the Humane Society of the United States, says that "there is only one diet that has ever been proven to reverse heart disease on majority basis is plant-based diet."[25] However, shifting to a meatless diet for health reasons alone may not give our resolve its stick-to-itiveness that only a deep compassion can.

Our Habits Are Not Us

We are all creatures of habits. And we are told that habits die very hard. Take the word "habit" itself, for example. If we remove "h," "a-bit" remains, and if we remove "a," "bit" remains. And finally, if we remove "b," "it" still remains! And the reason we are not able to see the truth in "it" is because of the crooked, lower case "*i*." In other words, our *little pride* (small "*i*") prevents us from seeing the *big truth* of our habits.

In any change efforts, therefore, it is important to remember that our habits are *not* us. Our habits are the by-products of the tradition we are born in and the sociocultural milieu we grow up in. And separating ourselves from our habits is the first step in growing out of them or overcoming them.

The second factor in overcoming some of our unwholesome habits is to be aware of what psychologists refer to as "confirmation bias." *It is very hard to agree with the truth that disagrees with us; it is even harder to disagree with the untruth that agrees with us*. To understand this, is to guard against our confirmation bias. Generally, the path we take in the formation of our habits is not informed by much logic or research. Once formed, we keep defending our habits impulsively, blissfully oblivious of our confirmation bias. Confirmation bias ensures that when reason is against us, we turn against reason. We do not like to end up on the side of being wrong, given our emotional investment in our habits. Confirmation bias also dictates the *sources* of our search for facts; it determines the information we *select* from those sources, its *interpretation*, and the *conclusions* we draw from the selected evidence. Once we become aware of the operation of our confirmation bias, we begin to become free from it noose.

Our eating habit is also just that, a habit. Mindful of our confirmation bias and armed with more research and awareness, we can make conscious choices about our food – choices that are good for us, good for the environment, and good for countless innocent creatures that get killed mercilessly every day for the sheer gratification of our taste buds. Real change is tough and, of course, seeing our defenses is easier than changing them. We need both engaged humility and patience to see through and overcome our pet habits and beliefs.

[25]VEGAN 2016 – The Film [PART 1] Retrieved November 17: https://www.youtube.com/watch?v=G5ufn_Gy_Ns (see also Greger 2015). In this informative book, Dr. Greger describes which foods to eat to prevent the leading causes of disease-related death and shows how a diet based on fruits, vegetables, tubers, whole grains, and legumes might even save your life.

The Ethics of Eating: A Case for Vegetarianism

> *Lama Tsering Gyaltsen, when asked what the one thing is we could do to further our spiritual path, said, "Eat less meat or give it up all together. It would reduce the amount of suffering in the world and would create a healthier body, mind, spirit."*[26]

Many people are attracted to vegetarianism primarily for health and ethical reasons. There are at least three main reasons for anyone to turn to a plant-based diet: health, sustainability, and compassion. Let's take the health reasons first, for they furnish the most natural motivation. Research shows that eating red and processed meat increase the risk of cancer and heart disease (Rocheleau 2015). A recent report from the WHO's International Agency for Research has indicated that bacon, ham, and sausages rank alongside cigarettes as a major cause of cancer, placing cured and processed meats in the same category as asbestos, alcohol, arsenic, and tobacco (Boseley 2015; also see Wu 2014). Processed meat refers to meat that has been salted, cured, fermented, smoked, or undergone other processes to enhance flavor or to improve preservation, according to the WHO. However, shifting to a meatless diet for health reasons alone may not give our resolve its stick-to-itiveness that only a deep compassion can.

Our dignity as humans should lie in protecting those who are weaker than us. Those who have more power ought to be more kind to those who are weak. All spiritual traditions teach us not to do to others what we don't want to be done to us. The compassionate basis of a vegetarian diet lies in the understanding that no living being wants to get hurt or to die, thus making harmlessness a universal value. Our self is the dearest of all to us. Love of self comes as a natural endowment, instinctually rooted self-preservation. This awareness can help foster "live and let live" way of life.

In today's factory farming system, animals have no legal protection from cruelty, which would be illegal if it were inflicted on our pets. Yet farmed animals are no less intelligent or capable of feeling pain than are the dogs and cats that we cherish as companions. Moreover, this cruelty to animals is not environmentally sustainable. In chapter two, we have seen that raising livestock for meat comes at a very high cost to the environment. In fact, recent research indicates that livestock industry produces more greenhouse gas emissions than all cars, planes, trains, and ships combined (Bailey et al. 2014). Let's be careful not to *upset* the very *setup* carelessly, in the name of progress.

Throughout the history, many great thinkers have recognized the salutary effect of a vegetarian diet on human temperament. For example, we have Einstein's testimony:

> Although I have been prevented by outward circumstances from observing a strictly vegetarian diet, I have long been an adherent to the cause in principle. Besides agreeing with the aims of vegetarianism for aesthetic and moral reasons, it is my view that a vegetarian manner of living by its purely physical effect on the human temperament would most beneficially influence the lot of mankind. (Calaprice 2005, p. 281)

[26]Lama Tsering Gyaltsen, speaking at the Omni Center for Peace, Justice and Ecology, Fayetteville, March 2, 2017. Source: Judi Neal, *Personal Communication*, March 25, 2017

Alan Watts, a British-born philosopher, writer, and speaker, best known as an interpreter of Eastern philosophy for a Western audience, when asked why he was a vegetarian, famously quipped: "I am a vegetarian because cows scream louder than carrots."[27] Life feeds on life, Alan Watts knew very well. But he was awakened to the deeper truth of existence – that all killing involves pain! And one should minimize the pain, as much as possible. To the list of great thinkers and immortals of pen who became vegetarian for reasons of morality, we can add such luminaries as Pythagoras, Plato, Leonardo da Vinci, Leo Tolstoy, Nikola Tesla, Gandhi, Albert Schweitzer, George Bernard Shaw, John Rawls, and Franz Kafka.

Dave Scott is a U.S. triathlete and the first six-time Ironman Triathlon Hawaii Champion (1980, 1982, 1983, 1984, 1986, and 1987) (Watson 2015). During peak training times, his highly regimented routine included cycling 75 miles, swimming 5000 m, and running up to 20 miles every single day. Widely considered to be one of the most difficult one-day sporting events in the world, an Ironman Triathlon format consists of a 2.4-mile (3.86 km) swim, a 112-mile (180.25 km) bicycle ride, and a marathon 26.2-mile (42.2 km) run, raced in that order and without a break within a strict time limit of 17 h. It is reported that in his bid for super-discipline, Dave Scott took his training regimen a few notches higher and used to *rinse his cottage cheese* with water to get extra fat off. What is even more remarkable is that, while training for triathlons, Dave Scott followed a *strict vegetarian diet*.[28]

Another great example of the power of a vegetarian diet is Hawaii legend Ruth E. Heidrich. After she was diagnosed with breast cancer, she switched to a completely vegan diet. With a strenuous exercise routine, a vegan diet, and an affirming mental outlook, Ruth not only overcame the cancer, she went on to become an award-winning, record-breaking triathlete (see Heidrich 2000; see also Heidrich 2013). Ruth has run six Ironman triathlons, over 100 triathlons, and 66 marathons. In 1999, she was named by *Living Fit* magazine as one of the ten fittest women in America. She still actively competes in marathons and triathlons, having won more than 900 trophies and medals since her diagnosis of breast cancer in 1982 at the age of 47.

Among the preeminent medical doctors who follow plant-based diet are Drs. McDougall, Michael Greger, Thomas Campbell, Neal Barnard, and Dr. Elmsworth.

There Are Alternatives

"Going vegan might save millions of humans, trillions of dollars, and maybe planet Earth," declare Drs. Thomas Campbell and Erin Campbell in their highly readable article, *Top 10 Plant-Based Research and News Stories of 2016*. They cite a study done by a group of researchers at Oxford University comparing the future effects of three different dietary scenarios out to the year 2050 in terms of their effects on

[27]Retrieved June 20, 2017, https://answers.yahoo.com/question/index?qid=20110220230000AAK4KO3

[28]See Dave Scott (triathlete) entry in *Wikipedia*. Retrieved March 28, 2017, https://en.wikipedia.org/wiki/Dave_Scott_(triathlete)

global human mortality, greenhouse gas emissions, and economic value of health and environmental benefits. The three dietary patterns were (1) a moderate pattern following dietary guidelines, (2) vegetarian, and (3) vegan.

Although global adoption of any of the three dietary scenarios would be beneficial, the more plant-based the diet, the greater the benefit, notes the study. Drawing upon the Oxford Study, Dr. Campbell and Campbell further provide the following revealing statistics:

> Global adoption of a vegan diet was projected to avoid 8.1 million deaths per year and reduce mortality by 10% for all causes by 2050. Vegan diets were projected to reduce food-related greenhouse gas emissions by 70% of those predicted in 2050. A vegan diet was projected to save $1067 billion USD per year in health-related costs (3.3% of the predicted global GDP) and $570 billion USD per year due to avoided environmental harm.[29]

In 2014, the Heinrich Böll Foundation, Berlin, Germany, and Friends of the Earth Europe, Brussels, and Belgium, jointly produced *Meat Atlas: Facts and figures about the animals we eat*. Atlas invites you to take a trip around the world. This report provides "insights into the global connections made when we eat meat." It rightly avers that "only informed, critical consumers can make the right decisions and demand the political changes needed." It represents one of the most balanced documents which highlights the problem of irresponsible livestock production and suggests positive, insightful policy level solutions, and alternatives.

Notes Barbara Unmüßig President, Heinrich Böll Foundation:

> In many countries, consumers are fed up with being deluded by the agribusiness. Instead of using public money to subsidize factory farms – as in the United States and European Union –, consumers want reasonable policies that promote ecologically, socially and ethically sound livestock production.[30]

The Meat Atlas presents the following lessons to learn about meat and the world:

1. Diet is not just a private matter: Each meal has very real effects on the lives of people around the world, on the environment, biodiversity, and the climate that are not taken into account when tucking into a piece of meat.
2. Water, forests, land use, climate, and biodiversity: The environment could easily be protected by eating less meat, produced in a different way.
3. The middle classes around the world eat too much meat: Not only in America and Europe but increasingly in China, India, and other emerging countries as well.
4. High meat consumption leads to industrialized agriculture: A few international corporations benefit and further expand their market power.

---

[29]Thomas Campbell and Erin Campbell, Top 10 Plant-Based Research and News Stories of 2016, New Letter published by T. Colin Campbell Center for Nutrition Studies. Retrieved July 24, 2017, http://nutritionstudies.org/top-10-plant-based-research-and-news-stories-2016/

[30]Cited in MEAT ATLAS (2014, p. 6).

5. Consumption is rising mainly because city dwellers are eating more meat. Population growth plays a minor role.
6. Compared to other agricultural sectors, poultry production has the strongest international links, is most dominated by large producers, and has the highest growth rates. Small-scale producers, the poultry, and the environment suffer.
7. Intensively produced meat is not healthy – through the use of antibiotics and hormones, as well as the overuse of agrochemicals in feed production.
8. Urban and small-scale rural livestock can make an important contribution to poverty reduction, gender equality, and a healthy diet – not only in developing countries.
9. Eating meat does not have to damage the climate and the environment. On the contrary, the appropriate use of agricultural land by animals may even have environmental benefits.
10. Alternatives exist. Many existing initiatives and certification schemes show what a different type of meat production might look like – one that respects environmental and health considerations provides appropriate conditions for animals.
11. Change is possible. Some say that meat consumption patterns cannot be changed. But a whole movement of people is now eating less meat or no meat at all. To them it is not a sacrifice; it is part of healthy living and a modern lifestyle (Ibid., pp. 8–9).

These lessons are full of engaged hope and provide essential alternatives to unsustainable factory farming. Their analysis is insightful and balanced and the solutions offered are eminently doable. Yes, alternatives exit and change is possible. It is our future and the choice is ours too.

Compasssion: Our Best Gift to the Universe

> *When a scholar named Chou Yu was cooking some eel to eat, he noticed that one of the eels bending in its body such that its head and tail were still in the boiling point liquid, but its body arched upward above the soup. It did not fall completely in until finally dying. Chou Yu found the occurrence a strange one, pulled out the eel, and cut it open. He found thousands of eggs inside. The eel had arched its belly out of the hot soup to protect its offspring. He cried at the sight, sighed with emotion, and swore never to eat eel.*[31]

As is clear from the foregoing, plant-based diet is good for us and good for our planet. The health hazards of eating meat and the high environmental costs of a meat-based diet are very well documented in the current scholarly and popular literature on health and nutrition. In the final reckoning, however, the decision to shift to a plant-based diet hinges on compassion. It depends upon understanding and honoring

[31] *Record of Protecting Life*. Retrieved March 21, 2017, http://centrebouddhique.fr/a-buddhist-perspective-on-vegetarianism/

the principle of *live and let live*. Many spiritual traditions recommend plant-based diet based on moral and compassionate grounds. Nonviolence, *ahiṁsā*, is the basis for the vegetarianism within Jainism, Hinduism, and Buddhism though it goes well beyond just being vegetarian. This core principle is derived from the Vedic injunctions *mā hiṁsyāt sarvabhutani* (do no harm to living creatures) or *hiṁsām na kuryāt* (do not cause injury). This recommendation is also repeated in the Upaniṣads, the Hindu books of wisdom. A commitment to a nonviolent way of life emanates from the profound understanding of the moral and metaphysical basis of life. It is only when one is able to perceive and "realize one's self in the Self of all" can one become nonviolent in the truest sense. The Christian dictum of "love thy enemy as thyself" – because our self is dearest to us – the practice of loving all, including our enemies, as "ourselves," pivots on the realizing the fundamental oneness of all life.

It is true that what we eat is largely governed by our personal beliefs and choices. These choices, being habit-driven, are not always easy to change, even if one is willing. The spirit is *willing*, says the Bible, but the flesh is *weak*. Observation and reflection make it clear that as human beings we are not the most rational creatures when it comes to forming our beliefs and making our choices. If life were rational, nobody would choose to smoke. For some, the decision to become vegetarian happens instantly. They read some study on the risks of eating meat or watch documentary footage of a factory farm, and meat is off their menu for good. For others, the decision may come in fits and starts.

It is an inevitable principle of life that life feeds on life. Our responsibility is to *minimize* it. As a Vedic verse puts it, "Life lives by living off another life" (*jīvo jīvena jīvati*). It is true that vegetarians, too, cause harm by killing plants or using animals to plough the fields, so inadvertently harming other beings in the process of raising crops. However, this seems minimal compared to the routine cruelty that is involved in raising, transporting, and slaughtering animals for food. For want of a nervous system, the plants cannot feel the pain, but the animals can. Like us, these animals can feel the pain and do not wish to be physically hurt or killed.

It is true that no one in reality can have a completely harmless existence. But that does not mean that we should abandon the core value of harmlessness. We must minimize the harm we cause to other creatures as far as possible. The Buddha said, "All tremble at violence; all fear death. Putting oneself in the place of another, one should not kill or cause another to kill" (Acharya Buddharahhita 1985, p. 43). Clearly no one is arguing that Eskimos and others who have no other means of sustenance should adopt a vegetarian diet. However, abstaining from eating meat is possible for nearly all of us, given the choices that the modern life accords. This is the minimum all of us can do.

Concluding Thoughts

According to one estimate, 150 billion marine and land animals are slaughtered every year worldwide by the meat, dairy, egg, and fish industries with cruelty that has no parallel anywhere, not even within the animal kingdom itself. At this rate, the entire human population of the world will be wiped out in less than 20 days! Again, 150 billion animals are ruthlessly killed every year for a sandwich and human greed and gluttony![32] How can we claim to be the "crown of creation?" Perhaps, "bane of creation" is more like it. If one realizes the terror of the situation, living just by the "golden rule" alone – the ethical compass most people use to gauge right from the wrong – meat will be off the table for good. All this suffering and misery is preventable. We can all change what we eat, if we want to. The choice, as always, is ours!

Of course, we cannot appeal to the tigers and lions in the jungle to become vegetarians.

Carnivorous animals are programmed as such by nature. This is not the case with the humans.

Gary Yourofsky, an American animal rights activist and a vegan superstar, is succinct:

> If you put a live bunny rabbit and an apple in the crib of a 2-year-old, let me know when the child eats the bunny rabbit and plays with the apple. *We are purely herbivorous*. We have no carnivorous or omnivorous instincts whatsoever. And physiologically if your jaw moves from side to side in grinding motion when you chew, you are hundred percent herbivorous. If you were a meat eater like lion, your jaw will only go up and down, rip and swallow, then you are a carnivorous. If you sweat through your pores to cool yourself, you are herbivorous.[33]

Besides, animals do not have the awareness to choose differently based on what is right and what is wrong. As humans, we have choices and can certainly choose to become vegetarian/vegan as a healthy decision both for ourselves and for the environment. We can also choose to become vegetarian/vegan out of love, kindness, and compassion. By way of spiritual rationale of vegetarian diet, we present the following excerpt based on author's meeting with a contemporary sage, Muni Narayana Prasad:

Q: Can a Self-Realized person be non-vegetarian?
Muni: A Self-Realized person realizes that the same Truth is in everyone and sees his or her own very self in others. Therefore, s/he cannot harm others, since that will be harming one's own self. Hence, the value of non-harming, *ahiṃsā*.
Q: So, it cannot be otherwise?
Muni: Yes! It is so.[34]

> Taittirīya Upaniṣad 2.1.2, an important wisdom text, says that all food is vegetarian.

Swami Dayananda Saraswati, a contemporary Vedānta teacher-scholar and a vociferous advocate of vegetarian diet, used to aver that though one may have a

[32]Gary Yourofsky, Best Speech You Will Ever Hear (Updated). YouTube video retrieved March 25, 2017, https://www.youtube.com/watch?v=_K36Zu0pA4U

[33]Gary Yourofsky, Vegan Activist destroys Ignorant Reporter. YouTube video retrieved on March 25, 2017, https://www.youtube.com/watch?v=xYP1GGdRMYo

[34]*Meetings with Remarkable People*: Muni Narayana Prasad. Unpublished Interview Transcript: December 22, 2015

nonvegetarian meal, food comes from vegetation and can therefore be vegetarian alone.[35] He maintains that vegetarianism is an expression of nonviolence, *ahiṁsā* (Swami Dayananda 2007, p. 41), and that plant-based food is the rational/ethical choice for human diet (Ibid., p. 44). Every biologist will agree that the primal food chain starts with plants. The reason why the Taittirīya Upaniṣad says food means plants is because only plants have the ability to make their own food through the process of photosynthesis using sunlight to make sugars from methane (C02) and water (H20). A principal difference between plants and animals is the plant's ability to manufacture its own food. All animals either eat plants or eat those animals that eat plants; they have the ability to only *digest* food and not *manufacture* food. Therefore, food means plant-based food only.[36] As we have seen in the foregoing pages, plant food is good for us, good for the planet, and good for all beings.

I remember an incident from my childhood when I fell very sick. My uncle took me to a nearby doctor. The doctor examined me and said to my uncle, "The child looks very week due to protracted fever. Give him some fish curry to eat. It is good for him." I do not know what possessed me at the time, I could not resist saying, "But, doctor, it is not good for the fish!" Once this author heard a sage explain, "I can live without fish. Why bother fish?" Exactly! Why bother the poor fish or a chicken or a cow. Of course, one can find a thousand reasons to rationalize and continue doing what we are doing in terms of one's eating habits. It has been observed that "when the reason is against man, man turns against reason." Choosing not to cause the suffering of other living creatures for the satisfaction of our taste buds and appetites is the minimal expression of compassion we all can offer. It is good for us and it good for the environment too. This indeed is one thing we all can do to make our planet more sustainable.

Any change at a personal or social level requires commitment, hard work, and patience. Sometimes we feel disheartened in wake of the enormity of the undertaking. We feel overwhelmed. We start entertaining thoughts such as "what difference does it make? I am just one person." The following story[37] illustrates that when we choose to follow the right course of action, it *does* make a difference.

[35]Swami Tattvabodhananda narrates the following explanation in this regard:

"In this context, I would like to share what Tattvavidanandaji told us this morning in the Taittirīya Upanishad class where the mantra "ओषधीभ्यो अन्नम्" came up for discussion. He acknowledged the objection that people raise about plants too having life and dismissed it saying that though plants too have life, the objection is misplaced. This is because, ओषधी is explained by our Rishis as "लपरिपाकान्तम्:" 'That whose life ends with its giving its respective ripened fruit.' A paddy crop, for example, dies after it gives us the rice grains. The crop is over after it gives its fruit, which in this case is rice grains. So is the case with wheat and most of the vegetables. Take tomatoes, or sugar cane or even bananas, for example. That is why farmers have to start afresh every year. Therefore, we are not killing to eat, but only eating that whose life is already over." Retrieved June 3, 2017, https://www.facebook.com/groups/KaUSO/

[36]See the discussion on vegetarianism: https://www.advaita-vision.org/vegetarianism-q-327/

[37]Originally written by Loren Eiseley (1907–1977), the story has appeared widely over the Web. This version was prepared by Catherine Ludgate on November 21, 2006.

You Can Make a Difference

Once upon a time, there was a wise man who used to go to the ocean to do his writing. He had a habit of walking on the beach before he began his work.

One day, as he was walking along the shore, he looked down the beach and saw a human figure moving like a dancer. He smiled to himself at the thought of someone who would dance to the day, and so, he walked faster to catch up.

As he got closer, he noticed that the figure was that of a young man, and that what he was doing was not dancing at all. The young man was reaching down to the shore, picking up small objects, and throwing them into the ocean.

He came closer still and called out "Good morning! May I ask what it is that you are doing?"

The young man paused, looked up, and replied, "Throwing starfish into the ocean."

"I must ask, then, why are you throwing starfish into the ocean?" said the somewhat startled wise man.

To this, the young man replied, "The sun is up and the tide is going out. If I don't throw them in, they'll die."

Upon hearing this, the wise man commented, "But, young man, do you not realize that there are miles and miles of beach and there are starfish all along every mile? You can't possibly make a difference!"

At this, the young man bent down, picked up yet another starfish, and threw it into the ocean. As it met the water, he said, "It made a difference for that one."

Cross-References

- ▶ Education in Human Values
- ▶ Expanding Sustainable Business Education Beyond Business Schools
- ▶ Just Conservation
- ▶ Responsible Investing and Corporate Social Responsibility for Engaged Sustainability
- ▶ The Spirit of Sustainability
- ▶ The Sustainability Summit
- ▶ To Eat or Not to Eat Meat
- ▶ Transformative Solutions for Sustainable Well-Being

References

Acharya Buddharahhita (Trans.). (1985). *Dhammapada: The Buddha's path of wisdom*. Kandy: Buddhist Publication Society.

Andrews, A. D. (1970). *Fit food for men*. Chicago: American Hygiene Society.

Bailey, R., Froggatt, A., & Wellesley, L. (2014). Livestock – Climate change's forgotten sector: Global public opinion on meat and dairy consumption. A research paper. London: Chatham House, the Royal Institute of International Affairs.

Berry, T. (1999). *The great work: Our way into the future*. New York: Harmony/Bell Tower.

Boseley, S. (2015). Processed meats rank alongside smoking as cancer causes – WHO. *The Guardian*, October 26, 2015. Retrieved July 20, 2017, http://www.theguardian.com/society/2015/oct/26/bacon-ham-sausages-processed-meats-cancer-risk-smoking-says-who

Calaprice, A. (2005). *The new quotable Einstein*. New Jersey: Princeton University Press; Enl. Commemorative Ed.

Carrington, D.(2014). Giving up beef will reduce carbon footprint more than cars, says expert. *The Guardian*, July 21, 2014. Retrieved July 25, 2017, http://www.theguardian.com/environment/2014/jul/21/giving-up-beef-reduce-carbon-footprint-more-than-cars?CMP=share_btn_fb

Cassidy, E., & Van Hoesen, S. (2015). Eating more veggies: A recipe for sustainability. February 20, 2015. Retrieved June 25, 2017, http://www.ewg.org/enviroblog/2015/02/eating-more-veggies-recipe-sustainability

Chemnitz, C., & Becheva, S. (Eds.). (2014). *Meat Atlas 2014: Facts and figures about the animals we eat*. Berlin/Brussels: Heinrich Böll Foundation/Friends of the Earth Europe. Retrieved July 3, 2017, https://www.foeeurope.org/sites/default/files/publications/foee_hbf_meatatlas_jan2014.pdf

David Tilman, G. & Clark, M. (2014). Global diets link environmental sustainability and human health. *Nature: International Weekly Journal of Science, 515*, 518–522. Retrieved July 6, 2017, http://www.nature.com/nature/journal/v515/n7528/full/nature13959.html

Fiala, N.(2009). How meat contributes to global warming. *Scientific American*, February 1, 2009. Retrieved July 25, 2017, http://www.scientificamerican.com/article/the-greenhouse-hamburger/

Foer, J. S. (2009). *Eating animals*. New York: Little, Brown and Company.

Goodland, R., & Anhang, J. (2009). What if the key actors in climate change are. . .cows, pigs, and chickens? *World Watch*, November/December 2009, p. 11. Full report. Retrieved June 30, 2017, https://www.worldwatch.org/files/pdf/Livestock%20and%20Climate%20Change.pdf

Greger, M. (2015). *How not to die: Discover the foods scientifically proven to prevent and reverse disease*. New York: Flatiron Books.

Heidrich, R. E. (2000). *A race for life*. New York: Lantern Books.

Heidrich, R. E. (2013). *Lifelong running: Overcome the 11 myths about running and live a healthier life*. New York: Lantern Books.

INFOGRAPHIC. (2016). The true environmental cost of eating meat. Retrieved July 28, 2017, http://inhabitat.com/infographic-the-true-environmental-cost-of-eating-meat/

MEAT ATLAS. (2014). Facts and figures about the animals we eat. Retrieved July 23, 2017, https://www.boell.de/sites/default/files/meat_atlas2014_kommentierbar.pdf

Messina, V., & Messina, M. (1996). *The vegetarian way: Total health for you and your family*. New York: Harmony.

Pavlenko, S. (2016). *Leo Tolstoy: A vegetarian's tale: Tolstoy's family vegetarian recipes adapted for the modern kitchen*. CreateSpace Independent Publishing Platform.

Pimentel, D., & Pimentel, M. (2003). Sustainability of meat-based and plant-based diets and the environment. *The American Journal of Clinical Nutrition*. Retrieved June 25, 2017, http://ajcn.nutrition.org/content/78/3/660S.full?sid=e4afbdc0-d16e-4324-99d1-3c567de50774

Ricard, M. (2016a). Why I am a vegetarian, October 17, 2016. Blog Entry. Retrieved July 28, 2017, http://www.matthieuricard.org/en/blog/posts/why-i-am-a-vegetarian

Ricard, M. (2016b). *A plea for the animals: The moral, philosophical, and evolutionary imperative to treat all beings with compassion*. Shambhala: Boston.

Ricard, M. (2017a). A dog for dinner? *Blog entry*, July 25, 2017. Emphasis added. Retrieved July 30, 2017, http://www.matthieuricard.org/en/blog/posts/a-dog-for-dinner

Ricard, M. (2017b). Climate challenges and the audacity of altruism. *Huffington Post*, May 4, 2017. Retrieved July 30, 2017, http://www.huffingtonpost.com/entry/climate-challenges-and-the-audacity-of-altruism_us_590b5462e4b0f71180724208

Robbins, J. (2010). *The food revolution: How your diet can help save your life and the world*. New York: Conari Press.

Robbins, J. (2012). *Diet for a new America: How your food choices affect your health, your happiness, and the future of life on earth*. Novato: H J Kramer.

Rocheleau, M. (2015). In wake of study on processed, red meats, what should you do? *Boston Globe*, October 26, 2015. Retrieved August 20, 2016, https://www.bostonglobe.com/metro/2015/10/26/study-says-eating-red-processed-meats-can-cause-cancer-what-should-you/gHfuGjmhYc3Gat0rDzllNK/story.html

Sabaté, J., & Soret, S. (2004). Sustainability of plant-based diets: Back to the future. *American Journal of Clinical Nutrition, 100*(Suppl. 1), 476–482. https://doi.org/10.3945/ajcn.113.071522.

Smith, C. (2014). New research says plant-based diet best for planet and people. *Our World*, brought to you by United Nations University. Retrieved July 9, 2017, https://ourworld.unu.edu/en/new-research-says-plant-based-diet-best-for-planet-and-people

Springmanna, M., Charles, H., Godfraya, J., Raynera, M., & Scarborougha, P. (2016). Analysis and valuation of the health and climate change co-benefits of dietary change. http://www.pnas.org/content/113/15/4146.full.pdf

Stashwick, S. (2016). Cut the beef – For health, for the environment and, er, for business? *GreenBiz Webcasts*, September 14, 2016. Retrieved June 5, 2017, https://www.greenbiz.com/article/cut-beef-health-environment-and-er-business

Steinfeld, H. et al. (2006). Livestock's long shadow: Environmental issues and options. Rome: Food and Agriculture Organisation of the United Nations. http://www.virtualcentre.org/en/library/key_pub/longshad/A0701E00.htm

Swami Dayananda. (2007). *The value of values*. Chennai, India: Arsha Vidya Research and Publication Trust.

Swāmī Mādhavānanda. (2008). *Brihadāraṇyaka Upaniṣad, with the commentary of Śaṅkarācārya*. Kolkata, India: Advaita Ashrama.

Swāmī Swāhānanda (Trans.). (1996). *Chāndogya Upaniṣad*. Mylapore, Madras: Ramakrishna Math.

Tallman, P. (2015). *The restore-our-planet diet: Food choices, our environment, and our health* (p. 19). CreateSpace Independent Publishing Platform.

Thean, T. (2011). How meat and dairy are hiking your carbon footprint. *TIME*, July 26, 2011. Retrieved July 22, 2017, http://science.time.com/2011/07/26/how-meat-and-dairy-are-hiking-your-carbon-footprint/

Tolstoy, L. (1909). The first step. In A. Maude (Trans.), *Essays and letters* (pp. 82–91). New York: H. Frowde. Retrieved: March 10, 2017, http://www.ivu.org/history/tolstoy/the_%20first_step.html

Watson, J. (2015). Ironman Dave Scott knows what will be on his tombstone. *The Times*, August 2, 2015. Retrieved March 25, 2017, http://www.shreveporttimes.com/story/sports/2015/07/31/ironman-dave-scott-knows–tombstone/30933751/

Worland, J. (2016). How a vegetarian diet could help save the planet. *Time*, March 21, 2016. [emphasis added]. Retrieved June 27, 2017, http://time.com/4266874/vegetarian-diet-climate-change/

Wu, S. (2014). Meat and cheese may be as bad as smoking: Eating animal proteins during middle age makes you a candidate for cancer. *USC News*, March 4, 2014. Retrieved June 22, 2017, https://news.usc.edu/59199/meat-and-cheese-may-be-as-bad-for-you-as-smoking/

Moving Forward with Social Responsibility

Shifting Gears from Why to How

Joan Marques

Contents

Abstract

The Social Responsibility movement has manifested itself through several triggers over the past few decades. These triggers contributed to humanity's collective attention about seriously considering the ramifications of its behaviors affecting society on a short-term and long-term basis, including the unacceptable tendency to engage in practices for mere selfish gain. Reflecting on the positive trend of growing awareness, the Social Responsibility framework, through which individuals and organizations fulfill their moral duty to ensure society's well-being, will be reviewed. Within that context, the guidance standards, ISO 26000, of the International Organization for Standardization (ISO), will be reviewed. All of the above sets the stage of humanity's concerted efforts to safeguard Planet Earth; hence, the "why." In order to explain the "how," this chapter evaluates the following core elements of a Socially Responsible framework: human rights, labor practices, the environment, fair operating practices, consumer issues, community involvement and development, and organizational governance. For each element, an exemplary corporate citizen will be reviewed. In its final stages, this chapter summarizes common leadership traits these corporate society members

J. Marques (✉)
Woodbury University, Burbank, CA, USA
e-mail: Joan.Marques@woodbury.edu

S. Dhiman, J. Marques (eds.), *Handbook of Engaged Sustainability*,
https://doi.org/10.1007/978-3-319-71312-0_4

display toward the future, such as better decision making and more responsible overall management; greater trust from the public due to a better reputation; increased competitive strength; more constructive relationships with stakeholders; stronger innovations due to the connection with a broader base of stakeholders; improved employee retention, morale, and loyalty due to an increased emphasis on worker safety and well-being; increased fairness in trade and elimination of corrupt practices; solid bottom line thanks to improved effectiveness and efficiency; greater longevity of the organization due to its sustainable approaches; and better social ties with individuals, civic, and commercial entities overall.

Keywords
Social responsibility · ISO · Human rights · Labor practices · The environment · Fair operating practices · Consumer issues · Community involvement · Organizational governance

Introduction

Conscious Global Movement

Social Responsibility (SR) is here to stay, and for good reasons! Whether pertaining to the micro (personal) or macro (organizational/global) level, and regardless of the environment on which one focuses, there is an undisputed movement to be detected toward enhanced awareness in the behavioral realm. "Despite the obvious differences, enterprises in business and science confront similar challenges, including the governance of large and disparate organizations, the inculcation and transmission of culture and values, the reconciliation of self-interest and societal interests, and the proper balance between self-regulation and external regulation" (Conley et al. 2015, p. 64). Conley et al. point out that the search for a meaningful concept of corporate social responsibility (CSR) has been going on for more than 25 years now, with the aim of applying this concept through robust self-regulatory regimes. While the efforts have accomplished some tangible good, the CSR movement has also been criticized as a self-serving public relations ploy with a main purpose of using easily manipulated self-regulation to head off coercive governmental regulation (Conley et al. 2015, p. 64).

Nonetheless, the triggers detected for the SR movement over the past few decades have contributed to humanity's collective attention that it is time to become serious about our social behavior, and that it is no longer acceptable to engage in practices for mere selfish gain without considering the short-term and long-term ramifications these practices have on the environment, as well as human and nonhuman stakeholders. One of the most critical of these triggers is the petrifying fact that our current global footprint exceeds our earth's capacity to regenerate by about 30%, and that more than 75% of the human race lives in countries where the national consumption has exceeded the country's bio-capacity (World Wide Fund for Nature 2008).

If humanity continues its contemporary lifestyle, we will need the equivalent of two planets around the 2030s (ASQ and Manpower Professionals 2011). Wilting and van Oorschot (2017) warn that the current global demand for food, wood, energy, and water has damaging consequences for global biodiversity, in general, but most profoundly for animal species. They stress that further policy action is needed to decrease biodiversity loss, as this destructive trend will otherwise continue if current trends in population and income growth are continued. Sadly, the continued decline in global terrestrial biodiversity will incite a concentration of losses in biodiverse but economically poor countries (Wilting and van Oorschot 2017).

Since many of such trends are corporate driven, and prior to briefly examining some concerns in this realm, it may be prudent to first present a definition of Corporate Social Responsibility. Blowfield and Frynas (2005) offer the following definition of CSR:

> "[A]n umbrella term for a variety of theories and practices all of which recognize the following: (a) that companies have a responsibility for their impact on society and the natural environment, sometimes beyond legal compliance and the liability of individuals; (b) that companies have a responsibility for the behavior of others with whom they do business (e.g., within supply chains); and (c) that business needs to manage its relationship with wider society, whether for reasons of commercial viability or to add value to society" (p. 503).

It seems that the focus of CSR is oftentimes rather skewed, especially in developing countries. Amaeshi et al. (2016) report that Corporate Social Responsibility in developing countries, particularly in Africa, focuses more on multinational corporations (MNCs) and less on small and medium-sized enterprises (SMEs), simply because MNCs are (a) considered more powerful and visible; (b) SMEs presumably lack influence and resources; and (c) SMEs are presumed to be more reactive (avoiding irresponsible behavior) and less or not proactive (social activism). Lund-Thomsen et al. (2016) add to this concern that, in developing countries, the role of SMEs is tainted by buyer-driven global value chains, which, on the one hand, promote the introduction of Western-style CSR policies in these developing countries and, on the other hand, undermine labor and environmental standards through cut-throat pricing policies and the threat of relocating orders to other low-cost producers elsewhere in the developing world. This, then, gives rise to the risk that "CSR initiatives in developing country clusters become either an exercise in economic and cultural imperialism or an attempt by local SMEs to greenwash their environmentally and social destructive activities" (p. 22).

It can easily be concluded that such double standards lead to severe discouragement, abuse, and neglect of important constituents in the SR movement. There are, after all, far more SMEs in the world than MNCs. No longer can we overlook or derail midsized and smaller partners in this critical stage.

Fortunately, there is also some good news: awareness is on the rise. A 2009 McKinsey report confirms that almost half of all investment professionals agree that the recent global economic depression has elevated the essence of governance program, two-thirds of chief financial officers agree that environmental, social, and governance programs create value for shareholders, and two-thirds of executives believe that environmental and governance programs will increase shareholder value (Valuing Corporate Social Responsibility 2009). The spread of CSR around the globe is increasingly gaining visibility. More than 90% of the largest companies in major developed markets, such as the United States and the United Kingdom, have adopted the CSR codes of conduct, one of this trend's most prominent tools. However, as is the case in any evolving trend, there is still considerable heterogeneity in approaches to CSR between firms in different countries (Preuss et al. 2016).

One of the common trends to be detected in CSR is employee volunteerism. In recent years, corporate leaders have increasingly gravitated to this trend, not only because it enhances awareness on responsible behavior, but also because it has turned out to be a proven way of successfully attracting and retaining employees, while also improving the corporation's reputation and performance. About 90% of Fortune 500 companies currently have employee volunteer programs, in which employees are supported and/or subsidized to engage in volunteer activities and community outreach on company (Cycyota et al. 2016). Indeed, with the abundance of connectivity and expanded infrastructural alternatives at our disposal, corporate leaders and quality professionals have become increasingly aware of the moral duty that is embedded in performing socially responsible, from considering the environment to selecting their suppliers; from managing their production to operating fairly in the market; and from treating their workforce well to honoring their direct and indirect stakeholders.

In a broad sense, SR is a framework through which individuals and larger entities, such as corporations, fulfill their moral duty to ensure society's well-being. Yet, while intentions may be admirable, perspectives and interpretations differ. There is still a widespread lack of clarity about what it means to be "socially responsible" and how it can be accomplished (Common Ground: Quality and Social Responsibility 2007). This confusion has led to widely diverging results, and had been the reason for the International Organization for Standardization (ISO), with assistance from experts of more than 75 countries (ASQ and Manpower Professionals 2011), to launch an international standard in November 2010, ISO 26000, in order to provide guidelines for SR toward global sustainable development. Within the ISO 26000 context, SR is described as the responsibility of an organization for the impacts of its decisions and activities on society and the environment, through transparent and ethical behavior that contributes to sustainable development, including well-being of society, acknowledges stakeholder expectations, complies with applicable laws and international behavioral norms, and is embedded in the organization's internal and external relationships (ISO 26000: Guidance on Social Responsibility 2010).

Moratis (2017) explains that the ISO 26000 standard includes several recommendations for enhancing the reliability of corporate CSR claims. Part of these suggestions can be retrieved in a dedicated clause called "Enhancing credibility regarding social responsibility." Yet, the standard also implicitly refers to strategies for enhancing CSR credibility in other parts of its text. Moratis (2017) further affirms that ISO 26000 considers stakeholder engagement a critical means of increasing confidence in the fact that the interests and intentions of all participants are considered.

ISO 26000: Shifting Gears from Why to How

As ISO 26000 has rooted itself into the day-to-day vocabularies and practices of quality professionals, the "why" of SR has become increasingly clear: This is a trend that will rather augment with time than ever diminish, because humanity as a whole is increasingly grasping the importance of safeguarding our planet and all its inhabitants for decades, centuries, and hopefully millennia, to come. However, the "how" has turned out to be a little thornier, particularly when evaluating a company's daily processes within a larger performance scope. After all, strategies, markets, resources, relationships, laws, environments, and stakeholders differ for every company and every industry.

One of the question marks that has emerged in past years is how to perceive SR in light of quality. Fortunately, the seasoned quality professional has figured that one out by now: quality and SR complement one another in such a sense, that SR has given more depth than ever to quality, since now not only output but also input and processes are considered in the equation (Maher 2014).

Investing in quality principles such as continual improvement, employee empowerment, and reduction of errors and waste contributes to the overall SR profile of an organization. The strong correlation between quality and SR can also be considered through the lens of current and future outcomes, with quality providing a conceptual approach and supporting tools for analyzing current behaviors and needs, and SR outlining a universal structure toward creating a sustainable future (Robinson et al. 2012).

However, the quality discussion is just the tip of the iceberg, for, as quality professionals dive deeper into the operational side of SR, they fully understand the bulwark of challenges a complete and sound SR implementation brings. Setting up an SR framework means including the core elements of human rights, labor practices, the environment, fair operating practices, consumer issues, community involvement and development, and organizational governance (American National Standard 2011). Each of these elements is essential yet intricate in itself. Let us find out how some corporations have dealt with these core elements.

Human Rights

These are the foundational rights for all human beings as emphasized by the international community in the International Bill of Human Rights and its core instruments. Human rights are generally divided into two categories: (1) pertaining to civil and political rights and (2) concerning economic, social, and cultural rights (American National Standard 2011). Both categories aim to ensure human dignity and proper quality of life. Organizations have the power to affect human rights by virtue of their negotiations, as well as their economic and social influence. Among the many factors organizations should consider in this light are, for example, due diligence and responsible risk management, avoiding condoning crimes or discrimination, resolving grievances and observing civil rights in order to positively, rather than negatively, impact human rights. Including human rights in the SR strategy is therefore critical.

How Organizations, Leaders, and Quality Professionals can Fulfill this Element
Fulfilling the human rights requirement within an organization can be complex, because, as indicated above, there are many angles to consider. However, once a culture of stakeholder inclusion and respect is incorporated into the structures, this element will no longer be a source of concern. Some useful ways to consider in fulfilling this element are

- Creating a member-rotating, interdepartmental committee with involvement of a Human Resources representative, to meet regularly and consider the organization's internal and external performance throughout its production cycle in observation of human rights. The committee should not only assess current activities, but also proposed ones, to make sure human rights are not affected. This committee could further review issues that have emerged, propose solutions, and build structures to prevent recurrence of problems in this area. Consideration of social, cultural, and political climates should not be excluded. The rotating aspect is needed to prevent the committee from becoming too comfortable in its ways and from developing groupthink.
- Ensuring a direct communication line with top management to solidify swift and efficient strategic action where this needs to be taken.
- Avoiding connections and partnerships with entities that engage in human rights abuse.
- Ensuring that the company does not fall into the pattern of placing profit over people: This is such an easy trap to fall into, so it will take conscious examination of internal processes, but also those entities that a company outsources parts of its productions to, in order to make sure that people are not underpaid, discriminated against, overworked, or being subjected to glass or pink ceilings.
- Depending on its size, the organization can instate a Chief Diversity Officer or Diversity Director to maintain a legal and sensible approach in equal treatments of internal and external stakeholders.

CSR Case Study: IKEA
IKEA, the Swedish furniture company with a global presence, has been known for quite some time for its socially responsible approach, not only in production, but in all aspects of its operation. "IKEA is a good example of a 'globally integrated enterprise', with a business model that integrates economic, environmental, and social perspectives as a basis for excellence, innovation, and sustainability" (Edvardsson and Enquist 2011, p. 536). While some might consider IKEA to be a product retailer, the company prefers to view itself as a service provider – because its focus is not necessarily on the furniture it sells, but rather on 'solutions to real-life problems' and making a contribution to a 'better life' for the majority of people (Edvardsson and Enquist 2011). This is directly linked to the vision of Ingvar Kamprad, IKEA's founder, "to create a better daily life for the many people" (Morsing and Roepstorff 2015, p. 400). In order to realize this vision, IKEA International's CSR policy focuses on diversity as a value to signify the company's contribution to the best of the many people in a number of ways (Morsing and

Roepstorff 2015). Thus, IKEA perceives its physical products more as a platform for a service experience, resulting in customer value. Included in this perception is also the notion of sustainability: IKEA's current CEO and President, Peter Agnefjäll, reports in the company's 2016 sustainability report that IKEA prides itself in designing products that are beautiful, functional, high-quality, affordable, and sustainable, so that it adheres to the well-being of its customers, the planet, and its own performance (IKEA Group Sustainability Report 2016).

Yet, there is more to IKEA than merely its product delivery and the way the company perceives this. For several decades now, IKEA has been active in supporting vulnerable groups to strengthen themselves and become economically and socially self-reliant. IKEA is very vocal about banning child labor and is strongly involved in UNICEF projects, but also in self-initiated endeavors. For example, in the early 1990s the company's representatives became aware of the serious impact of child labor trends in the carpet weaving belts of India. Merely funding schools was useless, as could be seen from the many sponsored but empty school buildings in those days. This led to IKEA's "Carpet Project" in 2000, a creative solution of enabling the child-laborer's low-caste mothers to form self-help groups and pay off the loan sharks, upon which they could place their children back in school to obtain a decent education and have a chance on a better future (Luce 2004).

In 2017, the IKEA Foundation announced its new three-year partnership with Reach for Change, a nonprofit organization that was launched in Sweden in 2010, focuses on improving children's lives through social innovation, and has a presence in about 18 countries on three continents, providing global support to more than 400 entrepreneurs. The partnership between the IKEA Foundation and Reach for Change intends to assist 120 Ethiopian entrepreneurs in their efforts to develop social enterprises aimed at improving the health, education, and protection of children in Ethiopia, which is Africa's second most populous country. To this end, the IKEA Foundation donated €3.3 million grant to roll out accelerator, incubator, and rapid scale programs and provide supports to Ethiopian social entrepreneurs through a social enterprise forum. More than 84,000 Ethiopian children will benefit from this partnership, which will thus provide a major boost to the country's social enterprise sector. More specifically, this grant aims to ensure better health, better school performance, and protection from trafficking and abuse for Ethiopian children.

A review of IKEA's 2014 and 2016 sustainability reports highlights a variety of ways in which the company maintains its socially responsible mindset. Seven decades of continuous improvement and focus on doing the right thing have led IKEA to become a leading global force in all elements of SR: From granting major donations to children, women, refugees, and other vulnerable groups, to developing a "People and Planet Positive strategy" in which customers, co-workers, suppliers, and other stakeholder groups are carefully considered within the company's ongoing developments. IKEA remains alert about the resources it uses. It has strong and clear policies on reducing waste, applying energy efficiency, and delivering energy-efficient products. The company invites co-workers' and customers' input in new product developments, encourages co-workers to engage in sustainability movements, complies with forestry standards, and ensures that suppliers do the same. In fact, IKEA has been working steadily toward deforestation, and has established partnerships with sustainability-focused global entities, while also focusing on product safety, reducing the company's carbon footprint, and inspiring co-workers to do the same. IKEA has not stopped with educating and involving its suppliers: The company is also working toward empowering people within the company's extended supply chain to create better lives for themselves. IKEA admits, learns, and improves from past mistakes (IKEA Group Sustainability Report 2014). Mindful about climate change, IKEA's Chief Sustainable Officer, Steve Howard, alludes to the Paris Agreement of climate change and the Sustainable Development Goals, thus expressing support toward a cleaner, fairer world. He also underscores IWAY, which is IKEA's code of conduct for suppliers, to ensure better quality of lives for the employees of its business partners. Another important point of progress is the company's resource management, with currently, 61% of its wood and 100% of its cotton coming from sustainable sources. In its support of a waste-free world, IKEA's leadership team promotes a circular economy in which waste becomes a resource for new products, and more products are made with recycled material, while the company also delivers over a million spare parts to enhance durability of its products, hence, less waste (IKEA Group Sustainability Report 2016).

In 2015, IKEA Switzerland, with its 3,000 co-workers, became the first company worldwide to achieve "LEAD," the premier certification of EDGE, a global standard for gender equality. EGDE assessed IKEA's Swiss operations on basis of equal pay for equivalent work, recruitment and promotion, leadership development training and mentoring, flexible working and company culture (Dunn 2015).

Labor Practices

Labor practices are inevitable in the realm of organizational performance: At the very inception of a company there is work to be done, whether the company's output is product or service based. Labor practices encompass more than just policies and practices of a company toward its current workforce. They comprise of all work-related relationships a company maintains, all the way to the subcontractor's levels. Within labor practices are also included the nature of work, the working conditions, grievance procedures, training and skills development, co-workers' health, work safety, schedules, remuneration, recognition, and involvement (American National Standard 2011).

How Organizations, Leaders, and Quality Professionals can Fulfill this Element

The first thought that enters most minds when thinking of labor practices is the Human Resource department. Indeed, this department should be the center of responsible labor practices, but it mostly does so from a legal standpoint. The topic of labor practices is too comprehensive and too important to end with mere compliance with local laws. Human beings are sensitive, and their sense of satisfaction and meaning often depends on more than just a correct legal environment. Here are some additional approaches to consider:

- In a socially responsible organization, every department manager should take on the responsibility to ensure the most optimal utilization of compensation, meaning, respect, development, opportunities, job security, reduced stress and conflict, and all other aspects that make labor practices rewarding.
- Department managers could consider meeting every quarter to share best practices, learn from each other's problem areas, help brainstorm toward constructive solutions, and plan work cycles tactically toward greater consistency and less volatility.
- Within each department, co-workers could form "meaning pods," which can consist of small groups of creative thinkers that suggest ways to make their work more interesting (within the established parameters, of course), and find inexpensive but meaningful ways to celebrate birthdays, anniversaries, and other joyful moments.
- Quality professionals could consider an annual get-together for external constituents such as suppliers or independent contractors, to kindle relationships, enhance mutual appreciation, and increase collaboration.
- The HR department, along with quality professionals, could invite external parties from either government or other involved entities, to discuss new trends and attention areas in regards to labor practices.
- In regards to health and safety, an interdepartmental team, as suggested in the Human Rights section, might be a useful body to detect, address, thus safeguard co-workers from hazards in that regard.
- Under labor practices also fall nondiscrimination, opportunities for people with disabilities, and training opportunities. While formally HR issues, socially responsible organizations can add value to these topics by making them part of regular topics of attention in an interdepartmental SR team.

CSR Case Study: Columbia Sportswear

Presenting Columbia Sportswear as an example of a company that engages in SR with special emphasis on labor practices was driven by the way the company is portrayed through multiple online channels. Before getting into the accolades, however, it should be stated that this company could evaluate the serious lack of diversity in its list of officers and directors, with 21 of the 22 officers being Caucasian and one Asian, and only four of these 22 officers being women, one of them being the Chairperson of the Board. These statistics give rise to a slight sense of discomfort as to why the company's integrative labor practices did not reach the top echelons of the organization.

Nonetheless, Columbia Sportswear has made its way into some praiseworthy initiatives, such as developing a set of Standards of Manufacturing Practices" (Standards of Manufacturing Practices 2015), to which all collaborating suppliers and subcontractors have to comply. The company regularly monitors implementation of these standards, in which it delineates (a) having inspection rights on working premises of suppliers to inspect working conditions, (b) abandonment of any type of forced labor, (c) compliance with local laws on minimum labor ages, (d) respectful treatment of co-workers, including safe and harassment-free work environments, (e) dismissal of any kind of age, race, gender, ability, status, or other type of discrimination, (f) recognizing co-workers' freedom of association toward improved circumstances, (g) compliance with legal wage laws as well as overtime rates, (h) humane and legally sound approaches in working hours and conditions, (i) observing decent safety, nourishment, and health conditions in the work premises, (j) compliance with applicable environmental laws, and (k) legal and ethical business conduct.

Aside from ensuring proper work environments in immediate and distant work environments, Columbia Sportswear is engaged in a number of uplifting labor-based projects, such as the HERproject, an organization that focuses on women empowerment all over the world. HERproject, operating in developing countries in South and Middle America, Africa, and Asia, supports low-income women through workplace-based programs, capacity building, and advocacy with business and government (Where HERProject Works N/A). Columbia Sportswear is also known to actively support the Skin Cancer Foundation and Mercy Corps, as well as organizations that focus on the wellness of children, especially in the community where its headquarters are located. Aside from the immediate human support aspect, Columbia Sportswear also contributes to large and small environmental organizations. Sports products that are out of the top quality range are donated to charity. Among the variety of socially responsible activities in which Columbia Sportswear engages, is its involvement in conservation and restoration of marine ecosystems. In 2010, Columbia Sportswear Company joined forces with the Ocean Foundation to protect and restore seagrass meadows. Seagrass is responsible for about 15% of total carbon storage in the ocean. Unfortunately, human activities have caused the loss of one-third of the world's seagrass meadows, and Columbia Sportswear Company was the first company in the outdoor industry to acknowledge the importance of seagrass to marine ecosystems, our economy, and quality of life. Along with the Ocean Foundation, Columbia enables the education of thousands of online visitors about the importance of seagrass beds in protecting our oceans and our planet, as well as how to safely travel through them. While the initial donation was not large, about $10,000, the initiative can be applauded given its awareness importance (Columbia Sportswear Company Teams Up … 2010).

As was apparent with IKEA, maintaining an SR framework is rather comprehensive in nature, and Columbia Sportswear also demonstrates that. The company maintains solid approaches toward human rights, labor practices, the environment, fair operating practices, consumer issues, community involvement and development, and organizational governance. As part of its environmental cognizance efforts, Columbia Sportswear has started to apply energy-conscious structures at its Portland headquarters, such as a solar electric system that reduces the organization's carbon emissions by 80 t per year (English 2013).

The Environment

Environmental concern has long passed the stage of being considered an overstatement. While, not even three decades ago, caring for the environment was perceived as emotional softness, governments of most countries around the world are now supervising the impact of organizations on the environment, and an increasing number of corporate leaders have started to acknowledge and embrace actions to comply and support legislation (Hopen 2010).

Depletion of natural resources has become a real challenge, which many of us experience on a daily basis. Corporations are facing immense moral dilemmas related to pollution, climate change, deprivation of human settlements, destruction of habitats, loss of species, and the collapse of entire ecosystems (American National Standard 2011). The continued growth of the human population is a major contributor to the above concerns, causing ever-increasing consumption, and, hence, even more environmental challenges. As trendsetters and employers of human populations worldwide, corporations have to take on a moral leadership role in reducing unsustainable patterns, not only with the aim to comply with legal regulations, but from the stance of creating improved standards toward environmental protection.

How Organizations, Leaders, and Quality Professionals can Fulfill this Element

Because we all depend on the environment, the element of environmental responsibility cannot merely be classified as one for organizations, leaders, and quality professionals only. This SR element should be shared with co-workers at every level: they should be encouraged to engage in projects that warrant a reduced toll on the environment at the company level, but also at the personal level. Just as IKEA is encouraging its co-workers to be mindful about their transportation means to and from work, and Columbia Sportswear installed a solar electric system that reduces the organization's carbon emissions, so can others also make environmental sensitivity part of the fabric of their organization. In regards to the environmental element, organizational leaders and quality professionals could therefore

- Organize workshops for co-workers and other interested stakeholders (customers, suppliers) to rethink their own carbon footprint. Including co-workers in the leadership of these workshops might be a good idea, as this will boost morale and increase a sense of ownership.
- Encourage co-workers to think of ways in which the company can be more environmentally responsible. A good way of moving into this direction is establishing an environmentally conscious task force, with members from various departments, but particularly the production units, and with a clear communication line to top management. The task force should not only focus on improvement of ongoing processes, but also on precautionary practices.
- Identifying environmentally-oriented projects in the community, and encouraging co-worker teams to participate for a few days every year at full pay by the organization. Co-workers should also be allowed to submit suggestions of projects they would like to support.

CSR Case Study: Interface

In 1973, Ray Anderson founded Interface, a company that would grow out to become the world's largest carpet tile provider. Little did he know, or even care at that time, that he would once be called "the greenest chief executive in America" and receive numerous awards for being a model environmentally conscious leader ($5 million commitment names … 2015). In the first two decades of its existence, Interface was highly profit oriented. Yes, Anderson complied with the legal prescriptions of corporate performance, but, as was customary in those days, was not really concerned about the environmental effects of his corporation's activities. It all changed, however, when in 1994, a team of co-workers started forwarding him questions from customers about Interface's environmental vision, which was completely absent at that time. As Anderson got confronted with these probing questions, he got confronted with a series of books, such as Paul Hawken's "The Ecology of Commerce" and Daniel Quinn's "Ishmael" about humanity's destructive effects on planet earth, and they provoked a complete paradigm shift within Anderson. Anderson realized the immense crime so many businesses commit to our environment without being punished, and decided to become the change he wanted to see in the world. He started his mission of making carpets sustainably, something that only gradually transformed from a prior "impossibility" to an achievable dream, because carpet production, by default, is highly destructive to the environment (Langer 2011).

For the next 17 years, Anderson operated on many fronts to enhance awareness for environmental sustainability: He worked internally, with his co-workers and his suppliers, but also externally through presentations, books, and articles, to encourage other CEOs in doing the same. He got encouraged to do all this after reading that the same source that caused the destruction through a "take-make-waste" approach (business), could also be the initiator of a restoration of the crisis in the biosphere (Anderson 2007). The internal project toward increased environmental respect was implemented by a task force and was called "Climbing Mount Sustainability." The plan consisted of seven focus points: (1) eliminating waste, (2) eliminating toxic substances from products, vehicles, and facilities, (3) operating facilities with renewable energy, (4) redesigning processes and products to solidify a more responsible production cycle, (5) enhancing efficiency to reduce waste and emissions, (6) creating a culture that integrates sustainable principles and engages all stakeholders therein, and (7) creating a new business model that demonstrates and supports the value of sustainability-based commerce (Anderson 2007).

Through the ups and downs in the sustainability journey, Interface has managed to move from using less than 1% of its raw materials from recycled and renewable sources, to 49% (Davis 2014). Not all efforts were rewarded. Some processes, which initially seemed exciting and progressive, turned out to be mere enlargers of the company's environmental footprint, and had to be discontinued. Similarly, some miracles surfaced, of which the Interface team had never expected to see the light. One year before his passing, Anderson wrote a reflection on the role of businesses and industries in environmental sustainability, in which he related the achievements of Interface Inc. for the zero environmental footprint goal. At that time, Interface had decreased greenhouse gas emission by 44% and cut water use by 80%. Anderson strongly believed that business is the major cause and solution for environmental degradation. In his article, Anderson asserted that he would be one of the business people who would continue to exert efforts for sustainability (Anderson 2010).

Ray Anderson passed away in 2011, but Interface's mission to become fully sustainable by 2020 is still fully in progress. Anderson's incessant efforts to raise awareness on CSR during the last two decades of his life have not gone unnoticed. In 2013, 2 years after his passing, US Green Building Council (USGBC) instated the annual Radical Industrialism Award in 2013 in honor of Ray Anderson, for having been a corporate sustainability pioneer who was influential in the development of LEED green building certification. The award is sponsored by the Ray C. Anderson Foundation, and is granted each year to a leader in the manufacturing sector whose commitment to and achievements in sustainability exemplify Ray's vision, integrating sustainability into the very heart of their company (Colgate-Palmolive 2015).

Fair Operating Practices

Regardless of the type of market in which an organization performs, there will be other entities to collaborate with: partners, suppliers, contractors, customers, competitors, and the like. A company that operates from a fair practices standpoint engages in ethical and respectful market participation, free from corruptive and other types of malicious behavior (American National Standard 2011). Behaving morally is not always easy: Any business participant can attest to that. It is much easier to cut corners, deceive, or engage in false promotion. Yet, in the long run, the courage of doing the right thing pays off by way of a strong reputation, a respected position in the industry, and, most of all, the satisfying awareness of having done the right thing.

How Organizations, Leaders, and Quality Professionals can Fulfill this Element

In order to maintain a clear view of the company's performance, it is critical to meet on a regular basis (quarterly or bi-annually) and review the major strategic decisions that have been made, those that are about to be made, and how they interact with the company's code of ethics. That being said, there has to be a code of ethics, which differs from the mission and vision statement. A code of ethics is a specific document that underscores the company's moral beliefs. It is a living document that has to be assessed annually and, where needed, updated. As will be highlighted in the case below, having a task force that does this annual code of ethics review is a good idea. The task force should preferably consist of organizational members from different levels and departments, and preferably alternated annually. It may also be prudent to instate a different committee to assess the company's SR attainment on an annual basis. Participation of or a direct communication line to top management is essential for these committees to perform effectively.

CSR Case Study: The SAS Institute

The SAS Institute is a privately held software service corporation, founded in 1976 by Jim Goodnight, and located in Cary, North Carolina. The company is the largest independent vendor in the business intelligence market, with sales subsidiaries in about 140 countries. What makes this company so special is the inception of an SR program, long before this term entered the buzz-word spectrum. SAS has been consistent in its efforts to achieve a three-prong leadership base: (1) being the best employer, partner, and vendor to its stakeholders, (2) being committed to high ethical standards in all its dealings, and (3) being proactive in discovering and developing ways to improve in all areas that matter.

It should therefore come as no surprise that this company holds an exceptional reputation and continues to score high on the lists of best companies to work for and most ethical companies to work with. SAS maintains durable relationships with its suppliers, both direct and indirect, and prides itself in high moral standards toward all constituents: colleagues, customers, suppliers, and competitors. As an example of its ongoing efforts in that regard, SAS has an anticorruption program that includes an online anticorruption course for all new co-workers and certain third parties. In a similar vein, the company educates direct stakeholders about bribery, gift giving, fair marketing, and donation policies.

SAS maintains a CSR Task Force, in which co-workers from different departments serve. The task force meets every other month and discusses everything that is important in light of the company's SR behavior. There is a direct communication line with top management to ensure effectiveness and swift action. The company ensures proper preparation of new employees, as well as some contract workers, by requiring some or all of the following training: Code of Ethics, Information Security, Export Controls Awareness, Respect in the Workplace, Global Anti-Corruption, and Privacy and Data Protection for Global Companies. Additionally, all employees and some contract workers have to take refresher ethics training (SAS Corporate Social Responsibility Report 2016).

Another useful approach from SAS is its active participation in public policy discussions, especially those related to the industry in which the company operates. This involvement keeps the SAS team on top of new developments, and positively affects the long-term decisions being made (SAS Corporate Social Responsibility Report 2016). The SAS Code of Ethics was recently updated as a result of its annual review process (SAS Code of Ethics 2016). SAS Institute holds the proud significance of being a pioneer in two critical fields: One is its source of service, analytics, and the other is on workplace culture. The attractiveness of the SAS Institute for prospective employees is the fact that the company ensures meaningful work through excellent leadership and a world-class work environment. Since Fortune started its "100 Best Companies to Work for," the SAS Institute has consistently been part of the list. The fact that the company's leaders understand the direct correlation between SAS Institute's constructive employee culture and its uninterrupted business success, manifested in 40 years of incessant revenue growth and profitability, may serve as an indication that this corporation is one to keep on the radar when responsible corporate behavior is of interest.

Consumer Issues

A stakeholder group that is particularly important for existence and growth of any corporation is its consumer base. While this seems to be clear, it has come to light time and again that this group is oftentimes gravely betrayed and exposed to products or services that reduce rather than enhance the quality of consumers' lives. The SR element pertaining to consumer issues focuses on safeguarding consumers' well-being through fair and factual marketing, protecting their health and safety, ensuring sustainable consumption, providing decent service and support, protecting their privacy, providing them access to services needed, and enhancing their awareness through education (What are the core subjects of social responsibility N/A).

How Organizations, Leaders, and Quality Professionals can Fulfill this Element

Consumer well-being is an issue that cannot be achieved in silos. It has to be a collective and continuous effort, simultaneously applied in multiple organizational layers. Some examples of these actions:

- Enabling communication mechanisms from product and service consumers with the company's leadership or quality team.
- Educating the marketing department on fair and honest information to consumers, and verifying this through assessment on a regular basis. This assessment can be done by the SR committee.
- Encouraging co-workers at all levels to interact with consumers, and create a communication line (possibly monthly sessions) in which co-workers are encouraged to share their viewpoints, ideas, and lessons learned.

CSR Case Study: Trader Joe's

Trader Joe's easily comes to mind when thinking of a company that invests efforts in its consumers. After all, this company is known for taking note of customers' suggestions and stocking stores with requested products (Couch 2013). In addition, Trader Joe's has been a trendsetter in placing allergy labels on products to warn consumers, and giving local products a boost even before they reach the national market. The company is not a health food store chain by name but has been highly responsive to consumers' requests for eco-friendlier, responsibly harvested food. Trader Joe's is also involved in taking care of the less fortunate in its communities: products that are no longer suitable for sale but still good for consumption are donated by each store's designated donor coordinator to food banks, food pantries, and soup kitchens for further distribution (Brown 2013).

A review of Trader Joe's CSR practices shows that the company has more of a localized than a general approach in this, because it heavily focuses on local communities for its SR-related activities. "Trader Joe's CSR practices are centered [on] Michael Porter's concept of creating value for both the business and the consumer" (Trader Joe's LA 346 2017). Even though this store consists of more than 350 stores nationwide, it is still privately held, and information is therefore not readily accessible. Yet, there are some great things about this company that are well-known within the community, for instance, that Trader Joe's does not charge slotting fees for product display on its shelves, unlike most grocery stores (Martin 2015). The savings are shared with the consumer. Co-workers' salaries are within satisfactory ranges, with decent benefits. Trader Joe's is also known to refrain from advertising, and only distributing its "fearless flyer" in local communities as a means of promoting its products. Keeping such expenses down also leads to savings shared with consumers. Aside from plain donations, the company also participates in an annual ladies' golf tournament, securing donations to three Florida food banks for each "birdie" (Trader Joe's LA 346 2017).

A 2013 consumer study conducted by Market Force, in which 6,645 men and women in the United States and Canada were interviewed, found that Trader Joe scored higher than any other grocery chain in regards to customer satisfaction. Participants to this consumer study submitted various reasons for their notion of satisfaction at Trader Joe's, varying from the stores' atmosphere and fast checkouts, to cleanliness, courteous staff, merchandise selection, and accurate pricing. In particular, it was the company's atmosphere and courteous staff that made the golden difference for Trader Joe's (Anderson 2013). DiSalvo (2015) adds some more reasons for Trader Joe's to be considered consumers' favorite: Sensible prices and fresh groceries, and the stores' design, which makes a happy-home-like impression with its combinations of cedar, brick, and bamboo. Just like other authors, DiSalvo also discusses the employees at Trader Joe's, whom he describes as a "certain kind of person": People who are truly engaged, and who perform cross-functionally: they are not limited to one task or one department. They really seem to care, and they exude friendliness. They do so, because they are treated well. They eat and drink in their workplace and are encouraged to try the new products, which turns them into passionate fans of those products, and therefore, authentic, infectious sales people (DiSalvo 2015). All of the above may explain why Trader Joe's, with its workforce of 8,000, is listed as number 16 on Forbes' 2016 list of "America's Best Employers."

Due to insufficient direct report from the company, we can only make a cautious assessment regarding Trader Joe's consumer issues strategies; however, there are still enough testimonials available to state that, in recent years, Trader Joe's has displayed an outstanding attitude toward consumer involvement, protection, and support.

Community Involvement and Development

Inasmuch as human beings have an interdependent relationship with their community, so too do organizations. Today, it is generally understood and accepted that companies are not only existing for the purpose of making money off their customers, but that they should also be involved in their communities (American National Standard 2011). An organization that participates in development and growth of its community is generally more respected and appreciated, hence, more patronized. Community involvement and development are therefore perceived as important elements of a company's SR. This element can focus on multiple aspects of community well-being, such as education, culture, health, employment, wealth creation, and social investment (What are the core subjects of social responsibility N/A).

How Organizations, Leaders, and Quality Professionals can Fulfill this Element

The suggestions presented below were listed earlier (see under "The Environment"), which demonstrates that multiple elements of a company's SR can be addressed in one organizational step. For instance,

- Encourage co-workers to think of projects in which the company can participate. Assigning a community-conscious task force, with members from various departments, would be highly constructive.
- Identifying developmental initiatives and projects in the community, and encouraging co-worker teams to participate for a few days every year at full pay by the organization. Co-workers should also be allowed to submit suggestions of projects they would like to support.

CSR Case Study: Patagonia

Providing co-workers the opportunity to engage in community involvement with full pay is exactly what Patagonia is doing. When, the Native Fish Society took on a project to secure the conservation of wild native fish in the Pacific Northwest, Patagonia co-workers volunteered in various ways, such as assisting scientists and speaking with visiting groups. The Patagonia co-workers received their full salary and benefits while volunteering in this community project (Scott 2012).

Patagonia, Inc. was founded in 1973, and it focuses predominantly on high-end outdoor clothing. The company's active membership in environmental movements is at least as well known as the products it sells on the consumer market. Performing in the apparel industry, Patagonia's leadership team seems to be well-aware of the sweatshop reputation this industry holds. The company's website provides ample information about its steps against the long, underpaid, and unhealthy work-hours people invest in this industry. Patagonia claims good pay, decent benefits, flex time, paid volunteerism, and subsequently low turnover for its immediate workforce of nearly two thousand people. Aware of the problems that can rise with subcontractors, the company carefully scrutinizes each new factory it considers for collaborative practices. For that purpose, Patagonia has instated two teams: a special Social/Environmental Responsibility team and a Quality team, both of which holding the power to veto a decision to work with a new factory.

Vividly involved in its industry, Patagonia is a major advocate for improved working conditions for garment workers, and pays additional premiums to workers in participating factories in order to help increase wages or engage in community development. Similarly, Patagonia advocates improved labor conditions in the supply chains within its industry. The company has also been instrumental in the Fair Labor Association's Fire Safety Initiative in 2013.

Agrawal (2017) underscores the importance of corporations to be in tune with the perceptions of Gen Z, and step up their CSR performance and authenticity, as this generation distinguishes itself from all its predecessors in its strong views on companies as members of society that should not only think of themselves but also of others. Agrawal presents some examples of great companies, who care for more than just the bottom line, and thereby includes Patagonia. Aside from operating as a premier outdoor brand, Patagonia distinguishes itself from competitors through considering the bigger picture. Customers of this company are outdoor types, who love nature, and want to know that their products come from companies that try to do the right thing. In Patagonia's case, its customers are very curious about what the company does for the planet. It therefore matters a lot that Patagonia pays fair wages to employees, engages in proper environmental conduct, and focuses on encouraging the restoration of the outdoors (Agrawal 2017).

Organizational Governance

While frequently listed as the first element in the SR range, we kept organizational governance as the final section in this chapter, because of its overarching nature. Organizational governance is the mental engine that drives an organization toward achieving its goals. It is the foundation of all decisions made in the organization, but also of all strategic changes that the organization decides to implement. The overall quality of life and success rate for any organization, and therefore also of its stakeholders, ultimately depends on the strategic decisions of its management team. These decisions pertain to (a) tolerable risks, (b) quality of output, (c) process consistency, and (d) SR approaches in an increasingly complex world (Robinson 2013). When a management system embeds quality at the root of all its practices (Duckworth 2014), good governance is practiced. The cultural, political, economic, and social contexts in which the organization operates, along with its size and type, are also great determinants of an organization's governance. Because of its immense influence on the way an organization is steered, organizational governance can be seen as the core of a company's SR implementation.

How Organizations, Leaders, and Quality Professionals can Fulfill this Element

Organizational governance, due to its strategic nature, has to begin with the leadership team. This is the team that determines the direction of the organization, so the initiative to formulate socially responsible goals is their primary task. Once the foundation is established, the leadership team could assign a governance team to supervise and update the organizations governance statements and performance.

CSR Case Study: The Starbucks Company

Starbucks' governance statement (Starbucks Corporation 2015) highlights a number of critical issues in regards to the company's SR performance, from the requirement to maintain diversity in background and perspectives to the maintenance of three board committees, one of them being the Nominating and Corporate Governance Committee. This committee has to make sure, among other things, that membership of all board and sub-committees alternates.

As is the case with all the companies presented in this chapter, the Starbucks Company has elevated SR compliance into a culture that is embedded in all its practices. Not only has Starbucks, as CEO Howard Schultz presents it, reinvented itself from a coffee-business serving people to a people-business serving coffee (Talpau and Boscor 2011); the company has displayed innovative thinking by conjuring up new consumer demands in what could be considered a declining industry. Starbucks has redefined coffee as a beverage, and has restructured the locations in which it is consumed (Rindova and Fombrun 2001), while simultaneously working on increasing SR performance. Some highlights in Starbucks' SR approaches are (a) there are full health insurance benefits and stock awards for part-time co-workers, (b) co-workers are treated with proper dignity and are granted ample career opportunities within the company, and (c) the company's impact on the community is carefully scrutinized (Outram 2014). The fact that Schultz retook control of the corporation after several years of having stepped down speaks volumes as well: contrary to the usual insatiable quest for expansion among companies, Schultz intervened in 2009 to slow down unbridled growth and help Starbucks refocus on doing what is right. In early 2017, Schultz stepped down again and handed over the reins of the company to Kevin Johnson.

As a way of community involvement, the Starbucks foundation has been supporting a campaign called "Create Jobs for USA," which focuses on funding job development in deprived areas (Saporito 2012). Like Trader Joe's, the Starbucks company refrains from expensive advertising campaigns, and has been stepping up its fair trade practices in recent years. The company heavily relies on synergistic partnerships, as stated in their annual report: "The first three stakeholder relationships discussed in their annual report are – in this order – partners (co-workers), customers, and coffee farmers" (Kleinrichert 2008). Starbucks is also a member of the Free Trade Coffee (FT) alliance, and even though the percentage of free trade coffee this company processes is still rather small, the effect of this membership has been a great stimulus to coffee growers and leads to rising numbers of members in the FT registry.

Starbucks is well aware of its customers' (mostly Gen Z members) preferences: it shifts the way drinks and food are treated, but even more, the way employees are treated (Agrawal 2017).

In an overview of Starbucks' SR strategy, Vandevelt (2015) identifies three pillars upon which the company performs: Community, Ethical Sourcing, and the Environment. In regards to community involvement, the first pillar, Starbucks develops stores that partner with local nonprofits, thus collaborating in offering services to meet local needs. Starbucks' nonprofit partners receive $0.05 to $0.15 per transaction. In addition, the company aims to hire about 10,000 veterans and military by 2018. Youth training opportunities are high on the company's priority list, so much so, that there is a Starbucks nonprofit foundation specializing in furthering community initiatives. In regards to Ethical Sourcing, the second pillar, the Starbucks Corporation commits itself to purchasing its products from ethical farmers and suppliers. In regards to the Environment, the third pillar, Starbucks perceives and promotes the earth as its most important business partner, thus ensuring an ongoing reduction process in environmental impact. Starbucks stores focus on recycling and conserving water and energy, and pursue strategies that address climate change on a global level. The fact that this three-pillar SR strategy works can be concluded from the company's overall customer ratings: about 93% of consumers perceive the Starbucks Corporation as positive. Due to a higher than average employee satisfaction rate, the turnover at Starbucks is relatively lower than that of its competitors (Vandevelt 2015).

Figure 1 below provides an impression of how an organization could implement the elements of CSR, discussed in this chapter. Important is Strategic Management's formulation and buy-in of a statement of governance and a code of ethics. These two critical documents will have to be developed in collaboration with an

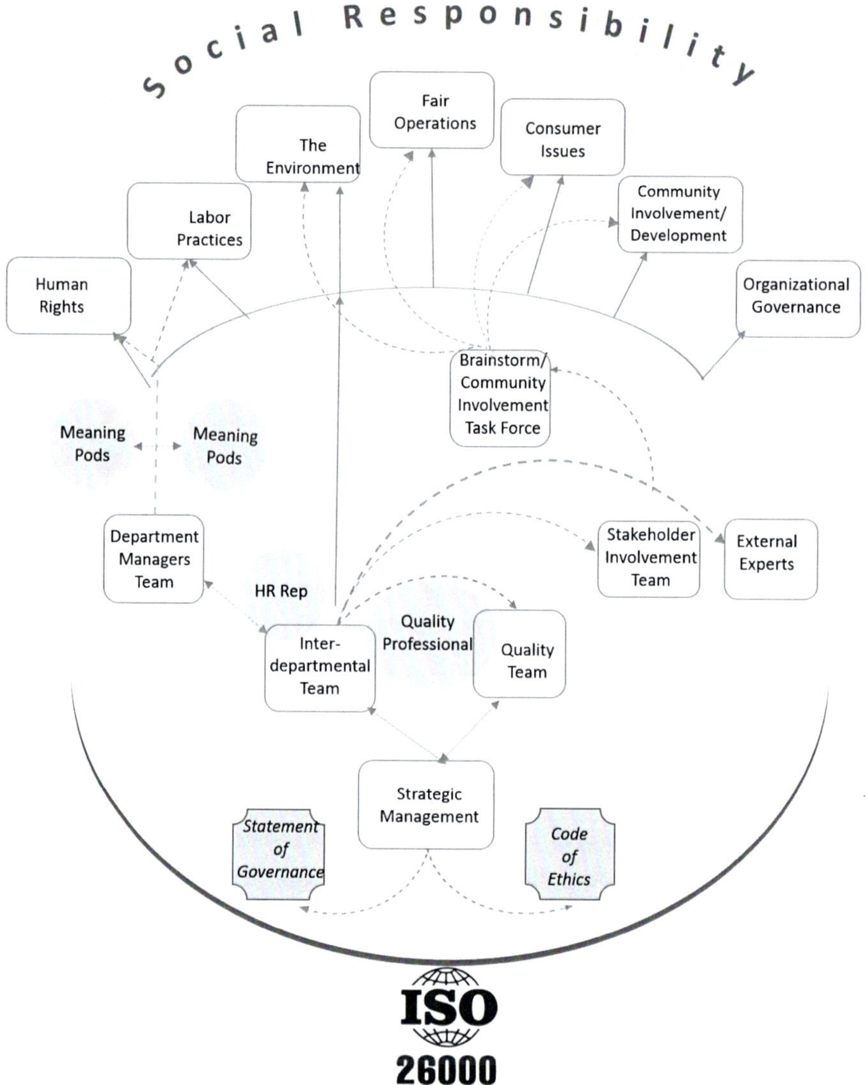

Fig. 1 Embedding SR into the organizational culture

interdepartmental SR team, as well as a quality team. The interdepartmental SR team should consist of department managers, so that there is broad support and understanding of the motivations and values to which the corporation will adhere. The interdepartmental SR team will serve as a liaison with the quality team, a stakeholder involvement team, and potential external experts, attracted to specific areas of CSR implementation. This interdepartmental SR team will carry general responsibility for the seven SR elements in the organization: Human Rights, Labor Practices, The

Environment, Fair Operations, Consumer Issues, Community Development and Involvement, and Organizational Governance. It might be useful for the company's ongoing CSR performance to also create meaning pods, supervised by the department managers team, and focused on continued review and fine-tuning of the meaningful performance of the organization. The department managers team may be the appropriate body to carefully co-monitor the company's human rights and labor practices performance. It is also prudent to have a representative of Human Resources and a Quality Professional involved in the entire CSR implementation structure, possibly participating in the interdepartmental SR team and the quality team.

Finally, it would be effective to instate a Brainstorm and Community Involvement Task Force, supervised by the stakeholder involvement team, and supporting efforts in the areas of The Environment, Fair Operations, Consumer Issues, and Community Development and Involvement.

Conclusion

As can be derived from this overview, companies that make SR part of their organizational culture usually emerge as leaders in their industries (Robinson et al. 2012). A bird's eye view on the implications of SR within the ISO 26000 scope, as demonstrated by the companies presented in this chapter, shows that a solid SR strategy leads to a number of aspects (ASQ and Manpower Professional 2011) that make up for perfect, accountable, and recurring business excellence:

- Better decision making and more responsible overall management.
  - IKEA's integration of economic, environmental, and social perspectives epitomizes this tendency, as can also be gathered from the way the company chooses to define itself: as a service provider rather than a furniture store.
  - Similarly, Columbia Sportswear's series of moral rules captured in its overall management construct, and expressed through its "Standards of Manufacturing Practices," demonstrate responsible management practices on a broad scale.
  - Interface's founder, Ray Anderson, also ensured responsible decision-making and better overall management when he steered the company onto the "Climbing Mount Sustainability" path, thereby including a plethora of stakeholders.
  - The SAS Institute demonstrates responsible decision-making and better overall management through its steadfast focus on being the best employer, partner, and vendor to its stakeholders; being committed to high ethical standards in all dealings, and being proactive in discovering and developing ways to improve in all areas that matter.
  - Trader Joe's responsible management and decision-making processes manifest themselves in the products it delivers, the way employees and customers are

treated, and the care this corporation gives to the less fortunate in its communities.
  - Patagonia, Inc. demonstrates this quality through its ability to see the bigger picture: having a very outdoors oriented customer base, Patagonia is aware of the importance of nature and focuses on encouraging the restoration of the outdoors.
  - The Starbucks Company has also displayed great overall management by focusing on major issues to assist the economy, varying from establishing synergistic partnerships, to offering employment for veterans, military, and disabled individuals, as well as its three performance pillars: Community, Ethical Sourcing, and the Environment.
- Greater trust from the public due to a better reputation.
  - IKEA has established longitudinal trust from the global customer community by living up to its founder's vision, "to create a better daily life for the many people" and thereby focusing on diversity as a value to signify the company's contribution to the best of the many people in a number of ways.
  - Columbia Sportswear has increased the trust from consumers through its engagement in labor-based projects, such as the HERproject, an organization that focuses on women empowerment all over the world, and its support to several other organizations that address concerning issues in society such as cancer, children's wellness, and the environment.
  - Interface continues to enhance the community's trust by implementing Ray Anderson's vision to produce as sustainably as possible, and trying to encourage others to do the same.
  - The SAS Institute holds an exceptional reputation and continues to score high on the lists of best companies to work for and most ethical companies to work with, due to its three-prong leadership base: (1) being the best employer, partner, and vendor to its stakeholders, (2) being committed to high ethical standards in all its dealings, and (3) being proactive in discovering and developing ways to improve in all areas that matter.
  - Trader Joe's has also achieved its reputation of trustworthiness through its responsiveness to consumers' requests for eco-friendlier, responsibly harvested food, and through taking care of the less fortunate in its communities.
  - Patagonia, being focused on outdoors apparel, has won trust of its customers and other stakeholders through its active membership in environmental movements and its careful scrutiny of each new factory it considers for collaborative practices.
  - The Starbucks Company has also won major trust through its community involvement, by establishing stores that partner with local nonprofits, thus collaborating in offering services to meet local needs.
- Increased competitive strength. The brief overview of the seven corporations discussed in this chapter demonstrates that doing the right thing, and going the extra mile toward safeguarding the important things for communities, the environment, and the earth as a whole, may bring some extra expenses and efforts along, but also yield greater rewards and admiration from global stakeholders.

- More constructive relationships with stakeholders.
  - IKEA perceives its physical products more as a platform for a service experience, resulting in customer value. The company has been active in supporting vulnerable groups to strengthen themselves and become economically and socially self-reliant. It is very vocal about banning child labor and is strongly involved in UNICEF projects, but also in self-initiated endeavors. The IKEA Foundation partners with Reach for Change, a nonprofit organization, on improving children's lives through social innovation.
  - Columbia Sportswear's earlier discussed "Standards of Manufacturing Practices" does the same, as it ultimately focuses on stakeholder well-being, and encompasses areas such as abandonment of any type of forced labor, compliance with local laws on minimum labor ages, respectful treatment of co-workers, including safe and harassment-free work environments, and dismissal of any kind of age, race, gender, ability, status, or other type of discrimination.
  - Interface takes care of its stakeholders through its seven focus points that consider the environment, health awareness, reduction of waste, increased efficiency, and sustainable business performance.
  - The SAS Institute maintains durable relationships with its suppliers, both direct and indirect, and prides itself in high moral standards toward all constituents: colleagues, customers, suppliers, and competitors. Its anti-corruption program includes an online anticorruption course for all new co-workers and certain third parties.
  - Trader Joe's places allergy labels on products to warn consumers, and gives local products a boost even before they reach the national market. It is highly responsive to consumers' requests for eco-friendlier, responsibly harvested food. It heavily focuses on local communities for its SR-related activities.
  - Patagonia provides co-workers the opportunity to engage in community involvement with full pay. The company's website provides ample information about its steps against the long, underpaid, and unhealthy work hours people invest in this industry. It is a major advocate for improved working conditions for garment workers, and pays additional premiums to workers in participating factories in order to help increase wages or engage in community development. Similarly, Patagonia advocates improved labor conditions in the supply chains within its industry.
  - The Starbucks Company grants full health insurance benefits and stock awards to part-time co-workers, treats co-workers with proper dignity, and carefully scrutinizes its impact on the community. The Starbucks foundation has been supporting a campaign called "Create Jobs for USA," which focuses on funding job development in deprived areas. Starbucks is also a member of the Free Trade Coffee (FT) alliance.
- Stronger innovations due to the connection with a broader base of stakeholders. The fact that the seven reviewed companies continue to perform successfully in an increasing competitive world, and manage to reinvent themselves time and again, demonstrates their potential to innovate through their broad stakeholder base.

- Improved employee retention, morale, and loyalty due to an increased emphasis on worker safety and well-being. This factor is one that also exudes in the reviewed company's approaches: From IKEA's encouragement to employees to take ownership in problem-solving, to the SAS Institute's HR practices that accommodate employees' needs optimally in order to enhance their job satisfaction, and from Trader Joe's employee motivation to take on any task, thus developing ownership, to Starbucks' conscious benefits and stockholder's options, even for part-time employees.
- Increased fairness in trade and elimination of corrupt practices. The care with which each of the seven reviewed corporations formulated their SR practices also demonstrates, sometimes verbatim, and at other times implied, that trade fairness and corruption elimination are critical in their business routines.
- Solid bottom line thanks to improved effectiveness and efficiency. Even though the seven reviewed corporations undoubtedly have their ups and downs in revenue influx, their rankings on various lists, along with their continued expansions and innovations, demonstrate that their bottom line is sound, particularly because most of these companies consider their bottom line within a broader scope than just financial.
- Greater longevity of the organization due to its sustainable approaches. Each of the reviewed corporation has a track record of successful existence that spans over several decades, which is evidence of their longevity, to be credited to their sustainable approaches.
- Better social ties with individuals, civic, and commercial entities overall. In several of the points above, the splendid social ties of the seven corporations reviewed have also been underscored.

Waiting until SR can be proven in hard, compartmentalized numbers of profits may be an erroneous mindset, just like the notion that SR is just another episode that will sooner or later fade away. "SR is here to stay, and organizations will either deal with it or SR will deal with them," a reality that may be the foundational reason why "more organizations are looking to move SR beyond simple public relations" (ASQs New SR Integration Guide 2014). Because of its comprehensive nature, it is not simple to delineate where exactly the benefits of SR lie. However, the overall statistics speak volumes: companies that seriously engage in SR find this culture to not only be beneficial to their bottom line, but also to their reputation, and to all their stakeholders. Not to be underestimated: their leaders also feel better about what they do, and, ultimately, that is the greatest reward one can desire in life.

Cross-References

▶ Collaboration for Regional Sustainable Circular Economy Innovation
▶ Designing Sustainability Reporting Systems to Maximize Dynamic Stakeholder Agility
▶ Empathy Driving Engaged Sustainability in Enterprises

- ▶ Ethical Decision-Making Under Social Uncertainty
- ▶ Expanding Sustainable Business Education Beyond Business Schools
- ▶ Moving Forward with Social Responsibility
- ▶ People, Planet, and Profit
- ▶ Responsible Investing and Corporate Social Responsibility for Engaged Sustainability
- ▶ Selfishness, Greed, and Apathy
- ▶ The Spirit of Sustainability
- ▶ The Theology of Sustainability Practice

References

$5 million commitment names Ray C. Anderson Center for Sustainable Business. (January 30, 2015). Georgia Tech News Center. Retrieved from http://www.news.gatech.edu/2015/01/30/5-million-commitment-names-ray-c-anderson-center-sustainable-business.

Agrawal, A. J. (April 23, 2017). How to use social responsibility to appeal to generation Z. Forbes. Retrieved from https://www.forbes.com/sites/ajagrawal/2017/04/23/how-to-use-social-responsibility-to-appeal-to-generation-z/#6cfa5024507a.

Amaeshi, K., Adegbite, E., Ogbechie, C., Idemudia, U., Kan, K., Issa, M., & Anakwue, O. (2016). Corporate social responsibility in SMEs: A shift from philanthropy to institutional works? *Journal of Business Ethics, 138*(2), 385–400.

American National Standard: Guidance on social responsibility. (January 17, 2011). American Society for Quality. Retrieved from http://asq.org/learn-about-quality/learn-about-standards/iso-26000/.

Anderson, R. (2007). Doing well by doing good. In D. Church (Ed.), *Einstein's business: Engaging soul, imagination, and excellence in the workplace*. Santa Rosa: Elite Books.

Anderson, R. C. (2010). Earth day, then and now. *Sustainability: The Journal of Record*, 73–74.

Anderson, G. (July 30, 2013). Why are trader Joe's customers the most satisfied in America? Forbes. Retrieved from https://www.forbes.com/sites/retailwire/2013/07/30/why-are-trader-joes-customers-the-most-satisfied-in-america/#7fa92eda1ea0.

ASQ and Manpower Professional. (2011). Social responsibility and the quality professional: The implications of ISO 26000. asq.org/2011/02/iso-26000/social-responsibility-and-the-quality-professional-the-implications-of-iso-26000.pdf.

ASQ's New SR Integration Guide Pathways to Social Responsibility: Successful Practices for Sustaining the Future (April 2014). Retrieved from http://asq.org/2014/04/social-responsibility/pathways-2014.pdf, pp. 20–21.

Blowfield, M., & Frynas, J. G. (2005). Editorial: Setting new agendas – Critical perspectives on corporate social responsibility in the developing world. *International Affairs, 81*(3), 489–513.

Brown, S. (22 April, 2013). Corporate social responsibility feature: Trader Joe's. Momentum telecom. Retrieved from http://www.momentumtelecom.com/corporate-social-responsibility-feature-trader-joes/.

Colgate-Palmolive. (November 30, 2015). Colgate-Palmolive receives sustainability leadership award at 2015 Greenbuild conference. Business Wire (English).

Columbia Sportswear Company Teams Up with The Ocean Foundation to Help Save Marine Habitat. (March 11, 2010). Columbia sportswear. Retrieved from http://www.csrwire.com/press_releases/29072-Columbia-Sportswear-Company-Teams-Up-With-The-Ocean-Foundation-To-Help-Save-Marine-Habitat-.

Common Ground: Quality and Social Responsibility. (2007). An ASQ white paper. Retrieved from www.asq.org/social-responsibility/pdf/social-responsibility-white-paper.pdf and http://docplayer.net/storage/57/41106478/1498973644/W7L5wotKRIodP1JjTCWGog/41106478.pdf.

Conley, J. M., Lázaro-Muñoz, G., Prince, A. E. R., Davis, A. M., & Cadigan, R. J. (2015). Scientific social responsibility: Lessons from the corporate social responsibility movement. *American Journal of Bioethics, 15*(12), 64–66.

Couch, C. (August 15, 2013). 10 companies with excellent customer service. *The Huffington Post*. Retrieved from http://www.huffingtonpost.com/2013/08/15/best-customer-service_n_3720052.html.

Cycyota, C. S., Ferrante, C. J., & Schroeder, J. M. (2016). Corporate social responsibility and employee volunteerism: What do the best companies do? *Business Horizons, 59*(3), 321–329.

Davis, M. (September 3, 2014). Radical industrialists: 20 years later, Interface looks back on ray Anderson's legacy. Greenbiz.com. Retrieved from http://www.greenbiz.com/blog/2014/09/03/20-years-later-interface-looks-back-ray-andersons-legacy.

DiSalvo, D. (February 19, 2015). What trader Joe's knows about making your brain happy. Forbes. Retrieved from https://www.forbes.com/sites/daviddisalvo/2015/02/19/what-trader-joes-knows-about-making-your-brain-happy/#3c91ba7f1213.

Duckworth, H. (April, 2014). Integrative social responsibility pathways to social responsibility: Successful practices for sustaining the future. Retrieved from http://asq.org/2014/04/social-responsibility/pathways-2014.pdf, pp. 4–5.

Dunn, J. (September 29, 2015). IKEA Switzerland is the first company worldwide to reach highest level of gender equality certification from EDGE. PR Newswire Europe Including UK Disclose.

Edvardsson, B., & Enquist, B. (2011). The service excellence and innovation model: Lessons from IKEA and other service frontiers. *Total Quality Management & Business Excellence, 22*(5), 535–551.

English, N. (December 6, 2013). The 14 athletic wear companies that are actually good for the world. Retrieved from http://greatist.com/fitness/athletic-wear-companies-social-good.

Hopen, D. (2010). The changing role and practices of successful leaders. *The Journal for Quality and Participation*, 4–9. Retrieved from http://rube.asq.org/quality-participation/2010/04/leadership/the-changing-role-and-practices-of-successful-leaders.pdf.

IKEA Group Sustainability Report, FY14. (2014). IKEA. Retrieved from http://www.ikea.com/ms/en_US/pdf/sustainability_report/sustainability_report_2014.pdf.

IKEA Group Sustainability Report, FY16. (2016). IKEA. Retrieved from http://www.ikea.com/ms/en_US/img/ad_content/IKEA_Group_Sustainability_Report_FY16.pdf.

ISO 26000: Guidance on Social Responsibility. (2010). Retrieved from https://www.iso.org/standard/42546.html.

Kleinrichert, D. (2008). Ethics, power and communities: Corporate social responsibility revisited. *Journal of Business Ethics, 78*(3), 482.

Langer, E. (August 10, 2011). Ray Anderson, 'greenest CEO in America,' dies at 77. *The Washington Post Obituaries*. Retrieved from http://www.washingtonpost.com/local/obituaries/ray-anderson-greenest-ceo-in-america-dies-at-77/2011/08/10/gIQAGoTU7I_story.html.

Luce, E. (15 September, 2004). Ikea's plan to tackle child labour: CORPORATE SOCIAL RESPONSIBILITY: The Swedish retailer is supporting schools in India's carpet belt in a commercially sound way. *Financial Times* [London (UK)], p. 14.

Lund-Thomsen, P., Lindgreen, A., & Vanhamme, J. (2016). Industrial clusters and corporate social responsibility in developing countries: What we know, what we do not know, and what we need to know. *Journal of Business Ethics, 133*(1), 9–24.

Maher, F. J. (2014). Integrating social responsibility with business strategy: A guide for quality professionals. Retrieved from asq.org/2014/10/social-responsibility/sr-integration-guide.html.

Martin, E. J. (August 24, 2015). Joe knows snacks. CSP magazine special issue. Retrieved from http://www.cspnet.com/print/csp-magazine/article/cover-story-joe-knows-snacks.

Moratis, L. (2017). The credibility of corporate CSR claims: A taxonomy based on ISO 26000 and a research agenda. *Total Quality Management & Business Excellence, 28*(1/2), 147–158.

Morsing, M. m., & Roepstorff, A. (2015). CSR as corporate political activity: Observations on IKEA's CSR identity-image dynamics. *Journal of Business Ethics, 128*(2), 395–409.

Outram, C. (2014). *Ten pitfalls of strategic failure*. Fountainebleau: INSEAD.

Preuss, L., Barkemeyer, R., & Glavas, A. (2016). Corporate social responsibility in developing country multinationals: Identifying company and country-level influences. *Business Ethics Quarterly, 26*(3), 347–378.

Reach for Change. (April 28, 2017). Reach for change and IKEA foundation: New partnership to impact 84,000 children in Ethiopia. Business Wire (English).

Rindova, V., & Fombrun, C. (2001). Entrepreneurial action in the creation of the specialty coffee niche. In C. B. Schoonhoven & E. Romanelli (Eds.), *The entrepreneurship dynamic: Origins of entrepreneurship and the evolution of industries*. Palo Alto: Stanford University Press.

Robinson, C. (2013). Integrating quality, social responsibility, and risk: Key principles and important tools. *The Journal for Quality and Participation, 35*(4), 24–29.

Robinson, C., Jacobsen, J., & Hopen, D. (Eds.) (2012). Quality and social responsibility: A key business strategy for enhancing competitive position. Retrieved from asq.org/2012/07/social-responsibility/a-key-business-strategy.pdf.

Saporito, P. (2012). Starbucks' big mug. *Time, 179*(25), 51–54.

SAS Code of Ethics. (2016). Retrieved from https://www.sas.com/content/dam/SAS/en_us/doc/other1/code-of-ethics.pdf.

SAS Corporate Social Responsibility Report. (2016). Retrieved from http://www.sas.com/content/dam/SAS/en_us/doc/other1/csr-107835.pdf.

Scott, R. (July 16, 2012). 3 stellar examples of corporate community involvement programs. Causecast Blog, Complete Corporate Philanthropy and Employee Volunteering Platform. Retrieved from http://www.causecast.com/blog/3-stellar-examples-of-corporate-community-involvement-programs.

Standards of Manufacturing Practices. (2015). Columbia Sportswear Company. Retrieved from http://demandware.edgesuite.net/aasn_prd/on/demandware.static/-/Sites-Columbia_US-Library/default/dwe783eab2/AboutUs/PDF/Standards_of_Manufacturing_Practices_08.pdf.

Starbucks Corporation Corporate Governance Principles and Practices for the Board of Directors. (June 23 2015). Retrieved from http://globalassets.starbucks.com/assets/c3a0ba7c48814d19953 46996151ae6e9.pdf.

Talpau, A., & Boscor, D. (2011). Customer-oriented marketing – A strategy that guarantees success: Starbucks and Mcdonald's. *Bulletin of the Transylvania University of Brasov. Economic Sciences Series V, 4*(1), 51–58.

Trader Joe's LA 346: Law & Ethics – Walk the Talk Team Project. (2017). Storify. Retrieved from https://storify.com/PumpkinBar/trader-joes.

Valuing Corporate Social Responsibility: McKinsey Global Survey Results. (February, 2009). McKinsey & Company. Retrieved from http://asq.org/newsroom/news-releases/2011/2011-0215-asq-manpower-sr-whitepaper.html.

Vandevelt, K. (September 24, 2015). Corporate social responsibility: How Starbucks is making an impact. WhyWhisper.com. Retrieved from http://www.whywhisper.co/the-blog/2015/9/24/corporate-social-responsibility-how-starbucks-is-making-an-impact.

What are the core subjects of social responsibility? (N/A). ASQ. Retrieved from http://asq.org/learn-about-quality/learn-about-standards/iso-26000/core-subjects.html.

Where HERproject works. (N/A). Retrieved from http://herproject.org/where-we-work.

Wilting, H. C., & van Oorschot, M. M. (2017). Quantifying biodiversity footprints of Dutch economic sectors: A global supply-chain analysis. *Journal of Cleaner Production, 15(6)*, 194–202.

World Wide Fund for Nature. (2008). The 2008 Living Planet Report. Retrieved from http://wwf.panda.org/about_our_earth/all_publications/living_planet_report_timeline/lpr_2008/.

Social Entrepreneurship

Where Sustainable Leading Meets Sustainable Living

Joan Marques

Contents

Abstract

Social entrepreneurs cleverly combine business techniques and private sector approaches in order to develop solutions to social, cultural, or environmental problems, and do so in a variety of organizations. The chapter will set the stage by first presenting some commonalities between conventional entrepreneurs and social entrepreneurs, such as their internal locus of control, their zest for innovation, their ambition, and their perseverance; and subsequently discussing the most obvious differences such as their foundational structures (wealth accumulation vs. making a difference) and their performance measurement (return on investments vs. return to society). Three different social entrepreneurial business models will briefly be discussed: (1) The Leveraged Nonprofit, a business model leveraging resources in order to respond to social needs; (2) The Hybrid

J. Marques (✉)
Woodbury University, Burbank, CA, USA
e-mail: Joan.Marques@woodbury.edu

S. Dhiman, J. Marques (eds.), *Handbook of Engaged Sustainability*,
https://doi.org/10.1007/978-3-319-71312-0_3

Nonprofit, a structure that can take on a variety of forms; and (3) The Social Business Venture, designed to create change through social means (Elkington and Hartigan 2008). The chapter will then present two exemplary social entrepreneurs: Muhamad Yunus (Grameen Bank) and Elon Musk (Tesla Electric cars, SolarCity, SpaceX, and Hyperloop). Both of these social entrepreneurs decided to address societal needs and solve social problems by changing the status quo, presenting and spreading a solution, and persuading their societies to move in a different direction. In its final section, the chapter will encourage readers to consider social entrepreneurial efforts toward a better world, by becoming society's change agents, and either righting a wrong, creating something right, or improving on something good.

Keywords

Social entrepreneurs · Internal locus of control · Nonprofit · Social problems · Sustainability · Social responsibility · Culture · Environment

Introduction

The term "entrepreneur" has been in use for at least two centuries now. Originating from French, the term implies qualities of leadership, initiative, and innovation. While the term "entrepreneur" in its early interpretation mainly pertained to new venture design, it currently finds broader use in innovative behavior across various realms. Yet, when we say that someone is an entrepreneur, most people still think of an individual who creates and runs one or more businesses, and taking the necessary financial risks that accompany this process. Conventional entrepreneurs start their ventures for many reasons, but most of the time these reasons are profit-driven, whether pertaining to the production of goods or offerings of services.

The term "social entrepreneur" seems to be about five decades old. According to El Ebrashi (2013), it was first mentioned in 1972 in Joseph Banks' book *The Sociology of Social Movements*. Social entrepreneurship practices developed in the 1980s with the establishment of Ashoka, the first organization to support social entrepreneurs in the world (Mastrangelo et al. 2017). Yet, it wasn't until we made our way into the twenty-first century, and human awareness increased, with concepts such as environmental abuse, global footprint, sustainability, and corporate social responsibility entering the business vocabulary, that the social entrepreneur got a prominent foothold in the world of business.

Multiple definitions of this concept were formulated, such as Martin and Osberg's (2007) description that social entrepreneurs are those that identify "a stable but inherently unjust equilibrium that causes the exclusion, marginalization, or suffering of a segment of humanity that lacks the financial means or political clout to achieve any transformative benefit on its own" (Martin and Osberg 2007, p. 35). Seelos and Mair (2007) describe social entrepreneurship as a phenomenon that encapsulates the individuals, the organizations, and the initiatives engaged in entrepreneurial activities with a social goal. In another resource, these same authors define social entrepreneurship as "entrepreneurship that creates new models for the provision of products and services that cater directly to the social needs underlying sustainable development goals such as the MDGs (Millennium Development Goals)" (p. 244).

Zahra et al. (2009) explain social entrepreneurship as a cluster of activities and processes aimed at "discovering, defining, and exploiting opportunities in order to enhance social wealth by creating new ventures or managing existing organizations in an innovative manner" (p. 519). Mastrangelo et al. (2017) describe social entrepreneurship as "an outstanding phenomenon that links entrepreneurship, social change, and economic development" (p. 437).

Based on a thorough investigation, Dart (2004) found that the legitimacy of social entrepreneurship is based on morality rather than pragmatism. Dart's research underscored that social entrepreneurship has emerged into a genuine and acceptable organizational form because it made solid sense to the core business mindset of using market-based tactics as the critical path toward successfully addressing social and ecological problems. Dey and Lehner (2017) add that social entrepreneurship is also in many ways "a response to the 'global crisis of value' instigated by the ascendancy of free-market liberalism as the dominant political ideology" (p. 753). They rightfully aver that social entrepreneurship, as a critical embodiment of progressive social change, has become a prominent topic in Management and Organization Theory, especially when Business Ethics are considered. While there are, of course, healthy skeptics, who underscore that the concept of social entrepreneurship can sometimes be abused by individuals and corporations to profile themselves as something they are not, the overall picture is a positive and hopeful one. In general, it can be stated that social entrepreneurship finds its origins in numerous developments and movements, which allots it a broad and solid base. Social entrepreneurship is a strong contributor in adding meaning to work and life. Indeed, "social entrepreneurs are increasingly important in filling unmet social needs. Yet, there is little understanding of their motivations and opportunity recognition in the domain of social entrepreneurship" (Yitshaki and Kropp 2016, p. 546).

(Social entrepreneurs cleverly combine business techniques and private sector approaches in order to develop solutions to social, cultural, or environmental problems, and do so in a variety of organizations.)

Common Factors

According to *BusinessDictionary.com*, an entrepreneur, in the collective sense of the word, is someone who exercises initiative by organizing a venture to take benefit of an opportunity and, as the decision maker, decides what, how, and how much of a good or service will be produced. The source further explains that entrepreneurs have to be willing to take risks, and will have to monitor and control their venture's activities. An entrepreneur can be a sole proprietor, a partner, or a person owning the majority of shares in an incorporated venture. All of the above remains the same, whether an entrepreneur chooses to be commercially or socially driven. *Business-Dictionary.com* also states that economist Joseph Alois Schumpeter (1883–1950) clarified that entrepreneurs are not necessarily motivated by profit but regard it as a standard for measuring achievement or success. This statement brings to light the fact that many entrepreneurs, regardless of their type of business, engage in their practice for fulfillment, satisfaction, accomplishment, and securing a livelihood.

Strauss (2015) describes entrepreneurs as having a greater capacity for pain and discomfort than most. "They can stay up later, work longer hours, stay more focused and, somehow, are able to set so much aside in deference to their dreams and visions" (par. 1).

In addition to the above common factors, it should also be understood that both, commercial and social entrepreneurs, have to possess a high internal locus of control. Testé (2017) explains, "The notion of locus of control (LOC) refers to an individual's perception of factors that account for the occurrence of outcomes. Specifically, internally oriented individuals believe outcomes are primarily related to internal factors (e.g., their own actions), whereas externally oriented individuals believe outcomes are influenced mostly by external factors (e.g., chance)" (p. 81). This implies that a person with an internal locus of control will harbor a high level of responsibility and confidence, which drives him or her toward leadership and the poise to start something new, where others don't even entertain that though, or if they do, refrain from following up due to fear for failure.

Another common factor between both types of entrepreneurs is their zest for innovation. Jiarong and Shouming (2016) found that entrepreneurs' degree of self-efficacy impacts their ability to be innovative. Self-efficacy refers to an individual's belief in his or her capacity to execute behaviors necessary to produce specific performance attainments. Defined that way, it becomes obvious that self-efficacy is closely aligned with one's internal locus of control. This positive relationship between self-efficacy and innovation performance highlights the interdependency of the qualities for the entrepreneur's degree of success and perseverance. Jiarong and Shouming (2016) found that a higher level of self-efficacy is related to improved innovation performance.

Entrepreneurs on either side of the spectrum have to harbor a certain degree of ambition. "Ambition is the fuel that sets our internal engine into gear, especially when it comes to professional performance" (Marques 2016, p. 65). "Ambitious people see opportunities where others see challenges, and they continuously try to expand their skills and connections in order to convert these opportunities into reality" (Marques 2016, p. 65). Ambitious people, particularly the entrepreneurial kind, seek suitable partners, are innovative, keep their ego under control, sharpen their focus, and, therefore, outperform many others (Marques 2016). It may be prudent to underscore here that ambition is a quality that needs to be monitored, because it can also lead to intolerance, micromanagement, increased dissatisfaction, and damaged trust or reputation, when escalated (Marques 2016).

Perseverance is yet another common factor within entrepreneurs, whether commercially or socially oriented. Perseverance enables people to go an extra mile when others give up, and this quality oftentimes turns out to be an asset in accomplishing a target. Perseverance brings a gamut of other skills along, such as patience, timing, communication, creativity, and emotional intelligence, in order to accommodate others, and convince them of the usefulness of pursuing the project at hand. Perseverance, like ambition, is a quality that also needs to be handled with mindfulness. A leader who exerts too much perseverance can be considered overly demanding, unreasonable, and unpleasant to work with, thus alienating good potential collaborators. Yet, when applied to a proper degree, perseverance is a great quality for all types of entrepreneurs.

Tobak (2016) describes entrepreneurship as a game of attrition: of having the determination, discipline, and cash to persevere. While he doesn't disregard vision, passion, savvy, and guts, he considers perseverance the critical glue that binds all other qualities together. He briefly reviews some entrepreneurs who became billionaires, such as Sergey Brin and Larry Page, the Google founders, who pitched their idea to numerous potential investors in Silicon Valley before finding one who wrote them a $100,000 check to start. And even then, it took years before the search engine really started to flourish. He also discusses Jeff Bezos, who founded Amazon.com in 1994, and displayed unremitting perseverance in the years thereafter to prove its advantage and effectiveness. Tobak (2016) finally considers Steve Jobs, who also felt that perseverance was the critical quality that separated successful entrepreneurs from unsuccessful ones. Perseverance means holding it all together, even when employees lose hope, and the media becomes mean.

The Differences

Bacq et al. (2016) affirm that commercial entrepreneurs are often perceived as driven by selfish reasons, even though they are praised for their economic- and performance-based contributions to society. Social entrepreneurs are perceived distinctive from their commercial counterparts based on their "motivations, intentions and relative importance of actively doing good for society (as opposed to being motivated by private gains)" (p. 703). Bacq et al. (2016) observe a positive interaction in dynamics between the two groups: because social entrepreneurs are so often applauded for their efforts in creating social wealth and bringing important social change in challenging times, commercial regular entrepreneurs become more aware of the need to also engage in socially responsible and morally sound behavior.

Mastrangelo et al. (2017) found that social entrepreneurs have the following traits in common: communication, teamwork, delegation, active listening, and attitude. Their study also found that particularly communication, active listening, and attitude can have a strong influence on employees' commitment toward involvement.

Yitshaki and Kropp (2016) found from a study among 30 social entrepreneurs in Israel, that there are different motivations that lead to social awareness and opportunity recognition among social entrepreneurs: a majority in their study (60%) were motivated by "pull factors," which included prosocial behaviors based on past or current life events, while a smaller number (40%) were motivated by push factors, including job dissatisfaction and a search for meaning. In some more detail, the "pull factors" varied from life influences in past or present that impacted the perspectives of these social entrepreneurs and compelled them to engage in their movement.

Some of their social entrepreneurial endeavors came straight from coping with problems, either in their own youth or through their children, that they felt needed larger attention in society. The issues varied from personal hardship, sadness, or embarrassment, leading to the decision to spare others the same experience, or at least, reduce their suffering. Sometimes a "pull factor" was also manifested through an example set by previous generations, and at other times, it was imbued with an ideological motivation that served as fuel for their initiative. There were also social entrepreneurs who ascribed their "pull" to a spiritual guidance, which arrived through certain eye-opening experiences (Yitshaki and Kropp 2016).

As for the "push factors," the reasons varied from nonpersonal, yet disturbing confrontations with injustice, to the quest for more meaning in work and life after holding an unsatisfying job (Yitshaki and Kropp 2016).

Even though there is a longheld belief that entrepreneurs positively contribute to society (e.g., through job creation) (Schumpeter 1934; Van Praag and Versloot 2007), what is assumed to distinguish social entrepreneurs from their commercial counterparts is their motivations, intentions, and relative importance of actively doing good for society (as opposed to being motivated by private gains) (Dacin et al. 2010; Zahra et al. 2009).

In a study that aimed to question the moral label generally ascribed to social entrepreneurs, Bacq et al. (2016) pinpoint the difference between social entrepreneurs and commercial entrepreneurs through the following two propositions:

1. Intention of blended value creation is not what distinguishes social entrepreneurs from commercial entrepreneurs: social entrepreneurs differentiate themselves by the primacy they attach to social value creation and their intention to pursue collective interests over and above economic value creation (p. 713).
2. Social entrepreneurs display a less favorable entrepreneurial profile compared to commercial entrepreneurs, as (a) they are less self-confident in their capabilities to start a business; (b) they are less likely to regard entrepreneurship as a desirable career choice; and (c) they devote less time and effort to their organizational social mission (p. 714).

Social Entrepreneurial Business Models

Social entrepreneurship, as may have become clear by now, is not a single concept. There are many ways in which social entrepreneurship can be exerted. Not only do the reasons for undertaking actions aiming at positive social change vary, but the operational models utilized are also divergent. In her article "It's About People, Not Profits," Hartigan (2006) distinguishes and explains three different social entrepreneurial business models, to be briefly reviewed below:

1. The Leveraged Nonprofit, a business model leveraging resources in order to respond to social needs. Social entrepreneurs engage in this model when they aim to adopt an innovation and set up a nonprofit for that purpose. The nonprofit

is an appealing model to gather support and commitment from various constituents, such as organizations – private and public, as well as individuals. While such a nonprofit organization is dependent upon donations and grants, it is fairly stable, thus sustainable, due to the support it gets from multiple societal layers. Some examples Hartigan (2006) mentioned for this model are, "Wendy Kopp's Teach for America, Jeroo Billimoria's Childhelpline International (India and beyond), and Javier Gonzalez Quintero's abcdespañol (Colombia and Latin America)" (p. 44).

2. The Hybrid Nonprofit, which is a structure that can take on a variety of forms. In this case, the entrepreneur creates a nonprofit that comprises a certain level of cost recovery, which is done through the sale of goods and services to a variety of associated entities, both public and private, along with target population groups. In order to keep up the transformational activities and continue to serve its oftentimes needy or dependent clients, the entrepreneur solicits other sources of resources from the public and/or donor organizations. Two examples Hartigan (2006) listed for this model are "Martin Fisher and Nick Moon's Kick Start (Kenya) and Rodrigo Baggio's Committee for the Democratization of Information (Brazil)" (p. 45).
3. The Social Business Venture, which is designed to create change through social means (Elkington and Hartigan 2008). In her 2006 article, Hartigan referred to this model as the "Hybrid for-Profits or Social Businesses," whereby she explained that, in this model, entrepreneurs create businesses aimed at driving transformational change. Even though there are profits generated, the main aim is not financial return maximization for shareholders but rather the expansion of the social venture in order to ensure a more effective reach of people in need. Wealth accumulation is therefore not a priority in this model, and revenues beyond costs are therefore reallocated to the business for increased support to the focus group. Entrepreneurs engaging in this model look for investors with an interest in combining financial and social returns. Hartigan presents the following examples of such social businesses, Issac Durojaiye's DMT Toilets (Nigeria), Nic Frances' Easy Being Green (Australia), Ibrahim Abouleish's SEKEM (Egypt), and Javier Hurtado's Irupana (Bolivia) are exemplars of such social businesses (p. 45).

Roper and Cheney (2005) also share their views on the business models of social entrepreneurship, and represent them as follows:

1. Private social entrepreneurship, in which the social entrepreneur enjoys the advantage of taking ownership in planning, profit and innovation of the venture. Private social enterprises have greater freedom to adopt popular contemporary business trends.
2. Social entrepreneurship in the not-for-profit sector, where ventures, often sprung from social-movement organizations, social advocacy groups, or community initiatives, compete for the scare funding resources from philanthropic entities. Even though their structures and focus points may differ widely, most of these ventures aim to allocate their funds in a socially laudable way such as

regenerating or expanding economic activity, advancing the public good, ensuring community ownership, and solidifying democratic structures.

3. Public-sector social entrepreneurship, which pertains to economic applications of business and market models to the public sphere, such as water, electricity, and waste management. Yet, public social entrepreneurship may bring substantial political and administrative constraints on the latter. Public organizations have a more difficult time adapting to changing circumstances and innovating owing to constitutional, executive, legislative considerations, as well as to sheer habit. In other words, the private sector allows for greater freedom and experimentation, as seen from this standpoint (Roper and Cheney 2005).

Social Entrepreneurship and Sustainability

The term sustainability holds at least two meanings for social entrepreneurial ventures. The first pertains to the literal interpretation of maintaining a sustainable business model: will the venture be able to continue performing? This is not always easy or clear, because in social enterprises, the financial influx oftentimes arrives from different sources (e.g., donors, benefactors, and philanthropists) than the outflow (mainly to challenged entities). This means that social entrepreneurs are incessantly and creatively searching for ways to sustain their operations, preferably without perpetually depending on illustrious, yet capricious donors (Dudnik 2010).

The other interpretation of sustainability pertains to the resilience of the social endeavor and its effects. It is very disheartening if hard-invested efforts get lost as soon as the social enterprise moves to the next project. Ensuring sustainability of actions is never guaranteed, but in order to enhance the likelihood of it to happen, social ventures will have to build in models that secure continuation (Dudnik 2010).

In an interesting article about sustainable social enterprises, Osberg and Martin (2015) describe the work of social ventures that address challenges too narrow in scope to trigger governmental involvement, yet important enough to make a critical difference. They point out that such ventures maintain a balance between attaining social goals, while at the same time meeting their financial pressures. They discuss social business entities such as GoodWeave, an Asia-based carpet company that ensures no child labor involved in their products, yet holds its own amid less scrupulous competitors, through a keen focus on financially sustainable operations. Treading a golden middle path between business and government assistance in their mission, some of these social entrepreneurial ventures engage in major environmental projects such as deforestation in the Amazon, requiring awareness campaigns amongst multiple stakeholders, safe landmine detonation activities in Asia, to ensure a healthier environment for upcoming generations, and drug inventory tracking systems to make up for the shortage of doctors and nurses in sub-Saharan Africa.

Social entrepreneurs focus on solving big problems in a simple, yet sustainable way, such as Daniel Snell's "Arrival Education," which is built on a reciprocity model: it takes young people from challenging backgrounds and gets them to help develop programs for businesses, who in turn develop them. Founded in 2003,

Arrival Education was first a part-time endeavor, until Snell dared to give up his corporate job in 2006 to fully devote himself to the cause. In developing his social concept, Snell discovered a unique interplay between problematic youth and business corporations, thus delivering valuable service in sustainable efforts, and making a living that way. Both corporations and schools love this organization because it reforms constituents in a positive way: corporate leaders utilize the training skills of Arrival Education's team, and schools do the same, benefiting by keeping challenged youngsters in school. Snell warns that doing the sustainable and socially responsible thing is not necessarily a big money maker. He underscores the common aspects of perseverance, less than average income, a shift of social environment, critical listening (and scrutinizing), and knowledge of the business that drives the social endeavor (Snell 2012).

Another interesting take on entrepreneurship, and most definitely what we have come to know as social entrepreneurship in recent decades, is the old view from Joseph Shumpeter (1934), who pointed out that entrepreneurial activities are actually creative destruction, because "sustainable entrepreneurs destroy existing conventional production methods, products, market structures and consumption patterns, and replace them with superior environmental and social products and services" (Schaltegger and Wagner 2011, p. 222).

Social Entrepreneurs in Action

In order to obtain a decent understanding of social entrepreneurship, we will review two exemplary, yet very different social entrepreneurs in the next section: Muhamad Yunus (founder, Grameen Bank), and Elon Musk (founder and co-founder of Tesla Electric cars, SolarCity, SpaceX, and the Hyperloop). Both of these social entrepreneurs are known for addressing societal needs and solving social problems by changing the status quo, presenting and spreading a solution, and persuading their societies to move in a different direction.

Muhammad Yunus

Muhammad Yunus is thus far the only business person who received the Nobel Peace Prize. This happened, along with his brainchild Grameen Bank, in 2006. This prestigious award was granted to Yunus because of his initiatives and decades of banking for the poor, and developing a microcredit system that was later adopted by many organizations in many nations. In his Nobel Peace Prize acceptance speech, Yunus pointed out that poverty is a threat to peace. He shared the grim reality that 94% of the world's income goes to 40% of the global population, while 60% of the people have to share only 6% of income among each other. Poverty is the absence of all human rights. If we want to build stable peace, we have to create means for the poor to rise above their deprived state (The World's Top 20 Public Intellectuals 2008).

The Nobel Foundation, in a brief biographical sketch, describes Yunus as "Banker to the Poor" (Muhammad Yunus Biographical 2006). Yunus was born in 1940 in Chittagong, a seaport city in Bangladesh. As a student at Dhaka University, he received a Fulbright scholarship to study in the US. He did so at Vanderbilt University in Tenessee, where he received his PhD in Economics in 1969. Upon graduating, he lectured at Middle Tennessee State University, but decided to return to Bangladesh, when the country became independent in 1971. He became affiliated to Chittagong University, where he taught Economics, and became head of the economics department. On his daily walks through the streets in the village outside the campus, he was shocked to see poverty all around. He saw hard-working people, who simply lacked the chance to get ahead in life, regardless of their actions. From his interactions with them, Yunus learned that these people were trapped in the clutches of money-lenders, who determined how much they wanted to pay for the products the poor people produced. This way, their poverty was sealed, while the money lenders had a guaranteed and abundant income flow. Deeply contemplating on his impressions, Yunus decided to organize a research project with his economics students, in order to find out how much money the poor people in the nearby village owed to the money lenders. The amount was a little more than $27. 00. Yunus then visited the local bank, where he received confirmation to what was already known: poor people could not get loans, because the general perception existed (and still exists) that poor people are not creditworthy, and will not be able to pay back their loans. Yunus decided to loan the poor people the money out of his own pocket, and found that, contrary to what the banks assumed, he received 100% of the loan amount back. Unfortunately, the local banks were unwilling to surrender their viewpoint and give the poor people a chance, regardless of the data Yunus presented. This disturbing fact led to an early, but prophetic statement from Yunus: "The question is not 'Are people credit-worthy', but rather, 'Are banks people-worthy?'" (Black 2012).

Fascinated by the debilitating state of being stuck in an obsolete paradigm, and appalled by the vicious cycle of perpetuating poverty due to a system that disfavored the poor, Yunus established the Grameen Bank in Bangladesh in 1983, particularly fueled by the conviction that credit is a fundamental human right. The recipients of the first loans were 42 women, all basket-weavers from the village of Jobra. While the loans were minimal, the impact was tremendous, leading to a major culture change, whereby even beggars were enabled to receive loans (Hall 2013). Yunus' aim was to assist poor people in escaping poverty by granting them loans on terms that were appropriate to them and by teaching them a few sound financial principles so they could help themselves (Hall 2013). In 1983, Grameen Bank received authorization to perform as an independent bank, and Yunus could start realizing his dream of reducing, and possibly 1 day even eradicating poverty in Bangladesh. Grameen Bank offered collateral-free, income-generating housing, student, and microenterprise loans to poor families (Vlock 2009). At first, Grameen bank would loan money with a heavy emphasis on men (98%), with only 2% of female lenders. Yet, Yunus quickly found out that women were more serious in utilizing the money toward actual progress for their families, and were prompter in paying back their loans. This led the bank to prioritize loans to women, leading to a current base of 98% female lenders (Esty 2011). Woman who wanted a loan needed to have the support of a team of others, who would become coresponsible for repaying the money. "Over the years, [Yunus'] Grameen Bank, now operating in more than 100 countries, has loaned nearly $7 billion in small sums to more than 7 million borrowers-97% of them women. Ninety-eight percent of the loans have been repaid" (The World's Top 20 Public Intellectuals 2008, p. 55). Through his Grameen Bank project, Yunus has helped elevate 50 million Bangladesh people out of poverty. Since the start of Grameen Bank, the concept has been replicated in more than 100 countries (Hall 2013), and Yunus is seen as a global hero, who has been awarded numerous honorary doctorates and other tokens of appreciation worldwide.

Over the last decade, however, Dr. Yunus has been under siege by political leaders in his home country, due to his once expressed interest in starting his own political party in order to eradicate corruption. Even though he never started the party, the leader who came to power after Yunus' statement, Sheikh Hasina, had been fighting his every step along the way. Hasina started a vivid campaign against Yunus, and even had him ousted as the President of Grameen Bank in 2010, under claims that he would be past retirement age, thus too old to hold such a position. Unless Hasina changes her mind, chances are that Grameen will become a government bank, which places the micro-lending institution in a dire place regarding its service to the poor. Many charitable organizations and political leaders worldwide condemned the Bangladeshi government's actions, but the attacks have continued, and tax probes have been launched against Yunus and his seven social-business firms, accusing them of evading millions of dollars in taxes (Hall 2013). The group of opponents to the Nobel Peace Prize winner in their country has grown to include Islamic extremists, since Dr. Yunus made a public statement in support of gay people's rights.

While the politically fueled turmoil continues in Bangladesh, Yunus remains active on the world scene, attending numerous conferences globally and conducting presentations. He may seem unaffected by the political and religious hate campaigns in his country at the personal level, but is concerned about the possibility of Grameen Bank falling in the hands of the government, because he feels that it has always been and should remain to be the bank of the poor people.

Microcredit, as any other phenomenon, has its advocates and adversaries. There have been critics, who labeled microcredit "a death trap" or a "damaging intervention." Other sources question the rules at Grameen specifically, where loans are never forgiven but only restructured (e.g., Adams and Raymond 2008). Nonetheless, the numbers of microloan applicants have steadily grown – and continue to grow – worldwide, with Bangladesh being one of the nations with the highest percentage of micro-borrowers. The fact that this concept entices many people all over the world can be concluded from the immense interest that remains for Yunus' speeches wherever he goes. Yunus claims that repayment of microcredit has consistently leveled around 97% in Bangladesh, and in New York, where there are now 14,000 women using microloans, the repayment in over 99% (Hall 2013). He adds that microcredit has to be monitored in such a way that borrowers don't fall too far into debt, as this can otherwise become a financial burden rather than a relief. Yet, he claims that conventional banking has far more flaws, as it is merely structured on making the affluent richer, and shunning small individuals.

In his advocacy for the social business model, Dr. Muhammad Yunus underscores that all humans are entrepreneurs by our sheer history (Cosic 2017). He makes a point in favor of the social model over the commercial way of doing business. To effectively spread the message, he founded "Yunus Social Business" in 2011, an entity that operates globally by establishing social business centers at universities around the world. Acknowledging that poverty is a globally manifested phenomenon, Yunus now targets upcoming generations with his concept, informing youngsters about social and commercial business, and leaving them the choice to decide where they will go (Cosic 2017). Having successfully changed the lending culture in Bangladesh since the 1970s, and encouraging women to become entrepreneurs, he now implements his vision on a larger scale. Yunus defies the notion that entrepreneurship is just something a handful of people can do. He looks at the history and points out that all our ancestors were entrepreneurs, and that our current system with a small group that owns almost all wealth in the world is relatively new in human history (Cosic 2017). Today there are about 160 million people, mostly women, who use microcredit, demonstrating their entrepreneurial skills on a daily basis. Witnessing the contemporary global trends of outsourcing, robotics that eliminates human labor, and accelerated technological development, Yunus maintains that entrepreneurship is the solution to the problems of our times. Those who select social business as their mode of operation will become the trendsetters. France has been rather proactive in this regard, as there have been quite some Franch companies that joined the social entrepreneurial movement early on, such as Danone, which produces dairy products, and Veolia, which bottles clean water. Both companies have joined forces with Grameen Bank in order to provide healthy products at a reasonable price to the poor in Bangladesh. Corporations from other nations have also picked up on the social business trend. For instance, McCain, an American food company, works with Grameen to assist farmers in Colombia. Yunus is particularly proud of the fact that the company leaders and shareholders are increasingly valuing these kinds of collaboration, even though they yield less financial returns. On the other hand, however, there is a major feel-good component in doing something tangible for children in poor countries, and this emotion seems to be valid and appreciated amongst major corporations, such as the ones just mentioned (Cosic 2017).

Indeed, Dr. Muhammad Yunus' social entrepreneurial efforts are praiseworthy. Even though he remains a human being, making his fair share of mistakes as he goes, he has accomplished an immense service to the eighty million people in Bangladesh (about half the population) that uses Grameen services, whether microfinance, savings, insurances, eye hospitals, mobile phones, solar energy, high nutrition yoghurt from Danone, or one of the many other facilities aimed at enabling people to have a chance at a decent quality of life and economic growth (Black 2012).

In any of the projects and collaborations Yunus established in Bangladesh, he demonstrated insight, willpower, initiative, and a desire to break through irrational and unfair boundaries. He utilized every opportunity he had, reached out to international businesses to expand on the services he could give to the people in Bangladesh, and created a web of innovation that knows no match. Some of his innovations were original, others adopted from places where he had seen them working. And while some people may contest the term "innovation" in those cases, there is no denial that Yunus has first-mover qualities: he delivered improvements that were not yet implemented in his country. Yet, the aim never changed: helping poor people get a better life. Yunus has stated on many forums that his aim is to place poverty in museums (Black 2012), so that future people will only know what it looked like when they visit such a place with historic artifacts. A strong element in implementing the changes he could bring to Bangladesh was his capacity to establish relationships with people from all walks of life: heads of state, business leaders, entrepreneurs, opinion formers, grassroots innovators, and citizen organizations (Black 2012). His immense network has been highly instrumental in keeping him out of jail in his country, when the corrupted regime of Hasina came into power. Yet, as well-connected and awarded as Yunus is, as untouched is he by all the honors. He spends just as much time with students when he presents his lectures, as he does with presidents and CEOs. He has not become an elitist.

Moses (2014) shares the following six lessons for young social entrepreneurs from Muhammad Yunus:

1. He focused on an unserved or underserved market: as stated earlier in this chapter, he addressed the need of the micropreneurs in the village outside the university where he was teaching, and tapped into an untapped opportunity, which was, providing loans to the poor in order to help them stabilize their lives.
2. He dreamed big: While the Grameen project started at a small level, he realized its potential and dared to push it forward in multiple directions that ultimately far exceeded microlending. Starting with a few hundred, his project affected many millions over the past decades.
3. He collaborated with others: Understanding that growth cannot be perpetuated without assistance from others, Yunus sought out multiple partners, such as financial donors, but also global corporations to help steer Grameen to its current levels.
4. He diversified his business construct: he did not fall into the myopic practice of holding on to only his core activities, but awaited the right moment to expand his trajectory into areas that would enrich the experience of those who sought out the services of his institution. He branched out into trusts, funds, software, cybernet, knitwear, mobile phone, and more.
5. He helped others: with his Yunus Social Business (USB), he now helps young social entrepreneurs to start their venues, through networking and idea development.
6. Be prepared for criticism: in spite of the obvious success of his ventures, the concept of microfinancing has been and remains to be criticized by various scholars and entities.

In his own words, Dr. Muhammad Yunus describes social business as cause-driven. While the money invested can be recouped, the intention is not to take dividend beyond that point. Even though there is no objection to profits, the main purpose is to achieve social aims. Also, the profit should stay with the company. The company covers the expenses and makes profit while working toward accomplishing the social goals (Yunus 2007). Yunus alerts us that social business is relatively new in the business realm, and that it should not be seen as the end or a competition to commercial business, but rather an enrichment to the available options. He actually underscores that it is possible to run conventional and social business at the same time. The key is the aim to solve a clear and present problem in society. Yunus asserts further that social entrepreneurship brings a whole new perspective: it forces you to think creatively, reach out for support, learn new paths and strategies, and see things the way you did not consider them before. Most importantly, continues Yunus, engaging in social entrepreneurship makes you a happier person, because you feel good about what you are doing, and it provides you a way to change the world for the better (Yunus 2007).

Elon Musk

Elon Musk grew up in South Africa and comes from an adventurous family, with entrepreneurial skills instilled from both his parents: his father was a serial entrepreneur, and his mother a fashion model (Sun 2016). He started university at the tender age of 17 and ended up with two Bachelor's degrees: one in Physics from the University of Pennsylvania's College of Arts and Sciences, and one in economics from the same university's high profile Wharton School of Business. He dropped out shortly after starting a PhD program at Stanford University in applied physics and materials science, as he was more interested in pursuing his entrepreneurial goals. He has been married twice, once to a novelist, whom he met in college, but whom he soon considered insufficient as a wife, telling her that he would have fired her if she was his employee. He has five children with his first wife (twins and triplets). His second marriage was to a British actress, whom he married twice. Yet, the tumultuous marriage still ended in divorce in 2016 (Sun 2016).

The suave billionaire was not always a popular person. Growing up, he was bullied, because he was younger and smaller than most other children in the class, and in addition, also introverted and quiet. At one point, he was so severely beaten by other children, that he had to be hospitalized, according to his father. Fortunately, he was removed from the school after the incident (Sun 2016).

The combination of intelligence and entrepreneurship was converted into multiple millions when he sold his first company, Zip2, to Compaq. The year was 1999, and Musk was only 27 years old at the time. From there on, he founded multiple other companies, such as X.com, which later became PayPal, and was sold to eBay in 2002 for the nice little sum of $1.5 billion. That same year, he started SpaceX, an endeavor that almost became his downfall, because of three failed rocket launches, an activity that gulps up millions like no other. With relatively few resources left, the fourth rocket was launched in 2008. This time, the launch was successful, and Musk was back in business: he received several multibillion-dollar contracts from NASA to assist in the resupply of the International Space Station (ISS), and restore astronaut travel to and from the space station (Sun 2016).

Musk started Tesla Motors and launched his first electric car in 2008. Therein he also experienced some setbacks, as there were multiple recalls needed to improve the product. In fact, Bloomberg reported in 2013 that Tesla was leaning near bankruptcy. Fortunately, the product improved, and so did sales. As of 2016, the third generation of Tesla cars was scheduled for hitting the market in 2017, with about 325,0000 orders in presales (Sun 2016).

As of today, Elon Musk has been labeled a business magnate, investor, engineer, and inventor, and he may not be the most apparent embodiment of a social entrepreneur, but when we review the most well-known companies he is involved in, there is one overwhelming common factor: they are all geared toward improving a social problem about which he feels major concern:

- Tesla electric cars address the depletion of fossil fuels and air pollution.
- Solar City delivers inexpensive, clean, pure energy from the sun, rather than the expensive electricity extracted from fossil fuels such as coal and natural gas.
- The Hyperloop has practically a similar goal: rapid mass transportation on renewable energy to avoid the use of burning fossils, thus reducing air pollution.
- Space X, to actively seek a new habitat for a gradually overpopulating human cohort outside of planet Earth.

Admitted, these projects have not made him any poorer: Musk's net worth in 2017 has been estimated at about 16.6 billion dollars. Nothing to sneeze about, especially when we consider that he is only in his forties. Musk is known for his valiant work ethic, consisting of long hours and little vacation time. He is also known as a resilient individual, with the laudable ability to bounce back in times of setbacks. All these factors contribute to his success as a versatile entrepreneur. Yet, what may be the most important element of his success is exactly that: versatility. Musk does not limit himself to only one activity. Within his focus of reinventing unsustainable trends in society, he branches out into multiple areas, which could classify him as an expert-generalist: one who studies in multiple fields, understands the deeper principles that connect those fields, and applies the principles to their core specialty (Simmons and Chew 2016). Simmons and Chew (2016) warn that an expert-generalist is not the same as a "jack of all trades, master of none; because an expert-generalist knows how to combine his or her expertise with newly encountered fields of interest, in order to come up with unique innovations and strategies. Elon Musk has always ensured to keep himself well-informed: he is an eager learner and has been an avid reader since he was a teenager, reading an average of two books a day, which easily adds up to about 60 per month! (Simmons and Chew 2016). It is this learning habit of Musk that enabled him to bring together acquired knowledge from different fields, and then diffuse it into multiple areas again. In his case, he brought together the foundational principles from artificial intelligence, technology, physics, and engineering into a variety of services, such as aerospace (SpaceX), automotive (Tesla), and trains (Hyperloop) (Simmons & Chew). More recently, he also created OpenAI, a nonprofit artificial intelligence (AI) research company, Neuralink, a neurotechnology startup company, to integrate the human brain with artificial intelligence, and The Boring Company, a company that will specialize in tunnels.

While there may not be many people who consider Elon Musk a social entrepreneur today, Schrang (2015) makes a strong point to underscore that we should start looking at Musk as such: an unconventional social entrepreneur, who sells an expensive product, Tesla, to people who mainly care about showing off their prize speed-machine, but who have – deliberately or not – reduced their carbon footprint by driving on electricity rather than fossil fuels. Musk's story may not come across as the usual heartwarming story of helping a deprived group in society, as we have seen in the case of Muhammad Yunus, but his actions are not any less socially driven. As stated earlier: whether that is his agenda or not, Elon Musk produces electric cars that cause far less pollution in our air than those that run on gasoline. Through his renewable energy company, he tries to make humanity less dependent on the unsustainable fossil fuels, one solar panel, and one Tesla car at the time.

Musk's approach, therefore, illustrates that we should consider adopting a broader view of what social entrepreneurship is. The ultimate purpose of an action, and what is attained thereby, should also be considered within this context. The fact that Musk caters to the affluent rather than the poor demands a paradigm shift in our perception of social entrepreneurship, but who says that the rich don't need as much help as the poor, albeit in entirely different regards? Musk's companies are aimed to sort impact by spawning a systematic revolutionary change in the way we use energy. He is changing a destructive status quo, which has been in place for far too long. This is why he fits the label of a social entrepreneur very well.

It is also good to know that money was not Elon Musk's main driver in creating his current corporations. He has stated on several forums over time that his aim, from his college days on, was to affect the future in a positive way (Krasny 2014). According to Musk, he gets motivated by observing things that don't work well, becoming saddened about the impact this trend might have on the future, and a desire to fix it toward a better future (Sun 2016). Thus, the fact that the positive change Musk envisions with his activities happens in a high-dollar industry should be seen as a side effect, but not the ultimate driver. His plan with Tesla is to follow the three-step process that seems to be tied to new technology: first a low volume, high expensive product, then a medium-priced, medium-volume iteration, and finally, a low-priced, high-volume cycle. The fact that Tesla cars are still rather pricey today should therefore not be a predictor for the future production cycles (Sun 2016).

Mack (2016) also proposes the perception of Elon Musk being a social entrepreneur, since, according to his article, all Musk's actions are motivated by his concern about climate change and the use of fossil fuels that greatly promote this occurrence. He underlines the fact that large-scale use of fossil fuels will inevitably lead to future harm. He has been working on a massive facility to make batteries that will store renewable energy, hoping that villages in poorer countries will be able to use low-cost energy, and thus progress at a greater pace. He admits that he cannot do it alone: according to his calculations, there are about 100 giga-factories, such as the one he has built, needed worldwide. He, therefore, does not shy away from sharing his strategies, as his hope is, that more companies will follow suit, which will result in an accelerated transition to more sustainable energy worldwide.

With all of the points made above, Elon Musk could be considered a decent example of a contemporary leader practicing "right view." He has the ability to not only consider the major challenges our world faces, but also actively and devotedly work on finding solutions to these seemingly insurmountable problems. Musk is not an advocate of incrementalism but has revolutionary views on redesigning the future. His visions are not years but decades ahead of their time (Vance 2012). As a business leader, Musk actively works on addressing the heavy toll we have thus far taken on our environment and on finding a solution for our evergrowing human population. In regard to the first, his Tesla automobiles that drive on electricity rather than unsustainable and environment contaminating fossil fuels speak volumes. In regard to the latter, he has been diligently evolving Space Exploration Technologies, known as SpaceX, the first-ever private company to deliver cargo to the Space Station, as mentioned before. While the current activities of SpaceX are rather lucrative, Musk focuses on the bigger picture: occupying Mars as humanity's second home (Vandermey 2013). Having been likened to Steve Jobs more than once, Musk's most outstanding leadership quality, similar to Jobs, is not invention but vision. Having acquired a design thinking mindset through an early interest in science and history, and holding degrees in physics as well as business (Vandermey 2013), Musk has developed the critical leadership skill of seeing past the here and now, and responding to needs that are still considered insoluble by most. For that very reason, Musk has been called one of the greatest optimists in history, definitely when considering an optimist within the scope of physicist David Deutsch, who describes optimists as people who believe that "any problem that does not contradict the law of physics can ultimately be solved" (Vandermey 2013, p. 90).

Musk has determined for himself what a better future for humanity should look like, and he has inspired an entire legion of workers to help him realize this vision. He worked hard on his dream even before there was any external confidence in his vision, and he himself had a little or no certainty that he would succeed. In an interview, Musk explained that he considers the California bullet train a setback rather than a sign of progress, because it will move people from Los Angeles to San Francisco at a speed of merely 120 miles per hour, something they can achieve on the freeway by car. He boldly expressed his disappointment in such a project in a leading hub in the USA to the California Governor, Jerry Brown, underscoring that we should not focus on the glory of a small group, but on the progress of an entire nation (Musk 2014). Musk's attitude is one of great example to leaders of today: dream big and constructive with no immediate focus on money, as this will ultimately come in with much more abundance when perceived as a consequence than as a primary goal. Musk's ideas focus on macro wellbeing, and he has found a way to communicate these ideas in cross-disciplinary ways, leading to increased interest from a broad base of thinkers (Vandermey 2013). Musk is not in the business of merely making a fortune, but in the movement of making an impressive yet sustainable difference for the earth and its inhabitants. He has thus far translated this passion in

Tesla, SpaceX and SolarCity, and possibly also the Hyperloop, all focused on a better quality of life on (or off) mother earth.

Comparison of the two Cases

Presenting the examples of Muhammad Yunus and Elon Musk brought a few insights, which may be useful to share, as we conclude this section of the chapter. There are some interesting commonalities and some fascinating differences to be detected between these two entrepreneurs (Table 1).

Why Considering Social Entrepreneurship?

There are many reasons why social entrepreneurship has become such a popular phenomenon these days. As a final note to this chapter, here are some of the reasons that can be useful in considering social entrepreneurship:

Social entrepreneurship is a fulfilling practice. As Yunus presented it: it may not always bring a lot of money, but it definitely brings happiness. When doing something one feels good about, one cannot help but feeling fulfilled, even though there will undoubtedly be many challenges to overcome.

Social entrepreneurship is appreciated in society, and increasingly finds support from philanthropic individuals and entities that seek to be involved in doing the right thing. If, therefore, one has a viable and clearly developed idea on improving the wellbeing of small or large communities, chances are that sponsors and collaborators will emerge.

If one has an urge to become an entrepreneur, why not combine this wonderful quest with a laudable purpose? The generation of millennials that is currently entering the workforce holds high ideals on morally sound practices. A social enterprise fits better in this scope than one that is started for mere commercial causes. As Guy Kawasaki once stated in a Stanford University clip: we should start businesses to either (1) create something new that will bring positive progress, (2) correct something wrong, or (3) discontinue something bad. Yunus and Musk have operated mainly in the areas of points (1) and (2) above: they created corporations that brought positive progress with the underlying mindset of correcting something wrong.

Social entrepreneurs are respected and appreciated, because they are bringing about positive change, which others may have perceived, yet not pursued. By pursuing projects that improve social circumstances, a social entrepreneur steps into a heroic position. Social entrepreneurship opens doors, brings valuable connections, and elevates awareness.

There is never a dull moment in social entrepreneurship. Not all moments may be equally uplifting, but there is so much to explore, and there are so many derivatives that will manifest themselves once one starts with a social entrepreneurial project. Just consider our two examples: Yunus started Grameen Bank, but eventually ventured out into yogurt, mobile phones, savings, insurances, eye hospitals, solar

Table 1 Case comparison Yunus versus Musk

Common factors	Differences
1. Both entrepreneurs have started their most successful projects on basis of other considerations that making money: They observed a problem in society and decided to fix it. Yunus was disturbed by the status quo of banks refusing to lend money to the poor, simply because poor people have no collateral, so he created a bank specially to lend money to this group. Musk was disturbed by the grim visions of a rapidly overpopulating human society, using fossil fuels that caused harm to the environment, so he started companies that deliver products in renewable energy.	1. Yunus and Musk cater to entirely different customer populations: Yunus focused on the poor and destitute in his country, and wanted to give them a chance to function as "normal citizens," with similar rights as most. Musk initially focuses on the affluent, who can afford participating in his projects, yet with a long-term vision to make his projects accessible to less affluent members of society as well.
2. Both Muhammad Yunus and Elon Musk are known to be innovative, creative, and very resilient. Yunus has been ousted from his brainchild for political reasons, but is now operating on a much larger scale, spreading and enabling awareness on social entrepreneurship in universities worldwide. Musk has been on the brink of bankruptcy with both, SpaceX and tesla, yet consistently found ways to pursue his dream, and succeeding against all odds.	2. Yunus is an apparent social entrepreneur, as we have come to understand the concept, while Musk, being a billionaire, is not too often considered within a social context. Yet, the differences are potentially mere dollar signs, since the underlying factors are for both still improvement of the quality of life for the living.
3. Both Yunus and Musk are highly intelligent and insightful people: Yunus holds a PhD in economics, while Musk holds two Bachelor's degrees in physics and economics. Interestingly, their areas of performance are very much aligned to their educational background: Yunus solved an economic problem in his home country and beyond, while Musk focuses on areas related to the intersection between physics and economics to solve the problems that disturb him.	3. Musk's first business endeavors, while innovative and useful (Zip2 and X.com) were not specifically geared toward solving global problems. His urge to tackle major concerns may have always been there, but he was able to actually create businesses around those concepts after making millions with the two early ventures. Microlending was Yunus' first entrepreneurial project.
4. Both men have endured crises in their personal lives, particularly marriage-related. While not mentioned before, Yunus was married and had a daughter when returning to Bangladesh. His then-wife could not get used to the Bangladeshi turmoil, and returned to the USA with their daughter. He later remarried and got another daughter. He only saw his oldest daughter, now an opera singer, again when she was an adult. Musk has gone through three marriages with two women.	4. Due to the nature of his social project, Yunus focused primarily on women, as they demonstrated a higher level of responsibility in using the funds to the advantage of their entire family. In Musk's case, his customer base is not defined, but because he currently caters more to the space industry and the affluent, one could conclude that his audience thus far has been slightly more male-dominated.
5. Both men have an urge to share their projects, in hopes that others will follow suit, as they understand that global changes can only be made if there are implementers of their initiatives in multiple nations of the world.	5. Yunus' microlending project started on a microscale: By lending $27.00 to poor micropreneurs in a Bangladeshi village. Musk's SpaceX, SolarCity, and tesla companies required major investments from the start.

energy, and a general support system for social ventures worldwide. Musk started SpaceX and Tesla, then ventured out into SolarCity, the Hyperloop, and more recently, OpenAI, Neuralink, and The Boring Company.

As human awareness keeps augmenting, due to our massive exposure to multiple cultures, nations, and their problems, the enormous web of social entrepreneurial options reveals itself to younger and more mature individuals with an urge to do well while doing good. This is a very hopeful and proud trend for humanity, and one we should cultivate and bring to full fruition when, how, and where we can.

Cross-References

- ▶ Education in Human Values
- ▶ Empathy Driving Engaged Sustainability in Enterprises
- ▶ Ethical Decision-Making Under Social Uncertainty
- ▶ Moving Forward with Social Responsibility
- ▶ People, Planet, and Profit
- ▶ Responsible Investing and Corporate Social Responsibility for Engaged Sustainability
- ▶ Selfishness, Greed, and Apathy
- ▶ The Spirit of Sustainability

References

Adams, J., & Raymond, F. (2008). Did Yunus deserve the Nobel peace prize: Microfinance or macrofarce? *Journal of Economic Issues (Association for Evolutionary Economics), 42*(2), 435–443.

Bacq, S. s., Hartog, C., & Hoogendoorn, B. (2016). Beyond the moral portrayal of social entrepreneurs: An empirical approach to who they are and what drives them. *Journal of Business Ethics, 133*(4), 703–718.

Black, L. (12 September 2012). Muhammad Yunus: The model social enterprise leader? *The Guardian*. Retrieved from https://www.theguardian.com/social-enterprise-network/2012/sep/12/muhammad-yunus-social-enterprise-leader.

Cosic, M. (28 March 2017). 'We are all entrepreneurs': Muhammad Yunus on changing the world, one microloan at a time. *The Guardian*. Retrieved from https://www.theguardian.com/sustainable-business/2017/mar/29/we-are-all-entrepreneurs-muhammad-yunus-on-changing-the-world-one-microloan-at-a-time.

Dacin, P. A., Dacin, M. T., & Matear, M. (2010). Social entrepreneurship: why we don't need a new theory and how we move forward from here. *Academy Of Management Perspectives, 24*(3), 37–57.

Dart, R. (2004). The legitimacy of social enterprise. *Non-Profit Management & Leadership, 14*, 411–424.

Dey, P., & Lehner, O. (2017). Registering ideology in the creation of social entrepreneurs: Intermediary organizations, 'ideal subject' and the promise of enjoyment. *Journal of Business Ethics, 142*, 753–767.

Dudnik, N. (October 18, 2010). Social entrepreneurs' tricky issues of sustainability and scale. *Harvard Business Review*. Retrieved from https://hbr.org/2010/10/social-entrepreneurs-tricky-is.

El Ebrashi, R. (2013). Social entrepreneurship theory and sustainable social impact. *Social Responsibility Journal, 9*(2), 188–209.

Elkington, J., & Hartigan, P. (2008). *The power of unreasonable people: How social entrepreneurs create markets that change the world*. Boston: Harvard Business Review Press.

Esty, K. (2011). Lessons from Muhammad Yunus and the Grameen Bank. *OD Practitioner, 43*(1), 24–28.

Hall, R. (26 October 2013). 'What did I do wrong?': Why the banker who helped millions of Bangladeshis out of poverty became his country's enemy number one. *Independent*. Retrieved from http://www.independent.co.uk/news/world/asia/what-did-i-do-wrong-why-the-banker-who-helped-millions-of-bangladeshis-out-of-poverty-became-his-8899838.html.

Hartigan, P. (2006). It's about people, not profits. *Business Strategy Review, 17*(4), 42–45.

Jiarong, Y., & Shouming, C. (2016). Gender moderates Firms' innovation performance and Entrepreneurs' self-efficacy and risk propensity. *Social Behavior & Personality: An International Journal [serial online], 44*(4), 679–692.

Krasny, J. (March 6 2014). The real reason Elon Musk became an entrepreneur. *Inc.com*. Retrieved from https://www.inc.com/jill-krasny/elon-musk-tesla-motors-stanford-business-school.html.

Mack, E. (Oct 30, 2016). How tesla and Elon Musk's 'Gigafactories' could save the world. *Forbes*. Retrieved from https://www.forbes.com/sites/ericmack/2016/10/30/how-tesla-and-elon-musk-could-save-the-world-with-gigafactories/#3aaa047a2de8.

Marques, J. (2016). *Leadership: Finding balance between ambition and acceptance*. New York: Routledge [Taylor & Francis].

Martin, R. L., & Osberg, S. (2007). Social entrepreneurship: the case for definition. *Stanford Social Innovation Review, 5*(2), 28–39.

Mastrangelo, L. M., Benitez, D. G., & Cruz-Ros, S. (2017). How social entrepreneurs can influence their Employees' commitment. *Journal of Promotion Management, 23*(3), 437–448.

Moses, N. V. (18 March 2014). 6 lessons for young social entrepreneurs from Muhammad Yunus. *Your Story*. Retrieved from https://m.yourstory.com/2014/03/6-lessons-young-social-entrepreneurs-muhammad-yunus/.

Muhammad Yunus Biographical. (2006). The nobel foundation. Retrieved from: https://www.nobelprize.org/nobel_prizes/peace/laureates/2006/yunus-bio.html.

Musk, E. (2014). I hope artificial intelligence is nice to us. *NPQ: New Perspectives Quarterly, 31*(1), 51–55.

Osberg, S. R., & Martin, R. L. (May 2015). Two keys to sustainable social Enterprise. Harvard business review. Retrieved from https://hbr.org/2015/05/two-keys-to-sustainable-social-enterprise.

Praag, van, C. M., & Versloot, P. H. (2007). What Is the value of entrepreneurship? a review of recent research. *Small Business Economics, 29*(4), 351–382.

Roper, J., & Cheney, G. (2005). Leadership, learning and human resource management: The meanings of social entrepreneurship today. *Corporate Governance, 5*(3), 95–104.

Schaltegger, S., & Wagner, M. (2011). Sustainable entrepreneurship and sustainability innovation: Categories and interactions. *Business Strategy and the Environment, 20*(4), 222–237.

Schrang, T. (2015). Elon musk, a social entrepreneur? What social entrepreneurs can learn from tesla. *Medium*. Retrieved from: https://medium.com/@Sastre_desastre/elon-musk-a-social-entrepreneur-da3e0eef895f.

Schumpeter, J. A. (1934). The theory of economic development. London: Oxford University Press.

Seelos, C., & Mair, J. (2007). Profitable business models and market creation in the context of deep poverty: A strategic view. *The Academy of Management Perspectives, 21*(4), 49–63.

Simmons, M., & Chew, I. (Aug 11, 2016). How to think like Elon Musk. *Fortune*. Retrieved from http://fortune.com/2016/08/11/how-to-think-like-elon-musk/.

Snell, D. (16 August 2012). Building sustainability into your social enterprise. *The Guardian*. Retrieved from https://www.theguardian.com/social-enterprise-network/2012/aug/16/funding-social-enterprise-bootstrap-growth.

Strauss, K. (Jan 21, 2015). What makes an entrepreneur? *Forbes*. Retrieved from https://www.forbes.com/sites/karstenstrauss/2015/01/21/what-makes-an-entrepreneur/#38be8e365238.

Sun, C. (April 3, 2016). Here's what you need to know about Elon Musk. *Entrepreneur.com*. Retrieved from: https://www.entrepreneur.com/slideshow/273580.

Testé, B. (2017). Control beliefs and dehumanization: Targets with an internal locus of control are perceived as being more human than external targets. *Swiss Journal of Psychology, 76*(2), 81–86.

The World's Top 20 Public Intellectuals. (2008). *Foreign Policy*, (167), p. 55.

Tobak, S. (January 25, 2016). What makes a successful entrepreneur? Perseverance. *Entrepreneur.* Retrieved from https://www.entrepreneur.com/article/269840.

Vance, A. (2012). Elon Musk, man of tomorrow. (cover story). *Business Week,* (4296), 73–79.

Vandermey, A. (2013). The shared genius of Elon Musk and Steve jobs. *Fortune, 168*(9), 98.

Vlock, D. (2009). Teaching Marx, Dickens, and Yunus to business students. *Pedagogy: Critical Approaches to Teaching Literature, Language, Composition, and Culture, 9*(3), 538–547.

Yitshaki, R., & Kropp, F. (2016). Motivations and opportunity recognition of social entrepreneurs. *Journal of Small Business Management, 54*(2), 546–565.

Yunus, M. (2007) Social business. Yunus Centre. Retrieved from http://www.muhammadyunus.org/index.php/social-business/social-business.

Zahra, S. A., Gedajlovic, E., Neubaum, D. E., & Shulman, J. E. (2009). A typology of social entrepreneurs: Motives, search processes and ethical challenges. *Journal of Business Venturing, 24*, 519–532.

Sustainable Decision-Making

Moving Beyond People, Planets, and Profits

Poonam Arora and Janet L. Rovenpor

Contents

Abstract

This chapter highlights the immense difficulty of achieving sustainability unless the language of business changes from an obsessive focus on profit, and leaders use new frameworks to view business problems. Traditional framing in business includes the worlds of accounting, warfare, sports, and games, which are at odds with humanistic and holistic approach essential to developing leaders with the courage and engagement needed to address the challenge of the sustainability of our planet. The chapter introduces the need for a new language for sustainable business, examines the unhappiness of the millennial generation with the current language, and explores how business-as-usual changes when a more holistic language is used. The chapter then examines how when the decision-making

P. Arora (✉) · J. L. Rovenpor
Department of Management, School of Business, Manhattan College, Riverdale, NY, USA
e-mail: poonam.arora@manhattan.edu; janet.rovenpor@manhattan.edu

S. Dhiman, J. Marques (eds.), *Handbook of Engaged Sustainability*,
https://doi.org/10.1007/978-3-319-71312-0_40

frame changes from one of war and losses to one of community and social learning, decision makers are more likely to reframe both their goals and the focus of their analysis to go beyond monetary outcomes to include socially and environmentally sustainable outcomes. It also analyzes the role of holistic language, motivation and appropriateness in moving organizational cultures from a single-minded profit focus to one of triple bottom line sustainability. The chapter ends with a road map and several examples of how organizations can change the way they frame sustainability problems, develop new goals that are holistic in nature and aligned with global sustainability goals, and implement metrics to keep track of real progress towards sustainability.

Keywords
Sustainability · Sustainable decision making · Triple bottom line · Framing · Holistic management · Poetry and business

Introduction

The Increasing Importance of Sustainability

Despite recent political rhetoric to the contrary, the scientific community continues to present evidence that environmental issues are urgent and climate change is a reality (https://www.nasa.gov/press-release/nasa-noaa-data-show-2016-warmest-year-on-record-globally). The Global Footprint Network, a nonprofit organization that keeps track of ecological overshoot, estimates that human beings use significantly greater ecological resources and services than nature can regenerate through its restorative processes. Our waters are overfished, forests are over-harvested, and more carbon dioxide is released into the atmosphere than can be absorbed. If the current state of affairs remains unchecked, human demand on the earth's ecosystems is expected to exceed what nature can regenerate by 75% by 2020 ("Overshoot Day," 2017).

The good news is that business leaders are paying attention. According to the *KPMG 2013 Survey of Corporate Responsibility Report* (which covers 4100 companies in 41 countries), the number of firms voluntarily reporting some noneconomic (i.e., environmental and social) measures to corporate sustainability rating agencies increased from approximately 10% to over 90% between 1993 and 2013. This increase was fueled in part by greater stakeholder expectation that reputable businesses be concerned with sustainability.

In 2013, almost all of the top 250 corporations on *Fortune's* Global 500 list provided some type of reporting on their corporate social responsibility (CSR) activities. Eighty-two percent of them referred to the Global Reporting Initiative (GRI) guidelines which emerged from work carried out by CERES, a Boston-based nongovernmental organization (NGO) focused on encouraging the adoption of sustainable business practices and solutions by economic entities. CERES was previously known as the Coalition for Environmentally Responsible Economies (see www.ceres.org for more information). Though GRI's original purpose (dating back to the 1990s) was to provide mechanisms to determine environmentally responsible conduct, over the years the framework and protocols have expanded to include environmental, social, and governance activities (ESG). In its latest iteration, the fourth generation of the guidelines (called G4) provides reporting principles and standard disclosures. They also identify the criteria that an organization should use to prepare its sustainability report, including evidence of economic, environmental, employee, shareholder, and stakeholder impact (see www.globalreporting.org for more information).

The bad news is that corporate efforts to really understand and implement "sustainability" and "sustainable development" in their operations are only half-hearted. Until managers can define what these concepts mean in different organizational decision-making contexts, they will be slow in developing real goals and strategies to address today's pressing environmental challenges. The underlying issue is that sustainable development requires a consideration of the ethics and ecology underlying the economics. Current business focus does not allow for the in-depth consideration of such an interconnected concept. Thus, in an effort to more clearly define sustainability, businesses are being encouraged to focus on the so-called triple-bottom line: economic well-being, environmental quality, and social justice. But Milne and Gray (2013, p. 24) note that "such conceptions are entity focused and reinforce the notion that business first not ecological systems must remain going concerns." Managers are far more comfortable with developing concrete and measurable indicators of financial performance than they are with viewing the world as a fragile ecosystem in which negligent actions in one part of the world have significant ripple effects for global economies, social systems, and natural environments.

Another major problem is that companies frequently use their reporting about sustainability as a surrogate for actually making progress toward being more sustainable (Milne and Gray 2013). Although concern with sustainability is usually accompanied by measurement across the triple-bottom line, simply reporting ESG outcomes is not evidence of positive overall impact on environment and society. Additionally, evidence shows that firms tend to report cherry-picked positive ESG activities while overlooking aspects of their operations that have negative impacts and connotations. Even worse, some managers engage in hypocrisy, erect organizational facades, and use impression management tactics to manage expectations, preserve corporate reputations, and give the appearance that they are working hard at social and environmental issues.

Progress can be made if conversation in boardrooms, drawing attention to existing environmental and social consequences of economic decisions, and introducing new levels of transparency and changes in legislation continue (Wilburn and Wilburn 2014). Greater cooperation among different CSR rating agencies also suggests a move toward understanding that economic growth, social well-being, and environmental quality are interdependent. Even more dramatic progress can be made when decision makers in organizations develop business strategies that embrace three distinct but inter-related worldviews: rationalism, naturalism, and humanism (Senge et al. 2007). Rationalism refers to the efficient use of resources (i.e., getting the most output with the least amount of input), naturalism considers human activity as part of a larger ecosystem in which species are interdependent, and humanism recognizes that individuals in today's society seek meaning and purpose. The integration of the three world views is not just a theoretical exercise, the Network for Business Sustainability has developed a framework for planning for businesses that integrates the three world views making it possible for business to move beyond talk to implementation of the triple bottom line.

Understanding the Role of Framing in Sustainable Decision Choices

Research in social cognition and behavioral decision-making strongly demonstrates that the way in which a problem is framed shapes opinions and influences the choices made by individuals. A frame is "a socially based, abstract, high-level knowledge structure that organizes certain information about the world into a coherent whole" (Huckin 2002, p. 354). Generally speaking, a frame consists of a set of words, metaphors, or symbols that are carefully chosen by a communicator over others in an effort to persuade and convince an audience to adopt the same world view. Framing is in itself just a technique that effectively uses language to create a context without any inherent morality. The resulting context, however, can and frequently does have moral implications. Thus, framing can be used to encourage individuals to engage in prosocial and altruistic behaviors or self-enhancing, community-eroding behaviors, and these behaviors can have a long-term or short-term impact. Framing, therefore, can also have temporal implications.

There are many interesting and powerful examples of how framing shapes attitudes and behaviors across multiple domains, ranging from politics and business to social interactions and psychology. In the area of environmental sustainability, fishermen can be persuaded to change the way they catch fish from pouring cyanide on or using dynamite to blast open coral reefs, to using nets and rods, based on whether they perceive the reef as a rock or resource to be exploited versus a living entity that supports a complex system of many species and needs to be treated with care so that the habitat can self-perpetuate. Spence and Pidgeon (2010) found that when information on climate change was framed positively as a gain (e.g., "By preventing further sea-level rises, we can prevent the inland migration of beaches and save up to 20% of coastal wetlands, maintaining the habitat availability for several species that breed or forage in low lying coastal areas") instead of negatively as a loss (e.g., "With further sea-level rises, beaches will migrate inland and threaten up to 20% of coastal wetlands, reducing the habitat availability for several species that breed or forage in low lying coastal areas"), subjects viewed climate change mitigation more favorably and perceived climate change impacts more severely.

In the domain of social cognition, Liberman et al. (2004) conducted a series of experiments using a version of the Prisoner's Dilemma (a classical game which examines the extent to which two participants will cooperate with each other for mutual benefit or compete against each other in pursuit of their own self-interest). In this particular simulation, half of the participants in the sample were told that there were going to play a game called the "Wall Street Game" while the other half were told that they were going to play a game called, the "Community Game." In both situations, the game was the same (except for the title). It was assumed that the label, the "Wall Street Game," would connote rugged individualism, competitiveness, and self-interest, whereas the label, the "Community Game," would signify interdependence, cooperation, and collective interest. The results showed that a greater percentage of subjects playing the "Community Game" cooperated, achieving outcomes that were collectively greater and more equitable, compared to subjects playing the "Wall Street Game" – merely changing the name changed the frame and the ensuing behavior.

In the political domain, George Lakoff (2016), a well-regarded cognitive linguist, has argued that President Donald Trump's use of simple, short phrases have a great subconscious appeal to listeners because they activate the neurons in the brain and increase the perception that an idea is more probable than it really is. When Trump uses repetition – "We're going to win, win, win" – an individual's neural circuitry gets stronger and stronger. Furthermore, when the term "radical Islamic terrorists" is said over and over, the link between "Islam" and "terrorism" is reinforced, consequently resulting in a public that fears Islam as terrorism. In fact, Lakoff argues that conservative politicians endorse a world view that associates them with "family values" and "fatherhood." This motivates their supporters to view opposing positions in high-valence issues, such as abortions and the human role in climate change as misguided and immoral, causing significant misunderstanding among counterparts and a breakdown in the open exchange of ideas.

Framing can also occur spontaneously where the individual imposes, frequently subconsciously, a certain frame upon a context. For example, ownership of an object has psychological consequences where owners spontaneously "endow" the object they own with more positive than negative attributes as well as assign greater monetary value to it (Thaler 1980). A frequently observed example of this is when a homeowner insists that the sales price for a much-loved home be set higher than the value assigned to it by the market due to its perceived "special characteristics."

Arora et al. (2015) show that agribusinesses in the Argentine pampas, one of the most fertile areas in the globe and a net producer of global food commodities, frame their goals for the land they own differently from their goals for land they rent. Although the economic goals for all the activities are the same for the agribusiness (to optimize profitability across the product portfolio), their focus in the farmed-land that they own is on the long term, and thus they are willing to invest in maintaining the quality of the land, employees-learning, and creating social capital in the local communities. This is in stark contrast to the approach in lands that they rent, where they perceive the rent-payment as a loss and are more likely to focus on making back the rent money spent by maximizing profit and minimizing any costs or investments. Arguably, the spontaneous framing made salient by the context impacts the overall sustainability of the agribusiness sector in Argentina – as a larger agricultural area is rented, the switch from the holistic ownership frame that prompts long-term investment in the quality of the land and community development is replaced by the short-term profit maximization frame of a renter.

The true impact of framing, therefore, is felt in its influence upon which choices appear salient and how options are perceived when solving problems and making decisions. In the example of the Argentine agribusinesses, any expenses incurred to maintain land that is owned are framed as investments and perceived as gains in the long run, while similar expenses are framed as costs and losses when the land is rented. Not surprisingly, when framing of the purpose of a business is defined in economic terms, noneconomic metrics support the environmental and social aspects of the triple bottom line as viewed as external to the true purpose of business; they are onerous costs to be avoided.

The Language of Business: The Need for a New Frame

Current Business Jargon Provides Narrow and Competitive Frame

As observed, language can be manipulated in order to promote the adoption of a particular point of view, which in turn establishes the context for decision-making. The contextual framing establishes priorities by highlighting who and what are most salient and therefore important in the decision process.

In his *Encyclical Letter*, dated May 24, 2015, Pope Francis issued an urgent appeal with regards to the care of our "common home" (Bergoglio 2015). The only way to really protect it was to "bring the whole human family together to seek sustainable and integral development." The pope called for a "new dialogue about how we are shaping the future of our planet. We need a conversation which includes everyone, since the environmental challenge we are undergoing, and its human roots, concern and affect us all." Multinational corporations were expected to join the challenging effort since they were responsible for unemployment, abandoned towns, depletion of natural resources, and pollution of rivers when, for financial reasons, they ceased operations in developing countries.

Unfortunately, the pope's new dialogue and holistic approach to better addressing one of the most significant societal problems we face is at odds with the language and single-minded pursuit of economic goals adopted by many of our most prominent leaders in the business community. Chief executive officers (CEOs) and business leaders continue to frame their companies' futures egocentrically, in terms of sports, competitive games, and military warfare. A quick survey of the business headlines in the *Wall Street Journal* during the first week of May 2017 showed the top five most frequently used words: win, lead, war, game, and earnings. The next five words were of similar vein (price, million, more, money, and accounting). Despite some talk about consideration of people and planet-related variables, business leaders have not seriously started to either use holistic thinking or to develop a robust set of metrics for "triple-bottom line" performance, which involves equal attention to planet, people, and profits. Books such as Sun Tzu's *The Art of War* are frequently found on the reading lists of business leaders, highlighting the attention paid to economic outcomes and the strong emphasis on "winning strategies" for the "game/war" of business. This chapter argues that such situational framing may serve to justify overly competitive and aggressive behaviors. If a business is "at war," then it is acceptable to cut corners, break promises, develop shoddy products, spy on competitors, mistreat employees, and harm the environment. If a business is "a game," it is similarly justifiable to cheat just a little to win for a game is rarely seen to have consequences that can be the difference between life and death. Given climate change, however, life and death may be exactly what's at stake.

The current language of business, for the most part, may inadvertently provide the very opposite of the holistic approach that business leaders and their employees must take to promote greater sustainability. Rationality, viewed as the bedrock of business, economic, and scientific thinking (Scott 2000), assumes that the optimal technique for decision making is deliberate calculation that involves an effortful cognitive process aimed to maximize self-interest (Smith 1991). Thus, the prescription in business is to approach decisions across contexts by using deliberate calculative strategies, and not surprisingly, the language used reflects as well as encourages a calculative mindset. Implicit support for this explicit understanding is reinforced by use of stories like Warren Buffett suggesting that the best college major is "Accounting – it is the language of business" (as cited in Buffett and Clark 2006). The analogy resonates deeply since business activities culminate in the corporate annual report and represent a way for them to "keep score," by disclosing revenues, profits, cash flow, assets, and debt. As important as these functions are to the viability of a business, it is equally imperative that businesses not be constrained to or defined by its mere economic sustainability – environmental and social sustainability are at least as, if not more, important to the overall success of the business.

Business and organizations are characterized by the use of language that keeps track of winners and losers, that encourages weighing costs and gains, and that provides a strong market perspective. Thus, in addition to the language of accounting, one finds a prevalence of the use of the language of warfare and sports (Camiciottoli 2007). A quantitative study of magazine and newspaper texts on the business strategy of mergers and acquisitions found that the "fighting metaphor" was predominant (Koller 2004). For years, military terminologies that is "market invasion," "price wars," and "guerilla warfare," were used in the marketing discipline (Laufer 2010). The war analogy provides the strategic language for marketing professionals, and as long as it dominates current thinking, it crowds out social concerns. Most metaphors used in business are "rhetorical" such as "we hit a home run" or "the bubble burst" which do not promote different ways of thinking and do not offer new insights (Von Ghyczy 2003).

When CEOs communicate with others during shareholder meetings, earnings calls, and televised interviews, they do more than just objectively report on their firm's performance. They weave a story and provide a context, a frame, in which they justify their cost cutting actions because they were faced with "a fiercely competitive environment" or they praise their less than spectacular earnings because they "persisted and successfully weathered an unprecedented economic, political and social storm." They reassure shareholders that they will continue to develop winning strategies in the years to come.

In a quarterly earnings call, when John Chambers, CEO of CISCO, turned to his successor, Chuck Robbins, he said, "You and I are on the 18th hole, we're already ahead by five strokes in a team play, and all of our competitors have hit their golf ball into the woods. By the time we're through, they're still going to be looking for their golf balls" (as quoted in Clark 2015). The analogy to golf creates the impression that business is a game to be won instead of a means to important ends for shareholders, employees, customers, and citizens, as well as a way to protect the environment so that their wants and needs can continue to be satisfied.

In his 2014 letter to shareholders, Jamie Dimon referred to his company, JP Morgan Chase, more than once as an "endgame winner" which had "well-fortified moats" to protect it from debilitating competition and unforeseen events. We find here a "winner takes all" attitude combined with overly defensive posturing. The language is intended to assure shareholders that JP Morgan has what it takes to succeed, but it also models appropriate behaviors that are aggressive and combative in nature. Such language, however, may not be sufficient for today's challenges because it is "a technical language, honed to its specific purpose but constrained in wider, more complex applications" (Doughty as cited in Windle 1994, p. 2).

There are, of course, some notable exceptions. In a reading of Salesforce.com's recent letter to shareholders, we see a broader focus on revenue growth, customer service, top rankings for one of the "top ten companies to work for" and one of the "most admired software" companies, as well as information on how the company is not just transforming business but also communities. It emphasizes employees' adherence to its core values of trust, innovation, growth, and equality. Salesforce.com's CEO, Marc Benioff, is a lead advocate for social change. He has been pressuring politicians to develop policies to close the gender gap in pay and to protect employees with religious beliefs from being fired by faith-based organizations (Langley 2016).

The framing of business in the language of accounting, warfare, and sports has several concerning consequences. If a business is "at war," then the stakes are high and it might be acceptable to cut corners, break promises, develop shoddy products, spy on competitors, mistreat employees, and harm the environment. If a business is just a "game," then a little bit of cheating can be forgiven as long as the referee does not notice.

Hamington (2009) cautioned that viewing business as a game could result in four potential harms:

1. Compartmentalizing morality
2. Truncating ethical content
3. Trivializing stakes
4. Privileging adversarial relationships.

Language that encourages a deliberate, calculative mindset, as is the case with metaphors of war and games, though grounded in rational theory, activates calculative assessment of even nonmonetary contexts. It has been shown to result in number of negative, including disengagement from work, declining ethics, lack of interpersonal trust, lower concern for others, and greater concern for oneself. In addition, such language may be viewed as the language of the past and thus not be fully embraced or even accepted by the Millennials – those entering or in relatively junior positions in the workforce.

The Millennials Disagree with the Language and Intent of Business-as-Usual

Millennials (those who are currently between the ages of 16 and 34 years) have expressed dissatisfaction with the core values for long-term success currently adopted by most companies, and yet future business leaders will come from among them. The Deloitte 2016 Millennial Survey (which surveys only 16 to 21 year olds rather than all Millennials) concluded that a significant leadership gap exists between the priorities Millennials would have if they led their organizations and where they believe their senior leadership teams are focused (Fuller 2016). As in previous surveys, Millennials continue to place far greater emphasis than current leaders on "employee wellbeing" and "employee growth and development." They would be less focused on "personal income/reward" or "short-term financial goals." Millennials would like businesses to devote effort towards improving the skills, income, and "satisfaction levels" of employees, creating jobs, and ensuring that their goods and services have a positive impact on users. Almost 9 in 10 Millennials believe that "the success of a business should be measured in terms of more than just its financial performance" (Fuller 2016).

Millennials are calling for a new language a business. In Deloitte's 2014 survey of young millennial (www.deloitte.com) aged 16 to 21 years, almost 50% strongly disagreed with what they felt were frequently used business phrases:

- It's a dog eat dog world, if I don't bite first, I'll be eaten
- Business is war
- The purpose of business is shareholder value
- You are always competing – against other businesses, other employees

Millennial disengagement is so high that two in three Millennials expect to leave their current places of employment by 2020 (Fuller 2016). When asked what positive impacts business had on society, businesses got high marks for "creating jobs and increasing prosperity." Millennials however gave corporate executives low marks for what they felt were the three other main challenges of the twenty-first century: climate change and environment, managing resource scarcity, and reducing inequality of incomes. Millennials do not seem to put much faith in the willingness of today's executives to protect their future or ensure that they will be able to thrive in a world of abundance. Their mistrust of corporations is turning them off from current organizational workplaces. They will not become cooperative partners of business unless executives work with them to develop a sense of shared values and mutual respect. As consumers, Millennials are starting to exert their purchasing power by shopping locally at farmer's markets, buying fair trade products, and cleaning out shelves stocked with organic goods.

Millennials also seek new educational opportunities that are not necessarily found in today's graduate programs in business. A small-scale study found that MBA applicants whose possess a balance of self-interest and social good are attracted to a graduate curriculum that views spiritual qualities (e.g., respect for others, transcendence and social justice) as an integral part of enhanced managerial capacities offered by a Catholic institution (More and Todarello 2013). Its MBAE program was designed to "capture and connect to the spiritual core of future students" (More and Todarello 2013, p. 23).

It is urgent for today's business leaders to assume greater responsibility towards society's future decision-makers by showing much greater concern for people, planet, and profits. They are at risk of causing low morale, skepticism, and poor performance if they do not keep up with the growing concerns and demands of the next generation of managers. Incremental changes and lip-service to the triple bottom line is no longer sufficient – there is a need for a new framing of the business context and it begins with a new language for business.

The Arts as a New Frame for Business

The Holistic Frame of Poetry and the Arts

The arts can provide insights into the new lens by which to view the mission and goals of modern day organizations and their significant relationships to society to ensure sustainability. Bartunek and Ragins (2015) called on scholars and scholar-practitioners to increase their understanding of (a) the types of art that can inspire their thinking and theorizing and (b) the ways in which the arts can open their minds to fresh ideas. They referred to the following relevant art experiences: poetry, fine art, crafts, film, documentaries, photography, dance, theater, music, architecture, and others. Arts-based methods can result in skills transfer; reflection through projection; illustration of the essence; and the release of subconscious ideas, experiences, and emotions through the process of creating something physically.

While all arts appear to be broadening, with the capacity to help leaders create the basis for long-term sustainable growth for their organizations without having to "win" at a cost to others, it is suggested that poetry, in particular, can help leaders create a more holistic approach towards both their roles and the cultures of their organizations by changing their immediate mindset and focus (Adler 2015). Because poems are multidimensional, they promote the ability to detect different modes of meaning and to deal with ambiguity and uncertainty, which are fundamental to decision making under conditions of climate change and environmental volatility. Because they are almost infinitely interpretable, poems help develop the ability to consider other viewpoints and to examine and revise current insights and perceptions. Because poems draw attention to human needs and motivations, they allow one to address ethical issues (Morgan 2013).

In the next section, research showing the positive impacts that poetry can have on attitudes and emotions is described.

Literary Neuroscience: A growing number of scientists are studying the brain wave activity of subjects as they read prose compared to poetry and as they read a poet's actual verses compared to a simplified translation. Using fMRI technology, Zeman et al. (2013) found that brain wave activities of volunteers differed when they read literary prose (e.g., an extract from a heating installation manual) compared to when they read poetry (e.g., sonnets). Poetry activated the posterior cingulate cortex and medial temporal lobes, which have been linked to introspection. Davis and his colleagues at Liverpool University (as reported in Henry 2013) conducted a study in which the brains of volunteers were scanned when they read four original lines by William Wordsworth and four easy to comprehend, translated lines:

She lived unknown, and few could know
When Lucy ceased to be;
But she is in her grace and oh,
The difference to me!
Versus:
She lived a lonely life in the country,
And nobody seems to know or care,
But she is dead,
And I feel her loss.

The original lines triggered greater brain activity in the left hemisphere (for language) and in the right hemisphere (for reflection, autobiographical memory and emotion) than the translated lines. Davis concluded that "Serious literature acts like a rocket-booster to the brain" and that it is better than "self-help books" in dealing with serious human situations (as quoted in Henry 2013).

Organizational Behavior: Research in this area suggests that poetry can help individuals connect deeply with their inner emotional self that, in turn, elicits empathy and the ability to connect at an emotional level with others. It can add value to a business student's self-awareness and artistic expression (Morris et al. 2005). Van Buskirk and London (2012) found that poetry helped students in an Organizational Behavior course to develop new insights and express themselves better as well as arrive at a deeper understanding of the course content. Morris et al. (2005) concurred and found that presenting students with poetry enhances emotional intelligence, which is so critical for effective management.

Romanowska et al. (2014) found that participants in an arts-based leadership program (which included contrasting phrases of poetry) showed less laissez-faire management, increased self-awareness, improved humility, and greater capacity to handle stress compared to participants in a conventional leadership program (which included lectures on organizational and leadership theories). Parker (2003) reported that writing and studying poetry helped managers improve their business writing. Poetry writing has been shown to encourage creative exploration and informed empathy; it can also be used by leaders as a tool to build trust, demonstrate empathy, communicate more effectively, and inspire others (Grisham 2006).

Strategy: Morgan (2013), a fiction writer, critic, and director of the graduate creative writing program at the University of Oxford, was hired by the BCG's Strategy Institute to study the relationship between poetry and strategic thinking. She linked the characteristics of poems to the development of certain managerial and leadership skills, concluding that poetry helps the reader become a "sharpener" instead of a "leveler." A sharpener is a person who can tolerate ambiguity, is ready to think and perform symbolically, and keeps in mind, simultaneously, various aspects of the whole. A leveler suppresses differences and emphasizes similarities, seeks perceptual stability, is anxious to categorize sensations, and is unwilling to give up a category once it has been established.

There have been many corporate executives who were also great poets, including Wallace Stevens (who was the vice president of the Hartford Accident and Indemnity Company) and TS Eliot (who worked for 10 years at Lloyd's Bank of London). Dana Gioia is an American poet and the chair of the National Endowment for the Arts. He was also a marketing director of General Foods. Gioia credits his ability to turn around the Jell-O product line from a $7 million loss to a $20 million profit in the 1980's to skills he developed as a poet. He remarked, "How did it happen? I looked at things differently. I made associative connections. I thought around and beyond and through the data that confronted me" (as quoted in Morgan 2013, p. 43).

In view of the many global challenges we face as a society, with economic instability and climate change at the top of the list, we need sharpeners who can see past the superficial rhetoric promoted by others that "market capitalism is the only economic system that works" or that "melting in the polar ice caps is caused by natural climate cycles." An individual who can tolerate ambiguity realizes it is not "either – or" but, more often, "both." Sharpeners need to take responsible action and stand firm – economic, environmental, and social goals are critical to the future of human well-being; the ice caps can be melting and freezing over at the same time due to both natural causes and man-made carbon emissions.

Leading a business towards sustainability is not an art or a science – but both. We need both short-term and long-term thinking to propel our organizations to success: We need leaders capable and willing to take responsibility for short- and long-term success. And perhaps, akin to Peter Senge's systems thinking, we need leaders who can comprehend the interrelationship among the parts in the whole and engage in holistic problem solving (Senge 1990). As Senge (1990, p. 69) reminded us, "Complexity can easily undermine confidence and responsibility – as in the frequent refrain, 'It's all too complex for me,' or 'There's nothing I can do. It's the system.' System thinking is the antidote to this sense of helplessness that many feel as we enter the 'age of interdependence' . . . by seeing wholes we learn how to foster health," and by thinking at the level of the system, we take responsibility for the whole.

Today's Business Leaders Use Poetry to Create Sustainable Organizations

Business leaders and students alike are often told to "think outside the box" to solve today's unforeseen problems. Poetry can be the mechanism that allows for nonobvious associative relationships, essential for out-of-box-thinking, and central to sustainable long-term success.

Jim Rogers was the influential chairman, president, and CEO of Duke Energy between 2006 and 2013. Robert Frost's poem, "The Road Not Taken," (see Box for complete poem) was very meaningful for him. The poem is about a traveler who has to choose between one of two paths. It is not clear which one he should take especially since he thinks one might be better than the other but he is not sure. It may be that one road is less traveled and it may be that both roads are actually the same. Only at the end of his journey will the traveler know in retrospect that he indeed did take the more promising road. This is analogous to a leader who has to choose among two courses of action for her company. It is not clear which one will lead to greater innovation, profits, or relative market share. In many cases, the data are ambiguous and will only predict so much. The leader must rely on her expertise, gut feeling, and intuition; both right-brain and left-brain thinking are required when faced with tremendous environmental uncertainty.

The Road Not Taken by Robert Frost

Two roads diverged in a yellow wood,
And sorry I could not travel both
And be one traveler, long I stood
And looked down as far as I could
To where it bent in the undergrowth;
Then took the other, as just as fair,
And having perhaps the better claim,
Because it was grassy and wanted wear;
Though as for that the passing there
Had worn them really about the same,
And both that morning equally lay
In leaves no step had trodden black.
Oh, I kept the first for another day!
Yet knowing how way leads on to way,
I doubted if I should ever come back.
I shall be telling this with a sigh
Somewhere ages and ages hence:
Two roads diverged in a wood, and I –
I took the one less traveled by,
And that has made all the difference.

Poem available in Public Domain. Obtained from: http://publicdomainpoems.com/theroadnottaken.html

Rogers reported that in 1988 he faced "two diverging roads" as CEO of PSI Energy (which later became Duke Energy). He could follow industry trends in which executives considered shareholders to be the only stakeholders of the corporation and focus on maximizing short-term profits or he could depart from this conventional approach and balance the competing needs of many stakeholders, including customers, employees, regulators, suppliers, partners, the environment, and future generations (Intrator and Scribner 2007). Rogers chose the latter, less traveled road. He called this road his "true north," following it helped his leadership team through good times and bad (Intrator and Scribner 2007, p. 110).

In a 2011 commencement speech at North Carolina State University, Rogers urged students to write their own best-selling book called, *No Limits* (Rogers 2011). He advised students to become strong central protagonists in their own story. They should take charge of the writing, which others may have started, and not be afraid to rewrite some of the book's chapters. There is a parallel here between the traveler choosing a road and a graduating student embarking on his or her next step. There are lots of directions to take, but the individual, be it a student or a manager, needs to figure it out for himself and chart his own course. It is fine to have doubts and to make mistakes. But the individual should try to find the path that will make "all the difference." For Rogers, this meant pursuing one's passion with conviction, putting the needs of others ahead of one's own and taking a holistic and sustainable approach to business, i.e., leading responsibly.

While engaged in strategic action, it is difficult for managers to know if they will be ultimately successful. Like the traveler in Frost's poem, they will only know the results when they look back at the past. As Rogers notes, "Bursting out beyond the limits involves a certain amount of risk. At the time, you never know for sure which were the right choices or the wrong ones, which were the good breaks or the bad ones in your life. You may only know when you look back on them years later. Just as you don't know what happens until you reach the end of a book" (Rogers 2011, para. 10).

To convince others of the urgent need to reverse global warming, Rogers quoted a line in Shakespeare's play, *Julius Caesar*, "There is a tide in the affairs of men." We need to get on board, ride the tide, and find alternative sources of energy. We need to work collaboratively with our diverse stakeholders to come up with a solution. Rogers also used poetic language to let us know that the journey ahead is long and difficult. He said, "We really have to have what I would call cathedral thinking, where we are looking out and saying we need to address this problem over many decades, in the same way the cathedrals of Europe took many decades to build. It's going to take many decades of both mitigation and adaptation to get to the right place on this planet" (as quoted in Zakaria 2007, p. 48).

Poetry can be a way of identifying the essence of a problem stripped of confusing jargon and difficult numeric calculations. It might seem strange that Dr. Gregory Johnson, an oceanographer, would write a series of 19 haiku poems, illustrated with watercolor paintings, focusing on the major report findings highlighted in the summary of the 2013 international report on climate change science issued by the UN Intergovernmental Panel on Climate Change (see Box for examples). Johnson served as a lead author for the chapter on ocean observations in the 1535-page report and was having difficulty synthesizing the vast amount of information written in detailed, technical language. The report consisted of such topics as "changes in the water cycle and cryosphere" and "radiative forcing from anthropogenic aerosols." One weekend, when he was ill and housebound, Johnson occupied his mind by writing haikus. They were subsequently published online and praised for summing up climate change in an "understandable and even moving, way" (Mooney as quoted in Doughton 2014, para. 5).

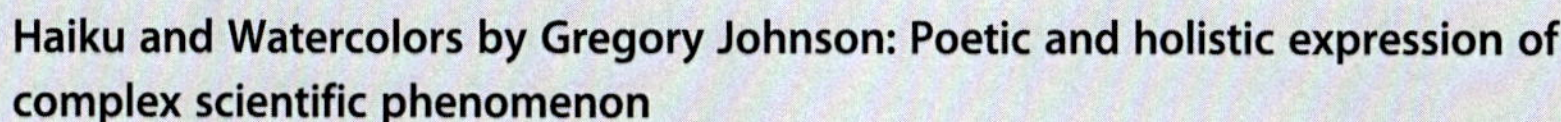

Haiku and Watercolors by Gregory Johnson: Poetic and holistic expression of complex scientific phenomenon

Abyss warms, coasts flood.
Air moistens – salt patterns shift.
Carbon sours oceans.
HISTORY, WATER

Our industry has
warmed oceans, air, lands – changed rains –
melted ice – raised seas.
ATTRIBUTION

(*continued*)

Wet will get wetter
and dry drier, since warm air . . .
carries more water.
WATER MEETS AIR

Used with the permission of Gregory C. Johnson and Sightline.
Obtained from: http://daily.sightline.org/2013/12/16/the-entire-ipcc-report-in-19-illustrated-haiku/

The grim UN report on climate change could cause one of two negative emotions for a concerned scientist: fear usually triggers an avoidance motivational system in which the individual perceives that risks are higher than they really are and that the best strategy is to flee; anger is associated with an approach motivational system in which the risks are perceived as being lower than they really are, so the individual stays and fights. Johnson, overwhelmed by the evidence that the earth was on a course that could not easily be reversed, worked hard to contain the dread and fear and started to write haiku and paint watercolors. These artistic endeavors calmed him down and enabled him to gain clarity and inspiration. He came up with evocative words and images to depict a future that no one wants – sour oceans, melting ice, snow retreat, frozen earth, and raised seas. Johnson did this for himself, but his friends and family liked his work so much that he put it in a booklet, which went viral, alerting others to the environmental challenges we face. He recognized that the natural environment is an important stakeholder of business.

Poetic devices, such as metaphors, have been considered a great tool for helping organizations identify themselves, create a sense of purpose, and revisit strategic decisions. One might wonder what cloud computing has in common with a Grecian urn. For Satya Nadella, the recently hired CEO of Microsoft and a veteran computer scientist, poetry is both a passion and a powerful device for communicating vision and the need for cooperation. As Nadella noted, there is nothing more haunting than the final lines in his favorite poem, Ode on a Grecian Urn, by WB Yeats: "Beauty is truth, truth beauty – that is all ye know on earth, and all ye need to know." He compares poetry to software code: "You're trying to take something that can be described in many, many sentences and pages of prose, but you can convert it into a couple lines of poetry and you still get the essence, so it's that compression" (as quoted in Bedigian 2014).

Nadella faces the challenging job of reinventing the software giant, Microsoft. Under Bill Gates and Steven Balmer, the company relied heavily on its Windows Operating System and Windows Office Suite as major sources of revenues. In the process, it failed to embrace open source code, mobile devices, and online software. As Microsoft's new CEO, Nadella needed to do two things quickly: differentiate himself from his predecessors and chart out an exciting future for the company. He did this by creating a new narrative for the company, filled with poetic and literary references. Stories and narratives are powerful tools that leaders can use as literary weapons.

In a memo sent out on his first day at work, Nadella paraphrased a quote from Oscar Wilde: "We need to believe in the impossible and remove the improbable." He continued to write, "This starts with clarity of purpose and sense of mission that will lead us to imagine the impossible and deliver it. We need to prioritize innovation that is centered on our core value of empowering users and organizations to 'do more" (http://news.microsoft.com/2014/02/04/satya-nadella-email-to-employees-on-first-day-as-ceo/).

Within 45 s of his first public appearance, Nadella quoted TS Eliot's poem, "Little Gidding":

> We shall not cease from exploration
> And the end of all our exploring
> Will be to arrive where we started
> And know the place for the first time.

The poem was used by Nadella to convey an important message: it was time for Microsoft to learn from the past, move forward, and reinvent itself. Leaders, like Nadella, use linguistic and nonlinguistic resources at their disposal in an attempt to communicate with and persuade others about complex issues with a clarity that is otherwise difficult to attain.

The holistic language used by leaders like Nadella nudges individuals to move from a narrow self-oriented perspective to a broader other-encompassing one, opening them up to greater insights and nonobvious connections between variables. The result is a re-framing of the problem in a language that is multidimensional, promoting the ability to detect different modes of meaning and to deal with ambiguity and uncertainty. This not only allows for consideration of other viewpoints, an examination and revision of current insights and perceptions, but also draws attention to human needs and motivations, allowing one to address issues of ethics and engagement.

In order for organizations to follow this new holistic approach, it is not sufficient for leaders to just take from poetry and change their language. The underlying motivational systems and decision-making processes also need to change to reflect the complex nature of current business problems, which require creativity and innovation along with discipline and perseverance against all odds. Traditional motivational mechanisms used in today's organizations are predominantly incentive based, leading to a calculative, task-oriented mind-set, rather than innovative holistic approaches. Although an in-depth review of motivation and decision making is beyond the scope of this chapter, the next section outlines an approach by which economic, environmental, and social goals can be made more salient in the decision process by changing motivational underpinnings.

The Logic of Appropriateness and Sustained Motivation

March (1994) suggests that decisions are made to be appropriate to the situation or context. Thus, a decision is the result of the decision maker's response to the question "what does a person like me do in a situation like this?" This question, also referred to as "the logic of appropriateness," contains three subquestions: (i) what defines the situation, (ii) who is the person in this situation? or what is his/her appropriate role? and (iii) given a person and a situation, what is the appropriate norm, decision rule, or choice?

The framing of the situation plays a major role in how it is defined by the decision maker and makes certain roles and goals more salient. For example, when a task is called a competition, decision-maker focus is on the choice that allows for an individual win, even at the cost of others. Changing the name of the same task to a team activity changes the focus to ensuring everyone involved in the task benefits from the outcome, i.e., there is no one single winner. Similarly, organizations can choose the context within which the decision is framed.

Decision-maker characteristics also influence the saliency of roles and goals. Internal characteristics and factors can be thought of as variables and processes relevant to the decision that are internal to the decision-maker, such as being risk averse (as opposed to risk-seeking), and pro-social (concerned about others impacted by the decision) or proself (concerned only with the outcomes for oneself). External factors, on the other hand, can be thought of as variables within the context of a decision that change how internal factors may be expressed. The expression of internal tendencies can be emphasized or attenuated by situational characteristics. In fact, the same person may choose options that are more holistic in a collective context, but may reverse the valuation when in an individualistic context and considering benefits only him/herself. This reversal of choices is spontaneous and automatic and thus is frame or context-dependent (Arora et al. 2012).

The last element in the appropriateness framework is the rule applied to the decision. These can be the result of fast, frugal processing as in a heuristic, or a more deliberate process, what Kahneman (2011) labeled as "System 1" and "System 2" modes. In both cases, the rules seen as "appropriate" will be influenced by both decision factors that are salient. Here too, organizations can choose which rules and heuristics they support through rewards and recognition. The logic of appropriateness suggests that decision makers are motivated to achieve congruence between their context, internal characteristics, and options chosen. When an organization motivates actions that are incongruent with its stated values, employees and customers feel a lack of fit, which has been shown to decrease productivity and create dissatisfaction as well as a lack of trust. For example, stating that authentic interactions with customers are valued while rewarding the number of customer complaints handled in a time-period (independent of the quality of the interaction) will result the latter being the focus and the former being ignored due to an incongruence between means and ends. The resulting nonfit is likely to reduce employee satisfaction and lead to greater distrust of the organizational leadership and culture.

It is not surprising then that Millennials, who see the main challenges for the twenty-first century as including complex wicked problems like climate change, resource scarcity, and income inequality find the current language of business of economic profits and income-generation confining, resulting in an ever-declining level of trust in businesses. These problems require creativity, empowerment, innovation, and effective interdisciplinary collaborations. The millennial response to the appropriateness question in their identity as business-people who are expected to maximize profits does not fit with their self-perceived identity as socially minded individuals who wish to be creative and have impact. This non fit however is not an unsurmountable problem – it can be ameliorated by creating organizations with cultures that subscribe to goals beyond economic profit, encourage holistic approaches to decision making, and focus on intrinsic rather than extrinsic motivation.

Ultman (1997) defines intrinsic motivation as the innerdirected force that causes an individual to engage with a task for the sake of the task itself. Intrinsic motivation is frequently accompanied by a high level of interest in the task itself, a propensity to exercise creativity in the task, and greater satisfaction from achieving high level of task-accomplishment rather than desired outcome. Other major elements of intrinsic motivation are its connections with meaningfulness of task and self-determination of action in the task. Empowerment of all members of an organization to be engaged in their tasks increases intrinsic motivation. Intrinsically motivated work is therefore simultaneously meaningful and challenging, allowing for introspection and creativity to achieve expertise and satisfaction in a job well-done.

It bears pointing out that both intrinsic and extrinsic motivation have a place in an organization's repertoire of motivational tools – using both allows for congruency between means and ends resulting in greater fit, thereby increasing both satisfaction and productivity. External rewards and punishments work best when the tasks are routine, consistent, with easily and objectively measured outcomes. Intrinsic motivation is a more effective tool when tasks are nonroutine, requiring creativity, collaboration, and cognitive effort with subjectively measured outcomes.

In sum, this chapter argues for a comprehensive approach by business leaders and managers to use the holistic language of the arts to frame business problems as social endeavors, change the metrics such that success is defined and measured evenly across the triple bottom line, and create motivational mechanisms that empower and support individual decision-makers to achieve fit resulting in greater satisfaction and productivity – a promising recipe for engaged decision making for a sustainable future. Although far from simple, the steps outlined above are achievable across industries and corporate boardrooms as illustrated in the next section.

Implementation of a New Language

The road ahead for businesses seeking to have a positive impact on planet, people, and profits is difficult, but not impossible. By reframing the issues, redefining company mission and goals, developing comprehensive metrics to evaluate progress and make continuous improvements, and allocating time and money to sustainable initiatives, today's leaders and managers can make valuing and supporting long-term sustainability a fundamental characteristic of their organizational cultures. Even more encouraging is the fact that companies operating in some of the most environmentally vulnerable industries – mining (e.g., AngloGold Ashanti in Ghana), timber (Weyerhaeuser Productos in Uruguay), and agriculture (e.g., Cargill in Vietnam) – are already successfully executing such changes.

What is most notable about these examples is their use of a "small wins" strategy. Working on projects to address specific problems in regions in which a corporation has a significant presence provides momentum and results as large problems are broken down and solved step by step. As noted by Amabile and Kramer (2011), "When we think about progress, we often imagine how good it feels to achieve a long-term goal or experience a major breakthrough. These big wins are great—but they are relatively rare." More often, change occurs incrementally. In less than 2 years, AngloGold Ashanti was able to reduce the number of reported cases of malaria in Ghana by 73% through an integrated disease control program; in the process, 127 permanent jobs as spray operators for local residents were created and a malaria control center was opened along with support from the Ghanaian government (Linnenluecke et al. 2014). The model was subsequently extended to mining areas in Tanzania and Guinea.

Weyerhaeuser Productos supervises projects in Uruguay to enhance land productivity, increase local employment, build and operate plants using biomass fuel, plant forests in areas where they had never been trees (a practice called "afforestation"), and promote strong worker safety programs. Cargill Vietnam has brought technology to cocoa farmers, developed an independent certification program to safeguard the environment, offers farmers a premium price for their crops, and has established 76 new schools in rural communities (Buchanan 2016). Both companies received the US State Department's Award for Corporate Excellence.

The progress evolving at AngloGold Ashanti, Weyerhaeuser Productos, and Cargill can only occur if there is strong commitment from top management to reframe the business and align corporate strategic goals with the eight UN Millennium Development Goals (see http://www.unmillenniumproject.org/goals/). The work done by Mark Cutifani, the CEO of Anglo American (a mining firm that formerly owned AngloGold Ashanti), is a worthwhile example for consideration. In collaboration with the Kellogg Innovation Network, Cutifani is helping to shift mining from an "isolated extractive" industry to a "resource development" industry (Cutifani and Bryant undated, p. 10). Remember how a change in language enabled fishermen to view the reef as a "living entity" and find safer ways to capture lobsters? In this case, a change in language is helping miners view the industry as less exploitative of the earth's valuable resources and more of a catalyst for the socioeconomic development and wellbeing of mining communities around the world. Executives are being encouraged to recognize that there is no one "silver bullet". ... Instead, companies need to recognize that a variety of actions will be required and that these need to be underpinned by a ***changed mindset*** that reevaluates the role of mining in the societies in which they operate (Cutifani and Bryant undated, p. 8; bold italics added).

The next step is to integrate the new frame into a company's mission statement and goals. Anglo American's mission is: "Together we create sustainable value that makes a real difference" (http://www.angloamerican.com/about-us/our-approach). With this holistically framed mission statement as a point of reference, the second step is to develop goals and measures that attest to the company's ability to "make a real difference." Goals and targets are set to ensure that the company does no harm to its workforce, minimizes harm to the environment, shares the benefits of mining with local communities and governments, and has an engaged productive workforce. These holistic goals are measured with metrics that are meaningful and reflect the focus of the company beyond just economics. It is important that the metrics and processes show high degrees of transparency, as recommended by the Global Reporting Initiative (GRI), by documenting positive and negative impacts of performance ("balance"), enabling shareholders to compare performance results across time ("comparability"), providing solid and detailed qualitative/quantitative information ("accuracy"), committing to reporting on performance on a regular schedule ("timeliness'), describing information on performance in an understandable way ("clarity"), and allowing information to be subject to examination ("reliability').

Anglo American's biggest gains have been in reducing new cases of occupational disease, increasing the number of HIV-positive employees in disease management programs, using less water in its operations, reducing the number of environmental incidents, and increasing procurement spending with black-owned and managed companies. Its code of conduct available from its website in English, Spanish, and Portuguese ensures that the important values of safety, care and respect, integrity, accountability, collaboration, and innovation infused within its organizational culture, thus signaling support for and valuation of the sustainability focus. This is the final step of continuous improvement towards sustainability. Clearly, such a transformation is not possible without considerable investment of resources and complete organizational commitment. When there is such commitment, coupled with holistic thinking, organizations transform, becoming beacons of sustainable practices.

The conceptual framework in Diagram 1 provides a road map for the kind of transformation described above, as envisioned by the authors. The lower-left quadrant in the diagram (Q1) refers to the practical misfit between the company's stated goals and what it actually seeks to achieve. For Anglo American, this began by understanding that the environmental degradation associated with mining and the social ills prevalent in the population around its local operations were inexorably linked to its ability to sustain its operations profitably over the long term. Thus, its current structures did not serve its true mission and intent.

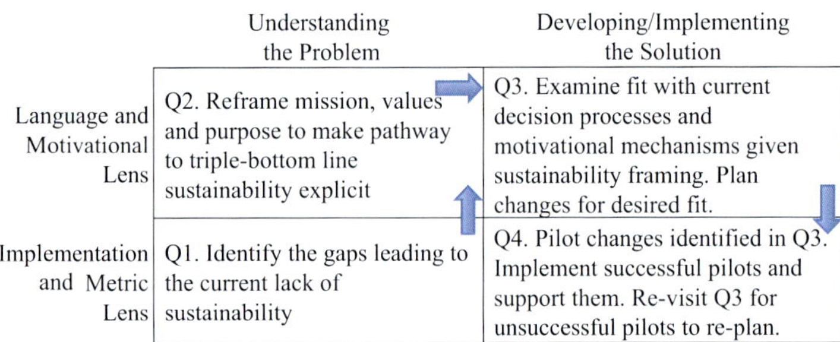

Understanding the Problem
Developing/Implementing the Solution
Language and Motivational Lens
Implementation and Metric Lens
Q2. Reframe mission, values and purpose to make pathway to triple-bottom line sustainability explicit
Q3. Examine fit with current decision processes and motivational mechanisms given sustainability framing. Plan changes for desired fit.
Q1. Identify the gaps leading to the current lack of sustainability
Q4. Pilot changes identified in Q3. Implement successful pilots and support them. Re-visit Q3 for unsuccessful pilots to re-plan.

Diagram 1: Conceptual Framework and Road Map for Transforming Organizational Cultures to Value and Support Long-term Sustainability

From here the organization moves to the top left quadrant (Q2) of the diagram, which brings it to the realm of holistic language and motivation by asking two questions:

A. What is the true intent of the organization – the legacy it hopes to leave behind, the impact it aspires to have in the immediate and distant future?
B. How can the true intent identified in (A) be best captured in a succinct and meaningful way? This requires a move away from current language of the business to using a new more holistic language to redefine goals, values, and mission.

Responses and materials developed in Q2 are just words unless they are transformed into metrics, processes, and organizational culture. This next step, outlined in the top-right quadrant (Q3), requires an understanding of how decision makers in the organization currently respond to the logic of appropriateness question (how are they framing the tradeoffs required by current decisions) versus how should they respond, and what processes will allow them to respond as they should, to that question. Gaps between what is expected and what may be observed, given current motivational mechanism, can be identified by asking how the current motivational systems align with desired responses. New motivational mechanisms may need to be designed to ensure that behavior changes throughout the organization, thereby transforming its culture.

Finally, in moving to the bottom right-hand quadrant (Q4), the organization begins to implement the plans designed in Q3 to achieve the desired changes. Implementation is best undertaken via small and/or pilot projects (as seen in the "small-wins strategy") that can be adequately supported. Successful endeavors can then be implemented across the organization. It is vitally important that at this stage all changes be well supported with appropriate resources (training, restructuring of work, jobs and roles, performance evaluations, and compensation). Without the availability of sufficient or appropriate resources, any efforts to move an organization on the road to sustainability are likely to fail. Since lasting change requires empowerment of those who carry out the day-to-day actions to change their behavior, it is crucial that those leading such an effort do not impose their views. Rather, they should act as mentors who build the bridges needed for the organization to undertake a meaningful journey through the four quadrants.

Conclusion

Businesses are expected to be organizations focused on creating shareholder wealth, wherein the decision-makers, using the current frames afforded by games and war, consider each choice, decision, or action as either resulting in a "win" or a "loss." A win enhances shareholder wealth, while a loss reduces it. Given the strong human tendency for loss aversion, and for losses to always loom larger than gains, it is quite logical then that businesses wish to win at all costs. The resulting calculative mindset, however, has been shown to lead to less ethical actions, reduced concerns with consequences of one's decision for others, and greater likelihood of cheating to ensure a win (Wang et al. 2014). The current language of business (whether thought of as accounting, war, or games) frames decisions as wins and encourages being "analytical" by focusing only on measurable outcomes, perpetuating the short-term unsustainable focus on shareholder wealth creation. Given climate change and impending resource scarcity due to environmental degradation, it is time for a new, more holistic language of business – one that has its roots in the arts and can broaden the perspectives of managers regarding their vital role in sustainability.

A new language for business leading to a refocus by corporations on genuine efforts towards greater sustainable development needs support from larger elements in society, especially from politicians and educators. The Trump administration has been criticized for deleting all mention of climate change and sustainability from its whitehouse.gov website in an attempt to communicate that there is no such concern. Similarly, in a state at high risk for beach erosion and flooding, officials working for the Florida Department of Environmental Protection were apparently banned from using the terms "climate change," "global warming," or sustainability" (Korten 2015). Politicians use a strategy in which denying a phenomenon rules out its very existence. When politicians downplay the significance of climate change by referring to it as an economic and technical issue without addressing its human and social implications, it becomes difficult for people to have a meaningful, inclusive discussion about mitigating the risks and adapting to the consequences of climate change.

Much is lacking within business schools as well. Giacalone (2004, p. 416) lamented, "Our fundamental business curriculum has no higher order ideals ... We teach students a simple pay-off matrix: Increase the company's wealth and improve the chances of increasing your own affluence and status. In our lesson plans, there is no selflessness, no objective for the non-financial, collective improvement of our world, and no generative aspiration to leave behind a better world for those who follow." Unfortunately, research has shown that materialistic values are related to lower levels of organizational citizenship behavior and higher levels of interpersonal deviance. By focusing exclusively on rationality and profit-based metrics, are business schools inadvertently teaching business students about self-interest, expediency and profit maximization at the cost of sustainable growth?

According to a Chinese proverb, "may we live in interesting times," and these are indeed interesting times that pose interesting issues of long-term sustainability of our planet. If they are to be addressed and resolved in a timely manner, there is an urgent need to transition away from traditional means of framing business, which have included the worlds of accounting, warfare, and games, resulting in a calculative mindset where decisions are made solely on the basis of operating expenses, revenues, and profits. Businesses should adopt a more holistic view of their raison d'être, one that focuses on triple bottom line performance and uses additional criteria, such as percentage of suppliers that have goals in place to reduce GHG emissions, yearly fixed donations to charitable causes, and percentage of directors elected by employees, to assess their contributions to society as a whole.

A reflection on self, a concern for others, and a consideration of the common good convert decision makers from game-players and warriors to humans who bring with them the capacity for analysis but also the capacity for concern for goals that go beyond monetary to include social and environmental goals. It creates the basis of holistic and sustainable decision making and is a fundamental step on the path towards achieving Pope Francis' goal of addressing today's complex problems with "a new and universal solidarity."

Cross-References

▶ Designing Sustainability Reporting Systems to Maximize Dynamic Stakeholder Agility
▶ Education in Human Values
▶ Responsible Investing and Corporate Social Responsibility for Engaged Sustainability
▶ The Spirit of Sustainability

References

Adler, N. (2015). Finding beauty in a fractured world: Art inspires leaders-leaders change the world. *Academy of Management Review, 40*(3), 480–494.

Amabile, T, & Kramer, S. J. (2011). The power of small wins. *Harvard Business Review*. https://hbr.org/2011/05/the-power-of-small-wins.

Arora, P., Peterson, N. D., Krantz, D. H., Hardisty, D. J., & Reddy, K. S. (2012). To cooperate or not to cooperate: Using new methodologies and frameworks to understand how affiliation influences cooperation in the present and future. *Journal of Economic Psychology, 33*(4), 842–885.

Arora, P., Bert, F., Podesta, G., & Krantz, D. (2015). Ownership effect in the wild: Influence of land ownership on economic, environmental & social goals and decisions in the argentine pampas. *Journal of Behavioral and Experimental Economics, 58*, 162–170.

Bartunek, J. M., & Ragins, B. R. (2015). Extending a provocative tradition: Book reviews and beyond at AMR. *Academy of Management Review, 40*(3), 474–479.

Bedigian, L. (2014). 5 unusual things you should know about Microsoft's new CEO, Satya Nadella. Nasdaq. Retrieved on 18 May 2015 from the World Wide Web: http://www.nasdaq.com/article/5-unusual-things-you-should-know-about-microsofts-new-ceo-satya-nadella-cm323615.

Bergoglio, J.M. (Pope Francis). (2015). http://w2.vatican.va/content/francesco/en/encyclicals/documents/papafrancesco_20150524_enciclica-laudato-si.html

Buchanan, M. (2016). These companies focus on rights to find success. https://share.america.gov/companies-focus-on-rights-to-find-success/

Buffett, M., & Clark, D. (2006). *The Tao of Warren Buffett*. New York: Scribner.

Camiciottoli, B. C. (2007). *The language of business studies lectures: A corpus-assisted analysis*. Amsterdam: John Benjamins BV.

Clark, D. (2015). A swan song, with ode to a deer. *The Wall Street Journal*, B1.

Cutifani, M, & Bryant, P. (undated). Reinventing mining: Creating sustainable value. http://www.kinglobal.org/uploads/5/2/1/6/52161657/pb_kin_dpf_final_12_4_5mb.pdf

Doughton, S. (2014). Seattle scientist distills 2,200-page report into haiku. *The Seattle Times*. Retrieved May 14, 2014 from the World Wide Web: http://seattletimes.com/html/localnews/2022657923_climatehaikuxml.html

Fuller, A. (2016). What can Millennials tell us about the future of business? Deloitte Digital. http://www2.deloitte.com/global/en/pages/about-deloitte/articles/gx-millennials-shifting-business-purpose.html#report

Giacalone, R. (2004). A transcendent business education for the 21st century. *Academy of Management Learning & Education, 3*(4), 415–420.

Grisham, T. (2006). Metaphor, poetry, storytelling and cross-cultural leadership. *Management Decision, 44*(4), 486–503.

Hamington, M. (2009). Business is not a game: The metaphoric fallacy. *Journal of Business Ethics, 86*, 473–484.

Henry, J. (2013). Shakespeare and Wordsworth boost the brain, new research reveals. *The Telegraph*. Retrieved June 3, 2014 from the World Wide Web http://www.telegraph.co.uk/news/science/science-news/9797617/Shakespeare-and-Wordsworth-boost-the-brain-new-research-reveals.html

Huckin, T. (2002). Textual silence and the discourse of homelessness. *Discourse & Society, 13*(3), 347–372.

Intrator, S. M., & Scribner, M. (Eds.). (2007). *Leading from within*. San Francisco: Jossey-Bass.

Kahneman, D. (2011). *Thinking, fast and slow*. New York: Farrar, Straus and Giroux.

Koller, V. (2004). *Metaphor and gender in business media discourse: A critical cognitive study*. NY: Palgrave Macmillan.

Korten, T. (2015). *Florida, officials ban term 'climate change*. Miami: Herald.

Lakoff, G. (2016). Understanding Trump. Blog at Wordpress.com. https://georgelakoff.com/2016/07/23/understanding-trump-2/

Langley, M. (2016). Tech CEO turns rabble rouser. *The Wall Street Journal*, A1.

Laufer, D. (2010). Marketing warfare strategies. In *Wiley International Encyclopedia of Marketing*, 1, Wiley.

Liberman, V., Samuels, S. M., & Ross, L. (2004). The name of the game: Predictive power of reputations versus situational labels in determining Prisoner's dilemma game moves. *Society for Personality and Social Psychology, 30*(9), 1175–1185.

Linnenluecke, M, Verreynne, M, de Villiers Scheepers, R, Gronum, S & Venter, C. (2014). Planning for a shared vision of a sustainable future. South Africa: Network for Business Sustainability. http://nbs.net/wp-content/uploads/NBS-SA-Shared-Vision-SR-Final.pdf.

March, J. (1994). *A primer on decision-making: How decisions happen*. New York: Free Press.

Milne, M., & Gray, R. (2013). W(h)ither ecology? The triple bottom line, the global reporting initiative, and corporate sustainability reporting. *Journal of Business Ethics, 118*(1), 13–29.

More, E., & Todarello, E. (2013). Business education and spirituality – The MBA with no greed. *Journal of Global Responsibility, 4*(1), 15–30.

Morgan, C. (2013). *What poetry brings to business*. Ann Arbor: University of Michigan Press.

Morris, J. A., Urbansky, J., & Fuller, J. (2005). Using poetry and the visual arts to develop emotional intelligence. *Journal of Management Education, 29*(6), 888–904.

Overshoot Day. (2017). Global Footprint Network. http://www.footprintnetwork.org/our-work/ecological-footprint/.

Parker, S. G. (2003). Rhyme and reason: What poetry has to say to business writers. *Harvard Management Communication Letter*, 3–4.

Rogers, J. (2011). Write your own book. *Electric Perspectives*. Retrieved June 3, 2014 from the World Wide Web: http://mydigimag.rrd.com/article/Another_Perspective/858768/84051/article.html.

Romanowska, J., Larsson, G., & Theorell, T. (2014). An art-based leadership intervention for enhancement of self-awareness, humility, and leader performance. *Journal of Personnel Psychology, 13*(2), 97–106.

Scott, J. (2000). Rational choice theory. In G. Browning, A. Haleli, & F. Webster (Eds.), *Understanding contemporary society*. Thousand Oaks: Sage Publishers.

Senge, P. M. (1990). *The fifth discipline*. New York: Doubleday.

Senge, P. M., Lichtenstein, B. B., Kauefer, H. B., & Carroll, J. S. (2007). Collaborating for systemic chance. *MIT Sloan Management Review, 48*(2), 44–53.

Smith, V. (1991). Rational choice: The contrast between economics and psychology. *Journal of Political Economy, 99*, 877–897.

Spence, A., & Pidgeon, N. (2010). Framing and communicating climate change: The effects of distance and outcome frame manipulations. *Global Environmental Change, 20*(4), 656–667.

Thaler, R. (1980). Toward a positive theory of consumer choice. *Journal of Economic Behavior & Organization, 1*(1), 39–60.

Utman, C. H. (1997). Performance effects of motivational state: A meta-analysis. *Personality and Social Psychology Review, 1*, 170–182.

Van Buskirk, W., & London, M. (2012). Poetry as deep intelligence: A qualitative approach for the organizational behavior classroom. *Journal of Management Education, 36*(5), 636–668.

Von Ghyczy, T. (2003). The fruitful flaws of strategic metaphors. *Harvard Business Review*. https://hbr.org/2003/09/the-fruitful-flaws-of-strategy-metaphors.

Wang, L., Zhong, C. B., & Murnighan, J. K. (2014). The social and ethical consequences of a calculative mindset. *Organizational Behavior and Human Decision Processes, 125*(1), 39–49.

Wilburn, K., & Wilburn, R. (2014). The double bottom line: Profit and social benefit. *Business Horizons, 57*(1), 11–20.

Windle, R. (1994). *The poetry of business life: An anthology*. San Francisco: Berrett-Koehler Publishers.

Zakaria, F. (2007). 'Cathedral thinking'; Energy's future. *Newsweek, 150*(8/9), 48. Retrieved June 3, 2014 from the World Wide Web: https://www.duke-energy.com/pdfs/Newsweek_14952.pdf.

Zeman, A. Z. J., Milton, F. N., Smith, A., & Rylance, R. (2013). By heart. An fMRI study of brain activation by poetry and prose. *Journal of Consciousness Studies, 20*(9–10), 132–158.

Transformative Solutions for Sustainable Well-Being

Designing Effective Strategies for Addressing Our Planetary Challenges

Annick De Witt

Contents

Abstract

Our severe environmental and social issues challenge us to think in new and innovative ways about the needed solutions. In this chapter, I argue we need to move beyond mere instrumental, linear, and reductionist approaches, toward more transformative, emergent, and aspirational approaches. Considering the nature of our global sustainability problems – often characterized as profoundly systemic, highly complex, and ultimately human-created – I use insights from three distinct academic fields in order to articulate a number of principles for effective sustainability strategies. *Sustainability science* urges us to engage with the intent or purpose of the system, thus shifting worldviews, mindsets, and

What sustainability science, complexity science, and positive psychology teach us about designing effective strategies for addressing our planetary challenges

A. De Witt (✉)
Copernicus Institute of Sustainable Development, Utrecht University, Utrecht, The Netherlands
e-mail: annick@annickdewitt.com; a.dewitt@uu.nl

S. Dhiman, J. Marques (eds.), *Handbook of Engaged Sustainability*,
https://doi.org/10.1007/978-3-319-71312-0_12

paradigms, or inviting reflection on them. From such a systems perspective "multi-problem-solvers" are preferred. *Complexity science* recommends drawing in collaboration with diverse stakeholders and viewpoints, so strategies are co-created and co-owned, as well as allowing 'emergence' through inviting experimentation and self-organization. *Positive psychology* proposes supporting people to strengthen intrinsic goals and values; behave in autonomous, volitional, or consensual ways; and be mindful. This perspective also emphasizes the importance of cultivating a positive, empowering sustainability narrative, which challenges the empirically faulty - yet highly persistent - idea that hedonism leads to happiness, and which demonstrates alternative, *eudaimonic* routes to well-being that are both fulfilling and sustainable. I then discuss a dietary change toward more plant-based diets as an example of such a transformative solution pathway. Arguing that highly potent pathways are often overlooked in the sustainability debate, I invite reflection on the nature and characteristics of the solutions we need, and the sustainable world we collectively envision.

Keywords
Sustainable well-being · Transformation · Climate change · Sustainability science · Complexity science · Positive psychology · Leverage points · Worldviews · Multi-problem-solvers · Emergence · Eudaimonia · Mindfulness · Sustainability narrative · Dietary change · Meat consumption

Introduction

Considering the many challenges in our world today, many people agree we are in dire need of powerful solutions that reach far, wide, and deep, and have the potential to shift the development trajectories humanity is on toward more sustainable and life-enhancing ways of being on this planet. The growing sentiment, both among the public and among experts, is that we need a "whole-systems" change, a profound, societal transformation toward ways of living that correspond more closely with ecological systems.

Yet our thinking about solutions is often not as bold, comprehensive, and aspirational as our problems require them to be. Dominant scientific discourses still address sustainability problems from largely disciplinary perspectives, often neglecting the profoundly interconnected, mutually reinforcing, systemic nature of our planetary problems (Spangenberg 2011). Frequently, there is a focus on "quick fixes" and instrumental solutions, rather than on the underpinning, ultimate drivers of current trajectories. For example, many of us may have heard about sustainability "solutions" with negative side effects more substantial than the benefits achieved (e.g., Sarewitz 2004). According to some authors, "much of what might be constituted as sustainability science fails to engage with the root causes of sustainability, and is therefore unlikely to substantially alter our current development trajectories" (Abson et al. 2016, p. 30). This complaint – that most of our problem-solving is oriented around relieving symptoms rather than addressing deeper causes – has been echoed by environmental philosophers, social thinkers, and sustainability advocates alike.

Another often-heard criticism is that many approaches are based on linear thinking, which doesn't account for the profound complexity and unpredictability that characterizes our contemporary challenges. According to some authors, technocratic responses of planning and technical problem-solving, which are based on assumptions of order, sufficient knowledge, and certainty, are fundamentally unable to address complex issues: "Dominant efforts to address our most serious challenges waste precious resources, time, and talent. These planning-based approaches – so common across government, civil society, and even business – represent a neo-Soviet paradigm, one that is spectacularly out of step with what we know about complexity, about systems, about networks, and about how change happens" (Hassan 2014, p. xiii). Since complex challenges are characterized by the absence of a shared problem definition as well as agreement on its solutions, they demand a shift in attitude from hierarchically imposed, predefined outcomes toward participatory processes in which a shared direction collectively emerges.

Moreover, rather than being driven by an inspirational vision of a sustainable future we all share, many approaches are characterized by a reductionist problem-solving mode, coming to expression in an almost exclusive focus on technology and regulation. The underlying conception of sustainability issues appears to be one of a series of problems that need to be overcome. Dominant narratives tend to focus on reducing harm or avoiding damage, rather than on a bright and beautiful future humans can aspire to. However, arguably, sustainability is about more than reducing unsustainability. For example, Ehrenfeld (2004, p. 4) defines sustainability as "the possibility that human and other forms of life will flourish on the earth forever." Although one may quarrel about the term forever, this definition underscores that sustainability is about *possibility* – about bringing forth and calling in a new reality; about creating something that does not exist yet but that we can envision and strive toward. It also underscores that sustainability is about *flourishing*, about a positive quality of life and the sense of well-being humans desire. In Ehrenfeld's words (2004, p. 4):

> Sustainability and unsustainability are not just two sides of the same coin. They are categorically different. Unsustainability is measurable; it can be managed, and incrementally reduced. But sustainability – the possibility of flourishing in the future – is aspirational.

As sustainability is thus about calling a new reality into being – a world in which humans flourish in accordance with the natural systems around them – our efforts toward sustainability require imagination and an inspiring vision of what that world could be. This is essential, as, in the words of Amartya Sen, "it is difficult to desire what one cannot imagine as a possibility."

Certainly, innovative solutions are already emerging that orient around such inspirational visions. Rather than offering technological fixes, these interventions often scaffold ways of living that are not only more sustainable but also more healthy and fulfilling. They are not focused on solving a single problem, or changing a particular behavior, but can often be characterized as "multi-problem-solvers."

Take for example *permaculture*, which is a system of social design principles centered around simulating or directly utilizing the patterns and features observed in natural ecosystems. Following *permaculture* principles, everyone could be living in settlements that are more like gardens than cargo containers, which purify air and water, generate energy, treat sewage, and produce food – at lower cost, while connecting humans with the natural systems surrounding them (Hemenway 2009). Permaculture thus has the potential to address a variety of sustainability issues while also supporting people to thrive through connecting with nature. Research has shown connecting with nature is profoundly beneficial for humans, in terms of their well-being, health, and relationships, among others (Green and Keltner 2017). It has also been shown to coincide with a sense of environmental responsibility (Hedlund-de Witt 2013a; Hedlund-de Witt et al. 2014).

However, while permaculture has a history of successful grassroots application, and while it has generated a broad, international movement with a relatively high public profile, it has received little systematic scrutiny in the scientific literature and has been largely ignored by the mainstream sustainability debate (Ferguson and Lovell 2015). Indeed, as also Abson et al. (2016, p. 33) claim, "shallower interventions are favoured in both science and policy." Or, as other voices express the same observation, we find ourselves "in a system that privileges reductionist approaches" (Van Beurden et al. 2011).

Therefore, it's important to *rethink* and *re-search* what more effective approaches to sustainability could look like: *What do solutions look like that address deeper causes, rather than merely relieving symptoms? What do solutions look like that collectively emerge, rather than being hierarchically imposed? What do solutions look like that orient around an inspirational vision of a sustainable future in which humans thrive, rather than just reducing unsustainability? How are these solution pathways structurally and characteristically different from more conventional approaches? What can we learn from existing examples of such approaches?* In a broad sense, these are the questions I aim to explore in this chapter. That is, this chapter is intended as contemplation on what our most powerful and life-enhancing solutions *could,* and perhaps *should,* look like.

However, in order to do this, it is essential to have some understanding of the nature of the problems we are trying to address. Three central characteristics are often mentioned in analyses of our global sustainability issues. These are (1) their interconnected, systemic natures; (2) their high complexity; and (3) their ultimate causation by human assumptions, attitudes, and behaviors.

To start with the first characteristic, our many, multifaceted, global problems – from climate change to the obesity epidemic; from challenges to water, food, and energy security to extreme political polarization; from abject poverty to financial systems that thrive on perverse incentives – are profoundly *systemic* in their nature, fundamentally interconnected and intertwined, often mutually reinforcing each other. In order to address any single issue effectively, the larger system of issues needs to be taken into account. Some authors argue for the need for addressing these multiple crises as a whole, as a "poly-crisis," to use philosopher Edgar Morin's and Kern (1999) term.

Secondly, our planetary issues are nearly impossible to solve due to their high complexity. Some scientists refer to them as "wicked problems" (Hulme 2009).

These problems typically have numerous causes and no clear solutions. They involve multiple stakeholders, who often disagree about the problem definition as well as its potential solutions. They require lots of people and organizations to shift their mindsets and behaviors and are characterized by psychological, cultural, behavioral, and systemic constraints. Due to incompleteness of data, complex webs of nonlinear cause-effect relationships, and the continuous evolution of involved systems, our understanding of these problems is limited, introducing high levels of uncertainty and unpredictability.

Thirdly, our global problems tend to be human-created, thus being social, cultural, and psychological in their origins. It's a simple fact we frequently seem to forget:

> Behind the world's most difficult problems are people – groups of people who don't get along together. You can blame crime, war, drugs, greed, poverty, capitalism, or the collective unconscious. The bottom line is that people cause our problems. (Mindell 1995, cited in Hassan 2014)

Since humans are generally their ultimate cause, actively engaging the human dimensions of assumptions, attitudes, and behaviors is essential in forging the most effective and transformative solutions. Some authors argue that a sustainable world "is not a world achieved solely by technological measures. . . . it is a world with more humanness, fairness, and awareness, as well as less focus on consumption" (Ericson et al. 2014, p. 74). According to some, addressing our global problems therefore demands a change in human psychology and worldviews, as much as requiring technological fixes and institutional changes (e.g., De Witt et al. 2016; Hedlund-de Witt 2013b; O'Brien 2010).

In light of these observations, I draw on insights from three distinct scientific fields, namely, sustainability science, complexity science, and positive psychology. These fields have attempted to theoretically understand as well as pragmatically approach, respectively, the profoundly systemic, the highly complex, and the human-created nature of the global issues that have come to define our contemporary world.

As I discuss more extensively in the next section, according to strands within sustainability science we need "deep leverage points" for initiating transformational change. That is, interventions or change strategies that engage with the *intent* (paradigms, worldviews, mindsets) and *design* of the system as a whole, rather than merely addressing parameters and feedbacks (e.g., Abson et al. 2016; Meadows 1999). This enables us to address ultimate drivers, thereby affecting and transforming systems as a whole. Inevitably, solutions able to engage with, and have transformational impact on, these deep drivers will affect a range of interconnected issues and may therefore be characterized as "multi-problem-solvers." Faced with mutually interacting crises, getting to "the heart" of the system implies contributing to multiple societal issues simultaneously.

In the section thereafter, I discuss insights from complexity science with respect to how to respond to highly complex challenges, such as most of our sustainability issues. Complexity thinking underscores that since complex problems are characterized by high levels of uncertainty and unpredictability, any decision-making strategy needs to be based on a fundamental assumption of not-knowing and (retrospective) recognition of emergent patterns, rather than on predictive analyses and the assumption of order (e.g., Kurtz and Snowden 2003). This underscores the need for more distributed and participative leadership styles, experimentation, collaboration among diverse stakeholders and viewpoints, self-organization, and *emergence*.

As I discuss in the section on positive psychology, insights from this field point in the direction of solutions that are inherently positive and rewarding, empowering *intrinsic* goals, values, and motivations (Ryan and Deci 2000; Ryan et al. 2008). Solution pathways should be aspirational and dovetail with the powerful, intrinsic human motivation for a better, more fulfilling life. Thus, sustainability should not just be about solving problems but also, and probably more so, about creating the conditions under which humans can thrive, in alignment with the earth community they are part of. Practically, this means solutions should not just advance ecological or social goals but *enhance human well-being*, in the eudaimonic sense, thereby overcoming the empirically faulty – yet still widespread – idea that sustainable lifestyles are based on self-sacrifice and the negation of one's own needs. Empirical research shows that, in many ways, the opposite is true.

See Table 1 for a conceptual overview of the central features of these different approaches to sustainability. After having laid out the tentative principles for

Table 1 A conceptual overview of central features of different approaches to sustainability. Discrete, localizing, time-bound issues may call for instrumental, linear, and reductionist solutions. However, such a problem-solving approach tends to be rendered ineffective when confronted with the highly systemic, complex, and human-created nature of many of our global sustainability issues, including climate change. For those issues, transformative, emergent, and aspirational approaches appear to be more commensurate with their problem set and thus tend to be more effective.

Our planetary issues are	Our mainstream solutions are	More commensurate, effective solutions would be
Systemic	*Instrumental* Limited in scope, disciplinary focus Focused on relieving symptoms	*Transformative* "Multi-problem-solvers," transdisciplinary Focused on addressing ultimate causes
Highly complex	*Linear* Paradigm of planning and control Focus on predefined, hierarchically imposed, concrete goals and outcomes	*Emergent* Paradigm of experimentation and self-organization Focus on collectively emerging, collaborative, innovative solution pathways
Human-created	*Reductionist* Reductionist "problem-solving" mode; negative, non-engaging narrative Almost exclusive focus on technology and regulation	*Aspirational* Integrative "enhancing thriveability" mode; positive, engaging narrative Focus includes human well-being and intrinsically motivated sustainable behavior

transformative, emergent, aspirational solutions, I use these principles in the section "Exploring Transformative Solutions for Sustainable Thriving" to analyze an example of what could potentially be such a transformative solution pathway, namely, a dietary change toward more plant-based diets. I end with a discussion and conclusion in the last section.

Sustainability Science: Pathways of Systemic Transformation through Engaging "Deep Leverage Points"

Sustainability science is an emerging field of research seeking to understand the fundamental character of interactions between nature and society. It studies the complex, dynamic, and continuously evolving interactions between natural and social systems and how they affect the challenge of sustainability – meeting the needs of present and future generations while substantially reducing poverty and conserving the planet's life support systems (Kates et al. 2001). As Weinstein (2010, p. 2) formulated the overarching research question of the field: "At multiple scales and over succeeding generations, how can the earth, its ecosystems, and its people interact toward the mutual benefit and sustenance of all?"

As opposed to the "value-free" stance of the natural sciences, sustainability is a normative ethically justified concept, describing a state of economy, society, and environment considered optimal. Sustainability science is therefore characterized more by its purpose than by a common set of methods or objects (Spangenberg 2011) and, as such, must be aimed at action and "real-world" solutions (e.g., Kaufmann 2009). However, since the concept of sustainable development does not articulate what needs to be sustained, developed, or how, it is fundamentally intersubjective and intercultural (Hedlund-de Witt 2014). This makes it essential to consider a plurality of worldviews that directly interact with people's notions of the quality of life they like to see sustained and/or developed (see also De Vries 2013).

While the field started out as an advanced form of complex systems analysis, over time, this descriptive-analytical, more disciplinary-based mode has been broadened with a more transformational and transdisciplinary approach (Spangenberg 2011; Wiek et al. 2012). In the words of De Vries: "A transdisciplinary approach is called for, in which the quantitative and the qualitative, the natural and the social and also theory and practice (or science and policy) are reconciled and creatively combined. Such an integrating and synthesising approach deserves the name *sustainability science*" (2013, p. 4, italics in original).

In the context of such more transformational approaches to sustainability, it may be useful to revisit Donella Meadows' (1999) well-known concept of *leverage points*. Leverage points are places in complex systems where a small shift may lead to fundamental changes in the system as a whole. In the most basic sense, a *system* is any group of interacting, interrelated, or interdependent parts that form a complex and unified whole, organized around a specific purpose (De Vries 2013). Planet earth is a system, as is every human being. A family is a system, as are the activities involving the production, processing, transportation, consumption, and disposal of food (i.e., the food system). System dynamics are highly complex because they encompass the interaction of global processes with ecological, social, cultural, and economic characteristics of particular places and sectors and integrate the effects of key processes across the full range of scales from local to global (Kates et al. 2001). Moreover, system dynamics operate not only within systems but also *between* them (e.g., changes in the global food system impact the earth system and vice versa).

In considering how to influence the behavior of a system, Meadows (1999) identified 12 leverage points ranging from more "shallow" leverage points – places where interventions are relatively easy to implement yet bring about little change to the overall functioning of the system – to "deep" leverage points, which are generally more difficult to alter but potentially result in transformational change (see Fig. 1, and also Abson et al. 2016). The deepest leverage points involve the *intent* characteristics of the system, which relates to the dominant values, norms, and goals, and the underpinning paradigms out of which the system arises. In other words, these deep leverage points address the *purpose* of the system, which defines it as a discrete entity and provides the integrity for holding the different parts together. Meadows (1999) in her famous essay:

> People who manage to intervene in systems at the level of paradigm hit a leverage point that totally transforms systems. You could say paradigms are harder to change than anything else about a system, and therefore this item should be lowest on the list, not the highest. But there's nothing physical or expensive or even slow about paradigm change. In a single individual it can happen in a millisecond. All it takes is a click in the mind, a new way of seeing. Of course individuals and societies do resist challenges to their paradigm harder than they resist any other kind of change.

From a systems perspective, seemingly "failing" systems may in fact understood to be well-functioning systems that are designed and have emerged out of a purpose society at large no longer subscribes to. Take, for example, our global food system. From the perspective of sustainability, food security, and human health, the current, dominant food system is generally considered highly problematic. Yet when we consider the purpose it has evolved out of, which can arguably be characterized in largely economic terms (e.g., maximization of profits, increase of yields, lowering of costs), the system may seem to be thriving rather than failing. Thinking about systemic change from an understanding of deep leverage points may therefore result in entirely different research questions being asked and different policy arrangements and intervention designs being pursued. As Abson et al. (2016, p. 33) state:

> . . . shallower interventions are favoured in both science and policy. For example, most high profile work on food security has focused on issues of food production (e.g. Foley et al. 2011). Such a focus emphasises material flows and buffer stocks, rather than deeper issues such as the rules, structures, values and goals that shape food systems. . . . questions such as "is the global food system oriented to provide food security for all?" and "if not, how can its intent be changed?" have rarely been asked by scientists. Yet it is these questions that address the more fundamental challenges, and provide input to thinking about deeper leverage points.

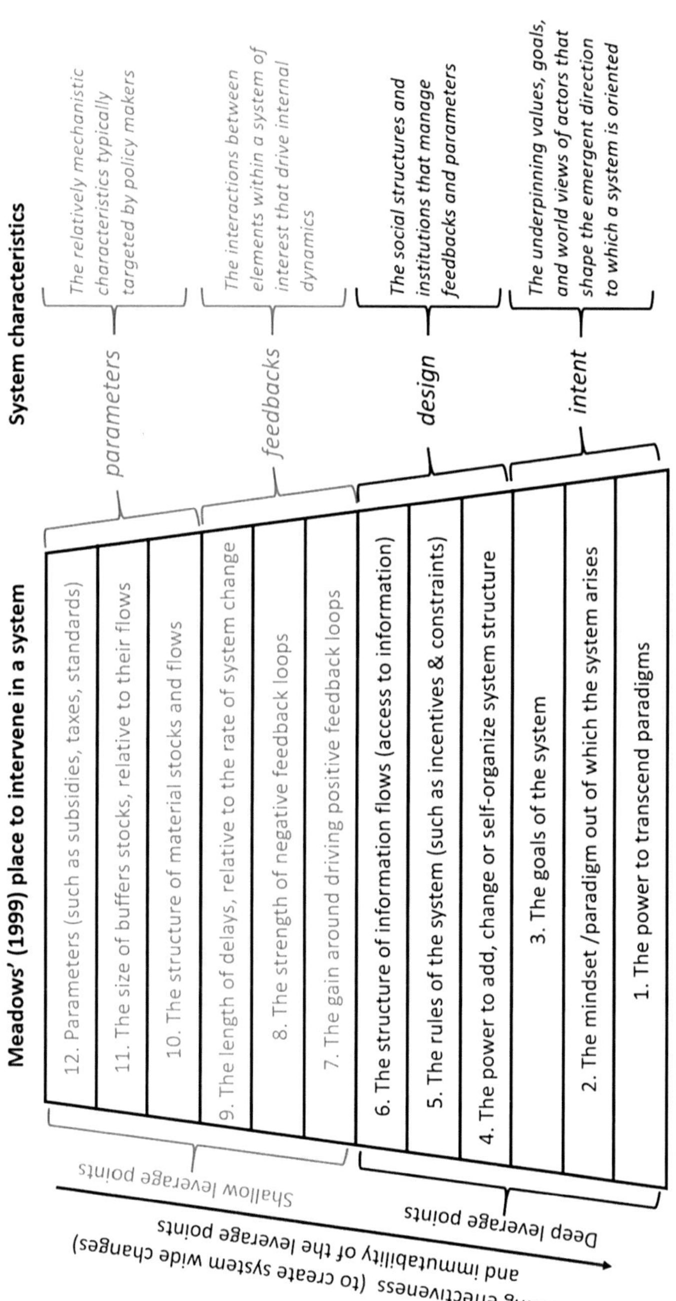

Fig. 1 From 12 leverage points to four system characteristics. (Figure adopted from Abson et al. 2016, p. 33, 1999, p. 11)

As these authors argue, "many sustainability interventions target highly tangible, but essentially weak, leverage points (i.e. using interventions that are easy, but have limited potential for transformational change). Thus, there is an urgent need to focus on less obvious but potentially far more powerful areas of intervention" (Abson et al. 2016, p. 30). This insight – that in order to get optimal leverage, we *need to focus on the deepest leverage points* – may be seen as key to designing effective interventions for sustainable transformation.

Next to the importance of engaging the intent of the system, a systems perspective reveals we need solutions that generate benefits *addressing multiple societal issues at the same time*. That is, we need "multi-problem-solvers." Because of the complex interdependencies in our earth system, issues tend to occur in relation to each other. What may seem like an isolated problem is often part of an interconnected network of issues. Changes in one system tend to impact changes in other systems. Effectively addressing any single issue therefore means we need to consider the larger systems this issue interacts with and address it in this broader context.

Moreover, because of the interconnected set of circular relationships that characterize systems, *feedback loops* need to be taken into account. Feedback occurs when outputs of a system are routed back as inputs as part of a chain of cause and effect that forms a circuit or loop, thereby creating the possibility of self-reinforcing cycles. When the direction of reinforcement is desired, we refer to "virtuous cycles"; when the direction is undesired, we refer to "vicious cycles." An example of a vicious cycle is the global rising of temperature due to climate change, resulting in the melting of permafrost and therefore the release of methane, a powerful greenhouse gas, which in its turn exacerbates global warming, to then result in more permafrost melt and thus more release of methane, and so on (Schuur et al. 2015).

Another key principle for sustainability interventions is therefore to design them in such a way that self-reinforcing feedback loops are generated in directions we desire (e.g., sustainable transformation). In practice, this means that ideally the benefits as generated by these interventions should mutually support, enhance, and reinforce each other, thereby creating virtuous cycles that accelerate their positive impact. Consciously designing our interventions with the transformative power of virtuous cycles in mind may be of essential importance in order to address the complexity and interconnected nature of our current "poly-crisis."

Moreover, for these solutions to be truly effective, they need to potentially have a global reach and impact. This means they need to be *transferable and scalable*, with all the systemic complexities and cultural challenges associated with that. Studies of transformational sustainability projects have revealed a tension between understanding the complexity and specificity of a case in its real-life context while also contributing to a common body of theoretical knowledge on generalization and generic principles. In the words of the researchers (Wiek et al. 2012, p. 19):

> Even if research led to actionable knowledge, deficits in applicability and synthesis undermined its value. A clear focus on small-scale (local) units and a close link to the actual implementation ... are favorable factors. Yet, they do not guarantee the transferability or

scalability of the solution options generated. Additional research efforts are needed to overcome applied research with one-case solutions towards the aspired transformational research that generates widely applicable solution options.

While widely applicable and scalable solutions are indispensable in a profoundly interconnected world facing systemic challenges of planetary proportions, the complexities involving such a transfer and implementation of solutions are not to be underestimated. That is, also the most potent solution options require decisions and political willingness as well as adaptation to the specific (e.g., economic, cultural, geographic, institutional) context of implementation. While I do not aim to cover these complexities here, transferability and scalability are important principles that need to be considered in our understanding of more effective and transformative solutions.

Another consideration for designing transformational sustainability solutions is the potential risks and unintended consequences or negative side effects involved with their implementation. As elaborated above, feedback loops can operate in both "virtuous" and "vicious" directions, and larger systemic consequences of local actions and interventions are frequently not (fully) overseen beforehand. In fact, as we have seen in the past decades, what some propose as viable sustainable solutions are by others seen as environmental problems in their own right, often due to latent risks or side effects.

For example, certain groups advocate for the further industrialization of agriculture, supported by genetic modification of crops, with the aim of sustainably feeding our growing world population. However, for others this pathway is itself a threat to the environment (e.g., Levidow et al. 2012). And while some see nuclear energy as a sustainable form of energy production, for others the waste products and risks associated with this technology are themselves considered environmental hazards. The production of biofuels is another well-known example of a highly controversial sustainability solution pathway, as latent risks and negative side effects may override its positive benefits.

This situation has led to many environmental issues and their potential solutions becoming highly controversial and politicized, with not only political and social actors deeply divided but scientists as well, arguably leading to increasing cultural polarization and gridlock and stalling effective international efforts and policies (Sarewitz 2004). Due to the systemic and complex nature of our sustainability challenges, solutions with few risks and negative side effects are preferable.

Complexity Science: Inviting Experimentation, Collaboration, and Emergence for Transforming Complex Systems

Contemplating the nature of our global challenges, many authors argue for distinguishing between fundamentally different types of problems, such as between "complex" and "complicated" ones (e.g., Kurtz and Snowden 2003; Snowden and Boone 2007), between "wicked" and "tame" (e.g., Grint 2010), and between "complex adaptive" and "technical" challenges (e.g., Hassan 2014). As all these authors argue, while the default approach is to address sustainability issues as complicated, technical, or tame problems, it may be more accurate and effective to address them as complex or wicked problems. According to Heifetz et al. (2009), "confusing complex, adaptive challenges with technical challenges is a classic error."

The origin for this line of thinking can be found in complexity science, which differentiates the qualities of *complex adaptive systems* from mechanical systems. A complex adaptive system is a system in which a perfect understanding of individual parts does not automatically translate into a perfect understanding of whole-systems behavior. In other words, a complex system is "more than the sum of its parts" (Snowden and Boone 2007). Its behavior is fundamentally different to, and cannot be predicted from, the behaviors of its constituent agents (e.g., we cannot predict behaviors of a crowd from individual behavior) (Van Beurden et al. 2011).

These systems are complex in the sense that they are dynamic networks of interactions. They are adaptive in the sense that individual and collective behaviors mutate, evolve, and self-organize in response to (collections of) events. Examples of complex adaptive systems include the biosphere, forests, reefs, stock markets, the immune system, humans, organizations, and communities (Van Beurden et al. 2011). Complexity science is therefore concerned with problems that are dynamic, unpredictable, multidimensional, and characterized by nonlinearity (in contrast with more straightforward problems, characterized by linear "cause-and-effect" relationships).

Since our sustainability problems are highly complex, we need to adequately account for and respond to that complexity in our intervention designs and leadership styles. Not doing so may create new, unforeseen issues, as mechanistic interventions may precipitate unexpected problems by stimulating latent feedback loops within the complex web of cause and effect (Van Beurden et al. 2011). Also, the "emergent," self-organizing characteristics of complex adaptive systems highlight the importance of context and the limitations of linear program delivery (Keast et al. 2004).

In the leadership and management literature, *the Cynefin framework* is used for addressing complex challenges. Especially when used as a sensemaking tool, this framework can help practitioners understand the level of complexity of issues, identify appropriate strategies, and avoid the pitfalls of applying reductionist approaches to complex situations (Van Beurden et al. 2011). *Cynefin*, pronounced ku-*nev*-in, is a Welsh word that signifies the multiple factors in our environment and experience that influence us in ways we can never fully understand. The framework has now been applied to knowledge and strategy management, research, policy making, and leadership training (Snowden and Boone 2007).

The Cynefin framework (see Fig. 2) sorts the issues facing leaders into five contexts defined by the nature of the relationship between cause and effect. *Known* (or simple) and *knowable* (or complicated) contexts assume an ordered universe, with perceptible relationships between cause and effect, while *complex* and *chaotic* contexts are unordered, with cause-and-effect relationships only appearing coherent

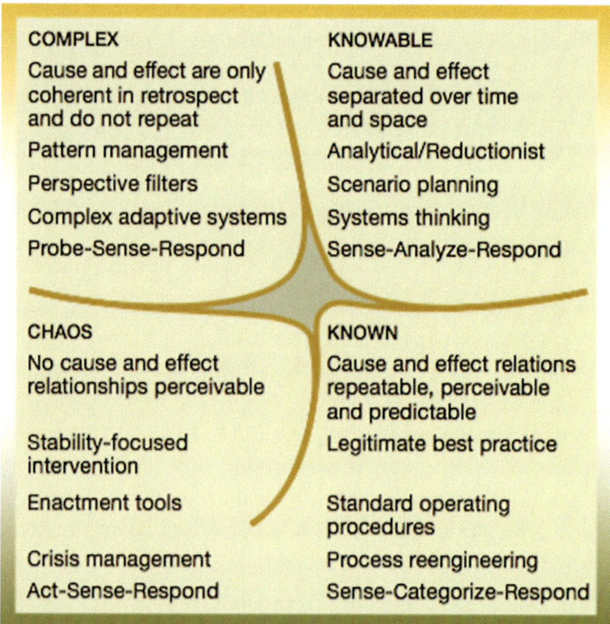

Fig. 2 The Cynefin domains. (Adopted from Kurtz and Snowden 2003, p. 468). Courtesy of International Business Machines Corporation, © (2003) International Business Machines Corporation

in retrospect. The ordered world is the world of fact-based management; the unordered world represents pattern-based management. The fifth context, situated in the middle of the figure, is referred to as *disorder* and applies when it is unclear which of the other four contexts is predominant (Kurtz and Snowden 2003).

Complicated problems or technical challenges are characterized by clear cause-and-effect relationships, though extensive analysis and expertise may be needed for understanding them. Although such issues are complicated, they can be analyzed and assumptions of order and rational choice generally hold (Kurtz and Snowden 2003). Some call this the domain of the "known unknowns," and in more general terms, the *knowable* domain. That is, we know what we don't know.

In contrast, *complex challenges* are characterized by intricate webs of nonlinear cause-and-effect relationships across multiple scales. They have unpredictable, emergent properties, high levels of uncertainty, and the potential for feedback loops that may rapidly precipitate catastrophe. Although coherent cause-and-effect relationships exist, they defy adequate categorization or analytic techniques (mainly due to the number and complexity of agents and interactions). Some call this the domain of the "unknown unknowns" or the complex or *unknowable* domain. That is, we don't know what we don't know.

Hassan (2014, p. 20) describes the difference between complicated and complex challenges as follows:

> An example of a technical challenge is sending a man to the moon. The problem is clearly defined and the solution unequivocal. Implementation may require solving many difficult problems, but the desired outcome is plainly understood and agreed upon. In contrast, multiple perceptions of both the problem and the solution are characteristic of complex systems. . . . Complex challenges are therefore dynamic and can change in unexpected ways over time, whereas technical challenges are relatively stable and static in comparison.

Therefore, rather than predicting patterns, complexity theory studies how patterns *emerge* through the interaction of agents, allowing one to make sense retrospectively. Generally, "emergence . . . refers to the arising of novel and coherent structures, patterns, and properties during the process of self-organization in complex systems. Emergent phenomena are conceptualized as occurring on the macro-level, in contrast to the micro-level components and processes out of which they arise" (Goldstein 1999, p. 49). That is, as complex adaptive systems are more than the sum of their parts, something new arises, emerges, out of the whole (macro-level), which could not be predicted by its parts (micro-level).

Since sustainability issues are mainly found in the complex domain, and require different, often counterintuitive, responses, I concentrate here particularly on that context. The decision model in complex contexts is to create *probes* to make patterns or potential patterns more visible before we take any action (Snowden and Boone 2007). That is, rather than analyzing with the aim to predict what will happen (as is appropriate in complicated contexts), we probe with the aim *to reveal what is emerging*. Probes are experimental interventions that allow one to see how the system responds to change and input. Examples are genuine engagement with communities; skilled facilitation to enable agreed priority areas; and pilot projects or prototypes that experiment with new courses of action and collaboration.

Understanding the nature of emergence naturally translates into a preference for more participative and distributed leadership styles and "bottom-up" approaches, which are grounded in the experience that "we don't know" and honor the innovative potential of allowing self-organization. This is in sharp contrast with more hierarchical "command and control" styles and "top-down" approaches. In the words of Grint (2010, p. 171):

> Tame [or complicated] problems might have individual solutions in the sense that an individual is likely to know how to deal with it. But since Wicked [or complex] Problems are partly defined by the absence of an answer on the part of the leader then it behooves the individual leader to engage the collective in an attempt to come to terms with the problem. In other words, Wicked Problems require the transfer of authority from individual to collective because only collective engagement can hope to address the problem. The uncertainty involved in Wicked Problems imply that leadership, as I am defining it, is not a science but an art – the art of engaging a community in facing up to complex collective problems.

Complex problems ask us to assume that no one has the solution in isolation and to acknowledge that the problem is of a systemic rather than of an individual nature, so that it cannot be caused by or solved by a single aspect of the system. A highly *collaborative* approach to group function is therefore desirable, and the more diverse the partners, representing a multiplicity of perspectives, the better a system can be engaged and appropriate probes developed (e.g., Snowden and Boone 2007; Van Beurden et al. 2011). This also means we need to involve stakeholders representing different worldviews, as they tend to understand the problem differently, and propose different solutions (De Witt 2015; Hedlund-de Witt 2014).

Analytic techniques appropriate to the ordered domains are inadequate here, while narrative-based sensemaking methods are particularly helpful (Snowden and Boone 2007). Or, to put it differently, addressing complex problems is more about asking the right questions and less about providing the right answers (Grint 2010). After probing, we need to "sense" which initiatives are useful in order to "respond" by amplifying and resourcing them (and disrupting those that are undesirable). The aim is to develop open-minded observation rather than hasty action based on preconceived ideas and the entrained patterns of past experience.

An important sidenote is that problems, once identified as sitting in a certain domain, do not necessarily stay there. In fact, examining transitions at boundaries between the domains is key to facilitating intervention work. An issue can shift across a boundary as a project progresses, or context changes. For example, aspects of a complex issue may shift into the ordered domains for scientific "unpacking" or for implementation of a "best practice" strategy (Van Beurden et al. 2011). Or, issues may shift from chaotic to complex, after some stability is established as a result of initial "rapid response" interventions. The task is thus to use strategies and approaches *appropriate* to the context or domain an issue manifests itself in.

Thus, there are a number of aspects we need to consider when applying these insights with the aim of developing effective sustainability interventions, strategies, and approaches. In the first place, our change strategies need to allow for self-organization and emergence, rather than attempting control in a top-down manner. In practice, this may, for example, mean that governmental policies primarily orient around facilitating sustainable trends, projects, and approaches emerging from society. Secondly, our change strategies need to be based on collaboration with a rich diversity of stakeholders and viewpoints, thereby overcoming the tendency to primarily engage people who share our worldviews, concerns, and priorities and avoid those who think differently.

These shifts – from technocratic, command- and control-type approaches toward approaches that actively invite collaboration, experimentation, and emergence – demand great flexibility, willingness, openness, and somewhat of a paradigm or attitude shift on the part of leaders, organizers, and policy makers. The leadership field attests to the importance of this shift, as a substantial amount of leadership books have been written in the past decade or two with the aim of facilitating leaders in making this transition toward more "participatory," "distributed," "servant," "adaptive," "co-creative," and "agile" forms of leadership.

Next to solution pathways needing to be more transformational and emergent, they also should be more "aspirational," which will be further discussed in the next section on the lessons emerging out of the field of positive psychology.

Positive Psychology: Cultivating Eudaimonia and Harnessing the Power of Intrinsic Motivation for Sustainable Well-Being

The happy man both lives well and does well. – Aristotle.

At present our global challenges may seem overwhelming or depressing to many people, with the immensity and formidability of these issues discouraging conscious action. Moreover, culturally undesirable impressions exist of environmentalists as living simple, barren lives, defined in a negative, self-sacrificing manner and characterized by a gloomy outlook on the world. More generally, the dominant sustainability narrative is often characterized by what we should or shouldn't do as well as the disastrous impact of what we are already doing – which is often not experienced to be inspiring or engaging. And while all of us search for "the good life" in our own ways, for many that good life seems to be in conflict with a life that is also sustainable and environmentally conscious. These factors may therefore put substantial limitations on the degree of public support for, and engagement with, sustainability issues at large, hampering constructive change on all levels.

In this section, I therefore explore the insights of positive psychology, a fairly young, yet rigorously empirical field of psychological inquiry and research, which focuses on the well-functioning, development, and *flourishing* of human beings, rather than on their illnesses, pathologies, and distortions (Seligman and Csikszentmihalyi 2000). The field departs from a conception of human nature as the progressive unfolding of innate potentials and an objective of facilitating positive, healthy growth. As we will see, empirical results from this field shatter the "myth" that "a good life isn't green, and that a green life isn't good." In fact, they demonstrate that pro-social and more sustainable attitudes and lifestyles coincide with human flourishing, while ecologically destructive attitudes and lifestyles correlate with a lack of psychological health and well-being. Perhaps, challenging the faulty perception that sustainable lifestyles are based on sobriety and self-sacrifice, while highlighting the positive link between such lifestyles and quality of life, may be needed to elicit a more widespread engagement with our ecological issues.

In their quest for understanding the nature of human flourishing, positive psychologists have developed a model of *eudaimonia*, based on extensive empirical testing and elaboration (Ryan et al. 2008). While the concept of eudaimonia originally stems from Aristotle, it is just as relevant in today's world, as it may suggest important alternative routes to individual and societal wellness and thereby may play a critical function with respect to economics, social policies, and sustainability interventions (e.g., Ryan et al. 2008).

Eudaimonia is generally defined as "living well," in the sense of living a complete human life, in which one realizes valued human potentials. Eudaimonic approaches to wellness are described by pursuing goals that are intrinsically valued and by processes that are characterized by autonomy and awareness. In contrast, "hedonic" approaches focus on "feeling good," interpreted as the occurrence of positive affect and the absence of negative affect. That is, whereas eudaimonic approaches focus on the *content* of one's life, and the *processes* involved in living well, hedonic conceptions focus on a specific *outcome*, namely, attainment of pleasure and avoidance of pain.

Rather than a psychological state or outcome, eudaimonia thus refers to *a way of living* – a way of living *that is focused on what is intrinsically worthwhile to human beings*. Intrinsic values are first-order values, meaning they are irreducible to other values and do not exist for the sake of other values. Extrinsic aspirations, on the other hand, are pursued as a means to other goals. For example, somebody may strive for wealth or fame, because of the freedom that wealth promises to provide or the admiration associated with fame. That is, goals like wealth, fame, image, and power are often sought because they are instrumental to other goals. In contrast, goals like personal growth, affiliation and intimacy, community contribution, and physical health tend to refer back to themselves; i.e., people strive for them because they enjoy these values in their own right.

This distinction has proven highly potent for an empirically based understanding of human well-being. Results revealed that the strength of intrinsic relative to extrinsic aspirations was positively related to a host of well-being indicators, including self-actualization, positive affect, and vitality, and negatively to indicators of ill-being, including depression, negative affect, anxiety, and physical symptoms (Ryan et al. 2008). Similar findings have been found cross-culturally (e.g., Grouzet et al. 2005). The researchers suggest that the reason intrinsic and extrinsic aspirations are differently related to psychological health and well-being is the degree to which they are linked to the satisfaction of the basic, psychological needs for *autonomy*, *competence*, and *relatedness* (e.g., Ryan and Deci 2000). These needs are understood to be innate and universal, essential for an individual's psychological health, and when satisfied allow optimal functioning and growth.

One may wonder why people would pursue extrinsic goals while it is demonstrably not supporting their overall health and happiness and is thus not in their self-interest. Psychologists understand these extrinsic goals as *substitutes* for underlying needs that were thwarted (Grouzet et al. 2005; Ryan 1995). That is, these goals have their salience because they attempt to serve something more basic and intrinsic, even though the person may not be conscious of the connection. Seeking fame because one desires to feel loved is an indirect (and, as the research shows, generally ineffective) route to the intrinsic value of love and connection. When that intrinsic need was left unsatisfied, i.e., due to factors in one's upbringing, people start to pursue more indirect, extrinsic pathways in an attempt to get their needs met. In that process, individuals frequently become preoccupied with second- and third-order values that are derivative and now disconnected from the deeper intrinsic needs that were unsatisfied (Ryan et al. 2008). In other words, pursuing extrinsic goals can be seen as a psychological *defense mechanism*, that is, a way of coping with a reality in which one's fundamental human needs were thwarted.

Although eudaimonia is not defined by a focus on specific outcomes, it is associated with numerous outcomes, which include hedonic happiness as typically assessed but also includes a fuller, more stable, and enduring type of happiness than that obtained when one's goals are more directly hedonistic. Among these enduring positive outcomes are a sense of meaning, subjective vitality, higher quality relationships, and better physical health indicators, especially with respect to symptoms related to stress (Ryan et al. 2008). Beyond these, researchers have started to map out additional kinds of well-being outcomes, showing that eudaimonic individuals (Ryan et al. 2008, pp. 162–163):

> … have high levels of inner peace, as well as frequent experiences of moral elevation and deep appreciation of life; feel connected not only with themselves but also with a greater whole that transcends them as individuals; have a sense of where they fit in to a bigger picture and are able to put things in perspective; and describe themselves as "feeling right" (as opposed to "feeling good," the state that hedonically oriented individuals seem to pursue).

It is therefore not surprising that living well as defined in the eudaimonic sense also contributes to social and environmental goals. In fact, multiple studies have found correlations between (eudaimonic) happiness and sustainable behavior (Brown and Kasser 2005; Corral Verdugo et al. 2011; Jacob et al. 2009). The emerging evidence suggests that people high in eudaimonia are likely to be more socially responsible and often derive a sense of well-being from contributing to the greater good. These studies attest that promotion of eudaimonic living may be better for a society as a whole, insofar as its members show more care, concern, and responsibility in their actions (Ryan et al. 2008).

This body of work thereby reveals that when humans flourish (in the eudaimonic sense), they tend to display more pro-social and pro-environmental behaviors and attitudes. The research thus suggests that, put simply, psychologically healthy and happy human beings tend to also be "good for the planet." As argued above, these findings are in sharp contrast with dominant assumptions in the environmental debate, in which "the good life" is frequently juxtaposed with living sustainably, and happiness and ecological care are portrayed as conflictual pursuits (see e.g., Brown and Kasser 2005). Sustainable behavior is often conceptualized to be emerging from negative emotions, such as fear, guilt, or shame and as having negative consequences, like discomfort, inconvenience, and sacrifice (Corral Verdugo 2012).

However, this association between unsustainability and happiness seems to stem from a hedonic approach to wellness. Indeed, attempts to maximize pleasure and avoid pain are often associated with selfishness, materialism, objectified sexuality, and ecological destructiveness. Empirical research confirms this idea. For example, Brown and Kasser (2005) found that people embracing the extrinsic goal of materialism consumed more and left bigger environmental footprints. Other studies found that people with materialistic, extrinsically oriented worldviews were more likely to display instrumental attitudes toward nature and less sustainable lifestyles, while people with more intrinsically oriented worldviews correlated positively with pro-environmental attitudes and lifestyles (Hedlund-de Witt et al. 2014).

The idea that altruistic and pro-social behaviors are based on a negation of one's self-interests is still widespread. However, this body of research shows that people pursuing hedonic lifestyles are, strictly speaking, doing this *against* their self-interest, as such lifestyles are associated with less happiness and even ill-being. In contrast, people pursuing eudaimonic approaches to happiness prove to be better equipped to take care of their own interests by fulfilling their basic, psychological needs.

Understanding hedonic attitudes and behaviors as psychological defense mechanisms opens up a more compassionate perspective, as well as new potential pathways for responding to them. Rather than "guilt-tripping" and blaming people for materialistic or social or ecologically destructive behaviors and attitudes, approaches that support people to find more constructive and effective ways of fulfilling their psychological needs may prove more helpful and rewarding. Moreover, challenging the faulty cultural perception that social and sustainable lifestyles are based on a sacrifice and negation of one self could have a substantial impact on broadening social involvement in sustainability.

The lessons we can draw from decades of research into human flourishing is thus that facilitating more eudaimonic lifestyles will powerfully serve our aim for sustainable transformation in a number of ways. As the research shows, eudaimonic individuals are more inclined to act in pro-social ways, which therefore generates a multitude of social and environmental benefits. Moreover, facilitating eudaimonic living implies the "unleashing" of human potential and creativity, which may result in major ripple effects and positive feedback loops in terms of new initiatives and innovations for a wide range of social and sustainability causes.

On the societal level, facilitating eudaimonic lifestyles means forging and enabling a new pathway toward sustainable well-being. Thereby it provides an important counterweight to the pathways almost incessantly promoted by the increasingly globally dominant, neoliberal, market-based paradigm, which tends to emphasize extrinsic values and motivations, with destructive consequences for the environment:

> Consumerism is prompted by continuous exposure to desire-creating advertisements, which often attempt to promote insecurity in order to create a sense of need. Moreover, the winner take all atmosphere associated with the values of a competitive market economy can crowd out altruism, sense of community, and other prosocial attitudes. It seems that the eudaimonic life is continuously threatened by the individualistic attitudes associated with such economies, whereas hedonic well-being has a much closer fit with the capitalist ethic. (Ryan et al. 2008, p. 165)

In fact, one could argue that capitalist marketing is specialized in selling the illusion of extrinsic goals and values as able to fulfill people's deeper, intrinsic needs. That is, commercials seduce us to think we are not just buying a car; we are buying freedom. We are not just buying soda; we are buying social connections. At the same time, policy makers, campaigners, and social scholars seem increasingly aware that competitive individualism and hedonic happiness for the masses will lead to, and ultimately be compromised by, an unsustainable environment, all the while not fulfilling its promise of enhancing happiness.

For sustainability interventions to be able to offer deep, transformational changes, they should thus *facilitate and enhance human well-being in the eudaimonic sense*. Optimally, interventions should support people to live well, grow, thrive, and unleash their potential. At the very least, they should not be in conflict with that. Practically, this means these solutions should be *inherently positive and rewarding* for people to engage in and thus advance their own life circumstances (health, well-being, community connections, et cetera), while also contributing to the larger whole. While the answer to how we can we use these principles in our intervention design is not easy to give, the research suggests to follow three rules of thumb (Ryan et al. 2008):

1. Support people to strengthen intrinsic goals and values such as health, (contribution to) community, affiliation and intimacy, personal growth.
2. Support people to behave in autonomous, volitional, or consensual ways, rather than relying on controlled and heteronomous ways.
3. Support people to be mindful and act with a sense of awareness.

Moreover, rather than inadvertently reinforcing the faulty idea that people need to choose between taking care of their own needs and the needs of the social and planetary whole, an inspiring sustainability narrative should be cultivated that spreads the uplifting and empowering message that sustainability is about making life better and more beautiful, and that authentic (eudaimonic) human well-being is at the heart of the more sustainable world we collectively aspire to.

And while all of this may still sound a bit abstract, in the following section I will explore an example of what these principles may look like in practice.

Exploring Transformative Solutions for Sustainable Thriving

Having explored some of the insights emerging from the fields of sustainability science, complexity science, and positive psychology, it is noteworthy that there is a certain resonance with respect to the solution-directions they are each pointing in. For example, while from a systems perspective the importance of engaging with the *intent* of the system is emphasized for engendering systemic transformation, positive psychology similarly points to the need for a paradigm shift, arguing for a transition from a hedonic to an eudaimonic conception of human wellness. Also, from the positive psychology research follows that policies and programs that facilitate more eudaimonic ways of living would comply with some of the key principles as formulated based on sustainability science. For example, such policies and programs would contribute to multiple societal issues simultaneously, ranging from subjective well-being to psychological and physical health and from pro-social attitudes and behaviors to positive social relationships. The shift in emphasis from *outcome* to *process* and *content* are emphasized both in positive psychology as well as in the leadership literature informed by complexity science.

On the basis of the last three sections, we can formulate a tentative list of principles for more commensurate, effective interventions and change strategies. Probably needless to say, this list is provisional and somewhat arbitrary and thus up for debate, further probing, and research. However, on the basis of the current insights, I argue that highly effective change strategies:

- Engage with the *intent* or *purpose* of the system as a whole, and therefore tend to shift mindsets and paradigms, or invite reflection on them.
- Generate benefits that systemically address multiple issues simultaneously.
- Are transferable and scalable.
- Carry limited risks and few negative side effects.
- Draw in collaboration with diverse stakeholders and viewpoints so solutions and strategies are co-created and co-owned.
- Invite *emergence*, through experimentation (probing) and self-organization.
- Facilitate emerging (bottom-up or societal) projects, ideas, and movements that move in desirable directions.
- Are inherently positive and rewarding for those who engage in them and thus enhance human well-being in the eudaimonic sense.
- Support people to strengthen *intrinsic goals and values*, such as health, (contribution to) community, affiliation and intimacy, personal growth.
- Support people to behave in autonomous, volitional, or consensual ways, rather than relying on controlled and heteronomous ways.
- Support people to be mindful and act with a sense of awareness.
- Cultivate an uplifting narrative that challenges the faulty idea that hedonism leads to happiness and demonstrates fulfilling ways to live a life that is both "good" and "green"; a narrative in which enhancing the ability to "flourish" or "thrive" is at the heart of sustainability.

In this section I briefly explore an example of what could potentially be such a transformative solution pathway, namely, a dietary change toward more plant-based diets. In this process I rely on existing literature, not intending to offer an exhaustive review of the research on this topic but rather to offer insight into what such a transformative approach may look like "in the real world."

Dietary Change Toward More Plant-Based Diets as Transformative Solution

An example of a potentially transformative solution is a dietary change toward more heavily plant-based diets, thereby moving away from animal-based diets. This is an extremely effective intervention with the potential to address multiple big challenges, including climate change and a host of other environmental problems, the obesity and health epidemic, and large-scale animal suffering, among others.

The food system is responsible for more than a quarter of all greenhouse gas emissions (Vermeulen et al. 2012), of which up to 80% are associated with livestock production (FAO 2006). Experts and policy makers increasingly agree that for the necessary transition to a low-carbon society, change in the Western diet, in particular the reduction of proteins sourced from animals, is a highly potent pathway (De Boer et al. 2016; Popp et al. 2011). A recent study found that a global transition toward low-meat diets could reduce the costs of climate change mitigation by as much as 50 percent by 2050 (Stehfest et al. 2009). Another study found that widespread adoption of a vegetarian diet would bring down emissions by 63% (Springmann et al. 2016). The "outstanding effectiveness" of this climate mitigation option is therefore widely recognized by climate experts (De Boer et al. 2016). Moreover, the reduction of meat consumption and production is effective in addressing a host of other environmental issues, including deforestation, air pollution, water depletion and pollution, and biodiversity loss (FAO 2006).

According to a 2015 *Chatham House Report*, people in industrialized countries consume on average around twice as much meat as experts deem healthy. In the USA the multiple is nearly three times. High consumption of red and processed meat and low consumption of fruits and vegetables are important diet-related risk factors contributing to early mortality in most regions, while over a billion people are overweight or obese (Lim 2012). Adoption of a healthy diet would therefore not only create healthier populations, but it would also generate over a quarter of the emission reductions needed by 2050 (Wellesley et al. 2015).

Moreover, such a dietary change would substantially lower health-care costs. A recent study found that the monetized value of improvements in health would be comparable with, or even exceed, the (already high monetary) value of environmental benefits, estimating the economic benefits of improving diets to be 1–31 trillion US dollars, which is equivalent to 0.4–13% of global gross domestic product in 2050 (Springmann et al. 2016). The researchers conclude their study saying that "there is a general consensus that dietary change across the globe can have multiple health, environmental, and economic benefits ... The size of the projected benefits ... should encourage researchers and policy makers to act to improve consumption patterns" (2016, p. 4150).

Next to health, economic, and environmental benefits, there are other advantages of shifting to a low-meat diet, as it would reduce large-scale animal suffering due to factory farming practices. Upon investigation, these practices are often considered deeply unethical because of the wide-ranging suffering they instill, as well as highly problematic in multiple regards, including (but not limited to) the emergence of increasingly antibiotic-resistant strains, large-scale animal diseases such as mad cow disease, and recalls of contaminated meat products (e.g., Eisnitz 1997; Pluhar 2010). Factory farming has been defined as "an industrialized system of producing meat, eggs, and milk in large-scale facilities where the animal is treated as a machine." However, our current understanding of animal intelligence and behavior underscores that animals cannot be considered, and therefore should not be treated as, machines. A wide range of moral theories agree in declaring current meat production practices to be morally unacceptable (Pluhar 2010).

A dietary change toward more plant-based diets also supports people to connect with nature through the consumption of natural, unprocessed vegetables. In fact, themes such as the wish to return to a more natural lifestyle, distancing from materialistic lifestyles, and reverting to a more meaningful moral life, as well as connectedness to nature, awareness, and purity are found to be motivators for people with more whole-food, organic, and vegetarian dietary styles (Schösler et al. 2012, 2013). Next to such *ethical* motivators, research has found strong emphasis on *aesthetic* reasons to eat more plant-based (and often more local or whole-food) diets, including taste, diversity, creativity, and experimentation in the kitchen ("vegetable gastronomy"), and enjoyable social relations due to food (i.e., connection with the local farmer) (Schösler 2012). Together these motivations represent a wide range of positive values that can be associated with plant-based diets (Schösler and Hedlund-de Witt 2012).

Clearly, a dietary change toward plant-based diets generates benefits that contribute to addressing multiple societal issues simultaneously. Moreover, these benefits may mutually reinforce each other. For example, as plant-based diets positively contribute to addressing the health and obesity epidemic, this in its turn positively feeds back to addressing climate change. Currently, the reverse relationship is being observed. That is, as humanity becomes more rotund, more resources are needed to cool, nourish, and transport that extra weight, a trend that accelerates climate change by requiring the consumption of more fossil fuels, which result in more greenhouse gas emissions. In fact, increasing population fatness could have the same implications for global food energy demands (with significant implications for food security, climate change, and sustainability) as an extra half a billion people living on the earth (Irfan 2012). This example thus shows how benefits in terms of weight loss and health reinforce other beneficial feedback loops, such as lower food and energy demands.

This intervention arguably also engages with the food system's *intent*, as it invites for a reflection on our relationship with the (natural) world around us, and our aspirations for the "good life," while offering practical pathways of different ways of relating to nature and animals in particular. In fact, food and consumption experts have argued that food habits are profoundly culturally engrained, with everyday consumption choices enmeshed in a web of non-instrumental motivations, values, emotions, self-conceptions, and cultural associations (Sorin 2010). A shift toward different consumption patterns therefore tends to coincide with a change in values and worldviews and vice versa.

Sociologist Colin Campbell (2007), for example, has signaled a dramatic change in popular beliefs and attitudes toward nature that has occurred over the past 30–50 years, coming to expression in the rise of the animal rights movement, the swing to vegetarianism and the consumption of whole and organic food, the holistic health movement, and the environmental movement itself. In Campbell's eyes, these are all different manifestations of the idea that some sort of spirit, life force, or higher value is present in all of nature, which therefore needs to be treated with respect or even reverence. This idea has positive implications for environmental behaviors, including attitudes toward meat eating. Animals are increasingly considered in terms of their well-being and rights and seen as sentient "fellow creatures" instead of merely "food" (Verdonk 2009).

This intervention can also be argued to be transferable and scalable, as, from a physical-geographical perspective, a transition to lower meat consumption can in principle be done almost anywhere. In fact, in most of the Western world, this transition means going back to consumption patterns that were the norm until the 1960s. However, changing deeply engrained cultural habits is challenging – a challenge that could be argued to be inherent in attempts to generate transformational change through accessing deep leverage points. Simultaneously, as Colin Campbell and others have argued, a cultural shift in this direction is already happening and thus can be facilitated and supported, rather than needing to be imposed in a top-down manner (Schösler and Hedlund-de Witt 2012). Moreover, a dietary change toward more plant-based diets could be scaled up through integration in educational programs and contexts, as well as in work places and institutions.

The risks and unintended consequences associated with this intervention are arguably low. While some nutritional deficiencies are associated with completely vegan diets, this risk can be mediated through an emphasis on lowering meat and increasing plant consumption, rather than emphasizing the need for a transition toward completely vegetarian or even vegan diets.

Also, the benefits attained by changing toward more plant-based diets will not only be experienced on the collective level but on the individual as well. This means this intervention generates benefits that are inherently positive and rewarding, as it enhances human health and can contribute to weight loss, is associated with a range of positive ethical and aesthetic values, from connectedness to nature to creativity and diversity in the kitchen, and reduces health-care costs. In that sense it clearly supports people to strengthen intrinsic goals and values, such as health, (contribution to) community, affiliation and intimacy, and personal growth.

Moreover, the intervention may support people to be more mindful and act with a greater sense of awareness, with respect to the origin and production of their food, their relationship to nature and animals, and their own sense of health and well-being. Although it depends on implementation and communication strategies, this intervention or solution pathway has the potential to support people to behave in autonomous, volitional, or consensual ways – that is, people should be supported to make the best choice for themselves, considering all factors at play, rather than being coerced or habituated a certain way (i.e., through meat tax or meat subsidies on the one hand or oppressive cultural norms and practices on the other).

Importantly, for such an intervention to be "aspirational," it should challenge often deeply engrained, cultural ideas that a fulfilling meal has to contain meat and portray compelling alternative meal options that are not only more climate-responsible and environment-friendly but also healthy, tasteful, and attractive in their own right. The international food awareness organization Proveg may offer an example of how such a more positive and inspiring narrative could look like, as their main emphasis is on the five, positive benefits attained from eating a predominantly plant-based diet, as they portray such a diet as "pro-health," "pro-animals," "pro-environment," "pro-justice," and "pro-taste" (see www.proveg.com).

Discussion and Conclusion

As argued in the introduction, we need to conceptualize sustainability in more positive, aspirational terms, potentially understanding it as "the possibility that human and other forms of life will flourish on the earth forever," rather than merely as a series of complex problems we need to overcome, or the reduction of unsustainability. Arguably, thinking about more life-enhancing, transformative solutions itself involves a *paradigm shift*, as it implies a move away from disciplinary, narrow, instrumental problem-solving and technological fixes to interventions that appreciate the profound interconnectedness between human well-being and the flourishing of larger natural systems.

Drawing on insights from sustainability science, complexity science, and positive psychology, I argue that more effective and transformative solutions are characterized by certain principles, which I summarize in the section on "Exploring Transformative Solutions for Sustainable Thriving". This list of principles is explicitly of a suggestive, rather than a conclusive, nature. My intention is that it may fulfill a role in stimulating our thinking, research, and debates about how to most optimally and constructively respond to our sustainability challenges.

The major insights from each of these three fields with respect to addressing our profoundly systemic, highly complex, and ultimately human-created planetary challenges can be summarized as follows.

Sustainability science urges us to orient toward whole-system solutions and "multi-problem-solvers" that creatively make use of the profoundly intertwined, systemic natures of our planetary challenges. In engaging "deep leverage points" such as the intent and design characteristics of the involved systems, change will be more transformational and impact multiple issues and systems at once. Design efforts should attempt to generate virtuous feedback loops, so that positive impacts are accelerated.

Complexity science teaches us that appropriate responses to unpredictable, multidimensional, and nonlinear challenges are characterized by more participative, distributed, and collective leadership styles as well as design and development processes that emphasize collaboration across diversity, experimentation, and the active allowance of emergence. From this perspective, hierarchical, planning-based, "command and control" approaches are fundamentally unable to address the complexity inherent in our sustainability challenges.

The empirical findings of *positive psychology* demonstrate that through designing our solutions in such a way that they enhance the human ability to thrive "thriveability", rather than just "problem-solve" environmental issues, we harness the powerful, intrinsic human motivation for a better life. We also reap the multiple, positive benefits that flourishing human beings offer to society or the world as a whole, including generally more social and sustainable behaviors and attitudes. Moreover, this direct connection to human flourishing could be the foundation for a more positive and empowering sustainability narrative that mobilizes large-scale engagement and support for sustainable transformation.

Subsequently I have used these principles to analyze a potential, real-world example of such a transformative solution pathway, focusing on the widely acknowledged (environmental and climate mitigation) potential of a dietary shift toward more heavily plant-based diets. Through discussing this example, I have attempted to illustrate what transformative solutions may look like, rather than promoting any practice or intervention in particular. Thus, a dietary change is merely an example (albeit a potentially powerful one) of how we can start to think in more bold ways about intervention design for sustainability, inviting a paradigm shift in our understanding of these issues, as well as of our thinking around the transition pathways that are needed in response.

Now I would like to turn to a few caveats and considerations with respect to the approach taken in this chapter.

In the first place, as stated in the introduction, transformative solutions are not easy nor necessarily quick, or obvious. In fact, one of their features is that despite their benefits and/or great potential for transformational change, they are frequently overlooked, downplayed, or dismissed. This makes sense as they fall outside the dominant modes of thinking, which is precisely what makes them so potent. That is, they invite for, and are based on, a paradigm shift, on a different set of assumptions and values, on a different worldview.

We see this clearly in the sustainability debate with respect to curbing meat consumption. While energy generation, transportation, and buildings have long been a target for governments, businesses, and campaigners looking to reduce emissions, the impact from food production has often been left out (Harvey 2016). Often, interference with what people eat is considered too personal, intimate, and/or culturally challenging. Especially when we consider the outstanding potential of a global meat reduction for addressing climate change, the focus seems to have been almost exclusively on technological solutions, particularly in the realm of direct energy production and use. Analyzed through the lens of a systems perspective, this means that the more obvious, direct parameters and feedbacks are used for affecting change, while the deeper intent and design of the system are overlooked, thereby ignoring a substantial potential for sustainable transformation.

Another consideration is that complex systems are, well, complex, which means that many influencing factors, feedback loops, and delays are often uncertain, or unknown. How the benefits of transformational interventions interact with each other will often be hard to predict. My aim in this chapter has therefore not been to make hard claims in that regard but rather to demonstrate that potentially we can start to use system dynamics to work *for* us. That is, while there is widespread acknowledgment of the existence and dangers of vicious cycles (e.g., in the context of climate change), we may be in need of a more comprehensive understanding and acknowledgment of how we may be well served by incorporating the potentiality of generating virtuous cycles into our thinking about sustainable solutions.

Thirdly, with my choice for the exemplary solution of a dietary change, I have opted for an intervention that is particularly low-tech and behavioral change oriented in its nature. This is not to suggest that solution pathways should always share that orientation. In fact, it would be naïve to overlook the vast potential for sustainable transformation from more technologically oriented solutions. However, it should also be acknowledged that for technological solutions to be optimally utilized, political will and support from the public at large are generally conditional. Increasingly even "technological optimists" admit that technology in itself, without facilitating changes in lifestyles and consumption patterns, will most likely fall short of offering the solutions we need – particularly in a world with an expanding population, and poorer groups gradually shifting to environmentally more burdensome lifestyle patterns. Since the sustainability debate has been dominated by technological solutions, highlighting the transformative potential of low-tech, low-risk, and highly accessible solutions that more directly engage the human dimensions is therefore arguably of great relevance.

Contemplating the insights of these three different and relatively young fields of research may enable and empower us to think in more comprehensive, bold, and inspirational ways about responses to the myriad crises and issues we are faced with. In fact, in our world today it seems like *nothing less* than profoundly transformative solutions will help us forge new pathways for more sustainable and fulfilling ways of being on our beautiful planet – pathways that are so enticing and full of potential that people will want to join and participate, simply because these new ways (of living) promise more authentic joy and fulfillment. In the words of Buckminster Fuller, an American architect, systems theorist, and inventor: "You never change things by fighting against the existing reality. To change something, build a new model that makes the old model obsolete."

And as he also said: "We are called to be architects of the future, not its victims." May it be so.

Cross-References

- ▶ Collaboration for Regional Sustainable Circular Economy Innovation
- ▶ Education in Human Values
- ▶ Environmental Stewardship
- ▶ Gourmet Products from Food Waste
- ▶ Social Entrepreneurship
- ▶ The Spirit of Sustainability
- ▶ The Sustainability Summit
- ▶ To Eat or Not to Eat Meat
- ▶ Utilizing Gamification to Promote Sustainable Practices

References

Abson, D. J., Fischer, J., Leventon, J., Newig, J., Schomerus, T., Vilsmaier, U., … Lang, D. J. (2016). Leverage points for sustainability transformation. *Ambio: A Journal of the Human Environment, 46*, 30–39.

Brown, K. W., & Kasser, T. (2005). Are psychological and ecological well-being compatible? The role of values, mindfulness, and lifestyle. *Social Indicators Research, 74*, 349–368.

Campbell, C. (2007). *The easternization of the west. A thematic account of cultural change in the modern era*. Boulder: Paradigm Publishers.

Corral Verdugo, V. (2012). The positive psychology of sustainability. *Environment, Development and Sustainability, 14*, 651.

Corral Verdugo, V., Mireles-Acosta, J., Tapia-Fonllem, C., & Fraijo-Sing, B. (2011). Happiness as correlate of sustainable behavior: A study of pro-ecological, frugal, equitable and altruistic actions that promote subjective wellbeing. *Research in Human Ecology, 18*(2), 95–104.

De Boer, J., De Witt, A., & Aiking, H. (2016). Help the climate, change your diet: A cross-sectional study on how to involve consumers in a transition to a low-carbon society. *Appetite, 98*, 19–27.

De Vries, B. J. M. (2013). *Sustainability science*. New York: Cambridge University Press.

De Witt, A. (2015). Climate change and the clash of worldviews. An exploration of how to move forward in a polarized debate. *Zygon: Journal of Religion and Science, 50*(4), 906–921.

De Witt, A., De Boer, J., Hedlund, N., & Osseweijer, P. (2016). A new tool to map the major worldviews in the Netherlands and USA, and explore how they relate to climate change. *Environmental Science and Policy, 63*, 101–112.

Ehrenfeld, J. R. (2004). Searching for sustainability: No quick fix. *The SoL Journal on Knowledge, Learning, and Change, 5*(8), 1–13.

Eisnitz, G. (1997). *Slaughterhouse: The shocking story of greed, neglect, and inhumane treatment inside the U.S. meat industry*. New York: Prometheus Books.

Ericson, T., Kjønstad, B. G., & Barstad, A. (2014). Mindfulness and sustainability. *Ecological Economics, 104*, 73–79.

FAO. (2006). *Livestock's long shadow. Environmental issues and options*. Rome: Food and Agriculture Organization of the United Nations.

Ferguson, R. S., & Lovell, S. T. (2015). Grassroots engagement with transition to sustainability: Diversity and modes of participation in the international permaculture movement. *Ecology and Society, 20*(4), 39.

Foley, J. A., Ramankutty, Navin, Brauman, Kate, A., Cassidy, Emily, S., Gerber, James, S., Johnston, Matt, ... Zaks, David, P. M. (2011). Solutions for a cultivated planet. *Nature, 478*, 337–342.

Goldstein, J. (1999). Emergence as a construct: History and issues. *Emergence, 1*(1), 49–72.

Green, K., & Keltner, D. (2017). What happens when we reconnect with nature. Research is discovering all the different ways that nature benefits our well-being, health, and relationships. *Greater Good Magazine*.

Grint, K. (2010). Wicked problems and clumsy solutions: The role of leadership. In S. Brooks & K. Grint (Eds.), *The new public leadership challenge*. London: Palgrave Macmillan.

Grouzet, F. M. E., Ahuvia, A., Kim, Y., Ryan, R. M., Schmuck, P., Kasser, T., ... Sheldon, K. M. (2005). The structure of goal contents across 15 cultures. *Journal of Personality and Social Psychology, 89*(5), 800–816.

Harvey, F. (2016). Eat less meat to avoid dangerous global warming, scientists say. *The Guardian*.

Hassan, Z. (2014). *The social labs revolution. A new approach to solving our most complex challenges*. San Francisco: Berrett-Koehler Publishers, Inc.

Hedlund-de Witt, A. (2013a). Pathways to environmental responsibility: A qualitative exploration of the spiritual dimension of nature experience. *Journal for the Study of Religion, Nature and Culture, 7*(2), 154–186.

Hedlund-de Witt, A. (2013b). Worldviews and their significance for the global sustainable development debate. *Environmental Ethics, 35*(2), 133–162.

Hedlund-de Witt, A. (2014). Rethinking sustainable development: Considering how different worldviews envision "development" and "quality of life". *Sustainability, 6*(11), 8310–8328.

Hedlund-de Witt, A., De Boer, J., & Boersema, J. J. (2014). Exploring inner and outer worlds: A quantitative study of worldviews, environmental attitudes, and sustainable lifestyles. *Journal of Environmental Psychology, 37*, 40–54.

Heifetz, R. A., Grashow, A., & Linsky, M. (2009). *The practice of adaptive leadership: Tools and tactics for changing your organization and the world*. Boston: Harvard Business Press.

Hemenway, T. (2009). *Gaia's garden. A guide to home-scale permaculture*. White River Junction: Chelsea Green Publishing.

Hulme, M. (2009). *Why we disagree about climate change: Understanding controversy, inaction and opportunity*. Cambridge: Cambridge University Press.

Irfan, U. (2012). Global shift to obesity packs serious climate consequences. Scientific American. https://www.scientificamerican.com/article/global-shift-obesity-packs-serious-climate-consequences

Jacob, J., Jovic, E., & Brinkerhoff, M. B. (2009). Personal and planetary well-being: Mindfulness meditation, pro-environmental behavior and personal quality of life in a survey from the social justice and ecological sustainability movement. *Social Indicators Research, 93*, 275–294.

Kates, R. W., Clark, W. C., Corell, R., Hall, J. M., Jaeger, C. C., Lowe, I., . . . Svedin, U. (2001). Sustainability science. *Science, 292*(5517), 641–642.

Kaufmann, J. (2009). Advancing sustainability science: Report on the international conference on sustainability science. *Sustainability Science, 4*(2), 233.

Keast, R. L., Mandell, M. P., Brown, K. A., & Woolcock, G. (2004). Network structures: Working differently and changing expectations. *Public Administration Review, 64*(3), 363–371.

Kurtz, C. F., & Snowden, D. J. (2003). The new dynamics of strategy: Sense-making in a complex and complicated world. *IBM Systems Journal, 42*(3), 462–483.

Levidow, L., Birch, K., & Papaioannou, T. (2012). Divergent paradigms of European agro-food innovation: The knowledge-based bio-economy (KBBE) as an R&D agenda. *Science, Technology & Human Values, 38*, 94–125.

Lim, S. S. (2012). A comparative risk assessment of burden of disease and injury attributable to 67 risk factors and risk factor clusters in 21 regions, 1990–2010: A systematic analysis for the global burden of disease study 2010. *The Lancet, 380*(9859), 2224–2260.

Meadows, D. (1999). *Leverage points: Places to intervene in a system*. Hartland: Sustainability Institute.

Morin, E., & Kern, A. B. (1999). *Homeland earth: A manifesto for the new millennium* (S. M. Kelly & R. Lapointe, Trans.). Cresskill: Hampton Press.

O'Brien, K. L. (2010). Responding to climate change: The need for an integral approach. In S. Esbjörn-Hargens (Ed.), *Integral theory in action. Applied, theoretical and constructive perspectives on the aqal model* (pp. 65–78). Albany: State University of New York Press.

Pluhar, E. B. (2010). Meat and morality: Alternatives to factory farming. *Journal of Agricultural and Environmental Ethics, 23*, 455–468.

Popp, A., Lotze-Campen, H., & Bodirsky, B. (2011). Food consumption, diet shifts and associated non-CO_2 greenhouse gases from agricultural production. *Global Environmental Change, 20*, 451–462.

Ryan, R. M. (1995). Psychological needs and the facilitation of integrative processes. *Journal of Personality, 63*(3), 397–427.

Ryan, R. M., & Deci, E. L. (2000). Self-determination theory and the facilitation of intrinsic motivation, social development and well-being. *American Psychologist, 55*, 68–78.

Ryan, R. M., Huta, V., & Deci, E. L. (2008). Living well: A self-determination theory perspective on eudaimonia. *Journal of Happiness Studies, 9*, 139–170.

Sarewitz, D. (2004). How science makes environmental controversies worse. *Environmental Science & Policy, 7*, 385–403.

Schösler, H. (2012). *Pleasure and purity. An exploration of the cultural potential to shift towards more sustainable food consumption patterns in the Netherlands*. Ph.D., VU University, Amsterdam.

Schösler, H., & Hedlund-de Witt, A. (2012). *Sustainable protein consumption and cultural innovation. What businesses, organizations, and governments can learn from sustainable food trends in Europe and the United States*. Amsterdam: Reprografie.

Schösler, H., De Boer, J., & Boersema, J. J. (2012). A theoretical framework to analyse sustainability relevant food choices from a cultural perspective: Caring for food and sustainability in a pluralistic society. In T. Potthast & S. Meisch (Eds.), *Climate change and sustainable development: Ethical perspectives on land use and food production* (pp. 335–341). Wageningen: Wageningen Academic Publishers.

Schösler, H., De Boer, J., & Boersema, J. J. (2013). The organic food philosophy: A qualitative exploration of the practices, values, and beliefs of Dutch organic consumers within a cultural-historical frame. *Journal of Agricultural and Environmental Ethics, 26*(2), 439–460.

Schuur, E. A. G., McGuire, A. D., Schädel, C., Grosse, G., Harden, J. W., Hayes, D. J., . . . Vonk, J. E. (2015). Climate change and the permafrost carbon feedback. *Nature, 520*, 171–179.
Seligman, M. E. P., & Csikszentmihalyi, M. (2000). Positive psychology. An introduction. *American Psychologist, 55*(1), 5–14.
Snowden, D. J., & Boone, M. E. (2007). A leader's framework for decision making. *Harvard Business Review, 85*, 68.
Sorin, D. (2010). Sustainability, self-identity and the sociology of consumption. *Sustainable Development, 18*, 172–181.
Spangenberg, J. H. (2011). Sustainability science: A review, an analysis and some empirical lessons. *Environmental Conservation, 38*(3), 275–287.
Springmann, M., Charles, H., Godfray, J., Rayner, M., & Scarborough, P. (2016). Analysis and valuation of the health and climate change cobenefits of dietary change. *Proceedings of the National Academy of Sciences, 113*(15), 4146–4151.
Stehfest, E., Bouwman, L., Vuuren, V., Detlef, P., Elzen, D., Michel, G. J., Eickhout, B., & Kabat, P. (2009). Climate benefits of changing diet. *Climatic Change, 95*(1–2), 83–102.
Van Beurden, E. K., Kia, A. M., Zask, A., Dietrich, U., & Rose, L. (2011). Making sense in a complex landscape: How the Cynefin framework from complex adaptive systems theory can inform health promotion practice. *Health Promotion International, 28*(1), 73–83.
Verdonk, D. (2009). *Het dierloze gerecht. Een vegetarische geschiedenis van Nederland [Animals to order. A vegetarian history of the Netherlands]*. Amsterdam: Boom.
Vermeulen, S. J., Campbell, B. M., & Ingram, J. S. I. (2012). Climate change and food systems. *Annual Review of Environment and Resources, 37*(1), 195–222.
Weinstein, M. P. (2010). Sustainability science: The emerging paradigm and the ecology of cities. *Sustainability: Science, Practice, & Policy, 6*(1), 1.
Wellesley, L., Froggatt, A., & Happer, C. (2015). *Changing climate, changing diets: Pathways to lower meat consumption*. London: Chatam House, The Royal Institute of International Affairs.
Wiek, A., Ness, B., Schweizer-Ries, P., Brand, F. S., & Farioli, F. (2012). From complex systems analysis to transformational change: A comparative appraisal of sustainability science projects. *Sustainability Science, 7*, 5–24.

The Spirit of Sustainability

The Fourth Dimension of the Bottom Line

Eugene Allevato

Contents

Abstract

Almost 30 years ago, the United Nations issued "Our Common Future" in search of a sustainable development path; however, little has changed in our society since then. The question is: have we gone too far away from nature? Have we undermined that we are part of nature? The present chapter will discuss that sustainability is not a trade, but a philosophy of life where spirituality plays an important role. Organizations and academia have utilized the triple bottom line as part of a sustainability balanced scorecard; however, sustainable development is predicted by interconnectedness and a worldview requiring a spiritually evolving experience that appreciates nature for its intrinsic value where humans are part of nature. The triple bottom line is expanded into a quadruple bottom line to integrate the spiritual realm as part of an effort to develop quality education for sustainability.

Keywords

Triple bottom line · Spirituality · Transformative learning · Critical pedagogy

E. Allevato (✉)
Woodbury University, Burbank, CA, USA
e-mail: Eugene.allevato@woodbury.edu

S. Dhiman, J. Marques (eds.), *Handbook of Engaged Sustainability*,
https://doi.org/10.1007/978-3-319-71312-0_17

Introduction

The world may be on the brink of catastrophe, and many humans still are oblivious to the reality of climate change. We still believe that in the last minute, something will happen to save us because humans are special. This reminds me of the 2008 American science fiction film "The Day The Earth Stood Still" I think this is a perfect example of human exemptionalism. In the movie, an alien came to destroy humans because humankind was a threat to the well-being of the Universe. But before starting the destruction the alien went to consult with another alien that came to Earth before and stayed for many years on a mission to try to understand humankind. That alien knew that humans would never change, but he still would rather stay and die together with humans because during this time on Earth he learned that there is something special about humans that is worth dying for. After the alien started the destruction process of humans in order to proceed with his mission to save the Universe, he saw the love of a woman for a child, then he decided to give humanity another chance. Another indication of human exemptionalism is that God is on our side. Humans are above the natural world and allowed to manipulate animal, vegetative, and mineral systems at their disposal. In fact, this disconnectedness of humans and the natural world may be the root cause of the human exemptionalism. We have forgotten that we humans are guests in this planet and at any moment if we go extinct, like many other species did in the past, the planet Earth will continue its existence. We need to realize that it is not the world that is on the brink of catastrophe, but human race. Based on evidence from radiometric age-dating of meteorite material, the age of the Earth is approximately 4.54 ± 0.05 billion years (Hedman 2007). In contrast, humans have only been around a few hundred thousand years, and further considering that civilization as we know it is about 6000 years, how can humans articulate that they are superior to nature and that nature is at their disposal?

We can argue that the conditions for human extinction have not been met yet, but we should be concerned that unsustainable human activity and exploitation of the natural world is compromising both natural ecosystems and human survival. Human extinction is quite plausible according to expert Nick Bostrom (2009); however, he argues that the current century or next centuries will be critical for humanity life expectancy due to technological powers and artificial superintelligence. Bostrom emphasizes the need of a clear and realistic understanding of the future in order to make sound decisions based on accurate scenarios of the future of humanity. However, the understanding of the future can be augmented based on the lessons learned from the past, and in this case environmental literacy is a crucial tool.

One of the most relevant contributions of environmental sociology is its critique of human exemptionalism, the belief that places humans outside the limits of the natural world (Catton and Dunlap 1978). As pointed out by Williams (2007), when sociologists address environmental problems, it is often assumed that humans have the ability to stand above the natural world and from an unrestricted view to rationally manage problems created by society. By placing human consciousness outside of nature, we fail to understand that consciousness is bounded by constraints that may or may not be overcome, but must be accepted in order to solve anthropogenic environmental problems. Williams claims in his endnotes, "One implication of conceiving human consciousness as limited is that we indeed may not be able to solve the problems we create." Even though this may be true, assuming that humans have the capacity for rationality, does not mean that we exercise or use this capacity. The question is why sustainability initiatives have not been more effective, as any rational analysis of the consequences of humans' behavior toward the environment would indicate the ludicrousness and foolishness of our actions. Beaver dams change the landscape dramatically like human dams but the difference is that they create many habitats for plants and other animals unlike humans that destroy habitat for the sake of their own species. We should expect humans to be rational, or at least to learn to behave, following the example of the symbiotic relationship between the beaver and its environment. It seems that humans are disconnected from nature, destitute from the symbiosisness that is found in nature as if humans were not part from the natural world. The beaver is often referred to as nature's own engineer. The beaver's ability to transform its environment to suit itself, and at the same time improve the habitat for fish species by improving water quality, reducing erosion, and reducing seasonal fluctuations in flow, is an example of nature interconnectedness. Humans need to learn to behave with an interconnected approach to the environment and do no harm. Therefore, better understanding of sustainable development may require strategies to educate society, possibly by exploring a worldview of appreciation for the natural world.

The definition of sustainability is by itself a challenge because of the diverse number of perspectives and interests as a consequence of the lack of a quality education strategy for sustainable development that takes into consideration attitudes toward sustainability with an intrinsic appreciation for nature. Both academia and businesses have been implementing a multitude of programs depending on their background and interests; however, they do not seem to focus on nature as an asset, and the outcome has been failure. In fact, many businesses have utilized the term sustainability as a buzz word to expand their profits and self-interest, this has been identified as green washing for public relations purposes, tainting any genuine effort toward sustainability. In order to provide genuine sustainable development, a closer look at the sustainability movement will be entertained in this study with perspectives from academia and business environments. The educational contributions of Socrates, John Dewey, and Paulo Freire provide compelling tools for transformative learning that may be utilized in quality education for sustainable development. This study explores the under-examined factors that may lead to pro-environmental behavior with the purpose to impact quality education for sustainability and recommendations for the design of appropriate curriculum.

Developing a responsible personal worldview is central to sustainable development, but achieving quality education to promote transformative learning for sustainability is thus far poorly understood. Most programs involving the education for sustainable development rely on changing behaviors rather than attitudes. The emphasis is on the scientific and utilitarian aspect of sustainability with negligible importance on the spiritual realm, emphasizing the intrinsic value of nature.

Campuses that have introduced sustainability projects have included building sustainable gardens and implementing energy-efficient upgrades with little or no focus on educating for genuine sustainable development by exploring students' values and beliefs. Even though green technology adoption may be the right thing to do, most schools are not targeting the root cause of the environmental crisis that fundamentally is human attitude. Instead, they are just providing palliative measures for a materialist perspective maintaining the status quo of the economic system. The lack of understanding is based on the fact that sustainability is not a trade that can be taught independently from the student philosophy and worldview, sustainability is in fact a philosophy of life.

Sustainable Development Implementation

History of Sustainable Development

As human population started to increase, the landscape of the planet shifted from a nomadic life style to a settled lifestyle with one place to cultivate the land and domesticate animals (Mebratu 1998). This transition was from a position of interdependence with the natural environment to an attitude of taming nature through science and technology to benefit humankind. Part of this process led to a devaluation of the natural world, assuming a perspective of human exemptionalism. However, beside the environmental degradation, it can be stated that ecological factors have been a major driving force behind social transformations affecting the planet and human life. It is true that the standard of living in the developing world has increased, but still the majority of people in developing countries are impoverished. Natural environment breakdown, such as habitat loss and degradation, is affecting 85% of all threatened species and indicates we are close to our limits (Brown et al. 1995). Additionally, Gottlieb (1996) discussed that as a consequence of the continuous growth of technology and the concentration of power, fuelled by religious and political ideologies, this has led to an increase in the belief that humankind is apart from the natural world.

At the other end to the concept of Limits to Growth, there are the deniers of climate change who argue that resource constraints can be overcome with little effort provided market-oriented policies. After the 1972 Stockholm Conference, there was a scientific consensus that the damage of human activities on the environment was unsustainable (Erkins and Jacob 1995). As a result, the term sustainable development was coined and the definition was framed by the report of WCED in 1987, *Our Common Future*.

The World Commission on the Environment and Development (WCED) as reported stated by Mrs. Gro Harlem Brundtland (1987) emphasizes the importance of instructing the young and changes of attitudes; however, it seems that the implementation does not follow the recommendations or the means to promote change were not effective as expected.

> We call for a common endeavor and for new norms of behavior at all levels and in the interests of all. The changes in attitudes, in social values, and in aspirations that the report urges will depend on vast campaigns of education, debate and public participation. To this end, we appeal to "citizens" groups, to non-governmental organizations, to educational institutions, and to the scientific community. They have all played indispensable roles in the creation of public awareness and political change in the past. They will play a crucial part in putting the world onto sustainable development paths, in laying the groundwork for our common future (Brundtland 1987).

This implies that sustainable development shall be implemented by a change of attitudes and behaviors through education, and it is, in essence, the responsibility of humanity, including individual citizens as well as organizations. Education is the key to sustainable development as long as it incorporates the development of critical thinking skills allowing people to make appropriate decisions. At this point, my question is how much weight will be given to the development of critical thinking of the public at large in order to avoid decision-making biases and manipulation of third party interests. It is true that debate and public participation is mentioned as a form of bringing awareness, but if the education process does not simultaneously include critical thinking development, the amount of information and effort will be useless. There is no question on the root-causes of unsustainability and accountability of our attitudes inherently in our educational programs for sustainable development that currently is focused on profiteering and consumerism tainted by a green façade. In addition, the education effort has to involve a mindset of transformation considering that sustainability is not a trade but mostly a worldview. Otherwise, the current general public understanding of environmental issues will allow the definition of sustainability to be manipulated based on rationalization of convenience and self-interest.

Unfortunately, so far the concept of sustainability was vastly utilized by organizations to capitalize on the goodwill of customers that were genuinely interested in helping to save the environment. The term "*green washing*" was coined by environmentalist Jay Westervelt in 1986 as a response to the behavior of some organizations in dressing up as environmentally friendly, portraying a green image of their products and practices, when in fact they have the sole motivation of profiting off the good faith of consumers. This was of great concern to me because these actions would not protect the natural environment. In fact, they would make things worse, by enhancing consumerism and escalating an irresponsible utilization of natural resources. Furthermore, green washing is unethical and allows businesses to mislead well-intentioned consumers who honestly want to help preserve nature, not for economic reasons, but based on a genuine sense of nature's intrinsic value.

Critical Analysis of Sustainable Development Implementation

The Influence of United Nations

The influence of major organizations and preconceived ideas on people's attitudes plays an important role. Although almost 30 years ago, the United Nations issued

"Our Common Future" in search of a sustainable development path; little has changed in our society since then in terms of genuine sustainability activities that actually have made any impact in the health of the planet because attitude change is not implicit in the current process of sustainable development. We are still debilitating ecosystems and pushing away other species that have the same right to inhabit the planet. We are still endorsing the attitude that nature was created for humans despite the fact that humans are part of nature and depend on the balance provided in the natural world to survive. Sustainability is still taught in schools from the perspective of a trade and application of technology that may reduce harm to the environment instead of promoting a worldview transformation based on spirituality and appreciation for nature, based on the premise that technology alone eventually will resolve all environmental challenges. This belief that human ingenuity will prevail any challenge because the human race is special is inaccurate and detrimental to a genuine sustainable development implementation.

Moreover, business organizations have utilized the triple bottom line as part of the sustainability balanced scorecard; however, sustainable development is predicted by interconnectedness and a worldview that appreciate nature for its intrinsic value where humans are part of nature. As currently described, the triple bottom line epitomizes the profit oriented view of sustainability. All three dimensions, namely, the environment, economic, and social justice are representing the perpetuation of the establishment fueled by consumerism and profiteering. The environment to represent the proper management and conservation of natural resources to maintain business growth, the economic aspect to ensure profit for the organization, and social justice to reduce poverty provide equity, access to social resources, and well-being with the hidden main objective of increasing the customer base and dependency of goods provided by corporations. The environment dimension should be viewed as a natural asset instead of a resource. In fact, the triple bottom line lacks a dimension representing spirituality and consequently should expand from triple into a quadruple bottom line to integrate the spiritual realm and more appropriately describe sustainability.

The Failure to Achieve Sustainable Development

Sustainable development has not been achieved yet, considering the complexity of our society from the political and economic activities. Most decisions toward sustainability are made with emphasis on economics and politics with little or no concern to the natural world from its intrinsic value, and for the most part the natural world is regarded as a resource instead of an asset. As a consequence of human exemptionalism, Pratarelli (2016) argues that unsustainable human activities and extensive exploitation of the environment is of concern since both global ecosystems and human security have already began to deteriorate. Pratarelli reminds us to consider that the state of global ecosystem with increasing threats of climate change whether attributed more or less to excessive human economic activity may support the assumption that the failure to achieve sustainability may be in our genes. Humans are wired to respond to short-term problems. If we want to try to understand why human behave unsustainably, we must account for the survival-based instinct to prosper in the short-term. LeBlanc claims that humans do not have a conservation ethic to keep them in ecological balance. "Literature addressing conservationist behavior among the worlds' remaining indigenous people reveals that the behavioral pattern of living sustainably is rare and not an inherently human behavior" (LeBlanc and Register 2003). LeBlanc concludes that the fundamental cause of warfare is correlated with the area's number of people, other living organisms, or crops that a region can support without environmental degradation. Because warfare and violent conflict have always existed throughout human history, humans have never lived in ecological balance with nature. The problem is that in the past, the population was relatively less than our current trend and ecological imbalances were localized and did not have a detrimental global impact. It is relevant to consider Bostrom (2009) claim that our capacity to learn from experience is not useful for predicting the future. From his perspective, predictability is a matter of degree, and different aspects of the future are predictable with varying degrees of reliability and precision.

Dobson and Bell (2006) discuss the relationship of self-interest and sustainable behavior when people are subject to a penalty-reward program to encourage environmentally sustainable behavior. In order to illustrate this scenario, Dobson shares the example of the Plastic Bag Environmental Levy (PBEL), implemented by the government of Ireland to encourage people to use reusable bags and to change attitudes toward litter and pollution. Dobson points out that it makes more sense that if our attitudes toward waste and pollutants change, our behavior will change, more likely than the vice-versa. However, the question remains: were people that changed their behavior motivated mostly by reward rather than by intrinsic attitudes? The answer remains unknown. Dobson says we do not know, as the subject matter is still to be investigated. It is important to consider that changing people's attitude to litter and pollute is much harder to assess, and to our knowledge no specific follow-up research on this issue has been done.

Further, Dobson stresses that every government should pay attention and seek to regulate and influence changing attitudes as well as behavior, since both are key to achieving objective sustainability. He admits, however, that this is easier said than done. If we want to focus on the sustainable development of an organization, it can be said that the level of commitment shall be similarly pursued from changes in both behaviors and attitudes. In this context – Andrew Dobson's claim that sustainable development is not possible without individual and organizational changes – it seems worth noting that monitoring top management and members of an organization may be valuable predictors of their level of sustainable development (Dobson and Bell 2006). In addition to the fact that change within individuals and the organization are prerequisites for sustainable development, it is more pertinent to secure changes in attitudes than just the behaviors (Dobson and Bell 2006). In this respect, the emphasis on designing an appropriate education program for sustainability to change attitudes becomes relevant. Additionally, the development of a metric system to evaluate the attitude of members in an organization, as well as the behavior specifically related to sustainability, is crucial to determining the main factors of sustainable development that will contribute to a successful implementation strategy.

The Triple Bottom Line

The triple bottom line concept was coined by John Elkington in 1994, based on the idea that companies should focus on three different bottom lines besides solely the traditional measure of profit. The triple bottom line consists of profit, people, and planet – the economic or financial, social, and environmental performance. The triple bottom line diagram is typically described as in Fig. 1.

Considering that what one measure is what one will pay attention to and control, the triple bottom line has an important role. Initially, the idea led to some success in light of corporate social responsibility agendas, remediation of climate change issues, and fair trade; however, it soon became clear that the transfer of production and social impact in countries such as China, India, and Brazil was devastating due to cheap labor, environmental costs, and natural resources exploitation. The main problem of the triple bottom line is that even though measuring profit is relatively easy, measuring the planet and people's performance is not as straightforward. What would be the cost of a forest? What is the cost of oil spillage or the cost of forcing children to work and depriving them of education? What is the cost associated with species extinction? How much is a plant species worth in terms of it genetic code and potential societal benefits, keeping in mind that nature's engineering took millions of years to develop? Those are immeasurable in terms of monetary value (Hindle 2008).

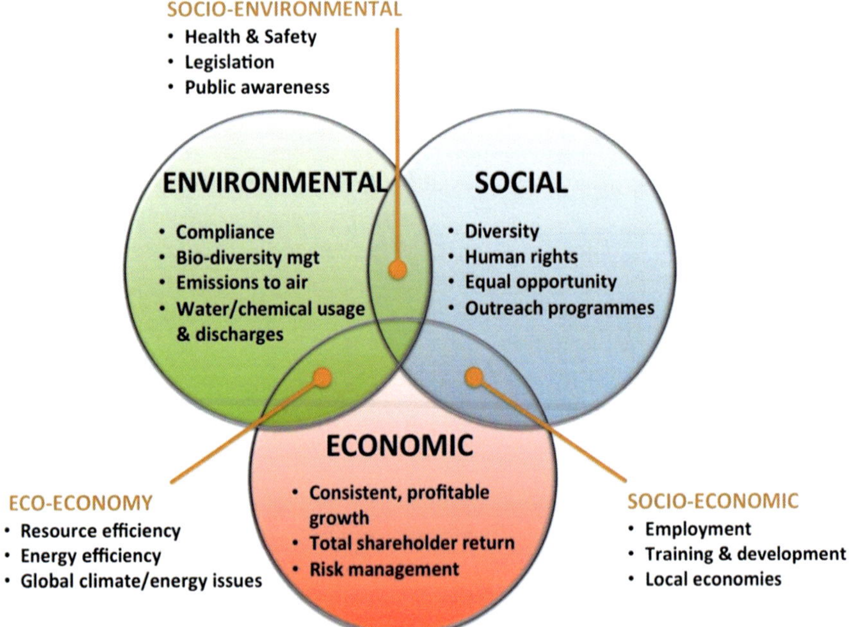

Fig. 1 Triple bottom line (Diagram retrieved on February 12, 2017, from http://www.sellingsustainabilitysolutions.com/what-is-sustainability/)

Opposite of the idea of utilizing an aggregation of different indicators, including economic, social, and environmental aspects, it seems that sustainable development cannot be truly pursued if the economic aspect is included, because the essence of sustainability cannot be monetized or subordinated to the logic of markets. We cannot consider sustainability as a commodity, but instead we should perceive sustainable development as a human responsibility to ensure that every species in the planet has access and the right to resources as needed for their survival, just as water is a human right under international law (United Nations 2014). The fact that optimizing a fabrication process is going to reduce costs should not be the emphasis, but rather reducing the carbon dioxide emissions and consequently benefiting society in terms of health and conserving biodiversity.

Ideally, individuals and organizations are interconnected entities sensible to the needs of other stakeholders that may not be just materialistic. Otto Scharmer criticizes that we have focused only on environmental and social dimensions and have neglected the spiritual or inner dimension causing unsustainable attitudes.

> This is highly visible and manifested in the act that we consume in our current economy 1.5-planet resources in a 1-planet world. The second divide is the social divide. It is manifested in the fact that 2.5 billion people live below the poverty line. But then there is a third divide, an inner or consciousness or spiritual divide, which has to do with me being disconnected to my own true sources of creativity and self (Zoeteman 2012, p. 321).

According to Scharmer, the spiritual aspect is disconnected from our behavior and consequently not often discussed by sustainable development stakeholders. The spiritual perspective, intertwined with nature, is part of the knowledge of indigenous people. Consequently, indigenous knowledge participation in the educational process of sustainability will help narrow the gap between spirituality and the natural world engrained by Western culture. A transformative education for sustainability is necessary in order to explicitly consider the spiritual dimension because Western culture is largely selfish and individualistic, resulting in anthropocentrism in contrast to eco-centrism. Additionally, technology solutions such as the proliferation of residential solar panel is not enough to remediate the environmental crisis. The challenging aspect of the environmental crisis is related to subjective topics such as control of population growth, attitude, behavior, and the ability to make technology utilization choices. I do not believe that modern science and technology should be banned in order to solve our problems, but I would like to emphasize that technology, as well as the economic system by itself, is not alone to blame for the current environmental crisis, as they are simply tools for humanity. We can only blame ourselves for the selflessness and lack of awareness. From my perspective, the economic system and environmental status quo are the result of peoples' lack of awareness and unfamiliarity of critical thinking, which could otherwise enable conditions that make sense based on solid information in any economic system utilized, because ultimately it is the people's actions of poor choices that lead to unsustainable activities.

Scharmer criticizes the departmentalization of academia and how non-governmental organizations (NGOs) are organized as another reason for unsuccessful attempts at sustainable development. I do agree with him in this aspect, but the tool behind sustainable development has to involve the individual's ability to employ a critical thinking approach as well as a spiritual or inner dimension. The individual's critical thinking process is fundamental because that will lead to awareness and behavior change within a social, economic, or political system. The analogy of a car or a gun is pertinent, as they can be utilized to bring comfort and protection when appropriately managed, but when they violate the common sense and established regulations, can cause aggression and harm. The critical thinking and empathy aspects are recognized by Scharmer as necessary for change of existing behavior. He explains,

> I believe that what is missing most are places and infrastructures that facilitate collective sense making. With infrastructure, I refer to places where people can come together, and then experience and make sense of the situation around them. These are basically places where systems can see itself. Sense making in society, but it is not a collective process. Making this shift in awareness is not a matter of developing new habits; it requires breaking through existing patterns of behavior. It requires the capacity and the experience of walking in the shoes of other stakeholder, and then making sense of this experience, sharing these stories, coming up with a system analysis, making sense of the larger situation that you face as an ecosystem together (Scharmer 2009).

Scharmer blames leadership decisions based on the larger systemic structure of our economy where political agenda have maximized shareholder value on the fact that many corporations, in part, have joined the sustainability wagon yet failed to follow a true sustainable development.

Mary et al. (2011) also explains the importance to developing a well-rounded and responsible citizen through education by exploring spiritual intelligence (SQ), despite emotional intelligence (EQ) and intellectual intelligence (IQ) in education. The need to instill human values in higher education to promote societal and environmental concerns is becoming recognized as fundamental in promoting change and to helping society embrace sustainability. Critical thinking or inquiry-based learning is an expression of intellectual quotient (IQ) where students have the opportunity to engage reflective practices in real world problems beyond the classroom. While emotional intelligence (EQ) is the capacity to empathize and recognize one's own feelings and those of others by developing self-awareness, social awareness, and social skills. Further research shows that spiritual intelligence (SQ) addresses problems of meaning and value, by placing our actions and lives in a wider and more meaningful context of passion that drives transformation. The spiritual intelligence will facilitate the dialog between the reason and emotion (Zohar and Marshall 2000). Whereas IQ, EQ, and SQ seem important to promote behavior change, they may be developed through education to instill genuine sustainable development.

Chilean economist Manfred Max-Neef has a negative perspective on human beings' ability to embrace sustainability assuming that people have limitless urges to consume and an obsession for materialistic possessions which produces more poverty and inequalities within society. He also claims that individual human beings are different and have consequently different needs and desires. He has classified fundamental human needs in subsistence, protection, affection, understanding, participation, recreation (including leisure, time to reflect, or idleness), creation, identity, and freedom (Max-Neef 2010). The model of promoting a shift from economic accounting has led and inspired other researchers to see indicators beyond existing metric including Gross Domestic Product (GDP) promoted by the Organization for Economic Co-operation and Development (OECD) and the United Nations (UN). Max-Neef's work has led to the creation of Human Development Index (HDI), which reflects many elements of sustainable development.

Attitudes are important indicators because they translate the mental stage of persons; in addition, attitudes are selective in terms of perception and memory interaction, in order to find information that agrees with chosen attitudes and avoid information that does not. Consequently, attitude predicts behavior, and attitude and behavior can be correlated. Attitudes may be classified based on past knowledge and weakly constructed on the spot. The strength of an attitude depends on factors such as attitude certainty, importance, accessibility, and ambivalence. It is crucial to identify strong attitudes because they impact behavior more than weak attitudes and are less susceptible to self-perception and remain stable over time. Attitudes of a person also have implications in utilitarianism, social adjustment, object appraisal, knowledge, value expression, and ego-defense. There are also implicit values that reflect moral behavior and long-term behavior that may serve to reveal attitude factors such as the willingness to invest in innovative technologies, to communicate openly on internal issues, or to engage with opposing actors.

In order to promote interconnectedness and a worldview that appreciate nature for its intrinsic value, it is suggested that the triple bottom line should be expanded into a quadruple bottom line to integrate the spiritual realm. The triple bottom line is not complete without a dimension that involves respect for nature and raise awareness that we are part of nature not above nature. It is true that we have the ability to transform our environment but we should do it by simulating the beaver's feat to suit itself and benefiting others. Quality education for sustainable development may be the main factor to facilitate and promote change in our attitudes.

Unfortunately, the triple bottom line has been utilized as an education tool in many schools of business. Even though the initial intention to create the triple bottom line concept may have been of good nature, there was definitely an oversight as what is required to have genuine attitude toward sustainability. As we discussed, humans have a tendency to believe that they are above and beyond nature due to cultural influence. This fact has to be accepted as the main factor leading to unsustainability toward materialistic and environmentally detrimental behavior. Humans have a predatory attitude toward the environment that without appropriate education may profoundly affect the natural world.

The Missing Dimension: The Fourth Dimension of the Bottom Line

Study conducted by Allevato (2017) based on interview of people's pro-environmental concern indicated that people's worldviews have been tainted by multiple definitions of sustainability, provided by various authority figures and international organizations. Most of these views are in opposition to a genuine and holistic definition of sustainability. The main results of this study suggest that both business students and green organization personnel hold predominant anthropocentric beliefs and lack of appreciation to the natural world. This finding endorses the need for an additional dimension to the triple bottom line, called systems thinking to bring awareness of the spirituality and interconnectedness inherent to the natural world that includes humans as part of the natural world. As described in Fig. 2, the fourth dimension is proposed in order to provide a better education tool to represent genuine sustainability. The quadruple bottom line for genuine sustainability also emphasize that the natural world should be regarded as a natural asset instead of natural resources to meet human needs.

It is clear that a different mindset toward sustainability is needed, besides those that revolve around technology and regulation. There is a need for a worldview that places humans within nature, instead of outside nature. The introduction of the systems thinking dimension forming the quadruple bottom line will lead to an understanding of interconnectedness and development of spiritual awareness, with regards to sustainability from the perspective of perceiving nature for its intrinsic value within a model of reciprocity. The sustainability education revered by many academic institutions following the triple bottom line for the last 30 years has so far proven to be inadequate. The need for a spiritual dimension within sustainability is

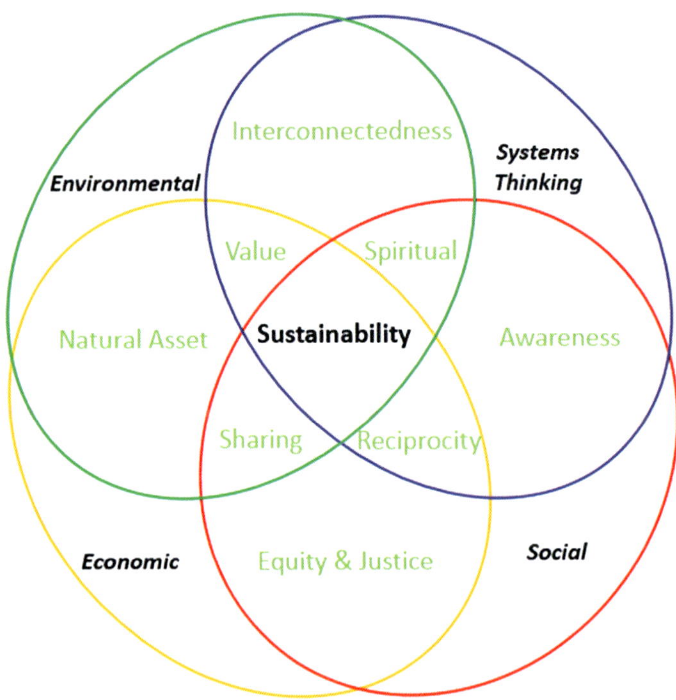

Fig. 2 Quadruple bottom line for genuine sustainability

slowly becoming recognized by other academics. During the 2015 International Conference for Sustainability Leadership in Croatia, I had the opportunity to interact with and learn of other researchers that were coming to the same conclusion that there is a need for the spiritual component in the education of sustainability

Sustainability is a complex and cross-disciplinary concept of personal and social transformation that involves both materialistic and spiritual components. The difficulty is that each individual has a tendency to perceive sustainability from their own perspective, based on their life experiences, professional training, and worldviews. It has to be understood that sustainability deals with living and nonliving things, and as such, if it involves living things, the spiritual component needs to be included. Consequently, a holistic education is crucial. Kopina et al. (2015) points out the importance of justice and equity between humans and other species in the discussion – ecological justice – based on a position of deep ecology and the recognition that all living things have rights; a context where entire species of plants and animals are considered to be much more than human property.

Unfortunately, the lack of a holistic approach to education, teamed with a worldview detached from the interconnectedness of all life, is based on maintaining the status quo. From my perspective, these are the main challenges to reaching genuine sustainable development. An example of the misrepresentation of genuine sustainable development is explained by the United Nations, in the document *Transforming our World: The 2030 Agenda for Sustainable Development* (UN 2015). This document contains distractions from the essence of sustainability – they state that sustainability is about ending of poverty and hunger, promoting sustainable consumption, managing natural resources, and ensuring that all human beings can enjoy prosperous and fulfilling lives where social and technological progress occurs in harmony with nature. These goals alone will not address the root cause of unsustainability. With a focus on poverty, violence, or economic development, little or almost nothing is mentioned in regards to the intrinsic value of nature or other species. There is no reference to the reality of a limit to growth or how to prevent overpopulation, a direct cause of poverty and lack of education, due to limited resources. A number of studies on attitudes toward sustainability have been undertaken in the last decade, but the focus has mostly been on economic development, including marketing trends and consumers' attitudes toward sustainability or green products, to provide to businesses guidance on better understanding their markets and optimizing sales. These studies often base the value of their results on the effect of economic indexes such as gross national product (Morel and Kwakye 2012; Harris Interactive Inc 2007). The misuse and multiple interpretations of the term sustainability as well as unanswered ethical questions with emphasis mostly on anthropocentric and economic perspectives has been pointed out by Desjardins (2007) as critical and subject of further investigation. I believe that the impasse on sustainable development is based on human exemptionalism and anthropocentrism as an evidence of a lack of the spiritual dimension within the definition of sustainability as portrayed by Brundtland report and triple bottom line. Consequently, the spiritual dimension should be added to the triple bottom line and emphasized as part of a curriculum strategy.

Micangeli et al. (2014) describes the importance of higher education in promoting changes in attitude through programs that do not only educate students for green careers but also gives them the tools to be problem solvers and agents of change. This implies a pedagogy specific for sustainability and developing competencies necessary to envisioning change and working with communities. But even though Micangeli has prepared an online questionnaire that listens to students to measure their learning preferences about themes of sustainability, little has been done to measure students' intrinsic appreciation of nature, and no proposals on how to impart such an appreciation on students has been presented.

In *Our Common Journey: A Transition toward Sustainability* (1999), the National Research Council acknowledges that anything that needs to be sustained will fall into one of three major areas: nature, life support systems, and community. But the authors make clear that the most common emphasis is on life support systems, where the life first to be supported is human. We recognize here, again, the anthropocentric perspective – it concerns human survival a priority. Furthermore, the emphasis of this human survival priority is based on natural resources; as useful for humans, ecologists call it "ecosystem services." I find this is a condescending attitude with which to view the natural world and other species. The way we disregard other species, plants, and all living things will result in not only biological species becoming endangered, but also push arguments about what should be sustained away from the natural world and onto human societies – with certain communities and cultures becoming differentiated, discriminated against, and deemed less important to humanity as a whole. As limits of resources worsen, the desensitization and differentiation toward other human beings will become more evident. Differential Human Life Value Perception is not a new concept – the history of humanity is rife with examples of discriminatory behavior based on citizenship and ethnicity. The experiments carried out on Jews in Nazi camps, abuse of African-Americans through The Tuskegee experiment, and exploitation of Guatemalans in The Guatemala Syphilis experiment are a few examples (Aggarwal 2012).

Humans are part of nature – it is presumptions to subjugate the entire natural world to our whims. The National Research Council (NRC) indicates the need to mobilize science and technology for a transition toward sustainability, but also recognizes the importance of other disciplines from natural and social sciences, engineering, and management. We also need to emphasize the importance of basic research, collaboration between industrial sectors, and policy aimed at utilization of technologies around the world. There is collaboration among an extended community of universities, businesses, and government agencies to address a specific set of social problems. But education for NRC does not include an understanding of the intrinsic value of nature or a transformative learning experience. It is mostly a development of science, solidifying the view that human ingenuity will eventually resolve all of humanity's issues – overpopulation, and abuse of the environment in disregard of limits to growth. Ignoring that there are limits to growth will lead to scenarios similar a 50-lane highway existing in China that resembles a parking lot due to the amount of vehicles.

Cristina Escrigas maintains that higher education institutions have always been committed to social service and the public good, but we should rethink universities' social relevance. "In times of major, worldwide change, it is essential to consider what higher education's role is and what it should be its contribution" (2008). The world today has two main areas of conflict: those that arise from the coexistence of people and those that arise from the relationship between people and the natural environment. In order to resolve these issues, education, individual willingness, and collective engagement could influence the world we live by altering how society is constructed.

> I think that what the world needs is for us to pay a little attention to it, in order to create paths towards a coexistence that is more sustainable than the present model. I think that it needs us to be more aware of the collective implications of our individual behavior. We have depersonalized life and societies. We have forgotten the individual behind the institutions, companies, parties, behind the system itself. (Escrigas 2008)

The need to promote coexistence between people and the environment is now on many political agendas. This aligns with the principles of eco-pedagogy and the concept of planetary citizenship as inspired by Paulo Freire, to bring humanity into the classroom by community engagement and learning the needs of the people. It is through dialectical thinking that students will learn not just from each other but also the value of a diversity of cultures and approaches, with the objective of effectively impacting political agendas.

Another point that Escrigas discusses, in favor of the implementation of eco-pedagogy, is the fact that the world of today is complex and interdisciplinary. So, compartmentalization of education is unlikely to be maintained in the context of globalization. The idea that higher education will train people who will reach positions of responsibility in society enhances the importance of quality education for sustainability, because they will be decision-makers in the future.

> Decision-making can be carried out using approaches that have a positive or negative effect on the overall progress of humanity and societies. Higher education therefore plays a decisive and fundamental role with respect to the contents of courses, as well as the values and the abilities that they incorporate (Escrigas 2008).

In her report, she explained that it is essential to rethink the academic contents of curricula in order to ensure values and deep understanding of human behavior and sustainable development as a process dependent on a social process.

> The need for mutual recognition, understanding and respect between different cultures and for diversity; the ability to deal with the expansion of technology, but to give it a human face and tackle its adverse effects and the ethical questions that it raises etc. All of the professions affect and interact with some of these items, or even with all of them. It is essential to break the hegemony of conformity of thought to be able to advance rapidly in a globalized society. Therefore, we should accept the complexity of reality and the interdependence of all areas of knowledge from a truly interdisciplinary approach to education (Escrigas 2008).

In this context a systems thinking dimension should be included in the triple bottom line, as it is fundamental in the implementation of quality education for sustainable development introducing a spiritual component that teaches inclusiveness and acceptance of different worldviews. If we create a classroom environment where everyone is involved, including members of the community, it will create an inspiring learning opportunity and thereby energize students and community members to work together for a common good. In this situation, the needs and interests of the community in harmony with the natural world will prevail instead of self-interests of the corporate complex.

As Demirel and Oner (2015) claims, usually the emphasis in designing curricula for business schools is based on a very career-oriented education; with little or no effort has been put on the "socialization" process. Considering that business schools should be the place where the fundamental values of the society are imparted to future managers and leaders, it seems that this objective has been overlooked. Learning to work together is the chief approach of eco-pedagogy, with focuses on improving not only as individuals, which is currently the common practice, but also as a collective, cohesive group toward the premise of genuine sustainable development.

Education Perspectives for Quality Sustainability

Eco-pedagogy Approach to Sustainability

To achieve genuine attitude toward sustainability, an education model that promotes a transformative worldview needs to be taken into consideration. Relying of educational philosophers such as Socrates (Jowett 1891), John Dewey, and Paulo Freire to develop instructional strategies of quality education for sustainability.

The Socratic method is fundamentally a school of thought that claims that knowledge can only be achieved by the student (Lam 2011). Instructors are supposed to guide students to answer their own questions on an exercise of dialectical exercise. This instructional philosophy is the basis of critical thinking development and behavior. Sustainability is not a trade and cannot be taught; it has to be earned through the ability to acquire understanding, self-awareness, and intrinsic motivation to appreciation to nature. In this case, the learner has to reach to the realization that sustainable activities are worth by themselves in order to have genuine environmental concern. The decision-making is a complicated and subjective process. Nevertheless, critical thinking is fundamental and a crucial skill for an accurate decision-making process.

John Dewey education model emphasize on the need to learn by doing with interaction with their environment in order to adapt and learn. In this context, the process of instruction is centered in the production of good habits of thinking, meaning that thinking is the method of an educative experience. In essence, education relies in having an experiential situation of continuous activity in which there is intrinsic motivation to pursue further knowledge and stimulate thought. This philosophy led to an interdisciplinary curriculum by connecting multiple subjects and involving community engagement. Students have to be involved in an activity that concerns sustainability. Dewey makes a distinction between what we experience, that is the thing we experience, and the act of experiencing by itself (Dewey 1916). This is the reasoning behind the idea that we can only develop intrinsic appreciation for nature as result of experiencing nature by itself.

Dewey specifies, in the context of his theory of inquiry, that many crucial skills including critical thinking, reasoning, as well as empathy, tolerance for the views of others, creativity, and imagination have the possibility of being properly recognized and developed. Dewey's education model provides an opportunity for a reflection of the natural world behavior what it is extremely important to the development and understanding of pro-environmental concern. By nature Dewey understands reality as both a biological and a social environment. Nature changes, renews, and grows, so the self also changes and develops. Whenever there are any tensions and disruptions, nature instinctively and automatically tries to overcome them. A human being is a part of its environment, helping to constitute nature itself; as a result, there are constant interactions between human beings and their surroundings. Growth within a social environment happens through the transmission of experience and through communication within a community. Humans live in a community surrounded by the things which they have in common; and communication is the way in which they come to possess things in common. Dewey stresses the importance of community to education and the idea of the inseparability of the inquiry process from its social environment (Czujko-Moszyk 2014).

Paulo Freire's methodology combining learning communities with eco-pedagogy specifically for higher education is an appropriate strategy to enhance nature appreciation through education and develop genuine sustainability. Reflective experiential learning opportunities of different cultures such as indigenous knowledge and worldview of the natural world are important to promote transformation learning opportunities. A central issue of Paulo Freire's pedagogy, based on "World Citizenship," is that it is directed at elementary and middle schools and involves members of the community in conjunction with students to resolve pertinent real world issues. The expansion of this concept to higher education is unique in nature and could be considered in other institutions as well.

Understanding sustainable development requires people's engagement on a thought process from an ethical and critical thinking perspective. In this respect, we cannot discuss sustainable development without considering the role of education, the freedom to become aware to act, and critical thinking as a tool of awareness, as it was proposed by the educator Paulo Freire in the Pedagogy of Freedom. Because the concept of sustainability is being incorporated into Western society as a new discourse, education is the starting point of sustainable development. Paulo Freire defines the term conscientization as the process of developing a critical awareness of one's social reality through reflection and action. Education, freedom, and critical thinking are fundamental in the conscientization (critical consciousness) of sustainable development as an intrinsic act by stakeholders. Freire says that we all acquire social myths which have a tendency to dominate our thinking, and so learning is a critical process which depends upon uncovering real problems and actual needs (2000). Conscientization is the framework of the cultural action for freedom where human beings act as agents who transform their world, instead of a passive copy of reality. So, transformation will not become a driver of sustainability within Western lifestyles unless people have the freedom to be educated to develop critical consciousness in order to act intrinsically toward sustainability and social change:

> The central thesis in Freire's work is the proposition, which was often stated with emphatic empiricism, to the effect that critical consciousness is the motor of cultural emancipation... Only dialogue, which requires critical thinking, is also capable of generating critical thinking. Without dialogue there is no communication, and without communication there can be no true education (Freire and Norton1990: 73–74)

While many businesses have genuinely invested in the sustainability cause, others have misused it as a new fad for turning a quick profit, and some have assumed the position that since everyone is doing it, why not? The idea of taking action without adequate analysis and reflection, as Paulo Freire has described in his *Pedagogy of the Oppressed* (1981), "action for action's sake" indicates and reinforces the need for quality education for sustainable development to be a prerequisite for transformation to occur. As Paulo Freire warns that the greatest danger that education faces is becoming a tool to sustain oppression, as the tendency is to harden any idea into a system, through a dominating bureaucracy that annihilates creativity. I believe that sustainability faces similar dangers in respect to becoming a tool of power for organizations that have the potential to affect professional and educational objectives.

Pedagogically, it is important to question why critical pedagogy is important for sustainable development. Sustainability has been tainted by multiple definitions, corrupted by self-interest, pressured from a culture of individualism, and self-centered by the anthropocentrism that abounds in American culture. Consequently, since the susceptibility to transformation is very small, and without drastic innovation and the understanding of the need for transformative learning strategies, curricular intervention is unlikely to happen in any real measure. There is a need to reflect, unlearn, and relearn in order to develop passion, consciousness of freedom of thought, and the ability to develop constructive action – characteristics of critical pedagogy that unfortunately are seldom explored in the current traditional education system. Almost 50 years ago, Paulo Freire, in his *Pedagogy of the Oppressed*, warned that humans were destroying systems of life in an unsustainable fashion. He said, "the oppressor consciousness tends to transform everything surrounding it into an object of its domination. The earth, property, production, the creations of people, people themselves, time—everything is reduced to the status of objects at its disposal" (Freire 1981).

Paulo Freire had a different consciousness of the world, with his pedagogy involving community participation in the learning process includes weekly meetings, in which students and community members participate in an attempt to "read the world" and solve issues that afflict society. This would allow instructors to learn students' understanding of their own schooling, their community as a whole, healthy activities, and other significant facts necessary to develop sustainable activities. Students are guided to identify unsustainable practices and find solutions within their means. Students are empowered and become committed to their ideas for change. In this context, sustainability is an opportunity to provide quality education to evaluate principles and values and to introduce a culture of responsibility for the environment. The theme of sustainability can be utilized to rebuild the current educational system, which is based on an exemptionalist view of the world (Gadotti 2010). The purpose of eco-pedagogy explores human relationships with the environment from the emotional and conscious level. As Paulo Freire said, "Without a proliferation of sustainable education, Earth will be perceived as nothing more than the space for our sustenance and for technological domination, the object of our research, essays, and sometimes of our contemplation" (Freire 1981).

Transformative Learning for Sustainability

Because humans have a malleable brain we are continuously changing as we are exposed to new experiences independent of age. Notwithstanding people in general can learn and gain ecological literacy toward sustainable development anytime depending on the education program. The question is what factors are important in promoting this mindset change?

According to Allevato (2017), the main factor behind positive pro-environmental concern is exposure to nature during childhood and special experiences during adulthood. Academic influence also play and important role in pro-environmental concern. The measures suggested by Allevato (2017) to be considered in the design of a curriculum that will promote worldview change toward genuine sustainability are summarized in Table 1, described as transformative learning core elements, or the learning elements that are essential in higher education for sustainable development. The fact that both the neglect of indigenous knowledge by Western civilization and lack of appreciation for nature has harming consequences to the health of Earth today as more real estate is taken from other species with devastating effects and should be considered in the education process for sustainability.

However, a curriculum adaptation to include sustainability aspects is not enough, as it should be understood that the instructional model and preparedness of the instructors is of extreme importance. Stubbs and Cocklin (2007) points out the importance of allowing students to examine contrasting assumptions underlying different worldviews because it helps them develop critical and reflective thinking skills. The aim is not to convert them to any particular viewpoint but rather to help them understand and articulate more than one side of the sustainability debate by developing reflective thinking skills. It is important to make clear that the sustainability framework is not static; instead, it is a continuum process of evolving understanding, positions, and perceptions that include intertwine interactions between humans and the environment.

For sure education has a role to play in changing behaviors and as a driving force to develop a sustainable future. However, in order for a curriculum to be effective, courses have to implement an interdisciplinary program in which interconnectedness

Table 1 Transformative learning factors and core elements

Transformative learning factors	Transformative learning core elements
Nature exposure (Childhood exposure)	**Individual experiential learning**
	Opportunity to explore
	Curiosity of the unknown
	Empathetic relationship between humans and environment
	Exposure to indigenous people
Academic experience (Freshman to senior)	**Dialectical thinking**
	Academic training and knowledge
	Freedom of thought
	Knowledge as tool of self-conscious and self-awareness
	Autonomous thinking
	Better thinking
Curriculum design (Liberal arts versus Business)	**Critical self-reflective skills**
	Exposure to interdisciplinary academic content
	Critical thinking
	Systems thinking design
	Innovation exercises
	Civic engagement
	Mainstream gap analysis
	Meaning perspective
	Value system and worldview assessment
	Ideas shape attitudes of society
	Personalized learning
	Project-based learning
	Sharing knowledge approach

is transparent across the curriculum and the theme of sustainability is approached by every course. Activities shall be structured with the intent to instill a vision of values and principles that can help frame a genuine mindset for sustainable development and de-emphasize human exemptionalism.

The role of education in promoting change to resolve environmental and social issues is discussed by Cohen after writing his book in 1995 *How many People Can the Earth Support?*, where he speculated that giving quality education to all children in the world would contribute to approaches to solving demographic, environmental, economic, and cultural problems risks. He describes that with respect to culture, education has the potential to reduce inequalities in helping resolve conflicts and to increase people's connection to different cultures. Education includes knowledge of nonhuman entities including living and nonliving components of Earth (Cohen and Malin 2010). Education instills values of care, reciprocity, and justice. The exposure to science and arts help to make good judgments, celebrate multiple perspectives, and understand that small differences can have large effects. The content of sciences and arts has civic implications about human uses of resources both living such as forests and nonliving such as minerals and other physical elements. With respect to population, education has the potential to reduce fertility rates and improve the health and survival of children and adults. With respect to economics, it has the potential to increase the productivity of workers and local capacity to use and develop technology. With respect to environment, it has the potential to improve environmental preservation and disease prevention (Cohen 1995).

Spirituality perceived from the natural world is part of the indigenous people knowledge, and their participation in the educational process of sustainability is of extreme importance to promote change and a transformative learning experience. A transformative education for sustainability will have to include the spiritual dimension because the Western culture is largely selfish and individualistic, resulting in anthropocentrism leading to exemptionalism. This adds to the challenges of successfully implementing sustainable development, emphasizing the need of quality education to help avoid unsustainable activities.

A transformation begins with an experience that contradicts existing attitudes and values including motivation to critically analyze the situation and reconsider their existing frame of reference. Lisa Quinn (2014) indicates that transformative learning may transpire from different sources – not just critical analysis and reasoning, but also through feelings, emotions, and spirituality, as she confirmed in her interview findings. The concept of transformative learning was developed by Mezirow, where learners change their beliefs, attitudes, and feelings by engaging in activities that require critical reflection of the experiences, leading to transformation and a new worldview. This transformation is the result of a life crisis or life transition from a disorienting dilemma or could also result from a series of incremental transformations over a period of time (Mezirow 1991).

The transdisciplinary approach seems a plausible way to bring students closer to nature and issues of sustainability. This approach has been suggested earlier and empirically tested in a classroom setting – Marques fuses systemic thinking from a scientific and spiritual perspective to bring awareness to students in order for them to become advocates of a new society of global citizens and a way of thinking that I consider fundamental to the survival of mankind as the basis of transformative learning (Eugene and Joan 2011). Students not only critically investigated course materials in respect to environmental science and spirituality, but also became facilitators in their own community, assisting in the development of good citizenship and enhancement of civic responsibility. As part of their final project students were asked to present their final findings in a community center. An educational system based on lectures and theoretical approaches is not enough to face today's complex issues including climate change and sustainable development. It is clear that community interaction as the means to raise awareness of environmental issues is very important in the curriculum design as a basis for bringing real world experiences into the classroom.

The New Curriculum and Pedagogical Framework

It is because we can change people's mindset that genuine attitudes toward sustainability can be taught given the appropriate learning tools and environment. The key ingredients to promote transformative learning environments are exposure to nature during either childhood or adulthood under special circumstances, as well as exposure to indigenous knowledge presented as major findings by Allevato (2017). In his doctoral dissertation, Allevato (2017) present a pair of studies exploring environmental concern of undergraduate students comparing different majors and year of enrollment as well as green organizations' personnel. The difference between major indicated that business students in particular have scored lower than liberal arts students what may indicate that a career oriented curriculum for the business school should be avoided in favor of a more transdisciplinary approach. Another important factor was the importance of childhood exposure to the natural world in their attitude toward sustainability. That work also found that even in adulthood an extraordinary event involving exposure to nature may trigger a worldview transformation toward a pro-environmental behavior. These findings have namely fomented the foundation of the new curriculum and instructional model explored by Allevato.

In this context, strategies to enhance nature appreciation through education and develop genuine sustainability are suggested, such as the inclusion of non-career-oriented classes from liberal arts in the traditional business school curriculum; the Socratic method based on argumentative dialectical instructional model to stimulate critical thinking and creating an inquiry mindset of questioning and letting students obtain the answers by themselves; John Dewey's philosophy of active and engaging students with meaningful and hands on topics; Paulo Freire's methodology combining learning communities with eco-pedagogy specifically for higher education to bring awareness of issues facing society; aspects of the Finnish communal education system by creating an environment of sharing information and cooperation; and reflective experiential learning opportunities of different cultures with infusion of indigenous knowledge into Western knowledge system to develop a worldview linked to the natural world.

Genuine values and appreciation for nature may be reached as students are exposed to nature and observe the interconnectedness between plants, microorganisms, and nonliving physical elements. Paulo Freire's educational model based on critical pedagogy applied to sustainability by exposing students to real world issues facing the community is spreading around the world. It was developed by the Paulo Freire Institute in the Municipality of Osasco, São Paulo, and begun in 2006. Based on the principles and values of sustainability, it focuses on the involvement of children and youth in exercising citizenship. This model could be utilized in higher education and in particular at business schools in order to transform exemptionalism into an appreciation for nature.

Antunes and Gadotti (2005) discuss the incorporation of the Earth's charter into education and eco-pedagogy with the purpose of educating planetary citizens as an ongoing process that does not stop in the classroom, but use the classroom in conjunction with the community as the first step toward life-long behavior of caring and appreciation for nature based on sustainable activities. Classic or traditional pedagogies are anthropocentric by principle. Eco-pedagogy is based upon a point of view that is more comprehensive, from man to planet, evolving from an anthropocentric vision to practicing planetary citizenship, with a new ethical and social point of reference. From a curriculum point of view, eco-pedagogy is a movement to educate and think globally, educate feelings toward nature, teach about the human condition in retrospect to the planet, develop a conscience for planetary inclusiveness, educate for collaboration and sharing, and educate for kindness and peacefulness. The concept of keeping Earth in mind is also shared by Orr (1994).

The conceptualization of eco-pedagogy is in its infancy, but in terms of a curriculum approach will incorporate values and principles in compliance with the nurturing and interconnectedness of the natural world. The failure of the academic and social institutions to see things from a holistic or interconnectedness perspective is well explained by David Orr (1993):

> The great ecological issues of our time have to do in one way or another with our failure to see things in their entirety. That failure occurs when minds are taught to think in boxes, and (are) not taught to transcend those boxes or to question overly much how they fit with other boxes (Orr 1993).

David further elaborates that another aspect of the failure becomes transparent in education's inability to join intellectual with affection and loyalty to the ecologies of particular places, which is to say a failure to bond minds and nature. Professionalized and specialized knowledge is not about loyalty to places or to the Earth, or even our senses, but rather about loyalty to the abstractions of a discipline.

The question of how can we bring curiosity for the search of the truth into our classrooms came to me as an inspiration from the work of educator Paulo Freire, in his *Pedagogy of the Oppressed*. I do not support the view that governments and economic systems are at fault for our ecological crisis. It is too easy to blame someone else, or institutions, for our missteps. Societal change has always been initiated by the people, and transformative education is a crucial element in providing the means to promote change. From my perspective, Paulo Freire's pedagogy is a resurrection of the Socratic maieutic method of inquiry, geared toward the immediate needs of the community. The climate crisis is, in this respect, a blessing for the awakening of society's consciousness; a chance to become more critical. A life of obscurity is contrary to the aims of happiness and harmonious life. Eco-pedagogy, as shown in the concept map at Fig. 3, is a framework involving dialectical eco-literacy and analytical skills to promote transformative learning, allowing students to change their worldview and to embrace the interconnectedness of life with positive skepticism fomented by an inquisitive mindset. Transformative learning is determined by an event of reflective inspiration leading to awakening of the mind and finding core meaning to life. We cannot blame government, economics, politics, science, or technology, because behind each organization or institution, there are people. Education based on reasoning and common sense creates a society that is conscientious of its actions. Education that evokes inquiry and dissatisfaction with uninvestigated information results in a society that will not allow injustices and irrational measures to take place. It is the mission of each educator to facilitate this transformation in our students. With the transformation into a critical mindset, empowerment becomes a natural consequence in our lives.

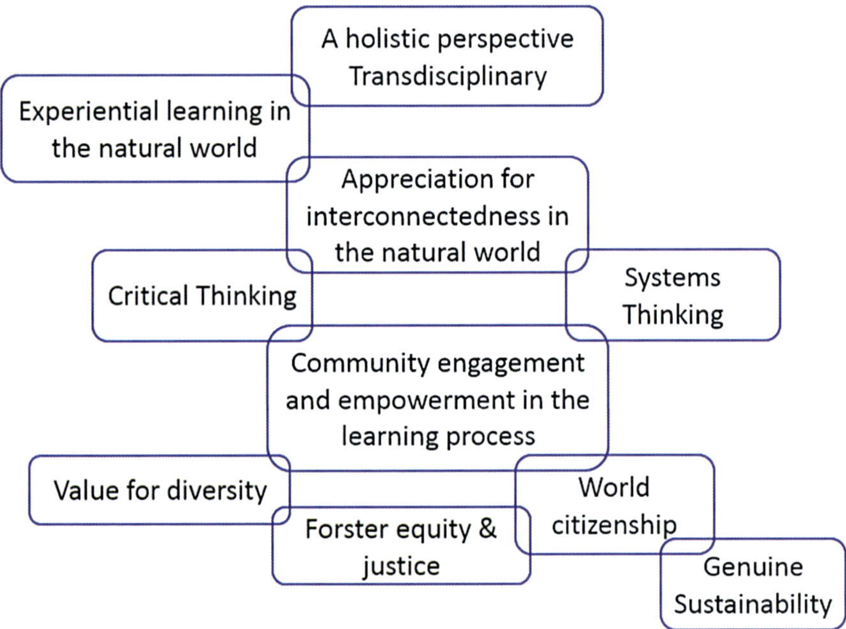

Fig. 3 Concept map for eco-pedagogy: integrating different aspects of effective pedagogy involving dialectical eco-literacy and analytical skills to promote transformative learning

Curricular reorientation cannot be based only on goals which the university sets. "Banking education" is the metaphor used by Paulo Freire to describe traditional education as a process of treating students as empty "containers" or passive recipients of information (Freire 1981). This is not aligned with the demands and goals of a quality education program built around the intrinsic value of nature for sustainability. Consequently, transdisciplinarity and eco-pedagogy is the foundation of transformative learning for quality education for sustainable development.

Conclusion and Recommendations

It is imperative that a new education model be implemented in order to promote change of people worldview that is in agreement with the concept of genuine attitudes toward sustainable development. One aspect of the transformative learning is the adoption of a fourth dimension to the triple bottom line to take into consideration systems thinking aspect of sustainability. This dimension gives the opportunity to explore subdimensions that include spirituality, interconnectedness, self-awareness of our attitudes, value of the natural word by recognizing that humans are part of nature not apart, and the notion of reciprocity to express a balance of mutual exchanges. In summary, our attitude needs to change from "we need natural resources, let's take care of nature" to "forget about our wants and needs, let's take care of nature," because nature is not meant to serve humans and it is an asset, not a resource.

According to Allevato (2017) findings from undergraduate students of different majors and staff of green organizations demonstrated that an educational strategy to inspire genuine sustainable development based on transformative learning is necessary due to their anthropocentric mindset. Analysis of current academic framework indicates that it is in accordance with the establishment status quo maintaining an unsustainable system where anthropocentric values are encouraged in detriment of appreciation and respect to nature. It became evident that business students had a different worldview and were less appreciative of nature, displaying stronger anthropocentric behavior than liberal arts students. If we want to change people's mindsets, we need a transformative education system. As was declared by educator Paulo Freire:

> Education either functions as an instrument which is used to facilitate integration of the younger generation into the logic of the present system and bring about conformity or it becomes the practice of freedom, the means by which men and women deal critically and creatively with reality and discover how to participate in the transformation of their world. (Freire 1981)

Because liberal arts students indeed scored higher on tests aimed at measuring pro-ecological opinions, it is important to compare the different disciplines and requirements for each major and try to find a common ground. Liberal arts students are exposed to a variety of disciplines from a transdisciplinary approach, rather than a career oriented for business. Additionally, sustainability topics or activities are rarely explicitly declared in the course descriptions. An alternative form of intervention, besides designing an entirely new curriculum with a focus on sustainability, would be to implement a different educational model. The approach that would be most effective is the utilization of educator Paulo Freire critical pedagogy and eco-pedagogy approaches, where dialectical thinking and environmental themes are brought to the classroom through engagement with the community providing elements for a transformative learning experience.

An important finding leading to the idea of transformative learning was the fact that early childhood exposure to nature is a pertinent factor in influencing pro-ecological behavior, as well as an opportunity for a unique experience or remarkable event during adulthood that would provide a turning point in someone's mind to relearn to appreciate nature. As a consequence of the realization that an individual's worldview is an important factor in determining if they would disclose an anthropocentric or eco-centric view. The triple bottom line does not represent the intrinsic value of nature but is a description of the status quo of unsustainability, a new dimension, the inclusion of systems thinking was proposed. The aim of this new dimenesion is to highlight the spiritual component of sustainability through aspects of interconnectedness (from the environment side) and awareness (from the social side). This inspired the design of the quadruple bottom line, as shown in Fig. 2 to be utilized as an instruction tool of transformative learning.

Since childhood exposure was found to be the main factor in determining pro-environmental concern, with subjects that had more exposure scoring more toward pro-environmental beliefs, contact with nature became a relevant consideration to be implemented in academia. The design of a curriculum for business schools focused on quality education for sustainable development has to include components of exposure to nature. In addition, the fact that even in adulthood the possibility of an event may cause a transformative experience leading to a different worldview toward pro-environmental supports the idea of a curriculum for sustainable development based on transformative pedagogy. Another important finding by Allevato (2017) is that the year of enrollment was a factor on environmental concerns. Indeed, the difference was significant between freshman and senior students, with senior students scoring higher than freshman. This suggests that schooling was somewhat beneficial in fostering pro-environmental concern, but the fact that even the score means for senior students were low demonstrates that improvement is still necessary possibly by emphasizing environmental literacy.

The inclusion of an additional dimension to the triple bottom line, namely, systems thinking, possibly can bring individual spirituality and systems consciousness as a fundamental tool to advert exemptionalism and anthropocentrism by tuning humans into the natural world again, to demonstrate the interconnectedness within nature. The challenges of climate change are in fact a wakeup call for our species to return to appreciate nature again, as we abandoned the interconnectedness with the natural world when we became "civilized." As population expands and we conquer other species' habitats and transform environments to accommodate our indulgence, we have turned our minds away from our essence and became subservient to our own predatory domination to nature. If measures are not taken to remediate this unsustainable trend, the devastation will not only eliminate other species, but humans as well. As we are biologically designed for survival and self-preservation, eventually we will turn against each other. Thus, the idea of the quadruple bottom line was presented, which includes the spiritual component.

In this study by no means are all problems of understanding human behavior toward sustainability well posed, let alone solved. I hope I have thrown some light on these problems and possibly a promising intervention through education as a contribution that I, as an instructor at a higher education institution, may be able to do during my career. Sustainability has a transdisciplinary character and cannot be approached from the environmental science and technological perspective alone, but also by the psychological and social sciences perspectives as well. As a last note, environmental knowledge does not translate into sustainable behavior because sustainability involves a change in mindset and worldview in order to affect behavior. Worldviews do not depend on knowledge, but knowledge may depend on worldviews.

In conclusion, transformative learning strategies may be applied in higher education to promote genuine attitudes toward sustainability. The uniqueness of the present study was in the identification of a lack of appreciation for the natural world identified both in business students as well as green organization personnel involved with sustainable development. Another important finding was that early childhood exposure to nature may be a predictor of pro-environmental behavior and behavioral intention. Interestingly enough, adulthood exposure to nature or an event of extreme wonder was also observed in this study, suggesting that transformative learning may occur in higher education. Consequently, contact with nature may provide an opportunity for transformative learning, and should be implemented in curriculum design and educational model in concurrence that common sense tells us that we need nature, nature does not need us.

Cross-References

- ▶ Education in Human Values
- ▶ Expanding Sustainable Business Education Beyond Business Schools
- ▶ People, Planet, and Profit
- ▶ Selfishness, Greed, and Apathy
- ▶ Sustainable Higher Education Teaching Approaches
- ▶ Teaching Circular Economy
- ▶ The Sustainability Summit
- ▶ Sustainable Decision-Making

References

Aggarwal, P. (2012). Commentary: Differential human life value perception, Guatemala experiment and bioethics. *Online Journal of Health Ethics, 8*(1).

Allevato, E. (2017). PhD thesis. Attitudes of green organizations' personnel towards genuine sustainable development.

Antunes, A., & Gadotti, M. (2005). A thematic essay which speaks to principle 14 on incorporating the values of the Earth Charter into education Eco-pedagogy as the Appropriate Pedagogy to the Earth Charter Process, The Earth Charter in Action, Part IV: Democracy, Nonviolence, and Peace. http://www.earthcharterinaction.com/invent/images/uploads/ENG-Antunes.pdf

Bostrom, N. (2009). The future of humanity. In J.-K. B. Olsen, E. Selinger, & S. Riis (Eds.), *New waves in philosophy of technology*. New York: Palgrave McMillan. http://www.nickbostrom.com/papers/future.pdf.

Brown, R. L., Lenssen, N., & Kane, H. (1995). *Vital signs: The trends that are shaping our future 1995–1996*. London: Earthscan Publications. https://www.researchgate.net/publication/265361134_The_Trends_That_Are_Shaping_Our_Future.

Brundtland, G. H. (1987). United Nations, report of the World Commission on Environment and Development: Our common future, Oslo, 20 Mar 1987. http://www.un-documents.net/our-common-future.pdf

Catton, W. R., Jr., & Dunlap, R. E. (1978). Environmental sociology: A new paradigm. *The American Sociologist, 13*, 41–49.

Cohen, J. E. 1995. How many people can the earth support? W. W. Norton & Company, New York. 532 p.

Cohen, J. & Malin, M.B., (2010), International perspectives on Goals of Universal Basic and Secondary Education, Routhledge Research in Education, Volume 22.

Czujko-Moszyk, E. (2014). John Dewey's community of inquiry. *Lingua AC Communitas*, ISSS 1230-3143, *24*, 129–141. http://www.lingua.amu.edu.pl/Lingua_24/10_Ewelina%20Czujko-Moszyk.pdf

Demirel, D., & Oner, M. (2015). Revisiting Schein (1965) study on the managerial values and attitudes of MBA students. *International Review of Management and Marketing, 5*(2), 73–84.

Desjardins, J. (2007). Sustainability: Business's new environmental obligation, portions of essay previously published in Business, Ethics and the Environment: Imagining a Sustainable Future. https://philosophia.uncg.edu/media/phi361-metivier/readings/DesJardins-Sustainability.pdf

Dewey, J. (1916). Democracy and education: An introduction to the philosophy of education. http://www.johndeweyphilosophy.com/books/democracy_and_education/Thinking_in_Education.html

Dobson, A., & Bell, D. (2006). *Environmental citizenship* (p. 258). Cambridge, MA: The MIT Press.

Elkington, J. (1994). Enter the triple bottom line. http://www.johnelkington.com/archive/TBL-elkington-chapter.pdf

Erkins, P., & Jacob, M. (1995). Environmental sustainability and the growth of GDP conditions for compatibility. In V. Baskar & A. Glyn (Eds.), *The north, the south and sustainable development.* Tokyo: United Nations University Press. http://archive.unu.edu/unupress/unupbooks/80901e/80901E06.htm.

Escrigas, C. (2008). "Putting the role of higher education on the agenda", Extracted from: The Global University Network for Innovation (GUNi) is an international network created in 1999 and supported by three partner institutions: UNESCO, the United Nations University (UNU) and the Catalan Association of Public Universities (ACUP). http://www.guninetwork.org/about-guni#sthash.ih4FuEjw.dpuf

Eugene, A., & Joan, M. (2011). Systemic thinking from a scientific and spiritual perspective: Toward a new paradigm and eco world order. *Journal of Global Responsibility, 2*(1), 23–45. https://doi.org/10.1108/20412561111128500.

Freire, P. (1981). *Pedagogy of the oppressed.* New York: Continuum.

Freire, P. (1998), *"Cultural Action for Freedom, Harvard Educational Review" 68*, N.4 Dec 1998.

Freire, P., & Norton, M. (1990). *We make the road by walking: Conversations on education and social change.* Philadelphia: Temple University Press. https://codkashacabka.files.wordpress.com/2013/07/we-make-the-road-by-walking-myles-and-paolo-freie-book.pdf.

Gadotti, M. (2010). Reorienting education practices towards sustainability. *Journal of Education for Sustainable Development, 4*(2), 203–211.

Gottlieb, R. S. (Ed.). (1996). *This sacred earth: Religion, nature, environment.* New York: Routledge.

Harris Interactive Inc. (2007). International survey highlights business attitudes towards sustainability. https://www.dowcorning.com/content/about/aboutmedia/SustainDoc.pdf

Hedman, M. (2007). Meteorites and the age of the solar system. In *The age of everything* (pp. 142–162). University of Chicago Press.

Hindle, T. (2008). Guide to management ideas and gurus. *The Economist.* https://bordeure.files.wordpress.com/2008/11/the-economist-guide-to-management-ideas-and-gurus.pdf

Jowett, E. (1891). The dialogues of Plato (428/27–348/47 BCE) translated by Benjamin Jowett E, texts prepared for this edition by Antonio Gonzalez Fernandez. http://pendientedemigracion.ucm.es/info/diciex/gente/agf/plato/The_Dialogues_of_Plato_v0.1.pdf

Kopina, H. et al. (2015). Sustainability: New strategic thinking for business. *Environment, Development and Sustainability.* https://doi.org/10.1007/s10668-015-9723-1. https://www.academia.edu/17509790/Sustainability_New_Strategic_Thinking_for_Business.

Lam, F. (2011). The socratic method as an approach to learning and its benefits. http://repository.cmu.edu/cgi/viewcontent.cgi?article=1126&context=hsshonors

LeBlanc, S. A., & Register, K. E. (2003). *Constant battles: Why we fight.* New York: St. Martin's Press.

Lisa Quinn, L (2014). *Social Action to Promote Clothing Sustainability: The role of transformative learning in the transition towards sustainability*, (Doctoral dissertation retrieved from https://umanitoba.ca/institutes/natural_resources/Left-Hand%20Column/theses/PhD%20Thesis%20Quinn%202014.pdf

Mary, K. A., Margaret, E., & Kavitha, N. V. (2011). Empowering young minds towards sustainable development. *Literacy Information and Computer Education Journal (LICEJ), 2*(1.) http://

infonomics-society.ie/wp-content/uploads/licej/published-papers/volume-2-2011/Empowering-Young-Minds-towards-Sustainable-Development.pdf.

Mebratu, D. (1998). Sustainability and sustainable development: Historical and conceptual review. *Environmental Impact Assessment Review, 18*, 493–452. http://citeseerx.ist.psu.edu/viewdoc/download?doi=10.1.1.474.8171&rep=rep1&type=pdf.

Mezirow, J. (1991). Transformative Dimensions of Adult Learning. San Francisco, CA: Jossey-Bass.

Micangeli, A., Naso, V., Michelangeli, E., Matrisciano, A., Farioli, F., & Belfiore, N. (2014). Attitudes toward sustainability and green economy issues related to some students learning their characteristics: A preliminary study. *Sustainability, 6*, 3484–3503. https://doi.org/10.3390/su6063484.

Morel, M., & Kwakye, F. (2012). Green marketing: Consumers' attitudes towards eco-friendly products and purchase intention in the fast moving consumer goods (FMCG) sector. Master thesis. http://www.diva-portal.org/smash/get/diva2:553342/fulltext01

National Research Council. (1999). Our common journey: A transition toward sustainability board on sustainable development. National Research Council, Policy Division, Board on Sustainable Development.

Neef, M. M. (2010). Chilean economist Manfred Max-Neef: US is becoming an "underdeveloping nation". http://www.democracynow.org/2010/9/22/chilean_economist_manfred_max_neef_us

Orr, D. W. (1993). The problem of disciplines/the discipline of problems. *Conservation Biology, 7*(1), 10–12.

Orr, D. W. (1994). Earth in mind: On education, environment, and the human prospect, Island Press, Washington.

Pratarelli, M. E. (2016). The failure to achieve sustainability may be in our genes. *Global Bioethics, 27*(2–4), 61–75. https://doi.org/10.1080/11287462.2016.1230989.

Scharmer, O. (2009). *Theory U: Leading from the future as it emerges*. San Francisco: Berrett-Koehler Publishers. Print; ELAIS: Creating platforms for leading and innovating on the scale of the whole system, paper prepared for FOSAD workshop Mount Grace Hotel, Magaliesburg, South Africa, Nov, 25 2009. Presencing Institute. https://www.presencing.com/resources/elias-creating-platforms-leading-and-innovating-scale-whole-system; Scharmer, O. Addressing the blind spot of our time. http://www.presencing.com/sites/default/files/page-files/Theory_U_Exec_Summary.pdf

Stubbs, W., & Cocklin, C. (2007). Teaching sustainability to business students: Shifting mindsets. *International Journal of Sustainability in Higher Education, 9*(3), 206–221. 2008. http://pages.ramapo.edu/~vasishth/Sustainability_Education/Stubbs+Teaching_Sust+Business_Students.pdf.

UN. (2015). The United Nations in the document Transforming our world: The 2030 agenda for sustainable development.

United Nations. (2014). Human right to water and sanitation. http://www.un.org/waterforlifedecade/human_right_to_water.shtml

WCED. (1987). Report of the World Commission on Environment and Development: Our common future. Retrieved November 5th, 2013, http://www.un-documents.net/wced-ocf.htm

Westervel, J. (1986). Westerveld coined the term "greenwash" in a 1986 essay examining practices of the hotel industry, "Beware of green marketing, warns Greenpeace exec". *ABS-CBN News*. 17 Sept 2008. Retrieved November 11, 2012. In 1986.

Williams, J. (2007). Thinking as natural: Another look at human exemptionalism. *Human Ecology Review, 14*(2.) http://www.humanecologyreview.org/pastissues/her142/williams.pdf.

Zoeteman, K. (2012). Sustainable development: taking responsibility for the whole. In K. Zoeteman (Ed.), *Sustainable development drivers: The role of leadership in government, business and NGO performance* (pp. 3–13). Cheltenham/Netherlands: Edward Elgar/Tilburg University.

Zohar, D., & Marshall, I. (2000). *SQ: Spiritual intelligence, the ultimate intelligence*. New York: Bloomsbury Press.

Part II

Systemic Paradigm Shifts About Sustainability

Just Conservation

In Defense of Environmentalism

Helen Kopnina

Contents

Abstract

Social scientists of conservation typically address sources of legitimacy of conservation policies in relation to local communities' or indigenous land rights, highlighting social inequality and environmental injustice. This chapter reflects on the underlying ethics of environmental justice in order to differentiate between various motivations of conservation and its critique. Conservation is discussed against the backdrop of two main ethical standpoints: preservation of natural resources for human use and protection of nature for its own sake. These motivations will be examined highlighting mainstream conservation and alternative deep ecology environmentalism. Based on this examination, this chapter untangles concerns with social and ecological justice in order to determine how

H. Kopnina (✉)
Faculty Social and Behavioural Sciences, Institute Cultural Anthropology and Development Sociology, Leiden University, Leiden, The Netherlands

Leiden University and The Hague University of Applied Science (HHS), Leiden, The Netherlands
e-mail: h.kopnina@hhs.nl

S. Dhiman, J. Marques (eds.), *Handbook of Engaged Sustainability*,
https://doi.org/10.1007/978-3-319-71312-0_5

environmental and human values overlap, conflict, and where the opportunity for reconciliation lies, building bridges between supporters of social justice and conservation.

Keywords
Anthropocentrism · Biodiversity conservation · Deep ecology · Ecological justice · Environmental justice · Environmentalism · Social justice · Sustainability

Introduction: "Just Conservation"

Environmental anthropology, political ecology, and social geography address sources of legitimacy of conservation policies as well as indigenous land rights in connection to conservation practice. In this chapter, conservation will be discussed in two main ways: conservation as preservation of natural resources for human use, associated with neoliberalism and utilitarianism, hereby referred to as neoliberal or mainstream conservation; and conservation as protection of nature for its own sake, associated with deep ecology and animal rights, hereby referred to as radical conservation (Shoreman-Ouimet and Kopnina 2016). This chapter will further distinguish between different types of "environmentalists," including the "mainstream environmentalists." The latter category is broken into subgroups based on the "nature" the environmentalists want to conserve: nature that is used for the sake of human welfare (instrumental value) or nature protected for its own sake (intrinsic value). These subgroups are then differentiated by their position on who needs justice: only less powerful people, nonhumans, entire habitats, or everyone; or those that are not concerned with justice at all. In this chapter, environmentalists and conservationists will be labelled in accordance to how the authors quoted refer to these groups.

Neoliberal conservation is described as a form of top-down environmental governance, which creates protected areas exploited for profit, and as critics argue, disadvantage local communities (e.g., Wilshusen et al. 2002; Brosius 2005; Büscher 2015). In this framing, large environmental nongovernmental organizations (ENGOs) are linked to a broader capitalist enterprise, which commodifies and profits from nature (e.g., Brockington 2002; Sullivan 2006). "Conservationist industry" (Wilshusen et al. 2002) is linked to industries such as timber and (eco)tourism, catering to political and corporate elites by appropriating natural resources through "green grabbing" (Igoe and Brockington 2007; Brockington et al. 2008). This type of conservation is often generalized by its critics to "environmentalism" in general and seen as imposed by postcolonial governments that exclude vulnerable communities in the process of capital accumulation (e.g., Kemf 1993; Escobar 1996; Brockington 2002; Goldberg 2010), particularly in developing countries (Kothari et al. 2013; Lyman et al. 2013; Rantala et al. 2013). This critique is mainly focused on social justice in conservation, namely equitable distribution of environmental benefits and burdens among the human groups (Gleeson and Low 1998; Kopnina

2014a). Martin et al. (2015) claim that conservation that succeeds in the "biological objectives" fails to address social injustices, introducing new forms of coercion or dispossession, and that "local cultural norms are forcefully displaced" (p. 166). This critical attitude to conservation as perpetuator of social injustice stems from the fields of political ecology and ecological anthropology. A much-quoted anthropologist Kottak (1999, p. 33) has implied that it is the job of anthropologists to prioritize human interests and not be "dazzled by ecological data." This position is summarized by the platform Just Conservation (http://www.justconservation.org/). In this view of "justice" in conservation, social and environmental justice are conflated as environmental justice refers to fairness in distribution of natural resources among people (Gleeson and Low 1998; Faber and McCarthy 2003; Gould and Lewis 2012; Gould et al. 2015).

The exclusive social justice perspective, in turn, has attracted some counter-critique from ecological justice proponents. In essence, this critique states that those that argue for the necessity to benefit local communities tend to present nature as a "warehouse for human use" (Miller et al. 2014, p. 509). Recently, however, a small number of anthropologists, along with the champions of animal rights, have leveled criticism against the humanist anthropocentric worldview for its presumption that only humans are morally considerable and that human rights trump those of nonhumans (Sodikoff 2011). Within the new wave of ecocentric anthropology (e.g., Kopnina and Shoreman-Ouimet 2011; Desmond 2013; Shoreman-Ouimet and Kopnina 2016), the hierarchical relationship between humans and nonhumans was critically examined, providing a very different ethical context for viewing conservation based on deep ecology and animal rights perspectives.

Opposing utilitarian view of conservation, the deep ecology perspective underlies the intrinsic value supporting close interconnections of cultural and biological systems (Naess 1973). As opposed to the utilitarian or "money green" environmentalism, the deep ecology conservation supports broader ecological justice, promoting environmental protection independent of human interests (Wissenburg 1993; Cafaro and Primack 2014; Kopnina 2014a; Cafaro 2015). While an instrumental motivation can produce environmentally positive outcomes in situations where both humans and environment are negatively affected, for example, when biodiversity is used by local people for eco-tourism or by pharmaceutical industry to develop medicines, anthropocentrism does not guarantee biodiversity protection which does not offer direct human benefits (Katz 1999; Kopnina 2012c; Bonnett 2015), nor safeguard animal welfare, let along animal rights (Singer 1977; Kopnina and Gjerris 2015; Kopnina 2016a, c). What allows pragmatic ethicists to rehabilitate anthropocentrism as a basis of utilitarian position outlined above is their rejection of the intrinsic value of nature (Noss 1992; Mathews 2016). By rejecting intrinsic value, a human environmental right or the right to use nature and animals subjugates all other needs of nonhumans to those of humanity (Bisgould 2008; Borràs 2016). This anthropocentric bias has been characteristic of much of conservation critique (Kopnina 2016a, b), ignoring ecological justice that encompasses justice between species (Baxter 2005; Schlosberg 2007; Higgins 2010; Kopnina 2014a; Cafaro 2015).

While the mutual accusations of conservation critics and supporters clearly speak of perceived flaws amidst conservationist efforts, they also illustrate the gap in agreement in what environmental justice entails. A significant problem in relation to the value judgments in regard to conservation and environmentalism is the lack of conceptual clarity and failure to agree on definitions – the problem that has dogged the pursuit of sustainable development (e.g., Faber and McCarthy 2003; Washington 2015). This chapter will reflect upon different strands of arguments, contending that it is important to recognize the ideological, political, and social forces active in shaping both the broad scope of environmentalism in order to differentiate goals in relation to conservation. The sections below outline various conservation perspectives that may help to build bridges between different types of justice.

Critique of Conservation

Several tropes emerged from the environmental justice critique of conservation, "things as classic as wilderness and as nouveau as carbon trading—as imagined categories invented as tools of the capitalist majority to wrest power away from the weak" (Wakild 2015, p. 43). This critique contains at least two threads of accusations. First, mainstream environmentalists supposedly collaborate in the very enterprise that they criticize – that of capitalism and corporatism (West and Brockington 2011). Indeed, the critics have pointed out that as protected areas approach designates certain areas as "wilderness," it restricts human habitation or resource use and seeks to separate humans from "nature" (Brockington 2002). In this critique of "fortress conservation," environmentalism is described as a view of the past as a "glorious unbroken landscape of biological diversity" (West and Brockington 2011, p. 2), with the "environment" represented as natural and benign and humans as other-than-natural and destructive. According to West and Brockington, this romantic view achieves separation between humans and nature "by seeking to value nature and by converting it to decidedly [word missing in the original] concepts such as money; and ideologically, through massive media campaigns that focus on blaming individuals for global environmental destruction" (Ibid).

The second accusation is the supposed proximity of ENGOs to the neocolonial capitalist enterprise (e.g., Chapin 2004; Adams and Hutton 2007; West 2008). West and Brockington (2011, p. 2) state that it is rare today to find ENGOs that challenge corporations or their logics and that environment has become another "vehicle for capitalist accumulation." Moreover, environmentalism is said to have gotten "snugly in bed with its old enemy, corporate capitalism" especially in developing countries (West and Brockington).

Responding to these concerns, some conservationists have pointed out that most of conservation is already targeted toward human welfare (Redford 2011; Doak et al. 2015). Based on research demonstrating that the top-down conservation is both inefficient and incompatible with local norms, values and beliefs (Chaudhuri 2012), participatory, bottom-up, community-based conservation (CBC) was proposed (Brechin et al. 2003; Brosius et al. 2005; Sullivan 2006).

In turn, critiques of CBC exposed community participation as a mechanism for masking persistent political power through the "creation of unwieldy projects aimed at top-down environmental management" (Brosius 1999, p. 50). It was also noted that CBC still garners a relatively low level of public acceptance (Hovik et al. 2010), resulting not only in poor environmental protection and poverty reduction, but triggering grassroots resistance (Horowitz 2012; Temudo 2012).

Arguing that the local communities' benefits should be considered as the primary objective of conservation, some commentators promoted, the "new conservation science" (NCS) stated that conservation should be for people's benefit, especially for vulnerable communities that live in or near protected areas (e.g., Kareiva et al. 2011; Marvier 2014). Utilitarian conservation can be characterized by maximizing human use of land, water, or minerals to safeguard natural resources or maintaining wetlands for duck hunting. In fact, Kareiva et al. (2011) have called for conservation to exclusively support programs for rural development targeted at human well-being. As many endangered species are not directly related to human welfare (Haring 2011), even protection of the critically endangered rhinoceros has led to acerbic critique of "politics of *hysteria* in conservation" (Büscher 2015).

The Noble Savage?

Defenders of indigenous rights contend that conservation threatens cultural survival. For example, in the case of Greenland, hunting contributes little to the national economy, but does constitute an important part of cultural identity of Icelandic traditional groups (Nuttall 2016). In this context, conservationists that seek to prohibit fishing or hunting are seen as threatening local culture. Illustrating this critique, both Einarsson (1993) and Kalland (2009) describe antiwhaling organizations as "culturally imperialistic," "intolerant," and even "militant" as they are opposed to the traditional way of life of whalers. In the case of the Inuit hunting, the animal rights activists are said to threat "cultural survival" of northern economies (Wenzel 2009; McElroy 2013; Nuttall 2016) (It is worth pointing out that those most concerned with social environmental justice are not necessarily the ones found in less powerful groups based on class, race/ethnicity, and national and global stratification systems. Also, the critiques of environmentalists are not uniform – while some are frustrated with political, corporate and ENGO elites masquerading as environmentalists, others are upset by the prioritization of environmentalism and animal rights over 'traditional lifestyles'. But there is also evidence that the less powerful have a difficult time getting their voice heard and can benefit from more powerful "established academics" helping to tell their story and help provide access to resources. And while the "noble savage" perception may be in play during some of the arguments, both by academics and by the indigenous, a noble savage to sustain the claim is not necessarily needed to sustain a claim for human rights. Often it is simply cultural protection, deemed necessary given the assimilationist, or worse, genocidal tendencies that have been present historically.).

The reification of "traditional cultures" as "noble" is not new. Roger Sandall describes "the noble savage" representation as "the romantic insistence on the superiority of the primitive" which is "increasingly grounded in a fictionalized picture of the past – a picture often created with the aid of well-meaning but misguided anthropologists" (Sandall 2000, p. 1). While social justice proponents insist that the local people should have special access rights – including hunting, in effect they reify the indigeneity ironically implying that these "traditional" societies should somehow be assigned different rights then more "modern" societies. Reminiscent of Sandall's (2000) critique, this view tends to disregard the drastic changes in local environment and the dwindling numbers of surviving wildlife. Social scientists failed to recognize that indigenous people are rarely isolated from global market forces (Poutney 2012, p. 215), thus stimulating the erroneous representation of the "noble savage" who lives "in harmony with nature" (Koot 2016).

Ironically, insistence on supporting special privileges (such as hunting endangered species) by the indigenous people presents "the natives" as fundamentally different from the rest of humanity. In fact, special rights, like other forms of positive discrimination, tend to reify disparities in power as much as they address them (Strang 2013). Strang (2013) inquires whether Aboriginal communities in Australia should have the "right" to extend their traditional practices, for example, to shoot rather than spear wallabies, to the point that the once plentiful wallaby population in Cape York has dwindled to critical levels. The possession of cars and rifles has enabled new forms of hunting within increasingly fragile habitat created by intensifying cattle farming. Should this be an Aboriginal choice? Should it be anyone's choice? How are we, as social scientists and as members of this planet, to "build bridges" between people inhabiting such different political, economic, and other cultural worlds, with different values, norms, and beliefs? Is that our responsibility as academics and/or practitioners?

Strang (2013) inquires whether anyone, advantaged or disadvantaged, has the right to prioritize their own interests at the expense of nonhumans. Besides, reification of indigenous rights to *exploit* nature is derived from the culturally and historically unique logic of industrial neoliberalism and is by no means "traditional." I need to emphasize, however, that bias towards reifying humanistic values might be present in much of academic work that considers itself liberal and inclusive. As academics, as well as nonacademic actors speak of justice, democracy, and equality, as some of the highest moral values – at least in the context of today's western morality, instructed by the heritage of enlightenment and humanism. Yet, ironically, all these noble values are not necessarily widely shared and may be in themselves a manifestation of a biased worldview based on these historically and culturally unique ideals of some Western liberal academics.

It also seems that proponents of traditional practices are very selective. As Western colonial governments have prohibited "barbaric practices" such as human sacrifice and headhunting, indigenous rights campaigners are not eager to revive them. Yet they seem to relegate animal killing to an "indigenous rights" domain.

It was argued that extreme cultural relativity, in which it is possible to ignore major abuses of human rights, can be seen as an abdication of moral responsibility

(Caplan 2004). If we extend this to nonhuman rights, then the key concern is our responsibility towards non-human species, independent of human interests. In fact, the "biological objectives" that Martin et al. (2015) dismiss, and the "politics of hysteria" that Büscher (2015) ridicules, includes desperate attempts to preserve the critically endangered species. The realization of the dire predicament of the biodiversity crisis highlights the need to consider ecological justice, or justice between species (Baxter 2005; Higgins 2010). Social justice advocates, while correctly identifying the larger destructive force of neoliberal capitalism that threaten both cultural and biological diversity, fail to recognize that these "forces conspire not just against the poor [who live near protected areas] but also against wild places [themselves]" (Wakild 2015, p. 52). The NCS position "restricts the focus of conservation to the advancement of human well-being, which it frequently conflates with narrow definitions of economic development, and thereby marginalizes efforts [...] to protect nature" (Doak et al. 2015, p. 30).

Alternative Environmentalism

There are many schools of thought within the broad label of "ecocentrism" – indeed conservation cannot be neatly construed in terms of insiders and outsiders (Igoe 2011, p. 334). A few generalizations about the core philosophy can be made. A tremendously diverse environmental movement is inspired in part by the work of transcendental writers such as Ralph Waldo Emerson and Henry David Thoreau, as well as the work of environmental philosophers Aldo Leopold (1949) and Arne Naess (1973). Both in the land ethics (Leopold 1949) and deep ecology (Naess 1973), humans are seen as part of nature. While anthropocentrism and shallow ecology typically sees environment in terms of human-centered interests, deep ecology recognizes that the attempt to ignore our dependence on environment and to establish a master-slave role has contributed to the alienation of man from himself (Naess 1973, p. 96).

In an ecocentric view, turning nature into a "natural resource" is inherently problematic (Rudy 2012). Ironically, it is this type of "economism that dominates human concerns in the West to override any conservationist concerns" (Bonnett 2013). This position of mastery over nature and resources resembles "ecological colonialism" (Eckersley 1998), a process in which environmental management becomes normative. By contrast, deep ecology argues that the natural world is a subtle balance of complex inter-relationships and sees human being as integral part of nature, recognizing the inherent worth of *all* living beings. The types of conservation organizations that promote conservation for the "sake of nature" are the Sierra Club, The Sea Shepard, and many "radical" environmentalists.

Recognizing this common victimhood of vulnerable human and nonhuman communities, deep ecology environmentalism neither attempts to separate humans from nature nor collaborates with the capitalist power holders (Merchant 1992; Taylor 2008). In fact, deep ecology is openly critical of neocolonial history that has displaced supposedly inferior humans and wild nature to the fringes of earthly landscapes and human mindscapes (Crist and Kopnina 2014). This displacement has historically served the "superior" human races that supposedly possessed the capacity for reason, morality, civilization, technology, and free will above animals or supposedly "inferior" races or minority groups (Crist and Kopnina 2014). This same displacement also made it permissible for nature to be exploited as a means for human betterment. In fact, deep ecology favors "diversity of human ways of life, of cultures, of occupations, of economies" (Naess 1973, p. 96) and thus embrace all. Deep ecologists support the fight against economic and cultural domination, and they are opposed to the annihilation of seals and whales as much as to that of human tribes or cultures (Ibid).

There is little doubt that radical environmentalists are far being complacent to the system of neoliberal oppression (Sunstein and Nussbaum 2004). In fact, many "radicals" are motivated by the belief that promotion of ecological justice is similar to the previous social liberation movements, such as liberation of slaves (Liddick 2006). Thus, radical environmentalists are not the ones who have "taken a back seat to corporate power" or "brought corporate leaders directly onto the boards of directors of their organizations" (West and Brockington). In fact, environmentalists who support ecological justice, animal rights, and biospheric egalitarianism are thwarted by those who allow anthropocentrically motivated utilitarian conservation to blossom. Indeed, alternative environmentalists are lesser heard, fewer in number, albeit an ecologically enlightened moral minority (Scarce 2011; Scruton 2012). In the United Kingdom and the United States, members of the Earth Liberation Front or Animal Liberation Front are considered to be terrorists (Liddick 2006).

Granted, mainstream environmental organizations do engage in strategic alliances with corporate partners (e.g., Van Huijstee and Glasbergen 2010; Van Huijstee et al. 2011; Kopnina 2016b). Perhaps many mainstream environmentalists are accepted in a wider society precisely because they have become "much more sensitive, well behaved, and well spoken" (Best and Nocella 2011). In this sense, deep ecologists and social scientists might agree that the "mainstream environmentalists" are part of the industrial neoliberal system.

The premise that "the conservation community needs to take justice issues seriously and that it cannot claim to be doing this until it has developed ways of assessing justice impacts of conservation interventions" (Martin et al. 2015) needs to be critically examined. What is intolerable is an injustice that threatens the very survival of all species but one. For deep ecologists, "the sense of ability to coexist and cooperate in complex relationships, rather than ability to kill, exploit, and suppress" (Naess 1973, p. 96) describes mutual justice. As Strang has argued (2013, p. 2):

> Discourses on justice for people often imply that the most disadvantaged groups should have special rights to redress long-term imbalances... However, if the result is only a short-term gain at the long-term expense of the non-human, this is in itself not a sustainable process for maintaining either social or environmental equity.

This opens up a question whether social scientists should promote cultural relativity to the degree that no universal human – or other – rights carry any weight.

Reflecting on the Arguments: Points of Conversion and Disagreement

While some of these conflicting perspectives attempt to combine social and ecological interests, both contain critiques of capitalism, but also of each other. At the most basic level, the primary distinction between these different schools of thought can be credited to the dualism between nature and culture. There is robust literature critiquing the nature and culture dichotomy, high-lightening interdependence and the human-nature continuum (e.g., Ingold 2006; Paterson 2006; Sullivan 2006; Shoreman-Ouimet and Kopnina 2016). Indeed, the nature-culture dichotomy is a significant contributor to the lack of progress toward ecological sustainability (Strang 2013). Due to the global reach of human impacts, there are few places left on Earth that could be considered to be "natural." Even if humans have never physically been there, our climate and other ecosystem-influencing impacts have. Also, there is nothing that can be considered to be human that is not shaped by the nonhuman world. Following this, the rights of humans and nonhumans cannot be thought of as wholly distinct, nor should one be valued over the other (Kopnina 2016c). Thus, collapsing the dichotomy and thinking of nature and human social systems as co-produced is more helpful than leaving the two areas as mutually exclusive (Moore 2015).

Yet the dichotomy needs to be addressed in legal terms as well (Kopnina 2016c). Deep ecology requires radical restructuring of societies in accordance with recognition of basic rights of individuals within the species, entire species, or even whole habitats or ecosphere (e.g., Naess 1973; Eckersley 1998). Classified under the banner of animal rights, or biospheric altruism that extends beyond "animals" and includes plants or entire habitats, the concept of ecological justice has recently become prominent in legal scholarship. As Sykes (2016, p. 75) has reflected, the emergent status of nonhuman protection as a matter of weight "both reflects and adds to a nascent consensus that a global conception of justice must include some notion of justice regarding animals." In recognition of the rights of nature, some countries have oriented their environmental protection systems around the premise that nature has inalienable rights, as do humans (Borràs 2016, p. 140).

However, there still appears to be a large gap between recognition of human rights and animal rights. This gap originates from what environmental sociologists William Catton and Riley Dunlap (1978) have termed the dominant "Human Exemptionalism Paradigm" or HEP. HEP is symptomatic of sociology's tendency to consider humans exempt from ecological influences and in seeing humans as morally superior.

Moral concern with social justice has increased during the postcolonial and post-world wars decades, expanding our concern for human lives *everywhere*. At present, few academics and liberal intellectuals would dispute the importance of human rights, environmental justice, racial and gender equality, and economic equality. Yet the daily subjectivism of animals and plants for the industrial food-production or pharmaceutical industry is often ignored (Crist 2012, p. 145). As witnessed by the massive scale abuse of animals in the industrial food production system (CAFOs), animal experimentation, and habitat destruction, it seems that our regard for the rights of other species has in fact decreased. While "raising the standard of living" everywhere is outwardly admirable, it is also a "euphemism for the global dissemination of consumer culture" (Crist 2012, pp. 141–142). Below, alternative types of environmentalism that defend the rights of other species are examined.

Explicating Standpoints

One of the main points of disagreement is the agency of blame and the value placed on the environment. Those conservationists concerned with protecting nature for its own sake typically addresses BOTH structural factors (e.g., population pressures) and the role of tenure, authority, and global markets. It is the exposure of structural factors that unleashes the most ardent counter-critique of social justice supporters. Martin et al. (2015) find that "local factors such as population growth and resource dependence" (p. 167) are irrelevant for analyzing conservation struggles. They state that the "narratives about population pressure, about local poor people being the main threats…. tend to exclude in depth political analysis of the role of tenure, authority, global markets, and the systemic implications of expanding capitalist relations" (Martin et al. 2015, p. 167). Fletcher et al. (2014) imply that overpopulation discourse is constructed by racists and elitists that blame vulnerable populations, while the real cause of environmental and other problems is industrial rapacious capitalism.

While identifying industrialism and top-down economic development a perpetrator of social inequalities and deepened ecological injustice is constructive, discounting demographic trends is short-sighted. For those that worry about the prospects of future generations, long-term perspective on population in relation to nature (even if only defined as a resource) needs to be considered (Hawkins 2012). If we assume that the well-meaning social justice proponents want *everybody* in this world to enjoy a decent standard of living, expansion of unsustainable consumptive practices that accompanies this process will necessarily cause greater pressure on the planet and thus hurt the future generations (e.g., Smail 2003). Thus, addressing population is not a condemnation of the poor, nor is it a call for coerced population control (Campbell 2012), as the straw-men arguments imply that "environmentalists" or other elites seek to do (Fletcher et al. 2014). Rather, it is a call to recognize the fact that there are many common factors contributing to global poverty, inequality, and environmental destruction, and that population growth exacerbates all of these (Wijkman and Rockström 2012).

The population pressure scales up all issues that might have been benevolent in "traditional" (preindustrial) settings, leading to fundamental incompatibility of agriculture with nature conservation in the context of the global demand for food (Henley 2011). This is why it is ironic that those who defend local communities' "ways of life" use the very vocabulary of the power-holders they criticize. Despite the claim that *all* biodiversity is needed to provide a safety-net for humanity (Rockström et al. 2009), much of biodiversity may be expandable from the utilitarian point of view (Crist 2012, 2013). It has been argued that since monocultures suffice in sustaining human material needs, conservation should be based on the intrinsic value of nature (e.g., Ehrenfeld 1988; McCauley 2006; Redford and Adams 2009). As Redford and Fearn (2007) note, a review of existing writings and available evidence suggests that there is no easy way for conservation professionals and organizations to defend conservation when it leads to forcible displacement of humans from areas that are to be protected, even if it is to stave off extinction of several species. Equally, however, it appears difficult to justify massive displacement and extermination of nonhumans in the name of justice (Crist 2013; Cafaro 2015). The common concern with suffering inflicted upon those displaced makes the question of displacement urgent for both social and ecological justice proponents.

Many environmentalists as well as social justice proponents could find a meeting point in placing the blame on political and corporate elites and more generally, the rich world's consumption. In fact, most environmental organizations subscribe to sustainable development framework and aim to address issues associated with human use of natural resources and ecosystem services, as well as aim to minimize negative environmental effects. An example of conversion of interests is Greenpeace campaign to minimize toxic pollutants or promote the use of "safe" energy (Greenpeace 2013).

There is an especially relevant thread of environmental justice literature that focuses on justice in distribution of environmental risks (such as pollution) and benefits (such as natural resources) to different human groups. The environmental justice literature maintains that people should not be disproportionately burdened by environmental threats or able to benefit from environmental goods because of race, ethnicity, gender, or economic status, or other characteristics (Fredericks 2015). Environmental justice examines the relationships between environmental toxins, risks, and the role of neoliberal policies in those injustices (e.g., Harrison 2014; Gould et al. 2015). The definition of environmental racism has arisen as research has shown that people of color and the poor are disproportionally burdened by environmental degradation (Faber and McCarthy 2003). Environmental justice advocates argue that everyone has the right to basic goods and services and should not be disadvantaged (Gould and Lewis 2012; Gould et al. 2015). Harrison (2014), for example, explicitly criticizes the social inequalities and relations of oppression that help produce environmental inequalities.

Whereas many environmental justice definitions focus on equal distribution of benefits of nature exploitation, Schlosberg (2007) highlights ecological justice by recognizing varied needs of human communities and nonhuman nature. If the considerations of "equal share" are extended to nonhumans, one can speak of inclusive justice or biospheric egalitarianism (Naess 1973; Baxter 2005; Schlosberg 2007). Environmental racism, in a more ecocentric interpretation, refers to human discrimination against nonhuman groups.

Many organizations work within established system trying to minimize environmental damage and accommodate human needs – something that the majority of engaged social scientists support. However, realization of our collective impact needs to move beyond blaming elites and toward the recognition of collective responsibility of us as a species towards millions of other living beings, from laboratory rats to battery chicken to captive elephants.

Are the accusations that mainstream environmentalists are "part of the system" (of neoliberal industrial capitalism) fair? The answer to this largely depends not only on what is meant by environmentalism but also on the degree of penetration of neoliberal ideology into social scientists' own rhetoric. The critics of conservation may be equally unjust in prioritizing human entitlement over the disadvantaged (and even critically threatened) non-human beings. The critiques of mainstream conservation implicitly lump together all environmentalism, while the "nonmainstream" environmentalists who speak for whales, seals, or lemurs as part of compassionate conservation (Bekoff 2013) remain marginalized. Most of the critics' writings on conservation, quoted above, contain generalizations implying that all conservationists are environmentalists and that they share similar values, lumping together different groups in a sweeping critique of supposedly socially unjust conservation and environmentalism. Are the critics of the mainstream conservation and environmentalism (which tends to be utilitarian and anthropocentric) prepared to accept deep ecology into the "mainstream"? If not, is it fair of them to "blame" mainstream environmentalism for being complacent? Are the critics themselves not part of the neoliberal system that treats "nature" as a "natural resources" under the guise of justice?

Attempting Reconciliation

There is a necessity is to find a way for conservation proponents and opponents, however diverse in their approaches and orientations, to work and thrive together. In fact, many ideas are already shared by proponents of community-based and conservation-for-the-sake-of nature approaches, such as the misgivings about the "dollar green" environmentalists. In distinguishing between different but at times overlapping schools of thought or motivations, the positions on the main issues can be outlined: (1) position on what/who is considered nature, (2) what kind of justice is sought and for who? and (3) how to best approach environmental problems that effect both human and nonhuman communities? Intuitively most easy way to reconcile positions is to include both human and nonhuman actors (thus, considering both humans and nonhumans as part of nature; seeking justice for both human and nonhuman beings, and approaching environmental problems that threaten both human and nonhuman survival). Notwithstanding many instances in which hard choices that may affect or exclude human and nonhuman groups have to be made (e.g., exploiting forest for timber plantations that may economically benefit local community versus preserving biodiversity of this same forest with strict controls on human use), including the interests of both humans and nonhumans and attempting to find ways to best balance them, remain an ideal that has been in many cases successfully implemented in practice.

Positive examples of reconciliatory approaches include valuation of nature for both its benefits to humans and as an intrinsic good, with ecosystem services including the services that all species provide each other (Lawrence and Abrutyn

2015). The common search for justice, addressing the conditions that have created domination and exploitation of the powerful classes over the oppressed ones, can be based both on appreciation of deep ecology and sensitivity to the local context and the human costs of conservation policies. Indeed, the outcomes in any conservation project are related to specific cultural, historical, and political circumstances (West et al. 2006).

The points of conversion in the case of polluting materials that effect both human health and the environment can be illustrated by the example of the use of DDT, an insecticide which was used to increase agricultural productivity. There were high death rates among nonhumans who were directly affected by the intake of DDT, but it was not until 1970 when DDT was prohibited after adverse health effects became apparent (Bateson 1972).

The underlying idea of law against ecocide is based on strategically powerful argumentation of human dependency on nature. Ecocide refers to "the extensive destruction, damage to or loss of ecosystem(s) of a given territory, whether by human agency or by other causes, to such an extent that peaceful enjoyment by the inhabitants of that territory has been severely diminished" (Higgins 2010). Crist (2012, p. 147) is more radical in her designation of ecocide as a form of genocide: "the mass violence against and extermination of nonhuman nations, negating not only their own existence but also their roles in Life's interconnected nexus and their future evolutionary unfolding."

Another point of conversion is the critique of neoliberalist or neo-colonial enterprise that commodifies nature and equates human progress with economic prosperity. A conservationist Redford (2011) and anthropologist Igoe (2011) both reflect that the differences between environmentalists' and social scientists' positions are not irreconcilable. Indeed, "there are openings for serious engagement by social scientists with conservationists and the broader conservation community" (Redford 2011, p. 329). We need to "foster more informed understandings of how best to promote conservation that is effective and equitable" (Igoe 2011, p. 334).

Inclusive justice can be conceived as a form of multiculturalism that most of us, academics writing about conservation, support (Kymlicka and Donaldson 2014). This multiculturalism is rooted in social justice, human rights, and citizenship, but also in ecological justice and animal rights, aiming to contest status hierarchies that have privileged hegemonic groups while stigmatizing minorities. This progressive conception, Kymlicka and Donaldson (2014) reflect, operates to illuminate unjust political and cultural hierarchies, to de-center hegemonic norms, and to hold the exercise of power morally accountable. Viewed this way, multiculturalism and animal rights are not in conflict, but flow naturally from the same deeper commitment to justice and moral accountability.

Exclusive attention to social and economic justice leads environmental myopia (Pluhar 1995; Kopnina 2012a, b; Wuerthner 2012) and even complacency in ecological genocide (Crist and Cafaro 2012; Crist 2013). The conservation struggle is not between conservation elites and poor communities, but between the larger forces of industrialism and "fickle but ravenous consumer desires" (Wakild 2015) and those that seek to protect the last remaining troves of cultural and natural diversity.

Conservation critique outlined above masks the fact that the very environmentalism is NOT an elitist or neoliberal invention but a truly transnational phenomenon (e.g., Dunlap and York 2008) with concern for environmental issues consistently observed in different cultures (Dietz et al. 2005; Milfont and Schultz 2016). Kelch (2016, p. 83), for example, has explored the impact of culture on animal advocacy, finding that there is "a universal facet composed of moral, ethical, empirical and other principles posited to be accepted across cultures." These universal principles can be constructed and utilized to advance the cause of animals worldwide (Peters 2016).

A number of studies on the nature and origins of environmentalism and conservation is instructive as it shows the breadth of movements and ideas (Milton 1993, 1996, 2002; Ingold 2006). Ecological justice position was elaborated from the deep ecology perspective, supported by the "species turn" (Haraway 2008; Ingold 2006) in social science. These perspectives could strengthen the theoretical framework of common justice that recognizes the artificiality of dualism, reintegrates the human and nonhuman, and thus enables reconciliation between the critical perspectives on these issues (Shoreman-Ouimet and Kopnina 2015, 2016). This theoretical frame has a potential not only to aid the mutual understanding of social and ecological justice proponents through articulating their differences, but may also promote their cooperation towards a more socially and ecologically just world. As one of the reviewers of the earlier version of this chapter has suggested, we need to ask: But what to do about those who do not support universal justice, many who seem to have the power historically and in the present to turn their standpoint into political, economic, and other cultural reality? Nobody has the easy answer and the lack of universal support for presumably inclusive values is an ongoing source of frustration for those that care about moral injustices. Yet it is the author's hope that we, as social scientists, environmentalists, or conservationists, continue to provide evidence that demonstrates the costs and benefits of different pathways, and that we should, as members of the Earth, advocate for those that best move us, based on what the science reveals, toward ecological sustainability.

Conclusion

Whether the two Norwegians Arne Naess (1912–2009) and Arne Kalland (1945–2012), one a fervent defender of all species, the other, of human and indigenous rights, could reconcile their views is unclear. Strang (2013) has raised a number of moral conditions to enable this reconciliation. First, one needs to recognize the provision of justice to those who can speak for themselves, in preference to those who cannot. Second, humans and nature are interdependent and that disruption for any of the participants has potentially major impacts on the others. Third, that the culture and nature dualism is theoretically inadequate. The author sees hope in accepting ecological justice in environmental justice debates. There are strategies for defending progressive causes, whether animal rights or human rights, against the danger of instrumentalization and cultural imperialism (Kymlicka and Donaldson 2014). As the (most of) humanity could be recently swayed to accept that slavery, racism, and sexism are morally wrong, so can the moral deficiency of anthropocentrism will be realized in time to extend justice to the most vulnerable beings.

Cross-References

▶ Environmental Stewardship

References

Adams, W. A., & Hutton, J. (2007). People, parks and poverty: Political ecology and biodiversity conservation. *Conservation and Society, 5*, 147–183.

Bateson, G. (1972). *Steps to an ecology state of mind* (pp. 494–499). Chicago: University of Chicago Press.

Baxter, B. (2005). *A theory of ecological justice*. New York: Routledge.

Bekoff, M. (Ed.). (2013). *Ignoring nature no more: The case for compassionate conservation*. Chicago: Chicago University Press.

Best, S., & Nocella, A. J. (2011). *The animal liberation front: A political and philosophical analysis*. New York: Lantern Books.

Bisgould, L. (2008). Power and irony: One tortured cat and many twisted angles to our moral schizophrenia about animals. *Animal Subjects: An Ethical Reader in a Posthuman World, 8*, 259.

Bonnett, M. (2013). Sustainable development, environmental education, and the significance of being in place. *Curriculum Journal*, 24(2), pp. 250–271.

Bonnett, M. (2015). Sustainability, the metaphysics of mastery and transcendent nature. In H. Kopnina & E. Shoreman-Ouimet (Eds.), *Sustainability: Key issues*. New York: Routledge Earthscan.

Borràs, S. (2016). New transitions from human rights to the environment to the rights of nature. *Transnational Environmental Law, 5*, 113–143.

Brechin, S. R., Wilshusen, P. R., Fortwrangler, C. L., & West, P. C. (Eds.). (2003). *Contested nature: Promoting international biodiversity conservation with social justice in the twenty-first century*. New York: State University of New York Press.

Brockington, D. (2002). *Fortress conservation. The preservation of the Mkomazi Game Reserve* (African issues series). Oxford: James Currey.

Brockington, D., Duffy, R., & Igoe, J. (2008). Nature unbound. In *Conservation, capitalism and the future of protected areas*. London: Earthscan.

Brosius, P. (1999). Green dots, pink hearts: Displacing politics from the Malaysian rain forest. *American Anthropologist, 101*(1), 36–57.

Brosius, P., Tsing, A., & Zerner, C. (Eds.). (2005). *Communities and conservation: Histories and politics of community-based natural resource management*. New York: Altamira.

Büscher, B. (2015). *"Rhino poaching is out of control!" Violence, heroes and the politics of Hysteria in online conservation*. Paper presented at the British International Studies Association. 16–19 June, London.

Cafaro, P. (2015). Three ways to think about the sixth mass extinction. *Biological Conservation, 192*, 387–393.

Cafaro, P., & Primack, R. (2014). Species extinction is a great moral wrong. *Biological Conservation, 170*, 1–2.

Campbell, M. (2012). Why the silence on population? In P. Cafaro & E. Crist (Eds.), *Life on the brink: Environmentalists confront overpopulation* (pp. 41–56). Atlanta: University of Georgia Press.

Caplan, P. (Ed.). (2004). *The ethics of anthropology: Debates and dilemmas*. New York: Routledge.

Catton, W. R., & Dunlap, R. E. (1978). Environmental sociology: A new paradigm. *American Sociologist, 13*, 41–49.

Chapin, M. (2004). A challenge to conservationists. *World Watch, 17*(6), 17–31.

Chaudhuri, T. (2012). Learning to protect: Environmental education in a South Indian tiger reserve. In Helen Kopnina (Ed.), *Anthropology of environmental education* (pp. 87–113). New York: Nova Science Publishers.

Crist, E. (2012). Abundant earth and population. In P. Cafaro & E. Crist (Eds.), *Life on the brink: Environmentalists confront overpopulation* (pp. 141–153). Athens: University of Georgia Press.

Crist, E. (2013). Ecocide and the extinction of animal minds. In M. Bekoff (Ed.), *Ignoring nature no more: The case for compassionate conservation* (pp. 45–53). London: Chicago University Press.

Crist, E., & Cafaro, P. (2012). Human population growth as if the rest of life mattered. In P. Cafaro & E. Crist (Eds.), *Life on the brink: Environmentalists confront overpopulation* (pp. 3–15). Athens: University of Georgia Press.

Crist, E., & Kopnina, H. (2014). Unsettling anthropocentrism. *Dialectical Anthropology, 38*, 387–396.

Desmond, J. (2013). Requiem for roadkill: Death and denial on America's roads. In H. Kopnina & E. Shoreman-Ouimet (Eds.), *Environmental anthropology: Future directions* (pp. 46–58). New York/Oxford: Routledge.

Dietz, T., Fitzgerald, A., & Shwom, R. (2005). Environmental values. *Annual Review Environmental Resources, 30*, 335–372.

Doak, D. F., Bakker, V. J., Goldstein, B. E., & Hale, B. (2015). What is the future of conservation? In G. Wuerthner, E. Crist, & T. Butler (Eds.), *Protecting the wild: Parks and wilderness, the foundation for conservation* (pp. 27–35). Washington, DC/London: The Island Press.

Dunlap, R. E., & York, R. (2008). The globalization of environmental concern and the limits of the postmaterialist values explanation: Evidence from four multinational surveys. *The Sociological Quarterly, 49*, 529–563.

Eckersley, R. (1998). Divining evolution and respecting evolution. In A. Light (Ed.), *Social ecology after Bookchin*. London: The Guilford Press.

Ehrenfeld, D. (1988). Why put a value on biodiversity? In E. O. Wilson (Ed.), *Biodiversity*. Washington, DC: National Academy Press.

Einarsson, N. (1993). All animals are equal but some are cetaceans: Conservation and culture conflict. In K. Milton (Ed.), *Environmentalism: The view from anthropology* (pp. 73–84). New York: Routledge.

Escobar, A. (1996). Constructing nature: Elements for a post-structuralist political ecology. In R. Peet & M. Watts (Eds.), *Liberation ecologies* (pp. 46–68). London: Routledge.

Faber, D., & McCarthy, D. (2003). Neo-liberalism, globalization and the struggle for ecological democracy: Linking sustainability and environmental justice. In J. Agyeman & R. D. Bullard (Eds.), *Just sustainabilities: Development in an unequal world* (pp. 38–63). New York: Routledge.

Fletcher, R., Breitlin, J., & Puleo, V. (2014). Barbarian hordes: The overpopulation scapegoat in international development discourse. *Third World Quarterly, 35*(7), 1195–1215.

Fredericks, S. E. (2015). Ethics in sustainability indexes. In H. Kopnina & E. Shoreman-Ouimet (Eds.), *Sustainability: Key issues* (pp. 73–87). New York: Routledge.

Gleeson, B., & Low, N. (1998). *Justice, society and nature: An exploration of political ecology*. London: Routledge.

Goldberg, J. (2010). The hunted: Did American conservationists go too far in Africa? *The New Yorker*, April 5.

Gould, K. A., & Lewis, T. L. (2012). The environmental injustice of green gentrification. In *The world in Brooklyn: Gentrification, immigration, and ethnic politics in a global city* (pp. 113–146). Plymouth: Lexington Books.

Gould, K. A., Pellow, D. N., & Schnaiberg, A. (2015). *Treadmill of production: Injustice and unsustainability in the global economy*. New York: Routledge.

Greenpeace. (2013). Eliminate toxic chemicals. http://www.greenpeace.org/international/en/campaigns/toxics/.

Haraway, D. (2008). *When species meet*. Minneapolis: University of Minnesota Press.

Haring, B. (2011). *Plastic pandas*. The Netherlands: Nijgh & Van Ditmar.

Harrison, J. L. (2014). Neoliberal environmental justice: Mainstream ideas of justice in political conflict over agricultural pesticides in the United States. *Environmental Politics, 23*(4), 650–669.

Hawkins, R. (2012). Perceiving overpopulation: Cannot we see what we're doing? In P. Cafaro & E. Crist (Eds.), *Life on the brink: Environmentalists confront overpopulation* (pp. 202–213). Atlanta: University of Georgia Press.

Henley, D. (2011). Swidden farming as an agent of environmental change: Ecological myth and historical reality in Indonesia. *Environment and History, 17*, 525–554.

Higgins, P. (2010). *Eradicating ecocide: Laws and governance to prevent the destruction of our planet* (pp. 62–63). London: Shepheard Walwyn Publishers Ltd.

Horowitz, L. S. (2012). Translation alignment: Actor-network theory and the power dynamics of environmental protest alliances in New Caledonia. *Antipode, 44*(3), 806–827.

Hovik, S., Sandström, C., & Zachrisson, A. (2010). Management of protected areas in Norway and Sweden: Challenges in combining central governance and local participation. *Journal of Environmental Policy & Planning, 12*(2), 159–177.

Igoe, J. (2011). Forum. Rereading conservation critique: A response to Redford. *Fauna & Flora International, Oryx, 45*(3), 333–334.

Igoe, J., & Brockington, D. (2007). Neoliberal conservation: A brief introduction. *Conservation and Society, 5*(4), 432–449.

Ingold, T. (2006). Against human nature: Evolutionary epistemology. *Language and Culture, 39*(3), 259–281.

Kalland, Arne. 2009. *Unveiling the whale: Discourses on whales and whaling* (Studies in environmental anthropology and ethnobiology series). New York: Berghahn Books.

Kareiva, P., Lalasz, R., & Marvier, M. (2011). Conservation in the anthropocene: Beyond solitude and fragility. *Breakthrough Journal*. Fall, 29–27.

Katz, E. (1999). Envisioning a de-Anthropocentrised world: Critical comments on Anthony Weston's 'the incomplete eco-philosopher'. *Ethics, Policy and Environment, 14*, 97–101.

Kelch, T. G. (2016). Towards universal principles for global animal advocacy. *Transnational Environmental Law, 5*, 81–111.

Kemf, E. (Ed.). (1993). *The law of the mother: Protecting indigenous peoples in protected areas*. San Francisco: Sierra Club Books.

Koot, S. (2016). Cultural ecotourism as an indigenous modernity: Namibian bushmen and two contradictions of capitalism. In H. Kopnina & E. Shoreman-Ouimet (Eds.), *Handbook of environmental anthropology* (pp. 315–326). New York: Routledge.

Kopnina, H. (2012a). Towards conservational anthropology: Addressing anthropocentric bias in anthropology. *Dialectical Anthropology, 36*(1), 127–146.

Kopnina, H. (2012b). Re-examining culture/conservation conflict: The view of anthropology of conservation through the lens of environmental ethics. *Journal of Integrative Environmental Sciences, 9*(1), 9–25.

Kopnina, H. (2012c). *Anthropology of environmental education*. New York: Nova Science Publishers.

Kopnina, H. (2014a). Environmental justice and biospheric egalitarianism: Reflecting on a normative-philosophical view of human-nature relationship. *Earth Perspectives, 1*, 8.

Kopnina, H. (2014b). Future scenarios and environmental education. *The Journal of Environmental Education, 45*(4), 217–231.

Kopnina, H. (2016a). Half the earth for people (or more)? Addressing ethical questions in conservation. *Biological Conservation, 203*(2016), 176–185.
Kopnina, H. (2016b). Animal cards, supermarket stunts and world wide fund for nature: Exploring the educational value of a business-ENGO partnership for sustainable consumption. *Journal of Consumer Culture, 16*(3), 926–994.
Kopnina, H. (2016c). Nobody likes dichotomies (but sometimes you need them). *Anthropological forum.* Special forum: Environmental and social justice? *The Ethics of the Anthropological Gaze, 26*(4), 415–429.
Kopnina, H., & Shoreman-Ouimet, E. (eds) (2011) Environmental anthropology today. Routledge, New York/Oxford.
Kopnina, H., & Gjerris, M. (2015). Are some animals more equal than others? Animal rights and deep ecology in environmental education. *Canadian Journal of Environmental Education, 20*, 109–123.
Kothari, A., Camill, P., & Brown, J. (2013). Conservation as if people also mattered: Policy and practice of community-based conservation. *Conservation and Society, 11*, 1–15.
Kottak, C. P. (1999). The new ecological anthropology. *American Anthropologist, 101*(1), 23–35.
Kymlicka, W., & Donaldson, S. (2014). Animal rights and aboriginal rights. In V. Black (Ed.), *Animal law in Canada*. Irwin: Toronto.
Lawrence, K. S., & Abrutyn, S. B. (2015). The degradation of nature and the growth of environmental concern: Toward a theory of the capture and limits of ecological value. *Human Ecology Review, 21*(1), 87.
Leopold, A. (1949/1987). *A sand county almanac and sketches here and there*. New York: Oxford University Press.
Liddick, D. R. (2006). *Eco-terrorism: Radical environmental and animal liberation movements*. Connecticut: Praeger Publishers.
Lyman, M. W., Danks, C., & Maureen, M. D. (2013). New England's community forests: Comparing a regional model to ICCAs. *Conservation and Society, 11*, 46–59.
Martin, A., Akol, A., & Gross-Camp, N. (2015). Towards an explicit justice framing of the social impacts of conservation. *Conservation and Society, 13*, 166–178.
Marvier, M. (2014). A call for ecumenical conservation. *Animal Conservation, 17*(6), 518–519.
Mathews, F. (2016). From biodiversity-based conservation to an ethic of bio-proportionality. *Biological Conservation, 200*, 140–148.
McCauley, D. (2006). Selling out on nature. *Nature, 443*, 27–28.
McElroy, A. (2013). Sedna's children: Inuit elders' perceptions of climate change and food security. In H. Kopnina & E. Shoreman-Ouimet (Eds.), *Environmental anthropology: Future directions* (pp. 145–171). New York/Oxford: Routledge.
Merchant, C. (1992). *Radical ecology: The search for a liveable world*. New York: Routledge.
Milfont, T., & Schultz, P. W. (2016). Culture and the natural environment. *Current Opinion in Psychology, 8*, 194–199.
Miller, B., Soulé, M. E., & Terborgh, J. (2014). 'New conservation' or surrender to development? *Animal Conservation, 17*(6), 509–515.
Milton, K. (1993). Introduction. In K. Milton (Ed.), *Environmentalism: The view from anthropology* (pp. 73–84). New York: Routledge.
Milton, K. (1996). *Environmentalism and cultural theory: Exploring the role of anthropology in environmental discourse*. New York: Routledge.
Milton, K. (2002). *Loving nature: Toward an ecology of emotion*. New York: Routledge.
Moore, J. W. (2015). *Capitalism in the web of life: Ecology and the accumulation of capital*. London/New York: Verso Books.
Naess, A. (1973). The shallow and the deep: Long-range ecology movement. A summary. *Inquiry, 16*, 95–99.
Noss, R. F. (1992). The wildlands project land conservation strategy. *Wild Earth, 1*, 9–25.

Nuttall, M. (2016). Climate, environment, and society in Northwest Greenland. In H. Kopnina & E. Shoreman-Ouimet (Eds.), *Handbook of environmental anthropology* (pp. 219–229). New York: Routledge.

Paterson, B. (2006). Ethics for wildlife conservation: Overcoming the human–nature dualism. *Bioscience, 56*(2), 144–150.

Peters, A. (2016). Global animal law: What it is and why we need it. *Transnational Environmental Law, 5*, 9–23. https://doi.org/10.1017/S2047102516000066.

Pluhar, E. B. (1995). *Beyond prejudice: The moral significance of human and nonhuman animals*. Durham: Duke University Press.

Pountney, J. (2012). Book review: Kalland, unveiling the whale. *Durham Anthropology Journal, 18*(1), 215–217.

Rantala, S., Vihemäki, H., Swallow, B. M., & Jambiya, G. (2013). Who gains and who loses from compensated displacement from protected areas? The case of Derema corridor, Tanzania. *Conservation and Society, 11*(2), 97–111.

Redford, K. H. (2011). Forum. Misreading the conservation landscape. *Fauna & Flora International, Oryx, 45*(3), 324–330.

Redford, K. H., & Adams, W. M. (2009). Payment for ecosystem services and the challenge of saving nature. *Conservation Biology, 23*, 785–787.

Redford, K.H., & Fearn, E. (Eds.). (2007). *Protected areas and human displacement: A conservation perspective. Wildlife Conservation Society* (Working Paper No 29).

Rockström, J., Steffen, W., Noone, K., et al. (2009). A safe operating space for humanity. *Nature, 461*, 472–475.

Rudy, K. (2012). If we could talk to the animals: On changing the (Post) human subject. In M DeMello (Ed.), *Speaking for animals: Animal autobiographical writing*. Durham, NC: Duke University Press

Sandall, R. (2000). *The culture cult: Designer tribalism and other essays*. Boulder: Westview Press.

Scarce, R. (2011). If a tree falls: A story of the earth liberation front interview: Rik scarce, author of 'Eco-Warriors'. http://www.pbs.org/pov/ifatreefalls/eco-warriors-rik-scarce-interview.php.

Schlosberg, D. (2007). *Defining environmental justice: Theories, movements, and nature*. Oxford: Oxford University Press.

Scruton, R. (2012). *How to think seriously about the planet: The case for an environmental conservatism*. Oxford: Oxford University Press.

Shoreman-Ouimet, E., & Kopnina, H. (2015). Reconciling social and ecological justice for the sake of conservation. *Biological Conservation, 184*, 320–326.

Shoreman-Ouimet, E., & Kopnina, H. (2016). *Conservation and culture: Beyond anthropocentrism*. New York: Routledge Earthscan.

Singer, P. (1977). *Animal liberation: A new ethics for our treatment of animals*. New York: Random House.

Smail, K. (2003). Remembering Malthus III: Implementing a global population reduction. *American Journal of Physical Anthropology, 123*(2), 295–300.

Sodikoff, G. (Ed.). (2011). *The anthropology of extinction: Essays on culture and species death*. Bloomington: Indiana University Press.

Strang, V. (2013). Notes for plenary debate – ASA-IUAES conference, Manchester, 5–10th Aug 2013. Motion: 'Justice for people must come before justice for the environment'. https://www.youtube.com/watch?v=oldnYTYMx-k.

Sullivan, S. (2006). The elephant in the room? Problematizing 'new' (neoliberal) biodiversity conservation. *Forum for Development Studies, 33*(1), 105–135.

Sunstein, C. R., & Nussbaum, M. C. (2004). *Animal rights: Current debates and new directions*. New York: Oxford University Press.

Sykes, K. (2016). Globalization and the animal turn: How international trade law contributes to global norms of animal protection. *Transnational Environmental Law, 5*, 55–79. https://doi.org/10.1017/S2047102516000054.

Taylor, B. R. (2008). The tributaries of radical environmentalism. *Journal for the Study of Radicalism, 2*(1), 61.

Temudo, M. P. (2012). "The white men bought the forests": Conservation and contestation in Guinea-Bissau, western Africa. *Conservation and Society, 10*, 354–366.

Van Huijstee, M., & Glasbergen, P. (2010). Business–NGO interactions in a multi-stakeholder context. *Business and Society Review, 115*(3), 249–284.

Van Huijstee, M., Pollock, L., Glasbergen, P., & Leroy, P. (2011). Challenges for NGOs partnering with corporations: WWF Netherlands and the environmental defense fund. *Environmental Values, 20*, 43–74.

Wakild, E. (2015). Parks, people, and perspectives: Historicizing conservation in Latin America. In G. Wuerthner, E. Crist, & T. Butler (Eds.), *Protecting the wild: Parks and wilderness, the foundation for conservation* (pp. 41–53). Washington, DC/London: The Island Press.

Washington, H. (2015). Is 'sustainability' the same as 'sustainable development'? In H. Kopnina & E. Shoreman-Ouimet (Eds.), *Sustainability: Key issues* (pp. 359–377). New York: Routledge.

Wenzel, G. W. (2009). Canadian Inuit subsistence and ecological instability – If the climate changes, must the Inuit? *Polar Research, 28*(1), 89–99.

West, P. (2008). Translation, value, and space: Theorizing an ethnographic and engaged environmental. *American Anthropologist, 4*, 632–642.

West, P., & Brockington, D. (2011). Introduction: Capitalism and the environment. *Environment and Society: Advances in Research, 3*(1), 1–3(3).

West, P., Igoe, J., & Brockington, D. (2006). Parks and people: The social impact of protected areas. *Annual Review of Anthropology, 35*, 251–277.

Wijkman, A., & Rockström, J. (2012). *Bankrupting nature: Denying our planetary boundaries*. New York: Routledge.

Wilshusen, P. R., Brechin, S. R., Fortwrangler, C. L., & West, P. C. (2002). Reinventing a square wheel: Critique of a resurgent "protection paradigm" in international biodiversity conservation. *Society and Natural Resources, 15*, 17–40.

Wissenburg, M. (1993). The idea of nature and the nature of distributive justice. In A. Dobson & P. Lucardie (Eds.), *The politics of nature: Explorations in green political theory*. London: Routledge.

Wuerthner, G. (2012). Population, fossil fuels and agriculture. In P. Cafaro & E. Crist (Eds.), *Life on the brink: Environmentalists confront overpopulation* (pp. 123–129). Atlanta: University of Georgia Press.

Ethical Decision-Making Under Social Uncertainty

An Introduction of Überethicality

Julia M. Puaschunder

Contents

Financial support of the Association for Social Economics, Austrian Academy of Science, Austrian Federal Ministry of Science, Austrian Office of Science and Technology at the Austrian Embassy to the United States of America, Bard Center for Environmental Policy, Research and Economy, Eugene Lang Liberal Arts College of The New School, Fritz Thyssen Foundation, George Washington University, ideas42, International Institute for Applied Systems Analysis, Janeway Center Fellowship, the New School for Social Research, New School University Senate, Prize Fellowship in the Inter-University Consortium of New York, Science and Technology Global Consortium, University of Kent, University of Vienna, Vernon Art and Science and the Vienna University of Economics and Business is gratefully acknowledged. The author declares no conflict of interest. All omissions, errors and misunderstandings in this piece are solely the author's.

J. M. Puaschunder (✉)
The New School, Department of Economics, Schwartz Center for Economic Policy Analysis, New York, NY, USA

Graduate School of Arts and Sciences, Columbia University, New York, NY, USA

Princeton University, Princeton, NJ, USA

Schwartz Center for Economics Policy Analysis, New York, NY, USA
e-mail: Julia.Puaschunder@newschool.edu; Julia.Puaschunder@columbia.edu; Julia.Puaschunder@princeton.edu

S. Dhiman, J. Marques (eds.), *Handbook of Engaged Sustainability*,
https://doi.org/10.1007/978-3-319-71312-0_34

Abstract

Prospect theory holds human to code gains or losses perspectives relative to an individual reference point to guide our actions. Monetary losses loom larger in human than the joy over gains – but does this hold for social status changes? Testing prospect theory for social status striving in the realm of socioeconomics helps understand the underlying mechanisms of social identity and social dominance theories. In two field experiments, social status prospects relative to an individual's reference point were found to influence social decision-making and action. Social status depletion was outlined in order to avoid repetition to drive social responsibility in the sustainability domain. Two field observations of environmentally conscientious recycling behavior and sustainable energy consumption at a North American university campus capture social status losses resulting in higher ethicality than social status gains. Ethicality as a socially appreciated, noble contribution to society may offer the prospect of social status gains resuscitation opportunities given the societal respect for altruism and pro-social acts. Social responsibility grants social status elevation opportunities. An Überethical filling of legal gaps or outperforming of regulatory obligations thereby is likely to occur after social status drops. Social status losses are identified as significant drivers of socially responsible environmental conscientiousness. Social forces thereby promise to become an effective means for accomplishing positive societal change.

Keywords

Elite education · Energy consumption · Ethicality · Libraries · Recycling · Prospect theory · Social identity · Social norms · Social status · Sustainability

Introduction

Decision-making research has been revolutionized by prospect theory. In laboratory experiments, prospect theory captures human to code outcome perspectives as gains or losses relative to an individual reference point, by which decisions are anchored. Prospect theory's core finding that monetary losses loom larger than gains has been generalized in many domains, yet not been tested for social status changes. Social status striving has been subject to social sciences' research for a long time, but until today, we have no clear picture of how social status prospects relative to an individual reference point may influence our decision-making and action. Understanding human cognition in the light of social status perspectives, however, could allow turning social status experiences into ethicality nudges. The perceived endowment through social status may drive social responsibility (Marques forthcoming; Puaschunder 2010; Ramananan forthcoming). Ethicality as a socially appreciated, noble societal contribution offers the prospect of social status gains given the societal respect for altruism and pro-social acts. An Überethical filling of current legal gaps or outperforming legal regulations grants additional social status elevation opportunities.

The following chapter provides an innovation application of social status theories in the sustainability domain. Building on prospect theory, two field observations of environmentally conscientious recycling behavior and sustainable energy consumption investigated if social status losses are more likely to be answered with ethicality than social status gains. Social status losses are found as significant drivers of socially responsible environmental conscientiousness. Testing prospect theory for social status striving advances socioeconomics and helps understanding the underlying mechanisms of social identity theories. Pegging social status to ethicality is an unprecedented approach to use social forces as a means for accomplishing positive societal change. Future studies may target at elucidating if ethicality in the wake of social status losses is more a cognitive, rational strategy or emotional compensation for feelings of unworthiness after social status drops.

Theoretical Framework

Social status is as old as human beings. Already ancient sources attribute rights and allocate assets based on status (DiTella et al. 2001). Status ranks individuals on socially valued individual characteristics and group membership (Ball and Eckel 1996; Hong and Bohnet 2004; Loch et al. 2000; Ridgeway and Walker 1995). At the same time, surprisingly scarce is the information on how individuals perceive status changes and how their social conscientiousness is related to social endowments. In general, social status upward prospects are seen as favorable – but the downside of social status losses is rather vaguely described, and no stringent framework exists on how status prospects impact human decision-making and actions.

One of the most influential theories explaining human decision-making under uncertainty is prospect theory (Kahneman and Tversky 1979). Prospect theory holds individuals' perceptions about prospective outcomes as individually evaluated changes from the status quo. Laboratory experiments find individual aggravation over losing monetary resources to be greater than the pleasure associated with gaining the same amount (Bazerman and Moore 2008). Originally prospect theory was captured for monetary gains and losses but replicated in various fields (Levy 1997). In the application of prospect theory, social comparisons have mildly been touched on – if we consider the impact social identities have on our day-to-day judgment, decision-making, and actions (Loewenstein et al. 1989). Understanding social status prospects' influence on individual behavior, however, could explain the underlying sociopsychological motives of decision-making in the social compound. More concretely, if certain social status prospects are found to be perceived as more or less favorable, they are prone to elicit certain behavior and may steer respective action. In individuals' constant striving for favorable social status enhancement, social status prospects could put people into a specific mindset that drives pro-social acts.

As a pro-social behavior, ethicality is socially honored. In the social compound, ethicality offers social status elevation prospects derived from respect for socially valued altruism. Ethicality as a noble act may thus grant social status elevating opportunities. In reverse, social status perspectives could be used to nudge people into pro-social behavior (Thaler and Sunstein 2008). The theory of nudging was introduced by Thaler and Sunstein (2008) drawing from psychology and behavioral economics to defend libertarian paternalism and create a favorable choice architecture that helps people to intuitively fall for a more health choice. Classical examples of how decisions can be influenced are anchoring, availability, representativeness, status quo, and herd behavior to create environments that aid people in making a choice that is beneficial on the long run and for the sake of common good. Applications of successful nudging range from food choice and health, over finance and retirement, to work discipline but also environmentalism.

In accordance with prospect theory holding that status losses loom larger than status gains and nudging theory, providing innovative examples of how individual decision-making can be influenced subliminally by group memberships, foremost social status losses may steer ethicality in the wish to regain social status based on a reference point relative to previously held status positions. If ethicality is related to social status gain perspectives, social status awareness could become a means to nurture a favorable climate within society. Social status endowments may thus be the core of socially responsible behavior; social status prospects the driver of the warm glow. In the light of ethicality being an implicit social status enhancement tool, social status losses are potentially answered by pro-social behavior. Social status manipulation could thereby serve as a nonmonetary nudge to foster ethicality in society (Thaler and Sunstein 2008). Prospect theory and nudging are proposed as means of social status enhancement with attention to regaining prior social status losses. Social status losses are portrayed to nudge people into pro-social action.

All cultures feature some form of social status displayed in commonly shared symbols. Social status attributions posit people in relation to each other in society (Huberman et al. 2004). As ascribed status can be improved throughout life, relative status positions are assigned in zero-sum games – thus one individual's status gain lowers another ones' status. Individuals implicitly weigh their social status based on the number of contestants in ranks above and below them (DiTella et al. 2001). In societal hierarchies, status is related to a diverse set of opportunities as different rules and availability of resources apply to variant social status positions (Young 2011).

As an intrinsic fundamental human characteristic, people are concerned about their social status in relevant domains, leveraging social status striving into a pivotal motivation factor in human life (e.g., Coleman 1990; Duesenberry 1949; Friedman 1953; Friedman and Savage 1948; Mazur and Lamb 1980; Ridgeway and Walker 1995; Weber 1978). Social status impacts on an individual's social identity and emotional state (Postlewaite 1998). Status gains and superiority are associated with positive emotions and well-being derived from positive interaction (Bird 2004; Galiani and Weinschelbaum 2007; Hong and Bohnet 2004). Individuals are psychologically satisfied when experiencing to be better off than others and feel uneasy when they see others doing better (Easterlin 1974; Hopkins and Kornienko 2004). Status losses are embarrassing and drive a desire to enhance one's self-image in the wake of experienced unhappiness and risk aversion (DiTella et al. 2001; Harbaugh 2006).

In the social compound, we favor positive status superiority of our groups compared to groups we do not belong to Tajfel (1978) and Tajfel and Turner (1986). Favorable group membership experiences are based on social opportunities (Meeker and Weitzel-O'Neill 1977; Ridgeway et al. 1985). Group members with high status have more control (Bales 1951; Berger and Zelditch 1985), receive more credit for success (Fan and Gruenfeld 1998), and enjoy higher degrees of well-being (Adler et al. 2000). In contrast, low-status group members are more likely to be neglected (Chance 1967; Savin-Williams 1979), more often blamed for failures (Weisband et al. 1995), and feel more negatively (Mazur 1973; Tiedens 2000).

Social status is directly related to social action and therefore provokes human behavior (Weber 1946/2009). Social status may bestow agents with subjective meaningfulness. The individual may perceive himself or herself as part of the larger and may therefore plan actions to be socially favorable. Actions may also arise naturally and subconsciously (Thaler and Sunstein 2008). The group has an anchoring guiding influence on the individual, which is described in very many domains ranging from sociology to history, politics, jurisprudence, ethics, and arts (Weber 1946/2009).

The emotional, social, and economic consequences of status striving and related decision-making appear as an open research question. Unsolved remains how social status pursuits drive motivation and behavior. No stringent decision-making pattern of social status prospects impacting on decision-making and human action can be given in the eye of social uncertainty. If social status endowments lead to more social conscientiousness or social status losses may loom larger than social endowment gains is an unsolved question. One of the most influential theories to predict human decision-making in the light of future uncertainty is prospect theory.

Prospect theory revolutionized decision-making sciences by capturing economic outcomes to be coded as gains or losses relative to a neutral reference point (Kahneman and Tversky 1979; Schkade and Kahneman 1998). In laboratory experiments, Kahneman and Tversky (1979) found individuals' perceptions about outcomes as evaluative changes from their current state. Based on deviations from the status quo, prospect theory depicts an S-shaped expected utility function for perceived monetary gains and losses with a convex value curve for gains and a concave value function for losses (Currim and Sarin 1989; Thaler 1999). A comparatively steeper loss than value function captures peoples' aggravation over losing money to be greater than their pleasure associated with acquiring the same amount of money (Kahneman and Tversky 1979). Individuals taking losses more serious than gains are motivated to preserve their status quo (Bazerman and Moore 2008). The status quo bias holds individuals to be risk averse in the domain of gains, while they are risk seeking when facing loss prospects (Jervis 1992; Kahneman et al. 1991; Kahneman and Tversky 2000; McDermott et al. 2008).

Overall, prospect theory has leveraged into one of the most influential social sciences paradigms (Kahneman and Tversky 1992). In manifold application of prospect theory, hardly any information exists on social status changes. Although social comparisons may directly affect decision-making under social uncertainty, decision-making in the eye of social status endowment prospect remains an underexplored scientific area. Until today, we have no information on the generalizability of monetary prospects on social status outcome perspectives and how decision-making is influenced by social status outlooks and endowments (Huberman et al. 2004).

While the idea that people care about their relative status is well acknowledged in social sciences, the behavioral consequences of social status prospects are rather unknown (Güth and Tietz 1990; Nowak and Sigmund 1998; Robson 1992; Tooby and Cosmides 1990; Wedekind 1998). The applicability of prospect theory on social status perspectives is untested (Loewenstein et al. 1989). Already in the original presentation of prospect theory, Kahneman and Tversky (1979) envisioned applications onto more "typical" situations of choice. Apart from prospective monetary outcomes, prospect theory was proposed to be investigated for probabilities, in which outcomes are not explicitly given and more likely to be based on skills and chances (Kahneman and Tversky 1979). Especially the status quo bias being theoretically isolated from motivational and social factors raised questions about social references, future aspirations, and recent gains and losses (Chernev 2009; Levy 1997, 2003; McDermott et al. 2008; Tversky and Kahneman 1991).

First comparisons of monetary and social utility in relation to prospect theory were started by Loewenstein et al. (1989). Social utility was defined as the level of satisfaction derived from outcomes of the self in comparison to others that were depicted as alternative or additional salient reference points. Regarding social comparisons and framing, Fox and Dayan (2004) found deviations from the original prospect theory in gain preferences over loss aversion.

Building on preliminary research that holds social status striving leading to investors' risk aversion (Roussanov 2009) and happiness being dependent on relative comparison results, Falkenstein (2006) argues that people become more risk seeking when facing negative social comparison outcomes (Axelrod 1984; Easterlin 1974; Siegel 2002). One application of prospect theory in social contexts showed that accommodations to losses tend to be slower than to gains and people incur excessive risks to recover from social status drops (Jervis 1992; Levy 1997). The view of social status striving as driver of risk-seeking behavior is also supported by descriptive research on wealth distribution and entrepreneurship (Cole et al. 1992). Naturally following experimental extensions of prospect theory could integrate the role of social comparisons for judgment and decision-making – as outlined by Festingers' (1954) social comparison and Adams' (1965) equity theory.

As most of our decision-making takes place in social and hierarchical contexts, applying prospect theory for social reference dependence appears as an interesting and necessary extension. In an attempt to explain the underlying sociopsychological motives of human decision-making under social uncertainty, investigating prospect theory in the social status domain will help understanding the behavioral consequences of social status prospects. In addition, becoming knowledgeable about emotions and behavior consequences related to prospective status changes could help unraveling the bounds of collective decision-making and overcome harmful collective decision-making outcomes – such as risk shifts, social stratification, and group polarization (Janis 1982). Finding how social status perspectives drive our actions and steer emotions will also enable us to create certain social status experiences that may instigate pro-social behavior. In particular, social status prospects could be used to drive ethicality.

Social status comprises of individual characteristics but also the amalgamated social status ascribed to groups one belongs to. Social identity captures that the mere belonging to a group contributes to an individual's status and self-esteem derived from the assigned respect toward membership groups (Tajfel and Turner 1979). As social identities are related to a certain social status, exposure to social identity cues creates situations of heightened social status awareness that influence the self-esteem. Social identity experiences put people in a specific mindset that affects their self-worth and determines their decision-making and actions.

In an implicit social hierarchy of social groups, individuals are in a constant struggle for status by orienting themselves onto higher social status groups they aspire to enter (Sidanius and Pratto 1999). Social group membership cues of groups with higher social status steer the wish to belonging to a group (Puaschunder 2015). The longing to gain access to higher social status groups and striving for their respect could enhance compliance on collectively shared goals.

Social status provides opportunities to distinguish among each other (Bourdieu 1979/1984). Social status prospects translated to economic means motivate people's actions. Cultural capital, such as elite education, may promote social mobility opportunities that satisfy a zest for social mobility improvement (Bourdieu 1979/1984; Ulluwishewa forthcoming). Different social classes may be established through education that constitutes and cultivates a certain taste within society. A certain habitus learned in educational institutions may be a foundation of different types of social classes (Bourdieu 1979/1984). Education may build a certain habitus that may then distinguish among social classes. A certain predisposition toward solving social situations may be acquired through education, which eventually leads toward class-based social groups handling social situations differently (Gil-Doménech and Berbegal-Mirabent forthcoming). Social choice preferences learned in educational institutions may be the foundation of social class differences. Predispositions to certain kinds of tastes – including the preference for ethicality – may be instilled in children through education, which creates a certain taste for social conscientiousness (Bourdieu 1979/1984). The beauty of Ivy League education may be to enter a certain social class and impede social upward mobility but also entail the positive externality to acquire social responsibility and taste for ethicality (Puaschunder 2016).

Research Questions

In these features, social identity experiences could be used to change peoples' actions according to social norms. In an implicit social contract, social norms trigger solidarity on common goals and cultivate virtues within society – foremost through emotional experiences. Groups bestow with self-worth elevating pride when members are complying with socially favorable goals and shame arises when individuals act socially irresponsible. Fear of social status losses breaks unfavorable antisocial habits. Through emotions, exposure to social identity and social norm cues may drive social responsibility. Emotions depending on social experiences could be used to drive pro-social action. Social forces could steer social norm compliance based on emotional experiences in the light of social status prospects. Pegging social identity to social norm cues may create a specific mindset that influences the judgment and decision-making of individuals who may then act in a socially favorable way. Through cognition and emotions, social status prospects and social norm cues may trigger social norm compliance contributing to ethicality.

Ethics capture social responsibility based on explicit and implicit social norms. Ethicality not only comprises people choosing to not do wrong or when people unconscientiously enter a slippery slope leading to unethicality (Bazerman and Chugh 2005; Bazerman and Tenbrunsel 2011; Tenbrunsel and Messick 2004). Ethicality also depicts when humans are outperforming legal requirements and policy recommendations in the search for doing more "good" than required. In this natural human drive to do "good" to others, humans are overdoing legal regulation while incurring costs and impose risks onto themselves. Similar to Zimbardo's heroic imagination (2011a, b) describing the voluntary service to others that involves a risk to physical comfort, social stature, or quality of life, this kind of Überethicality captures the voluntarily filling of legal gaps or outperforming of public policy goals that impose costs and risks onto the individual. In closing current legal gaps, the evolutionary-based natural law of Überethicality is forerunning legal codifications if considering laws to be the expression of our shared nature and amalgamated sum of societal norms over time (Cicero in Keyes 1966).

Ethicality offers potential implicit or explicit strategies to express and enhance social status in the social compound. In general, the natural behavioral drive of Überethicality ignites without material gain prospect. To draw an extreme example, Mother Theresa was monetarily unfortunate yet obtained highest social status for her Überethical pro-social work. In addition, financial investors who gained a fortune by rational market calculus are often prone to feel they have to return to society by philanthropy – for instance, Warren Buffett and George Soros dedicate extraordinary amounts of time and money to promote ethical causes.

Apart from monetary considerations, Überethicality derives from social incentives in the social compound. Status may play a key role. Based on Maslow's (1943) hierarchy of needs, one can only be Überethical if having reached a certain social status. Not having to worry about food and shelter frees mental capacities to address higher societal, ethical needs and future-oriented filling current legal gaps. As ethicality is perceived as noble act that grants others' respect, individuals may use ethical decisions as a conspicuous social status symbol in the social compound. Beyond governmental regulation and legal obligations, the nobleness of Überethicality may bestow individuals with social status elevation prospects. The foresightedness of fulfilling future requirements also implies leadership advantages (Young 2011). Given the natural respect for the voluntary willingness to incur risks for the sake of pro bono-outcomes as well as leadership advantages attributed to proactively tackling ethical problems that may likely cover future regulation, Überethicality is thus an implicit social status elevation means apart from any monetary gains.

Under the assumption of individual self-esteem being dependent on social status and human constantly wishing to maintain or gain positive social status, Überethicality is seen as a social status pedestal. How social status striving can drive socially responsible behavior will be studied in an elite education setting. Social preferences as a context-dependent phenomenon will be captured at an educational hub that individuals enter with the hope to improve their social standing and cultural capital. Proactive ethicality can be used to claim or regain social status. In accordance with prospect theory holding that losses loom larger than gains based on an individual reference point, especially the prospect of losing prior social status may trigger individuals' wish to compensate social status losses by gaining status through ethical acts. This phenomenon will be captured at points in life, when young people are eager to enter a higher social class and acquire the social skills to successfully transfer to a higher class. The idea to use social status prospects as a means to elevate pro-social behavior is a novel and an innovative nudge to drive social conscientiousness in the sustainability domain (Thaler and Sunstein 2008).

In order to investigate whether social status is related to ethicality in a way prospect theory would suggest, two studies tested if social status prospects lead to a rise in ethicality and if – in accordance with prospect theory and Überethicality assumptions – social status losses are associated with higher degrees of ethicality than social status gain prospects at an elite educational institution.

In accordance with prospect theory and Überethicality notions, study 1 tested if social status losses are more likely to be answered with ethicality than social status gain perspectives. People were nudged into ethicality in the prospect of losing social reputation and access to social capital in the educational sector and sustainability domain.

Empirical Results

A field experiment was staged at dormitories of a North American elite university during the 2011 summer that scrutinized if (1) the mere presence of social identity in combination with social norm cues leads to a rise in ethicality and if (2) social status losses heighten ethicality more than social status gain prospects.

The empirical evidence collected at a university environment was chosen for the unique setting of young individuals joining the summer school eager to enter a higher social class and acquire the social skills to successfully transfer to a better social standing. Academic circles and university settings are opinion leaders and educators of future leaders and can therefore be considered as a valid social laboratory. The data collected itself was in the sustainability domain featuring application in recycling and energy light consumption with widespread applicability to very many different outlets in the public and private sectors.

Four similar on-campus summer school residences were selected for creating different status experiences. During a summer school at a North American university campus, the summer school students, comprising of high school students, lived in the regular university campus dormitories. Four dormitories were chosen to stage field experiments featuring an observation of the summer school students' recycling behavior. In total, 711 summer school students lived in the selected dormitories, whose recycling choices were observed in 4 independent dormitories. The female and male students – mostly from the United States – were summer school attendees, mostly in their last year of high school, who took a 7-week-long program to specialize in classes in the fields of "arts, humanities, and social sciences"; "business," "computer science, math, and engineering"; "foreign language and literatures"; "sciences"; and "writing and journalism." The university chosen was an Ivy League, and therefore a certain distinct social status bias occurs.

Per dormitory, an average of 178 students' environmentally conscientious behavior was recorded every working day of the observation period. All dormitories were home to summer school students with approximately the same schedule – for instance, summer school students faced the same arrival period, study and exam periods, and move-out times. Extraneous influences that potentially could lead to a higher amount of disposals – like cardboard boxes from moving in or out – were therefore assumed to be constant for all dormitories.

In order to capture the effect of social identity pegged to social norm cues on socially responsible environmental ethicality (Hypothesis 1), the impact of a 3-week exposure to university logos and "Sustainability" initiative logos on summer school students' recycling behavior was observed.

For 6 weeks during the 2011 summer school, the residing summer school students faced different environments. Some of them were exposed to logos of the university (Test condition 1 representing social identity status striving), others to "Sustainability" initiative logos of the university (Test condition 2 representing social, ethical norms), others to both logos concurrently (Test condition 3 representing social identity status striving pegged to social, ethical norms), and even others did not see university logos or "Sustainability" logos of the university in dormitory recycling areas at all (Control condition 1, neutral).

Four posters were placed in each of the selected test dormitories around the recycling bins and buckets. Two of the 8.3 × 11.7-inche-sized posters displayed the university logo featuring the emblem and letters of the university slogan on a red shield and below the university name on a white bandage. Two other posters exhibited the "Sustainability" initiative logo featuring the described emblem and letters of the slogan a green shield and on the right-hand side the words "Sustainability" and the name of the university. Both logos were printed in color on individual posters filling approximately two thirds of the poster. All hardcover printouts of the university logo or "Sustainability" initiative logos were secured by a waterproof, lucent shield. The posters were placed on the wall next to recycling bins and buckets with easily removable tape. A similar poster location in all dormitory recycling areas was chosen. No posters were placed on doors, fences, entry posts, gates, poles, utility, or sidewalks.

In the test dormitory, four copies of the university logo or "Sustainability" logos per dormitory were installed concurrently around the recycling bins and buckets. In order to have comparison groups, the recycling behavior in three other, independent university dormitories of a North American campus was observed during the entire 6-week observatory period. In two of the other test dormitories, four copies of the university logo or "Sustainability" initiative logos were placed around the recycling bins and buckets. The control dormitory remained without logo installment. The time of the logo exposure was balanced between dormitories – one dormitory remained without any logos in the first 3 weeks and two featured no logos in the last 3 weeks in order to control for general temporal biases.

In the first 3 weeks, 153 students of test dormitory 1 were exposed to university and "Sustainability" logos. Test dormitory 2, which hosted 161 summer school students, did not feature any logo exposure prior to a planned future "Sustainability" logo exposure. Test dormitory 3 exposed university logos to 147 students. The control dormitory, in which 250 summer school students lived, had no logos.

In the beginning of week 4 of the field observation, the conditions changed for all the test dormitories. The 153 students of the test dormitory 1, in which students were exposed to university and "Sustainability" logos, faced an environment without any logo exposure. Test dormitory 2, hosting 161 summer school students – that did not feature any logo before – now showed "Sustainability" logos. In test dormitory 3 – with prior university logos exposed to 147 students – the logos were removed. The control dormitory, in which 250 summer school students lived, remained without logo exposure.

The recycling behavior of 2011 North American dormitories' residents was observed by weighting recycled disposals. Recycled waste was measured every day during the regular disposal collection in the four respective dormitories during the summer school. The recycled disposals were weighted on a regular scale. The weight of different recycling buckets was recorded manually on a paper spreadsheet and the data transferred onto an Excel computer spreadsheet later each day of the data collection.

The recycling behavior during times of exposure versus non-exposure to logos was compared within the test dormitory. In addition, the effect of logo installment versus non-logo exposure was captured between dormitories. The effect is determined by a significant change in the recycled disposals' weight.

The combined presence of university logos and "Sustainability" initiative logos heightened common goal compliance. After a time of exposure to the combined logos, in the phase of the removed social status and social norm insignia, pro-social behavior increased significantly ($t(_{27}) = -2.\,042; p < 0.032$), which was not the case in any of the other dormitories. Figure 1 holds the recycled weight during university and "Sustainability" logo exposure recycled weight monitored within the test dormitory over time. Logo versus non-logo exposure impacts on recycled disposals'

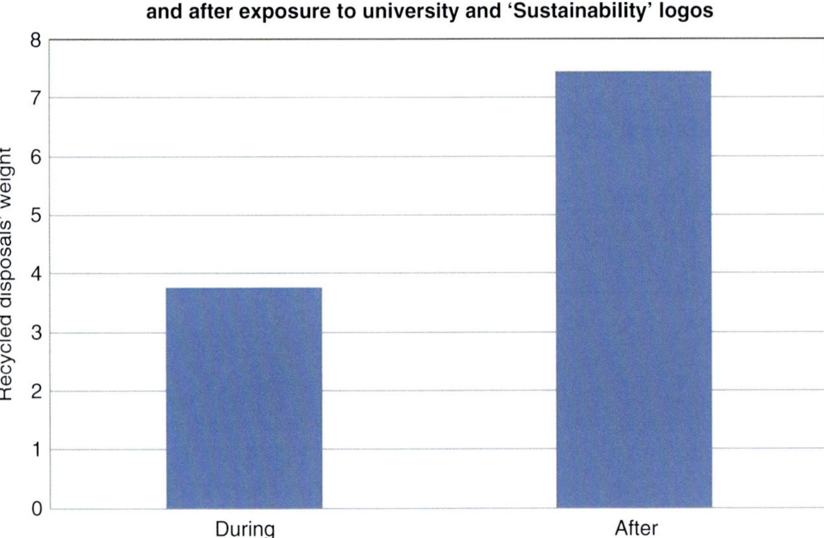

Fig. 1 Within test dormitory comparison of recycled weight during university and "Sustainability" logo exposure and afterward

weight within the test dormitory. Removed university in combination with "Sustainability" logos heighten recycling compliance significantly.

Figure 2 holds the recycled weight during university and "Sustainability" logo exposure recycled weight monitored within the test dormitory over time. Logo versus non-logo exposure impacts on recycled disposals' weight within the test dormitory. Removed university in combination with "Sustainability" logos heighten recycling compliance significantly.

Logo versus non-logo exposure impacts on the environmental ethicality measured by recycled disposals' weight between the test dormitories and the control dormitory. Removed university in combination with "Sustainability" logos heighten recycling compliance significantly when comparing between dormitories (one-way ANOVA $F_{(7,104)} = 5.914, p<0.000$). Recycling compliance was measured based on the recycled weight per resident during exposure and after exposure to cues.

Study 1 provides evidence for ethicality being a context-dependent phenomenon nudgeable by social forces. Social status prospects can steer ethicality. Social identity and social status striving in combination with social norms can be used to improve day-to-day environmental protection behavior. Building on prospect theory, social status losses more likely trigger social norm compliance than prospective social status gains. Another study was designed to test if the effect of social status losses nudging people into social conscientiousness holds for environmentally conscientious energy consumption.

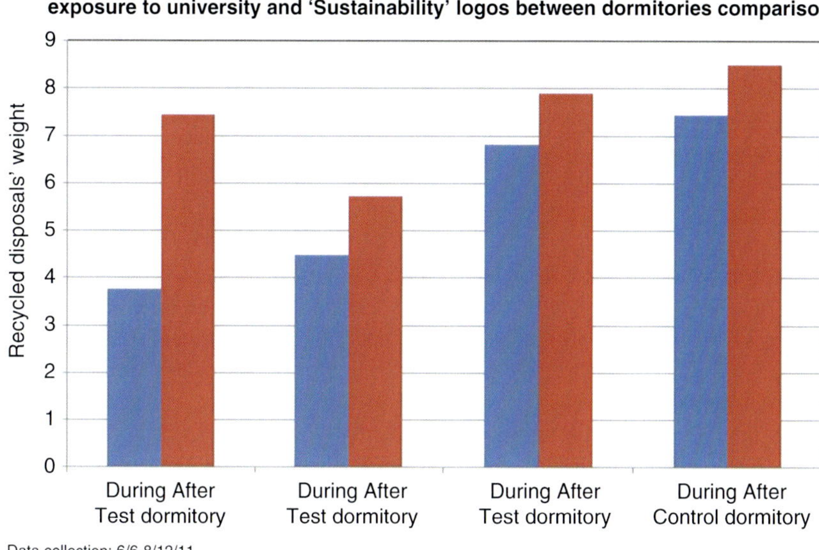

Fig. 2 Between test dormitory comparisons of recycled weight during university and "Sustainability" logo exposure and afterward

An additional field experiment at two North American university libraries targeted at investigating if (1) the mere presence of social identity in combination with social norm cues leads to a rise in ethicality and if (2) social status losses heighten ethicality more than social status gain perspectives.

The field experiment comprised of a field observation at two North American university libraries featuring user-controllable light switches at reading desks and carrels. In a quasi-public location of two university libraries that are accessible by all general students, researchers, and affiliates of the university, data was collected during 6 weeks. For 6 weeks, the energy light consumption was monitored and recorded manually.

In the first 2 weeks, the field experiment featured a baseline observation of energy light consumption without any installation followed by a 2-week exposure to social status symbols and "Sustainability" initiative logos featuring slogans followed by 2 weeks with removed logo exposure.

In order to capture the effect of social identity and social norm cues on socially responsible energy light consumption (Hypothesis 1) in a field observation, at one university, library students were exposed to university logos (Test condition 1 representing social identity status striving) and "Sustainability" (Test condition 2 representing social, ethical norms) at the respective university logos plus instructions to "Please avoid losing energy by switching off lights after use" (Test condition 3 representing social identity status striving pegged to social, ethical norms) for 2 weeks. The postcard-sized (5,1 × 3,1 inches) stickers displayed the university logo and/or "Sustainability" initiative banners stating "Please avoid losing energy by switching off lights after use" in color filling approximately two thirds of the card.

The cards were formed into cardboard tents. The tents were placed onto reading desks and carrels on three floors of one of the chosen university libraries. Per reading desk, either one or two tents were placed so that a person sitting at a carrel would see one tent. At the desks, three tents were installed so that approximately one to two people sitting down and using the reading desk would directly face one tent and could potentially also see two other tents. On the first floor of the test library, approximately 32 university logos were put on the reading desks and carrels. At the second floor of the test library, approximately 47 cardboard tents displaying the "Sustainability" at the respective university logo were placed on reading desks and carrels under scrutiny so that one reading desk user could see one tent. On the third floor of the test library, approximately 120 cardboard tents displaying the university logo and the "Sustainability" at the respective university logo were shown on the reading desks and carrels in similar locations per reading desk and carrel so that approximately one person could see one tent. All tents were maintained throughout the observation period and replaced on a daily basis if missing or damaged. As a control group, the energy light consumption of another library of the university campus was observed independently during the study. The control library (Control condition 1, neutral) remained without any logo or instruction placement.

The energy light consumption was observed several times daily in the respective libraries for 6 weeks. During the prospective observation time, the energy light consumption was recoded manually on a paper spreadsheet and transferred on an electronic computer spreadsheet on a daily basis. The library visitors were not informed about the energy light consumption measurement to avoid study participation biases. All information on energy light consumption was recorded in such a manner that the consumers were never identifiable insofar as linking the consumption to the individual library visitors, who were potentially university college and graduate students, postdoctoral researchers, scholars, and professors. All observation data was collected and analyzed anonymously. During the energy meter data collection, no contact with students or affiliates studying or working in the library was sought. At no point, there was any disruptive research activities that could influence or affect the study activities of library visitors.

The energy light consumption during times of exposure versus non-exposure to social status insignia and social norms instructions was compared over time within the test library and between the test and the control library. The recorded energy light consumption conscientiousness was measured by creating an index of abandoned burning lights divided by used burning lights per floor of the observed test library and the entire control library.

As Fig. 3 exhibits, significantly improved energy consumption conscientiousness was found for the floors featuring the university logo ($t = 3.127$, $df = 106$, $p < 0.002$) for the 2-week period after the logo had been removed. Improved energy consumption conscientiousness was recorded in the wake of removed university logos and social norm instructions. Significantly improved energy consumption conscientiousness was

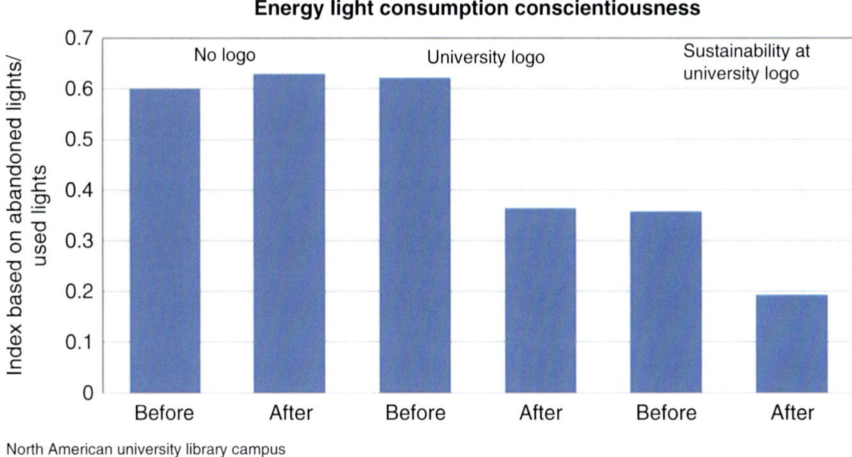

Fig. 3 Between test and control library comparison of energy light consumption before and after logo exposure

exhibited for the floors featuring the "Sustainability" at the university logo ($t = 2.898$, $df = 85$, $p < 0.005$) for the 2-week period after the logo had been removed. When comparing the energy light consumption conscientiousness between the test library and the control library, there was a significant effect between libraries (ANOVA, $F_{(3,\ 578)} = 7.009$, $p < 0.000$) after removed logos ($p < 0.001$). When comparing the test library floor featuring university logos and the control library, there was a significant increase of energy light consumption conscientiousness after the removal of status symbols in the test library (ANOVA, $F_{(7)} = 6.\ 695$, $p < 0.000$) after removed logos ($p < 0.025$). When comparing the test library floor featuring "Sustainability" signs and slogans between the control library, there was a significant increase of energy light consumption conscientiousness after the removal of social norm cues and instructions in the test library (ANOVA, $F_{(7)} = 6.695$, $p < 0.025$) after removed logos ($p < 0.000$). While there was a significant change in the light energy consumption in the test library, there was no significant change in the control library without any logos.

Overall, the presented studies provide evidence for ethicality being a context-dependent phenomenon. In the field of behavioral law and economics, connecting social status to ethicality contributes to the upcoming trend of sociology entering behavioral economics. Capturing social forces as the core of collective decision-making on social responsibility at the same time spearheads the idea of ethicality as a natural behavioral law. The presented results serve as evidence for situational cues as the driver of pro-social decision-making.

On a practical basis, pegging social status to ethicality is a novel idea to use social forces as a means for accomplishing positive societal change (Thaler and Sunstein 2008). Unraveling social status striving affecting ethical decision-making thereby helps fostering socially responsible goals beyond stringent legal enforcement and governmental policy control. Deriving information on circumstances under which decision-makers are likely to exhibit social responsibility grants recommendations on how social forces can be used to stimulate socially responsible outcomes.

Discussion

The presented connection of social status and social conscientiousness will help promoting ethicality. In a plethora of research on ethical downfalls and negative consequences of accidental ethical decision-making failures, Überethicality is a powerful contribution shedding light on how to strengthen our inner ethical forces. Understanding ethical decision-making under social uncertainty allows fostering a moral dimension of social life. Exploring further how psychological ownership and social status attributions can drive socially conscientious behavior offers cost-effective, easily implementable ethicality nudges to steer and promote socially responsible acts on a daily basis (Thaler and Sunstein 2008).

The current research also serves as a way to better understand the socio-dynamics of environmental ethicality. The application of social identity on environmental ethicality is an unprecedented approach to prepare for major challenges ahead for humankind in the light of climate change and natural resource constraints (Dhiman forthcoming). Finding circumstances under which decision-makers are environmentally conscientious and potentially identifying social forces that trigger environmental ethicality are aimed at modeling day-to-day environmental protection decision-making and find easily manipulated, real-world relevant sustainability incentives. On a grand scale, relating individual experiences to future common goals will allow public policy specialists in designing contexts that advance societal welfare and sustainable prosperity. Modeling the full realm of social decision-making promises nudges to steer civic virtues and motivate citizens to contribute to common goals.

The results can be considered transferable to other situations and social environments. Future research on social status losses as drivers of ethicality in the corporate domain appears interesting. Further, age variant differences in social group membership relevance could be additional research areas evolving out of these preliminary results. The results could also be transferred into other social domains, such as community service, ethical consumption, and charity giving. All these extensions would help solidify the case of a connection between social status and ethicality as well as social responsibility arising out of social status loss experiences.

Exploring ethicality in the light of social status drops is novel, and the results lead to very many different applications in the public and private sector domains. At current, we do not have any further information whether this effect is more likely to be driven by a rational calculus or rather an emotional whim. Although we find social status losses to be answered with ethicality, we do not know if ethicality spikes in the eye of social status losses are caused by a more future-oriented rational calculus or more likely to be elicited by an unconscious wish to release from a feeling of unworthiness in order to compensate for unpleasant past social status losses. Unanswered remains in this finding whether social status-related ethicality derives from a future-oriented cognitive strategy or a more subconscious emotional compensation mechanism in the aftermath of social status drops. Ethicality in the light of social status losses could be a rational calculus to gain a competitive advantage. In the wake of social status losses, individuals may strategically use pro-social acts to enhance their status by egoistic altruism (Becker 1976). At the same time, social status losses could also be accompanied by feelings of unworthiness and ethicality be a means to compensate for hurtful losses. As such, ethicality may serve as a way to relieve from a feeling of unworthiness in the wake of former social status drops.

Conclusion

Future follow-up studies may therefore investigate if social status loss-related ethicality is more likely to be associated with a rational profit maximization calculus or emotional loss compensation act. Exploring if social status-driven ethicality is more of a cognitive-rational strategy than subconscious, emotional social status loss compensation will help understanding the underlying mechanisms of social identity and societal responsibility.

Future planned research could add laboratory experiments to back our findings in the field. Finding similar effects in a controlled environment, for instance, by having students recycle under different conditions as they leave an experimental session or partake in some other type of ethical behavior, could back the hereby presented results. Extrapolating the results onto the organizational level would allow professionals to develop practical remedies on how to strengthen their inner moral muscle in a work setting. The distinct sample provides a peculiar snapshot of elite education, which leaves the generalizability of the results onto other religious groups, gender, age, and classes an open research question. On the micro level, it would also be interesting to see how the actual social environment plays a role and if the influence of social contexts changes throughout life. Additional research projects could address intercultural nuances of social status membership effects. Future investigations about different ethical dilemmas could provide additional practical applications of social status as a driver of ethicality following the greater goal of finding strategies on how to ensure a pro-social and sustainable humankind.

Cross-References

▶ Ecopreneurship for Sustainable Development
▶ Environmental Intrapreneurship for Engaged Sustainability
▶ Environmental Stewardship

References

Adams, J. S. (1965). Inequity in social exchange. In K. Berkowitz (Ed.), *Advances in experimental social psychology* (pp. 267–299). New York: Academic Press.

Adler, N. E., Epel, E. S., Castellazzo, G., & Ickovics, J. R. (2000). Relationship of subjective and objective social status with psychological and physiological functioning: Preliminary data in healthy, white women. *Health Psychology, 19*(6), 586–592.

Axelrod, R. (1984). *The evolution of cooperation*. New York: Basic Books.

Bales, R. F. (1951). Channels of communication in small groups. *American Sociological Review, 16*, 461–468.

Ball, S. B., & Eckel, C. C. (1996). Buying status: Experimental evidence on status in negotiation. *Psychology & Marketing, 13*(4), 381–405.

Bazerman, M. H., & Chugh, D. (2005). Bounded awareness: Focusing failures in negotiation. In L. Thompson (Ed.), *Frontiers of social psychology: Negotiation* (pp. 7–26). New York: Psychological Press.

Bazerman, M. H., & Moore, D. A. (2008). *Judgment in managerial decision-making*. New York: Wiley.

Bazerman, M. H., & Tenbrunsel, A. E. (2011). *Blind spots: Why we fail to do what is right and what to do about it*. Princeton: Princeton University Press.

Becker, G. (1976). Altruism, egoism, and genetic fitness: Economics and sociobiology. *Journal of Economic Literature, 14*, 817–826.

Berger, J. M., & Zelditch, M. J. (1985). *Status, rewards, and influence*. San Francisco: Jossey Bass.

Bird, C. (2004). Status, identity, and respect. *Political Theory, 32*(2), 207–232.

Bourdieu, P. (1979/1984). *Distinction: A social critique of the judgement of taste*. London: Routledge.

Chance, E. (1967). Group psychotherapy in community mental health programs. *American Journal of Orthopsychiatry, 37*(5), 920–925.

Chernev, A. (2009). *Goal orientation and consumer preference for the status quo*. Retrievable at http://www.ssrn.com/.

Cole, H., Mailath, G., & Postlewaite, A. (1992). Social norms, savings behavior, and growth. *Journal of Political Economy, 100*(6), 1092–1125.

Coleman, J. S. (1990). *Foundations of social theory*. Cambridge: Harvard University Press.

Currim, I. S., & Sarin, R. K. (1989). Prospect versus utility. *Management Science, 35*(1), 22–41.

Dhiman, S. (forthcoming). To eat or not to eat meat: Striking at the root of global warming! In S. Dhiman & J. Marques (Eds.), *Handbook of engaged sustainability*. Springer.

DiTella, R., Haisken-De New, J., & MacCulloch, R. (2001). *Happiness adaptation to income and to status in an individual panel*. Working paper, Harvard Business School.

Duesenberry, J. S. (1949). *Income, saving and the theory of consumer behavior*. Cambridge: Harvard University Press.

Easterlin, R. A. (1974). *Does economic growth improve the human lot? Some empirical evidence*. New York: Academic Press.

Falkenstein, E. (2006). *Why risk and return are uncorrelated: A relative status approach*. Working paper, Eden Prairie, Telluride Asset Management.

Fan, E. T., & Gruenfeld, D. H. (1998). When needs outweigh desires: The effects of resource interdependence and reward interdependence on group problem solving. *Basic and Applied Social Psychology, 20*, 45–56.

Festinger, L. (1954). A theory of social comparison processes. *Human Relations, 7*, 117–140.

Fox, S., & Dayan, K. (2004). Framing and risky choice as influenced by comparison of one's achievements with others: The case of investment in the stock exchange. *Journal of Business and Psychology, 18*(3), 301–321.

Friedman, M. (1953). Choice, chance and the personal distribution of income. *Journal of Political Economy, 61*(4), 277–290.

Friedman, M., & Savage, L. J. (1948). The utility analysis of choices involving risk. *Journal of Political Economy, 56*(4), 279–304.

Galiani, S., & Weinschelbaum, F. (2007). *Social status and corruption*. Retrievable at http://www.idec.gr/iier/new/corruption%20conference/Social%20Status%20and%20Corruption%20-%20Federico%20Weinschelbaum.pdf.

Gil-Doménech, D., & Berbegal-Mirabent, J. (forthcoming). People + planet + profit: Training sustainable entrepreneurs at the university level. In S. Dhiman & J. Marques (Eds.), *Handbook of engaged sustainability*. Springer.

Güth, W., & Tietz, R. (1990). Ultimatum bargaining behavior: A survey and comparison of experimental results. *Journal of Economic Psychology, 11*, 417–449.

Harbaugh, R. (2006). *Prospect theory or skill signaling?* Retrievable at http://papers.ssrn.com/sol3/papers.cfm?abstract_id=311409.

Hong, K., & Bohnet, I. (2004). *Status and distrust: The relevance of inequality and betrayal aversion*. RWP04–041. Working paper, Harvard Kennedy School.

Hopkins, E., & Kornienko, T. (2004). Running to keep the same place: Consumer choice as a game of status. *American Economic Review, 94*(4), 1085–1107.

Huberman, B. A., Loch, C., & Önçüler, A. (2004). Status as a valued resource. *Social Psychology Quarterly, 6*(1), 103–114.

Janis, I. L. (1982). *Groupthink*. Boston: Houghton Mifflin.

Jervis, R. (1992). Political implications of loss aversion. *Political Psychology, 13*(2), 187–204.

Kahneman, D., Knetsch, J. L., & Thaler, R. H. (1991). The endowment effect, loss aversion, and status quo bias. *Journal of Economic Perspectives, 5*(12), 1325–1347.

Kahneman, D., & Tversky, A. (1979). Prospect theory: An analysis of decision under risk. *Econometrica, 47*(2), 263–292.

Kahneman, D., & Tversky, A. (1992). Advances in prospect theory: Cumulative representation of uncertainty. *Journal of Risk and Uncertainty, 5*, 297–324.

Kahneman, D., & Tversky, A. (2000). *Choices, values, and frames*. Cambridge: Cambridge University Press.

Keyes, C. W. (1966). *Cicero, M. T. De republica, De legibus*. The Loeb Classical Library. Cambridge: Harvard University Press.

Levy, J. S. (1997). Prospect theory, rational choice, and international relations. *International Studies Quarterly, 41*(1), 87–112.

Levy, J. S. (2003). Applications of prospect theory to political science. *Decision Theory, 135*(2), 215–241.

Loch, C. H., Huberman, B. A., & Stout, S. K. (2000). Status competition and performance in work groups. *Journal of Economic Behavior and Organization, 43*, 35–55.

Loewenstein, G. F., Thompson, L., & Bazerman, M. H. (1989). Social utility and decision-making in interpersonal contexts. *Interpersonal Relations and Group Processes, 57*(3), 426–441.

Marques, J. (forthcoming). Moving forward with social responsibility: Shifting gears from why to how. In S. Dhiman & J. Marques (Eds.), *Handbook of engaged sustainability*. Springer.

Maslow, A. H. (1943). A theory of human motivation. *Psychological Review, 50*, 370–396.

Mazur, A. (1973). A cross-species comparison of status in small established groups. *American Sociological Review, 38*(5), 513–530.

Mazur, A., & Lamb, T. A. (1980). Testosterone, status and mood in human males. *Hormones and Behavior, 14*(3), 236–246.

McDermott, R., Fowler, J. H., & Smirnov, O. (2008). On the evolutionary origin of prospect theory preferences. *Journal of Politics, 70*, 335–350.

Meeker, B. F., & Weitzel-O'Neill, P. A. (1977). Sex roles and interpersonal behavior in task-oriented groups. *American Sociological Review, 42*, 91–105.

Nowak, M. A., & Sigmund, K. (1998). Evolution of indirect reciprocity by image scoring. *Nature, 393*, 573–577.

Postlewaite, A. (1998). The social basis of interdependent preferences. *European Economic Review, 42*(3–5), 779–800.

Puaschunder, J. M. (2010). *On corporate and financial social responsibility*. Dissertation, University of Vienna.

Puaschunder, J. M. (2015). *Meritocracy and intergenerational mobility*. The Worldly Philosopher Blog. Schwartz Center for Economic Policy Analysis. Retrievable at http://www.economicpolicyresearch.org/index.php/the-worldly-philosopher/1545-meritocracy-builds-equality.

Puaschunder, J. M. (2016). The beauty of ivy: When inequality meets equality. *Global Journal of Management and Business Research: Economics and Commerce, 16*(3), 1–11.

Ramananan, R. (forthcoming). Responsible investing and corporate social responsibility for engaged sustainability: Managing pitfalls of economics without equity. In S. Dhiman & J. Marques (Eds.), *Handbook of engaged sustainability*. Springer.

Ridgeway, C. L., Berger, J., & Smith, L. R. (1985). Nonverbal cues and status: An expectation states approach. *American Journal of Sociology, 90*(5), 955–978.

Ridgeway, C. L., & Walker, H. A. (1995). Status structures. In K. Cook, G. Fine, & J. House (Eds.), *Sociological perspectives on social psychology* (pp. 281–310). New York: Allyn & Bacon.

Robson, A. J. (1992). Status, the distribution of wealth, private and social attitudes to risk. *Econometrica, 60*(4), 837–857.

Roussanov, N. (2009). *Diversification and its discontents: Idiosyncratic and entrepreneurial risk in the quest for social status*. Working paper, Wharton School, University of Pennsylvania.

Savin-Williams, R. C. (1979). Dominance hierarchies in groups of early adolescents. *Child Development, 50*(4), 923–935.

Schkade, D. A., & Kahneman, D. (1998). Does living in California make people happy? A focusing illusion in judgments of life satisfaction. *Psychological Science, 9*(5), 340–346.

Sidanius, J., & Pratto, F. (1999). *Social dominance: An intergroup theory of social hierarchy and oppression*. New York: Cambridge University Press.

Siegel, J. J. (2002). *Stocks for the long run*. New York: McGraw Hill.

Tajfel, H. (1978). *Differentiation between social groups: Studies in the social psychology of intergroup relations*. London: Academic Press.

Tajfel, H., & Turner, J. C. (1979). An integrative theory of intergroup conflict. In W. G. Austin & S. Worchel (Eds.), *The social psychology of intergroup relations* (pp. 94–109). Monterey: Brooks-Cole.

Tajfel, H., & Turner, J. C. (1986). The social identity theory of inter-group behavior. In S. Worchel & L. W. Austin (Eds.), *Psychology of intergroup relations* (pp. 7–24). Chicago: Nelson-Hall.

Tenbrunsel, A. E., & Messick, D. M. (2004). Ethical fading: The role of self-deception in unethical behavior. *Social Justice Research, 17*(2), 223–236.

Thaler, R. H. (1999). Mental accounting matters. *Journal of Behavioral Decision-making, 12*, 183–206.

Thaler, R. H., & Sunstein, C. S. (2008). *Nudge: Improving decision-making about health, wealth and happiness*. New Haven: Yale University Press.

Tiedens, L. Z. (2000). Powerful emotions: The vicious cycle of social status positions and emotions. In N. M. Ashkanasy & C. E. Haertel (Eds.), *Emotions in the workplace: Research, theory, and practice* (pp. 72–81). Westport: Greenwood.

Tooby, J., & Cosmides, L. (1990). The past explains the present: Emotional adaptations and the structure of ancestral environments. *Ethology and Sociobiology, 11*, 375–424.

Tversky, A., & Kahneman, D. (1991). Loss aversion in riskless choice: A reference dependent model. *Quarterly Journal of Economics, 41*, 1039–1041.

Ulluwishewa, R. (forthcoming). Education in human values: Planting the seed of sustainability in young minds. In S. Dhiman & J. Marques (Eds.), *Handbook of engaged sustainability*. Springer.

Weber, M. (1946/2009). *Essays in sociology*. Oxford: Oxford University Press.

Weber, M. (1978). *Economy and society.* Berkeley: University of California Press.

Wedekind, C. (1998). Enhanced: Give and ye shall be recognized. *Science, 280*, 2070–2071.

Weisband, S. P., Schneider, S. K., & Connolly, T. (1995). Computer-mediated communication and social information: Status salience and status differences. *Academy of Management Journal, 38*(4), 1124–1151.

Young, I. M. (2011). *Responsibility for justice.* Oxford: Oxford University Press.

Zimbardo, Ph. (2011a). *My journey from evil to heroism.* Speech delivered at Webster University. 31 May 2011. Vienna.

Zimbardo, Ph. (2011b). *Evil no! Heroes yes!* Speech delivered at Harvard Law School. 26 Oct 2011. Cambridge.

Oceans and Impasses of "Sustainable Development"

Will McConnell

Contents

Abstract

One of the most powerful concepts to emerge in the multiple discursive practices that together form the conceptual and politico-pragmatic terrain of "sustainable development" is "environmental sustainability." In the emerging discourse of environmental awareness, this concept reveals its polyvalence as a speech act largely through the absences of meaning. The history of the term as now understood has its foundations in the Brundtland Report (Our common future, 1987. http://www.un-documents.net/our-common-future.pdf); despite the strength of this document in founding a global awareness, the definitions offered therein have created a foundation for fundamentally opposed drives toward understanding environmental sustainability, largely due to the privileging of neoclassical economics in land-based models for sustainable development. From ocean-based perspectives, the privileging of land-based, economically-driven modeling of sustainable development produces weak forms of "environmental" sustainability even in the best examples of current analytic approaches; the central concern of the discourse of environmental sustainability currently is to disarticulate the current paradigm through the tensions inherent in the discursive fields across which sustainability discourse produces its hegemonic (economic)

W. McConnell (✉)
Woodbury University, Burbank, CA, USA
e-mail: will.mcconnell@woodbury.edu

S. Dhiman, J. Marques (eds.), *Handbook of Engaged Sustainability*,
https://doi.org/10.1007/978-3-319-71312-0_47

properties. Major drivers of environmental pressures on the ocean originate outside of ocean systems – on land. In its "unsettling" nature, however, its divergence from land-based temporalities and spatial properties, the ocean offers multiple points at which the discourse of (land-based) sustainability collapses within its own tensions to reveal horizons of representation that offer a glimpse into the full complexity, and excitement, of reenvisioning an entire discourse of the (contested) future.

Keywords
Ecosystems · Sustainable development · Environmental sustainability · Brundtland Report · Neoclassical economics

Introduction

One of the most powerful concepts to emerge in the multiple discursive practices that together form the conceptual and politico-pragmatic terrain of "sustainability" is "environmental sustainability." In the emerging discourse of environmental awareness, this concept reveals its polyvalence as a speech act largely through the absences of meaning. The term engenders a seemingly calculated series of imprecisions: in its attempt to marry an emerging environmental awareness with traditional models of economic growth, the concept contains within it a powerful concatenation of historically layered meanings which continue to structure impasses to action developed through a "best practices" approach to re-situating a relation between "human" and "nonhuman," "animate" and "inanimate," "subject" and "object" of production, and discursive field or horizon of meaning and the relations – conceptual and practical or operable – that come to define a more "sustainable" approach to the dynamics of human-nonhuman interaction.

The concept privileges a passivizing of nature while at the same time places under erasure the perpetuation of economic growth as predicated on a (naturalized) consumerist ethos. In the language game of environmental sustainability, language denoting acts predicated on an ethos of "environmental development," for example, if it appears at all, appears in the discourse as the negative – literally, actively negative-ized – dimension of economic productivity. More often than not, the operability of "development" is unidirectional, presenting an attendant, passivized language of "conservation" and "protection," implying that the best strategy – perhaps, the only strategy – is to be "designed" in a shoring-up of "whatever remains" of nature after human activity has produced mass destruction of ecosystems, that is, after a species, the human species, has effectively destroyed the life systems essential for its own survival. "The environment," within the discourse, is barely visible and, when visible, is widely disfigured.

Similarly, this set of linguistic strategies for identifying "solutions" to the effects of human-based destruction of nature occupies the language of "best practices" throughout key models designed for measurement of environmental damage and the economic consequences of that damage. The models can be seen, then, as the negation of nature's visibility as an analytic object – with a concomitant assertion that the perspective governing the discourse is one of cost measurement as solutions – as itself environmental sustainability.

Thus, this discourse rendering nature throughout as a kind of absence – a presence that is negated – in actuality renders "nature" into invisibility. In such a discourse, progress toward "sustainability" is constructed as, at best, the containment of destruction and the "managed" transformation of mountains, oceans, and organisms into the stasis of systems of waste in the production and transaction of consumer object. Uncoding "nature" as the human or built environment, produces a conceptualization of nature and the natural world as the object not merely of protection, but also of design – design of a more symbiotic relation with the entirety of earth's processes, living, and nonliving. Such a conceptualization would re-interpret human waste streams as productive rather than destructive. This form of becoming, of bringing nature into the human world as a "new" object – of study, engagement, and practice – is only now emerging in the current, globalized discourse of environmental sustainability. However, such directions for interpretation, analysis, and modeling are crucial to develop *as* the discourse itself. As a foundation for new forms of modeling future scenarios – a hallmark of the discourse – working actively to produce this shift in the discourse of sustainable development would work more concertedly toward the reversal rather than the containment of damage.

Given these foundational tensions in the conceptualization of orienting models of thought in sustainability discourse, it is not surprising that, despite widely agreed upon and scientifically based, if dire predictions of social and economic collapse due to the anthropogenic nature of issues arising from global warming, the discursive practices of "globalization" have not produced any transnational grounding in "best practices" of envrionmental sustainability globally. Agreements, policies, and treaties emerge and proliferate; goals based on a human equality surge and recede; local action is linked to national, international, and planetary action; bewilderingly, however, significant reductions in environmental damage do not result in the reframing gestures inherent in "sustainable development" or in environmental sustainability discourse more generally. Baker and Eckerberg (2010) note that one of the difficulties inherent in shifting behaviors remains the "profound lack of knowledge about the complex and dynamic interactions between society, economic development, technology and nature."

Nowhere is the perceptual elision supported by, contained in, "sustainable development" more apparent than in comparisons of land-based and ocean-based modeling scenarios of environmental impacts in such discursive locations as the emerging (behavioral) science of ocean management, the theory and practices of "sustainable development" articulated as "marine protected areas," and, more generally speaking, in the unsettling nature of the largely undertheorized space/temporality of "the ocean." All the more remarkable is the relative newness of ocean research, about which the FGIM authors observed: "humans have been using the ocean for millennia, [but] it is only in the past 120 years or so that serious exploration of the seven tenths of the planet covered by the sea has been in progress" (42). In its sheer impenetrability as an object readily available to human thought, the ocean reveals the practico-operative impasses of "sustainable development" or "environmental sustainability" as a means toward articulating "best practices" in a globalized conceptual framework. Farther away still in the linguistic-conceptual horizon of intelligibility is the development of a framework of best practices in the discursive sphere of a future-oriented relation to "the ocean"– indeed, in a relation to "nature" itself, human as well as nonhuman, animate, and inanimate objects of (human) consumption. An encounter with the ocean reveals not only significant limitations in current "environmental sustainability" theory and behavior(s) but also questions current operations of "markets," global and local, as well as a host of other concepts held in place by the economics of marketplace dynamics in human activity: "work," "leisure," "production," "consumption," surplus value," "wealth creation," etc. A comparison of the multidisciplinary linguistic field through which ocean assessment is practiced and the language in which land-based valuation is understood can expose productive tensions that reveal fault lines in the discourse of environmental sustainability. Such tensions can lead to a new vision of (human) relations to both the land and the ocean – a vision not merely to reduce environmental damage, or mitigate and adapt to the results of thinking damage a priori of economics, but also to reverse the wide-scale production of destructive patterns of consumption – to mobilize a form of global awareness from within the construct that environmental damage is not merely a consequence but a preconception of production itself.

If such a critical encounter with the past/present of our own linguistic edifice reveals ideologically encoded tensions inherent in democratic policies, concepts, and practices (of human-to-human and human-to-environment sustainabilities), such an encounter with the present of our existing linguistic field(s) of environmental sustainability can also lead to a reenvisioning of our relation to the seas and other material processes that sustain us. Scientists across the planet have made clear, in evidence-based findings, that the human journey, in order to continue, must include a rapid transformation of thought, action, and behaviors. What was once unthinkable is now a given of daily life: humans are destroying the very systems that sustain life, all life, in the acts that are understood as the pursuit of "the good life." Despite regularly occurring predictions by scientists across the globe – perhaps due to the enervation that comes from the regularities of witnessing anthropocentric production of extinction event conditions – the destruction of the earth continues to accelerate. Our ability to understand environmental discourse itself *as action that sustains* rather than resumption that measures is crucial. The first step to understanding clearly the pathways from existing impasses out of the linguistic, politico-social, and cultural dimensions of environmental damage is to reevaluate what we consider "understanding" the earth (and, by necessary inclusion, ourselves) in the first instance. If our enlightenment values have led us to conceptualize the present world "as is," then the discursive fields of those values must be deconstructed, such that a new perspective on can be made actionable, a paradigm shift in seeing into the production of "the environment" as ready-to-hand damage must be understood as actionable across the globe. Ideas and ideals that continue to inhere in the discursive fields of "sustainability" must be unseated in order to move quickly beyond the current impasses in our understanding of the enlightenment project of democratizing social, material, and environmental processes. The current repository of work toward global democratizing principles is the United Nations (UN) program to envision the building of global equalities from within enlightenment values. More specifically, for "environmental sustainability," the UN-driven Brundtland Report (1987) can situate the current tensions that inhere in environmental sustainability discourse.

Sustainable Development and the Brundtland Report (1987)

In the United States, the late 1960s and 1970s were marked by an intensification of concern about pollution. At the same time, researchers realized that this concern was marked by a complex, if predominantly misunderstood, interplay between humankind, global resources, and social and physical environments; part of this realization was that humans' interactions with "nature" produced a highly contextualized experience of the natural world: no one experience of nature is likely to exhaust the range of relations possible between humans and the natural world. The implications of this insight about the social and political context in which nature could be understood gave rise, in an emerging global context, to the concept and practices of "sustainable development"; although the concept had been in use, to a limited degree, in the 1950s and 1960s, the current set of terminologies and conceptual relations across the discourse of "sustainable development" were presented more coherently, initially, by the International Union for the Conservation of Nature and Natural Resources (IUCN) in 1980.

Seven years later, the term was expanded upon by the United Nations Environment Programme, in the Brundtland Report, commonly known as *Our Common Future* (1987). The term has been in wide circulation since then, its seeming ubiquity all but obscuring its beginnings as a means to articulate a different relation between "the environment" and "the economy." The Brundtland Report presents a case for the integration of environmental policies and development strategies, in a direct critique of the mid-1980s operative structuring of an opposition between "the environment" and "the economy." From this perspective, the Brundtland Report can be understood as an early critique of the thinking that environmental damage must be an inevitable consequence of economic growth; further, the authors of the Brundtland Report subtly challenged the "rich versus poor" paradigmatic thinking that attends the accumulation of, acceleration of, "wealth" in global economic processes. As the authors of the Brundtland Report framed the idea (Brundtland 1987), "[a]fter a decade and a half of a standstill or even deterioration in global co-operation...the time has come for higher expectations, for common goals pursued together...Environmental degradation, first seen as mainly a problem of the rich nations and a side effect of industrial wealth, has become a survival issue for developing nations." The Brundtland Report authors then gestured toward reframing the call for sociopolitical change through repositioning the environment in what they suggested would be seen, from a future perspective, as a historicized political and environmental stasis: "the 'environment' is where we all live; and 'development' is what we all do in attempting to improve our lot within that abode. The two are inseparable." The problem with this formulation is that the abode of the environment is always already ready-to-hand as a resource toward improvement. To deny this principle is to deny the inalienable right of all citizens of the earth to acquire the very materiality of excess production: to convert the inert in their surroundings into "wealth."

This linguistic formulation is an attempt to situate (economic) development on an equal conceptual footing with "the environment." Further, the authors of the report were clear in their call for social change and prescient in their recognition that specific elements of global society must change: "Many of the development paths of the industrialized nations are clearly unsustainable." The central problem in articulating how (and when) a global equality might appear, of course, is that that nascent, perhaps emergent, economy of (global) equality requires considerable contributions, and internal programs of socioeconomic restructuring, from and within the very countries whose "globalized" economic systems have produced worldwide social and economic inequities. The Brundtland Report authors, aware of the potentially revolutionary context of a "new" paradigm for thinking the organization of meaning through nationalist, humanist, and enlightenment paradigms, as well as in human-environment relations, attempted to minimize a direct challenge to the existing power relations inherent in the globalization of both the environment and the economy across the 1980s and 1990s.

One way to understand the performative dimensions of the Brundtland Report's application of language is suggested by the definition of the key term "paradigm" in Thomas Kuhn's *Structure of Scientific Revolutions* (1970). The Brundtland Report language gestured toward Kuhn's precise definition of change via "paradigm" shift, suggesting the same historical processes as occurred in scientific revolutions would begin to materialize in global social justice discourse via the (sub) or tertiary discourse of environmental sustainability. In this sense, the Brundtland Report could be contextualized as a future-oriented interplay of paradigmatic shifts in language; that is, future practitioners, putting the "new" language into play, would then make emerge, as a process of future history, a different social, political, and environmental understanding and set of practices. As Kuhn articulated his key concept (for identifying how change occurs in specific science research contexts), a paradigm shift must meet the following two conditions: the emergent conceptual apparatus must be "sufficiently unprecedented to attract an enduring group of adherents away from competing modes of...activity" and "sufficiently open-ended to leave all sorts of problems for the redefined group of practitioners to resolve" (1970, 10). Thus, "history" as environmentally oriented becomes, explicitly, a formation of processes through democratized linguistic practices, speech acts on the world stage that form the world as a (future-oriented) process of collective rebuilding.

Interestingly, in Kuhn's explication of his thinking on paradigmatic forms of change, "[a]chievements that share these two characteristics" [quoted above] also share an orientation beyond problem-solving, and conform to rules (in effect, limitations of effectivity) that map or trace processes that become 'solution-objects': "if it is to classify as a puzzle, a problem must be characterized by more than an assured solution. There must also be rules that limit both the nature of the acceptable solutions and the steps by which they are to be obtained" (Kuhn 39). Thus, the Brundtland Report would attempt to deconstruct enough of the relation of key terminology by which the existing problems produced in the current (linguistic) system of representation could remain intact while at the same time reposition the existing language's axiomatic properties enough to destabilize the relation subtending underlying conceptual structures that produce key concepts in the discourse of (human) equality. That is, while the meaning of these key words in the discursive field remained intact enough so as not to disrupt the formulation of the now global problems of (human-to-human) inequality, the eradication of these structural/systemic problems is presented as a puzzle for future generations to reformulate through a repositioning of key relations across concepts. In the Brundtland Report, the environment becomes an *object (petit) a*: seemingly at the new center of the desire for a future-oriented discourse which could enable a new paradigm, the environment becomes a "partial object," a "transitional object." The "environment," paradoxically, becomes structured, consistently throughout the report as throughout the wider environmental (global) sustainability discourse, as a remnant: "environment" is that which becomes left behind in the resumption of reified forms of economic theory that infuse the language of (human) equality within the prevailing global sensibility of constructed and rigorously maintained disproportions in the economic and social development of a "global" equality. Thus, in the report, "environment" functions as Slavoj Žižek's *petit objet a*, in the socio-psychic operation of global environmental awareness: at once what cannot be accounted for in the system of meaning and signification and that which simultaneously produces the entire system of language and meaning, locating and guaranteeing the chain of meanings within it. As the *petit objet a* of this discursive field, the constellation of intelligibility in which all meaning occurs, "environmental sustainability," as a global/environmental understanding, functions as "the lack, the remainder of the Real that sets in motion the symbolic movement of interpretation, the hole at the center of the symbolic order, the mere appearance of some secret to be explained, interpreted" (Žižek 1996).

The most significant example of this operation of language and representation in the Brundtland Report is the anchoring gesture of oppositional thinking in the current paradigm: "economy" versus "environment." No longer should the "environment" be understood as in "natural" opposition to "the economy," but rather, the long operative opposition between "economy" and "environment" can now be historicized as a partial understanding of often misunderstood complexities requiring a paradigmatic shift. Thus, the document articulated not a new paradigm, but the grounds upon which the new paradigm would emerge, in concert with a more rigorous approach to democratic principles of participation in articulating global econo-environmental relations across existing national interests. The report attempted to articulate a deeper structure by which new "rules" for transforming the deep structures of existing social inequities could be mapped out as nation states moved toward a shared vision of global econo-environmental discourse. Through such a perspective, much of the language of the Brundtland Report could avoid proscriptive declarations and, instead, focus on the construction of a linguistic and practical mitigation in the present, positioning the construction of a descriptive critical language as an emergent paradigm, such that, in the existing situation of "first world" relations to "third and second world" countries, economic and political threats to the status quo could be mitigated in a language denuded of direct challenge to the conceptual, linguistic, and practical infrastructures of inequality represented by the current globalization of production in the "global economy." Balancing the creation of a new horizon of intelligibility for a global equality with the need to maintain sociopolitical relations, the report also highlighted "interlocking crises" as a paradigm for change; as the authors suggested, the "older" paradigm that limited the understanding of complex human relations across concepts and practices had begun to reframe itself: "Until recently, the planet was a large world in which human activities and their effects were neatly compartmentalized within nations, within sectors (energy, agriculture, trade), and within broad areas of concern (environment, economics, social). These compartments have begun to dissolve."

If, in the Brundtland Report, the "compartmentalization" of "economy" and "environment" were in need of rethinking, two additional key concepts for articulating the compartmentalized understanding of the earth also required reworking: "needs" and "limitations." In the Brundtland Report, "needs" carry the sociopolitical weight of the "needs of the poor," which the authors argue consistently must be prioritized over other forms of need. In part, this argument was developed around the assumption that poorer communities represented not only the most vulnerable in democratic (and other) forms of society but also that these communities were the most likely to be negatively impacted by the effects or impacts, globally and locally, of climate change due to global warming. In this articulation of interactive concepts of "need" and "limitations," these communities often are interpreted as both a cause and a consequence of unsustainable behavior.

If the rethinking of "needs" is relatively straightforward, the concept of "limitations," as presented in the Brundtland Report, is remarkably nuanced, if under-articulated, referring neither to humans nor to the environment but to the complex interchange of these via the intermediary of human technological development and deployment of emerging technologies across an array of aging technologies. As suggested in careful reframing of "limitations" in the Brundtland Report, Baker et al. (2010) note that one of the difficulties inherent in shifting human behaviors remains the "profound lack of knowledge about the complex and dynamic interactions between society, economic development, technology and nature." The Brundtland Report was an (early) attempt to reframe this set of concepts – and the often occluded relationships constructed across them – as a point of embarkation for an international discussion that not only repositioned "human" and "material world" relations but also offered pathways for reorganizing the relation between economic development and human (democratic) formulations of (global) equality.

Noting that the "mainspring of economic growth is new technology," the Brundtland authors focused not on the ability of technology to mitigate environmental degradation but instead gestured toward the net impact of technology in a consumerist paradigm. "[W]hile this technology offers the potential for slowing the dangerously rapid consumption of finite resources, it also entails high risks, including new forms of pollution and the introduction to the planet of new variations of life forms that could change evolutionary pathways. Meanwhile, the industries most heavily reliant on environmental resources and most heavily polluting are growing most rapidly in the developing world…" Thus, the report subtly acknowledged that "technology" is an unreliable, often unpredictable, source of environmental protection and/or conservation. The authors stopped short of stating, or even implying, that the earth itself was finite – and as a consequence, the authors also refrained from questioning deeply what remains the governing economic paradigm of our time: that economic growth itself, the growth of production of excess "wealth" *as* the economy, was unlimited. Instead, the authors clearly shied away from implying that this component of the economic model of growth is itself unsustainable: "The concept of sustainable development does imply limits – not absolute limits but limitations imposed by the present state of technology and social organization on environmental resources and by the ability of the biosphere to absorb the effects of human activities."

Social Effects and Market Behaviors

Not mentioned in the report is the seemingly contradictory impact of technologies that reduce pollution at point of production, only to contribute higher levels of pollution via purchasing behavior in the market, further along the chain of a product's life cycle. For example, the paradoxical consequence of reducing the volume of plastic in each single-use water bottle has led not to a decrease in plastic entering oceans but to an increase: companies selling water in single-use plastic bottles that decrease the volume of plastic per bottle earn the reputation of "sustainable behavior" in the marketplace; as market consumption increases due, in part, to this reputation, the company produces and distributes more plastic overall, thereby increasing rather than decreasing the levels of plastic produced, sold, and discarded. Since the mid-1980s, much of this plastic has found its way into the ocean, a seemingly intractable problem with to which this essay will return. But this is a problem only because we have been unwilling and to date unable to develop our capacity for seeing the mutually supporting edifice, the attitudinal-behavioral links, between our paradigms for economic growth and the consumerist ethos that maintains ecological degradation at unprecedented levels historically.

If, in the model espoused by the Brundtland Report, the global economy could continue to be understood as ever-expanding and societies' growth rates were included in that expansionist rhetoric, as the conceptual model of "sustainable development" developed across the 1980s and 1990s, the earth was increasingly recognized as being the "limitation" to this model and rhetoric – the governing concept for understanding the human place in and on the earth. The two concepts of "need" and "limitations," taken together as found in the Brundtland Report, articulate an intragenerational vision of social justice and interpersonal equality, such that resources available to one generation may not be exhausted in and by the model of "economic growth," simply because this use of resources might be technologically feasible or socially desirable in the present.

In the time that has passed since the Brundtland Report, there has emerged no agreement on the meaning and valence of "sustainable development." As early as 1989, analyses by Pezzey (1989) and others (i.e., the Pearce Report (1989)) collated pages of definitions for sustainable development, signaling the ambiguity and inconsistency with which the term came to be applied to the problems caused, and societies built, by producing environmental damage all but ensured in the status quo of deploying "sustainable development" as a strategy for understanding the uneasy, untenable links between environmental sustainability and economic growth. Thus, Baker et al. (2012) called for a repositioning of "sustainable development" as a political construct, capable of illuminating the stasis between attitudes and behavior. "Viewing sustainable development as a social and political construct makes it possible to move beyond the search for a unitary and precise definition and to focus instead on the objectives underlying the original formulation of each of the two concepts 'sustainable' and 'development'" (6). Throughout these debates that followed the Brundtland Report, researchers have noted that "sustainable development" is a concept, or set of concepts and practices that are similar to overarching, largely heuristic abstractions such as "democracy," "liberty," and "social justice" (Lafferty 1995; Jacobs 1999; O'Riordan 1985).

But while these latter concepts are part of a democratic process for coming to terms with "best practices" in socioeconomic arguments and the reordering of global resources that accompany production and consumption, they rarely remain clearly defined practices in themselves. This has been equally true of "sustainable development" – arguably, with the notable exception that this concept has a reified base or structure that subtends, polices, and directs much of the conceptual terrain of environmental sustainability discourse. Positioned in this way, the application of "sustainable development" as a theoretical concept then ushered into the discourse of sustainability the application of case studies for multiple practices and policies taking place globally, as approaches to, rather than the attainment of, "sustainable development." The polyvalence in meaning produces multiple, often contradictory, definitions, which produces fissures in relations between actions, processes, and, ultimately, often facile models and measurements of "sustainable development." Paradoxically, metrics of measurement in the discourse often are more "sustainable" than the results themselves. If this linguistic multiplicity, this particular form of linguistic openness, has necessitated research approaches governed by case study methodologies, it has also embedded a troubling fragmentation of what specific achievable targets might be designed for either reducing (or, more importantly, reversing) widespread consumption and production behaviors that, somewhat perversely, reproduce everyday practices of environmental damage. Thus, to date, from nearly any set of concepts for measuring "progress" toward a more clearly defined definition of "environmental sustainability" – i.e., a definition that articulates not merely cessation or mitigations of damage but instead produces a reversal of economic damage reversal – the perpetuation of practices of wide-scale, global damage constitutes the collective vision that is "environmental sustainability."

Nationalism, exercised on the global scale, appears only to exacerbate this form of discursive fragmentation.

As Griffin (2013) and Dryzek (2005) assert, this polyvalence in meaning for the term has produced a proliferation of definitions operating along specific trajectories of meaning and concomitant practices; for example, sustainable development "is imbued with notions of economic growth and managerial techniques for governing" (Griffin 2013, 35). As Dryzek (2005) concludes, sustainable development "involves a rhetoric of reassurance. We can have it all: economic growth, environmental conservation, social justice...no painful changes are necessary" (205). As Griffin (2013) notes, in the European Union (EU) context, "official EU strategies such as the Sixth Environmental Action Plan ostensibly position the environment 'center stage' in the discourse. But in reality, things are different. In *practice*, economic development policy is likely to continue putting a premium on old-style approaches to growth based on economic expansion at all costs, especially during economic stagnation" (41). This interpretation has been borne out by multiple analyses; for example, in 2011, the European Commission itself, tasked with reporting out on the performance of existing political and social approaches, observed that "the decoupling of resource use from economic growth has not led to a decrease in overall resources use" (European Commission 2011). Like the situation of decreasing the production of plastic in single-use containers of water mentioned above, the strategy – and ethos behind the strategy – of "producing more with less" has led to a proliferation of examples in which the net impact has been an increase, rather than a decrease, in pollution and environmental degradation. As Griffin (2013) observes, "market rationality may be incompatible with inter-generational equity...although each unit of what is being produced in Europe might be more 'sustainable,' development as economic growth means that *more* of these units are being produced and consumed, thus tending to nullify any environmental gains" (42). In the European context as in the North American marketplace, "conventional sustainable development discourse perpetuates the commodification of our environment" ("Review of Utopian Themes," 2013).

As early as 1995, Escobar noted the tendency for economic terminology to supersede environmentally oriented meanings in the discourse of environmental sustainability, with what has been called "weak sustainability" (Bebbington 2000, 19). According to Bebbington and Escobar, conventional discourse forces meaning into patterns of deeply embedded economic relations, such that terms for environmental processes are reduced to specific relations within human (meaning) and economic relations, such as occurs in terminology like "natural capital" and "ecosystem services." As David Pearce (1992) noted, "What economic valuation does is to measure *human preferences* for or against changes in the state of environments. It does not 'value the environment.' Indeed, it is not clear exactly what 'valuing the environment' would mean" (7). Throughout the Brundtland Report, for example, this language reoccurs, ostensibly to direct human thought to understand the benefits to humans in understanding the many benefits that the environment supplies "naturally"; however, this language actually is integrated into the human economy of assessing or directing the "meaning" of nature into the "value" of goods and services that accrue in "enabling" the material world *as* object in human-ized systems of meaning and production.

Escobar calls this cathexis of language a "regime of representation" (1995, 10). The echo and redirection of Marxist language are no accident here, as a mode of identifying reifications of meaning and economic exchange that, even in Marx's powerful project for rethinking social relations via denuding the sociocultural meanings of unilateral economic circulation, under-represents a fundamental shortcoming in understanding the reality of human existence. Schemas of "unlimited production" in economic paradigms – the fundamental assumption of neoclassic economics – present both the failure of democratically driven reform as well as leads directly to the most significant re-democratizing project in human history. This project is the design of systems of thought, language, and exchange – economic and environmental relations – that can ensure not only the availability of environmental resources for future generations but also can ensure the survival of the human species itself. For all of its sociocultural and political power, the enlightenment project should increasingly be understood as having exhausted both the material world and the drive toward the increasing "perfection" of democratic forms of exchange and (national and personal) identify formation.

From Land to Ocean: Patterns of "Cost" and Linguistic Dynamics of Stasis

Nowhere is this pattern of ideologically encoded meanings more charged – between ongoing development of policy and practices of "response" and in the continued, large-scale and complex forces of degradation at work in the ocean. How does this seemingly counterintuitive logic, in which reductions in resource use lead directly to increases in the multiplicity of forms of pollution – play out *as* the ocean?

The answer to such a question is remarkably straightforward: the best location from which to begin to approach the ocean is the land. As Àlvarez-Romero et al. (2011), Beck (2003), Stoms et al., and others observe of conservation efforts, "[d]espite shared conceptual roots, conservation planning in terrestrial and marine realms have largely proceeded as if the ecological systems were unconnected" (382). Àlvarez-Romero et al. (2011) note that this "lack of integration" is particularly troubling due to the asymmetrical patterns of influence: "physical and ecological connections" are often not conceptualized, either in systems of (economic) or environmental interpretation. Thus, as with other areas of the ocean, areas designated as "marine protected" (MPAs) are susceptible to damage originating outside their boundaries. In part, this is due to the fact that much of the pollution in the ocean originates on land; thus, somewhat paradoxically, understanding environmental sustainability *as* the ocean begins on land. Multiple researchers have articulated the need to move quickly in this reconceptualization process; in June 2011, for example, the European Union International Programme on the State of the Ocean (IPSO) released its "State of the Oceans Report." Much like the Intergovernmental Panel on Climate Change (IPCC), the IPSO compiles and analyzes the latest scientific evidence about the condition of the ocean by using a considerable body of disciplinary approaches and research contributions. The report identified seven key concerns: among the most significant was the alarm they raised about "the speeds of many negative changes" to the ocean, which, by 2011, were already recognized as "tracking the worst-case scenarios from the IPCC and other predictions." Further, the IPSO report noted that "the magnitude of the cumulative impacts on the ocean is greater than previously understood" (State 2011).

The report had a (then) stunning conclusion: current uses of the ocean "are not sustainable," and the situation "demands change in how we view, manage, govern, and use marine ecosystems" (State 2011). Such change must begin with a paradigm shift, first and foremost, *on land*. The IPSO report's authors alluded to this in more generalized language, but the implication was clear: the need for wide-reaching and rapid changes in understanding and behavior were crucial. IPSO recommended a more holistic approach to understanding the interplay of land-based activity impacting ocean environments, one that could address "all activities that impinge marine ecosystems" (State 2011).

More recently, the same problem – land-based, or non-source, pollution *as* always already the ocean and the urgency of the call for change – was echoed in *The First Global Integrated Marine Assessment* (FGIM 2017) or, as it is also known, *The First World Ocean Assessment*. The significance of this effort is underscored by the long timeline of its conceptualization and development and also suggests the equally significant temporal dimensions of enacting human change – even when faced with having contributed to immanent conditions of ocean collapse. The global movement toward this "first" global-scale assessment had begun many years earlier, in the *United Nations Convention on the Law of the Sea* (UNCLOS 1980); UNCLOS mapped out a new language for thinking globally about ocean provision systems (the terrain of the ocean understood across human activities in legal terminology). Perhaps the most complex global agreement ever reached, the document took 9 years of negotiations until member nations signed on in 1982; then, the document came into force 12 years later, in 1994.

Although ostensibly a continuation of land-based juridico-political interpretation, this document also articulated a fundamental paradigm shift from land-based approaches, a linguistic foundation that continues to inform ocean-based approaches across the multiple approaches that constitute this body of work. It was explicitly based in multilateral dialogue, with identification of specific terminology to capture the complexity of ocean geographies and scales of temporality that differed from understandings of land-based time, spanned seabed as well as open ocean "resources" and territories, and created a linguistic-conceptual mapping upon which to build ocean research protocols (i.e., "high seas" refer to marine areas beyond any national jurisdiction; "area" refers to the ocean floor beyond any national jurisdiction; "exclusive economic zone" refers to the marine area within a 200 mile nautical mile contour line around a country, etc.). The document articulated a juridico-political, globalized framework for understanding the oceans across underwater infrastructures, environmental protection(s), pollution, navigation, dispute settlement, management of resources, law, and jurisdiction.

The approach was not without its critics (and significant weaknesses for moving from land- to ocean-based paradigms). As late as 2017, the European Commission on International Governance concluded: "as of 2017, 64% of marine waters that are in areas beyond those defined, in the UNCLOS document, are beyond national jurisdiction; there are over 300 UN related entities involved in international ocean governance – but these entities are not governed by an overarching body...[t]he current International Ocean Governance framework has gaps and shortcomings" (European Union International Ocean Governance 2017). The United States still has not joined the UNCLOS agreement, suggestive of just how far behind the United States is lagging in its paradigm for understanding sustainability in approaches to the ocean.

The authors of FGIM (2017), while not singling out the United States, made the limitations of the current approaches clear. As they noted, "each of [the] many players tends to have a limited view of the ocean that is focused on their own sectoral interests." Further, echoing the European Union's work across the 2000s, the assessment report makes clear the need for a "sound framework in which to work"; it was out of this recognition that, some 16 years ago, the World Summit on Sustainable Development (out of which came the assessment report) recommended "a regular process for global reporting and assessment of the state of the marine environment, including socioeconomic aspects" (2).

The report also recognized that the weaknesses inherent in the current land-based approaches have continued to disrupt representation of the ocean within modeling efforts and, thus, had perpetuated misrecognition of the ocean itself in representations of land-based measurements of damage and economic valuation. As the authors noted,

> The current sectorial approaches to ocean management do not take into account the interconnectedness of resources and activities surrounding the sea, including the most basic measurements that could capture the overall impact on marine ecosystems and coastal communities. This results in a lack of efficiency in the remarkably limited resources already allocated to the ocean's governance, as well as poor coordination of approaches. (FGIM 2)

In unveiling significant weaknesses in land-based approaches, as well as in moving toward an ability to see more clearly assessment practices and analytic approaches that could more accurately capture oceans as an independent, if interdependent phenomenon, the FGIM report would attempt to shift this paradigm. In the foreword by the Secretary-General of the United Nations, Ban Ki-Moon pointed to the global democratic effort of the evidence-based approach: "hundreds of scientists from many countries, representing various disciplines...indicate that the oceans' carrying capacity is near or at its limit. It is clear that urgent action on a global scale is needed to protect the world's oceans" (Foreword FGIM). The report provides an "important scientific basis for the consideration of ocean issues by governments, intergovernmental processes, and all policy-makers" involved in ocean affairs. The scope of the report's ambitions breaks multiple limits in previous assessments – not least of which was its attempt to form a common language for multiple nations, widely differing sociocultural and economic processes and purposes, and ocean processes themselves.

The authors of FGIM expressed the need to evaluate – perhaps even to conflate – land-based and ocean-based issues. "[M]ajor drivers of the pressures producing change in the ocean are to be found outside the marine environment. In particular, most of the major drivers of anthropocentric climate change are land-based" (39). The conclusions the report drew from this situation were multiple and offered a stunning, if simply articulated, critique of the ways in which land-based models for understanding climate change (as *impacts* on *human* resources) presented severe limitations on approaches to understanding ocean stresses – and therefore, how to work to eliminate them. Somewhat trenchantly, the authors concluded, "Thus, as far as social and economic aspects of the marine environment are concerned, many of the most significant drivers are outside the scope of the present Assessment. . .[t]he present Assessment of the marine environment cannot therefore reach conclusions on some of the main drivers affecting the marine environment without stepping well outside the marine environment and the competencies of those carrying out the Assessment." The report mobilizes the more generalized language of the IPSO (2011) report and recontextualizes the conceptual mapping through creating a productive tension between "land-based" and "ocean-based" research activity: the work ahead "will require the consideration of the full range of factors relating to human activities affecting the ocean" (39).

The linguistic echo cathected into two simple phrases in the two reports (IPSO's reorganization of approaches to include all activities that impinge ocean systems and FGIM's "consideration of the full range" of [human] factors) contains an imbedded, complex critique of assumptions – and therefore, findings – governing land-based research, *outside of its relation to* oceans; similarly, however, the critique also includes the assumptions governing ocean-based approaches, given the inability to amalgamate land-based research effectively into modeling of ocean-exclusive damage. Thus, without articulating directly the path forward out of this impasse, the FGIM report, by way of this critique, expresses an awareness that a new ground for constructing knowledge, as for understanding both the ocean and land as contiguous, interdependent systems, must be developed. The FGIM authors drove this conclusion home:

> [e]ven within the scope of what has been requested, it has not proved possible to come to conclusions on one important aspect: a quantitative picture of the extent of many of the non-marketed ecosystem services provided by the ocean. Quantitative information is simply insufficient to enable an assessment of the way in which different regions of the world benefit from those services. Nor do current data-collection programmes appear to make robust regional assessments of ocean ecosystem services likely in the near future. (39)

As both reports made very clear, at stake in the critique is a crucial, and missing, interface between these two components – land-based perspectives repositioned with and through ocean-based approaches – for sustaining nonhuman as well as human life.

Embedded in these two simple statements, however, is a powerful, paradigmatic change, and the horizon of a critique in which the limitations of land-based modeling of climate change would reveal themselves in the encounter with ocean systems. The enlargement, perhaps implosion of current assumptions underlying modeling human activity in relation to environmental processes is suggested here, as is the awareness of the ground of a significant paradigm change thrust upon us. Action to save the ocean is governed by the double bind of a critical descriptive language that restricts land-based paradigms to narrow foci in understanding processes incidental to the creation of damage – in the ocean but also on land. This language also privileges land-based processes of extraction, production, distribution, and consumption and places under erasure any number of processes, of both land and ocean, for ensuring the maintenance of this first chain of systems. To address the asymetricality in the conceptual interface between land and oceans, a review of the structuring in the majority of land-based issues is necessary. Arguably, there is no better place to begin than to turn to the country that bears the largest responsibility, currently, for the situation of global as well as national "confusion" in information about, or direct obfuscation of knowledge building within, environmental reporting: the United States. How do some of the most significant assessment and reporting efforts in this country interpret dynamics between land and sea?

The United States Government Accountability Office (GAO) is likely largely unknown – even within the United States. The office produces studies on topics requested by the United States Congress, and in September 2017, the GAO submitted *Climate Change: Information on Potential Economic Effects Could Help Guide Federal Efforts to Reduce Fiscal Exposure*. As the title of the report implies, the analysis delivered to congress maintained the governing paradigm of relations between "economy" and "environment." Although the GAO examined 30 studies, as well as interviewed 26 experts whose knowledge helped the GAO evaluate the strength and limitations of these studies, the report relies for its findings primarily on two national-scale studies that examine the potential economic effects of climate change in the United States: the *American Climate Prospectus* (2014) and *Climate Change Impacts and Risks Analysis* (2015). The GAO authors used these two reports because "[o]nly recently have studies analyzed the economic effects of climate change using frameworks that can compare effects across different sectors and regions within the United States on a national scale" (19). So recent, in fact, that these two studies are, to date, the *only two* conducted on a national (US) scale (7). Another widespread weakness of current environmental discourse is the series of significant knowledge and research gaps that continue to exist; these significant knowledge gaps disrupt the "future orientation" of the discourse in significant ways. Given IPCC scientists' projected, rapid acceleration of environmental damage, and concomitant economic and social impacts, it is remarkable that the United States has so produced so few studies that can range effectively across sectors and regions nationally. Yet, the linguistic structuring of a "future orientation" remains a hallmark of global environmental sustainability discourse.

Interestingly, the two studies, taken together, mirror the pattern of temporal structures that characterized meanings produced in and by the discourse of "environmental sustainability": while *Prospectus* provided information on the probability "of a set of economically important climate change impacts comparable across sectors," the Environmental Protection Agency's *Climate Change Impacts* assessed potential benefits to the United States of global action on climate change. Thus, although both reports provided the temporal protention and retention linguistic structures characteristic of the *Brundtland Report* (future orientation as a projection of changes in a historical pattern of actions), neither *Prospectus* nor *Climate Change Impacts* offered alternative actions as a basis for a polity of concerned actants coming to "best practices" in reorienting a past toward a more environmentally sustainable future.

The GAO report found that "the federal government has not undertaken strategic government-wide planning to manage climate risks by using information on the potential economic effects of climate change to identify significant risks and craft appropriate federal responses"; further, despite the number of years that have produced significant global knowledge about global warming and its effects on environments and humans, "the federal government could take an *initial step* [italics added] in establishing government-wide priorities to manage such risks." The GAO worded carefully this set of consistently voiced admonitions throughout the report: couched in the language of an "opportunity" for the federal government, the GAO report actually issues an occluded warning. Framed in the language of economic loss now and in the future, the report drew upon the internal reporting of the President Obama's proposal for fiscal year 2017 and studies across the preceding decade: "the federal government has incurred direct costs of more than $350 billion because of extreme weather and fire events, including $205 billion for domestic disaster response and relief; $90 billion for crop and flood insurance; $34 billion for wildland fire management; and $28 billion for maintenance and repairs to federal facilities and federally managed lands, infrastructure, and waterways" (1). Although future-oriented estimates, these "costs" are expected to continue to increase beyond current valuations of (financial) cost, as environmental scientists predict that extreme weather events will continue to build in frequency, duration, and severity beyond the capacity of current modeling systems to measure accurately. Chief of the National Oceanic and Atmospheric Administration's (NOAA) Monitoring Branch National Centers for Environmental Information, Deke Arndt, noted that "the Earth has moved into a new climate regime" when climate change broke algorithms used to monitor the accuracy of data for the month of November 2017 in Utqiaġvik, Alaska. "[The Arctic] is changing faster than anywhere else in the world" (Murphy 2017). As an outlier of patterns of climate change, the Arctic sea ice decline was depicted visually in the National Climate Report (2014 see Fig. 1).

It is no mere irony that the government's (own) "accountability" office refuses to engage in direct speech, instead opting for linguistic structures that suggest invitation to action rather than construct a more forceful, direct language to invoke, or perhaps compel, federal participation. If the past could do little to structure direct action in the present, the framing of "costs" as predominantly economic certainties of a projected future could: "recurring costs the federal government incurred as a result of climate change could increase by $12 billion to $35 billion per year by mid-century

Arctic Sea Ice Decline

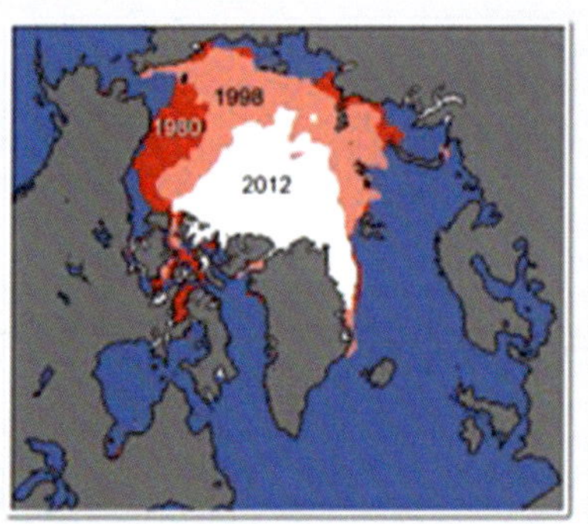

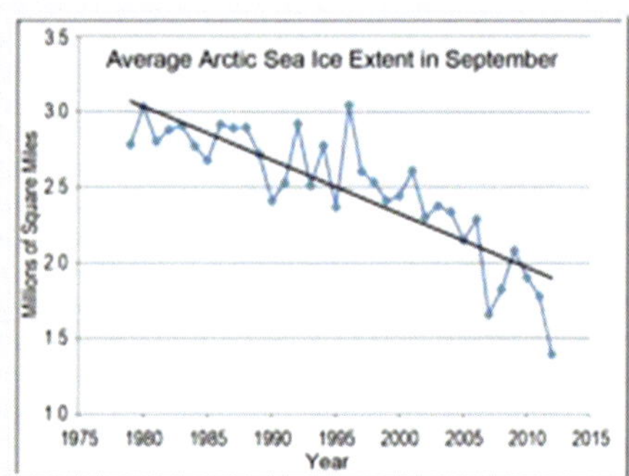

Fig. 1 Arctic sea ice decline

and by $34 billion to $112 billion per year by late century" (1). In addition, this linguistic structure frames "costs" in particular ways across the environmental sustainability discourse's discursive structure of future orientation(s). In this construct of the future, "costs" are constructs of "mitigation" as post-environmental damage response rather than investments in "pre-damage" or preventative strategies.

Thus, the network of meanings in the discourse actually discourages propensities to support awareness of other opportunities, such as direct action driven toward the avoidance of damage itself. For example, the direct (economic) action of imposing a carbon tax of $15 per ton of emissions would allow the United States to meet its goals as agreed upon by former President Obama in the Paris Accord (2017): among other targets is the reduction of 1.6 billion tons of carbon by 2030. According to the World Bank, a number of developed countries have adopted such a schema of direct action for a specific, measurable result within the future-orientation framing that is a hallmark of the discourse of environmental sustainability. These countries include Ukraine, France, Japan, Ireland, and 35 others. However, the Trump administration has consistently refused to impose a version of the carbon tax and has also adopted a rhetoric of withdrawing the United States both from its existing obligations and from any future negotiations – effectively withdrawing from the Paris agreement altogether. A similar pattern is observable in marine protected areas in the Trump administration's initiatives: "Interior Secretary Ryan Zinke has proposed that three ocean monuments be opened up to commercial fishing, shrinking the protected areas to undetermined sizes" (Conley 2018).

Perhaps more telling for its deliberate occlusions or manipulations of linguistic constructs available in the current meaning structures in environmental sustainability discourse, on President Trump's inauguration day, the URL to the White House climate page disappeared; in its place was a page titled "An American First Energy Policy Plan." A multitude of other proposed policy shifts in the United States energy sector, which drives the structure of economic thought, have now been proposed: under the rhetoric of the "American First Energy Policy Plan," the United States government would subsidize power plants that can store more than 90 days of fuel on site – effectively subsidizing current coal producing facilities. This support for coal production counters the current global trend: from January 2016 to January 2017, coal plants worldwide registered a "48% drop in pre-construction activity, a 62% drop in construction starts, a 19% drop in ongoing construction, and a 29% drop in completed projects" (Shearer et al. 2017).

Similarly, the EPA reopened a review of rules requiring auto manufacturers to reach 54.5 miles per gallon of fuel by 2025. In 2012, the EPA estimated that enacting the new standards would eliminate 140 million tons of greenhouse gases from reaching the atmosphere by 2025 (Temple 2017).

The Clean Power Plan (2015), former President Obama's project to retire coal plants and encourage investment in cleaner energy generation, targeted the single largest source of US global warming emissions. The plan required the electricity sector, by 2030, to cut carbon production to 32% below 2005 levels. In effect, this would reduce carbon emissions by as much as 267 million tons by 2025. According to a study conducted by the Union of Concerned Scientists (2015), the benefits of the EPA's Clean Power Plan far outweigh the costs. "An in-depth analysis of the final rule by the EPA found that the combined climate and health benefits of the Clean Power Plan will. . .deliver billions of dollars in net benefits each year, estimated at $26 billion to $45 billion in 2030" (Union 2015). Although accessible as an archived webpage, the study itself has since been removed from direct accessibility on the EPA webpages, as the webpages are in the process of being "updated."

In the report on the Clean Power Plan's climate benefits, the EPA used "the social cost of carbon, an official monetary estimate of the costs imposed by climate change impacts, such as property damage from increased flood risk" (Union 2017). However, a number of economists consider the social cost of carbon measurement model an underestimation of the true costs of climate change, suggesting that the measurement of benefits produced by the environment should be greater than that provided by the EPA's study – and conversely, that costs associated with the forms of damage being produced are also underestimated. In a study that appeared in *Nature* (2014), scientists and economists working across existing cost-modeling approaches reported that "an interagency working group for the United States government used three leading models to estimate that a tonne of carbon dioxide emitted now will cause future harms worth US$37 in today's dollars. This 'social cost of carbon' represents the money saved from avoided damage, owing to policies that reduce emissions of carbon dioxide" (Revesez et al. 2014). In the same article, the climate scientists, economics, and legal experts asserted that "because the models omit some major risks associated with climate change, such as social unrest and disruptions to economic growth, they are probably understating future harms" (Revesez et al. 2014).

They note that future costs could be higher, for four main reasons: first, "the impacts of historic temperature changes suggest that societies and economies may be more vulnerable than current models predict." Weather variability, for example, is already impacting crop yields of valuable foodstuffs such as almonds and cherries in the central growing basin of California. According to the National Climate

Assessment Report (NCA 2014), almonds grown in the region account for 95% of all almonds consumed in the United States (693). Globally, almonds produced in the San Joaquin and Sacramento valleys – a 400-mile tract of land from Tehama county to Kern County – account for 82% of all almonds consumed (Pierson 2017). Almonds were the state's number one agricultural export in 2012. Patterns of increasing temperatures due to anthropocentric warming are already documented in the region; the production of these and many other types of crops is predicted to be severely impacted, and costs of growing these crops will continue to impact profitability margins. As global warming increases, drought conditions in an already arid region, unaccounted for costs in the future, will include not only the loss of significant industry that supports this entire region but also the need to shift entire infrastructures for producing these crops entirely out of California, as the viability of production regimes continues to decline. Authors of the NCA report (2014) suggest that, if this is true for almond production, it is also true for a host of other crops grown in the same region: "[t]he Southwest produces more than half the nation's high-value specialty crops…[among them are] apricots, almonds, artichokes, figs, kiwis, raisins, olives, cling peaches, dried plums, persimmons, pistachios, olives [sic], and walnuts" (NCA 2014).

The warning signs of "unanticipated," wholescale disruptions in the productivity of the region are already here: in 2014, the year in which the NCA third report was released, unstable water supplies resulted in smaller almonds (Pierson 2017). Climate change scientists are predicting significantly more serious impacts by 2050. As the authors of *Climate Change and Infrastructure, Urban Systems, and Vulnerabilities: Technical Report for the US Department of Energy in Support of the National Climate Assessment* (2014) note, "[d]isruptions of services in one infrastructure will almost always result in disruptions in one or more other infrastructures, especially in urban systems, triggering serious cross-sectoral cascading infrastructure system failures in some locations" (Wilbanks, xviii). Such scenarios suggest unaccounted for costs in human terms: as industries "dry up," simultaneous mass migrations are likely. No current models for the assessment and measurement of these categories of costs exist, and these costs are likely to produce large-scale social and economic upheaval in differing, difficult to measure patterns across much of the United States by 2050, if current trends in policies and environmental behaviors continue unabated. As the *Climate Change and Infrastructure* report authors note, "in past assessments, cross-sectoral issues related to infrastructures and urban systems have not received a great deal of attention; in fact, in some cases the existing knowledge base on cross-sectoral interactions and interdependencies, at least as represented in published research literatures, appears to be quite limited" (1).

Second, models "omit damages to labor productivity…productivity growth, and to the value of capital stock, including buildings and infrastructure." Such losses are not static; that is, these losses, like the losses indicated above, are likely to produce economic feedback effects (cumulative and exponential forms of loss that are difficult to measure). As the writers of the article note, "[a]lso not taken into account are the risks of climate-induced wars, coups, or societal collapses and the resulting economic crises" (Revesez et al. 2014). Within the borders of the United

States, such political unrest seems unlikely at present; however, even if this does not represent accurately the future of the "future orientation" exercised through current environmental sustainability discourse in the United States, it is not at all inconceivable that such events will begin to dominate global politics, thereby producing immeasurable effects in the flow of consumer goods to and from production and consumer streams upon which the United States economy and its cultural social life rely.

The third issue with current models for estimating future costs of climate change: "models assume that the value that people attach to ecosystems will remain constant." However, related to the second reason above, if a commodity becomes increasingly scarce, its cost in the marketplace rises, based on long-held principles observed in theories of supply and demand in economics. In times of flooding, dry land's value increases; similarly, in regions in which wildfire is predicted, land values fall. As global warming degrades ecosystems, costs of future damage from climate change "will rise faster than the models predict." Currently, these models do not account for such variables, or unprecedented variability – what is likely to be understood, in the future, as the *volatility of commodities* – in a systemically less stable environment.

The fourth issue with current modeling of more traditional concepts of "cost": "the US analysis assumes a constant discount rate to translate future harms into today's money." Economists have demonstrated that discount rates that decline should be used, an approach that would "yield a higher present value to the long-term impacts of climate change." And, experts from across disciplinary specializations underscore the limitations of their studies: as the authors of *Climate Change and Infrastructure* (2014) conclude, although their report "breaks new ground. . .some of its assessment findings are rather speculative, more in the nature of propositions for further study than specific conclusions offered with a high level of confidence." The authors note that this direction for study represents "a welcome start" (1) toward a more definitive articulation of policy directions; however, the current administration's occluding of language and meaning structures forces these and other researchers to, at best, stall efforts toward this refining of the economic components of the discourse of environmental sustainability and, at worst, eliminate the trends of promising progress altogether toward research that works more concertedly toward redefinitions of meaning structures in environmental sustainability discourse. Such impacts are already visible, as reported by National Public Radio, in a study of "self-censoring" among research-based scientists. "Grants about 'climate change' are down 40% this year [2017]" As Hersher notes, "[c]limate change research is an inherently interdisciplinary field and shared terminology allows people to collaborate" (Hersher 2017).

Beyond the destructive power of explicit production of politically motivated linguistic exclusions in the discursive fields of environmental sustainability, however, the central weakness of all of these models is that the IPCC has estimated that, without effective strategies of mitigation, warming is likely to reach 4 °C by the end of this century – beyond current human experience. Current reports from that body – the most significant source of global information for a global planet (and the information all nations now use to adjust policy and spur technologically based research efforts) – indicate that, by the end of this century, the globe is tracking global warming models predicting the higher estimate of 4 °C (rather than the 2 °C commonly accepted as the maximum threshold for "safe" levels of warming). All current modeling of future costs, however, is predicated on the 2 °C threshold. Studies are beginning to emerge that suggest a scientific consensus is building toward acceptance that additional greenhouse gas emissions reductions are necessary to achieve "safer" warming levels across this century. For example, as Brown and Caldeira (2017) report, "we find that the observationally informed warming projection for the end of the twenty-first century for the steepest radiative forcing scenario is about 15% warmer (+0.5 °C) with a reduction of about a third in the two-standard-deviation spread (−1.2 °C) relative to the raw model projections reported by the Intergovernmental Panel on Climate Change. Our results suggest that achieving any given global temperature stabilization target will require steeper greenhouse gas emissions reductions than previously calculated" (*Nature*).

Yet another example of how the language of the discursive field of "environmental sustainability" is now functioning (in the United States): recently, $3 billion of the Advanced Research Projects Agency-Energy department's budget has been cut. This is perhaps the deepest shift in environmental sustainability discourse within the United States, given that the impact on any ability to create breakthroughs in the future is immeasurable – and this also denudes the US researchers of their ability to participate in cutting-edge global discoveries, as well as allows more action-oriented nations access to economic leadership in global markets for technologies that will shape the future. Similarly, the EPA's 2018 budget was cut by 31%.

In the meantime, any movement toward a globally or more national and localized, organized approach to understanding a different economy of production, as well as a different linguistic economy for "environmental sustainability," continues to recede into an obscured past, despite the discourse's linguistic foundational framework of a "future orientation." In this linguistic economy, "the environment" as an object of intervention becomes erased in the linguistic inoperability of "economic" approximations toward measurement; thus, environmental damage comes under the erasure of an economic determination at the level of the discursive field of meanings produced in and by "environmental sustainability." In this horizon of intelligibility, "environmental damage" becomes the more positive speech act of "mitigation efforts," or, in language adopted by the GAO report, "climate change adaptation" (2). Mobilization of this kind of speech act in the language of environmental sustainability enacts entire substructures of ideologically encoded meaning; these meanings structure the appearance of positive, definitive responses – a future-oriented series of acts – rather than signify a present series of acts compelled to stasis in a constellation of meanings consigned to a past orientation. These "past" meaning structures thereby re-enabled in and as the present become acts the government (or its citizens) *can* or *will* undertake when signs of environmental extremes appear in the future; however, this interpretative field ensures that such decisions to "mitigate" or "adapt" in the present actually *produce* a future in which the shaping of future damage, rather than the avoidance of damage now, becomes the future. That is, these acts, currently coded as conditional options, in actuality guarantee that such responses *must* be undertaken in the future. The framing of this series of strategies contextualizes a predictable and widespread pattern of extreme events in "localized," geographically "isolated" situations of "post-damage."

Little subtlety in the current administration of the US condensation and fissuring of linguistic structures of "environmental sustainability" is being exercised, however. In the first year of the Trump administration in the United States, staff at the United States Department of Agriculture (USDA) have been advised to avoid the use of the conceptual tools signifying complex concepts in the sustainability discourse – the now readily identifiable meanings of such significant concepts as "climate change," "climate change adaptation," "reduce greenhouse gasses," and "sequester carbon" in favor of, respectively, "weather extremes," "resilience to weather extremes," "build soil organic matter, increase nutrient use efficiency," and "build soil organic matter" (Milman 2017). The interpretative emphasis is directional: each of these "substitutions" advantages a meaning that appears as a positive rather than a negative impact in the environment. Remarkably, Bianca Moebius-Clune, director of soil health at the agency, noted that "we won't change the modeling, just how we talk about it" (Milman 2017). Jimmy Bramblett, deputy chief for programs at the National Resources Conservation Service (NRCS), noted in an email to senior employees, "[i]t has become clear one of the previous administration's priorities is not consistent with that of the incoming administration. Namely, that priority is climate change" (Milman 2017). Further eroding the underlying conceptual network across the language of environmental sustainability, Jonathan Thompson, senior ecologist at Harvard Forest, recognized that "[s]cientists I know are increasingly using terms like 'global change,' 'environmental change,' and 'extreme weather', rather than explicitly saying 'climate change'" ("Climate Scientists Watch Their Words"). This is a linguistic performative fracturing, or fissuring, of conceptual mapping that subtends widely supported, longer-term research into global warming –and citizens' abilities both to perceive the underlying unity of scientific consensus on global warming and its effects and to understand how to orient their own interpretations in the future orientation of environmental sustainability discourse itself.

Although adopting a form of linguistic exclusion in the actual use and applicability of the term "climate change" in significant scientific reporting venues in the United States is impactful enough, a similar operation is at work more generally in the shift from "global warming" to "climate change" in the environmental sustainability discourse. While recently widely adopted in the discourse globally, this shift in language creates specific forms of ambiguity in interpretation, allowing for imprecise thinking and, more negatively, opportunistic mobilization of these forms of meaning ambiguity. In the United States, for example, in the dismissal of dimensions of meaning in "climate change," an opportunistic mobilization of meaning is evident in the Environmental Protection Agency (EPA) Director Scott Pruitt's recent pronouncement that "Our activity contributes to the climate changing to a certain degree. Now measuring that with precision. . .I think is more challenging than is let on at times." Despite the admitted complexity of, and imprecision in, measuring *the rates* at which CO2 is building in the atmosphere, there remains no legitimate challenge to the anthropomorphic nature of current rates of climate change.

Scientists studying global warming across the globe have asserted, through the latest (Fifth Assessment Report of 2014) reports issued by the IPCC, that with 95% confidence rates, human activities have been the dominant cause of global warming. Similarly, advantaging this same field of linguistic ambiguities inherent in "climate change," Pruitt asserted, "I think there's assumptions made that because the climate is warming, that that is necessarily a bad thing." (Exclusive Interview). Further, Pruitt mobilizes the neutrality of meaning inherent in "climate change" to articulate his own "future-oriented" framing – unanchored from the more precise terminology governing the reporting of scientific findings in environmental sustainability discourse. Pruitt introduces a moral language and judgment into the linguistic "space" made available in this linguistic structural ambiguity:

> Is it an existential threat—is it something that is unsustainable, or what kind of effect or harm is this going to have? We know that humans have most flourished during times of, what, warming trends? Do we know what the ideal surface temperature should be in the year 2100 or year 2018? It's fairly arrogant for us to think we know exactly what it should be in 2100. (Exclusive Interview)

As Richard Lazarus, Harvard University professor of environmental law, observed, "[Pruitt] is much more organized, much more focused than the other Cabinet-level officials. . .Just the growing number of environmental rollbacks in this time frame is astounding" (Irfan 2018). In the overt series of "futures" mobilized as a characteristic feature of the discourse, evidence-based structures in assertions predominate. Yet, as Pruitt's (EPA-based) power in the environmental economy of ideas attests, the discourse is governed by a paradox, an inversion of the normative, predominating power of hegemonic concepts and established linguistic networks: despite the hegemonic nature, in environmental sustainability discourse, of "evidence-based" findings, more powerful still is the exercise of doubt as declension, rather than "proof" *as* review and the production of knowledge *in* processes of (global) consensus. Yet another characteristic of the discourse is the creation of global-wide and nation-wide review processes for knowledge generation. The IPCC, for example, since its inception in 1988, has assembled thousands of scientists across the globe "to prepare, based on available scientific information, assessments on all aspects of climate change and its impacts, with a view of formulating realistic response strategies" (History). For the initial meetings to approve the outline of the Sixth Assessment Report (AR6), 195 member governments discussed the draft and agreed on a final outline (Intergovernmental Panel on Climate Change 2017).

The reports are produced across 5-year intervals and represent 5-year-long processes for the testing of ideas across scores of disciplinary researchers which comprise the content of IPCC report structures. This global network of researchers from scores of disciplinary perspectives is, perhaps, a new form of hegemony in discourse regulation and formation: a hegemony borne of open processes of (self-) regulation, rigorous review, multidisciplinary participation, and scientifically based forms of evidence construction. As a descriptor of this process for coming to knowledge claims in the public sphere, "arrogance" seems, at best, a mis-apperception of a deeply democratic form of global participation in understanding, and characterizing, "the future."

The GAO report, released September 2017 – 5 months before Pruitt articulated his broadly based ideas above – suggested another crucial characteristic of research in "climate change": the complexity of contributory disciplinary methodologies and the intricate interdisciplinary linguistic base that drives the practice of research and language construction across an array of disciplinary perspectives and conceptual tools. The transdisciplinary conceptual navigation(s) involved in research that challenges existing communication strategies and paradigmatic disciplinary adhesions also produces powerful forms of discovery, critique, multidisciplinary and global forms of review, interdisciplinary language building and concept sharing, and scientific consensus. Describing a small if dominant quotient of approaches (3) in the discursive fields of environmental sustainability discourse, the authors noted structures of "sequentiality" in modeling processes. The three of multiple models and processes for building knowledge across multiple research processes suggest the processes of linguistic and conceptual telescoping –hegemonic disciplinary clustering – *as* climate change research. Chosen for representation (and thus, conceptual activation) in the report are "climate models," "individual sectoral economic models," and "economy-wide models" (9–12). GAO's authors noted: "Methods used to estimate the potential economic effects of climate change in the United States are complex because, according to literature we reviewed and many experts we interviewed, they use different types of complicated climate and economic models that are linked together in a sequential framework that uses the results of one model as input to another." (9) As is suggested in the IPCC processes outlined in brief above, the three specific types of models chosen for representation suggest that the GAO report's insight into one characteristic of environmental sustainability discourse is, in reality, its own structural and conceptual limitation.

In addition to the potential for undermining the perception of the scientific integrity of current work on climate change, such "substitutions" and elisions across the language games of environmental sustainability realign the discourse by occluding current, rapid, and ongoing contributions of insights by way of the superimposition of a language of "facticity" and techno-economic conceptions of valuation that police structures of meaning as representations of discursive proficiencies. If meaning structures that inhere to "climate change" are rendered ambiguities in such linguistic shifts – deliberate disarticulations and occlusion of agreed upon language to produce an array of imprecise meaning structures introduced into the linguistic networks that construct patterns of imbricated, complex concepts *as* thought – the construction of global movements toward an understanding of a more sustainable present and future is likely to be similarly impacted.

Such language meaning structures occlude multiple opportunities inherent in shifting these meaning structures altogether into "environmental sustainability" opportunities that encourage rethinking across the meanings generated by an "environmental sustainability" anchored in, and largely subsumed by, the reproduction of existing economic realities. Thus, any repositioning of identity politics across relations of consumption, patterns of production, and assumptions behind concatenations of "leisure," "productivity," and "environmental awareness" may deliver the eruption of meaning into the stasis produced by the dynamics of linguistic exchange in current environmental sustainability discourse. In its "unsettling" nature, the ocean offers multiple points at which the discourse of (land-based) sustainability collapses within its own tensions to reveal horizons of representation that offer a glimpse into the full complexity, and excitement, of re-envisioning the future of an entire discourse of the (contested) future.

Two of the most promising possibilities for rethinking paradigms of damage, and the behaviors that reproduce this damage as a daily occurrence, emanate from the ocean(s). The authors of FGIM note that, although "there are widespread gaps in the skills needed to assess the ocean" such as the lack of "integration of environmental, social, and economic" interactions that continue to produce damage in and for the ocean, as well as gaps in capacity building and resources needed for "successful application of knowledge," the research on the ocean has produced the most powerful critique of land-based methodologies *and* called explicitly for a paradigm shift that can accommodate both land- and ocean-based modeling. Noting that three possible foci the FGIM authors could have adopted each have strengths and weaknesses –ecosystems services, habitats, and human-based pressures on the marine environment – the authors chose the first to focus their assessment. Why? Largely for its ability to interrupt the current hegemony of economics in land-based approaches to understanding sustainable development through destabilizing "environmental" sustainability. The authors clarified the modeling weakness that disfigures both the understanding of land and oceans and ensures that damage remains a daily, escalating effect of human behavior:

> …benefits and costs have been hidden within the 'natural system', and are not accounted for financially; such hidden costs and benefits are considered 'externalities' by neoclassical economists. While the neoclassical economic toolbox includes non-market valuation approaches, an ecosystem services approach emphasizes that 'price' is not equal to 'value' and highlights human well-being, as a normative goal. The emergence and evolution of the ecosystems services concept offers an explicit attempt to better capture and reflect these hidden or unaccounted benefits and associated costs when the natural 'production' system is negatively affected by human activities. (Part III, 1)

In contrast to the land-based assessment models, the "Millennium Ecosystem Assessment defines an ecosystem as a dynamic complex of plant, animal and micro-organism communities and their non-living environment interacting as a functional unit"; ecosystem services are "the benefits that humans obtain from ecosystems" (Part III, 2). Thus, the ocean-based modeling reflects, to a far greater degree than current land-based models, both nonmarket valuation of costs (and benefits) and non-animate contributors as continuous with one "eco-service" system. The shift of human's predominance in an ecosystems approach is subtle but clear: humans become on facet of a multilayered series of systems for creating well-being.

The second promising approach involves less of a paradigm shift than an intensification of neoclassical economics via a different viewpoint on "waste." The effort here is to shift one of the most serious consequences for the ocean though limitations in the neoclassical economics of "cost" in land-based pollution: the production of plastics. "Cost" here is reconceptualized as loss to the economy via unnecessary waste. The extent of the problem produces far more dire consequences than is currently understood: as 78% of earth's surface, if oceans collapse, the land is sure to follow. As the much larger mass, oceans regulate earth's climate; similarly, ocean processes provide the strongest carbon sink on earth, sequestering 23% of all carbon currently being produced by anthropocentric means. Finally, ocean processes produce the oxygen of one of every two breaths every human on earth takes – every day. But as the McKinsey Center for Business and the Environment recently released its much-anticipated report, *Stemming the Tide: Land-Based Strategies for a Plastic-Free Ocean* (*Stemming* 2017). The report updated the sheer volumes of plastic entering the sea, "estimated at eight million metric tons a year – greatly exceeding any previous estimates" (6).

Without a new paradigm for intervention in the status quo of this form of damage production, the global quantity of plastics entering the ocean is likely to exceed "250 metric tons by 2025" (6). Similarly, the Ellen MacArthur Foundation found that, by 2025, in a "business-as-usual scenario, the ocean is expected to contain 1 t of plastic for every 3 t of fish"; by 2050, the ocean would contain more plastics than fish (*New Plastics Economy* 2016). The recent attention toward producing more rigorous studies on plastics in the ocean, although deeply unsettling, represents a significant advance in the knowledge building and modeling capacity necessary for intervention strategies. The study analyzed different regions and proposed region-specific strategies of intervention. Applying the economic modeling to analyses of plastic "waste," the study remained within the parameters of the neoclassical economic models for understanding "the environment"; however, the study also suggests the power of the model in applications of this nature. Through the neoclassical lens, the study found that, by 2020, improving infrastructure and plugging post-collection gaps can reduce annual leakage by 50% (*Stemming* 34). In the medium term (2025), the development of commercially viable waste treatment can reduce annual leakage by an additional 16%. And in the long term (2035), a number of additional measures applied to the problem of leakage from multiple plastic streams of production could curtail rebound effects of leakage based on the propensity to create additional plastics to meet additional demand (34–36).

More promising still, however, is the paradigm shift signaled in the World Economic Forum's *New Plastics Economy: Rethinking the Future of Plastics* (2016). The paradigm pursued in the report is governed by the following question: "how can collaboration along the extended global plastic packaging production and after-use value chain, as well as with governments and NGOs, achieve systemic change to overcome stalemates in today's plastics economy in order to move to a more circular model?" (3). By "circular model," the authors "overarching vision of the New Plastics Economy is that plastics never become waste; rather they reenter the economy as valuable technical or biological nutrients" (7). The logic underpinning both of these (overlapping) approaches to plastics can be applied to others streams of waste. Although these do not represent a whole-scale paradigm shift, these approaches do represent a significant step forward in reversing the paths of destruction increasingly evident in the ocean today. Paradoxically, for the ocean to continue to provide humans with significant benefits – for ocean and human habitats to survive – land-based interpretations of value, and the models upon which these are built, must be understood as indivisible. The paradigm shift necessary is simple to phrase, if difficult to enact as a daily experience: everything we do becomes the ocean.

References

Álvarez-Romero, J. G., Pressey, R. L., Ban, N. C., Vance-Borland, K., Willer, C., Klein, C. J., & Gaines S. D. (2011). Integrated Land-Sea Conservation Planning: The Missing Links. Annual Review of Ecology, *Evolution, and Systematics 42*(1):381–409.

Baker, S., & Eckerberg, K. (Eds.). (2010). *In pursuit of sustainable development: New governance practices at the sub-national level in Europe*. London/New York: Routledge.

Baker, S., Kousis, M., Richardson, D., & Young, S. (2012). *The politics of sustainable development: Theory, policy and practice within the EU* (pp. 14–15). London: Routledge.

Bebbington, J. (2000). *Sustainable development: A review of the international development, business and accounting literature*. Aberdeen papers in accountancy, finance and management working paper 00–17.

Beck, M. W. (2003). The sea around: conservation planning in marine regions. In C. Groves (Ed.), *Drafting a conservation blueprint: A practitioner's guide to planning for biodiversity* (pp. 319–44). Washington, DC: Island.

Brown, P. T., & Calderia, K. (2017). Greater future global warming inferred from Earth's recent energy budget. *Nature, 552*, 45–50. https://www.nature.com/articles/nature24672. Accessed 4 Jan 2018.

Brundtland Report. (1987). Our common future. http://www.un-documents.net/our-common-future.pdf. Accessed 2 Aug 2017.

Conley, J. (2018). With oceans under greatest threat ever, Trump administration urges even less protection for marine life. *Common Dreams*. https://www.commondreams.org/news/2018/01/02/oceans-under-greatest-threat-ever-trump-administration-urges-even-less-protection. Accessed 2 Jan 2018.

Dryzek, J. S. (2005). *The politics of the earth: Environmental discourses*. Oxford: Oxford University Press.

Escobar, A. (1995). *Encountering development: The making and unmaking of the third world*. Princeton: Princeton University Press.

European Commission. (2011). Final assessment of the 6th Environmental Action Programme shows little progress in environmental policy – but with shortfalls in implementation. European Commission 31 August 2011. http://europa.eu/rapid/press-release_IP-11-996_en.htm. Accessed 1 Aug 2017.

State of the Oceans. (2011). European Union International Programme on the State of the Oceans. http://www.stateoftheocean.org/pdfs/1906_IPSO-LONG.pdf. Accessed 4 Oct 2011.

European Union International Ocean Governance. (2017). https://ec.europa.eu/maritimeaffairs/sites/maritimeaffairs/files/docs/body/2015-international-ocean-governance_en.pdf. Accessed 3 Feb 2018.

Exclusive Interview: EPA Chief Scott Pruitt goes one-on-one with News 3. https://www.youtube.com/watch?time_continue=281&v=h-Pq782F9zA. Accessed 13 Feb 2018.

First Global Integrated Marine Assessment (World Ocean Assessment I). http://www.un.org/Depts/los/global_reporting/WOA_RPROC/WOACompilation.pdf. Accessed 3 Dec 2017.

Griffin, L. (2013). Governance for sustainability in the European union – A post-political project. In P. M. Barnes & T. C. Hoerber (Eds.), *Sustainable development and governance in Europe* (pp. 34–47). New York: Routledge.

Hersher, R. (2017, November 29). Climate scientists watch their words, hoping to stave off funding cuts. *NPR*. https://www.npr.org/sections/thetwo-way/2017/11/29/564043596/climate-scientists-watch-their-words-hoping-to-stave-off-funding-cuts. Accessed 30 Nov 2017.

Intergovernmental Panel on Climate Change: Fifth Assessment Report (AR5). https://www.ipcc.ch/report/ar5/. Accessed 4 Nov 2014.

Intergovernmental Panel on Climate Change (IPCC). History. http://www.ipcc.ch/organization/organization_history.shtml. Accessed 23 Nov 2017.

Irfan, U. (2018). Scott Pruitt is slowly strangling the EPA. Vox. Accessed 3 Jan 2018. https://www.vox.com/energy-and-environment/2018/1/29/16684952/epa-scott-pruitt-director-regulations. Accessed 30 Jan 2018.

Jacobs, M. (1999). Sustainable development: A contested concept. In A. Dobson (Ed.), *Fairness and futurity: Essays on environmental sustainability and social justice* (pp. 21–45). New York: Oxford University Press.

Kuhn, T. (1970). *Structure of scientific revolutions. Enlarged* (2nd ed.). Chicago: Chicago University Press.

Lafferty, W. M. (1995). The implementation of sustainable development in the European Union. In: J. Lovenduski & J. Stanyer (Eds.), *Contemporary political studies. Proceedings of the Political Studies Association* (Vol. 1). Belfast: PSA.

Milman, O. (2017, August 7). US federal department is censuring use of term 'climate change', emails reveal. https://www.theguardian.com/environment/2017/aug/07/usda-climate-change-language-censorship-emails. *The Guardian*. Accessed 7 Aug 2017.

Murphy, P. P. (2017). Climate change just broke a NOAA algorithm. CNN. http://edition.cnn.com/2017/12/14/weather/arctic-temperatures-break-noaa-algorithm-trnd/. Accessed 14 Dec2017.

National Climate Assessment (NCA). (2014). Southwest: Threats to Agriculture. https://nca2014.globalchange.gov/report/regions/southwest. Accessed 7 May 2014.

Ocean Conservancy and McKinsey Center for Business and Environment. (2017). Stemming the tide: Land-based strategies for a plastic-free ocean. https://oceanconservancy.org/wp-content/uploads/2017/04/full-report-stemming-the.pdf. Accessed 3 May 2017.

O'Riordan, T. (1985). Environmental Issues. *Progress in human geography* 9(3):401–414. http://journals.sagepub.com/doi/abs/10.1177/030913258500900305. Accessed 4 Dec 2015.

Pearce, D., Markandya, A., & Barbier E. B. (1989). Blueprint for a green economy. London, Earthscan.

Pearce, D. (1992). Green economics. *Environmental Values, 1*(1), 3–13. http://www.environmentandsociety.org/sites/default/files/key_docs/pearce_1_1.pdf. Accessed 2 Jan 2017.

Pezzey, J. (1989). Economic analysis of sustainable growth and sustainable development. Environmental Department Working Paper No. 15. Washington D.C: World Bank.

Pierson, D. (2017). California farms lead the way in almond production. *Los Angeles Times*. http://www.latimes.com/business/la-fi-california-almonds-20140112-story.html. Accessed 2 Aug 2017.

Revesz, R. L., & Howard, P. L., et al. (2014, April). Global Warming: Improve economic models of climate change. *Nature*. https://www.nature.com/news/global-warming-improve-economic-models-of-climate-change-1.14991. Accessed 4 Apr 2014.

Shearer, C., & Ghio, N., et al. (2017). Boom and bust 2017: Tracking the global coal pipeline. CoalSwarm, Greenpeace USA, Sierra Club. https://endcoal.org/wp-content/uploads/2017/03/BoomBust2017-English-Final.pdf. Accessed 10 Apr 2017.

Temple, J. (2017). Trump's five biggest energy blunders in 2017. *MIT Technology Review*. https://www.technologyreview.com/s/609631/trumps-five-biggest-energy-blunders-in-2017/. Accessed 27 Dec 2017.

United Nations Convention on the Law of the Sea. (2017). http://www.un.org/depts/los/convention_agreements/texts/unclos/unclos_e.pdf. Accessed 3 Aug 2017.

Union of Concerned Scientists. (2015). How much will the clean power plan cost? https://www.ucsusa.org/global-warming/reduce-emissions/how-much-will-clean-power-plan-cost?utm_source=fb&utm_medium=fb&utm_campaign=fb&s_src=socnet&s_subsrc=facebook#.Wo4FiBPwZE4. Accessed 8 Aug 2015.

Union of Concerned Scientists. (2017). The social cost of carbon underscores an obvious fact: climate change is costly. https://blog.ucsusa.org/rachel-cleetus/the-social-cost-of-carbon-underscores-an-obvious-fact-climate-change-is-costly. Accessed 28 Feb 2017.

Wilbanks, T. J., & Fernandez, S. J. (Eds.). (2014). Climate change and infrastructure, urban systems, and vulnerabilities: Technical report for the US department of energy in support of the national climate assessment. https://www.eenews.net/assets/2014/03/06/document_cw_01.pdf. Accessed 3 Jan 2015.

World Economic Forum. (2016). *The new plastics economy: Rethinking the future of plastics*. https://newplasticseconomy.org/publications/report-2016. Accessed 15 Jan 2016.

Žižek, S. (1996). *Love they symptom as thyself.* https://www.youtube.com/watch?v=vM12ddwbHOE. Accessed 13 Sept 2014.

Environmental Stewardship

Pathways to Community Cohesion and Cultivating Meaningful Engagement

Sachi Arakawa, Sonya Sachdeva, and Vivek Shandas

Contents

Abstract

What pathways do people take on the journey to stewardship and what rewards do they reap? Numerous studies emphasize the underlying values, whether moral, spiritual, or religious, which provide the foundation for engaging in environmental behavior. Yet, many cases of stewardship are founded not on lofty environmental ideals but on pragmatic, localized ambitions. As cities work to rectify historical inequities in access to environmental assets like trees and green spaces in low-income communities of color across the United States, it is important to understand the cultural and socioeconomic dimensions of environmental stewardship and the distinct pathways to stewardship. Understanding these can lead to policy and programmatic changes, helping city foresters and environmental advocacy groups better engage and serve marginalized communities. In this

S. Arakawa (✉) · V. Shandas
Portland State University, Portland, OR, USA
e-mail: sachi@pdx.edu; vshandas@pdx.edu

S. Sachdeva
Northern Research Station, USDA Forest Service, Evanston, IL, USA
e-mail: sonyasachdeva@fs.fed.us

S. Dhiman, J. Marques (eds.), *Handbook of Engaged Sustainability*,
https://doi.org/10.1007/978-3-319-71312-0_37

review, we use several cases from our work and others' to illustrate the possible barriers to engaging low-income communities and communities of color in environmental stewardship, how notions of identity, power, and agency impact the ways in which underserved communities respond to environmental issues, and finally, what paths stewards take in finding meaning in their work.

Keywords
Stewardship · Environmental values · Environmental justice · Motivation · Community resilience

Large-scale environmental problems, like the depletion and overconsumption of natural resources, continue to impact humanity on a global scale, yet state resources for interventions, from the neighborhood to the national level, continue to shrink. There is a growing recognition that sustainability outcomes, particularly in cities, cannot be achieved by top-down management alone but instead must be integrated with human systems, with people leading local, grass-roots stewardship efforts (Romolini et al. 2012). Social and cultural issues are an essential piece of environmentally sustainable solutions and development, and yet they remain the most nebulous and perhaps least understood dimensions (Chiesura and de Groot 2003). These issues can range from large and abstract (e.g., how are our relationships with nature formed?) to highly granular (e.g., what parks are most frequently used in a neighborhood and by whom?). Fundamentally, researchers have tried to understand the socio-cultural factors that influence attitudes toward sustainability, and ultimately, understand what brings people "to the table" of environmental conservation, i.e. the pathways to stewardship. We hope this chapter can be meaningful for policymakers, volunteer, and stewardship organizations in designing effective and engaging environmental programs by describing a) our current understanding of environmental stewardship, and b) the socio-political structures and processes that mediate broad participation in stewardship activities.

Environmental stewardship, as an area of research, has gained prominence relatively recently, emerging only in the past four decades. It is now recognized as a notable confluence of sociocultural norms and ecological sustainability which is crystallized into practice and conservation strategy (Worrell and Appleby 2000). The practice of stewardship not only creates more environmentally responsible citizens, but also expands the capacity of management organizations by redistributing the responsibility of environmental monitoring and maintenance to the average citizen (Berry 2006; Buell 2009). Citizen-based environmental stewardship programs are increasingly relied upon by government agencies to expand capacity for restoration efforts, greening initiatives, and other improvements to urban ecosystems (Romolini et al. 2012; Baker 2014). The term "stewardship" tends to get applied broadly and can be difficult to pinpoint for analysis.

Stewardship has also been lauded as a way to build community resilience, to increase civic engagement, and to create partnerships between government and community (Baker 2014; Romolini et al. 2012; Fisher et al. 2011). Citizen stewardship initiatives are increasingly relied upon by governments that do not have the resources to improve social or environmental conditions (Romolini et al. 2012; Baker 2014), and citizen participation in environmental conservation has come to be considered a fundamental part of democracy (Shandas and Messer 2008). Programs are intended to connect individuals to their environment through learning and meaningful action, but some critics argue that current approaches tend to oversimplify complex sociocultural context and overlook some moral and ethical dimensions of stewardship behaviors (Baker 2014). Other studies suggest precisely the opposite – that stewardship is not only shaped by personal motivations but involves both organizational goals and outcomes (i.e., measurable environmental improvements) and also process-related goals such as engaging community members, doing outreach, and collaborating with like-minded others (Romolini et al. 2012).

Social Constructions of Environmental Problems

The existing literature suggests that how we define stewardship varies depending on the context, from meanings derived from modern environmentalist theory to definitions created by agency organizations. Approaches that provide a contextual basis for stewardship are needed, recognizing that engagement in such efforts are tied to social and political histories that enable the capacity to participate. This chapter draws from social constructionist theory where "environmental problems" are seen as social problems in that they are, as Taylor suggests, "socially constructed claims defined through collective processes" (Taylor 2000, p. 509). From this perspective, knowledge and folk theories of environmental issues are created by the interactions of individuals within society, as groups perceive, identify, and define environmental problems collectively through shared meanings and interpretations (Taylor 2000). This means that the framing of environmental issues is particularly relevant to how they are perceived, how people will mobilize around them, and who engages in stewardship efforts (Carmichael 2017).

Therefore, our understanding of environmental stewardship begins with the acknowledgment that individual perceptions, institutional dynamics, and historical injustices recursively create the places we inhabit. Such biophysical and social interactions have complex interpretations at multiple scales (e.g., individual and collective, past, present, future, and across space). Importantly, Clarke and Agyeman (2011) point out that environmental conditions are not equal in all communities. In places where the environment is heavily degraded, or environmental resources are lacking, adjacent communities often receive societal blame for not doing more to conserve their environment. Such claims miss an essential aspect that not everyone has inherited a democratized or shared sense of environmental responsibility in this way. In fact, racial and economic inequalities impact the way that historically marginalized communities experience and frame environmental issues, and whether or not they chose to participate in traditional environmental stewardship activities.

As a result, different approaches to stewardship activities in communities of color, for example, may stem from perceiving a lack of agency, and loss of trust in government and government processes (Carmichael 2017; Clarke and Agyeman

2011). Examining the multiple spaces of identity, power, and agency in which minority and low-income communities respond to environmental issues can help us to understand and break down the barriers to environmental stewardship. The "environmental sustainability" framework has in the past prioritized environmental conservation over social inclusion or equity, and only more recently embraced the concept of "green justice" or "just sustainability," with an eye toward social justice concerns (Clarke and Agyeman 2011; Agyeman and Evans 2003). Increasingly clear is a need to examine dimensions of fairness, equality, and justice when attempting to get buy-in for efforts in environmental conservation (Agyeman and Evans 2003). Yet, we are still understanding effective approaches to engage communities with these complex topics, especially when engaging with those who distrust and may bear animosity toward government actions and/or formal organizations.

To address these topics, we use a case-based approach to examine real-life examples of environmental stewardship efforts, discuss their impact, and reflect on how policy and governance can be changed toward a more inclusive stewardship model. While a fundamental goal of the present chapter is to illustrate alternative engagement models that address fairness, equality, and justice, we also build on previous theoretical work that helps to frame the practical guidance that underscores our approach. We start with definitions of stewardship, discuss why the concept can be problematic and describe the various pathways and barriers to contemporary stewardship efforts. We close by offering cases where organizations are finding novel successes in expanding and deepening participation in environmental conservation by focusing on community cohesion and helping people find meaning in the work they do.

Defining Environmental Stewardship

As use of the term has proliferated, so have the many definitions of environmental stewardship. In particular, we try to define how the concept of stewardship might be distinct from conservation behaviors or land management more generally. For some, the terminology can be traced back to the Old Testament (Worrell and Appleby 2000). These origins can be viewed as somewhat problematic as they bootstrap onto more traditional, anthropocentric Christian notions of stewardship which might not place an intrinsic value on nature or nonhuman life (Devall and Sessions 1985; Palmer 2006; Routley 1973; Van Dyke 1996). For instance, the seminal work by Lynne White (1967) claims that far from connecting humans to their environment, traditional Christian religious ecologies served to reify the boundaries between humans and the natural world. In recent years, however, this perspective has been challenged by many who have pointed out that a) the religious interpretation of stewardship is not limited to the Christian faith, and can be found in indigenous communities and most major religious traditions around the world and b) the Christian interpretation of stewardship need not be anthropocentric (Worrell and Appleby 2000). For instance, a growing "green" movement within the Evangelical Christian community suggests that engaging in climate change adaptation, mitigation, and more environmentally sustainable behavior overall (Roberts 2011) can be aligned with a Christian stewardship ethic.

A more secular interpretation of environmental stewardship is also often invoked within the literature. To some, stewardship bears a moral core, providing a path for people to characterize their relationship toward nature as one of right or wrong actions or an ethic of care (Welchman 1999). The idea of values and ethics as motivating behavior has also been empirically validated. In a study with environmental stewards in the Seattle, Washington area, Romolini and colleagues (2012) used a cognitive mapping approach to arrive at a functionalist definition for stewardship. Their analyses revealed that volunteers defined stewardship as rooted in a core set of values and ethics, including **environmental values** such as reducing the human impact on the environment, and a sense of moral obligation. Interestingly, however, stewardship, in the volunteers' minds, was not only defined by personal motivations but involved both organizational goals and outcomes (i.e., measurable environmental improvements) and process-related goals such as engaging community members, doing outreach and collaborating with like-minded others (Fig. 1).

There is also a strong idea of grassroots action within most definitions of stewardship, where a local group of people take initiative to conserve or protect some environmental good. In moving beyond the biblical sense of stewardship, the environmental good need not have instrumental or utilitarian value (e.g., timber is useful for building houses), but rather the management of natural resources can be based in spiritual, religious, cultural, or aesthetic value (e.g., the majesty of our forests). Since the early 1990s, grass-roots stewardship organizations have proliferated at all geographic scales (local, regional, national) to promote conservation of

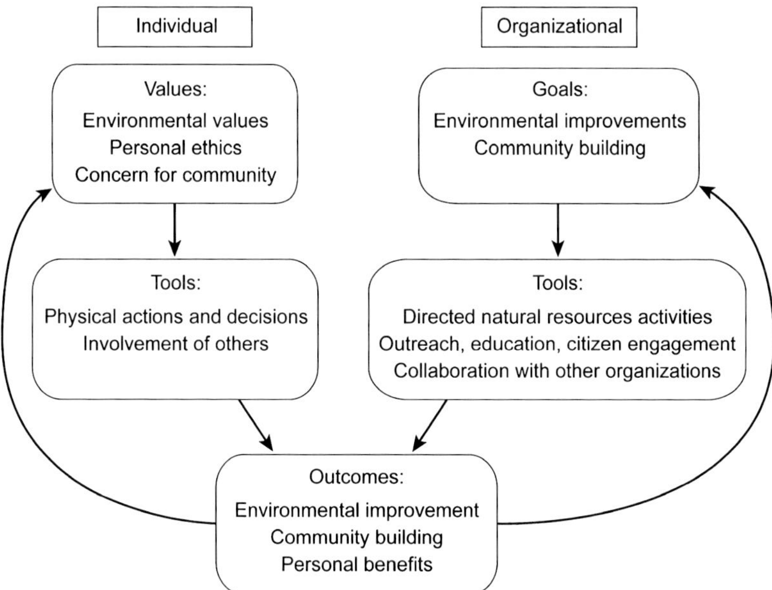

Fig. 1 Romolini et al.'s (2012) framework for environmental stewardship based on cognitive maps of Seattle area environmental stewards

wilderness areas, rivers or particular species of plants or animals (Carr 2002). Government stewardship programmes have also followed suit such as the Urban and Community Forestry program managed by the US Forest Service. This program supports local tree planting and care initiatives, among other projects.

As these efforts proliferate, it is becoming more widely acknowledged that stewardship is not solely a human-driven effort, where humans are the only agents of change in the environment. Rather, the significance of coupled human-nature systems in shaping human perception and bounding stewardship efforts is now well-established within the literature (Carr 2002). That is, we note that the structure and shape of the natural environment or landscape has an impact on human behavior and is also an agent of change in impacting local stewardship. Furthermore, while the abundance or scarcity of a particular resource or the local socio-ecological context will necessarily shape human behavior so too will the overall access to or level of engagement with natural landscapes.

And in fact the emphasis on the local environment is another true hallmark of stewardship, which relative to advocacy or other forms of environmental conservation efforts, is less about engaging in lobbying efforts and more likely to have an action- or protection-oriented role within a clearly defined geographic region (Carr 2002; Leopold 1949). Indeed, environmental stewardship appears to contain a strong sense of place, and many stewards draw on their connection to their hometowns and other personally meaningful sites to motivate their work. For instance, tree-planting rituals and trees themselves became symbols of socio-ecological resilience in post-Katrina New Orleans where a desire to restore their neighborhood motivated stewards' tree planting, which in turn helped to sustain their sense of place (Tidball 2014). These efforts in New Orleans, and other like them across the country, make the connection between place attachment and management explicit, indicating a stewardship ethic based on an implicit sense of ownership and agency (Svendsen and Campbell 2008). These implicit feelings of ownership can have many positive stewardship outcomes – for instance, a sense of ownership in a rented plot in a community garden may increase commitment toward managing that plot over time and foster long-term engagement (Kaplan 1985; Stone 2009; Teig et al. 2009).

As these examples illustrate, stewards' sense of place or ownership intermingles with concepts such as belongingness and identity, and is not restricted by geographical constraints nor confined by temporal bounds. That is, long-term residents of a given neighborhood or city may not necessarily become engaged environmental stewards, while recent immigrants to a place with a strong sense of community resilience may. For example, Johnson (1998) discusses the long history of enslavement and oppression faced by African American communities in the South and how that might lead to a decreased attachment to local wildland areas. On the other hand, Krasny and Tidball (2017) note that local immigrant populations are among the most avid community gardeners, bringing their agricultural backgrounds to their new hometowns while concurrently building stronger social community networks.

Historic Socioeconomic Inequities

The influence of place identity and belongingness on stewardship efforts suggests that a discussion of perceptions of environmental stewardship must be situated within larger sociocultural contexts, including the historical context of race. For instance, the American environmental movement before the 1960s was often populated by the middle and upper middle class and emphasized outdoor recreation, an activity that limited its participants to those who had access to transportation, gear, leisure time, and public spaces. This was often alienating to African Americans and other racial or ethnic minorities who could not participate in these activities (Taylor 1998). African American post-slavery were not often welcomed on public lands such as parks and natural areas, and in fact were threatened with violence at times, as in the case of a young African American in Chicago who was stoned to death while swimming in Lake Michigan in 1919 (Taylor 1998). And while time and social progress have perhaps healed some wounds, there can still be a troubled relationship between some African American communities and environmentalism (Finney 2014; Parker and McDonough 1999). Studies suggest that African Americans perceive the environmental movement and perhaps environmental activities in general as "a white thing" (Elmendorf et al. 2005; Taylor 1998).

These culturally-bound and defined frames (i.e., perceiving environmental activities as a "white thing") are therefore a critical factor in understanding pathways and barriers to environmental stewardship (Taylor 1998; Carmichael 2017). Framing is characterized by Taylor as "a scheme of interpretations that guides the way in which ideological meanings and beliefs are packaged by movement advocates and presented to would-be supporters" (Taylor 2000, p. 511). The early American environmental movement, inspired by the Romantics/Transcendentalist ideology of the purity of nature, framed itself as advancing an agenda of land conservation and protection of public lands, issues that did not resonate with many communities of color including African Americans, who were largely denied access to these amenities (Taylor 2000). The civil rights movement's framing was much more relevant, timely, and empowering to African Americans (Taylor 1998). Consequently, the environmental movement had little overlap with the civil rights movement or other social justice movements of the 1960s and 1970s and failed to engage the majority of African Americans (or any cultural minority group for that matter).

Framing continues to be an issue with environmental organizations looking to reach a broader audience (Carmichael 2017). Thus, understanding the frames that motivate certain types of actions is an important and not well understood issue for most environmental organizations looking to inspire and engage stewards. As the next section will highlight, there are many different aspects of stewardship which appeal in different ways to people. Frames have the power to highlight these varied angles and appeal to distinct motivations of environmental stewards. In short, there are many pathways to stewardship and here we highlight the different motivations which might lead people down these paths.

What Motivates Stewards?

It is first worth noting that motivations to engage in environmental stewardship share much in common with motives for civic engagement in other forms. For instance, all of these behaviors have been shown to have a strong values-based component, which could be either intrinsic or extrinsic, and many theories have been proposed about what types of values can motivate prosocial action, and under which contexts. In one descriptive framework, Batson (1994) highlights four distinct motivations: self-interest or egoism, altruism, collectivism, and principled motivation. Often seen as a special case of prosocial action, environmental stewardship has been found to be motivated by a similar set of values and beliefs (Turaga et al. 2010). Stern's (2000) value-belief-norm theory has been especially influential in understanding environmental activism and behavior. In this model, values (i.e., egoism and altruism), similar in form to Batson's four motivation model, predict environmental beliefs, leading to a set of personal norms which translate into distinct types of action. However, environmental stewardship is also distinct in some notable ways from other forms of prosocial action. For instance, many environmental stewards share a belief that nature and all life are worthy of protection and that human beings should aim to reduce our impacts on them (Romolini et al. 2012). This has led researchers to propose the existence of an ecocentric or biospheric value orientation (Schultz and Zelezny 1998; Schultz 2001; Stern and Dietz 1994) in which the decision to engage in environmental behavior or stewardship is predicated on the perceived consequences to the environment, rather than to the self (egoism) or society (altruism) (de Groot and Steg 2008). Biospheric values are predictive of a variety of pro-environmental behaviors and attitudes, from green consumer choices (Sachdeva et al. 2015) to perceived connectedness between nature and the self (Martin and Czellar 2016), whereas ecocentric values have been shown to be negative predictors of environmental behavior (Kollmuss and Aygeman 2002). In self-reports, helping the environment, a proxy for biospheric concern, is a common response when environmental stewards are asked why they engage in long-term volunteerism (Bramston et al. 2011; Liarakou et al. 2011).

The academic attention given to the link between environmental values and behavior belies other types of motivations that stewardship may entail. That is not to say that values related to the self, community, and earth are not meaningful but rather that there may be alternative pathways to stewardship which have not been studied as extensively. As a point of intervention for volunteer groups or environmental organizations, a focus on motivators separate from ideological values might be beneficial in engaging the broader public. Although values can form the basis of environmental concern, they are often more immutable, subject to broader cultural or macrosocial factors and may even run to counter to sustainability goals (Markowitz and Shariff 2012). Here, we describe two other motivations that may foster stewardship – engaging in conservation efforts for fairly pragmatic reasons such as building life skills and on the opposite end of the spectrum, stewardship that is based on finding spiritual meaning. We also find that a sense of fostering community ties and building cohesion is a common thread through both of these distinct motivations.

Stewardship for Pragmatism

Engagement in environmental stewardship requires long-term commitment like many other social causes but can also be much more concrete and visceral than other forms of volunteerism. It often requires volunteers to engage in difficult manual labor, learning new skills and techniques at a particular action site. However, despite these challenges, environmental stewardship may be especially rewarding because stewards may see a visible or tangible product of their labors, and therefore more direct evidence of their efforts (Grese et al. 2000). For many of these stewards, the rapidly changing climate or extreme weather disturbances may be strong concerns, yet the abstract, distant quality of concepts such as global warming may make action difficult and worry paramount (Fritze et al. 2008). Localized, community-based environmental stewardship programs can be a way to cope with these anxieties and help people regain a sense of personal agency (Westphal 1999).

The Community Watershed Stewardship Program in Portland is a case in point. Watershed management and planning is not the most prototypical domain of community-based environmental stewardship; rather watershed management is usually viewed as a regulatory matter. However, as Shandas and Messer (2008) point out, community involvement in the management of local watersheds in the Portland area not only yielded important insights for complex urban resource management efforts, but it also increased civic engagement, local knowledge, and feelings of ownership of shared water resources citywide. This program was able to bring together multiple stakeholders, including universities, government agencies, and community groups from the design to implementation phases of the project. Ownership of the project was fostered by involving both community members and technical experts (e.g., hydrologists, ecologists, and botanists) from the onset, and continued participation in the project led to an increased awareness among community members of how their actions affect local water quality and availability.

Along with providing a concrete, localized means to address larger global environmental concerns, participating in environmental stewardship may be beneficial in other ways as well. Several studies have shown that intrinsic desires to learn about the environment and gain knowledge about ecological processes are important motivations in nurturing and maintaining environmental volunteerism (Ryan et al. 2001; Grese et al. 2000; Bramston et al. 2011). Volunteers who feel as though they are gaining expertise, fostering personal growth, and learning skills that could be transferable to other facets of their life experience more satisfaction and are perhaps, consequently, more likely to remain involved in such projects. These educational interests can be fulfilled by integrating programs such as species identification walks (i.e., plants, birds, etc.) during primary stewardship activities. Or, as in the case of the nonprofit organization Keep Indianapolis Beautiful (KIB) in Indianapolis, Indiana, youth education and workforce development can be used to encourage conservation and stewardship behavior in young people.

This model also affords the opportunity to increase diversity in the field of environmental management. KIB employs approximately 70 high school-aged students and 20 adults in a summer program, watering about 5,000 trees a week, as well as mulching, staking, and pruning. The program includes weekly enrichment activities ranging from outdoor recreation to vocational skills for "green-collar" jobs and networking opportunities. Partnerships with research institutions also present opportunities to nurture an early interest in environmental management. KIB and the Bloomington Urban Forestry Research Group (BUFRG) at Indiana University developed such a partnership which resulted in members of the Youth Tree Team being trained in data collection methods by BFURG researchers (Faris 2017). The data contributed to a study of planted-tree survival and growth in urban neighborhoods. Mari Aviles is an alumnus of KIB's Youth Tree Team who now works for the organization as a community arborist. For Mari, the Youth Tree Team was a direct connection to a real job in the environmental field. She says that she believes that changing the perceptions of who environmentalism is for will go a long way toward getting buy-in from marginalized communities (Aviles 2017). Currently, the members and volunteers of environmental organizations are predominantly white (Taylor 2014). Having more minorities working in green-collar jobs especially in positions of power gives people of color recognition and representation where historically they had neither. Note, however, that members of KIB are paid for their work and so motivations for these stewards may be slightly different for those who engage in stewardship on a volunteer basis.

Recruitment for new staff in environmental work often occurs informally through word of mouth, creating an insular network that makes it difficult for communities of color, the working class, or anyone outside of traditional environmental networks to access these jobs (Taylor 2014). Workforce development in environmental conservation and management is one direct way to bring a broader spectrum of people into the environmental field. For The Greening of Detroit (TGD), a not-for-profit organization in Detroit, Michigan, training a workforce in green jobs like forestry, landscaping, and conservation is seen as one step toward a more sustainable economy for the struggling city. Currently, the unemployment rate in many Detroit neighborhoods is still well above the national average (Hay 2017). Detroiters are in desperate need of jobs, and not just in the same industries that have been declining in the city for decades (auto, manufacturing). TGD's apprenticeship program has trained more than 350 people through their adult workforce training cohort since its inception in 2012. This program is designed to train recently-paroled people in landscape maintenance and arboriculture. Participants attend an 8-week training session that takes place in the classroom and in the field, and are paid a small stipend to help them make ends meet during the session. Training can take place in the nursery, city parks (especially in underserved areas where parks have fallen into disrepair), and in tree plantings on public rights-of-way. After they get their certificate, program participants are placed in a job. TGD reports the program as having a 100% placement rate. TGD hires some as crew leaders; others land jobs in local landscaping, or construction and maintenance companies. Most make $12–13/hour to start, which is considered a living wage in Detroit (Hay 2017).

Stewardship for Meaning

In addition to increasing personal efficacy and ecological knowledge, spiritual fulfillment may also be an important motivator for environmental stewardship. In recent years, psychologists have focused on the feeling of awe, as a special case of a spiritual connection to nature (Vining and Merrick 2012). Awe, defined as an overwhelming feeling of wonder, of something bigger than the horizons of one's everyday life, has been linked to spirituality, certainly, but has also been found to foster prosocial behavior and connection to other people (Rudd et al. 2012). Furthermore, many people report feeling a sense of awe or wonder as they realize the vastness of nature, and their relatively small role within it (Piff et al. 2015). Perhaps it is these sensations which help foster a spiritual bond between humans and nature – whether in the case of indigenous traditions all over the world highlighting the interconnectedness of humans and nature (Berkes 1999, 2017) or the reverence of natural resources within most major religious groups (Gottlieb 2006; Sachdeva 2016).

Humans' spiritual connection to nature can result in positive environmental action as well. Take, for example, Sadhana, a progressive Hindu organization active in the New York area. Through a stewardship project called Project Prithvi, Sadhana organizes monthly cleanups at Jamaica Bay on Long Island. Jamaica Bay is part of the Gateway National Recreation Area administered by the National Park Service and is used by many local residents for a variety of purposes, including recreation. However, it is also used by some members of the local Hindu, Indo-Caribbean community as a sacred site where the sandy shores of the bay are used to conduct religious rituals and, as is customary in many Hindu rituals, the flowing water is used to make offerings. We have found, in our ongoing collaboration with Sadhana, that the bay is often used as a stand-in for the Ganges River in India, a site of utmost religious importance within Hindu practice.

In addressing the concerns of local fishers, ecological groups, and the National Park Service that these practices might introduce contaminants into Jamaica Bay's delicate ecosystem, Sadhana is employing the Hindu concept of ahimsa, or non-violence, in raising ecological awareness among the Indo-Caribbean community. As co-founder Aminta Kilawan puts it, "the goal is not to tell 'the worshippers' to not pray there, but to do so respectfully [of local ecological needs]." Moreover, Sadhana is also committed to bridging understanding between the local worshipping community and the National Park Service and others for whom these practices may seem foreign and unnecessary. By building an open dialogue between all parties, Project Prithvi aims to promote a spiritually based sense of environmental stewardship while ensuring that religious practices continue to be sustained at a site of such profound sociocultural meaning.

It is precisely this sense of meaning, whether place-based or not, that might be important in promoting other environmental stewardship initiatives as well. In the case of volunteer oyster gardeners in New York City, Krasny, Crestol, and colleagues (2014) find that volunteers' memories and meanings attached to oysters as a keystone species for New York and their socio-ecological associations or memories with local places are strong motivating factors. They go on to explicitly state that this form of meaning attachment is distinct from an altruistic or biospheric value orientation because it is personal and linked to volunteers' self-identity. Finding meaning from working with particular species or in particular regions is not specific to oyster gardening. Sometimes, a connection to megafauna can be emphasized to engage stewards, as in the case of sea turtles as described in Campbell and Smith's (2005, 2006) work. People have also been shown to form these special connections with plants and trees. In the Chicago area, for example, Westphal (1993) notes that oak trees are a particularly iconic species with volunteers reporting vivid childhood memories about particular oak trees in and around their neighborhoods.

The generation of meaningful landscapes can take many forms. In another Portland example, the use of trees in urban neighborhoods is helping to shed light on forgotten histories. Canopy Story, a project that provides a platform to uncover personal histories using neighborhood trees, is integrating complex spatial analytical technologies with the craft of localized, place-based storytelling. Using remote sensing-generated maps to pinpoint the location of every tree in Portland, participants can use an online link to identify and share stories about an individual tree. Although several projects have shown that people do form meaningful attachments to trees in their neighborhood (LaFrance 2015; Westphal 1993), Canopy Story allows us to systematically analyze the types of meaning people attach to trees. Through this project, we can assess whether people are more likely to tell stories about their childhood, spiritual awakening, or falling in love with nature and then link those stories to places and neighborhoods. Although this project is still in its nascent stages, Canopy Story may become a tool for managers and volunteer groups to gain valuable information about the most meaningful trees in a given area. As a management tool, Canopy Story will not only have implications for management outcomes, i.e., keystone trees may need to be cared for differently, but also that these trees can become symbols for stewardship efforts, and help engage people in taking care of trees in their neighborhoods. The path to becoming an engaged steward is varied and an intricate mix of personal and ideological motivations. One aspect that appears to be a common thread among all of the examples and types of motivations described in this section is the desire to improve social cohesion. The KIB example from Indiana and the Community Watershed Program in Portland demonstrate that people develop ownership and pride in their communities through stewardship efforts. Similarly, one of Sadhana's explicit goals for Project Prithvi is to make communication between the Indo-Caribbean community, the National Park Service, and other local residents, as fluid as possible. As these examples illustrate, people value the idea of connecting with one another and building interpersonal networks as part of their journey to becoming environmental stewards. Much like caring for the environment, social fulfillment and the opportunity to be connected to committed others are powerful, if not specifically environmental, drivers of stewardship. To reiterate our stance from earlier in this section, we do not claim that values or concerns for the environment are not important precursors to engaging in environmental stewardship. They most certainly are, and many volunteers do get involved in stewardship activities as a means of restoring their connection to nature, or because they believe it is the right thing to do. However, personal, environmental values are not the sole, nor even the most predictive, factors in leading to sustained environmental stewardship. Others, such as community building, social resilience, personal efficacy, spirituality, and a sense of purpose, also play important roles.

Overcoming Barriers to Stewardship

Citizen stewardship initiatives are increasingly relied upon by governments who do not have the resources to improve social or environmental conditions (Romolini et al. 2012; Baker 2014). In recent years we have seen a proliferation of programs and events like the New York's MillionTreesNYC and Portland's Tree Inventory Project promoting environmental stewardship as a means to increase the capacity of government managers (Fischer et al. 2011). Despite the popularity of this model, evidence shows that citizen stewardship efforts have often been unsuccessful at engaging marginalized and underrepresented groups, stemming partly from lack of local resident buy-in to these types of stewardship activities (Clarke and Agyeman 2011).

Case studies further suggest a current trend toward distrust of politicians and public processes, reflecting a growing discontent with political processes and their outcomes in marginalized communities. In a survey conducted in the Puget Sound region of Seattle (Shandas 2007), respondents were asked about organizations from which they trust information. The results revealed that when participating in environmental projects, respondents most trusted friends, family, and neighbors (23%), professional associations (17%) (e.g., Adopt-a-Stream Foundation, Washington Trout, etc.), and university scientists (20%). Least trusted were government agencies, at the county, city, state, and federal levels. Although the focus on friends, family, and neighbors is consistent throughout the world, the lack of trust suggests a major challenge for those agencies in engaging with communities, though exceptions are also likely. Perhaps unsurprisingly, government agencies are increasingly contracting with nonprofit organizations to coordinate stewardship efforts (Shandas and Messer 2008; Chaskin et al. 2001; Romolini et al. 2012).

In Detroit, Michigan, The Greening of Detroit (TGD) experienced community backlash in some areas where they had planted trees, including vandalism and illegal removal of trees. TGD received "no-tree requests" (NTR) from 24% of residents in neighborhoods where they planned to plant between 2011 and 2014 (Carmichael 2017). Outreach by TGD revealed that despite understanding the benefits of trees, many people did not want them planted in their neighborhoods. People voiced concerns about maintenance issues once the trees were planted, and they felt like the community had been left out of the planning process. Situated in the context of the history of urban forestry in Detroit, these concerns are easily understood. Detroit's tree canopy has suffered from sustained tree loss and damage for over 50 years, thanks to pests, lack of funding, and neglect. Longtime residents remember the devastation of Dutch elm disease in the mid-twentieth century and more recently emerald ash borer, both of which wiped out millions of trees, and the latter from which the city has still not recovered. Residents were left with dead trees lining their streets, and a city government that lacked the resources to remove them, let alone maintain the living trees. As a result of these historical issues, trees may symbolize hardship, neglect, and failure of government agencies to properly manage environmental assets to many people (Carmichael 2017).

Not all Outreach Needs to be "Environmental" or Science-Based

In recent decades, perhaps due to increasingly scarce financial resources for environmental initiatives, managers have tended to make decisions about resource management based on quantifiable characteristics (putting a dollar value on environmental assets like trees), sometimes at the expense of important sociocultural values associated with nature like spiritual significance, beauty, and cultural identity and heritage (Chan et al. 2012; Schroeder 2011). This might mean the prioritization of ecosystem services, such as the carbon sequestration or air purification impacts of a tree planting program, with less emphasis on cultural values and long-term social impacts of environmental changes. Cultural environmental services were first defined by the Millennium Ecosystem Assessment as "the nonmaterial benefits people obtain from ecosystems through spiritual enrichment, cognitive development, reflection, recreation, and aesthetic experience including, e.g., knowledge systems, social relations, and aesthetic values" (Millennium Ecosystem Assessment 2005). The value of cultural, environmental services may be difficult to capture in a monetary valuation, but their value regarding stewardship and community investment in conservation efforts should not be underestimated. By better understanding and representing these values and including them in environmental decision-making processes, we will have a more robust representation of the value of environmental assets (Chan et al. 2012). Bringing things like storytelling (Sanderock 2000), placemaking, and environmental heritage and identity into planning may yield higher rates of stewardship and stronger and longer citizen engagement in stewardship.

Incorporating strong, already existing social structures into stewardship efforts may be one way of promoting and sustaining public engagement. For instance, a church is often the center of social life and networks, especially in communities of color (Taylor 1998). Migration and Me is a not-for-profit program in Chicago that describes itself as "focused on conservation and stewardship that engages people of faith in sharing their personal migration stories, connecting their stories to the migration of other species" (Migration and Me 2017). Migration and Me asks people to share their personal stories of migration and relate those experiences to animal migration, in this case, the migration of monarch butterflies. They work with dozens of places of worship, and across faiths. The goal of the organization is not only to help people understand their narratives of identity and place but also to have them connect their own religious lives with the natural world. The hope, ultimately, is that reinforcing the connection between community members' religious perspectives and nature will help them become stewards in their communities. In essence, Migration and Me builds on existing strong community institutions to weave new environmental understandings into the social network of the community.

Reframing environmental issues with a social justice lens is another way to reach disenfranchised communities.

For example, in Portland, Oregon, the city's Bureau of Environmental Services (BES) is partnering with the social justice organization, Asian Pacific Network of Oregon (APANO), and the environmental justice group OPAL to help the city get information about new tree planting efforts to residents in the Jade District, an area surrounding 82nd and Division in SE Portland which is about 40% communities of color and is a landing zone for many new immigrants to Portland. The Jade District is an area that has been identified as having low tree canopy relative to other parts of the city (City of Portland 2016b), and BES has committed to planting 100 new trees in the neighborhood. This is an area where the traditional outreach and educational efforts to engage the public, based on ecocentric appeals highlighting the environmental benefits of tree planting and stewardship, have not been as effective as city managers expected (Karps 2017). By partnering with APANO, which has strong ties with the Asian community in east Portland, city officials hope they will be better able to reach the community. Brining to light environmental inequities (such as the lack of tree canopy in neighborhoods with high percentages of low income communities of color) has been pointed to as a way to mobilize communities of color toward stewardship (Taylor 1998), yet policy and practice often favors environmental outcomes and "eco-lifestyle" projects rather than environmental justice concerns (Lubitow and Miller 2013). This new approach by the City of Portland to partner with a coalition of social justice organizations to reach environmental goals shows that city governments are beginning to understand the importance of framing when doing environmental work. According to Jennifer Carps, Tree Program Coordinator with the Bureau of Environmental Services, "a big part of my work is to win hearts and minds" (Karps 2017). Reframing is one way of reaching the hearts and minds of those who may not respond to more ecocentric messaging around environmental issues.

When designing environmental interventions or thinking about environmental activities, institutional agents (whether governmental or private) and individuals may often diverge, partly because of from issues of scale. Typically, an individual makes decisions based on household level needs, whereas government agencies or environmental organizations will necessarily think about environmental issues on a much larger-scale, whether that is a neighborhood, city, or forest. Differences in scope and scale of this sort may cause a disconnect between the priorities and goals of individual versus institutional actors.

For the Greening of Detroit (TGD), this disconnect was identified in interviews with both TGD staff and residents who lived in neighborhoods where TGD was doing tree plantings. TGD staff thought of tree plantings primarily regarding city-scale ecosystem services like carbon sequestration and stormwater runoff mitigation, while residents thought of tree planting in terms of the individual trees planted adjacent to their property which they would have to see and care for as long as they occupied their home. One of the primary complaints of residents was that they were not given a choice about what species of tree to plant. TGD felt that it was not problematic for them to choose the type of tree that was planted because they would select the trees that would maximize diversity and other city/neighborhood-level ecosystem services (Carmichael 2017). However, 22% of residents said that they might not accept a tree if they did not have a choice of the tree type that was planted. So, even though TGD's tree choices might maximize environmental benefits, their success would be minimized if they did not work with residents to find a mutually agreeable solution (Carmichael 2017).

The degree of community power in decision-making processes can also influence an individual's perceptions of environmental issues. Meaningful participation for residents can help to develop and reproduce shared narratives within the community. A community-based approach can improve dialogue between residents and environmental managers, and increase participation in conservation activities because individuals feel that they have agency (Shandas and Messer 2008). This is illustrated by an example from TGD's Neighborhood Nurseries program. The idea behind Neighborhood Nurseries was for The Greening of Detroit to plant sapling trees in empty lots in several neighborhoods around Detroit, then train neighborhood residents to care for trees as they grow, and finally help people to replant the trees in the neighborhood once they were mature. While this program had success in several neighborhoods where the community was onboard with the program, it failed in others where the community was not adequately informed or empowered.

For example, one of the nurseries that was a great success story was in the neighborhood of Grandmont-Rosedale on the northwest side of Detroit. The city of Detroit has made a substantial investment in this community, and its residents earn a higher income than the citywide average. All of the trees in the Grandmont-Rosedale nursery were cared for and eventually planted. However, trees in nurseries in two other neighborhoods were vandalized. A possible explanation is that people in the community did not accept or acknowledge ownership of these trees; rather they believed that the city was simply planting trees without permission. Another issue was the choice of lot for the nursery. People were using the lot as a soccer field and social space, and they understandably viewed the tree plantings as an attempt by the city to take over their space. TGD had done community outreach before undertaking this project, but in this case, their efforts fell short. The engagement and buy-in were not occurring with the right people, and TGD was not on the same page with the residents they were trying to serve. This was a case where their message and mission did not get out to the community effectively, with the result that the planted trees were not properly cared for (Hay 2017). Environmental, ecocentric outreach is not always an effective way to achieve good environmental outcomes, In this case, rather than ecocentric outreach, what was needed was social outreach, to understand the needs and desires of the community using the space where the trees were being planted.

In Chicago's Pilsen neighborhood, a lower west side community that is made up primarily of working class, immigrant communities, the Pilsen Alliance works to build grassroots movements that advance an environmental justice agenda. The nonprofit organization spent more than a decade fighting to shut down coal-based power plants after pollution, and air quality issues became major issues for public health. Pilsen Alliance leader Byron Sigcho says that though Pilsen residents do want environmental improvements, and have an interest in stewardship and volunteerism, the priority of most organizations working in the neighborhood is affordable housing and keeping working-class residents in their homes as the neighborhood improves (and gets more expensive). As in the case of the Jade District in Portland, Oregon, the Pilsen neighborhood is fearful of displacement as gentrification comes to their neighborhood. A major concern is environmental gentrification, where environmental improvements in a neighborhood lead to displacement of working-class residents. Leadership at the Pilsen Alliance believe that special interest groups, planners, and developers want to seize the opportunity to move into the improving neighborhood and take the benefits for themselves, without giving back to the existing community. These beliefs within the community may cause a resistance to engage in city-sponsored environmental improvement events (Sigcho 2017).

The examples cited above suggest that environmental managers may sometimes assume that trees are a universal salve to fix a wide range of problems in a neighborhood, with insufficient attention paid to the means through which tree planting programs are implemented on the ground. Programs that are implemented without the input of residents can lead to wasted time and resources for the city, broken trust with the community, and fears of resident displacement due to gentrification. Another problematic assumption is that people who do not want to participate in environmental stewardship simply have not been educated (Carmichael 2017; Sigcho 2017). Framing the problem in this way leads to a top-down model where environmental managers attempt to convince someone to participate in stewardship by explaining the benefits of trees to them, but not expanding the decision-making power of residents in the planning process (Carmichael 2017).

A top-down approach to urban environmental issues undervalues the perspectives, opinions, and wisdom of the community. Experiential and expert knowledge come from two distinct ways of knowing (Carmichael 2017), which are each valuable in environmental planning processes. If community trust is gained, basic needs of residents are met, and their voices are heard, then stewardship can be a powerful and long-lasting model to increase participation in environmental activities, and increase the capacity of government agencies with a fleet of educated volunteers.

New Paths to Stewardship

The examples presented in our case studies shed light on how we can better engage a diverse audience in stewardship efforts. Perhaps the most important thing to take away from this discussion is the fact that when discussing the environment, we are not *just* talking about the environment. Using a social constructionist framework, we understand that individual interpretations of environmental issues are mediated through a set of institutional and collective perceptions (normative notions) that form attitudes about stewardship. To understand an individual's pathway to stewardship, we must understand the social dynamics at play that frame their perceptions about environmental issues. Foundational to this process is acknowledging the history of marginalization of certain populations and the effect that racial oppression has on attitudes toward stewardship. Social justice and environmentalism have only recently become aligned, and there will be growing pains. However, reframing environmental issues through a social justice lens can bring new voices to the environmental field, and engage a broader audience in environmental stewardship work. Though the current chapter does not touch upon pathways or barriers to stewardship in developing countries, or the challenges faced by communities outside of the United States, it seems likely that similar concerns about justice are going to be salient within those contexts as well (Adger 2001; Mertz et al. 2009; Thomas and Twyman 2005).

We also see from our case studies that not all environmental outreach needs to be technical or science based. Examples like Migration and Me, Sadhana, and Canopy Stories show us that stewards are often made through storytelling or by making a personal or even religious connection to nature. Conversely, in the case of The Greening of Detroit, we see that presenting people with numbers and facts about the technical benefits of trees was not effective at changing residents' minds about whether or not they wanted to plant them. The personal connection to nature, often tied to a sense of awe or spirituality in nature's presence, is a crucial driver of stewardship.

Inevitably, there are some problems that are so systemic and deep-rooted that stewardship alone may not be able to effectively address them. In these cases, the top-down strategy may become more costly because the trees fail due to lack of care, or as in the case of the case study with The Greening of Detroit discussed in this chapter, because of direct vandalism. Such issues include gentrification and displacement concerns around environmental improvements. In places like Chicago's Pilsen neighborhood, local government may need to rebuild trust with the community and take care of essential issues like housing stability before buy-in for stewardship efforts can be expected from the community into stewardship efforts. While environmental issues are certainly critical, it is important to acknowledge and understand that in some communities basic needs are also not being met. To lift up the people in a community may lay the groundwork for future stewards of the land.

Conclusion

With mounting environmental pressures comes the need to engage local and global communities in conservation efforts. Environmental stewardship efforts are increasingly being recognized as a valuable means of engaging communities across geographic scales in conservation efforts. As this chapter has demonstrated, however, substantial barriers remain in inspiring people to join these efforts, particularly within historically marginalized communities where people may have different relationships with their local parks, forests, and the environment, more generally. A primary aim of this chapter is to highlight some of these barriers so that managers, policymakers, and volunteer organizations can recognize hurdles and ultimately prevail in building stewardship efforts. Identifying alternative pathways that can engage people in conservation efforts may be key to success. Some of these pathways can be found by reconceptualizing stewardship from a value-based effort to one that builds community resilience, provides people with valuable life and workplace skills, and may help them feel pride and ownership in their neighborhoods. These alternative pathways to stewardship may help create more effective environmental stewardship programs – effective both in terms of promoting sustainable outcomes and fostering long-term engagement on the part of local stewards.

Cross-References

► Community Engagement in Energy Transition
► Education in Human Values
► Empathy Driving Engaged Sustainability in Enterprises

References

Adger, W. N. (2001) Scales of governance and environmental justice for adaptation and mitigation of climate change. *Journal of International Development 13*(7), 921–931

Agyeman, J., & Evans, T. (2003). Toward just sustainability in urban communities: Building equity rights with sustainable solutions. *The Annals of the American Academy of Political and Social Science, 590*, 35–53. Rethinking Sustainable Development.

Aviles, M. (2017). Interview with Mari Aviles of the Greening of Detroit. Interview by Sachi Arakawa.

Baker, D. (2014). *Environmental stewardship in New York City Parks and Natural Areas: Assessing barriers, creating opportunities, and proposing a new way forward.* Ph.D. dissertation, Yale School of Forestry and Environmental Studies.

Batson, C. D. (1994). Why act for the public good? Four answers. *Personality and Social Psychology Bulletin, 20*(5), 603–610. https://doi.org/10.1177/0146167294205016.

Berkes, F. (2017). *Sacred ecology.* New York: Routledge.

Berkes, F. (1999). Sacred ecology: traditional ecological knowledge and resource management. Taylor & Francis, Philadelphia.

Berry, R. J. (2006). Environmental stewardship: critical perspectives, past and present. T&T Clark, London.

Bramston, P., Pretty, G., & Zammit, C. (2011). Assessing environmental stewardship motivation. *Environment and Behavior, 43*(6), 776–788. https://doi.org/10.1177/0013916510382875.

Brundtland, G. H. (1987). *Our common future: Report of the World Commission on Environment and Development.* New York: Oxford University Press.

Buell, L. (2009). The future of environmental criticism: environmental crisis and literary imagination. Blackwell, London.

Campbell, L. M., & Smith, C. (2005). *Volunteering for sea turtles? Characteristics and motives of volunteers working with the Caribbean Conservation Corporation in Tortuguero* (pp. 169–194). Costa Rica: MAST.

Campbell, L. M., & Smith, C. (2006). What makes them pay? Values of volunteer tourists working for sea turtle conservation. *Environmental Management, 38*(1), 84–98. https://doi.org/10.1007/s00267-005-0188-0.

Carmichael, C. (2017). *The trouble with trees? Social and political dynamics of greening efforts in Detroit, Michigan*. Ph.D. dissertation, Michigan State University.

Carr, A. (2002). *Grass roots and green tape: Principles and practices of environmental stewardship*. Annandale: Federation Press.

Chan, K., et al. (2012). Where are cultural and social in ecosystem services? A framework for constructive engagement. *Bioscience, 62*, 744–756.

Chaskin, R., Brown, P., Venkatesh, S., & Vidal, A. (2001). *Building community capacity*. Edison, NY: Aldine de Gruyter.

Chiesura, A., & de Groot, R. (2003). Critical natural capital: a socio-cultural perspective. *Ecological Economics, 44*(2), 219–231. https://doi.org/10.1016/S0921-8009(02)00275-6.

City of Portland. (2016a). Parks and Recreation. *Portland's Tree Canopy Present and Future*. https://www.portlandoregon.gov/parks/article/509398

City of Portland. (2016b). Bureau of planning and sustainability. In *Portland Comprehensive Plan 2035*. https://www.portlandoregon.gov/bps/57352.

Clarke, L., & Agyeman, J. (2011). Shifting the balance in environmental governance: Ethnicity, environmental citizenship and discourses of responsibility. *Antipode, 43*(5), 1773–1800.

de Groot, J. I. M., & Steg, L. (2008). Value orientations to explain beliefs related to environmental significant behavior: How to measure egoistic, altruistic, and biospheric value orientations. *Environment and Behavior, 40*(3), 330–354. https://doi.org/10.1177/0013916506297831.

Devall, B., & Sessions, G. (1985). Deep ecology. In L. P. Pojman, P. Pojman, & K. McShane (Eds.), *Environmental ethics: Readings in theory and application*. Cengage Learning. Salt Lake City, UT: Gibbs Smith

Elmendorf, W. F., Willits, F. K., & Sasidharan, V. (2005). Urban park and forest participation and landscape preference: A review of the relevant literature. *Journal of Arboriculture, 31*(6), 311.

Faris, N. (2017). Interview with Nate Faris of Keep Indianapolis Beautiful. Interview by Sachi Arakawa.

Finney, C. (2014). *Black faces, white spaces: Reimagining the relationship of African Americans to the great outdoors*. Chapel Hill: UNC Press Books.

Fischer, D., Connolly, J., Erika, S., Campbell, L. (2011). *Who volunteers to steward the urban forest in New York City? An analysis of participants in MillionTreesNYC planting events*. Environmental stewardship project white paper #1. National Science Foundation.

Fritze, J. G., Blashki, G. A., Burke, S., & Wiseman, J. (2008). Hope, despair and transformation: Climate change and the promotion of mental health and wellbeing. *International Journal of Mental Health Systems, 2*, 13. https://doi.org/10.1186/1752-4458-2-13.

Gottlieb, R. S. (2006). *A greener faith: Religious environmentalism and our planet's future*. Oxford: Oxford University Press.

Grese, R. E., Kaplan, R., Ryan, R. L., & Buxton, J. (2000). Psychological benefits of volunteering in stewardship programs. In R. B. Hull (Ed.), *Restoring nature: Perspectives from the social sciences and humanities* (pp. 265–280). Washington, DC: Island Press.

Hay, D. (2017). Interview with Dean Hay of the Greening of Detroit. Interview by Sachi Arakawa.

Johnson, C. Y. (1998). A consideration of collective memory in African American attachment to wildland recreation places. *Human Ecology Review, 5*(1), 5–15.

Kaplan, R. (1985). Nature at the doorstep: Residential satisfaction and the nearby environment. *Journal of Architectural and Planning Research, 2*(2), 115–127.

Karps, J. (2017). Interview with Jennifer Karps of the Portland Bureau of Environmental Services. Interview by Sachi Arakawa.

Kollmuss, A., & Agyeman, J. (2002). Mind the gap: Why do people act environmentally and what are the barriers to pro-environmental behavior? *Environmental Education Research, 8*(3), 239–260. https://doi.org/10.1080/13504620220145401.

Krasny, M. E., & Tidball, K. G. (2017). Community gardens as contexts for science, stewardship, and civic action learning. In J. Blum (Ed.), *Urban horticulture: Ecology, landscape, and agriculture*. New York: Apple Academic Press.

Krasny, M. E., Crestol, S. R., Tidball, K. G., & Stedman, R. C. (2014). New York City's oyster gardeners: Memories and meanings as motivations for volunteer environmental stewardship. *Landscape and Urban Planning, 132*, 16–25. https://doi.org/10.1016/j.landurbplan.2014.08.003.

LaFrance, A. (2015, July 10). When you give a tree an email address. *The Atlantic*. Retrieved from https://www.theatlantic.com/technology/archive/2015/07/when-you-give-a-tree-an-email-address/398210/.

Leopold, A. (1949). *A Sand County almanac: With other essays on conservation from Round River*. New York: Oxford University Press.

Liarakou, G., Kostelou, E., & Gavrilakis, C. (2011). Environmental volunteers: Factors influencing their involvement in environmental action. *Environmental Education Research, 17*(5), 651–673. https://doi.org/10.1080/13504622.2011.572159.

Lubitow, A., & Miller, T. (2013). Contesting sustainability: Bikes, race, and politics in Portlandia. *Environmental Justice, 6*(4), 121–126.

Markowitz, E. M., & Shariff, A. F. (2012). Climate change and moral judgement. *Nature Climate Change, 2*(4), 243. https://doi.org/10.1038/nclimate1378.

Martin, C., & Czellar, S. (2016). The extended inclusion of nature in self scale. *Journal of Environmental Psychology, 47*, 181–194. https://doi.org/10.1016/j.jenvp.2016.05.006.

Mertz, O., Halsnæs, K., Olesen, J. E. Rasmussen, K. (2009). Adaptation to Climate Change in Developing Countries. Environmental Management 43 (5):743-752

Migration and Me. (2017). *Faith in place*. www.faithinplace.org/our-programs/migration-me.

Millennium Ecosystem Assessment. (2005). *Ecosystems and human well-being: Synthesis*. Island Press. https://www.millenniumassessment.org/en/index.html.

Palmer, C. (2006). Stewardship: A case study in environmental ethics. In R. J. Berry (Ed.), *Environmental stewardship*. London, New York: A&C Black.

Parker, J. D., & McDonough, M. H. (1999). Environmentalism of African Americans: An analysis of the subculture and barriers theories. *Environment and Behavior, 31*(2), 155–177.

Piff, P. K., Dietze, P., Feinberg, M., Stancato, D. M., & Keltner, D. (2015). Awe, the small self, and prosocial behavior. *Journal of Personality and Social Psychology, 108*(6), 883–899. https://doi.org/10.1037/pspi0000018.

Roberts, M. (2011). Evangelicals and climate change. In D. Gerten & S. Bergmann (Eds.), *Religion in environmental and climate change: Suffering, values, lifestyles*. London: Bloomsbury Publishing.

Romolini, M., Brinkley, W., & Wolf, K. (2012). *What is urban environmental stewardship? Constructing practitioner-derived framework*. Portland: United States Department of Agriculture, US Forest Service: Pacific Northwest Research Station.

Routley, R. (1973, April 1). Is there a need for a new, an environmental ethic. https://doi.org/10.5840/wcp151973136.

Rudd, M., Vohs, K. D., & Aaker, J. (2012). Awe Expands People's Perception of Time, Alters Decision Making, and Enhances Well-Being. *Psychological Science, 23*(10), 1130–1136. https://doi.org/10.1177/0956797612438731.

Ryan, R. L., Kaplan, R., & Grese, R. E. (2001). Predicting volunteer commitment in environmental stewardship programmes. *Journal of Environmental Planning and Management, 44*(5), 629–648. https://doi.org/10.1080/09640560120079948.

Sachdeva, S. (2016). Religious identity, beliefs, and views about climate change. https://doi.org/10.1093/acrefore/9780190228620.013.335.

Sachdeva, S., Jordan, J., & Mazar, N. (2015). Green consumerism: Moral motivations to a sustainable future. *Current Opinion in Psychology, 6*, 60–65. https://doi.org/10.1016/j.copsyc.2015.03.029.

Sanderock, L. (2000). When strangers become neighbours: Managing cities of difference. *Planning Theory and Practice, 1*(1), 13–30.

Schroeder, H. W. (2011). Does beauty still matter? Experiential and utilitarian values of urban trees. In: Trees, people and the built environment. Proceedings of the Urban Trees Research

Conference; 2011 April 13–14. Edgbaston, Birmingham, UK. Institute of Chartered Foresters: 159–165.
Schultz, P. W. (2001). The structure of environmental concern: Concern for self, other people, and the biosphere. *Journal of Environmental Psychology, 21*(4), 327–339. https://doi.org/10.1006/jevp.2001.0227.
Schultz, P. W., & Zelezny, L. C. (1998). Values and proenvironmental behavior: A five-country survey. *Journal of Cross-Cultural Psychology, 29*(4), 540–558. https://doi.org/10.1177/0022022198294003.
Shandas, V. (2007). An empirical study of streamside landowners' interest in riparian conservation. *Journal of the American Planning Association 73*(2).
Shandas, V., & Messer, W. B. (2008). Fostering green communities through civic engagement: Community-based environmental stewardship in the Portland area. *Journal of the American Planning Association, 74*(4), 408–418. https://doi.org/10.1080/01944360802291265.
Sigcho, B. (2017). Pilsen Alliance Interview. Interview by Sachi Arakawa.
Stern, P. C. (2000). New environmental theories: Toward a coherent theory of environmentally significant behavior. *Journal of Social Issues, 56*(3), 407–424. https://doi.org/10.1111/0022-4537.00175.
Stern, P. C., & Dietz, T. (1994). The value basis of environmental concern. *Journal of Social Issues, 50*(3), 65–84. https://doi.org/10.1111/j.1540-4560.1994.tb02420.x.
Stone, E. (2009). The benefits of community-managed open space: Community gardening in New York City. In L. Campbell & A. Wiesen (Eds.), *Restorative commons: Creating health and well-being through urban landscapes*. Gen. Tech. Rep. NRS-P-39. Newtown Square, PA: U.S. Department of Agriculture, Forest Service, Northern Research Station. p. 278.
Svendsen, E., & Campbell, L. (2008). Urban ecological stewardship: Understanding the structure, function and network of community-based urban land management. *Cities and the Environment (CATE), 1*(1). Retrieved from http://digitalcommons.lmu.edu/cate/vol1/iss1/4.
Taylor, D. (1998). Blacks and the environment: Toward an explanation of the concern and action gap between blacks and whites. *Environment and Behavior, 21*(2), 175–205.
Taylor, D. (2000). The rise of the environmental justice paradigm. *American Behavioral Scientist, 43*(4), 508–580.
Taylor, D. (2014). *The state of diversity in environmental organizations*. Ann Arbor. Green 2.0. http://vaipl.org/wp-content/uploads/2014/10/ExecutiveSummary-Diverse-Green.pdf.
Teig, E., Amulya, J., Bardwell, L., Buchenau, M., Marshall, J. A., & Litt, J. S. (2009). Collective efficacy in Denver, Colorado: Strengthening neighborhoods and health through community gardens. *Health & Place, 15*(4), 1115–1122. https://doi.org/10.1016/j.healthplace.2009.06.003.
Thomas, D. S. G., & Twyman, C. (2005). Equity and justice in climate change adaptation amongst natural-resource-dependent societies. *Global Environmental Change, 15*(2), 115–124
Tidball, K. G. (2014). Trees and rebirth: Social-ecological symbols and rituals in the resilience of Post-Katrina New Orleans. In *Greening in the red zone* (pp. 257–296). Dordrecht: Springer. https://doi.org/10.1007/978-90-481-9947-1_20.
Turaga, R. M. R., Howarth, R. B., & Borsuk, M. E. (2010). Pro-environmental behavior. *Annals of the New York Academy of Sciences, 1185*(1), 211–224. https://doi.org/10.1111/j.1749-6632.2009.05163.x.
Van Dyke, F. (1996). *Redeeming creation: The Biblical basis for environmental stewardship*. Downers Grove: InterVarsity Press.
Vining, J., & Merrick, M. S. (2012). Environmental Epiphanies: Theoretical Foundations and Practical Applications. The Oxford Handbook of Environmental and Conservation Psychology. https://doi.org/10.1093/oxfordhb/9780199733026.013.0026.
Welchman, J. (1999). The virtues of stewardship. *Environmental Ethics*. https://doi.org/10.5840/enviroethics19992146.
Westphal, L. M. (1993). Why trees? Urban forestry volunteers values and motivations. In *Managing urban and high-use recreation settings* (pp. 19–23). St. Paul: USDA North Central Forest Experiment Station.

Westphal, L. M. (1999). *Growing power? Social benefits from urban greening projects*. Chicago, IL: University of Illinois at Chicago.
White, L. (1967). The historical roots of our ecologic crisis. *Science, 155*(3767), 1203–1207.
Worrell, R., & Appleby, M. C. (2000). Stewardship of Natural Resources: Definition, Ethical and Practical Aspects. *Journal of Agricultural and Environmental Ethics, 12*(3), 263–277. https://doi.org/10.1023/A:1009534214698.

The Theology of Sustainability Practice

How Cities Create Hope

Peter Newman

Contents

Abstract

Reflecting on a lifetime of sustainability practice as an academic, politician, public servant, and community activist, I have drawn on how theology has provided the roots of engagement in tackling the issues of change. Understanding the role of cities in theological history enables us to see how the global and local, the personal and the political, are linked in the journey we need to take towards sustainability. Key themes will be how nature and cities are intertwined, the role of prophets, the competing visions of a future city that have guided urban planners for centuries, and the role of activism and good work as a source of hope in creating the city of the future.

Keywords

Theology · Sustainability · Bible · Hope · Cities · Prophets · Romanticism · Activism

P. Newman (✉)
Curtin University Sustainability Policy (CUSP) Institute, School of Design and Built Environment, Curtin University, Perth, Australia
e-mail: p.newman@curtin.edu.au

S. Dhiman, J. Marques (eds.), *Handbook of Engaged Sustainability*,
https://doi.org/10.1007/978-3-319-71312-0_6

Introduction

I have just completed a book called "Never Again: Reflections on Environmental Responsibility After Roe 8" (Gaynor et al. 2017). It is an edited collection of reflections from 30 academics involved in a local action that stopped a freeway as part of a state election where the issue had become the main moral choice. It was a deeply divisive issue with huge media coverage of demonstrators versus police, of bulldozers versus cherished local bush. The newly elected government has since stopped the freeway, put the money into an extended public transport system, and converted the road reserve into an iconic green corridor. The previous government had begun to bulldoze the highly significant bush and wetlands so it could "show greenies strength" – how a government could stand up to environmental activists and win votes. Such values damage the soul of the city. They had to be stopped and were stopped.

The new book does not discuss the theology of such a political choice, but it certainly was a theological issue. It was a deeply spiritual choice made by thousands of people to demonstrate and for 176 to be arrested. It has had its impact directly on the soul of my city, and its future is very different now we have dealt with that issue at such a fundamental level.

My theology was shaped 40 years ago and this freeway was one of the first issues that attracted my activism back then. It was a very long-term struggle, as are most deeper issues in our cities.

Political activism needs to be based on theology in my view and this chapter is how I came to understand what it was that shaped my ideas and motivations.

How Nature and Cities Are Intertwined: A Theological Perspective

In the 1970s, as I was studying environmental science I began to read the work of Jaques Ellul, a French sociology professor who wrote 58 books and over 1000 articles in two different styles that he said were in a constant "dialectic." One style was secular sociology and the other was theology. He was also an activist having been a member of the French resistance in the Second World War (awarded a Yad Vashem medal for saving the lives of Jews) and was an elected local government councilor in Bordeaux.

Ellul wrote the Technological Society (his most well-known work) where he showed how pervasive the belief in a technological solution or technique had become in all areas of life and thinking (Ellul 1964). The partner theological book was called The Meaning of the City (Ellul 1970) which showed the theological base for understanding cities, humanities' greatest achievement, and our central problem to helping us solve the sustainability issues of our time at a deeper spiritual level.

Ellul went through the Bible from start to finish showing how cities are understood. The Old Testament views of nature and cities give us a foundation for sustainability, and as these books serve Islam, Judaism, and Christianity, it is possible to see how a large proportion of humanity are at least theoretically influenced (I am not an expert on other religions but I know enough to have seen similar approaches to environmental care and sustainability, see UNEP (2016).).

The Genesis perspective shows a world that has been created over epochs of time (not just literal "days") with an unformed chaos then becoming a place for nature and finally humans. It is declared to be "good" with humans living in the garden provided by nature and given a role "to tend and care for it." But humans choose to pursue the "knowledge of good evil" when given a choice and were expelled from the garden. Ellul and others see this as the choice to create settlements around agriculture and technology rather than the millennia of time that humans had been hunter-gatherers depending totally upon nature and an innocence that is still reflected in many traditional tribal groups. However, virtually all hunter-gatherers have made and continue to make the choice to live in cities or settlements of some kind rather than moving with the flows of nature as we did over deep time.

The choice to pursue a less nature-immersed life and build cities is soon seen to lead to disasters symbolized by the Tower of Babel in Genesis where the inability to communicate properly leads to violence and disaster. From here on, the history of cities emerges as a rise and fall based on the soul of the city and the ability or inability of urban citizens to respond to their prophets.

The Role of Prophets
"And the people bowed and prayed
To the neon god they made
And the sign flashed out its warning
In the words that it was forming
And the sign said 'The words of the prophets
Are written on the subway walls
And tenement halls
And whispered in the sounds of silence.'"
Paul Simon and Art Garfunkel, The Sounds of Silence

In the middle and last volumes of the Old Testament, there are many stories of how fragile is the civilization created within cities. Archeologists and ancient history academics can piece together why cities collapse. Megiddo, in Israel (a city after which the word Armageddon was developed), is a large archeological site where the remains of 22 cities have been located one on top of the other. The work of Jared Diamond (2005) called "Collapse" shows that the fundamental cause is that the people in a city choose not to change, they do not respond to the prophets of their time. This is a spiritual issue for cities. Thus, it is possible to see that spirituality seeps into all of life especially our cities which are the human creation that can be arrogant or humble about how we create our future; they can choose a future that can face up to the evils of corruption, inequality, frivolity, and destruction of nature or they can let it destroy them.

The Old Testament prophets first of all called this out publicly; they did not idly stand by: *"Don't listen to those who say all is well, who give us smooth words and seductive visions"* (Isaiah 30 10) but accepted there *was* a problem and that it was caused by the *"arrogance of ruthless men"* (Isaiah 13″). But they went beyond blaming individuals; they talked about it being something to do with the "soul of the city." They called this a choice between "Zion" and "Babylon," two city types that represented the choices of life or death, the city of hope or the city of fear and despair. Any city could be seen as making this choice at particular times in their history. The soul of the city was linked to a range of social injustices and to how the city impacted on the natural environment. Jeremiah said *"your wrong doing has upset nature's order."* Why? What wrong doing? Because they do not *"grant justice to the poor"* (Jer 5 25–28) and because as Isaiah said:

> Though in their pride and arrogance they say,
> The bricks are fallen but we will build in hewn stone,
> The sycamores are hacked down,
> But we will use cedars instead. Isaiah 9:10.

Technological pride and belief in the absolute ability of the market to replace any resource was not apparently invented by us moderns after all. So, in the end judgement falls on such arrogance and the evidence of such happening is in the history books. Ephesus, the last large modern western city to collapse and be abandoned (around 1000 AD), is said to have collapsed from overcutting its forests which silted up the river and made it no longer a navigable port. The greatest city in ancient times was Babylon. Isaiah, writing in the eighth century BC predicted that the towering city of Babylon would collapse as it did in 140 BC from over-exploiting of its forests which diverted the Euphrates River and destroyed their agricultural base. Isaiah dreamed of the day when this would happen to Babylon:

> The whole world has rest and is at peace it breaks into cries of joy,
> the pines themselves and the cedars of Lebanon exult over you:
> Since you have been laid low they say no man comes up to fell us. (Is 10 18, 19).

The Competing Visions of the Future

The city, as seen in this theological perspective, was not seen as fundamentally wrong; people were told not to "leave Babylon" until it was about to collapse. They needed to try and make it work. It was only wrong if it did not make the necessary changes shown by its prophets.

Cities will always involve a choice to avoid this collapse – first morally and then in its economy and social fabric. People will choose in cities to create more and more knowledge and more and more complex societies with more and more opportunity, but they will always be under the possibility of collapse depending on the soul of the city (Newman 2007).

We should accept that cities have a future, but they need to constantly be reclaiming their future, choosing the direction that enables them to be inclusive, productive, and sustainable. If they are not doing this then they need to adapt or else they open themselves to the possibilities of collapsing. This tension between futures for a city is the way that our western spiritual traditions saw the future. Such competing visions can continue to inform us.

The last book in the Bible pitches two scenarios of the future being two city-types which stand in tension, as throughout history. Once again one is called Zion, the City of God, and the other is called Babylon, the City of Man. Zion is not paradise, a remade perfect nature, it is a city that is built by human science and craftsmanship; it is pictured as a city of jewels or diamonds – which appear to be human-made achievements. Diamonds are scientifically discovered, crafted with great skill, and valued for generations as a symbol of hope. This city of diamonds is also a city in harmony with nature; a tree of life and a river of life flow through the city.

The other city, Babylon (not the historical city of Babylon but any city that does what it did and collapsed) is a city full of frivolous consumption, repression of people, and degradation of nature. There are no diamonds here, just fear and despair. This city is under judgement and it will collapse.

We do dream of more rural utopian pasts and try to help make them continue, but history tells us that the process is one way towards more cities. Cities provide opportunities, but they also are fragile. History, like that told by Jared Diamond, shows that the cities we create from our knowledge can collapse or can adapt and thrive. This theological perspective shows that we can even create a legacy of human diamonds, if we make the right choices in our cities and do the daily work needed to make it happen.

Cities were first built about 13,000 years ago. This occurred after the ice age though variations in climate did impact so when the world became colder and drier again, many of the first Middle Eastern settlements did collapse as their ecological base was ruined. These societies then spread west and east from the Fertile Crescent, and as the climate warmed, created the settlements and agricultural areas in Europe and Asia that we now know. At the same time, great cities were being built in Africa and South America, and even in North America there is evidence of large settlements pre-1492, as people created new economies of commerce and trade.

Although cities have collapsed on every continent as they depleted their soils or were unable to manage their settlements or were destroyed by invaders, they did not go back to the "garden" (the mystical Garden of Eden where humans first lived) and become hunter gatherers again. The broad sweep of history shows that cities tend to be rebuilt and have endured and grown, and the place of nature has changed in that journey. But they are still very fragile.

In more recent times the world watched in amazement as the city of New Orleans collapsed due to an extreme climate event. The way that all civilizing constraints disappeared so quickly as people tried desperately to find food and safety shocked us all. History though suggests that we should not be shocked – the potential is there in any city. However, we do then tend to rebuild such cities and sometimes try to learn from the lessons.

Does this mean that sustainability is somehow impossible, that nature will always struggle to have a place with human civilization? No, but first let me demonstrate that there are pitfalls.

Isn't Rural Life Better for us?

There have been attempts in the past to ruralize cities. Pol Pot and Mao were two recent leaders who believed in a rural-oriented revolution that would replace our cities. There is some attraction to most of us, perhaps a memory of our distant past, but history shows that such rural idealization of cities does not last and does little for rural production which soon collapses under the weight of incompetent urban peasants. In both China and Cambodia massive loss of life followed. Cities rapidly rebuild after such experiments. Phnom Penh and Beijing are not lacking in people after their failed attempts to empty them.

The idea that resource vulnerability and climate change will disperse our cities into rural settlements or even rural suburbs, with only those surviving being those who can farm a small block is a vision sometimes promulgated by the Permaculture movement (Mollison and Holmgren 1978). Permaculture is a set of design rules that help make cities more sustainable that are all eminently sensible and helpful. But sometimes it develops into an anti-urban ideology that tries to say cities cannot work and should be abandoned in favor of small rural villages that are self-sufficient in food.

This is not likely to be a realistic or helpful scenario. We are going to have come to terms with a new kind of agriculture and a new kind of city, but we are not likely to reverse 13,000 years of urban history. Nor would we want to as all the evidence suggests that such small rural holdings and garden suburbs on the fringes of our cities are the least sustainable parts of our cities, especially in transport, and do nothing much for regional productivity (Newman and Kenworthy 1999, 2015). There is a growing sense that rural settlements and peri-urban areas could perhaps be the best place for these ideas to be practiced but there is little hope for a uniform dismissal of the urban area as a place to live and work with shared economic opportunities not available outside cities.

The reason we rebuild and adapt cities is that our choices for returning to nature are very limited and, to most people, are not acceptable. Not only do we not want to become totally dependent on foraging or hunting for our food, we mostly do not want the responsibility for food production at all. The attraction of doing other things, which are only possible if we are freed from food production, drives us to cities. Thus, history has also shown multiple failures of rural utopian visions which are set up to ensure that people create their own food rather than being in cities or towns dependent on food. History is dotted with these experiments but those that succeed are usually either very short-lived or are heavily dependent on an urban area nearby. Locked out of Eden, we seem destined to be more and more urban.

So, do we just put up with cities knowing that it would be much better for us if we lived away from them? This is largely the issue of Romanticism and how we deal with the problems in cities.

Romanticism or Activism

There is a difference between Romanticism and a healthy view of cities and their potential. Romanticism sees nature as the source of all truth and beauty and the artifices of human beings only spoil this. The Romantic poets came out of the early industrial revolution; they saw Babylon being built there. Coleridge and Wordsworth fled the cities and wrote about the purifying effect of nature. The city was seen as deeply alienating with no possibility of being reformed.

Others at that time focused on what could be done about the cities. The famous song now sung at soccer games in the UK "Jerusalem" is based on William Blake's 1808 poem. He says:

> And was Jerusalem builded here,
> Among these dark Satanic Mills?

Blake could instead just see Babylon and was part of a movement trying to reform the cities of England. The original poem by Blake is prefaced by a quotation from the Bible: "Would to God that all the Lords people were prophets." (Numbers 11: 29). So, Blake's famous hymn says:

> I will not cease from Mental Fight,
> Nor shall my Sword sleep in my hand:
> Till we have built Jerusalem,
> In England's green and pleasant Land.

Commentators on Blake such as Christopher Rowland, Oxford Professor of Theology, suggest that: "The words of the poem stress the importance of people taking responsibility for change and building a better society 'in England's green and pleasant land'" (Rowland 2007). UK's cities today are much cleaner, indeed green and pleasant, than during the industrial revolution centuries. The UK is one of the leaders in the transition to renewable energy which is remaking the future world for cities everywhere. The last coal fired power station in the UK will close within a decade (Newman et al. 2017).

The alternative to a Romantic view of cities is thus a prophetic or activists view. Part of this is to ensure that cities respect nature and have nature as part of their everyday living and working environments. We call this today "biophilic urbanism," and it is not anti-urban but is anti-artificiality and the arrogance that asserts we do not need nature (Beatley 2011; Newman et al. 2017). Bringing nature into cities is not Romanticism, it is being practical and creative about human life as well as the world of nature that now lives in close partnership with us.

A pioneering woman in Western Australia, Georgiana Molloy, came out from the UK in 1829. She followed the Romantic poets of Wordsworth and Coleridge in the Lake District before sailing half way around the world to start a new life in a strange land. She brought their Romantic ideas with her to Western Australia and set up a campsite where she mostly had contact with local aboriginal people. She learned much from them and discovered the astonishing biodiversity of South Western Australia. She became a brilliant botanist but died at 37 without proper health care and the benefits of city life and said she was "utterly wretched" at the end despite towns nearby offering the potential for health services that may have saved her (Barry 2016). Is it not possible to have the best of urbanism with the best of nature?

The Sustainable City

The work I have been doing as an academic and an activist come together around the concept of sustainability and cities. The book that Jeff Kenworthy and I produced in 1999 introduced the concept of sustainability into the academic and activist agenda of cities (Newman and Kenworthy 1999). It is a good agenda to enable us to look at how cities are faring in the complex world of surviving into the longer-term future, of choosing a future of hope rather than a future of fear and despair.

How the sustainable city agenda helps to shape our choices for the future can be framed in different ways. Bruntland's definition of sustainable development emphasized ensuring we did not shut off the options of future generations. Others discuss integrated economic, social, and environmental thinking. In our approach, we emphasize the metabolism of cities and how it can be integrated with the human livability of cities. We defined sustainability in cities as how cities face up to reducing their input of resources and output of wastes while improving their livability (the social and economic opportunities for individuals and communities).

We collected data on cities from around the world and highlighted where they needed to change, especially in the highly car-dependent cities of America and Australia. But most of all (in terms of bringing about change), we told the stories of those cities that had done good things to improve their sustainability. We shared the diamonds of hope (Newman and Kenworthy 1999, 2015; Newman and Jennings 2008; Matan and Newman 2016; Newman et al. 2017).

Since then we have expanded the concept to understand how cities have three kinds of cities within their spatial urban area: walking cities, transit cities, and automobile cities – and these have very different sustainability outcomes built into their fabrics (Newman et al. 2016; Thomson and Newman 2016, 2017). More recently we have shown that sustainability is now happening much faster even while the need, especially climate change, is also growing faster (Green and Newman 2017; Newman et al. 2017). We have therefore begun pushing for an even more demanding concept: the Regenerative City (Newman et al. 2017). This means that cities should not just be reducing their resource consumption and waste but actually they need to be regenerating the atmosphere and the city's bioregional environment as well as creating better opportunities for people. As the stakes of *not* changing grow, the need for cities to regenerate their impacts is now becoming the real agenda for long-time survival.

Is this possible? I don't know but I remain hopeful. Hope is getting harder but its still hope. Why? Because the past 40 years has shown that when I am hopeful with what I write and what I do in politics, what I say in the media about the campaigns that matter, I believe that the diamonds of hope begin to be discovered, polished, and turned into legacies that are going to last for generations.

The rebuilding of suburbia is one of the great challenges of our time. Their advertized rural bliss is increasingly looking like the failed experiments of other rural utopias as they dissolve into long commutes and diminished services. As car-dependent transport impacts rise, the vulnerability of car-dependent suburbs will increase until they really will begin to collapse. Unless of course, we begin a massive process of rebuilding cities in the suburbs around new rail and bus lines, and redeveloping closer to centers (Newman and Kenworthy 2015).

This is our urban agenda. These ideas are in all the plans for most cities and there are many examples where it is working. But it will take so much more before our cities have truly adapted to this challenge. The wonders of science and all the best of human ingenuity and creativity will be needed to make our cities work, to come to grips with removing fossil fuels, adapting to climate change and responding to the needs of the Sustainable Development Goals. This is the focus of many commentators on sustainable development from a theological perspective including academics such as Vogt (2010) and the Encyclical Letter Laudato si' of Pope Francis on care for our common home (Pope Francis 2015). The opportunities of renewable energy combined with smart systems and battery storage are providing great hope that our cities can create this clean energy future (Green and Newman 2017), but the political power of the fossil fuel industry and the conservatism of professional practice will be a serious challenge.

The role of urban activists in catalyzing this change, providing the diamonds of hope that build the city, will be very important as communities must face these issues with confidence but not hubris. Ours is a tradition of making cities work for thousands of years, so we should not lose hope now.

The Role of Activism as a Source of Hope in Creating the City of the Future

The image of hope set out in the theological traditions outlined above is that we need to creatively accept that cities are our historical and spiritual home, and we need to work on them. They are subject to all of human frailty and can collapse if we casually accept a tradition of inability to change. The diamonds or jewels that make the city of the future are the legacies of winning the detailed battles of everyday work pushed by the need to be better, to be more creative, more positive, and more helpful. Activists are needed to help show that agenda and help to demonstrate it.

There is a film that sets out in a wonderful story of hope how New York was saved from several freeways and modernist high-rise housing projects that were designed to "clean the slate" in the old fabric of the city. The film is *Citizen Jane: Battle for the City* (2016). It is the story of how Jane Jacobs took on the top-down, elitist and corrupt processes of the slum clearance and highway engineer Robert Moses. The loss of Greenwich Village and many other highly vibrant and sustainable areas of the city were saved by a series of women led by the young journalist Jane Jacobs. The diamonds of hope can be seen very easily now and the inspiration for other cities is considerable.

A similar story can be told about Jan Gehl, a Danish urban designer who has worked in over 50 cities across the globe showing them what was wrong with their old central cities and making them walkable again; the stories are all different but they show a professional who is an activist, a prophet of our time, who did not pull back from telling cities what they needed to do to reclaim their cities from the automobile (Matan and Newman 2016).

Cities have a soul; they are a combination of the historical values won by generations before and by those who now set the priorities and practices today. The soul of a city is created daily by the activities of government, business, and civil society. If a city loses its way and then begins to deny its soul, it can collapse. If a city, however, is confronted by a set of issues that reveal its soul is being compromised and it fights these tendencies and wins, the diamonds of hope are built into a new and more spiritually strong city.

This is what I believe I saw in the Roe 8 issue in my own city in the last year and perhaps a range of other issues over the past 40 or so years that I have struggled with. I see similar victories happening in other cities and write about them so others can gain some hope in their struggles. I see other cities that just deny they have issues and begin the slow but inexorable decline and collapse that is historically quite possible.

Conclusions

There does seem to be evidence that human society is becoming more and more urban, while at the same time our cities seem to be more and more precarious. But this tension is not new. Our urban civilization has always been a mixture of both trends. The ancients could see this and certainly the prophets in the Bible suggest this. Cities became the dwelling place of humanity, but their potential to collapse could never be forgotten. The prophets saw their role as reminding people of this possibility, even about the problem of resources running out and how the environment of cities could never just be assumed. However, the theology described here is not based on despair and fear of the future, it is based on hope. For everyone the possibility is to create a diamond of hope that can contribute to the city which has long-term significance. Such is the basis of sustainability practice.

Reflection Questions: Through this pedagogical feature, the chapter will invite the readers/participants to reflect on their experience of engaged sustainability transpired by the case study and in terms of the concept(s) presented in the chapter.

1. What can you say about the "soul" of your city? What are the features that characterize its approach to the future? Is it a city of fear and despair or a city of hope?
2. Examine an issue in which you have been engaged personally. Can you see a diamond of hope that you helped create? How have you communicated its significance to the next generation of activists?
3. How important is it to call out the issues of your city and seek responses to a prophetic voice?
4. Can the world manage if it becomes even more urban in the future?

References

Barry, B. (2016). *Georgiana Molloy: The mind that shines*. London: Pan Macmillan.

Beatley, T. (2011). *Biophilic cities*. Washington, DC: Island Press.

Diamond, J. M. (2005). *Collapse: How societies choose to fail or succeed*. New York: Viking.

Ellul J (1964) *The technological society* (J. Wilkinson Trans.). New York: Knopf.

Ellul J (1970) *The meaning of the city* (D. Pardee Trans.). Grand Rapids: Eerdmans.

Francis, P. (2015). Encyclical Letter Laudato si' *On care for our common home*. The Vatican. https://w2.vatican.va/content/francesco/en/encyclicals.index.html.

Gaynor, A., Newman, P., & Jennings, P. (2017). *Never again: Reflections on environmental responsibility after roe 8*. Perth: University of Western Australia Press.

Green, J., & Newman, P. (2017). Citizen utilities: The emerging power paradigm. *Energy Policy, 105*, 283–293. http://www.sciencedirect.com/science/article/pii/S0301421517300800.

Matt Tyrnauer (Director). (2016). Citizen Jane: Battle for the city. http://www.imdb.com/title/tt3699354/.

Matan, A., & Newman, P. (2016). *People cities: The life and legacy of Jan Gehl*. Washington, DC: Island Press.

Mollison, B., & Holmgren, D. (1978). *Permaculture one: A perennial agriculture for human settlements*. Melbourne: Transworld Publishers.

Newman, P. (2007). Beyond peak oil: Will our cities collapse? *Journal of Urban Technology, 14*(2), 15–30.

Newman, P., & Jennings, I. (2008). *Cities as sustainable ecosystems: Principles and practices*. Washington, DC: Island Press.

Newman, P. W. G., & Kenworthy, J. R. (1999). *Sustainability and cities*. Washington, DC: Island Press.

Newman, P., & Kenworthy, J. (2015). *The end of automobile dependence: How cities are moving beyond car-based planning*. Washington, DC: Island Press.

Newman, P., Kosonen, L., & Kenworthy, J. (2016). Theory of urban fabrics: Planning the walking, transit and automobile cities for reduced automobile dependence. *Town Planning Reviews, 87*(4), 429–458. https://doi.org/10.3828/tpr.2016.28

Newman, P., Beatley, T., & Boyer, H. (2017). *Resilient cities: Overcoming Fossil Fuel dependence*. Washington, DC: Island Press.

Rowland, C. (2007). William Blake: a visionary for our time. openDemocracy.net.

Thomson, G., & Newman, P. (2016). Geoengineering in the Anthropocene through regenerative urbanism. *Geosciences, 6*(4), 46. https://doi.org/10.3390/geosciences6040046

Thomson, G., & Newman, P. (2017). Urban fabrics and urban metabolism: From sustainable to regenerative cities. *Resources, Conservation and Recycling*. https://doi.org/10.1016/j.resconrec.2017.01.010

UNEP. (2016). *Environment, religion and culture in the context of the 2030 agenda for sustainable development*. Nairobi: United Nations Environment Programme.

Vogt, M. (2010). Climate Justice from a Christian point of view: Challenges for a new definition of wealth, Vortrag bei der Konferenz "Religion in Global Environmental and Climate Change: Sufferings, Values, Lifestyles", Potsdam. http://www.kaththeol.unimuenchen.de/lehrstuehle/christl_sozialethik/personen/1vogt/texte_vogt/cj_potsdam.pdf.

Bio-economy at the Crossroads of Sustainable Development

José G. Vargas-Hernández, Karina Pallagst, and Patricia Hammer

Contents

J. G. Vargas-Hernández (✉)
University Center for Economic and Managerial Sciences, University of Guadalajara, Guadalajara, Jalisco, Mexico

Núcleo Universitario Los Belenes, Zapopan, Jalisco, Mexico
e-mail: Jvargas2006@gmail.com; josevargas@cucea.udg.mx

K. Pallagst · P. Hammer
IPS Department International Planning Systems, Faculty of Spatial and Environmental Planning, Technische Universität Kaiserslautern, Kaiserslautern, Germany
e-mail: karina.pallagst@ru.uni-kl.de; patricia.hammer@ru.uni-kl.de

S. Dhiman, J. Marques (eds.), *Handbook of Engaged Sustainability*,
https://doi.org/10.1007/978-3-319-71312-0_52

Abstract

This chapter aims to review, analyze, and systematize the knowledge created on bio-economy to develop a conceptual and theoretical framework based on the transdisciplinary study of biology and socioeconomy to be used in further research. It begins from the questioning of what are the benefits that bio-economy has compared to the neoclassical economy. The methods employed are the critical analytic, descriptive, and deductive-inductive and suggest holistic and transdisciplinary approaches. As a result, the core of the chapter presents the principles under which this new scientific paradigm in sustainable development can continue creating more scientific knowledge to be used in the formulation and implementation of strategic choices for the bioproduction, biodistribution, and bioconsumption process.

Keywords

Bio-economy · Green economy · Knowledge-based economy · Sustainable development · Strategic choices

Introduction

Any nation of the world faces major environmental, economic, and social challenges to be addressed to change for sustainable development and to better the way to live and work. Bio-economy is a “greener” alternative that has impacts on natural and environmental resources, food, soil, land, and livelihoods. Bio-economy has a relevant impact in important bio-products such as textiles, cosmetics, bioenergy, biofuels, building products, and other by-products and biopower. Bio-economy serves a market of environmentally sustainable satisfactors, products, and services, and to keep in pace in the long term to become global, it requires more research and development. Bio-economy is globally influencing biotechnological research and development, business models, and market structure. (Bundesministerium für Ernährung und Landwirtschaft (BMEL) 2016).

Bio-economics is considered as a more advanced scientific development than economics because it relies on the evolutionary process of humanity and nature. The advance of economic science extends to consider biological evolution, biology, and thermodynamics as important foundations of the economic process. The bio-economy connects and expands economics and biology to anchor in its empirical prediction to give it the power of regeneration and sustainability to the activities of the socioeconomic and biological systems.

This chapter analyzes the recent developments on bio-economy. It reviews the conceptualization of bio-economy, green economy, and ecological economics to analyze the deficiencies of classic economy leading to present the bio-economics as the new epistemological paradigm inextricable linked to sustainable development. From this framework, there are derived at the core of the subject and object of study some principles offered as the basis for further research: circular economy, sustainable development, holistic and transdisciplinary approaches, innovation culture and capacity creation, knowledge-based economy, global ethics, social capital, and culture of peace.

Finally, the chapter considers the importance of formulation and implementation of the bio-economy as a strategy to enable and ensure results in terms of contributions for the sustainable development of renewable resources. Also some concluding remarks are offered.

Conceptualization of Bio-economy

The concept of bio-economy is relatively new to name those economic activities derived from the biosciences advances and surge in the scientific knowledge in biotechnology, genetics, genomics, etc., to achieve practical applications from biological processes. The term *bio-economics* was coined by Georgescu-Roegen to explain the biological origin of the economic process and thus spotlight the problem of mankind's existence with a limited store of *accessible* resources, unevenly located and unequally appropriated (Georgescu-Roegen 1977, p. 361). Bio-economy is the sustainable production and conversion of biomass into a range of goods and services, among others food, health, fiber and industrial products, and energy. The term "bio-economic" is used to indicate both economic and biophysical components (Knowler 2002).

Bio-economy is a concept related to the economic activities derived from utilizing natural and biological resources or bioprocesses to produce bio-products. Bio-economy is an aggregated set of economic operations and activities related to biological products to capture economic value, growth, and welfare benefits for human development. The concept of a bio-economy refers to that economy where the basic components of materials, chemicals, and energy come from renewable biological resources, such as plant and animal sources. This type of economy can meet many of the requirements for sustainability from environmental and social aspects as it is designed and implemented intelligently.

Returning to the definition of the concept of bio-economics, the OECD suggests that it can be understood as the aggregate set of economic operations in a society that uses the latent value of biological products and processes to capture new benefits of growth and well-being for citizens and nations. This first definition, from the year 2006, essentially includes the same idea with regard to the means to achieve growth and prosperity, as is clear from the description of the OECD from 2009.

Turning to a European vision, the EC defines the bio-economy in its policy package as the production of renewable biological resources and the conversion of these resources and waste streams into value-added products such as food, feed, bio-based products, and bioenergy (European Commission 2015a, b, c). The bio-economy encompasses agriculture, forestry, fisheries, food, and biotechnology, as well as a wide range of industrial sectors, ranging from energy and chemical production to construction and transportation.

The concept of bio-economics has several meanings: It is the efficient management of human resources (Clark 2010) and is considered as an analogy of biology to explain economic theory. The concept of bio-economics is the biological analysis of economic relations, which as dynamic systems can be effected, for example, through frames of reference, mental schemes, and instruments of thought analysis.

The concept of bio-economy is under debate (Levidow et al. 2012). A bio-economy can be defined as an economy in which the basic building blocks for materials, chemicals, and energy are derived from renewable biological resources (McCormick and Kautto 2013). Bio-economy is defined as the global industrial transition of sustainably utilizing renewable aquatic and terrestrial biomass resources in energy, intermediate, and final products for economic, environmental, social, and national security benefits (USDA 2015). The bio-economy is defined as an economy based on the sustainable production and conversion of renewable biomass into a range of bio-based products, chemicals, and energy (de Besi and McCormick 2017).

Awareness on the concept of the bio-economy as potential in biosciences and biotechnologyis important to understand the implication for human beings.

While the concept of bio-economics began initially from the science of biotechnology, then it has expanded to incorporate other ideas. Biotechnology can be understood as the science of using living things to produce goods and services. Therefore, it involves manipulating and modifying organisms to create new and practical applications for primary production, health, and industry.

Another meaning of the concept is the one that considers the bio-economy as the influence that has the satisfaction of the human needs like the biological conditions in the economic behavior.

Bio-economy as Green Economy

Bio-economy also called green economy can be defined as one that has low carbon emissions, uses resources efficiently, and is socially inclusive (PNUMA 2016). The bio-economy sector is a warrantor for a green economy.

The "bio-economy" agenda has developed the concept "green economy" that emerged at the 2012 Rio+20 Summit and is promoted by the United Nations Environment Programme to pursue natural renewable resources bioenergy and sustainable biological products (Socaciu 2014). The agenda for bio-economy has been pushed by large corporations and developed states linked to green economy and the knowledge-based bio-economy (Hall and Zacune 2012). The Rio+20 agenda "Towards a Green Economy" gave rise to the bio-economic strategy for sustainable development and economic growth (European Commission 2012a, b, c). The European Union promotes among its member states the commitment to the agenda of "green economy," bioenergy (Paul 2013), and agrofuels as an alternative to fossil fuels (Hall and Zacune 2012).

Bio-economics biologicalizes the natural environment and the biological resources of business activities as intrinsic and inextricable elements of mutualistic coevolution. The bio-economy integrates the green economy that seeks to respond to the challenges of food security and clean bioenergy.

Bio-economy as Ecological Economics

However, bio-economics is also called ecological economics. For some authors the bio-economy is the ecological economy that reconciles the economy and the ecology, that at the same time looks for the economic efficiency, and that takes care of the natural resources that are essential for the humanity. The bio-economic approach calls for a change in values on the use of available resources and energy to be conserved for the use of future generations. The bio-economy is the basis of the business of ecological projects, agricultural, etc. Bio-economics studies the biological origin of economic process and the human activities associated with a limited stock of available and accessible resources that are unevenly located and unequally appropriated (Mayumi 2001).

Ecological disasters in recent years as a result of the subordination of the laws of nature to the laws of the market economy have demonstrated the destructive capacity of the forces of nature, to the point that has created an awareness by the environmental care. Some analysts consider that the bio-economy approach is destructive and should be restructured from an agroecological perspective.

The Deficiencies of Neoclassical Economics

The economy as a science oriented by the competition has serious methodological deficiencies that the bio-economy tries to overcome through biocentric processes and of bio-economic balance centered in the cooperation for the sustainable development and the conservation of the nature and the environment in their interactions with the humanity. From a holistic perspective, the interactions between biological and socioeconomic systems result in the field of bio-economics. The interactions of the natural biological and economic processes that give rise to the bio-economy have an impact on the complex and uncertain phenomena of the biosphere.

The physiocrats subordinated reproduction and economic transformation to nature and classical economics to capital, later to the goal of market equilibrium and the financial system with economic neoliberalism. The heterodox economy is more oriented to the human aspect with approaches such as humanist socialism that resists the clashes of the liberal economy and analyzes the contradictions of economic liberalism with the argument that the labor force is not a commodity that separates from the human being.

The market economy reduces natural, social, and moral values but cannot regulate the behavior of nature. In the First Report of the Club of Rome (Club de Roma 1972) which warns about the limits of economic growth and the

Brundtland Report (Brundtland Gro Harlem 1987) that reports threats to the mechanisms that regulate nature, the issues for sustainable development to maintain the balance between the interdependence of the economy, the nature, and the biosphere emerge.

Bio-economy: A New Economic Epistemology Paradigm

Bio-economics as a discipline is a new paradigm in environmental economic development (Soedigdo et al. 2014). The emergence of bio-economics as a different science to economics and biology represents a new paradigm in scientific disciplinary evolution for the study and analysis of the bio-economic causes of the environmental impacts of human actions (Mohammadian 2000). The interactions of biological and economic systems are integrated in the bio-economy from a new perspective that represents a paradigm shift. As an interdisciplinary science, bio-economics synthesizes sciences based on the empiricism of biology and the humanism of economics, resulting in a new, more holistic paradigm because it attempts to explain the interactivity of biological systems and nature with economics and the social.

Bio-economics represents a paradigmatic change in the evolution of disciplines whose main task is to investigate the problems that arise from the impact of human enterprise on the environment. These problems are not only due to purely biological causes or to purely economic causes. Bio-economics has been the result and in turn is provoking paradigmatic change in the epistemology of research for the generation of innovative theoretical and empirical knowledge with applications that are highly significant for their contributions. The paradigm focused on bio-economic development is based on the holistic perspective that results from the interaction of biological, socio-economic, and environmental development to explain the coevolution of human development that guarantees the provision of resources for present and future generations.

The bio-economic development paradigm (Mohammadian 2000) is an alternative to prevent the accelerated depletion of natural biological resources, mitigate environmental degradation, and reduce economic inequality and social inequity. The community participation and paradigm have been changing with the bio-economy process (Henry and Roche 2013). Sustainability preserves the identity and integrity of bio-economic systems by enhancing self-sufficiency and equity and respect for biospheric equilibria as the criteria for regulation of trade and economic activity in opposition to the neoclassical economic paradigm endorsing global trade to stimulate economic growth.

The new paradigm complements the tangible and intangible resources, the objective with the subjective, and the feelings with the facts with a sustainable and sustained growth orientation, with a material and economic operational benefit. The new paradigm of production and market has been called cognitive capitalism and bio-economics, in a word, bio-capitalism (Fumagalli 2007). Bio-economics makes aware of the interactions of biological and socioeconomic systems, to reach a conciliation and to adopt a holistic research. The three disciplines of the economy-environment have offered a change of perspective and a non-paradigm shift as bio-economics does.

Principles of Bio-economy

Circular Economy

Bio-economy links economic growth with environmental sustainability under the guiding principle of a circular economy. Bio-based circular economy is a bio-economy for a sustainable policy agenda which can solve the challenges of climate change, eco-technologies, agriculture and food security, blue growth, etc. For example, the wood-based bio-economy is a bio-based circular economy that originates from processing and recycling timber, pulpwood, and other forest products utilized for material and energy sources. Bio-economy innovation is shaping the new social economy combining elements of circular economy and social inclusion.

Sustainable Development

Bio-economy is a transitional step toward sustainable development. Bio-economy promotes sustainable development by protecting biodiversity and reducing dependence on fossil natural resources. The White Paper The European Bio-economy in 2030 sustains that bio-economy has to face challenges such as the sustainable management of natural resources, sustainable production, public health, mitigating climate change, integrating and balancing social developments, and global sustainable development.

The neoclassical model of the economy supports neoliberal capitalism, while the bio-economically centered sustainable development model encourages people to engage in cooperative socioeconomic activities that benefit in the long run. The bio-economy reduces dependence on fossil natural resources, prevents biodiversity loss, and creates economic growth and employment in line with the principles of sustainable development.

The bio-economy provides benefits for the increasing sustainable development by economic growth and outputs, replacing fossil fuels with renewable natural resources, biodiversity conservation, and increased energy self-sufficiency. Some contributions of bio-economics to sustainable development are in global food security, renewable raw materials, mitigation of climate change, etc. The utilization of bio-economic resources to promote sustainable development of marine fisheries resources requires empirical research on fisheries biology and economics. Bio-economic analysis of fisheries is scarce despite that there are some reports on economics and biological and resource management (Habib et al. 2014).

The bio-economic analysis is based on calculation of reference points (Thunberg et al. 1998). Mathematical bio-economics is used to study the management of renewable resources, such as differential equations to model resource processes based on guiding principles to determine profits (Clark 1990). One of these principles is the tragedy of the commons. Indicators for bio-economy sustainable development are required to be defined (Fritsche and Iriarte 2014).

The bio-economy agenda has already presented several limitations in all principles of ensuring food security, harvest limited to the capacity of regeneration, use of biomass for the highest value, and reducing, recycling, and reusing waste. The European Union imports aromas expanding the use of bio-economy, despite that its policies are being questioned and criticized. A good example is agrofuels, which are not actually renewable resources and, therefore, have failed to deliver the promises.

One of the main challenges of the bio-economy is to change the current production, distribution, and consumption systems to those that are more environmentally friendly and provide sustainable development for the conservation of natural resources, while meeting human current needs.

Holistic Approach

According to Mohammadian (2005), bio-economics is a holistic interdisciplinary science, also called biological economics, and is a new epistemology for the investigation of the interrelationships between biological and socioeconomic systems beyond the different approaches of environmental economics, economics of natural resources, and the ecological economy. Bio-economics aims to narrow biology as an empirical science and economics as a social science.

Bio-economics aims to achieve a harmony between the holistic balance of economic, social-human, and natural and environmental conditions. Thus conceived, this bio-economic balance of holistic character gives rise to the happiness of a life balanced in all the orders. Achieving the balance between the price and the quantity of the economy that is short term with the quality and the bio-economic value that is maintained in the long run is also to maintain a holistic perspective that balances selfishness and altruism.

Bio-economic education should focus on a holistic and transdisciplinary learning process that results from the scientific experimentation of new cognitive processes and synthesis of the interactions and interrelations of biological, economic, social, and environmental processes. To achieve this new mentality, the educational process must undergo a scientific-academic revolution through the synthesis of theories of biology, economics, and cognition, to promote an integrated education in the form of a bio-economic educational process. Such an educational process is holistic and interdisciplinary and helps to dismantle reductionist education to promote the synthesis culture in addition to facilitating the art of learning to learn.

Transdisciplinary Approach

Modern scientific research during the last centuries has suffered from the premises of simplicity, objectivity, and positivism. In addition, its Cartesian reductionist methodology has divided biospheric and socioeconomic systems into separate compartments, to be investigated by various disciplines, forgetting the unity of human life with its biological basis. It is evaluating interactive problems with an outdated mindset, evolved for a world it has left behind decades ago. Bio-economics is born as a response to the incremental advances of the other economics-environment disciplines through which the pathologies of capitalism and its industrial system have been investigated individually and separately.

The analysis from the bio-economy is based on the transdisciplinary interrelations between the human, natural, and economic areas (Nicolescu 1996). In addition, the reflections of Georgescu-Roegen (GR) helped to establish links between different disciplines and to address the economic issue beyond money, managed to introduce the biophysical dimension, so that the currents of industrial ecology, agroecology, or urban ecology have in the proposals of GR a precedent (Carpintero 2006).

Bio-economy is a transdisciplinary approach that encompasses biology, economics, physics, genetics, forestry, marine sciences, etc. Bio-economics as a transdisciplinary science proposes the integration of biophysical subsystems of nature and environmental, biological, and bio-spherical with economic, social, and human subsystems with long-term orientation to ensure the happiness of future generations. Bio-economics is a transdisciplinary science that seeks to study and analyze biological and socioeconomic systems from a perspective of cooperation and solidarity for the conservation of physical, natural, biological, and social capital. Bio-economics proposes among its transdisciplinary principles for the creation of bio-economic capital through biology and economics as the integrated solution to sustainable bio-economic development.

Heuristics is a method used by the bio-economy that is based on transdisciplinary and holism for the generation of knowledge through the absence or reordering of existing information. The functional method in research starts from emerging properties in the interaction of problems such as climate change in biological and socioeconomic systems. The transdisciplinarity of the principles of bio-economics and the knowledge economy raises its character of postmodernity.

Sustainable bio-economy needs transdisciplinary capabilities and expertise. The bio-economy is not only a transdisciplinary branch but also contributes to change the mentality of people, passing from a greedy, guided by the power that grants the money to a person who uses resources in a rational way, thinking about the future and the conservation of the planet.

Innovation Culture and Capacity Development

The bio-economic theory is based on the value of complex innovations based on an ethical – economic – humanist balance. Business culture to foster innovation is essential in bio-economy aligned with the needs of sustainable development and with the collaboration of all the stakeholders involved on strategic alliances, or other forms of cooperation between different sectors are essential. Education, research, and innovation skills are critical factors for development of bio-economy. However, the basis for the bio-economy comes from previous strategic agendas of the European Commission (2012a, b, c; 2015a, b, c) (EC), including the 1993 White Paper which highlighted the need for nonphysical investments based on knowledge and the role of biotechnology in innovation and growth or the 2000 Lisbon Agenda by stimulating the knowledge-based economy to ensure competitiveness and economic growth.

It can be pointed out convincingly that discoveries, ideas, and innovations play an important role in economic growth. Bio-economy is driven by material innovations. Bio-economy innovation emerges from research and bio-economy. The bio-economy web is created to integrate all the sectors involved in sustainable growth, including all bio-products, food and nonfood, and facing the challenges to become a major driver of innovation.

Small- and medium-sized enterprises (SMEs) are drivers of innovation in bio-economy, thereby driving SMEs in bio-economy. SMEs and groundbreaking innovation emerge in the bio-economy. Not only large corporations but also SMEs in all sectors are important for the development and implementation of bio-economy because they may be able to develop the capability to absorb the knowledge, produce in accordance and marketing innovations of the new bio-products. Bio-economy offers opportunities to promote innovations and market access and ensure operations and functioning solutions.

A transition toward bio-economy requires the innovation and development in biotechnology. The transition toward a bio-economy founded on biomass has been designed by different strategies but on a common direction based on technological research and innovation on biotechnology (de Besi and McCormick 2017). Biotechnology R&D requires bio-economy innovation to attend an increasing market demanding sustainable products and services.

Bio-economy initiatives contribute to shape a sustainable economy away from fossil resource dependency, prompting governments and firms to develop strategies aimed to conduct research and innovation. Bio-economy makes possible decarbonization which in turn supports innovations. Thus, biotechnology innovation drives bio-economy innovation and technology transfer from scientific institutions to industry through cooperation and networks with private firms and academic, social, and governmental institutions.

Bio-economy obtains biological natural resources from agriculture, forestry, and fishery providing innovations that are vital for biotechnology growth. Sustainable bio-economy objectives are based on research and innovation into breeding, feeding, health, and housing of livestock and fish. Some key influence factors identified for the wood-based bio-economy development are the biomass availability/forest structure, globalization and global economic development, energy and climate policies, supply and demand for wood, willingness to pay for bio-based products, and innovations along the value chain of wood (De Besi and McCormick 2015). Bio-economy innovation is driven by other influence factors such as the demand for biomass that may be reducing due to utilization of waste materials.

Urban green spaces are potentially development environments for sustainable bio-economy innovations in products and services that provide solutions in air quality, water, energy, health, waste management, etc. Interactions between urban green spaces and communities derive bio-economy benefits in areas such as renewable energies, food, health, etc. Bio-economy is already a major driver of rural and coastal development playing a key role in societal innovation.

Sustainable bio-economy activities of new business require investments in innovation and equity financing for growth of bio-economy services and products entering into global markets. Bio-economy business growth may be government funded, cooperation and public finance for research innovation and improve the bio-economy base products and services market position. An innovation system supported by institutional infrastructure and supportive mechanisms embedded within the waste-based bio-economy enable transition from traditional market to bio-economy-based market.

Sustainable bio-economy innovations require capabilities and competence development in research and expertise in product and services development. Development of bio-economy business expertise in some sectors still in their infancy is driven by the market attractiveness, risk investments for innovation and growth, and target of demand. Further research on bio-economy innovation is needed. The circulate nature of bio-economy processes influences the economics of energy use and raw materials as it is required to build a capacity for future development based on research, innovation, and market conditions.

Research, development, and innovation on diverse sustainable technologies, facilities, processes, and skilled workforce maximize environmental sustainable bio-economy. Innovations along the value chain to reduce costs are necessary. Institutional and technological innovations are essential for bio-economy product and process. A deficient innovation system may lead to weak innovative firms. Innovative innovation by the increase of bio-economy-based raw materials, increases the diversification and variety supply for the bio-economy. The development of the bio-economy integrates alternative forms of agriculture and social innovation to achieve a more distributed bio-economy with relationships between firms and community's involvement.

Certain conclusions are drawn from the strategic analysis and recommendations for the future of bio-economy based on research, development, and innovation. The bio-economy transition requires changes at the level of system involving social, political, government, and industry actors in the strategy formulation and implementation on research and innovation, land use, biomass, social change, and governance.

Bio-economy research and innovation strategy across the landscape resources meeting the scientific criteria is needed. Future opportunities of bio-economy applications for markets are the duties of research policies shaping the innovation and creativity as a framework conditions for the bio-economy locations.

Knowledge-Based Economy

To date, the progress of the bio-economy has been made possible by a recent search for knowledge and technical expertise to use biological processes for practical applications. The human enterprise based on the principles of bio-economics and novel ideas is really the knowledge economy. The dominant view of the bio-economy considered as a hard science based on knowledge and technological innovation has a life sciences perspective that is based on the argument that global value chains increase the efficiency and productivity of natural agriculture bio-products.

The bio-economy is based on the knowledge economy and green economy for an efficient management of renewable biological resources and the production of bio-products and agrofuels. The new bio-economic sustainable growth theory is an alternative theory to neoclassical growth theory that emphasizes knowledge and ideas as the most relevant production factor.

The European Commission (2012a, b, c) adopted the agenda "Innovating for Sustainable Growth: A Bio-economy for Europe" in 2012 under the life sciences strategy to address environmental, biological, and natural bio-products and bioenergy. However, the life sciences approach to knowledge-based bio-economy (KBBE) dominates bio-economy research more than the agroecology perspective. Life sciences is one of the dynamic sectors of bio-economy. The agroecological perspective of the knowledge bio-economy stands as an alternative under the premise that organic farming with fewer levels of supply chains gives more value to producers (Levidow et al. 2012).

The OECD further states that the bio-economy involves three elements: the use of advanced knowledge of genes and cellular processes to design and develop new processes and products, the use of renewable biomass and efficient bioprocesses to stimulate sustainable production, and the integration of knowledge and applications of biotechnology in a range of sectors. According to the White Paper, the implementation of bio-economy needs coherent and integrated policy for a sectorial and multidisciplinary investment on bio-economy research, knowledge development and innovation, entrepreneurship in bio-economy, and development of skilled workforce in bio-economy. However, it is highly debated and contested the top-down target-oriented approach to implement knowledge-based bio-economy as a programmatic framework.

The bio-economy pivots on the basis of the new knowledge economy with objectives of economic growth and competitiveness in close relation with the strategies of public policies that emerge from the neoliberal narrative. The bio-economy agenda is sustained by the knowledge-based bio-economy (KBBE) as an approach that emerged from the life sciences research focusing on bioagriculture. The dominant version of bio-economics is the life sciences approach which commodifies both nature and knowledge reinforcing dysfunctional behaviors and patterns instead of solving their causes.

Bio-economy is the knowledge-based bioproduction and use of renewable resources to make bio-products, processes, and services. The knowledge-based bio-economy develops biological processes technologically to provide a wide range of bio-products from renewable raw materials. The knowledge-based bio-economy promotes strategic research and innovation to support the transition from an oil-based to a bio-based economy that enables economic prosperity with ecological and social compatibility. Efficient and sustainable biological resources rest on bio-economy knowledge-based innovation aimed to business and large corporations to concentrate control over natural resources, production processes, and distribution chains.

Knowledge-based bio-economy supports research and innovation for the transition from oil-based economy to bio-based economy to foster ecological and social compatibility in a wide range of sectors. Bio-economy, bio-based economy, and knowledge-based bio-economy (KBBE) are similar concepts used to describe the transition from a fossil fuel-based economy toward an economy based on renewable biomass for bio-products and bioenergy (Schmidt et al. 2012; Birch and Tyfield 2012; Staffas et al. 2013).

Knowledge-based bio-economy is translating into economic growth by bio-based research and innovation; accelerating bio-based products and services, energy, health, food, IT sector, agriculture, and manufacturing; and strengthening competitiveness. The bio-economy is expanding knowledge on biosciences and biotechnologies based on nano-molecular structures to genetic science. The knowledge-based bio-economy is scattered among the industrial sectors and nations. Knowledge-based research in the area of the bio-economy has been conducted and substantially covered in agriculture and food policy.

Bio-based innovations promoted by a knowledge-based research strategy are centered around competitive bio-economy sectors, such as sustainable agricultural production, the global food security, healthy and safe foods, the development of biomass-based energy carriers, and the industrial application of renewable resources. Bio-economy involves the use of knowledge of genes and cell processes to develop new products and services.

Biotechnology benefits from bio-economy knowledge spillover contributing to a significant increase of economic growth. Bio-economics can maximize economic efficiency and social and human benefits from a sustainable use and maintenance of nature and environment through knowledge spillovers and commercial applications.

For the knowledge-based bio-economy to create sustainable development, it is required to include sectoral components at all levels, that is, at local, national, regional, and global level. The economic, ecological, and social contributions of the bio-economy can be realized through knowledge of the biological processes and systems and their interactions with the ecosystem, as well as the associated biotechnology and social implications. The knowledge-based bio-economy combines economic growth and prosperity with environmental sustainability and compatibility. The knowledge-based bio-economy contributes to global responsibility in the present and to appropriate foresight for future generations.

Bio-economy knowledge can be taken for granted by the emerging information channels. The bio-economy transformation process driven by technological knowledge and supported by structures of innovation networks is taking commercial advantages in global markets. Integrated bio-economy combines entrepreneurship with corporate responsibility and identity with other stakeholders and firms sharing knowledge networks. Incorporation of renewable natural resources in bio-economy programs aimed for creation of business and new jobs requires the expertise of research multidisciplinary models to generate bio-economy knowledge in creating business.

Bio-economy knowledge can be transferred from research centers to different sectors, including to business in different organizational forms among which can be joint ventures for development and innovation. Innovative developments in bio-economy technologies and products are stimulated for investments on research and development and knowledge transfer and exchanges, combined with a created demand of bio-products which are important components of bio-economy policy (Grubler et al. 2001; Borrás and Edquist 2013; Foxon et al. 2005).

The knowledge-based bio-economy accommodates results between technology, the economy, and ecology. Knowledge-based bio-economy competencies expertise is developed by research infrastructure extended in bio-economy research centers and information and communication technologies such as the one supported by platform technologies and institutional funding.

The knowledge-based bio-economy requires the implementation of complex interrelationships of training and research agents committed to the creation of sustainable bio-based economy in accordance to demographic and socioeconomic changes. Bio-economy requires well-trained and educated talent with multi-disciplinary and interdisciplinary qualifications.

Opportunities for a knowledge-based bio-economy, biomass-based raw materials, and global food security are enormous in growing markets with the increasing demand of bio-products and services. The bio-economy is a knowledge-based science requiring further collaborative intra- and interdisciplinary research and development to increase the innovation capable to elicit and solve the global societal challenges.

Global Ethics

Bio-economics represents a fundamental change in ideology in everything that is related to socioeconomic, biological, and ethical activities. Business, industry, and society can contribute to and benefit from bio-economics by considering the socio-economic and ethical implications. Homo bio-economicus is based on biological principles of conservation, recycling, regeneration, and respect, as well as on the socioeconomic principles of equity and equality and under the ethical principle of exploiting natural and biological resources but living off of its income management provided.

From a different perspective, bio-economics complements analysis of the neoclassical economics of socioeconomic, biological, and ethical activities. The principles of biological sustainability, environmental integrity, social equity, economic equality, and global ethics sufficiency of the bio-economy underpin socioeconomic activities.

The bio-economy supports economic activities based on the principle of an ethical practice to benefit all participants, thereby reducing transaction costs. The science of bio-economics is considered a postmodern science that considers the participation of all interest groups around the concepts of sustainability, quality, value, ethics, equality, social justice, fraternity, and compassion. Ideas and knowledge as intangible resources of the bio-economy emanate from heuristics as a tool for sustainable economic growth that is based on biological, environmental, economic, social, and ethical resources.

The implications of bio-economy for economy, society, and ethics have to be addressed such as long-term health care, food, etc. The framework of bio-economy interdisciplinary research projects contributes to discussions on social, legal, and ethical issues. Sustainable bio-economy global demand of bio-products and bio-services requires to be framed by ethical consumer behavior and legislation.

Social Capital and Culture of Peace

The business bio-economic activities have as components in bio-economic capital and social capital. Bio-economics is based on the cultural, noneconomic, and intangible dimensions of social capital such as trust, cooperation, solidarity, reciprocity, etc., values that validate the humanist and transformative vision of biology and socioeconomic activities of bioproduction with use value (Mohammadian 2003). Social capital facilitates the relations of solidarity and compassionate cooperation that contribute to the formation of a more generous and altruistic society.

For organizations and companies with a bio-economic orientation, social capital, biological capital, and financial and monetary capital have the same importance in their scale of values. In other words, a bio-economic enterprise attaches as much importance to social capital and biological capital as to money capital (Mohammadian 2005). At the center of social capital are trust and cooperation, also the main ingredients of a culture of peace.

The bio-economy is based on solid principles of trust and cooperation, fraternity, justice, and compassion to achieve the creation of bio-economic value and economic growth through the internalization of costs that produce the externality attached to the care of the biological foundation. The new institutional economy promotes the cultural principles on which the bio-economy is based, so that they coincide in the sustainable benefits obtained through just, cooperative, fraternal, reciprocal, and trustworthy socioeconomic transactions. For example, the noneconomic components of bio-economic accounting enable the processes of transformation of economic activities with capitalist profit purposes to carry out more humanistic activities that harmonize the relations of cooperation and competition.

Bio-economy has an impact on the quality and level of employment and on building societal trust. The "homo oeconomicus" is a being that seeks efficiency, focused on competition, and is predatory, greedy, and without any human feeling, while "homo bio-economicus" is a being that is sensitive to human needs and nature; centered on relations of cooperation and trust; in harmony with himself, with those around him, and with nature; and with a culture of sufficiency and conservation and with a sense of the values of solidarity and fraternity (Mohammadian 2000, 2003) promoter of sustainability. Bio-economics is considered as a postmodern science in terms of the values of cooperation, trust, and empathy on which it stands.

The institutionalization of bio-economics promotes the relationships of trust and cooperation between stakeholders and other groups of interest. To build consumer trust on bio-products and bio-economy, it is required to implement high global standards of reliable quality, safety, efficiency, transparency, and acceptability with the support from industry, society, policymakers, and regulatory agencies.

The sustainable bio-economy encourages relationships of cooperation between different economic sectors and improves the well-being of society. Homo bio-economicus, unlike homo oeconomicus, is a being satisfied with the resources that he has access to and that has self-fulfillment in his life and is sensitive to human needs and realities, in harmony, and with an attitude of cooperation and care with environmental, economic, social, and political situations. Bio-economy is sustained as the third path of economy because it is considered to be between the classical equilibrium economy and use value and the new global economy of complexity and exchange value and therefore benefits both (Mohammadian 2003) as models of competition and cooperation (Mohammadian 2000).

Bio-economy organizations are using innovative business models based on relationships of cooperation to create new entrepreneurial structures to foster research and development and the applications. All types of organizations involved in bio-economy activities and practices have different entrepreneurship structures to accommodate the participation and cooperation of all the stakeholders to achieve the goals. Sustainable bio-economy services associated with industrial bio-products contribute to develop efficient bio-economy value chains, create new bio-economy business, and encourage relationships of cooperation and partnerships.

As such, development cooperation between business units, regions, and nations demands sustainable bio-economy development activities in sustainable use and expertise of natural resources. Advancement requires active participation in relationships of international cooperation programs to strengthen support of bio-economy activities. A good example of sustainable bio-economy implementation is the cooperation with financial providers across sectoral boundaries. Bio-economy technologies, products, and services can be standardized and certified to follow the rules of global markets by means of international cooperation and allocation. Technology transfer in bio-economy in science delivers economic and social benefits if scientific results are delivered through new forms of cooperation and strategic alliances between firms, science institutions, communities, and government.

Although, in fact, there is much to do and develop, the projects of bio-economy involve a more democratic attitude supportive of a culture of peace, such as the specific cases of free green spaces.

Bio-economy as a Strategy

Bio-economy strategy aims on self-sufficiency in energy and raw materials and securing availability of biomass and low-carbon and resource-efficient society. A bio-economy strategy is aimed to identify energy, environment, water, food, health, social, etc. challenges and act upon the critical bio-economy research from using waste materials to gain market value and growth. A bio-economy strategy links bio-economy-based renewable resources with sustainability by ensuring sustainable production and use of biomass (Pfau et al. 2014).

The knowledge-based bio-economy is a sustainable economic strategy for the creation of sustainable natural capital (Birch et al. 2010) that is oriented toward the formulation of sustainability policies, in a way that attempts to link bio-economics projects with the knowledge-based economy and technological innovation for the design of public policies and institutional practices (Franco et al. 2011).

Bio-economics is presented as a win-win strategy. Bio-economy has been identified as a strategy and defined as encompassing the production of renewable biological resources and the conversion of these resources and waste streams into value-added products, such as food, feed, bio-based products, and bioenergy. Its sectors and industries have strong innovation potential due to their use of science and enabling and industrial technologies, along with local and tacit knowledge (European Commission 2012a, p. 3). The European Union's strategy adopted in 2012 was announced as innovation in the service of sustainable growth: a bio-economy for Europe promotes sustainable production and consumption capabilities based on life sciences to address the challenges of the environment, the use of energy, and food security.

The bio-economy as a greening strategy is advancing amid the debates on the so-called green economy proposed by the United Nations Environment Program (PNUMA) to promote bioproduction. Bio-economy as strategy has solutions to environmental concerns and sustainable economic growth moving forward to renewable biological resources in an increasing market share (European Bio-economy Panel and SCAR 2014; European Commission 2015a, b, c). The European Union recommends the bio-economy to the Rio+20 agenda "Towards a Green Economy" as a strategy to promote the sustainable development of environmental and natural resources efficiently, thereby making economic growth compatible and the sustainable use of biological and environmental resources (European Commission 2015a, b, c).

To target specific markets, sustainable bio-economy must study and analyze the development and functioning of the environment and development of sectors and identify global trends and challenges, opportunities and changes in drivers of marketing of products and services, and the specific needs to satisfy. The study of the bio-economy raises the need for a behavioral change of production systems called to play a strategic role in providing a response to a set of global challenges and uses the system approach to analyze related fields such as green economics and energy efficiency and productivity processes taking into account climate variations and increasing restrictions on the quality of biodiversity and natural resources such as air, water, and soil. This analysis is necessary to formulate and implement a strategy and policy actions. Strategy formulation formulates priorities for future bio-economy development.

A favorable operating environment, strategic choices, and policy actions are drivers for business bio-economy development, growth, and achieving competitiveness in a setting to operate. Strategic choices and implemented sustainable bio-economy policy actions to produce efficient goods and services targeted to satisfy the needs of consumers in the marketplace and aimed to support the growth of bio-economy business. Strategy choice implementation and policy actions supported by agents in the bio-economy sector may be able to set up a program aimed for resource efficiency and low-carbon society.

The bio-economy as part of a comprehensive global strategy including the renewable energies is able to solve the problems of sustainable development (Piotrowski et al. 2015). A sustainable bio-economy strategy is aimed to generate bio-economy business growth from added value bio-economy products and services while securing the nature's ecosystems and providing sustainable global solutions for saving nature's diversity, global warming, air pollution, consumption standards, etc. Sustainable bio-economy strategy aims for economic growth generation from high added value products and services in the competitive bio-economy business secured by natural ecosystems. Global bio-economy competitive strategy builds innovative firms supported by science and high-technology research and structural conditions.

Productivity and performance of the bio-economy in sectors internationally competitive result in growing demand for innovative products, processes, and services.

Strategies for bio-economy development focus in priority areas related to research, development, and innovation of new biotechnological and industrial bio-products; relationships between firms, communities, research institutions, and government to optimize the functioning of biomass use; and funding the development of bio-based activities. Strategies on import of biomass as a raw material is required in some industries are related to the development certification process to ensure sustainability and not have negative economic, social, and environmental impacts. Availability of biomass for the present and future of bio-economy is dependent on strategies on land use, agriculture, forest, and the other natural and biological resource diversity. Also important are technological development and economic and legal incentives (Welfe et al. 2014).

Bio-economy strategies based on societal awareness and fostering research and innovation are aimed to the transition toward the development of a bio-based economy. To achieve a bio-economy transition, the development of some strategies and actions is required such as to invest in research and development, to train human resources for innovative and completive skills and capabilities, and to engage stakeholders in interactive processes of collaboration with bio-products to meet the market. A transition strategy toward a sustainable bio-economy has to face many factors with significant uncertainty, such as consumption patterns, climate change, sustainability risks, etc.

Some other important bio-economy strategies are designed more specifically to foster biotechnology, using biomass and waste residuals, and promote collaboration relationships between different sectors and stakeholders and firms and research institutions, funding support, etc. A bio-economy research strategy should strengthen the interactions between the different stakeholders. Participation of stakeholders representing the bio-economy is vital to strategy formulation and implementation. The bio-economy research strategy is sustained by participation in research and development in collaborative projects between firms and science institutions aimed to achieve market leadership.

Research, development, and innovation are the basis of the regional and industrial strategy for bio-economy development. Regional strategies for bio-economy development are different and contextualized in its own strengths, capacities, and capabilities (Paterman 2014). Deployment of bio-economics specialization strategies can foster regional economic development and support regional bio-economy clusters. Research on bio-economy strategies at regional level is very limited and provides little knowledge on the bio-economy development.

Regional strategies can improve coordination and communication between groups and industry's platforms (BioPro Baden-Wuerttenberg 2014) and facilitate the role of local governments in supporting firms and research institutes to create bio-economy partnerships and developing standards for public procurement and labels for bio-based products, recognized by consumers (Carrez 2014). Regional collaboration between the different actors in the different regions has strategies for bio-economy development sustained by biotechnology, health, and life sciences. Regional bio-economy strategies are supported by the provision of funding and subsidies for research, development, innovation, communication, diffusion, and commercialization of bio-based activities and bio-products.

The development of a network system to support bio-economy business is relevant for the strategy implementation to enter a new international market in such areas as food, health, energy, etc. The global strategy of the bio-economy as an agenda of multinational corporations that are linked to local governments is based on processes of innovation and the sustainable development of the agricultural sector and the highly questioned natural resources. The corporate-led bio-economy agenda is a global strategy linking public and private sectors based on innovation and sustainable development. Strategic alliances for collaboration between public and private sectors are critical instruments for information sharing and the research and expansion of bio-economics and to encourage new programs of bioproduction.

A knowledge-based strategy intends to foster skills particularly for innovative start-ups and SMEs. Companies across bio-economy sector should encourage professional participation to gain experience in the field through different strategies and actions such as lifelong learning.

The benefits of bio-economy have to be communicated to consumers who may shift consumption away from fossil-based products and toward sustainable bio-products (FORMAS 2012). An external evaluation of the research bio-economy strategy assesses the positive effects and the impact on groundwork progress for the development of knowledge-based competitive bio-economy activities. The bio-economic model is an integrated approach for evaluation of fishery management strategies (Anderson and Seijo 2010; Armstrong and Sumaila 2000; Clark 1995).

A strategic plan on bio-economy is required to map future potential and directions, matching economic and social needs and formulating policy agendas. The strategist of the bio-economy should be to gradually integrate small-scale ecological practices as an alternative to the expanding biomass production (ETC Group 2015) with increasing independence of fossil fuels. However, some analysts consider that the bio-economy agenda has limitations and suggests that strategies should include more elements of agroecology. The bio-economy strategy becomes essential to the concept of circular economy in such activities as separation of biowaste collection and use, process of biodegradation, sustainable bioproduction, etc. This bio-economy strategy should be clear and transparent, engaging and involving all business, research and education institutions, community and social organizations, and all levels of government institutions.

A sustainable bio-economic strategy requires an operating environment with access to sustainable biomass for the creation of new bio-economy business based on competence. Transition toward sustainable bio-economy needs strategy design and implementation for industrial biotechnological transformation from fossil fuel resources to biomass-based production and the required consumer behavior changes (Birch 2015).

These conditions can contribute to the creation of bio-economy business for sustainable development. Sustainable bio-economy development can be accelerated by systematic strategic choices and policy actions for achieving economic growth and well-being. Society structure must enable sustainable strategic choices and policy actions for bio-economy development targeted to offer alternative well-being services and products to consumer choices.

The bio-economy strategy should be more inclusive of ecological activities and techniques. The strategies of bio-economy based on agro-industrial solutions need to expand to include new forms of adding value and agricultural knowledge involving social and community innovation to biological resources (Birch and Tyfield 2012; Schmidt et al. 2012; Levidow et al. 2012).

A research strategy should be based in a natural cycle of a bio based economy supported by a biological, ecology and technology systems in such a way that provide incentives to work and cooperate interdisciplinary. This strategy should develop the knowledge-based and innovation internationally competitive bio-economy.

Active dialogue, participation and relationships of cooperation among citizens, firms, new social movements and governments are required to support bio-economic initiatives embraced by public policies.

Concluding Remarks

The high dependence of today's economic development supported on fossil-based resources increases environmental sustainability concerns of production systems and food security. This situation justifies the urgent need for a transition from neoclassical economy type of development based on fossil resources toward a more sustainable development supported by the bio-economy based on biological resources and biological products. Sustainable development based on bio-economy optimizes the allocation of natural and biological renewable resources while increasing the environmental, food security, energy, and health concerns.

The theoretical framework of bio-economics for sustainable development was created as an alternative to neoclassical economic development model. The economic development model based on neoclassical economics is being criticized for not addressing issues such as resource scarcity, environment, social institutions, social policy, social organization, entropy, and bio-economics of economic activities.

The bio-economy sustainable development model focuses more on quality than on quantity as opposed to neoclassical economic development model. However, also the sustainable development model based on bio-economy is being under scrutiny. Some critical points addressed here regarding the sustainable development model are to consider natural resources as infinite goods which leads to an overexploitation and require massive inputs and the production method impact on agricultural land and the environment. This interpretation, which deals with similarities between economic and biological systems, has been harshly criticized. The explosive wave of bio-economics entails as policies the improvement in the quality of fuels and renewable energies; it is highly questioned and criticized for not delivering its promises.

This chapter has identified some principles of bio-economy which are critical issues affecting the prospects for the bio-economy-based sustainable development. Therefore, the chapter concludes that bio-economy is at the crossroads of the sustainable development paradigm.

Cross-References

- ▶ Business Youth for Engaged Sustainability to Achieve the United Nations 17 Sustainable Development Goals (SDGs)
- ▶ Collaboration for Regional Sustainable Circular Economy Innovation
- ▶ Ecopreneurship for Sustainable Development
- ▶ Environmental Intrapreneurship For Engaged Sustainability
- ▶ Environmental Stewardship
- ▶ Low-Carbon Economies (LCEs)
- ▶ Responsible Investing and Environmental Economics
- ▶ Sustainable Decision-Making

References

Anderson, L. G., & Seijo, J. C. (2010). *Bio-economics of fisheries management*. Ames: Wiley. 2010.

Armstrong, C. W., & Sumaila, U. R. (2000). Cannibalism and the optimal sharing of theNorth-East Atlantic cod stock: A bio-economic model. *Journal of Bio-Economics, 2*(2), 99–115. 2000.

BioPro Baden-Wuerttenberg. (2014). Bio-economy: Baden-Württemberg Path towards a Sustainable Future. Available online: http://www.biopro.de/biopro/downloads/index.html?lang=en. Accessed 27 Mar 2014.

Birch, K. (2015). York University (2015), Toronto, ON, Canada. Phone interview and email correspondence. Personal communication, 7 Jan 2015.

Birch, K., & Tyfield, D. (2012). Theorizing the bio-economy: Biovalue, biocapital, bio-economics or what? *Science, Technology & Human Values, 2012*(38), 299–327.

Birch, K., Levidow, L., & Papaioannou, T. (2010). Sustainable capital? The neoliberalization of nature and knowledge in the European knowledge-based bio-economy. *Sustainability, 2*, 2898–2918.

Borrás, S., & Edquist, C. (2013). The choice of innovation policy instruments. *Technological Forecasting and Social Change, 80*, 1513–1522.

Brundtland, G.H. (1987). *Our common future*. Oxford University Press.

Bundesministerium für Ernährung und Landwirtschaft (BMEL). (2016). Nationale Politikstrategie Bioökonomie—Nachwachsende Ressourcen und Biotechnologische Verfahren als Basis für Ernährung, Industrie und Energie. Available online: https://www.bmbf.de/files/BioOekonomiestrategie.pdf. Accessed 18 Jan 2016. (In German).

Carpintero, Ó. (2006). *La bio-economía de Georgescu-Roegen*. Madrid: Montesinos.

Carrez, D. (2014). Clever consult, Brussels, Belgium. Phone interview and email correspondance. Personal communication, 9 Apr 2014.

Clark, C. W. (1985). Modelling and fisheries management. New York: John Wiley & Sons.

Clark, C. W. (1990). Mathematical Bioeconomics: The Optimal Management of Renewable Resources. New York: John Wiley & Sons.

Clark, C. W. (2010). Mathematical bio-economics: The mathematics of conservation (3rd ed.). New York: John Wiley & Sons.

Club de Rome. (1972). Halte à la croissance?- traduction française Fayard.

De Besi, M., & McCormick, K. (2015). Towards a bio-economy in Europe: National, regional and industrial strategies. *Sustainability, 2015*(7), 10461–10478.

de Besi, M. & McCormick, K. (2017). Towards a bio-economy in Europe: National, regional and industrial strategies International Institute for Industrial Environmental Economics (IIIEE). Lund University.

ETC Group. (2015). The Bio-economy. http://etcgroup.org/issues/bio-economy.

European Bio-economy Panel and SCAR. (2014). What next for the European bio-economy? Available from: http://ec.europa.eu/research/bio-economy/pdf/where-next-for-european-bio-economy-report-0809102014_en.pdf.

European Commission. (2012a). Innovation for sustainable growth: A bio-economy for Europe. Available from: http://ec.europa.eu/research/bio-economy/index.cfm?pg=policy&lib=strategy.

European Commission. (2012b). *Innovating for sustainable growth: A bio-economy for Europe* (p. 2012). Luxembourg: Publication Office of the European Office.

European Commission. (2012c). Innovation for sustainable growth: A bio-economy for Europe. Available from: http://ec.europa.eu/research/bio-economy/index.cfm?pg=policy&lib=strategy

European Commission. (2015a). Bio-economy strategy. http://ec.europa.eu/research/bio-economy/policy/strategy_en.htm.

European Commission. (2015b). Bio-economy strategy. Available from: http://ec.europa.eu/research/bio-economy/policy/strategy_en.htm.

European Commission. (2015c). What is the bio-economy. http://ec.europa.eu/research/bio-economy/index.cfm.

FORMAS. (2012). *Swedish research and innovation strategy for a bio-based economy*. Stockholm: The Swedish Research Council for Environment, Agricultural Sciences and Spatial Planning (FORMAS).

Foxon, T. J., Gross, R., Chase, A., Howes, J., Arnall, A., & Anderson, D. (2005). UK innovation systems for new and renewable energy technologies: Drivers, barriers and systems failures. *Energy Policy, 2005*(33), 2123–2137.

Franco, J., Goldfarb, L., Fig, D., Levidow, L. & Oreszczyn, S. M. (2011). Agricultural innovation: Sustaining what agriculture? For what European bio-economy? Project-wide final report. CREPE.

Fritsche, U. R., & Iriarte, L. (2014). Sustainability criteria and indicators for the bio-based economy in Europe: State of discussion and way forward. *Energies, 2014*(7), 6825–6836.

Fumagalli, A. (2007). *Bioeconomia e capitalismo cognitivo: Verso un nuovo paradigma di accumulazione*. Roma: Carocci.

Georgescu-Roegen, N. (1977). Inequality, limits and growth from a bio-economic viewpoint. *Review of Social Economy, XXXV*(3), 361–375.

Grubler, A., Aguayo, F., Gallagher, K., Hekkert, M., Jiang, K., Mytelka, L., Neij, L., Nemet, G., & Wilson, C. (2001). Policies for the energy technology innovation system (ETIS). In *Global energy assessment – Toward a sustainable future* (pp. 1665–1744). Cambridge, UK/New York/ Laxenburg: Cambridge University Press/The International Institute for Applied Systems Analysis, 2012.

Habib, A., Ullah, H., Ngoc Duy, N. (2014). *Bio-economics of commercial marine fisheries of bay of Bengal: Status and direction*. Ahasan Thanasis Stengos.

Hall, R., & Zacune, J. (2012). *Bio-economies: The EU's real 'Green Economy' agenda?* London: World Development Movement and the Transnational Institute.

Henry, M., & Roche, M. (2013). Valuing lively materialities: Bio-economic assembling in the making of new meat futures New Zealand. *Geographer, 69*, 197–207.

Knowler, D. A. (2002). Review of selected bio-economic models with environmental influences in fisheries. *Journal of Bioeconomics, 4*(2), 163–181.

Levidow, L.. Open University, Buckinghamshire, UK. Phone interview. Personal communication.

Levidow, L., Birch, K., & Papaioannou, T. (2012). EU agri-innovation policy: Two contending visions of the bio-economy. *Critical Policy Studies, 6*(1), 40–65.

Mayumi, K. (2001). *The origins of ecological economics. The bio-economics of Georgescu-Roegen Routledge Research in Environmental Economics*. London: Routledge.

McCormick, K., & Kautto, N. (2013). The bio-economy in Europe: An overview. *Sustainability, 2013*(5), 2589–2608.

Mohammadian, M. (2000). *Bio-economics: Biological economics. Interdisciplinary study of biology, economics and education*. Madrid: Entrelíneas Editores.

Mohammadian, M. (2003). What is bio-economics: Biological economics. *Journal of Interdisciplinary Economics, 14*(4), 319–337. Guest Editor: Special Issue Dedicated to Bio-economics.

Mohammadian, M. (2005). La bio-economía: un nuevo paradigma socioeconómico para el siglo XXI. *Encuentros Multidisciplinares, 7*(19), 57–70.

Nicolescu B. (1996). La Transdisciplinarité Manifeste. Monaco: Ed. Du Rocher.

Patermann, C. (2014). Member of the bio-economy council, Germany. Phone interview and email correspondence. Personal communication, 24 Mar 2014.

Paul, H. (2013). A Foreseeable Disaster: The European Union's agroenergy policies and the global land and water grab, 19. The Transnational Institute, FDCL and Econexus. Available from: https://www.tni.org/en/briefing/foreseeable-disaster

Pfau, S., Hagens, J., Dankbaar, B., & Smits, A. (2014). Visions of sustainability in bio-economy research. *Sustainability, 2014*(6), 1222–1249.

Piotrowski, S., Carus, M., & Essel, R. (2015). Global bio-economy in the conflict between biomass supply and demand. *Nova Pap, 2015*(7), 1–14.

PNUMA. (2016). Programa de las Naciones Unidas para el Medio Ambiente. (2016). Hacia una economía verde: Guía para el desarrollo sostenible y la erradicación de la pobreza. 21 de noviembre de 2016, de Programa de las Naciones Unidas para el Medio Ambiente Sitio web: http://www.pnuma.org/eficienciarecursos/economia.php.

Schmidt, O., Padel, S., & Levidow, L. (2012). The bio-economy concept and knowledge base in a public goods and farmer perspective. *Bio-based and Applied Economics, 2012*(1), 47–63.

Socaciu, C. (2014). Bio-economy and green economy: European strategies, action plans and impact on life quality. *Bulletin UASVM Food Science and Technology, 71*(1), 1–10.

Soedigdo, D., Harysakti, A., & Usop, T. B. (2014). The elements driving local wisdom on the architecture Nusantara. *Journal of Perpect Architecture, 9*(1), 37–47.

Staffas, L., Gustavsson, M., & McCormick, K. (2013). Strategies and policies for the bio-economy and bio-based economy: An analysis of official national approaches. *Sustainability, 2013*(5), 2751–2769.

Thunberg, E. M., Helser, T., & Mayo, R. (1998). Bio-economic analysis of alternative selection patterns in the United States Atlantic silver hake fishery. *Marine Resource Economics, 13*, 51–74. 1998.

USDA. (2015). *An economic impact analysis of the U.S. biobased products industry – A report to the congress of the United States of America*. Washington: USDA.

Welfe, A., Gilbert, P., & Thornley, P. (2014). Increasing biomass resource availability through supply chain analysis. *Biomass and Bioenergy, 2014*(70), 249–266.

Environmental Intrapreneurship for Engaged Sustainability

Challenges and Pitfalls

Manjula S. Salimath

Contents

The author wishes to acknowledge helpful comments shared on a preliminary version during presentation at the 2014 *World Conference on Entrepreneurship* in Dublin.

M. S. Salimath (✉)
Department of Management, College of Business, University of North Texas, Denton, TX, USA
e-mail: Manjula.Salimath@unt.edu

S. Dhiman, J. Marques (eds.), *Handbook of Engaged Sustainability*,
https://doi.org/10.1007/978-3-319-71312-0_41

Abstract

The chapter explores the philosophy of environmental sustainability from an organizational standpoint. How do organizations push the frontiers of intrapreneurship in their proactive exploration and exploitation of environmental opportunity? Patagonia Inc. is among the most visible and early adopters of environmental intrapreneurship. This company has been engaged in resolving environmental degradation, a key issue that has challenged the business world. The chapter evaluates eight specific measures or programs that are indicative of active environmental intrapreneurship. These relatively new and radical programs range from supply chain logics, technical design standards, funding models, and cooperative networks and alliances with key stakeholders. From an inductive case-based approach, a two-by-two framework of environmental intrapreneurship initiatives is offered that are either intra- or inter-organizational as well as the presence or absence of collaborations. Practical outcomes include deriving guidelines for companies that may benefit from knowing the challenges and pitfalls of engaging in environmental intrapreneurship. Theoretical outcomes include the stimulation of research to build better explanations of engaged sustainability. Pedagogic outcomes include the potential for learning and teaching using decisional heuristics of the environmental intrapreneurship framework.

Keywords

Sustainability · Intrapreneurship · Environmental Sustainability · Stakeholder · Triple Bottom Line · Patagonia Inc. · Challenges · Pitfalls

Introduction

Sustainability includes the triple concerns of people (society), planet (environment), and profits (company) and has a generational aspect as well. According to the Brundtland Commission report (formally known as the World Commission on Environment and Development), sustainability is economic activity that "meets the needs of the present without compromising the ability of future generations to meet their own needs" (United Nations 1987). Sustainability is a relevant and pressing topic that has affected the way organizations approach their business topics, stimulating theory, research, and practice.

However, environmental concerns and its incorporation into the fabric of business priorities and considerations are a fairly recent phenomenon (Haigh and Griffiths 2009). In the past, the typical business, from a traditional standpoint, only recognized the primacy of the investor or owner of the business, who was considered the most important (and sometimes only) stakeholder. Thus, the purpose of the business was to appease the interest of the owner/investor by providing returns for their investment in the form of profits. If one were to consider the reason for businesses continued existence, following this logic, it would continue to exist as long as it satisfied the needs of its owner/investor. Soon enough, and in this vein, the primary identity of a business became the ability to generate profits. Even the field of strategic management began to espouse that the holy grail of all business was the pursuit of competitive advantage (Porter 1980; Christensen and Fahey 1984; Kay 1994). This competitive advantage referred to the ability to generate "above average" returns to the business owner/investor and referred almost exclusively to financial returns or profits. This trend continued even as businesses chose other legal forms such as becoming incorporated or publicly traded firms with many investors/owners. The pressures of public companies to appease their shareholders via quarterly or yearly dividends are a good example of how financial considerations and returns on investment remained at the forefront of all businesses, regardless of size or legal entity.

Over time and as other social and environmental values began to be recognized as being equally important as the financial considerations for firms, the idea of the triple bottom line (Spreckley 1981; Elkington 1997) gradually became more salient. Activists, ideological leaders, agencies, and sometimes even governmental bodies started to pay more attention to the actions of firms and their negative effects on the society and environment. In the most egregious cases, such as the Chernobyl disaster, Bhopal gas tragedy, Fukushima accident, or the British Petroleum oil spill, worldwide attention was drawn to the unfortunate fact that a neglect of environmental concerns by any company can prove disastrous to the society and ecosystem in which it is embedded. This was quite apart from and in addition to the financial losses suffered by the perpetrating company. These major catastrophes are among the worst industrial disasters of the world, causing enormous environmental damage by radiation, contamination, and pollution on a scale that was previously unknown, with far-reaching effects that continued over several generations. Greater oversight, tighter regulations, and new policies and fines typically ensued after much public outrage and protests were leveraged against the companies that caused such disasters.

In response to this negative attention, companies were pressured to be more cognizant of the consequences of their activities and its impact not just on financial measures, such as profitability, but the implications for environmental and societal well-being and health. However, such responses by companies after the fact or following policy restrictions are reactive and often forced and not being done of their own volition. This is a reflection of firm behavior that can be classified as a reactive response. On the other hand, some companies make environmental considerations a part and parcel of their operations, are proactive in ensuring that they do not cause any environmental harm, and may even go further by activating responses that would trigger positive effects on the environment. This is a reflection of firm behavior that can be classified as a proactive response that was voluntarily undertaken. Thus while there are the unfortunate companies on one extreme that are causing environmental harm, there are fortunately some firms that lie on the opposite spectrum because they proactively pursue environmental goals of their own volition, even in the absence of any mandated requirement.

Understanding that firm responses to environmental concerns may be either involuntary and reactive or voluntary and proactive has important implications. When firms react to external pressures to show concern for environmental aspects, they are doing the legal minimum that is being forced onto them. An example would be observing the legal limit on the emission of harmful gases. In contrast, when firms are proactive and voluntarily try to reduce environmental harm, they tend to impose a stricter standard on themselves which goes over and beyond what has been mandated. An example of this would be a company that tries to achieve zero emission of harmful gases by constantly measuring and monitoring its activities and taking remedial measures to prevent crossing its self-imposed standards. This type of proactive environmentalism generally delivers greater positives and is much more beneficial than the reactive type.

The motivations that explain why firms can choose whether to be proactive or reactive have been the subject of much research (Ricks 2005; Groza et al. 2011). Firms pursuing proactive approaches were found to do so prior to any negative information (Du et al. 2007), while reactive strategy usually occurs to reduce harm to the firm image after some negative information was reported (Murray and Vogel 1997; Wagner et al. 2009). As the motivations differ, proactive responses are viewed positively as they are considered to be altruistic (Becker-Olsen et al. 2006), while reactive responses were viewed negatively by consumers (Ricks 2005; Becker-Olsen et al. 2006; Lee et al. 2009).

It is often said that firms need some level of discretionary power to be able to make the choice of how they wish to react to environmental concerns. For instance, having high discretionary power means that firms have the autonomy and freedom to decide their course of action (Salancik and Pfeffer 1974). Not all firms have the same level of decision-making autonomy as there is heterogeneity in the locus of decision-making within different organizational structures. In a public corporation, for example, decisions are made by the top management team, and a single person is unlikely to have much say in the strategic planning. However, in most entrepreneurial firms, since the entrepreneur is the owner and manager, all or nearly a hundred percent of the decision-making autonomy is vested in this single person. So the entrepreneurs have high discretionary power and can choose to take a strategic path single-handedly without having to convince an entire team to agree. As such, entrepreneurs have the flexibility and freedom to pursue proactive environmental strategies if they wish to without having to encounter much opposition or obstacles in their own firms.

Entrepreneurship refers to the recognition and exploitation of opportunities that exist outside the firm and usually occurs in new ventures and start-ups, while intrapreneurship refers to the recognition and exploitation of opportunities within established firms. The term environmental intrapreneurship refers more specifically to the exploitation of environmental opportunities within established firms. In the next section, a brief literature review of intrapreneurship and entrepreneurship is provided as an illustrative background of these concepts.

Entrepreneurship and Intrapreneurship

The term intrapreneur was first found in a white paper written by Pinchot and Pinchot (1978) and was later picked up by other scholars and academicians, leading to a new field of inquiry over the past four decades since its original conceptualization. In 1983 Miller highlighted entrepreneurship at the enterprise level, offering new avenues for exploration. In 1985, Pinchot defined intrapreneurs as "Those who take hands-on responsibility for creating innovation of any kind within an organization. They may be the creators or inventors but are always the dreamers who figure out how to turn an idea into a profitable reality" (p. ix). Nielsen et al. (1985) defined intrapreneurship as "the development within a large organization of internal markets and relatively small and independent units designed to create, internally test market, and expand improved and/or innovative staff services, technologies or methods within the organization. This is different from the large organization entrepreneurship/venture units whose purpose is to develop profitable positions in external markets" (p. 181). In 1992, the word intrapreneur was added to the American Heritage Dictionary of the English Language where it was defined as "a person within a large corporation who takes direct responsibility for turning an idea into a profitable finished product through assertive risk-taking and innovation." In 1994, Carrie expanded the notion from large corporations to small and medium enterprises. Later scholars continued to refine the concept (Parker 2011; Gundogdu 2012; Antoncic and Hisrich 2001, 2003; Sharma and Chrisman 1999), stimulating research and clarifying its usage.

The term entrepreneur is derived from the French word *entreprendre* which means to begin something or to undertake. It was first used by Cantillon in Cantillon 1759 who explained that entrepreneurs are specialists in taking risk. Since then, due to the concerted efforts of scholars, an increased consensus is seen on the concept of entrepreneurship "as the process of uncovering and developing an opportunity to create value through innovation and seizing that opportunity without regard to either resources (human and capital) or the location of the entrepreneur—in a new or existing company" (Churchill 1992, p. 586).

Though both entrepreneurs and intrapreneurs exploit opportunity and create economic value, the former start new ventures outside an existing organization, while the latter develops a new venture within an existing organization. Morris and Kuratko (2002) highlight further differences between entrepreneurship and intrapreneurship. While entrepreneurs take the full risk and responsibility for their start-up ventures, own the intellectual rights, and have to rely finding their own resources, for intrepreneurs the situation is quite the opposite. In the case of intrapreneurs, the risk is borne by the company which also owns the concept and intellectual property, and the resources are provided by the company, prompting more risky decisions made by intrapreneurs. Intrapreneurship is considered to be a multidimensional construct with eight distinct dimensions, i.e., new ventures, new businesses, product/service innovativeness, process innovativeness, self-renewal, risk-taking, proactiveness, and competitive aggressiveness (Antoncic and Hisrich 2003).

As business environments get increasingly competitive and challenging, organizational survival depends on several factors. Key among them is the ability to stay current and manage the complexity of changing demands by focusing on innovating products, processes, and services. Intrapreneurship has been recognized as an important catalyst for innovation in organizations (Zhao 2005). Over the years, intrapreneurship has metamorphosized into being considered as an effective organizational strategy (Baruah and Ward 2015). Since intrapreneurship may be utilized as a tool to achieve a company's objective, it has also been viewed as a firm level capability. In particular, intrapreneurship can help companies achieve sustainability goals via process innovation (Hastuti et al. 2016).

Pinchot (1985) suggests that large firms try to promote intrapreneurship by making structural changes such as flattening hierarchy and delegating authority to units. Operational autonomy is important to allow employees to focus on new ideas without excessive constraints that may hamper the development and execution of innovations. When autonomous processes are enabled, employee initiatives and intrapreneurship emerge from lower levels to shape the businesses overall strategy and direction (Hart 1991). The role of product champions is critical as it allows these initiatives to be protected from dismissal, helps in securing adequate funding, and also generates positive interest from the rest of the company while simultaneously shielding it from organizational norms that may cause its rejection (Burgelman 1983; Kanter 1983; Peters and Waterman 1982). They also favored creating autonomy by bypassing usual procedures and bending rules (Shane 1994). Having the necessary organizational infrastructure, processes, and incentives facilitates the practice of intrapreneurship.

Intrapreneurship enables companies to be innovative and meet the challenges of competition. Companies that do not actively engage in innovation are likely to fail (Deeb 2015). Some examples of failed firms which chose to rest on past successes instead of pursuing continuing innovation are Pan Am in the airline industry, Montgomery Ward in the apparel retail store sector, and Borders in the book and music retail segment. Among the success stories in these same industries are Southwest Airlines, J.C. Penney, and Barnes & Noble bookstores. Several well-known products and businesses emerged from intrapreneurship encouraged within big firms such as iPhone of Apple, Post-it notes inside of 3 M, Saturn inside General Motors, Gmail and driverless cars inside Google, and Digital Light Processing technology inside Texas Instruments. Some of the popular and much publicized champions of intrapreneurship are Richard Branson from Virgin, Sergey Brin and Larry Page from Google, and Steve Jobs from Apple. Not surprisingly, Pinchot (1985) recognized that "The Future is intrapreneurial." The solution for organizational complexities is effectively offered by intrapreneurship (Baruah and Ward 2015). It has been shown that paradigm-breaking companies invest and nurture in intrapreneurship (Mohanty 2006).

Thus intrapreneurship is important for organizations, but even more so when the focus is on achieving environmental standards and sustainability goals, because the benefits go beyond the firm and impact the larger society, over an extended period of time. In the following sections, a detailed company example is presented to bring to life the various ways in which a firm can pursue environmental intrapreneurship. The chapter then moves on to displaying the common challenges and pitfalls that befall companies that wish to pursue environmental intrapreneurship as well as some pedagogical implications. The concluding section discusses implications and offers directions for future research in environmental intrapreneurship.

As described earlier, environmental intrapreneurship is highly beneficial to the environment and is among the ideal firm behaviors. The company that is highlighted is a particularly well-known firm, Patagonia Inc. This company is appropriate because it allows us to consider a key issue that has challenged the business world – environmental degradation – and because this organization (i.e., Patagonia Inc.) has been very intrapreneurial in taking specific steps to resolving this issue (Salimath 2014). To assist with the analysis, a qualitative case-based methodology is leveraged in order to explore an organization that has been one of the most visible and early adopters of environmental intrapreneurship. Consequently, this chapter evaluates eight specific measures or programs that are indicative of active environmental intrapreneurship. These relatively new and radical programs range from supply chain logics, technical design standards, funding models, and cooperative networks and alliances with key stakeholders. From this inductive case-based approach emerges a framework of environmental initiatives that are based within or between firms and are either collaborative or not. This discussion helps to shed light on specific organizations that are pushing the frontiers of intrapreneurship in their proactive exploration and exploitation of environmental opportunity. The practical outcome of this approach is to derive prescriptions and guidelines for other intrapreneurial companies that may benefit from knowing the challenges and pitfalls of engaging in environmental intrapreneurship.

In contrast to entrepreneurship which is typically the focus of smaller or new ventures, intrapreneurship is concerned with the entrepreneurial activities of established firms which may include large corporations. Thus regardless of size, when innovative processes, projects, or activities are carried out in established firms, it is considered to be intrapreneurship or entrepreneurship within organizations (Miller 1983; Antoncic and Hisrich 2001). The pursuit of intrapreneurship is important as it allows established firms to develop new initiatives and pursue opportunities that have positive consequences for firm longevity and economic development. Research in intrapreneurship has generally converged on three distinct domains – the characteristics of the individual entrepreneur (Souder 1981), the creation of new corporations (Hlavacek and Thompson 1973), and the characteristics of entrepreneurial organizations (Hanan 1976). Since the 1980s, scholars have shown that intrapreneurship has important beneficial effects on the performance of organizations and is a worthy topic of inquiry (McKinney and McKinney 1989).

A specific term, "environmental intrapreneurship," is chosen to describe intrapreneurship efforts that are focused on environmental issues. This would include activities, projects, and processes that established firms carry out to safeguard and protect the health of the natural environment which may be adversely impacted by business practices. Thus environmental issues are a subset of larger social issues and corporate social responsibility. As prior scholars have noted, intrapreneurship is a multidimensional construct (Felício et al. 2012), and as such, it is necessary to clarify which specific dimension of intrapreneurship is the focus of our study.

This chapter examines intrapreneurship, in a rather novel company context, à la Patagonia, or in the style of the company Patagonia. A case-based methodology helps to address how companies may choose to adopt, implement, and lead environmental efforts within their own organizations. While regulatory efforts by governments and institutions have imposed some level of mandatory compliance from businesses in terms of environmental responsibility, those companies that respond in a reactive fashion often do so to avoid penalties or sanctions and may try to maintain minimal legal standards of environmental responsibility. Not surprisingly, those companies that take a more proactive approach to environmental responsibility tend to go over and beyond mandated standards and are often environmental leaders and stewards.

Importance of Environmental Intrapreneurship

The issue of environmental degradation and the ensuing need to adopt protective mechanisms to restore the vitality of the planet's resources have begun to enter mainstream business thinking. Recent environmental disasters resulting from poorly managed businesses such as the British Petroleum oil spill in the Gulf of Mexico, Fukushima nuclear disaster in Japan, Bhopal Gas tragedy in India, and the Chernobyl disaster in the Soviet Union indicate that this has become a global phenomenon that can impact the lives and livelihood of people, as well as the success and failure of businesses. Often these disasters were caused because of inadequate control and monitoring systems within the organizations or sometimes a failure to acknowledge the severity of the impact of such negligence. As the above unfortunate incidents indicate, when these companies reacted later to stem further environmental damage (e.g., chemical or nuclear radiation, oil contamination, etc.), they were unable to effectively curtail or counteract the massive implications of their initial negligence toward environmental issues.

This leads to the need for proactive organizational attempts that would prevent the possibility of such environmental disasters. Some companies are pioneers, taking on environmental intrapreneurship to achieve specific goals, by effectively recognizing and exploiting environmental opportunity. Surprisingly, there is a paucity of research on environmental intrapreneurship, and this investigation attempts to fill the gap by considering and assessing such an ignored sample of business.

Yet despite these research gaps that are evident in the extant scholarship, on a practical level, it is possible to find that some businesses have moved ahead and paved the way for successful interpretations of environmental intrapreneurship. For example, Patagonia Inc. is well known for its environmental stewardship. For a fascinating look at the history of the company, from its early beginnings to a global giant and environmental leader, see the part biography offered by the founder of Patagonia (Chouinard 2005). More recently, the founder, along with the co-editor of its Footprint Chronicles, shares some valuable "actionable" steps to help businesses become socially responsible (Chouinard and Stanley 2012). Indeed, the emergence of Patagonia's social responsibility did not occur without several missteps. In fact, it appears that they view their sustainability efforts to be an ongoing and never-ending journey, and they humbly declare that they are not a "model" of social responsibility. Furthermore, in an effort to engage other businesses on the sustainability journey they have started on, they have created a checklist that can be used or modified to help other businesses that seek to become socially responsible organizations. The checklists are publicly available and cover five dimensions of critical stakeholders: business health, workers, customers, community, and nature.

Methodological Approach

The aim of the chapter is to examine environmental intrapreneurship in a different context – i.e., as it is reflected in the mode (à la) of a particularly well-known firm, Patagonia Inc (Patagoni Inc. company profile 2014). Thus an evaluation of eight specific measures or programs that are indicative of active environmental intrapreneurship is undertaken (Patagonia Company Initiatives 2014). These relatively new and radical programs range from supply chain logics, technical design standards, funding models, and cooperative networks and alliances with key stakeholders. Following recent calls, we use a case study using qualitative methodology. For our particular case, grounded methodology techniques are both suitable and appropriate (Strauss and Corbin 1990, 1998). This allows for fine-grained analysis and is suitable for examining phenomenon from an exploratory perspective.

From this inductive case-based approach, a framework of environmental initiatives is offered that are either intra- or inter-organizational as well as the presence or absence of collaborations. Theoretically, this investigation will shed light on specific organizations that are pushing the frontiers of intrapreneurship in their proactive exploration and exploitation of environmental opportunity. The practical outcome of this research is to derive prescriptions and guidelines for other intrapreneurial companies that may benefit from knowing the challenges and pitfalls of engaging in environmental intrapreneurship. In addition there are some pedagogical applications of environmental intrapreneurship and engaged sustainability that are also briefly presented.

Sample Case

The company selected is Patagonia Inc. which operates in the outdoor apparel and accessories business. According to the company profile, "Patagonia has scaled the peak of the outdoor apparel and accessories business. The company designs and markets rugged clothing and accessories to mountain climbers, skiers, surfers, and other extreme sport enthusiasts and environmentalists who are willing to pay for the Patagonia brand and its environmental ethic. Besides its signature Patagonia line (for outdoor gear and apparel), the company also sells items from sister companies Lotus Designs (paddling gear), Water Girl (women's sportswear), and Great Pacific Iron Works (retails outdoor gear and apparel). Patagonia, owned by Lost Arrow Corporation, sells these items, as well as luggage, through specialty retailers, a catalog, website, and its own stores" (http://biz.yahoo.com/ic/47/47281.html). It is a private company founded in 1972 by Yvon Chouinard and is headquartered in Ventura, California, USA.

Patagonia started initially as a small company that made tools for climbers and eventually grew larger and a bit more varied in its products over the years. However, it is endearing to note that Alpinism remains at the core of the global business. While it does make clothes for climbing, it also does so for other related sports such as skiing, snowboarding, surfing, fly fishing, paddling, and trail running. Interestingly these are silent sports that do not require a motor or the cheering crowds of other sports. Instead, the reward in these sports comes in the form of connection between humans and nature.

Patagonia's mission statement is quite direct and straightforward as to its intent and goals. Their mission states quite simply that

> We strive to make the best product, cause no unnecessary harm, (make our products with the least impact possible), and inspire and implement solutions to the environmental crisis (we do our best to accomplish this by conducting our business in a way that incorporates environmental considerations as well as contributing money and in-kind donations to a wide variety of environmental non-profits that are dedicated to protecting the earth's many wonderful things).

Patagonia's values reflect a minimalist style that is also seen in its bias for simplicity and utility in its product design. Given that their targeted sports typically occur in natural settings that are wild and beautiful, it also reflects a love of the planet and an ensuing fight to save environmental health of the planet. To reverse ill effects and environmental decline, this company has started numerous initiatives and donates time, resources, and service to other environmental groups around the world.

Sample Inquiry Areas

The following eight specific initiatives implemented by Patagonia Inc. were examined and assessed. All information was derived from secondary sources. The sources of data utilized were publicly available information from company sites and portals (http://www.patagonia.com/us/environmentalism), articles, and reports.

The Footprint Chronicles®

This initiative uses transparency about the supply chain to help reduce adverse social and environmental impacts, on an industrial scale. Right on their homepage, this initiative allows consumers to know where a particular product was made and its journey through the supply chain. There are videos and interviews with suppliers, growers, and others who play a role in creating the product at various stages. It also allows consumers to interact and share their thoughts and feedback to the company. "Our goal is to reduce the adverse social and environmental impacts of our products and to make sure they are produced under safe, fair, legal and humane working conditions throughout the supply chain" (http://www.patagonia.com).

Bluesign® Standard

This initiative works with Bluesign® Technologies, based in Switzerland, to reduce resource consumption and harm from dyes and finishes in fabrics that they use. Bluesign Technologies AG provides independent auditing of textile mills, manufacturing processes and emissions, and assessing each component's ecotoxicological impact. Textile mills can become certified "System Partners" if they commit to adopting Bluesign's recommendations and allowing verification through audit processes in five key production process areas: resource productivity, consumer safety, water emissions, air emissions, and occupational health and safety. It also includes screening of chemicals and allocation to one of the three categories: blue, safe to use; gray, special handling required; and black, forbidden under the standard. The factories using black chemicals are assisted in eliminating them and finding other alternatives for use. Consumers can be assured apparels with the Bluesign label are those that are the most environmentally friendly and socially conscious version. Patagonia was the first to adopt the Bluesign standard.

1% For The Planet®

This initiative is about forming an alliance of businesses committed to leveraging their resources to create a healthier planet. In 2002, in order to "encourage more businesses to donate 1% of sales to environmental groups," the founder of Patagonia (Yvon Chouinard) and the founder of Blue Ribbon Flies (Craig Mathews) started 1% for the planet. The mission is to "build, support and activate an alliance of businesses financially committed to creating a healthy planet." Proceeds go to "sustainability-oriented nonprofits" that companies have chosen to support. The nonprofit organization's track record, credibility and impact, and contributions received are verified. The following categories are supported: alternative transportation, climate change, energy and resource extraction, environmental education, environmental law and justice, environment and human health, food, land, pollution, water, and wildlife. This initiative had met with success, and it has been credited to have contributed more than $100 million to preserving the environment in 2012.

Conservation Alliance

This initiative encourages other companies in the outdoor industry to financially support environmental organizations and to become more involved in environmental work. It was cofounded in 1989 by Patagonia along with three peer companies, REI, The North Face, and Kelty, in order to increase outdoor industry support for conservation efforts to protect threatened wild places. The Conservation Alliance collects annual dues from companies in the outdoor industry. A 100 percent of the dues are then directly contributed to grassroots conservation organizations that are subjected to a strict grant proposal review. The alliance now boasts 180 member companies and has contributed more than $10 million.

Patagonia itself contributes generously to the Conservation Alliance with $100,000 annually. The alliance funds are said to have helped conservation of wild spaces significantly, having saved over 50 million acres of wildlands and protected over 2000 miles of rivers. This initiative has been very successful in its impact. With more than 200 member companies, the alliance expects to contribute $1.8 million in 2017.

Common Threads Initiative

This initiative started in 2011 (Patagonia 2011), created a partnership with their customers to keep clothing out of the landfill, and is guided by the five Rs: reduce, repair, reuse, recycle, and reimagine. In their own words, each of the five Rs are described succinctly below:

- *Reduce*: "We make useful gear that lasts a long time. You don't buy what you don't need."
- *Repair*: "We help you repair your Patagonia gear. You pledge to fix what's broken."
- *Reuse*: "We help to find a home for Patagonia gear you no longer need. You sell or pass it on."
- *Recycle*: "We will take back your Patagonia gear that is worn out. You pledge to keep your stuff out of the landfill and incinerator."
- *Reimagine*: "Together we reimagine a world where we take only what nature can replace."

Patagonia developed several promises to their consumers with the notion their consumers will reciprocate with a sustainable "pledge." The Common Threads Initiative also encourages five "Rs" between Patagonia and their consumers:

- Promise to create long-lasting products as long as consumers do not buy what is not needed (reduce).
- Promise to repair products as long as consumers fix what is broken (repair).
- Promise to "find a home" for Patagonia products as long as consumers continue to pass on their unwanted gear (reuse).
- Promise to take worn out products and dispose in a sustainable manner as long as consumers do not irresponsibly throw belongings in landfills (recycle).
- Together help encourage a world where only renewable resources are used (reimagine).

This initiative has been quite successful. Since 2004, Patagonia has recycled approximately 164,000 pounds of products. These recycled products are further transformed into a broad range of items that include cozies, dog coats, tablet cases, and hand planes for bodysurfers (Patagonia 2015).

Environmental Grants Program

This initiative formalized support of environmental activism by committing at the greater of t 1% of sales or 10% of pretax profits. The funding is provided to protect habitat, wilderness, and biodiversity programs that lack corporate sponsors. Being a privately held company gives Patagonia to be selective and has the freedom to fund programs that are not mainstream but may be off the more commonly known and well-trodden paths.

Patagonia therefore chooses to fund (a) only environmental work, (b) organizations that identify root causes of problems with a commitment to long-term change, and (c) focuses on organizations that create a strong base of citizen support. Patagonia believes that building grassroots momentum is the most direct path to real change, so they like to support provocative direct-action agendas of small grassroots activist organizations. They believe that the individual battle of frontline communities to protect a specific natural environment or species is the most effective in raising more complicated issues – especially biodiversity and ecosystem protection – in the public eye. Their grant guidelines specify fairly clearly those avenues that they do not fund such as green building projects (http://www.patagonia.com/grant-guidelines.html).

Enviro Internships

This initiative created an Employee Internship Program which allows employees to leave their jobs for up to 2 months to work for an environmental group of their choice, without losing salary or benefits. Patagonia employees enjoy the outdoors and are also passionate about preserving it, and this initiative allows them to work for an environmental group of their choosing without foregoing their paychecks. Having a free Patagonia employee intern work for any small, grassroots group in local communities is very beneficial. When interns return, their experiences inspire and reinvigorate commitment to Patagonia's environmental mission. Patagonia reports that their employees clocked in almost 10,000 volunteer hours for 43 organizations in a single year.

Conservación Patagonica

This initiative aims to protect and restore Patagonia's wildland ecosystems, biodiversity, and healthy communities through creating national parks. Conservación

Patagónica's mission is "to create national parks in Patagonia that save and restore wildlands and wildlife, inspire care for the natural world, and generate healthy economic opportunities for local communities." It was founded by Kristine Tompkins, former CEO of Patagonia Inc. in 2000. The first focus was on the creation of Monte León National Park in Argentina, which was also the country's first coastal national park. In 2004, a second project was undertaken – the creation of Patagonia National Park in Chile's Aysen Region. The Patagonia National Park Project consists of four major program areas: buying land, restoring biodiversity, building public access, and engaging communities. This initiative also was met with much success.

A Framework of Environmental Intrapreneurship

Following examination of the eight programs and initiatives related to environmental intrapreneurship, these initiatives are classified into internal and external. Internal initiatives are primarily concerned with restructuring or remodeling internal structures, models, and processes to achieve specific goals such as environmental intrapreneurship. External initiatives are primarily concerned with collaborative and network alliances with partners outside the organization to achieve specific goals such as environmental intrapreneurship. Between these two ends of the continuum, there lies a midrange of initiatives that involve both internal agents and collaborative mechanisms (such as partnerships with employees) to achieve specific goals such as environmental intrapreneurship. Thus a two-by-two framework of environmental initiatives is offered that is either intra- or inter-organizational as well as the presence or absence of collaborations (see Fig. 1). Note that these are broad classifications and that

Fig. 1 Framework of environmental intrapreneurship initiatives

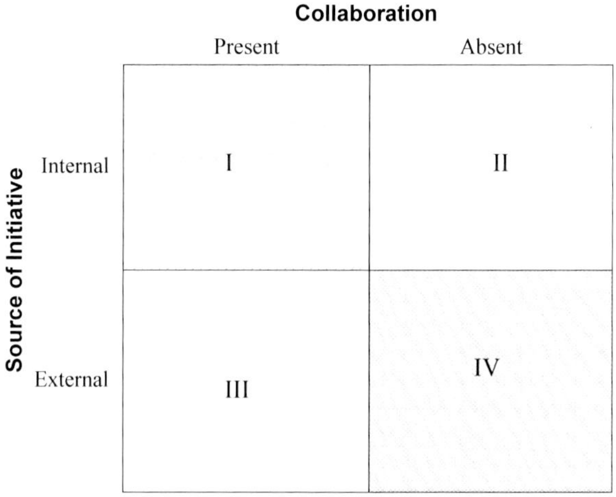

Source of Initiative \ Collaboration	Present	Absent
Internal	I	II
External	III	IV

collaborations generally refer to the extent to which they are considered essential for the initiative to be launched. Of course, any external collaboration will surely enhance its success but may not be essential for the initiative to be launched initially. For example, Conservación Patagonica efforts are targeted to occur outside the organization (building national parks) and were started as an initiative by Patagonia Inc. (as a US nonprofit incorporated in California) though it benefits from external collaborative support from other conservationists, agencies, and entities which help its success. Thus as one probes deeper into how each initiative may unfold over time and its many connections to other stakeholders, it is likely that the quadrants are not exactly watertight and that there is some overlap in how the initiatives may include elements inside and outside the organization as well as the presence or absence of collaborations. Other considerations of the complexity of the quadrants, such as its dynamic essence, are also discussed in later sections. Nevertheless, the parsimonious manner in which the framework allows organizations to make a preliminary classification of potential or existing sustainability initiatives is a valuable and useful organizational schema.

Below is the broad classification of the eight initiatives under each quadrant – presented as an example of an approach that can apply to classifying other organizational initiatives in different companies.

Quadrant I: Internal Source and Collaborative

In this quadrant, environmental intrapreneurship initiatives stem from within the organization; however, there is collaboration with other entities and firms in order to achieve the objectives.

Examples: The Footprint Chronicles® and Bluesign® Standard.

Quadrant II: Internal Source and Non-collaborative

In this quadrant, environmental intrapreneurship initiatives stem from within the organization; however, there is no collaboration with other entities and firms in order to achieve the objectives.

Examples: Environmental Grants Program and Enviro Internships.

Quadrant III: External Source and Collaborative

In this quadrant, environmental intrapreneurship initiatives stem from outside the organization; however, there is collaboration with other entities and firms in order to achieve the objectives. Examples: 1% For The Planet®, Conservation Alliance, and Common Threads Initiative.

Quadrant IV: External Source and Non-collaborative

In this quadrant, environmental intrapreneurship initiatives exist outside the organization; however, there is no collaboration with other entities and firms in order to achieve the objectives.

Example: Conservación Patagonica.

It must be noted that Patagonia Inc. has several sustainability-related efforts and projects besides the above that are either newly developed or continuing. This is not surprising, and quite expected, as the company continues to reinvent its role in ensuring sustainability in a variety of areas such as packaging and merchandizing, traceable down, responsible wool, supply chain, etc. The eight initiatives we have introduced here are sufficiently well known and broad in their impact on sustainability. For more information on the other initiatives by this company and others, one may visit the company websites and reports for details.

Limitations and Applications of the Framework

An exploratory and case-based approach was presented in order to facilitate the understanding of the ways in which environmental intrapreneurship may be best understood. For companies trying to get started on the path to becoming environmentally responsible, a wide range of options and choices are present which may inadvertently pose challenges of making the right choice. Thus, at a minimum, our framework helps firms to figure out what type of initiative to select based on two rather simple dimensions – source of the initiative and need for collaboration. However, given that the focus rested primarily on a single case and one that is fairly unique, there may be generalizability issues that arise. Hence future scholars are urged to analyze initiatives of a wider spectrum of organizations to enhance our two-by-two framework of environmental intrapreneurship.

In addition, it must be noted that the boundaries of the four quadrants are not fixed but have a dynamic quality. For example, initiatives that were initially internal or non-collaborative may, at a later stage, become external or collaborative in order to achieve the company objectives in a better fashion. The initiatives can likewise be scaled up or down depending on the firm's resources and goals. It is hoped that this helps to stimulate future research on the evolving nature of environmental intrapreneurship initiatives.

Finally, from a pedagogic perspective, this framework is valuable as it can be used in the classroom to teach students in a general business class to categorize various environmental initiatives that companies have already engaged or committed to. For more specialized courses, such as strategic management classes or entrepreneurship classes, students can use the framework to plan sustainable strategies or exercise entrepreneurial choice in deciding whether to go for collaborative approaches to environmental intrapreneurship.

Discussion of Challenges and Pitfalls of Environmental Intrapreneurship

The Patagonia Inc. company is among the most striking example of engaged sustainability and is also among the early adopters of sustainability as an integral part of its mission and identity. This company has taken a proactive approach and has embraced sustainability goals through environmental intrapreneurship. Most interestingly, this company has not only engaged in sustainability, but it has also been able to engage the larger community of stakeholders in this endeavor.

The role of the leader or champion in initiating and continuing sustainability efforts cannot be understated. Patagonia Inc.'s founder Chouinard was instrumental in shaping the company goals and was passionate about preservation and sustainability of the environment. This helped him to create a company ethos that breathed and lived on the sustainability mission. From its inception, and throughout the company's history, the sustainability motif spread and encompassed all products, processes, sourcing of inputs, advertising, and marketing. Indeed, almost all value chain activities were touched and influenced by sustainability. Almost as a necessary corollary to Patagonia Inc.'s focus on sustainability, other entities that were part of the value chain also began to be persuaded or influenced to likewise engage in sustainability. The various environmental initiatives that were collaborative in nature, no doubt, helped collaborating firms and entities to incorporate sustainability and environmentalism. Thus a ripple effect of engaged sustainability occurred, starting from the firm and spreading to suppliers, distributors, and end users. Perhaps the greatest challenge most companies face is that they lack a champion who is genuinely interested in furthering sustainability efforts.

Though the showcased company Patagonia Inc. has made tremendous inroads and has been recognized as a pioneer as far as environmental intrapreneurship is concerned, success is neither guaranteed nor is the path smooth and straightforward for all companies seeking to pursue environmental goals. It would be naïve to assume that there would be no obstacles or pitfalls on such a journey. Consequently, this section alerts firms to some of the potential challenges and pitfalls that may befall them.

Most companies jump on the sustainability bandwagon to earn quick brownie points to carry ideological favor with customers and the public by investing in environmentally friendly processes, products, and services. Unfortunately, they go after the low-hanging fruit which is relatively easier and simpler to do in a company and which generates significant return on investment. Some examples of low-hanging fruit include recycling or changing their energy consumption to renewable forms. Usually firms initially do those tasks that earn them money in the short term and are often guilty of "green washing." Some will make a huge deal of their environmentally friendly activities and then back down after making some money. For example, the fashion industry gets a lot of media attention for their promotion of a small percentage (1%) of organic cotton in their fabric selections. When it is time to do real hard work which involves long-term payout, they may no longer be interested in pursuing it. Thus sustainability initiatives have to be part of the vision and long-term strategic plan of the company. To be effective, a sustainability initiative has to have "buy in" from important stakeholders throughout the organization. A commitment to the organization's sustainability goals must be nurtured at all levels.

Among the classic sustainability pitfalls are the fairly obvious ones such as creating an environmental disaster by failing to have precautions or safety checks, having no budget set up to cover expenditures, giving no autonomy to sustainability staff to make decisions, having an immature supply chain, presenting conflicting messages from the top to the rest of the organization, over analysis, or forgetting that the business has still to make money in order to survive. Companies are also surprisingly finding out that they are far from immune from the problems found in their supply chains.

Another pitfall is that firms may quickly find that just because their initiatives hold great potential; it is not sufficient in gaining success right away. Potential is necessary, but efforts should be made to ensure that the market for the sustainable product is also cultivated. Sometimes a company may feel that they are not seeing a market response to their initiatives and are disappointed enough to give up. This can also occur when initiatives may have not yet reached the scale needed to sufficiently transform or change consumption patterns (Bocken 2017). In such situations, it is recommended that using social marketing type of techniques may hold some promise in encouraging sustainable consumption.

Failure to adequately perform assessment and measurement can prove to be problematic and is a pitfall that could be addressed during the planning stages. Assessment helps companies to realize the challenges they face in achieving sustainability goals and the pitfalls that they should try to avoid in the future. Several measurement tools exist that may be developed either specifically for the company and by the company itself or they may be independently developed by third parties that are external to the company. One of the most popular tools is an externally developed tool by B Lab, a nonprofit organization that has a team of experts from business and academia. The B impact assessment tool allows companies to measure the impact of their social and environmental impact. It is a free tool that has been utilized by over 40,000 companies to measure not only financial prowess but their ability to create value for their customers, employees, community, and the environment. The B impact score if positive indicates that the company is doing something positive for the environment and society. Scores range from 0 to 200 and are used by a range of companies (small to large) that may be interested in improving their environmental impact or getting certified as B corporations. B corporations pursue the triple bottom line of financial, environmental, and social goals simultaneously. The process is actively managed by B Lab as they also undertake verification of the accuracy of the reports submitted by companies.

There are other reporting platforms such as the Global Reporting Initiative (GRI) or IRIS that help by offering specific ways to report impact metrics to allow comparability across reports. For instance, they may define the best way to report a company's emissions, to allow for comparability with other company reports easily. IRIS is managed by the Global Impact Investing Network (GIIN), a nonprofit organization dedicated to increasing the scale and effectiveness of impact investing. The GIIN offers IRIS as a free public good to support transparency, credibility, and accountability in impact measurement practices across the impact investing industry. IRIS provides added value in the following ways: IRIS is the catalog of generally accepted performance metrics that leading impact investors used to measure social, environmental, and financial success, evaluate deals, and grow the sector's credibility. Thus the GRI or IRIS information tells you if the company is following best practice in reporting its emissions, while the B impact assessment tells consumers and activists whether the company has lowered or increased its emissions relative to others so that they may choose to support the business or not.

Assessment and accurate measurement are essential because it helps to reveal the areas that companies are investing their efforts in as well as the outcomes of their efforts. It also indicates what areas need more attention and resources and has room for improvement. Likewise, it shows what areas are proving to be counterproductive and may be disinvested. For example, better "facilitation" of employee involvement in greening initiatives by the company itself is much more effective than merely giving employees a day off toward those ends without providing them with specific avenues of action. Thus having some company-specific program for a day (e.g., cleaning debris and planting trees around the storage area) that is undertaken by the company itself would be more directional in aiding these efforts.

However, such company initiatives must also be assessed so that its impact can be evaluated. For example, Patagonia Inc. offers several opportunities for employees to participate in environmental or social activism, but without assessment, it is tough to know how many employees participated or to what degree. Similarly, companies like Dell, Kraft, and Campbell Soup Company are also pursuing sustainability initiatives that would benefit greatly from an assessment of their relative success or failures and important lessons for the future.

Interestingly, interpreting the assessment is not always straightforward, and simplistic conclusions about impact may be erroneous. It is important to understand that just because a particular program did not succeed at the first instance does not mean that its impact is nonexistent. This is a bit counterintuitive to understand. For example, the failure of Puma's biodegradable range did not ring the death knell of eco-fashion, as other companies producing biodegradable clothes are thriving. Companies that start new environmental programs seem to be instrumental in promoting a sustainability caused by creating initial momentum, which gets picked up and nurtured by other firms. Even if the initiating firm experienced nonsuccess, the cause is ignited and continues to shine in other locations and companies, often becoming a movement that has industry-wide effects.

The role of activists and vigilantes cannot be underestimated or can it be conveniently ignored either. Often due to their watchful efforts, many hard-to-find information about company violations of environmental or sustainability regulations become public knowledge. For example, Volkswagen slowly poisoned the environment by manipulating the emission controls in its engines to allow their cars to emit 40 times the legal limit of nitrogen oxide. The detection of this undoubtedly spurred greater regulation of the auto industry and automakers on the pollution that their products cause.

Due to the highly embedded nature of sustainability, it is important to note that a concerted effort by multiple stakeholders is necessary to achieve any threshold of sustainability. Collaborating with channel partners, suppliers, distributors, developers, and customers will ensure that the sustainability goals are adequately met at all stages of the product's life cycle. Otherwise it would be difficult to ensure that a product is sustainable because though sustainable sourcing and manufacturing were followed, the packaging and delivery were not subject to the same sustainability metrics. Consequently, teaming with other agencies and companies would lead to multi-stakeholder involvement that may have greater support from the society as well as greater momentum, energy, and visibility for sustainability goals.

One encouraging trend related to multi-stakeholder inclusion and transparency is that some companies are sharing their tier 1 and tier 2 supplier information. They not only measure their supplier's sustainability performance but are also willing to disclose this to industry peers who are struggling with supply chain sustainability challenges. This information was previously either not sought by firms or was considered proprietary and was closely held without being openly disclosed to the public.

Achieving a better balance between what is sent to the marketplace and what impact it has on the natural world is something that many firms struggle with. Having a best practice guide that connects the companies to others in the field who actually practice the various options would be quite a valuable learning experience of much benefit. The best practice guides can be generated by companies chronicling their experiences as well as by industry experts that can provide benchmarks for various stages or steps in the sustainability journey.

Merely having goals or even strategies appears insufficient unless there are corresponding structural changes with the organization to facilitate sustainability efforts. Reports show that firms also make governance changes to make sure that the sustainability goals are embedded in structural and procedural processes within the organization. Thus several companies have a board of directors formally oversee sustainability performance, many hold management accountable for sustainability performance, some have linked executive compensation to sustainability goals, and yet others have well-written corporate policies for sustainability.

Furthermore, having a sustainability strategy that is a stand-alone version does not augur well for its success. That is, instead of having a separate sustainability strategy, it is important to have sustainability integrated with the firm's competitive strategy. This would help with garnering ongoing funding from budgets allocated to strategic implementation. Companies such as UPS, Starbucks, Centrica, and Hitachi are infusing their strategies with sustainability by taking care to ensure that the ties between strategy and sustainability are natural, by connecting sustainability to new strategic opportunities, and integrating it with competitive strategy. However this can generate an interesting pitfall – if sustainability is successfully integrated with strategy, then employees have a harder time trying to articulate, identify, or know what the sustainability initiative is or how it differs from what another firm may be doing.

Implications and Future Research

To create a culture of environmental intrapreneurship, corporations should hire employees that are innovative and are not afraid of risk-taking and making decisions on their own. Big companies should also make it known that failure will not be punished and that it is ok to make mistakes. As an employee, it is important to find and stay in companies that will encourage you to pitch your environmental ideas to management and should be proactive in seeking opportunities to advance their innovations.

There are several barriers to achieving successful intrapreneurship in organizations (Buekens 2014) which are applicable to environmental intrapreneurship as well. He suggests that the analytic techniques for centralizing decision-making and scientific management techniques are suitable for making existing things better in an organization but destroy the ability to do something truly innovative and thus do not foster intrapreneurship. Having an environment at work that gives freedom and time to pursue new ideas helps environmental intrapreneurship to take root in the organization. Many companies such as Google give their employees a certain percentage of time during the workday where the sole task is to engage in new, creative, and innovative ideas that originate from the employees themselves. When organizations are rigid and do not allow such flexibility to explore new avenues, intrapreneurial employees get frustrated with the lack of opportunity given to them, and often leave, taking their ideas outside the company. Thus companies lose valuable talent and the potential to earn profit from employee innovations.

Regardless of size, organizations must encourage environmental intrapreneurship not only to attain a competitive advantage but also to meet sustainability goals. At a practitioner level, businesses such as Patagonia Inc. have taken proactive steps to pursue environmental intrapreneurship and have forged ahead creating their own methods and processes to do so. However not much is known about the factors, conditions, antecedents, and consequences of environmental intrapreneurship, and more scholarship and inquiry in this vibrant and emerging area are needed.

Future research can explore the antecedents of sustainability at the firm level such as the capabilities of environmental intrapreneurship. It may be helpful to know the boundaries and assumptions under which environmental entrepreneurship can be effective in organizations and whether the size, type, and industry context of firms would make a difference. It would be interesting to know if any cross-national variation exists in the practice and effectiveness of environmental intrapreneurship. Developing and developed economies of the world face different environmental challenges, and it would be useful to understand how country-specific environmental policies would unfold and influence environmental intrapreneurship in companies in these economies. Future questions that can be explored include longitudinal assessments of environmental intrapreneurship within the same organization and its evolution over the organizations life cycle. The impact of external pressures in the form of regulatory policy and environmental standards on companies would be yet another fruitful area for future research.

In conclusion, environmental intrapreneurship is a fruitful way for companies to engage in sustainability; however, care and caution must be undertaken to ensure that the common challenges are met and that the company does not fall victim to the typical pitfalls during the journey to engaged sustainability. The Patagonia Inc. case showcases some interesting ways to engage with other stakeholders to deliver sustainability goals. The two-by-two framework of environmental intrapreneurship may assist companies in deciphering where and how to plan its sustainability activities and prove useful in the educational efforts in the classroom and in the research efforts of scholars to build a theory of engaged sustainability.

Cross-References

- ▶ Ecopreneurship for Sustainable Development
- ▶ Responsible Investing and Corporate Social Responsibility for Engaged Sustainability
- ▶ Social Entrepreneurship

References

Antoncic, B., & Hisrich, R. D. (2001). Intrapreneurship: Construct refinement and crosscultural validation. *Journal of Business Venturing, 16*(5), 495–527.

Antoncic, B., & Hisrich, R. D. (2003). Clarifying the intrapreneurship concept. *Journal of Small Business and Enterprise Development, 10*(1), 7. Retrieved from https://libproxy.library.unt.edu/login?url=https://search.proquest.com/docview/219320555?accountid=7113

Baruah, B., & Ward, A. (2015). Metamorphosis of intrapreneurship as an effective organizational strategy. *International Entrepreneurship and Management Journal, 11*(4), 811–822. https://doi.org/10.1007/s11365-014-0318-3

Becker-Olsen, K. L., Cudmore, B. A., & Hill, R. P. (2006). The impact of perceived corporate social responsibility on consumer behavior. *Journal of Business Research, 59*(1), 46–53.

Bocken, N. (2017). Business-led sustainable consumption initiatives: Impacts and lessons learned. *Journal of Management Development, 36*(1), 81–96.

Buekens, W. (2014). Fostering Intrapreneurship: The challenge for a new game leadership. *Procedia Economics and Finance, 16*, 580–586. https://doi.org/10.1016/S2212-5671(14)00843-0

Burgelman, R. A. (1983). A process model of internal corporate venturing in the diversified major firm. *Administrative Science Quarterly, 28*(2), 223–244.

Cantillon, P. (1759). *The analysis of trade and commerce*. London: Macmillan.

Chouinard, Y. (2005). *Let my people go surfing: The education of a reluctant businessman*. New York: The Penguin Press.

Chouinard, Y., & Stanley, V. (2012). *The responsible company: What We've learned from Patagonia's first 40 years*. Ventura: Patagonia Books.

Christensen, K. and Fahey, L. (1984). Building distinctive competences into competitive advantage. *Strategic Planning Management*, February, pp. 113–123.

Churchill, N. C. (1992). Research issues in entrepreneurship. In D. L. Sexton & J. D. Kasarda (Eds.), *The state of the art of entrepreneurship* (pp. 579–596). Boston: PWS Kent.

Deeb, G. (2015). Big companies that embrace intrapreneurship will thrive. *entrepreneur.com*, March 19, https://www.entrepreneur.com/article/243884#

Du, S., Bhattacharya, C. B., & Sen, S. (2007). Reaping relational rewards from corporate social responsibility: The role of competitive positioning. *International Journal of Research in Marketing, 24*(3), 224–241.

Elkington, J. (1997). *Cannibals with forks: The triple bottom line of 21st century business*. Oxford: Capstone Publishers.

Felício, J. A., Rodrigues, R., & Caldeirinha, V. R. (2012). The effect of intrapreneurship on corporate performance. *Management Decision, 50*(10), 1717–1738.

Groza, M. D., Pronschinske, M. R., & Walker, M. (2011). Perceived organizational motives and consumer responses to proactive and reactive CSR. *Journal of Business Ethics, 102*(4), 639–652.

Gundogdu, M. C. (2012). Re-thinking entrepreneurship, Intrapreneurship, and innovation: A multi-concept perspective. *Procedia – Social and Behavioral Sciences, 41*, 296–303.

Haigh, N., & Griffiths, A. (2009). The natural environment as a primary stakeholder: The case of climate change. *Business, Strategy and the Environment, 18*, 347–359.

Hanan, M. (1976). Venturing corporations – Think small to stay strong. *Harvard Business Review, 54*(31), 139–148.

Hart, S. L. (1991). *Intentionality and autonomy in strategy-making process: Modes, archetypes, and firm performance* (Vol. 7). Greenwich, CT: JAI Press.

Hastuti, W. A., Talib, A. N. B., Wong, K. Y., & Mardani, A. (2016). The role of intrapreneurship for sustainable innovation through process innovation in small and medium-sized enterprises: A conceptual framework. *International Journal of Economics and Financial Issues, 6*(3). Retrieved from https://libproxy.library.unt.edu/login?url=https://search.proquest.com/docview/1809611972?accountid=7113.

Hlavacek, J. D., & Thompson, V. A. (1973). Bureaucracy and new product innovation. *Academy of Management Journal, 16*(3), 361–372.

Kanter, R. M. (1983). *The change masters: Innovation and entrepreneurship in the American corporation*. New York: Simon & Schuster.

Kay, J. (1994). *Foundations of corporate success*. Oxford: Oxford University Press.

Lee, H., Park, T., Moon, H. K., Yang, Y., & Kim, C. (2009). Corporate philanthropy, attitude towards corporations, and purchase intentions: A South Korea study. *Journal of Business Research, 62*(October), 939–946.

McKinney, G., & McKinney, M. (1989). Forget the corporate umbrella—Entrepreneurs shine in the rain. *Sloan Management Review, 30*(4), 77–82.

Miller, D. (1983). The correlates of entrepreneurship in three types of firms. *Management Science, 29*, 770–791.

Mohanty, R. P. (2006). Intrapreneurial levers in cultivating value-innovative mental space in Indian corporations. *Vikalpa, 31*(1), 99–106.

Morris, M. H., & Kuratko, D. F. (2002). *Corporate entrepreneurship: Entrepreneurial development within organizations*. Cincinnati: South-Western Publishing.

Murray, K. B., & Vogel, C. M. (1997). Using a hierarchy-of-effects approach to gauge the effectiveness of corporate social responsibility to generate goodwill toward the firm: Financial versus nonfinancial impacts. *Journal of Business Research, 38*(2), 141–159.

Nielsen, R. P., Peters, M. P., & Hisrich, R. D. (1985). Intrapreneurship strategy for internal markets—Corporate, non-profit and government institution cases. *Strategic Management Journal, 6*, 181–189. https://doi.org/10.1002/smj.4250060207

Parker, S. C. (2011). Intrapreneurship or entrepreneurship? *Journal of Business Venturing, 26*(1), 19–34.

Patagonia. (2015). Environmental and social initiatives. Retrieved from http://www.patagonia.com/on/demandware.static/Sites-patagonia-us-Site/Library-Sites-PatagoniaShared/en_US/PDF-US/patagonia-enviro-initiatives-2015.pdf.

Patagonia Company Initiatives. (2014). http://www.patagonia.com/us/environmentalism. Last accessed 17 April 2014.

Patagonia Inc. company profile. (2014). http://biz.yahoo.com/ic/47/47281.html. Last accessed 17 April 2014.

Patagonia. (2011, September 7). Introducing the common threads initiative: Reduce, repair, reuse, recycle, reimagine. Retrieved from http://www.patagonia.com/blog/2011/09/introducing-the-common-threads-initiative/.

Peters, T., & Waterman, R. (1982). *In search of excellence*. New York: Harper & Row.

Pinchot, G., & Pinchot, E. (1978). *Intra-Corporate Entrepreneurship*. Tarrytown: Tarrytown School for Entrepreneurs.

Pinchot, G. (1985). *Intrapreneuring: Why you don't have to leave the corporation to become an entrepreneur*. New York: Harper & Row.

Porter, M. E. (1980). *Competitive strategy*. New York: Free Press.

Ricks, J. M., Jr. (2005). An assessment of strategic corporate philanthropy on perceptions of brand equity variables. *Journal of Consumer Marketing, 22*(3), 121–134.

Salancik, G. R., & Pfeffer, J. (1974). The bases and use of power in organizational decision making: The case of a university. *Administrative Science Quarterly, 19*(4), 453–473.

Salimath, M. S. (2014). Environmental Intrapreneurship à la Patagonia. Proceedings, *World Conference on Entrepreneurship*, International Council for Small Business, Dublin, Ireland.

Shane, S. A. (1994). Are champions different from non-champions? *Journal of Business Venturing, 9*, 397–421.

Sharma, P., & Chrisman, J. J. (1999). Toward a reconciliation of the definitional issues in the field of corporate entrepreneurship. *Entrepreneurship Theory and Practice, 23*(3), 11–27. Retrieved from https://libproxy.library.unt.edu/login?url=https://search.proquest.com/docview/213809448?accountid=7113

Souder, W. E. (1981). Encouraging entrepreneurship in the large corporations. *Research Management, 14*(3), 18–22.

Spreckley, F. (1981). *Social audit – A management tool for co-operative working*. Penarth: Beechwood College.

Strauss, A., & Corbin, J. (1990). *Basics of qualitative research: Grounded theory procedures and techniques*. Newbury Park: Sage.

Strauss, A., & Corbin, J. (1998). *Basics of qualitative research: Techniques and procedures for developing grounded theory* (2nd ed.). Thousand Oaks: Sage.

United Nations General Assembly. (1987). Report of the World Commission on Environment and Development: our common future, toward sustainable development, Chapter 2, Paragraph 1.

Wagner, T., Lutz, R. J., & Weitz, B. A. (2009). Corporate hypocrisy: Overcoming the threat of inconsistent corporate social responsibility perceptions. *Journal of Marketing, 73*(6), 77–91.

Zhao, F. (2005). Exploring the synergy between entrepreneurship and innovation. *International Journal of Entrepreneurial Behavior and Research, 11*(1), 25–41. https://doi.org/10.1108/13552550510580825

Part III

Training the Mind, Education is the Heart of Sustainability

The Sustainability Summit

Using Appreciative Inquiry to Engage the Whole System

Dennis Heaton

Contents

Abstract

Engagement plays an essential role in advancing sustainability. Engagement elicits participation and collaboration from the diverse parties who all together impact sustainability. It is through engaged listening that an organization understands the value it is creating or destroying as seen from the perspectives of multiple stakeholders. For this reason, how effectively the organization creates

D. Heaton (✉)
Maharishi University of Management, Fairfield, IA, USA
e-mail: dheaton@mum.edu; dheaton21@gmail.com

S. Dhiman, J. Marques (eds.), *Handbook of Engaged Sustainability*,
https://doi.org/10.1007/978-3-319-71312-0_33

stakeholder engagement is a measure of sustainable business practice and is a key indicator in the responsible management reporting standards of Integrating Accounting and of the Global Reporting Initiative.

Appreciative Inquiry (AI) is a process through which companies and communities have effectively and efficiently engaged a broad representation of stakeholders in a collaborative visioning and design process to advance sustainability. Rather than investigating and solving problems, AI mobilizes change through positive inquiry into strengths and aspirations. A sustainability summit is a large group process which convenes diverse stakeholders to apply AI to cocreate a sustainable future.

This chapter provides practical guidelines for preparing and conducting a sustainability summit using AI. Two case studies are presented to demonstrate the successful application of AI for transformative change toward sustainability – Fairmount Minerals and the City of Cleveland, Ohio.

Systems move in the direction of what they consistently inquire about. Therefore, the questions used in Appreciative Inquiry shape what the system will become. Examples of discussion questions for the discovery, dream, design, and destiny stages of an AI sustainability summit are provided.

Keywords

Stakeholders · Engagement · Sustainability · Appreciative Inquiry · Organizational change · Large group planning · Positive

Introduction

How does an organization effectively transition toward greater sustainability? In his book *Leading Change Toward Sustainability: A Change-Management Guide for Business, Government and Civil Society*, Doppelt (2010) outlines a number of key factors. These include:

- Changing the goals of the system by crafting an ideal vision and guiding principles for sustainability
- Organizing deep, wide, and powerful sustainability transition teams
- Tirelessly communicating the need, vision, and strategies for sustainability

The Appreciative Inquiry (AI) sustainability summit process described in this chapter can enable any organization to address each of these factors in a swift and powerful way. By getting the whole system in the room to communicate in dyads, small groups, and large groups, it amplifies communication. By asking "Dream" questions to envision an ideal future state, it not only clarifies goals but taps into passionate commitment. And by enabling cross-group members to self-organize to create prototypes and action plans, it quickly builds powerful teams. One word that captures the essence of the summit process is engagement – in terms of how diverse participants are brought together, in terms of the empowered participative exercises in the summit event, and in terms of continuing team action on initiatives launched in the summit.

This chapter begins by highlighting the role of stakeholder engagement in socially and environmentally responsible management. It then introduces the five principles of AI and AI's four-D cycle of Discovery, Dream, Design, and Destiny. The use of AI for sustainability summits is illustrated by two cases. Practice exercises and reflection questions in this chapter guide readers to apply the content in their own settings.

The Importance of Stakeholder Engagement

According to stakeholder theory (Freeman 1984), stakeholder engagement creates social capital, which can lead to prospering. Stakeholders typically include investors, customers, employees, organized labor, media, local communities, government media, universities, and nongovernmental advocacy groups. Through formal and informal dialogues and engagement practices, managers connect stakeholders to strategies to achieve common goals (Andriof and Waddock 2002).

Engaging with diverse stakeholders enables an organization to understand how it creates or destroys value as perceived from various perspectives. Through two-way communication, the organization gains knowledge of opportunities to innovatively fulfill stakeholder interests as well as understanding of the reputational and financial risks around stakeholder sustainability concerns.

Global Reporting Initiative (GRI) is an international standard for sustainability reporting that is followed by the majority of large corporations. The GRI principle for stakeholder inclusiveness calls for each reporting organization to identify its stakeholders, their expectations and interests, and the firm's responses regarding each stakeholder issue. Stakeholders are entities or individuals who are affected by the reporting organization's activities, products, or services; or whose actions can affect the ability of the organization to implement its strategies (GRI.org 2016, p. 8). The GRI principles contend that engagement and accountability increase trust and result in learning for the organization.

Stakeholder engagement has also become an essential element of Integrated Reporting (n.d.), which links financial reporting to sustainability issues which are of concern to stakeholders. See also the chapter ▶ "Designing Sustainability Reporting Systems to Maximize Dynamic Stakeholder Agility" in this volume. Integrated Reporting recognizes and reports on not just financial capital but also manufactured, intellectual, human, social and relationship, and natural capital:

> The capitals are stocks of value that are increased, decreased or transformed through the activities and outputs of the organization. For example, an organization's financial capital is increased when it makes a profit, and the quality of its human capital is improved when employees become better trained. (International Integrated Reporting Council n.d., p. 11)

> An example of Integrating Reporting is the South African firm Telkom. Telkom reports its performance one each capital. Telkom's Integrated Report explains: "At Telkom we believe that relationship capital is about engaging with the people connected with us in every way possible, receiving their input, listening to them, informing them and taking action to put

> things right if we need to" (Telkom 2015, p. 88). Telkom's Integrated Report includes a table identifying stakeholders and the material issues they raised. Telkom also described the variety of communication channels it employs to engage with stakeholders, including: "face-to-face meetings, telephonic and electronic communication, websites, electronic and paper-based employee and customer newsletters, brochures, advertising, employee and customer forums and customer and investor roadshows" (p. 88).

As illustrated by the above example from Telkom's Integrated Report, dialogue with stakeholders can be facilitated through a number of communication channels. Each contact with a customer, supplier, employee, or other stakeholder provides an opportunity to strengthen organizational learning and social capital. One particular channel for engaging with a large number of stakeholders in a concentrated period of time is a sustainability summit employing the Appreciative Inquiry (AI) approach to organizational change. We will first give some background about AI and then describe its use to advance engaged sustainability for companies and communities.

Appreciative Inquiry for Stakeholder Engagement

AI is an approach to organizational development that focuses on the organization's strengths rather than its problems. The authors of *The Appreciative Inquiry Handbook* define Appreciative Inquiry as "the cooperative co-evolutionary search for the best in people, their organizations, and the world around them" (Cooperrider et al. 2008, p. 3). The AI process helps an organization to create images of its future and in so doing, create hope, energy, and commitment to change among the groups of people working to achieve that future. AI seeks out the very best of "what is" to help ignite the collective imagination of "what might be." The aim is to generate new knowledge that expands conceptions of what is possible and to facilitate team action to translate images of possibility into reality (Christian Reform World Relief Committee 1997). The AI process is said to release six freedoms: to be positive, to act with support, to choose to contribute, to dream in community, to be heard, and to be known in relationship (Cooperrider and Whitney 2005, p. 60).

A key element for successful application of AI is to engage voices of the whole system for richer conversations and the possibility of more innovation solutions. The whole system means all levels, all functions, and key stakeholders. Cooperrider (2012) explains:

> When it comes to enterprise innovation and integration, there is nothing that brings out the best in human systems — faster, more consistently, and more effectively — than the power of "the whole." Flowing from the tradition of strengths-based management, the AI Summit says that in a multi-stakeholder world it is not about (isolated) strengths per se, but about configurations, combinations, and interfaces.

The word *appreciate* emphasizes the positive focus of AI (Cooperrider and Godwin 2011). AI mobilizes collective inquiry around positive questions, to appreciate and strengthen a system's capacity to apprehend, anticipate, and heighten positive potential. The word *appreciate* has an additional meaning – which is to increase in value. An underlying principle of AI is the recognition that what we put our attention on will grow and evolve as a result of our positive, collective inquiry. The word *inquiry* means to ask questions, to search for understanding. An AI summit is a task-focused assembly to build a vision and plan of action for advancing a practical positive purpose. Through large group inquiry and dialog around a series of prepared inquiry questions, creative ideas, and actions are unleashed.

AI's positive approach is consistent within findings in the domain of positive organizational scholarship (Cameron and Spreitzer 2011) and positive psychology. Research supports the view that positive emotions broaden one's repertoire for thinking and acting, build optimism and resilience, and elevate relationships (Fredrickson et al. 2003; Haidt 2003). AI starts by identifying and celebrating a systems positive core, as opposed to traditional organizational development which would start by identifying and analyzing problems. The key tool in an AI intervention is the unconditional positive questions which turn attention to positive potential, in contrast to deficit-oriented approaches to problem identification and problem solving. Barrett and Cooperrider (n.d) provided the following examples of positive question:

- What possibilities exist that we have not yet considered?
- What solutions would have use both win?

What is the smallest change that could make the biggest impact? and contrasting negative questions:

- What is biggest problem here?
- Why do you blow it so often?
- Why do we still have these problems?

What makes AI different from other OD methodologies is that every question is positive. According to Laszlo and Zhexembayeva (2011) focusing on gaps and weaknesses, such as what is wrong with the environmental and social performance in our business, can engender blame and discouragement. To make progress in the direction of sustainability, "What we need now is an ability to appreciate – to grow – everything that is right, inquiring into successes, however small" (Laszlo and Zhexembayeva 2011, p. 141).

Over the past 30 years, Appreciative Inquiry has been applied around the world with transformative impacts. Examples of organizations that have used AI Summits include corporations like Hewlett Packard, Fairmont Minerals, Green Mountain Coffee Roasters, the BBC government agencies like the US Navy and the Environmental Protection Agency. AI has also been used in local governments, national organizations, and in worldwide initiatives like the United Nations and United Religions.

The UN Global Compact Summit (2004) was an AI event attended by 500 global business leaders with the task of building momentum for adoption of the Global Compact principles for socially and environmental responsibility and anticorruption. By now, 13,000 business and nonbusiness organizations from 170 countries have committed to the principles of the Global Compact and 680 universities around the world (including my Maharishi University of Management, USA) have signed on to advance the principles of the Global Compact through teaching Principles of Responsible Management Education (PRME 2017).

The following discussion questions from the UN Global Compact Summit illustrate how AI focuses attention on positive achievements and aspirations:

- Please recount one specific story that describes an innovative practice in global corporate citizenship. What do you believe is the most significant moment of choice, discovery, or outcome in the story? How can this innovation or practice be scaled up for greater impact?
- Describe one innovation that illustrates the type of global citizenship you want to see in the future.

At the conclusion of that summit, UN Secretary General, Kofi Annan wrote to David Cooperrider, "I would like to commend Appreciative Inquiry and to thank you for introducing it to the United Nations. Without this, it would have been very difficult, perhaps even impossible, to constructively engage so many leaders of business, civil society, and government" (Annan 2004).

Five Principles of AI

AI is an organization development (OD) process and approach to change management that grows out of social constructionist thought and that is situated within what is now known as positive organizational scholarship. It is a practice which makes extensive use of storytelling for narrative-rich communication. The following are five foundational principles of the AI approach (Cooperrider et al. 2008; Bushe 2013):

1. The **constructionist principle** recognizes that what we take to be organizational realities are constructed from the language and discourse of day-to-day interactions of the people in that organization. Words create worlds. The constructionist principle recognizes the habitual styles of thought, and preconscious assumptions come to define organizations. The questions asked in an AI intervention can liberate new possibilities of thinking and thus new possibilities for organizational realities.
2. The **principle of simultaneity** proposes that as we inquire into human systems we change them. Inquiry is intervention; organizations change in the direction of the questions they consistently ask. The moment we ask a question, we begin to create change. The careful framing of positive dialogue questions is fundamental to the course of an AI intervention.
3. The **poetic principle** proposes that organizational life is coauthored by organizational members through the stories they tell each other each day. There is a choice to dwell on stories of stress and failure or to bring attention to stories of moments of creativity, engagement, and success. AI makes use of this principle in

the Dream questions which ask members to envision and communicate vivid depictions of positive future states.

4. The **anticipatory principle** acknowledges that what we do today is guided by our image of the future. Images inspire action. Appreciative Inquiry elicits artful creation of positive imagery to inspire present actions toward an envisioned future. In the words of Cooperrider et al. (2008, p. 8): "Human systems are forever projecting ahead of themselves a horizon of expectation that brings the future powerfully into the present as a mobilizing agent."
5. The **positive principle** declares that momentum for sustainable change requires positive affect and social bonding. Research in positive psychology has shown that positive emotions like hope, excitement, and joy increase creativity and openness to new ideas and people (Cameron 2012). Positive questions lead to positive change. The use of positive questions in AI – like "tell a story about a time you felt most empowered in this company" leads to positive results, in contrast to the results from inquiry into "what is the cause of low morale."

Based on the assumptions that organizations are socially constructed and limited only by human imagination, AI create processes of inquiry that enable organizations to appreciate, envision, and actualize their potential to function at their best for organizational members and all those affected by the organization.

The Four-D Cycle of Discovery, Dream, Design, and Destiny

An AI intervention is a facilitated group process which generally progresses through four phases – discovery, dream, design, and destiny model. This first phase, "Discover," is about learning: discovering the best of the moments and memories in the history of an organization and its people. The second phase, "Dream," builds on these exceptional life moments to envision what the organization could be in the future. In the "Design" phase, elements of the collective future are developed into prototypes by teams of summit participants, leading to agreeing on each person's continuing role in the phase called "Destiny" (Fig. 1).

Each AI event is focused on an affirmative topic which comes out of a preliminary organizing stage. The affirmative topic tells what the system wants more of. It established the task for the summit. The definition of the topic or overall generative focus for the inquiry is sometimes referred to by a fifth D word – Define.

Some examples of affirmative topics for AI interventions are increased customer satisfaction, operational quality, teamwork, or high employee engagement. Topics for various actual AI events cited by Cooperrider and Barrett (n.d.) include: outstanding arrival experiences, sustainable value creation, creating an epidemic of health, empowering and enlightened leadership, and ecologically restorative production. In applying AI to engaged sustainability, a group may choose an affirmative topic about net positive environmental impacts or thriving societal well-being throughout the supply chain.

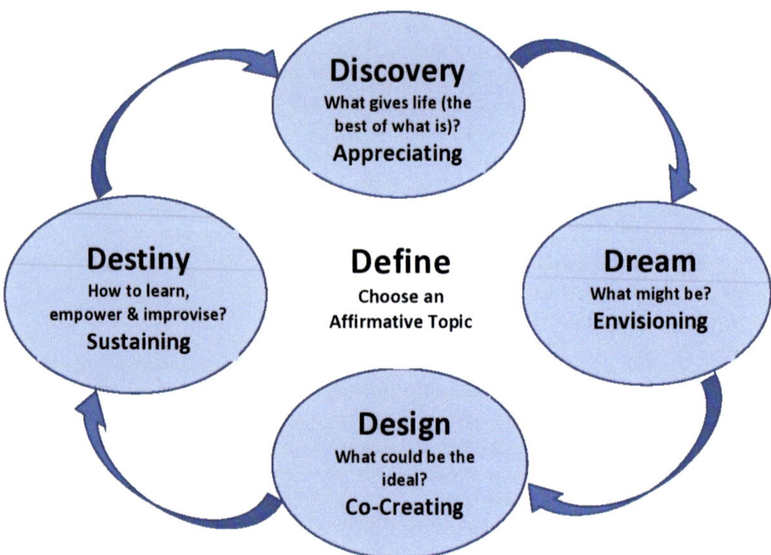

Fig. 1 Appreciative Inquiry 4-D Cycle (Based on Cooperrider and Whitney 2005)

The critical value of preframing a clear and compelling task or affirmative purpose is illustrated by the success of the Green Mountain Roasters AI Summit convened around the affirmative topic of "Coming Together to Create Phenomenal Growth and a New Model of Sustainability for Doing Good and Doing Well." As reported by Copperrider (2012, p. 106):

> At that time, the company's stock price was at $11 a share. A few short years later, it was trading at over $100 a share, and Green Mountain was selected as the most ethical corporation in the world two years in a row — unprecedented at that time.

Figure 1 identifies four phases in AI that occur after the "affirmative topic" is chosen. The **Discovery** phase using dialog to remember strengths and moments of excellence for the organization and for individual participants. Participants interview each other – taking turns as interviewer and interviewee – about their own "best of" stories. If the affirmative topic is about positive environmental impact, an example question could be "tell about something you feel proud that our organization has done for the environment." Telling and listening to meaningful, personal stories is considered central to creating widespread engagement and building relationships in the early stage of the change process. After dyad interviews, groups of dyads join together to look for key themes that underlie the times when the organization was most alive and at its best.

Valuing the best of "what is" leads naturally to searching further into what could be. The Discovery phase holds up and carries forward on an element of continuity as an organization journeys toward a desired future. On the basis of affirming past strengths and successes, people have more confidence for inventing possible futures.

During the **Dream** phase, participants are asked to share their images of a desired future. Participants are asked to imagine their group, organization, or community at its best in relation to the affirmative topic. A typical question for the Dream phase asks participants to imagine that you wake up 10 years in the future; describe a picture of what is happening. Another question can be: "I will be most proud of our organization in 2028 when I see . . .." After sharing these stories about the future, groups aim to identify common aspirations of system members and to symbolize their collective dream in a graphic representation.

Design follows Dream. A shared image of what might be inspires members to design innovative ways to help move the organization closer to the ideal. Participants are asked to coconstruct concrete proposals for the new organizational state that has been envisioned. The group work in this phase can include brainstorming actionable ideas and then rapidly building prototypes to show the summit what the innovation could look like. These mock-ups of innovations provide further momentum for implementation planning in the Destiny phase. Design targets are expressed in positive terms, are written in the present tense as if they are already happening, and represent a realistic stretch.

The **Destiny** phase is about sustaining progress to deliver the design. Outcomes of this final event of the summit consist of agreements about the who, what, and when of short-term, medium-term, and long-term action plans and participant pledges to specific actions.

The 4-D cycle can be utilized in a variety of schedules. A common format for an appreciative inquiry summit is a multiday retreat to quickly experience progress through discover, dream, design, and destiny. This process not only generates innovative designs but also builds the motivation and social structure to carry through on transformative ideas.

Resources for learning more about Appreciative Inquiry include *The Appreciative Inquiry Handbook* (Cooperrider et al. 2008) and *The Appreciative Inquiry Summit: A Practitioner's Guide for Leading Large-Group Change* (Ludema et al. 2003). A reference for framing questions for each part of the 4-D cycle is *Encyclopedia of Positive Questions: Using Appreciative Inquiry to Bring Out the Best in Your Organization* (Whitney et al. 2013). An online repository of downloadable materials and case examples can be found at the AI Commons (https://appreciativeinquiry.champlain.edu/).

Sustainability Summits Using Appreciative Inquiry

The focus in an AI summit is not on inquiring about what has gone wrong in environmental and social performance, but on bringing to life the best of what can be right. AI is a social technology to capture wholeness by engaging diverse elements of a community in order to design and implementing holistic solutions for sustainability. Senge et al. (2010) count "collaborating across boundaries" as one of the essential capabilities for realizing sustainability. They argue that changing the sustainable impacts of our human activities cannot be separated from changing how we work together. Isolated parties cannot construct holistic solutions without joining and communicating with other parties who can understand and affect other parts of the system. This is because the ecological, economic, and social context of business is too complex for any one person or even one organization to have full knowledge. Systems' solution comes not only from the inclusion of diverse stakeholders but from the chemistry of the combine strengths in collaborative interaction. A related chapter about collaboration in this volume is by Buch et al., ▶ "Collaboration for Regional Sustainable Circular Economy Innovation."

The AI summit process is a way to design initiatives to create sustainable value by getting the whole system in the room. Sustainable value means positive value for shareholders and stakeholders, while avoiding risks associated with negative social and environmental impacts. The summit process can enable organizations to identify and develop initiatives that embed sustainability into business strategy at the levels of risk, process, product, markets, firm reputation, and changing the business context (Laszlo and Zhexembayeva 2011). When sustainability is strategically embedded, it a source of innovation, differentiation, and competitive strategy rather than any added-on cost or constraint.

An AI summit convenes a cross-section of the organization and its stakeholders in a participative planning and design process. An advantage of this large group planning process is that it can harness the power of the whole: AI is suited to the multistakeholder nature of sustainability issues not just because it is a strength-based approach to change but because it is a process to realize configurations, combinations, and interfaces among multiple strengths. The engagement in such summits entails more diversity and less hierarchy than is usual in a working meeting, and a chance for each person and stakeholder group to be heard and to learn other ways of looking at the task at hand. Because the whole system is engaged together at the summit, it is possible to make rapid decisions and to make commitments to action in an open way that everyone can support. This summit process is illustrated below by one business example – Fairmount Minerals – and one example of sustainable community development – the City of Cleveland, Ohio.

Fairmount Minerals

In 2005, Fairmount Minerals convened a three-day sustainable development appreciative inquiry summit which provides a model for what other organizations can achieve through the social technology of AI. This example of the power of AI is featured in an article by Cooperrider (2012). Fairmount Minerals produces sand products for the oil and gas, water filtration, foundry, glass, sports, recreation, and building industries. It is held up as a model of embedded sustainability and positive organization development (Cooperrider 2012). The case of Fairmont Minerals is highlighted in the book *Sustainable Value* (Laszlo 2008) to illustrate "doing well and doing good," that is, achieving economic progress at the same time that one creates positive social and environmental impacts.

The affirmative purpose of the first sustainable design summit of Fairmount Minerals was to be a sustainable enterprise responsible for the triple bottom line of people, prosperity, and profit. The AI summit theme was SiO_2, for Sustainability in Our Organization. SiO_2 is also the chemical symbol for silicon dioxide, the main ingredient in much of what Fairmount does. The summit was attended for three full days by more than 300 internal and external stakeholders such as senior management, employees, customers, global supply chain partners, NGOs, and community citizens. The intent of the summit was to come away with prototypes of new products initiatives that would combine sustainability and business value, taking into account the life cycle of materials from sourcing to reuse or return to the earth.

The objectives of Fairmount Minerals sustainable development appreciative inquiry summit were laid out in Participant Workbook for this event, which has been published in Cooperrider et al. (2008):

- Discover Fairmount's core strengths and value past
- Dream and envision the shared future we want to create as a sustainable company
- Design sustainable development into our products, services, operations, organization, and personal lives
- Deploy actions for value creation initiatives to move us in the direction of our vision of sustainable development

Cooperrider (2012, p. 107) provides a vivid description of the how the event would appear to one of Fairmount Minerals customers when he arrives for the summit:

> You enter a Grand Ballroom. It is teeming with 350 people from the sand company. There is no central podium or microphone. As many as 50 round tables fill the room – each has a microphone, a flip-chart, and packets of materials including the summit's purpose, three-day agenda, and a pre-summit strategy analysis and fact base. As an external stakeholder of the company, you've been invited to roll up your sleeves and participate in a real-time strategy session devoted to the future. You sit down at your assigned roundtable and you are struck by the complex configuration of individuals: the CEO of the company; a sand loader operator; a marketing specialist; a potential solar energy supplier (external); a product designer; a corporate lawyer, an information technology (IT) professional, and a middle manager from operations. Soon the "whole-system-in-the-room" summit begins.

The program for the event involved some minilectures to convey concepts and to showcase benchmark examples. But most of the program was participatory and interactive. The program mixed individual reflection, paired interviews, discussion in small groups, and presentations from each small group to the whole summit.

The Discovery stage asked participants to share the best of the past of Fairmount Minerals – the positive core of strengths to build on while exploring possibilities for the future. The Dream stage asked participants to communicate a concrete image of what could be 10 years in the future. In the Design phase, opportunity areas coming out the Dream stage were identified and each participant chose to move to one opportunity area. The work of the Design phase was not just to come up with an aspiration or possibility, but to build a prototype of the product design. Examples of design groups in this AI summit included a team designing a new business re-using spent sand that had already been used in industrial processes. Another team worked on the design of a low-cost sand filter product to purify drinking water in impoverished areas. Still another group worked on the less tangible product of a draft of corporate sustainability principles.

The Design session involved a deep dive brainstorming process which generated ideas for 1 h, which were recorded and displayed on sticky notes. Then each group took the best brainstorming ideas for rapid prototyping in a five-hour working session. This meant creating a visual model, storyboard, or skit to put the emerging innovation into concrete form. As described by Cooperrider et al. (2008), completing this kind of visual mock-up gives the project impetus to help move it forward.

In the Destiny stage, which was called Deployment at Fairmount Minerals, participants built action plans to put prototypes into practice. This planning process provided opportunities for participants to commit themselves in terms of what actions they would take, whom they would collaborate with, and what would be measurable milestones of achievement. Coming right after the Design process in which they had worked on prototyping initiatives of their own choice, empowered participants come to the Destiny stage with voluntary enthusiasm to continue to make manifest realities from the ideas created during the summit.

The 2005 Sustainability Summit for Fairmount Minerals produced substantial results in process and product innovations. Prototypes of new products, the discovery of new markets, and the design of renewable energy facilities — were quickly executed by an energized workforce. Between 2005 and 2007 revenues almost doubled, while earnings from took a gigantic leap to more than 40% per year (Cooperrider 2012). This financial growth in large measure grew out of new sustainably designed products and operational efficiencies. A further consequence was that the firm was the 2006 winner of Corporate Stewardship Award for Small/Midsize Business from the US Chamber of Commerce Foundation. This award is given to a company that "understands the linkages between its operations and society, and conducts its business in a way that creates "shared value:" both financially and responsibly over the long term (US Chamber of Commerce Foundation 2006).

Cooperrider (2012) has explained that a key factor in the success of the sustainability summit at Fairmount Minerals was the element of wholeness. By bringing diverse parties together in the summit setting, synergistic connections were made across and between groups. An example is the connections which were created to find a use for spent sand which was discarded after its use in factories was connected to renewable fuel for Fairmount's trucks. Someone identified a profitable use for spent sand to help grow higher yields of biomass. This led to Fairmount Minerals developing partnerships with local farmers to grow lower cost, green biofuels to power Fairmount's heavy truck fleet. This a new multimillion dollar business opportunity came out of the summit process in which diverse contributors could invent together to advance the companies sustainability goals.

Sustainability breakthroughs coming out of this summit helped double Fairmount's already superior double-digit growth rates, and set it on a pathway of differentiation unheard of in its industry. The company's reputation for doing good and doing well has helped the company compete for license to operate in local communities.

Examples of Positive Question for Fairmount Minerals Sustainable Summit

An important part of preparation for an AI summit to prepare "unconditionally positive questions" (Cooperrider et al. 2008, p. 3) for the 4-D cycle, tailored to the affirmative purpose of the summit. Table 1 lists some questions, abbreviated from the Participant Workbook for the Fairmount Minerals 2005 Sustainable Development Appreciative Inquiry Summit.

Practice Exercise 1

Think about preparing an Appreciative Inquiry sustainability summit for your organization or your community.

1. Identify the affirmation topic for your summit.
2. Try to think of a catchy, creative phrase to identify the aim of your summit, like Fairmount Minerals' "SiO_2" or Cleveland's "Green City on a Blue Lake."
3. Write questions for each of the 4-D phases of your summit. Look at model questions in Table 1 and Table 2 or in other reference sources cited in this chapter. Tailor the questions to your organization and your affirmative topic.
4. Think of speakers who can inspire your summit participants with practical ideas. This should include influential leaders from within your own

Table 1 Sample Appreciative Inquiry questions based Fairmount Minerals Participant Handbook (source Cooperrider et al. 2008).

Discover	Please share one example in which doing good and doing well became a source of innovation and success for Fairmount rather than simply a cost. Think of a high point from your experience in this organization – a time you felt most engaged or really proud. Please share the story and your feelings and insights. List the high points of your stakeholder group and select your proudest proud and strongest strengths as a group.
Dream	Put yourself 10 years in the future. Visualize the company you want us to be. How did this come about? What makes his vision exciting to you? How does this vision help us unlock missed opportunities to generate value – New processes, products, services, cost savings, enhanced reputation, and competitive advantage?
Design	Brainstorm as many as ideas as possible to move in the direction of our future visions and dreams, assuming anything imaginable is possible. Each group takes three to five most promising areas from the brainstormer and builds a prototype of an initiative that has strategic value for the company. Prepare a three- to five-minute presentation to the whole group and be ready to make the business case for your sustainability initiative.
Destiny	Create an action plan: What, when, and who? What are your personal commitments and to-do's? How will you evaluate success of your prototype at 30, 60, 90 days?

Table 2 Sample Appreciative Inquiry questions based on City of Cleveland (2009)

Discover	Describe an experience that stands out when you felt most engaged, alive, and effective, and worked with others to build a better organization, community, or society. Tell one story of sustainable innovation that in your view helps exemplify some of Cleveland's strengths, assets, and valued areas of leadership. How did it benefit the community?
Dream	When you awake it is 2019, ten years from now. And while you were asleep Cleveland has created a green city on a blue lake. What do you see happening in business and industry? What do you see happening in terms of people and relationships? What do you see ecologically and in our relationships with nature and natural resources? How about our community, our schools, and public life? Our reputation nationally and globally? Taken as a whole, what's happening that is new? Better? What has changed? Can you describe some of the more compelling, exciting innovations? Finish this sentence: I will be most proud of Cleveland and our region when. . .." What are three of the smallest things that could come out of the summit that might have the largest impact on our aim of building an economic engine to empower a green city on a blue lake?
Design	Brainstorm as many ideas as possible related to your group's action initiative – ideas that can move Cleveland's economy in the direction of our future visions and dreams. Select one priority idea and build a visual prototype such as a model, a story, a news story, and a skit.
Destiny	Build an action plan to find the quickest, most effective ways of launching the rapid prototype or pilot. Closing reflections and pledges: The collective action initiatives that I will help make happen include . . ..

organization, and may include sustainability success stories from other organizations or communities.

5. Identify potential experienced facilitators for your AI summit.

Summit Event for the City of Cleveland, Ohio

The City of Cleveland utilized an Appreciative Inquiry summit to launch Sustainable Cleveland 2019 (SC2019). The city continues to hosts a summit each year to reconvene the whole system, measure progress, and adjust course. Since the first summit, an alignment of civic and business focus toward sustainability has produce new jobs, policies, green space, transportation access, increased local food production, more renewable energy sources, and even new entrepreneurial organizations.

The original three-day Appreciative Inquiry Summit for the City of Cleveland took place from August 12–14, 2009. This event, called "Sustainable Cleveland 2019: Building an Economic Engine to Empower a Green City on a Blue Lake," has been described in published articles by Glavas et al. (2010) and by Meyer-Emerick (2012). Andrew Watterson, who led the City of Cleveland's sustainability efforts for more than 6 years, has written about appreciative inquiry for the blog "Greenbiz" (Watterson 2012).

Cleveland is the home of Case Western Reserve University, where Appreciative Inquiry has been developed and widely applied by David Cooperrider and colleagues. Cleveland had adopted many sustainable policies and practices prior to 2009. Mayor Frank Jackson approached the Fowler Center at Case Western Reserve University to help do an Appreciative Inquiry summit to engage all sectors and walks of life in planning the future of the city based on a sustainable economy.

Beginning in February 2009, a summit design team started preparing for the August summit event. The design team consisted of 41 people from government, universities, metropolitan school district, businesses, financial institutions, nonprofit organizations, sustainability advocates, and a mix of stakeholders who would reflect the demographics of the 1.6 million people in the Greater Cleveland region. The design team worked on: the objectives, who should attend, marketing to community groups, organizing logistics, doing research needed prior to the summit, and preparing the speaker program, appreciative inquiry questions, and participant workbook for the summit.

The summit was held in August 2009 at the Cleveland Convention Center and attended by 700 participants. There was no admission charge, but participants were asked to commit to attend all 3 days of the summit to complete the AI 4-D process. The program included outside speakers – including Peter Senge, Ray Andersen, and representatives from green city projects from New York and Atlanta – to inspire community members to envision what transformations are possible. Dr. Cooperrider and the Fowler Center facilitated the four stages of the AI process over the three-day summit.

Examples of Positive Question for Cleveland Sustainable Summit

Organizations move in the direction of the questions we ask. Table 2 presents examples of questions copied or edited from the participant workbook for the 2009 Cleveland summit to engage collective energy in the direction of sustainable community development.

Summary of Initiatives Generated by Sustainable Cleveland 2019 Summit in 2009

Nearly 700 people participated in Sustainable Cleveland 2019, a three-day summit designed to generate an action plan to transform Cleveland's economy into a sustainable economy and the city into a green city on a blue lake. On the afternoon of day two, participants at the summit were asked to vote with their feet and join one of 20 subject-specific work groups. Participants were then asked, "How might we cultivate or advance..." the specific topic. Participants answered with 28 concepts. Table 3 summarizes the 28 initiatives compiled from the transcription of notes, flip charts, and work sheets created by participants during the Summit as reported by Green City Blue Lake (2009). These initiatives illustrate the types of ideas and plans that can come of a three-day sustainability summit in which community members come together with the goal of realizing sustainable development for the future of their city.

The Cleveland sustainability summit process also put into place structures to support continuing deployment after the summit (Glavas et al. 2010). At the Summit,

Table 3 Action areas for Sustainable Cleveland, from Green City Blue Lake (2009)

Advanced energy generation	1. Create the first utility-scale freshwater wind farm in the world, positioning Cleveland in the potentially booming offshore wind industry in the USA. 2. Create the "Solarail," the most sustainable transit line in North America. 3. Create a 400 MW combined heat and power plant and an industrial water management collection and reuse system.
Advanced energy research and commercialization	4. Create the Laboratories for Advanced Energy Commercialization (LAEC) for funding and incubation of advanced energy products and services.
Advanced manufacturing and materials	5. Create a global Center for Sustainable Design and Manufacturing to create industrial symbiosis between businesses and bring university research to market.
Communications campaign and branding	6. Build a solid communication plan that will provide a common language, spread awareness, spur conversations about sustainability, help market Cleveland as the green city on a blue lake.
Engage 1.6 million	7. Create a campaign that engages the entire region. The campaign will create a messaging campaign based on "The Year of _________," around a sustainability focus like recycling, local food, transportation, etc.
Green building	8. Promote green building. Retrofit 100% buildings for 50% energy reductions, 50% of buildings to be retrofitted to net zero, 25% of buildings to be retrofitted to be energy producers.
Health	9. Transform our community from a focus on sick-care to a focus on well-care.
Local foods	10. Support the development of a sustainable regional food system with six possible local food projects.
Metrics of success	11. Develop measures and incentives to quantify the sustainable success of transportation, business growth, waste, energy, and education; and report out to the community.
Postsummit momentum	12. Develop a postsummit structure, with input from community leaders and organizations, to help implement all of the initiatives from the summit
Public compact	13. Create a public compact – The Cleveland promise – so individuals and businesses can sign on for specific sustainability actions.
Social capital:	14. Create a protocol to foster collaborations between government, business, education, community, and other institutions to create a more sustainable community.
Social entrepreneurship in the urban Core	15. Develop, CLEVA, an online platform that would connect citizens interested in investing in Cleveland to opportunities to invest in socially and environmentally responsible Cleveland businesses.

(*continued*)

Table 3 (continued)

Strategic partnerships and learning	16. Create an international BETA site to foster collaboration with sister cities around the world and to bring sister cities to the 2010 sustainability summit in Cleveland. 17. Create a green neighborhood alliance/campus by fostering strategic partnerships between schools, community gardens, local businesses, libraries, and police forces. Students in a local school would grow produce in an on-site urban garden, sell it to a local business, and use the produce in school meals. 18. Create the Great Lakes sustainable cities Ning network to facilitate intercity sharing of best practices, coordinate with other local social networking groups.
Sustainable business innovation and incubation	19. Raise $100 million through individual donations and community support to create the regional sustainability fund, a self-sustaining fund that would make grants, loans, and investments in new large-scale sustainable ventures. 20. Create the global sustainability cooperative, a physical structure that will act as a hub for sustainable technology businesses, an incubator program, ecofriendly consumer product retailers, and locally grown food outlets like restaurants and gourmet food stores. It would act as a catalyst to bring emerging sustainable technologies to market as well as spark innovations through the connections and collaborations formed by the companies within the facility. The global sustainability cooperative would not simply be a group of companies within a building, but an integral part of the sustainability movement in Northeast Ohio by connecting and inspiring innovations in the sustainability industry. 21. Create the Cleveland Sustainable Business Growth Initiative to foster sustainable economic growth by educating, connecting, and inspiring businesses in Cleveland, recognizing businesses for their sustainable achievements.
Transportation	22. Promote sustainable transportation by focusing on public transportation, rail, bicycles, and pedestrian friendly development.
Vacant land use and green space	23. Develop available vacant parcels into contiguous, revitalized land areas, in coordination with other local projects such as Towpath Trail, Re-imagining Cleveland, and Cleveland Job Corps Center.
Waste to profit	24. Create the Wasteipedia Center: a processing transfer station that takes waste from residential, commercial and industrial sectors with the goal of zero waste by 2019. The center would be a repository for reusable materials, divert and track at least 2019 tons of waste per year. 25. The Regional Zero Waste Collaborative: Designed to support the development and implementation of policies and practices that create a zero waste economy in Northeast Ohio by 2019. 26. Waste-to-energy center.

(*continued*)

Table 3 (continued)

Water	27. Create Cleveland Water "Works" that will work toward the goals of clean water, education, and lakefront development.
World class sustainable education	28. Create a living system for education including the development of a green campus for students from early education through university.

Table 4 Yearly Themes for Sustainable Cleveland (based on Sustainable Cleveland 2019, 2010)

Year	Theme	
2011	Energy efficiency	
2012	Local foods	FRESH and LOCAL
2013	Advanced and renewable resources	
2014	Zero waste	zero waste
2015	Clean water	
2016	Sustainable mobility	
2017	Vibrant green space	
2018	Vital neighborhoods	
2019	People	

each group created its own internal self-governing structure. For those groups that needed nurturing or additional hands-on support, the Mayor's office and the Fowler Center provided support. There was also top-down support through the City's Office of Sustainability and an advisory group of 60 key community leaders. Finally, there was a steering committee, comprised of representatives of each of the initiative groups, that met to communicate the progress on each initiative.

Table 4 illustrates the yearly themes that were established for carrying out the 10-year vision for a green city on a blue lake (Sustainable Cleveland 2019, 2010).

At the time of writing this chapter, Cleveland is celebrating the Year of Vibrant Green Space "to better connect people to place, nature, and one another" (http://www.sustainablecleveland.org/). Yearly summits continue to be held in Cleveland. For example, the 9th annual Sustainable Cleveland Summit was held on September 27–28, 2017 at Cleveland Public Auditorium.

Reflection Questions for City of Cleveland Case and the Fairmount Minerals Case

1. How did your own organization go about establishing a vision, forming teams, and communicating about sustainability goal and projects?
2. Compare your organization's strategies to the experiences described in the two cases in this chapter.
3. Do you think a sustainability summit could be helpful for your organization? In what way?
4. How could the strategies that your organization has used work together with an AI summit?

Wholeness and Sustainability

Sustainability involves complex issues which impact many different constituencies. Isolated parties cannot construct holistic solutions without joining and communicating with other parties who can understand and affect other parts of the system. Thus Senge et al. (2010) count collaborating across boundaries as one of the essential capabilities for realizing sustainability. Appreciative Inquiry is a social technology to capture wholeness by engaging diverse elements of a community to collaboratively create desired sustainable futures.

The realization of wholeness in human awareness is also the aim of what Laszlo et al. (2012) refer to as reflective practices. Their research on spirituality and sustainability, conducted at the same Fowler Center which has championed AI, concluded that "sense of connectedness based on reflective experiences has enabled individuals to engage in qualitatively more powerful ways of thinking and acting aimed at flourishing" (Laszlo et al. 2012, p. 35). Such reflective practices, by facilitating the experience of lively wholeness in consciousness, can be complementary to the social technology of AI in bringing about transformative organizations and a sustainable world. Heaton (2016) has described the Transcendental Meditation Program (Transcendental Meditation® is a protected trademark and used in the U.S. under license or with permission.) as one such reflective practice to help achieve the aim of thought and action in harmony with the intelligence of nature. Maharishi Mahesh Yogi has explained that the Transcendental Meditation Program enables the mind to settle inward to experience transcendental pure consciousness, which is wholeness of natural law in its most settled state. In Maharishi's words, this holistic intelligence of nature "sustains existence and promotes the evolution of everything in the universe, automatically maintaining the well-coordinated relationship of everything with everything else" (Maharishi Mahesh Yogi 1995, p. 8).

The aim of sustainability is that human beings are coordinated with each other and with the environment is such a way that life is sustained and evolving for all. The systems within the natural world are holistic phenomenon in which interconnected parts are functioning together to maintain conditions for live and growth. Systems of thought such as biomimicry (Benyus 1997), biosphere rules (Unruh 2008), and the Natural Step (Bradbury and Clair 1999) guide managers to consider how to perform as nature would perform. As individuals grow to higher degrees of consciousness they do grow in capacity to think and act like nature. As Brown (2011) has found consciousness development unfolds capacities for sustainability, including the capacities to take a systems view on reality, to integrate ideas, to create long-term visions with profound purposes, and to build truly collaborative relationships. Such growth of consciousness, together with the interaction tools provided by AI, are complementary to each other in the realization of ever more wholeness and flourishing through engaged sustainability.

Engaged Sustainability Lessons

The AI summit processes introduced in this chapter, harness the power of wholeness to bring out the best in organizations and communities as they aspire to contribute to a sustainable world. The exceptional potential of AI for sustainability summits is that it applies a strengths-based approach at the macromanagement level (Cooperrider 2012). That means that AI works at the level of the whole system to facilitate how strengths can be joined together to design and implement holistic solutions.

Glavas et al. (2010) contend that the AI Summit process in Cleveland was successful due to four criteria that can be applied in any system, whether it is a city or a company. First, the summit had representation of the whole system – all sectors and demographics of the 1.6 million people in the Cleveland region. When representatives of the whole system participate in the summit process together, then decisions are transparent and change can be carried out immediately. Cooperrider likewise emphasizes the chemistry of the whole, pointing out that "improbable configurations that can combine strengths to create magic" (Cooperrider 2012, p. 111).

The second criterion of success according to Glavas et al. is that preparation for the Cleveland summit involved using a strength-based approach to inquire into the best in the people, organizations, and the region, which enabled the summit process to move in the direction of hope and aspiration. The lesson for others is that the leadership group for a summit event needs to understand strengths-based thinking and research and needs to have their own personal experience with AI methods.

Third, once the first summit began, the Mayor and other sustainability champions "got out of the way" and empowered the participants to interactively create and run with initiatives. Fourth, about 200 people were consulted in various aspects of the Summit. This widespread engagement built the social capital which helped make the project successful. Glavas et al. summarize (2010, p. 33):

> The greatest lessons learned from the process were in how it happened. We don't think that traditional management models will work in building a green economy. We don't have time.

Change needs to happen quickly, and it needs to grow exponentially. The severity of the issues we face around food, water, energy and waste and toxicity can't wait for slow, incremental change. In this environment, we need processes that engage the whole, are strengths-based and require managers to let go of control.

Traditional management models tend to be controlled from the top – the C-level officers who assign responsibility and monitor performance. In traditional management, the task of working on sustainability may be delegated to designated personnel, projects or teams within the organization. By contrast, a well-prepared summit aims to engage every summit participant in envisioning, designing, communicating, and taking action. In the summit process ideas emerge quickly and teams commit to follow-up plans. The breadth and speed of this process is more than can be planned and controlled by senior executives. Executives who ask their organizations to engage in AI as a process for change must be prepared to let go of some top-down control and empower and recognize participants for multi-prong change initiatives. Then change can happen quickly, and raise morale, effort, achievements and fulfillment for the organization.

The fourth principle for success highlighted by Glavas et al. (2010) is building social capital. They explain that prior to this summit Cleveland was an example of fragmented clans, lack of collaboration, and prevailing attitudes of skepticism. But a tipping point took place during the design team meeting in February. As many participants noted, it was when organizations actually came together to do something instead of just talking about doing something that the energy shifted. Gradually, Design Team members began engaging others until about 200 people became involved in various aspects of the Summit. While the Fowler Center at Case Western Reserve University took primary responsibility for planning and logistics, many parties helped with aspects of the Summit, such as providing input on stakeholders who should be invited, giving input into the pre-summit research, helping with communication, etc. All of this engagement helped build vital social capital.

While the experiences of Fairmount Minerals and the City of Cleveland with using Appreciative Inquiry for a sustainability summit do provide encouraging models for other communities, Meyer-Emerick (2012) offers two cautions. One is that considerable research and groundwork needs to be done prior to the summit to identify and harness existing resources and the potential within their community. This requires extensive engagement with key parties prior to the actual summit. And the second is that the process must be led by experienced AI Summit facilitators who can keep conversations positive and focused and who can target manage and steward the information generated by groups through each phase of the 4-D process. With the benefit of these lessons, take advantage of the power of the Appreciative Inquiry Sustainability Summit to achieve engaged sustainability in your organization or community.

Practice Exercise 2

The first exercise in this chapter asked you to draft some potential Appreciative Inquiry questions for each phase of a sustainability summit for your organization or your community. Here are some other important questions for you to work on in order to prepare a successful summit:

1. Who are all the stakeholders that you want to be included so that the wholeness of the system can be harnessed for shifts toward greater sustainability? How many total people would you like to participate?
2. How can you bring all voices – the "whole system" – into the room?
3. Who in your organization and stakeholder network has had experiences with Appreciative Inquiry?
4. What individuals and organizations will be approached to participate in a core organizing committee?
5. What are the roles and responsibilities for the organizing committee in order to lay the ground for a successful summit?
6. Write a draft invitation to the summit event. Define the affirmative topic for the summit and describe the role of participants who will attend the summit.
7. Draft some possibilities regarding where to hold your summit, for how long, when?
8. What are the some action steps (with target dates and names of committed, responsible parties) to be taken toward making a sustainability summit happen for your business or community?

Chapter-End Reflection Questions

1. What tools does your organization use to identify its stakeholders, their expectations, and interests?
2. Does your organization report on your stakeholders' concerns and on your responses regarding each stakeholder issue? Does your organization utilize some reporting standards such as Global Reporting Initiative or Integrated Reporting?
3. Appreciative Inquiry argues that organizational change is stimulated by appreciating and joining strengths, rather than diagnosing weaknesses. Do you agree? What have been your experiences with change efforts which focus on problems compared to your experiences with change efforts which focus on sources of pride?
4. Why is engagement of the whole system a formula for developing and implementing sustainability initiatives?
5. What is the role of C-level officers in a sustainability summit? Does this summit process call on C-level officers to do something different from what might be their accustomed roles?

Cross-References

▶ Collaboration for Regional Sustainable Circular Economy Innovation
▶ Community Engagement in Energy Transition
▶ Designing Sustainability Reporting Systems to Maximize Dynamic Stakeholder Agility

References

Andriof, J., & Waddock, S. (2002). Unfolding stakeholder engagement. In J. Andriof, S. Waddock, B. Husted, & S. Sutherland Rahman (Eds.), *Unfolding stakeholder thinking: Theory, responsibility and engagement* (pp. 19–42). Sheffield: Greenleaf Publishing.

Annan, K. (2004). Testimonial of UN Secretary General, Kofi Annan. Retrieved from https://appreciativeinquiry.champlain.edu/testimonials/united-nations-secretary-general-kofi-annan/

Barrett, F., & Cooperrider, D. (n.d.). AI workshop slides. Retrieved from https://appreciativeinquiry.champlain.edu/educational-material/ai-workshop-slides/

Benyus, J. (1997). *Biomicry: Innovation inspired by nature*. New York: William Morrow.

Bradbury, H., & Clair, J. A. (1999). Promoting sustainable organizations with Sweden's natural step. *Academy of Management Executive, 13*(4), 63–74.

Brown, B. C. (2011). Conscious leadership for sustainability: A study of how leaders and change agents with post conventional consciousness design and engage in complex change initiatives. Dissertation, Fielding Graduate University, Santa Barbara, California.

Bushe, G. R. (2013). The appreciative inquiry model. In H. Kesser (Ed.), *Encyclopedia of management theory* (pp. 41–44). San Francisco: Sage Publications.

Cameron, K. (2012). *Positive leadership: Strategies for extraordinary performance*. Oakland: Berrett-Koehler Publishers.

Cameron, K. S., & Spreitzer, G. M. (Eds.). (2011). *The Oxford handbook of positive organizational scholarship*. Oxford: Oxford University Press.

Christian Reform World Relief Committee. (1997). Using appreciative inquiry to build capacity. In S. Johnson & J. Ludema (Eds.), *Partnering to build and measure organizational capacity*. Grand Rapids: Christian Reformed World Relief Committee.

City of Cleveland. (2009). Sustainable Cleveland 2019 participant workbook.

Cooperrider, D. L. (2012). The concentration effect of strengths: How the whole system "AI" summit brings out the best in human enterprise. *Organizational Dynamics, 41*(2), 106–117. https://doi.org/10.1016/j.orgdyn.2012.01.004.

Cooperrider, D. L., & Godwin, L. N. (2011). Positive organization development: Innovation-inspired change in an economy and ecology of strengths. In K. Cameron & G. Spreitzer (Eds.), *Oxford handbook of positive organizational scholarship* (pp. 737–750). Oxford: Oxford University Press.

Cooperrider, D. L., & Whitney, D. (2005). *Appreciative inquiry handbook: The first in a series of AI workbooks for leaders of change*. Oakland: Berrett-Koehler Publishers.

Cooperrider, D. L., Whitney, D. K., & Stavros, J. M. (2008). *Appreciative inquiry handbook: For leaders of change* (2nd ed.). Oakland: Berrett-Koehler Publishers.

Doppelt, B. (2010). *Leading change toward sustainability: A change-management guide for business, government and civil society*. Sheffield: Greenleaf Publishing.

Fredrickson, B. L., Tugade, M. M., Waugh, C. E., & Larkin, G. R. (2003). What good are positive emotions in crisis? A prospective study of resilience and emotions following the terrorist attacks on the United States on September 11th, 2001. *Journal of Personality and Social Psychology, 84*(2), 365.

Freeman, R. E. (1984). *Strategic management: A stakeholder perspective*. Boston: Pitman.

Glavas, A., Senge, P., & Cooperrider, D. L. (2010). Building a Green City on a blue Lake: A model for building a local sustainable economy. *People & Strategy, 33*(1), 26–33.

Global Reporting Initiative. (2016). Consolidated set of GRI Sustainability Reporting Standards 2016. Retrieved from https://www.globalreporting.org/standards/gri-standards-download-center/

Green City Blue Lake. (2009). Summary of initiatives generated by Sustainable Cleveland 2019. Retrieved from http://www.greencitybluelake.org/Summaries%20of%20Proposed%20Sustainable%20Cleveland%202019%20Initiatives.pdf

Haidt, J. (2003). Elevation and the positive psychology of morality. In C. L. Corey & J. Haidt (Eds.), *Flourishing: Positive psychology and the life well-lived* (p. 275). Washington, DC: American Psychological Association.

Heaton, D. (2016). Higher consciousness for sustainability-as-flourishing. In S. Dhiman & J. Margues (Eds.), *Spirituality and sustainability: New horizons and exemplary approaches* (p. 121). New York: Springer.

International Integrated Reporting Council. (n.d.). International Integrated Reporting Framework. Retrieved from http://integratedreporting.org/resource/international-ir-framework/

Laszlo, C. (2008). *Sustainable value*. Stanford: Stanford University Press.

Laszlo, C., & Zhexembayeva, N. (2011). *Embedded sustainability: The next big competitive advantage*. Stanford: Stanford University Press.

Laszlo, C., Brown, J. S., Sherman, D., Barros, I., Boland, B., Ehrenfeld, J., ..., Werder, P. (2012). Flourishing: A vision for business and the world. *The Journal of Corporate Citizenship, 46*, 1–51.

Ludema, J. D., Whitney, D., Mohr, B. J., & Griffin, T. J. (2003). *The appreciative inquiry summit: A practitioner's guide for leading large-group change*. Oakland: Berrett-Koehler Publishers.

Maharishi, M. Y. (1995). *Maharishi University of Management: Wholeness on the move*. India: Age of Enlightenment Publications.

Meyer-Emerick, N. (2012). Sustainable Cleveland 2019: Designing a green economic future using the appreciative inquiry summit process. *Public Works Management & Policy, 17*(1), 52–67. https://doi.org/10.1177/1087724X11429043.

PRME (Principles for Responsible Management Education). (2017). About us: Six principles. Retrieved from http://www.unprme.org/about-prme/the-six-principles.php

Senge, P., Smith, B., Kruschwitz, N., Laur, J., & Schley, S. (2010). *The necessary revolution: Working together to create a sustainable world*. New York: Crown.

Sustainable Cleveland 2019 (2010). Action Areas. Retrieved from https://www.sustainablecleveland.org/action_areas

Telkom. (2015). Integrated Report 2015. Retrieved from http://www.telkom.co.za/ir/apps_static/ir/pdf/financial/pdf/Telkom%20IR%202015%20Final.PDF

United Nations. (2004). The Global Compact leaders summit participant worksheet. Retrieved from https://appreciativeinquiry.champlain.edu/wp-content/uploads/2016/01/workbook-June-2004.pdf

Unruh, G. C. (2008). The biosphere rules. *Harvard Business Review, 86*(2), 111–117.

US Chamber of Commerce Foundation. (2006). 2006 Citizens Awards. Retrieved from https://www.uschamberfoundation.org/article/2006-citizens-awards

Watterson, A. (2012). Creating positive change through Appreciative Inquiry. https://www.greenbiz.com/blog/2012/08/01/creating-positive-change-through-appreciative-inquiry

Whitney, D., Trosten-Bloom, A., Cooperrider, D., & Kaplin, B. S. (2013). *Encyclopedia of positive questions: Using appreciative inquiry to bring out the best in your organization*. Brunswick: Crown Custom Publishing.

Empathy Driving Engaged Sustainability in Enterprises

Rooting Human Actions in Systems Thinking

Ritamoni Boro and K. Sankaran

Contents

Abstract

This chapter suggests a vision for human activities in enterprises through the lens of engaged sustainability. We put together some important bits and pieces of insights and wisdoms that some seers among us have been able to envision. In the process we discuss empathy, emotional intelligence, social intelligence, and ecological intelligence and apply these ideas to learning organizations.

If human needs and wants were to be expressed and fulfilled only in material terms, humanity will not be able to recover the beleaguered Earth these have inflicted. If it is going to be business as usual, the unintended consequences which

R. Boro · K. Sankaran (✉)
Justice K. S. Hegde Institute of Management, Nitte University, Nitte, India
e-mail: ritaboro@nitte.edu.in; sankaran@nitte.edu.in

S. Dhiman, J. Marques (eds.), *Handbook of Engaged Sustainability*,
https://doi.org/10.1007/978-3-319-71312-0_20

human actions have caused would destroy everyone. Here we attempt to understand the underlying empathetic nature of human beings that seem to "naturally" point towards giving rise to saner voices and enduring solutions.

We take the view that recovering from the impediments we face in shaping a sustainable future will not be possible if we continue to ignore the "softer" aspects of personal nature and superior interpersonal interaction humans are privy to. This chapter aims to understand whether achieving such feat can be made possible by grooming our empathetic skills in a more mindful and conscious manner. To that end, the concept of learning organizations has been applied to enhance the developmental skills across enterprises. We suggest that all enterprises, institutions, and organizations, irrespective of whether governmental, corporate, or voluntary, could embrace systemic changes to pay respect to the biospheric systems in a manner that would reverse the damage done while ensuring that the needs of future generations are respected and upheld.

Keywords
Empathy · Engaged sustainability · Learning organizations · Ecological intelligence · Social intelligence · Emotional intelligence

Introduction

> It is no measure of health to be well adjusted to a profoundly sick society. – Jiddu Krishnamurti

Engaged sustainability is a choice. This adage is hard to believe in these modern times, as there is too much noise out there menacingly pointing to the view that human beings are totally helpless in the face of the destruction done to the planet. Perhaps one wouldn't even be aware about the destruction of the Earth if cities weren't filled with 24-hour smog screen making breathing difficult. Or, one wouldn't even start being concerned if one didn't have to rush for medical help because one drank water contaminated by the waste dumped in the water bodies by giant factories that claim benefits to denizens on the "other" half of the planet.

Herein lies the problem. It is a disease called "lack of sensitivity." It is lack of sensitivity to one's fellow humans and other living beings. It is lack of sensitivity to the environment that supports one's livelihood. It is lack of sensitivity to even one's own needs. It is lack of sensitivity to anything that hasn't been seen, touched, tasted, smelt, or felt through one's own senses. It would seem that "out of sight, out of mind" is the primary *modus operandi*, and the bigger problem is that this disease isn't being suffered by only one person or a few individuals. This is being suffered by almost all of humanity as a whole. It wouldn't be wrong to call it the invisible plague, a disease so rooted in the deepest of our psyche that we simply can't become aware of it through ordinary means (Tucker and Williams 2007; Gore 2006).

However, all isn't lost. There is in fact a very easy way to bring sustainability into one's life. In fact, this way will pave the way for making sustainability the very basis of what helps one thrive, and not just survive. Engaged sustainability can very well become one of the best-kept open secret in achieving higher qualities of fulfillment in all spheres while addressing one's wants and needs – physical, social, emotional, aspirational, experiential, spiritual, etc. For this there is just one shift in focus that we have to make. If "out of sight, out of mind" is what the root of all sustainability issues is, then, we just need to "bring everything into sight, to engage the mind with sensitivity." We need to seek radical transparency and become conscious of how we conduct our lives and activities in the process of living it (Bonnett 2017).

It might seem like a gigantic undertaking. How much effort would it take to dig through all the hundreds of years of collective outcomes of destruction that humanity has instigated? But imagine for once, what would it feel like to know that when one wakes up in the morning, the free air that sustains life in our bodies is of the purest quality? One doesn't need to be afraid that the water being fed to our children is not going to result in malfunctioning kidneys or deformed bones. What would it feel like to be completely confident that the buildings in which we spend a significant part of our waking time, our workplaces, are not contributing to the depletion of the very resources that sustains our living environment? Or for that matter, the various facilities that we use to make our work easier or prepare the packaged food that we eat aren't turning these very life-giving resources into life-threatening nightmares we can't escape (Nadeau 2006).

Yes, there is a way to overcome this helplessness. In fact, it is a gift of nature, which most probably we haven't used to the fullest extent possible. It is a qualitative inherent trait that all human beings have called "empathy." Empathy isn't a solution. It is a psychological process which contributes to the growth of a human being (Wispe 1991). When used appropriately, empathy works at the very source of all the sustainability-related problems preventing the need for even questioning whether engaged sustainability as a way of living is possible or not.

This chapter is an undertaking to encourage readers to use their inborn skill of empathy to curve out a more conscious way of living and turn that very way of empathetic living into an endeavor that rehabilitates Earth into a safe and joyful place to live in.

Section "Introduction" would look at empathy and focus on understanding how empathy is the most natural skill one is born with as a human being and how exercising the skill of empathy drives the natural learning and growth of a human being (Lipton 2005). This part will also look at how empathy and its relationship with emotional intelligence help one become connected to other people and gain better understanding about the cultural environments that sustain us. And most important of all, how can empathy enhance awareness regarding the different human creations that we deem necessary for survival? It is also important to inquire as to how it can help us make more informed and conscious decisions? Are what we deem as necessary in our lives beneficial in reality, or do they contain hidden threats?

In section "What is Empathy?," following Goleman and Ekman (2007), we define empathy and describe its three variants: cognitive, emotional, and compassionate. We suggest that when these three variants come together what emerge are endurable thoughts and actions that are truly empathetic. This is followed by connecting empathy with emotional intelligence in section "The Connection Between Empathy and Ecological Intelligence." Here we describe emotional intelligence in terms of self-awareness and self-management (Goleman 1995). Emotional intelligence is then connected with social intelligence and finally to ecological intelligence. A certain holistic substantive awareness of how the Earth actually supports life is required as part of ecological intelligence. Therefore we describe what the Earth's biosphere supports as pointed out by Rockström et al. (2009): nine life-support systems that comprehensively view the dynamic nature of life on Earth.

In section "Operationalizing Collective Ecological Well-Being," we examine ways and means to operationalize ecological well-being as members of organizations. It is a task of mammoth proportions to bring every human being together as one enterprise to achieve one common goal of designing an environment conducive of sustaining human survival. To address the issue of whether enforcing sustainability can be made operational on a systemic basis, we'll take the model of learning organizations (Senge 1990) and analyze its five disciplines – systems thinking, mental models, shared vision, team learning, and personal mastery – as applied towards building a sustainability-oriented enterprise. Finally section "Conclusion" provides a conclusion.

What Is Empathy?

According to Merriam-Webster dictionary, empathy is "The action of understanding, being aware of, being sensitive to, and vicariously experiencing the feelings, thoughts, and experience of another of either the past or present without having the feelings, thoughts, and experience fully communicated in an objectively explicit manner."

Let us seek an example. Many of us have faced situations where, when we meet a friend or colleague of ours, we are immediately able to guess their emotional condition without the person having uttered a single word (Lakin 2006). Most of the time, we would be correct. It might be discomfort due to an illness, or sadness at having faced some unfortunate happenings in their lives, or even excitement from having encountered some fortunate ones. And then there's the other example of a mother and her baby. A baby doesn't know how to speak, but somehow a mother is able to guess the different moods of the child from its activities and address situations appropriately. Maybe the baby is hungry. Or, maybe it needs a change of diapers. Perhaps it just needs to "hear" the voice of its mother because it is feeling lonely. The mother is able to sense the needs of the baby instinctively. This trait of being able to sense the feelings from another person without that person having to mention anything is empathy.

When a baby first learns to speak or identify objects, it does so by imitating its mother, which is, indeed, not a simple process (Pinker 2002). It tries to imitate the speech patterns and recognizes the significance of the objects it comes in contact with regularly – for example, a milk bottle will mean that its hunger will be appeased. Even as adults, if someone smiles at us, generally there is an automatic response on our side to return that smile. In both these cases, it is observed that there isn't always a need to consciously execute an action physically from our side. It is our motor nerves which automatically get activated and start the process of executing the act on an unconscious level.

The two-way process undertaken by the individuals concerned described above (whether the baby and the mother on the one hand or the two adult on the other) enables them to function optimally in specific situations. Of course, as human beings there isn't always a necessity to respond to a stimulus in predetermined fashion. Individuals have the option to consciously choose which actions to respond to/imitate and which not to. Goleman and Ekman (2007), the world-renowned experts on emotions and individuals' ability to respond to them in others, classify empathy as three types – cognitive empathy, emotional empathy, and compassionate empathy.

"Cognitive empathy" is the one that directly arises out of one's observational inputs. This empathy is also sometimes called perspective-taking in more professional situations. Through inferences derived based on one's observations, the individual is able to know how the other person might be feeling and what he or she might be thinking at that particular point in time and place. A person with sharpened cognitive empathic skills might be highly proficient in negotiation or motivating others. A study conducted at the University of Birmingham reports that managers with highly developed perspective were able to motivate their team members in giving their best efforts (Goleman 2006).

However, there is a dark side to cognitive empathy. Ultimately, cognitive empathy is an inherent skill. It is a tool to be precise. How it is used depends on the user. Social workers or caregivers may use this skill to motivate patients and the needy to alleviate their pain of suffering. At the same time, an unscrupulous conman might use this very ability to rob the vulnerable of their belongings. Nefarious politicians would use this same skill to garner votes from misled citizens without fulfilling any of the promises made.

While cognitive empathy is about simply "knowing," the next type of empathy, also called "emotional empathy," is about "feeling." This type of empathy entails someone being well-attuned to another person's inner emotional world. In this case, one can physically feel what the other person is feeling as though they have been infected by the other's emotion. This empathy may be the one most related to our learning and development. Goleman (2004) states that "Most of us have assumed that the kind of academic learning that goes on in school has little or nothing to do with one's emotions or social environment. Now neuroscience is telling us exactly the opposite. The emotional centers of the brain are intricately interwoven with the neurocortical areas involved in cognitive learning. When a child trying to learn is caught up in a distressing emotion, the centers for learning are temporarily hampered. The child's attention becomes preoccupied with whatever may be the source of the trouble. Because attention is itself a limited capacity, the child has that much less ability to hear, understand, or remember what a teacher or a book is saying. In short, there is a direct link between emotions and learning" (p: vii).

Learning doesn't stop with childhood. Even when people have moved on to their workplaces, one can see the presence of emotional empathy in all spheres of interactions. In an ideal working environment, employees are focused, fully attentive, motivated, and engaged and enjoy their work.

However, the biggest downside of emotional empathy surfaces when one is not capable of managing one's own distressing emotions. This can give rise to psychological exhaustion that leads to burnout and depression. Those in the fields of medicine have a way to cultivate purposeful detachment as prevention against such burnout. However, there is a risk of that very detachment to turn into indifference and apathy, instead of well-calibrated caregiving.

The two types of empathy we discussed above may be seen as "knowing" empathy and 'feeling" empathy, respectively. If a person wants to employ these skills as tools for progressive development, he or she needs to be highly aware of the repercussions of these two empathetic skills and make appropriate decisions. Herein comes the third form of empathy, known as "compassionate empathy." This form of empathy involves not just being able to know and feel what another feels but also taking action so as to alleviate their concerns if required. Also, compassionate empathy involves knowing when not to get overwhelmed by the emotional contagion and back off if the situation demands it. The basis of this empathy is in being conscious or aware of not just others' thoughts and feelings but also of own thoughts and feelings and how the self and the other interact emotionally with each other. So we see that in order to effectively exercise one's freedom of choice and action in life, one needs to have a highly developed awareness of one's ability to manage one's own thoughts and emotions. However, how does one distinguish between observations made consciously and the ones made unconsciously? This is where the field of emotional intelligence comes in, which is discussed in the next section.

The Connection Between Empathy and Ecological Intelligence

Albert Einstein famously stated that "Our task must be to free ourselves by widening our circle of compassion to embrace all living creatures and the whole of nature and its beauty." Empathy is the beginning of ecological intelligence as pointed in Fig. 1. Let us look at each of the connections one by one.

Empathy and Emotional Intelligence

Emotional intelligence is the process of understanding one's own emotions and handling them in a healthy manner. There are primarily two parts to emotional intelligence – self-awareness and self-management (Goleman 1995). These are discussed below.

Self-Awareness

Self-awareness is about being able to recognize feelings as they happen. It is the foundation of developing emotional intelligence. As one monitors the arising and passing flow of feelings, one can gain self-understanding as well as insights into the

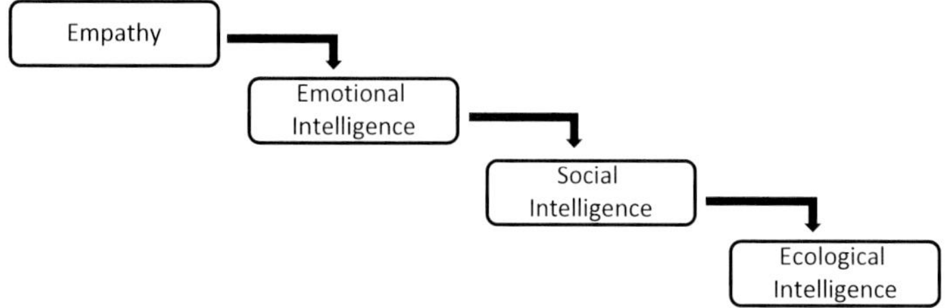

Fig. 1 Sequence of intelligences on application

current situations. This solidifies confidence in one's own feelings and enables one to make better choices in life. Aware of one's own moods and emotions, the person gains much clarity as to his/her own personality traits and what drives oneself. The person is better able to perceive his/her strengths and limitations, thereby enabling him or her to always have a positive outlook of life, which in turn maintains one's psychological health.

Correspondingly, the more open one is to own emotions, the more skilled the individual becomes in reading others' feelings. People with high self-awareness are able to be mindfully manage not just their own emotions but also appropriately address others' emotions.

On the other hand, an inability on our part to mindfully notice our true feelings will lead us to be overwhelmed by them. It would start feeling like nothing goes right in life. Psychological health will deteriorate leading to anxiety, melancholy, depression, and other such mental malaise.

Not just will such people be confused about their own emotions; they will also not be able to know what others around them are feeling. They won't be able to recognize the nonverbal, or even verbal, cues when someone is feeling sad or angry and may appear insensitive or cruel to others around them.

Self-Management

Self-management is about handling feelings and emotions appropriately as the situation necessitates. This is the second building block of emotional intelligence. Self-management required one to be aware of one's emotions first. Emotions have a strange tendency to form self-reinforcing loops (Checkland 1981). A positive emotion like joy would in turn bring in more feelings of joy, whereas a negative emotion like anger will keep flaming the fire higher and higher. In case of positive emotions, one might be willing to let the loop continue forever. After all, why should one change something that's already going the good way? However, too much of a good thing is also dangerous. When one is in a good mood, there is no desire to change the status quo, which may cause the individual to become unable to adapt to changes in life circumstances. As for the negative emotional loop, it is quite apparent that things aren't going to end well anyway – neither for the individual nor for the people around them.

Self-management helps one to break these loops of emotions and gain control over one's lives. It also helps people to adapt easily to life changes and recover from life's setbacks and upsets. This also means that they are able to adjust and build rapport with the people around with more ease and compassion. Between the divide of this self and the other lies empathy. Empathy is what enables us to extend the awareness of our own emotions to the awareness of others' emotions. It is what enables us in extending the self-management of our emotions to the management of emotions in others. This brings us to the field of social intelligence, which is an application of emotional intelligence with respect to other people.

Social Intelligence

There is a saying handed down by some wise man "Walk a little in my shoes. See what I see, hear what I hear, feel what I feel. Then maybe you'll understand why I am the way I am." Previously, it has already been mentioned how human beings learn and grow by imitating the people around them (Pinker 2002). The more number of times the imitation is repeated, the more number of times the synaptic impulses are fired in the brain for the same trait. Consequently, the stronger the emotional imprint becomes, the more strongly one mimics another person and the more accurate their senses regarding what the other person is feeling will become (Goleman 2006).

When a person empathizes or attunes himself or herself to someone, it is very natural to feel along with that person. That means, for example, even if one is in a good mood, the former is likely to be infected by the latter's bad mood too. This in turn will create the loop of bad mood between the two people concerned. Further this will give rise to all sorts of psychological malaise. This in itself is a very good reason to understand how to manage others' emotions in relationships.

There's also the responsibility of not transferring one's own negative emotions to the other person – who may or may not be well versed in the skill of self-management. Thus, how people connect with others has unimaginable and significantly huge consequences. Hence, people who are around us play a big role in changing the inner chemistry of our entire being – physiological and emotional. The natural question that arises in the context of our inquiry here is "What about the rest of the environment? How do they impact a human being's growth?" This brings us to the field of ecological intelligence which is the application of emotional intelligence and social intelligence with respect to how we relate to the environment which nurtures us.

Empathy and Ecological Intelligence

Johan Rockström, director of the Stockholm Environment Institute in Sweden, along with Nobel Laureate Paul Crutzen and NASA climate scientist James Hansen and others identified nine life-support systems essential for human survival (Rockström et al. 2009). These systems include biochemical cycles such as carbon and water and physical circulation systems such as the climate and the oceans.

The caveat is that as long as these nine support systems are in balance, human development can be considered to be securely operating in the safe zone. However, if these systems cross the limits, then only catastrophic consequences await humanity. As of now, seven out of these nine support systems are operating at the border or are already in the danger zone. The good news is that as long as immediate measures are being taken to reverse the situation, all will be well. So, how does empathy help in generating ecological intelligence?

Earlier we had, quoting Merriam-Webster dictionary, suggested that empathy has to do with feelings towards *fellow human beings*. A second definition of it by the same source expands the idea as the "the imaginative projection of a subjective state into an object so that the object appears to be infused with it." Earlier we had discussed empathy in terms of understanding the emotions of other people. Now we extend it to the Earth which will enable us to comprehend Earth's natural state of being and variations we engineer on it. No one person can make an impact on the global crises. However, the entire human community, one person at a time, can. We can extend the idea of social intelligence and ecological intelligence to heal the Earth and create sustainability for all of Earth's creations.

Before proceeding towards addressing the endeavor of redesigning organizations to be sustainability oriented, there is a need for being conscious of the nine major life-support systems present in the biosphere of the Earth's ecosystem as identified by Johan Rockström et al. (2009). A short introduction to the same is provided in Table 1.

In his book on ecological intelligence, Goleman (2010) reiterates the need for developing a systems thinking if we want to address the burning issue of how human communities can be made more sensitive to the biosphere that nurtures them and thereby find solutions which are more sustainable for human survival on Earth. Only when there is a better understanding of the subsystems of the Earth's ecosystem, the human communities, and the nature of relationship as dependent and nurturer between the two is established, will there be a breakthrough to finding solutions which are truly sustainable for generations to come.

Enterprise as a Nucleus for Collective Ecological Well-Being

In various kinds of organizations, whether governmental, corporate, voluntary, etc. there are already training programs in place to impart leadership skills and instill the employees with the capacities of self-awareness, self-management, social awareness, and relationship management. Thus, the workplaces are already well-equipped to dive right into this project. They can start with wherever they are, whichever stage of maturity there are. For example, they may start with identification of the different types of resources used for the functioning of the entire work building. Then perhaps measure their contribution in impacting the environment, following which look for more environment-friendly options, thus developing a more sustainable game plan.

The next section will look at the one such method of developing a game plan and how empathy can be a key driver in helping enterprises become more systems oriented and resource-friendly by following such a blueprint.

Table 1 Nine life-support systems of Earth

Sl. No.	Life-support systems	Impact
1	Climate change	The increase in concentration of atmospheric CO_2 has led to an imbalance in the energy systems in the biosphere. This has caused disruptions in the regional climate patterns, giving rise to phenomena like loss of polar ice caps, loss of glacial freshwater supplies, weakening of carbon sinks, etc.
2	Ocean acidification	Increase in carbonate ion concentration has led to the conversion of coral reefs to algal-dominated systems. This has induced the extinction of marine life and dissolution of coral reefs which provide support to the ocean beds
3	Stratospheric ozone depletion	The thinning of the ozone layers in the stratosphere is no longer able to stop incoming UVB radiation from outer space, thus leading to severe and irreversible effects on human health and ecosystems
4	Interference with the global phosphorus and nitrogen cycles	Increase in inflow of phosphorus to the oceans has led to depletion of dissolved oxygen impacting survival of marine life. Removal of nitrogen from the atmosphere has affected the "overall resilience of ecosystems via acidification of terrestrial ecosystems and eutrophication of coastal and freshwater systems"
5	Rate of biodiversity loss	Sometimes, due to one of the factors mentioned above or several of the factors combined together, there is an impact on the functioning of the continental and ocean basin scales resulting in a massive loss of biodiversity. This in turn impacts on several other biospheric systems like "C storage, freshwater, N and P cycles, land systems"
6	Global freshwater use	Consumption of freshwater may effect regional climate patterns (e.g., monsoon behavior). Originally, the natural cycles of the atmosphere would be enough to replenish the water consumed. However, due to the increase in usage of freshwater at a faster rate and higher volumes, the amount replenished naturally can no longer meet the required criteria for systemic balance. This has affected "moisture feedback, biomass production, carbon uptake by terrestrial systems and reducing biodiversity"
7	Land system change	The conversion of massive patches of land across the globe for crop cultivation purposes has triggered irreversible changes in the regional biodiversity and thereby indirectly affected the carbon storage and ecosystem resilience in those particular regions

(*continued*)

Table 1 (continued)

Sl. No.	Life-support systems	Impact
8	Aerosol loading	Natural aerosol concentration occurring as a result of the cyclical processes of nature acts as a regulator for Earth's radiation by scattering incoming radiation back to space. It indirectly influences cloud reflectivity and persistence by acting as regulator for temperature fluctuations. However, human-induced aerosol loading has short-circuited this regulatory phenomenon leading to erratic climatic changes like disruption of monsoon systems and production of hazardous phenomenon such as smog, etc. which affects human health
9	Chemical pollution	Chemical pollutions here include common examples like carbon emissions, uncontrolled plastic disposal, heavy metal usage, nuclear wastes, etc. Unregulated usage of chemicals may lead to dire results pertaining to human health, as well as impact the overall functioning of the ecosystem

Operationalizing Collective Ecological Well-Being

Change is hard to adopt for humans. A fluid and more systemic transitioning endeavor will need to be implemented to reinforce change conducive to the enterprise. One such concept is promoted by Peter Senge (1990) in his book *The Fifth Discipline*. He calls it "the learning organization" – a highly dynamic system that incorporates sensitivity to change and at the same time enables the organization to flourish while addressing those changes.

There are many models for management of organizations. Learning organization is a good means to ensure systems thinking and collective learning. We use this as a metaphor to illustrate that workplace achievement (both at the individual and organizational levels) can be attained while maintaining ecological sustainability as the key purpose.

The concept of a learning organization is essentially a route map of managing the learning needs of an organization, thereby making the members of the organization more sensitive towards change as well as be agile while addressing them. This concept comprises of the following five disciplines – systems thinking, mental models, shared vision, team learning, and personal mastery. This concept is now established in various fields as obvious from the collection of case studies in the book *The Fifth Discipline Fieldbook* (Senge et al. 1994). These cases provide ample evidence that the framework of learning organizations can be applied as a basis for developing and adopting engaged sustainability practices instead of rediscovering the wheel.

Systems Thinking: Thinking Beyond Boundaries

> If I had an hour to solve a problem I'd spend 55 minutes thinking about the problem and 5 minutes thinking about solutions. – Albert Einstein

Peter Senge (1990) ends his book *Fifth Discipline* with the profound notion of *wholeness*. The roots of this idea can also be traced back to the century-old scriptures or indigenous cultural philosophies – wise words handed down by the ancestors on how to live a harmonious life. The core idea behind this concept is that the universe is not made up of separate parts stacked together. Instead, it is a group of small individual subsystems which flow into each other to form bigger self-sufficient systems. These systems are efficient loops of self-correcting processes that balance and adjust themselves to accommodate any change being triggered due to factors occurring either internally or externally. This ensures that even a tiny change in one part of a subsystem immediately gets reflected in the other subsystems as well as the greater system of which they are a part.

Systems thinking is the overall discipline of looking at organizations as a system in order to be able to understand the connections between causes and effects, the patterns of recurring results, and see the bigger picture that emerges. This makes the organization agile and quick in adapting to changes as the situation demands. This way of observation also brings into view the hidden undercurrents that might be hampering an organization in moving towards its goal.

In the current times, as people are becoming more conscious of their environment, several policies are being enforced to ensure eco-friendly measures be taken by the organizations. Where earlier enterprises were only concerned about profit-making, now there are new sustainability concepts like "triple bottom line" (people, planet, and profits) being enacted throughout the organization. This is a change which has opened the doors to a lot of contemplations and discussions on how to make our enterprises more eco-friendly. However, the realization that emerges is that addressing the direct impacts of the concerned enterprise on the surrounding environment turns to be only a stopgap measure. The repercussions of the choices made and the resources consumed are more far-flung than originally thought to be.

Hence, it becomes necessary that the entire living environment be mindfully observed in order to examine how an individual interacts with it to the minutest detail possible. Let's take, for example, one day in the life of an employee. The workday for the person begins the moment the person enters the premises of the workplace. Most probably, the first interaction of the employee will be with the elevator pressing the button to take him or her to the relevant floor of the building. An elevator is that magical artifact that transports one from one floor to another without having to expend muscular energy and be tired from it. Are we aware how much electricity is being exhausted in keeping such a magical instrument operating 24 hours a day? Then we reach our desks. In the middle of the desks, there will be most probably a computer. On one side, there might be an intercom and on the other might be a printer. A common man would definitely not inquire about the raw materials going into the making of a computer or a printer or an intercom, not to mention the electricity required to run the same. What about the milky white A4 papers that need to be printed on? And then there's the pen with which notes are taken and signatures given. Also, there's the large quantity of refills to ensure that the pen keeps on writing. How many trees might have been felled to produce the tables that hold our work paraphernalia? Not to mention whatever chemicals the carpenters use, like colors, turpentine, etc., to achieve the highly polished look. Sometimes the material might not be wood but metal or plastic and tons of associated chemicals. The ergonomically designed chairs too are definitely to be included in our zone of awareness.

And then, we move on to the cafeteria. Most organization will have a food-serving area and several coffee and tea vending machines every floor. Where did the food ingredients come from? Are they locally made or have they traveled across seven continents in aluminum-foiled packets, on specially consigned vehicles to reach the building? And then there's the building itself. Has any measure been taken to offset the harm caused to the soil while erecting this building? What about the air-conditioning systems that ensure an easy working environment in extreme temperatures?

No, it is definitely tough to survive without these modern facilities with today's lifestyle. And no, no one is being demanded to give up this lifestyle either. But what is asked for here is a bit more awareness: consciousness of things that people use up and then discard into the surroundings. One person might exhale only a few grams of carbon dioxide per day. But when tens of trillions of organisms do so, the grams add up to tens and thousands of tons. Remember, carbon dioxide does not recognize geopolitical boundaries. With the decrease of tree cover, the carbon dioxide has nowhere to go but dissolve into the ocean water. Which in turn raises the acidity level of the oceans and triggers reduction of marine life forms. And when one form of life goes extinct in a particular ecosphere, the balance of the system disrupts giving rise to a host of other issues, like increase in bacteria that may be harmful to human life. And so, the circle of biological cohabitation continues in never-ending loop of cause and effect.

We may not be able to give up modern amenities, but by being a little bit more aware, we definitely can plant a tree. Tens of millions of people, belonging to different generations across long spans, each planting one tree, will start playing their role of being a net carbon dioxide sink, thus reducing ocean water acidification and stopping the rest of the trigger reactions in the biospheric chain.

This is just one of the many examples of the need to become aware of the nonliving entities which make life on Earth possible. Earth isn't the one which is in danger of survival. It is us human who are going to become extinct should the life-support systems of Earth go out of human tolerance levels.

As can be seen from the above, it can be a very scary notion to behold that from the very food that we eat, to the products that we use, and every item (cars, mobiles, computers, air-conditioners, water heaters, etc.) in between that have been deemed as necessities for modern life – personal or professional – adds another death knell to continued survival for future generations. It can be very overwhelming to question everything about our current lifestyles in order to restore even a part of the biosphere.

However, change need not be so painful. Change on a global scale can definitely not be done alone. However, as a community, working one step at a time towards the goal of making our enterprises resource-friendly is achievable. Here, the other four learning disciplines that characterize a learning organization come into play – mental models, shared vision, team learning, and personal mastery.

A systems effort propagates the necessity for reexamination of the thoughts and assumptions underlying the practices undertaken by organizations. The emphasis would be on identifying and differentiating between the assumptions which would work and which no longer serve the organization's purpose. This brings us to the second learning disciple mental models.

Mental Models: Five Practices of Ecologically Engaged Literacy

> The ability to perceive or think differently is more important than the knowledge gained. – David Bohm

According to Senge et al. (1994), "the assumptions held by individuals and organizations are called mental models," and they represent the ability to be "reflecting upon, continually clarifying, and improving our internal pictures of the world, and seeing how they shape our actions and decisions" (p:6). As the wise would say, the ideas that gave rise to the problems of today won't be able to provide the solutions that in the first place created them. To gain fresh ideas, a shift in thinking is necessary.

According to the theory behind learning organizations, once the current mental models have been revisited, it becomes immensely easier for the members to embrace change. A change in mental models allows people to anticipate change and respond to them faster. Thereby they are poised to adapt to changes seamlessly without any major hiccups to the system. "Change and learning may not exactly be synonymous, but they are inextricably linked" (Senge et al. 1994, p:6).

History stands witness to rising civilizations and changing vistas. From nomadism to building prosperous civilizations, from stone tools to artificial intelligence, from agriculture-based societies to intelligence-oriented societies, every change is driven by some sort of idea figured out by humans. These ideas in turn gave rise to changing demands in the way an individual conducts his or her life – be it farming, hunting, socializing, etc.

Consumerism today may have its roots in Industrial Revolution, and even further back in the dynastic revolutions, no one can say for sure. But one reality we all agree to is that human beings reaching out to their needs and novelty is the biggest causes of change, so much so that humans have even started changing the very contours of the biosphere that supports their lives. To regulate human choices so as to not cross the limits of these boundaries has become the urgent need. It is time to reexamine our needs and demands from the very core.

As argued earlier in the chapter, human beings learn and grow by imitating other humans in their immediate living environment. This is one of the reasons why people of a particular cultural origin behave in one particular way, and the members of another community behave in another way. It is the assumptions handed down through culture that give rise to the particular manner in which one conducts one's life. One the way, the underlying assumptions may get modified due to cataclysmic factors such as natural disasters, foreign invasions, and so forth. The issue arises when these very same assumptions are maintained unconsciously for generations on end without reviewing the consequences periodically. Industrial Revolution was considered an unqualified boon for humans until recently. Even the common man could now live at a higher standard of living with the help of machines. But today, this very demand for better and better machines and scientific revolutions is creating havoc to the environment.

Hence, it is very important that the goals for change be set at a community level – be it a commercial company, a politically recognized country, or even an association of several countries. This is because, as it has been established before, the biosphere does not recognize any geopolitical boundaries. It just follows its own system of physical laws giving rise to certain natural interactions, which may or may not be conducive to the survival of living organisms (humans being one of these).

Goleman et al. (2012) mention five practices that can allow people to strengthen their capacity to live sustainably. His work is aimed at grooming students to become future leaders who are armed with enough knowledge and skills to become well balanced in their emotional, social, and ecological dealings with the living environment. And what works for future generations of leaders definitely works for the current generations of leaders as well. The only way to inspire action is by living it ourselves. These five practices integrate emotional, social, and ecological intelligence. These are:

1. Developing Empathy for All Forms of Life: The first idea encourages one to develop a sense of compassion for other forms of life. "By shifting from our society's dominant mindset (which considers humans to be separate from and superior to the rest of life on Earth) to a view that recognizes humans as being members of the web of life," it becomes easier to "broaden their care and concern to include a more inclusive network of relationships" (Goleman et al. 2012, p:12).
2. Embracing Sustainability as a Community Practice: This idea emerges from "knowing that organisms do not exist in isolation" (Goleman et al. 2012, p:12). The collective ability to survive and thrive is determined by the quality of the network of relationships within any living community – irrespective of whether that relationship is between humans or between humans and other living organisms. This inspires one to "consider the role of interconnectedness within their communities and see the value in strengthening those relationships by thinking and acting cooperatively" (Goleman et al. 2012, p:12).
3. Making the Invisible Visible: The third idea enables one in "recognizing the myriad effects of human behavior on other people and the environment" (Goleman et al. 2012, p:12). Making the invisible visible through observation and awareness will make us understand "the far-reaching implications of human behavior and enables us to act in more life-affirming ways" (Goleman et al. 2012, p:12).

4. Anticipating Unintended Consequences: This is a two-pronged method to address the dual necessities of "cultivating a way of living that defends rather than destroys the web of life," and at the same time "build resiliency by supporting the capacity of natural and social communities to rebound from unintended consequences" (Goleman et al. 2012, p:12).
5. Understanding How Nature Sustains Life: Earth has been in existence for billions of years. It has stood testimony to countless species of living organisms that had once existed and then gone extinct. Hence, to learn how to thrive in the Earth's environment, it is urgently necessary to examine "the Earth's processes, and learn strategies that are applicable to designing human endeavors" (Goleman et al. 2012, p:12).

Continuously reviewing the mental models that drive the needs and demands of our society will ensure that we never get lulled into false sense of security that everything is okay. As we become more aware of systemic structures, like the ones that exist in the biosphere (the nine life-support systems discussed earlier), and feel compelled to modify them, having a set goal towards formulating an actionable plan calls for brainstorming and execution at a collective level. This is where the next discipline "shared vision" comes in.

Shared Vision: Bringing the Ecosphere Down to Earth

> The way to get things done is not to mind who gets the credit for doing them. – Benjamin Jowett

According to Senge et al. (1994), shared vision is "a sense of commitment in a group, by developing shared images of the future we seek to create, and the principles and guiding practices by which we hope to get there" (p:6). The premise of shared vision is not just about working towards a set goal with the help of a team. It is more about building a dynamic process, a way of following the philosophies and practices revised during the mental modeling phase. It is not just about a single or a short-term achievement. It is about moving beyond the premise of "achieving" itself and establishing a way of working which encourages and embraces all types of changes and learning requirements.

The most successful visions of any organization are those that are usually built on the basis of the individual visions of employees at all levels of the organization. This gives a sense of identity to the entire group and provides a point of focus for everyone. Can we envisage organizations where the energy to deliver gets concentrated on the "one goal" of sustainability.

The current times spectate a huge number of cutting-edge technological changes in every sphere of life. Earlier, when technology had just debuted into human society, only the wealthy elites had access to enjoy the latest gadgets and the life of ease it provided. However, times are changing. With increased disposable income and better standards of living, everyone starting from a teenager to multimillionaire tycoons have access to the latest communication gadgets and lifestyle facilities. Technology is shaping the way people engage in their day-to-day lives. Where earlier change could be anticipated at the turn of every two to three decades, now last year's technology has already become obsolete. It is now important to proactively create rather than merely react to change. It calls a tremendous amount of concentrated effort for a big organization to remain nimble and dance with the changes.

So, the question arises, is it possible to create an organizational culture whose every strategy and plan is focused on rehabilitating the Earth's ecology first and then fulfilling the rest of the organizational aspirations? There is no single answer yet. Everybody, who has taken it upon themselves to be a contributor to this Earth-friendly movement, is trying. And perhaps to a certain extent, it can be said that they have been successful.

Awareness regarding eco-friendliness and engaged sustainability is slowly trickling down into the minds of everyone. Policies are being proposed at the government level to ensure consumption is being handled in a responsible way. Nonprofit organizations have dedicated their lives to spread the message of green living. Individuals are becoming more mindful about their lifestyles and considering one-person activities like indoor gardening, minimization of personal assets, recycling, planting of trees, and so on.

Such activities not only encourage us but also give us hope that people might be more agreeable to take to a shared vision of aligning our needs to the planet. The best thing is that it would be the ideal kind of shared vision, where everyone contributes, out of their own volition, and not as a one-upmanship game.

It is okay to start small. The only requirement is that the appropriate questions need to be asked. Goleman et al. (2012) urges us to answer author and farmer Berry (2009) who argues that in order to be more empathetic towards nature, instead of thinking about the ecosphere, an individual must start by looking at their own community first. He urges us to "think local." The following eight questions are a part of the curriculum Berry proposes in order to challenge young leaders to strengthen empathy towards their own communities. The community here could also mean organizational communities:

1. What has happened here?
2. What should have happened here?
3. What is here now? What is left of the original natural endowment of this place? What has been lost? What has been added?
4. What is the nature or genius of this place?
5. What will nature permit us to do here without permanent damage or loss?
6. What will nature help us to do here?
7. What can we do to mend the damage that we have done?
8. What are the limits: Of the nature of this place? Of our own intelligence and ability?

Questioning what has never been questioned before is bound to raise a lot of confusion and uncertainty about what action to take next. And being humans, the first tendency would be to snap back from our familiar comfort zones. To address this, having members come together in small groups and start a conversation about the unknown will direct the process in a positive direction. Thus, we come to the next discipline Senge (1990), "team learning."

Team Learning: Cultivating Ecoliterate Learning Communities

If you want to go quickly, go alone. If you want to go far, go together. – Old African proverb

The discipline of team learning of Senge et al. (1994) consists of developing "transforming conversational and collective thinking skills, so that groups of people can reliably develop intelligence and ability greater than the sum of individual members' talents" (p:4). This is where empathy plays its role as a key driver of civilizational transformations. A single person would not be able to carry a stone hundred times their size. But a committed and energized workforce of a thousand can build the mighty pyramid. One lone individual will find it hard to build a rocket. But a dedicated group of scientists can even place a human on the moon. One alone can't go against the forces of nature. But the citizens of an entire country can at least take precaution against the hurricane Katrina and ensure some measure of safety. Having the skills and attitude to successfully undertake stupendous collective human endeavor is what team learning is all about. Proactive and conscious collective learning ensures that no blind spot has been left untouched. A dialogue that begins with a sense of camaraderie and shared visionary energy is bound to be carried forward into the future.

With respect to shifting to living in a sustainable manner, perhaps the biggest obstacle is our identity as a part of some community – a religious identity, a cultural identity, a national identity, identity with some lifestyle belief, etc. These are very sensitive issues, and dealing with them calls for a high level of maturity. Most often sensitive topics are generally not logically dealt with. It's almost like one is being questioned about one's way of life itself. No one likes to be told that the way he or she is living previously is incorrect or unjustified. As Brown (2012) suggests, this becomes an affront to the person's past self which is the basis of who he or she now is. It altogether feels like a rejection of the very self. While sensitive conversations such as these require tremendous self-awareness as well as social awareness, being empathetic would open up hearts and minds to tough conversations and, finally, to changes.

The following are the five practices that Goleman et al. (2012) suggest to focus on while connecting with people:

1. Don't communicate from a place of anger
2. Reach people on the human level through stories
3. Foster dialogue instead of debate
4. Speak from the heart
5. Make ecological connections clear to others

Every one of us is a significant part of the systems we work within, and the most significant leverage may come from changing our own orientation and self-image. This requires work in developing our own personal vision and learning to see the world from not just a reactive point of view but also a creative and interdependent perspective. Thus, we come to the final discipline of learning organizations, personal mastery.

Personal Mastery: Giving Voice to a Way of Life

> To practice a discipline is to be a lifelong learner. You 'never arrive.' The more you learn, the more acutely aware you become of your ignorance. – Peter Senge

In the context of learning organizations, personal mastery is all about "learning to expand our personal capacity to create the results we most desire, and creating an organizational environment which encourages all its members to develop themselves toward the goals and purposes they choose" (Senge et al. 1994, p:6).

Team learning definitely zooms in the microscope to all our individual personality quirks that make us ready, or not ready, for change. Even with such doubts if we agree to go ahead with changes, what could be the further steps? How can we ensure a little bit of a less painful process of growth? Maybe it is by changing habits to be more eco-friendly. Or may be, even by overhauling our entire life to accommodate the change.

This is where personal mastery comes in. The premise behind this idea is that only when we take responsibility and ownership of change at the personal level can change be permanent. This requires us to become more self-aware about our own thoughts and ideas, the habits that groom us into behaving the way that we do, or lead the lives that we live. And becoming self-aware makes it easier to exercise self-management skills. And we come back a full circle to emotional intelligence. With further help from empathy, we also become better at managing relationships with other individuals as well as the living environment, i.e., our social intelligence and ecological intelligence also grow.

Just imagine, when tens of billions of people become highly empathetic towards one another and the other organisms in our ecosystem, doesn't the task of ensuring engaged sustainability become much simpler? It wouldn't be a big stretch to say that by choosing better options for our own selves, we make the world a better place. And all this can be done simply by strengthening our natural skill of empathy.

Conclusion

By using the framework of learning organizations as a method for helping organizations become more sustainability oriented, there is one single lesson that prominently stands out. Small changes can accumulate and make the big overwhelming goals easier to achieve.

Three insights emerge from this chapter:

1. The Earth's biospherical systems and the human social systems are inherently interconnected. The changes made in the human systems can trigger unimaginable consequences in the biospheric system disrupting the entire existential balance of the other living organisms. This in turn comes back a full circle hindering the survival of humanity itself. The implication here is that, if humanity wants to ensure its survival, they have no other choice but to adopt ways and means of interacting with the planet in a manner that keeps the greater life-support systems of the planet in balance.
2. The endeavor to survive, and also to thrive under new ways of conducting human affairs, calls for awakening the different talents inherent in the human psyche. First, improve the faculty of consciousness by becoming more self-aware and be adept at self-management by training the capacity for emotional intelligence. Second, extend this to social intelligence that enhances this awareness to the other members of the human society in order to understand what causes this system to work for or against the bigger life-nurturing biosphere. Third, include the other living members of the biosphere in order to recognize mindful practices of conducting life which does not deprive any living organism (including humans) of its right to life on Earth.
3. To develop consciousness (or awareness or mindfulness), there is just one prerequisite; switch on the innate ability to empathize. Strengthen the empathetic muscle in order to strengthen an individual's capacity to handle his or her own as well as others' emotions and manage them towards a positive direction. This will ensure that one's own growth as well as humanity's collective growth is conducted in a manner that balances Earth's life-support systems.

Thus, it can be seen that engaged sustainability is a choice. And this choice is the natural choice if the innate ability of humans towards empathy is strengthened. Empathy is what makes human beings the most powerful living being on this Earth, the only beings who have the power to make a choice to include all life and nonlife forms on Earth to coexist in a sustainable manner.

Cross-References

▶ Agent-Based Change in Facilitating Sustainability Transitions
▶ Bio-economy at the Crossroads of Sustainable Development
▶ Business Youth for Engaged Sustainability to Achieve the United Nations 17 Sustainable Development Goals (SDGs)
▶ Collaboration for Regional Sustainable Circular Economy Innovation
▶ Ecopreneurship for Sustainable Development
▶ Education in Human Values
▶ Environmental Intrapreneurship for Engaged Sustainability

- Environmental Stewardship
- Ethical Decision-Making Under Social Uncertainty
- Expanding Sustainable Business Education Beyond Business Schools
- From Environmental Awareness to Sustainable Practices
- Just Conservation
- Moving Forward with Social Responsibility
- People, Planet, and Profit
- Selfishness, Greed, and Apathy
- Smart Cities
- Social Entrepreneurship
- Sustainable Decision-Making
- Teaching Circular Economy
- The Spirit of Sustainability
- The Sustainability Summit
- The Theology of Sustainability Practice
- To Be or Not to Be (Green)
- Transformative Solutions for Sustainable Well-Being
- Utilizing Gamification to Promote Sustainable Practices

References

Berry, W. (2009). *Bringing it to the table: On farming and food*. Berkeley: Counterpoint.

Bonnett, M. (2017). Sustainability and human being: Towards the center of authentic education. In B. Jickling & S. Sterling (Eds.), *Post-sustainability and environmental education: Remaking education for the future*. Cham: Palgravee Mcmillan.

Brown, B. (2012). *Daring greatly: How the courage to be vulnerable transforms the way we live, love, parent and lead*. New York: Gotham Books.

Checkland, P. (1981). *Systems thinking, systems practice*. New York: Wiley.

Goleman, D. (1995). *Emotional intelligence: Why it can matter more than IQ*. New York: Bantam Books.

Goleman, D. (2004). In Forward to Zins, J. E., Weissberg, R. P., Wang, M. C., Walberg, H. J. (Eds.), *Building academic success on social and emotional learning: What does the research say?* New York: Teachers College Press.

Goleman, D. (2006). *Social intelligence: The new science of human relationships*. New York: Bantam Books.

Goleman, D. (2010). *Ecological intelligence: The coming age of radical transparency*. Camberwell: Penguin.

Goleman, D., & Ekman, P. (2007). Three kinds of empathy: Cognitive, emotional compassionate. http://www.danielgoleman.info/three-kinds-of-empathy-cognitive-emotional-compassionate/

Goleman, D., Bennett, L., & Barlow, Z. (2012). *Ecoliterate: How educators are cultivating emotional, social, and ecological intelligence*. San Francisco: Jossey-Bass. (Amazon Kindle e-book).

Gore, A. (2006). *An inconvenient truth: The planetary emergency of global warming and what we can do about it*. New York: Rodale.

Lakin, J. L. (2006). Automatic cognitive processes and nonverbal communication. In V. Manusov & M. L. Patterson (Eds.), *The sage handbook of nonverbal communication*. Thousand Oaks: Sage Publishing.

Lipton, B. (2005). *Biology of belief: Unleashing the power of consciousness, matter and miracles.* New York: Hayhouse.

Nadeau, R. L. (2006). *The environmental endgame: Mainstream economics, ecological disaster and human survival.* New Brunswick: Rutgers University Press.

Pinker, S. (2002). *The black slate: The modern denial of human nature.* NewYork: Viking Penguin.

Rockström, J., Steffen, W., Noone, K., Persson, Å, Chapin, F. S., Lambin, E., Lenton, T. M., Scheffer, M., Folke, C., Schellnhuber, H., Nykvist, B., De Wit, C. A., Hughes, T., van der Leeuw, S., Rodhe, H., Sörlin, S., Snyder, P.K., Costanza, R., Svedin, U., Falkenmark, M., Karlberg, L., Corell, R. W., Fabry, V. J., Hansen, J., Walker, B., Liverman, D., Richardson, K., Crutzen, P., Foley, J. (2009, November 18). *Planetary boundaries: Exploring the safe operating space for humanity.* Retrieved September 27, 2017, from https://www.ecologyandsociety.org/vol14/iss2/art32/.

Senge, P. M. (1990). *The fifth discipline: The art and practice of the learning organization.* New York: Doubleday.

Senge, P. M., Kleiner, A., Roberts, C., Ross, R. B., & Smith, B. J. (1994). *The fifth discipline fieldbook: Strategies and tools for building a learning organization.* New York: Currency, Doubleday.

Tucker, M. E., & Williams, D. R. (2007). *Buddhism and ecology: The interconnection of dharma and deeds.* Cambridge: International Society for Science and Religion.

Wispe, L. (1991). *The psychology of sympathy.* New York: Plenum Press.

Education in Human Values

Planting the Seed of Sustainability in Young Minds

Rohana Ulluwishewa

Contents

Abstract

Most conventional strategies adopted for achieving sustainability are designed to bring about external changes in the form of technological, institutional, and infrastructural changes. However, past experiences in development show without inner changes – changes in our values, which influence our attitudes and behavior – external changes are unable to achieve sustainability. Materialistic values such as money, material possessions, power, social status, recognition, fame, and reputation etc., dominant in our modern society cause our relationships with fellow human beings and with nature to be self-centered and exploitative, and our behavior unsustainable. To achieve sustainability, these external changes need to be supplemented with inner changes to bring a shift from materialistic values to human values: love, truth, right conduct, peace, and nonviolence, leading to relationships that are selfless, loving, and nonexploitative. Thus, human values can play a significant role in achieving sustainability. Recent discoveries in quantum physics and neuroscience have revealed that these human values are intrinsic to human beings and are hardwired in our brains.

R. Ulluwishewa (✉)
(Former) Massey University, Palmerston North, New Zealand
e-mail: ulluwishewa@xtra.co.nz

S. Dhiman, J. Marques (eds.), *Handbook of Engaged Sustainability*,
https://doi.org/10.1007/978-3-319-71312-0_23

Education in human values is a program designed to bring out these values and guide our behavior and attitudes. Drawing on empirical evidence from the value-based water education implemented by the United Nations Human Settlements Programme in six African countries, this chapter highlights the potential capacity of education in human values to plant the seed of sustainability in young minds.

Keywords
Sustainability · Intrinsic values · Extrinsic values · Human values · Materialistic values · Education · Water management · Universal consciousness · Quantum-neuroscience

Introduction

Sustainability is variously defined depending on the context in which the term is used. Following the concept and the definition of sustainability presented by the Brundtland Report (1987) there is a general agreement that sustainability entails meeting the needs of the present generation without compromising the ability of future generations to meet their needs. Sustainability is also seen in three domains: (1) environmental sustainability: the ability of the environment to maintain its capacity to support the human society indefinitely, (2) economic sustainability: the ability of an economic system to maintain its production indefinitely, and (3) social sustainability: the ability of a social system to function at the defined level of social well-being indefinitely. Sustainability of environment, economy, and society is necessary for human society to be sustainable. Therefore, sustainability generally refers to the ability of the humankind to sustain itself indefinitely. It is now widely evident that the ability of the humankind to sustain itself is under threat. There is a growing concern about the capacity of the environment to support our survival. Some of the major concerns are global warming, loss of biodiversity, erosion of soil, and pollution and contamination of water and air. Beside these environmental concerns, there are also social and economic concerns. These include: overconsumption of Earth's resources by the rich whose greed for material wealth and power results in the unequal distribution of resources, poverty, hunger, malnutrition, poor health, divisions and conflicts in social and political fields, and the breakdown of the family institution.

Most strategies that have been designed and adopted by development agencies and policy makers to achieve sustainability have almost exclusively aimed at bringing external changes: introduction of technologies which improve resource efficiency and substitution of nonrenewable resources with renewable resources; development of environment-friendly infrastructural facilities for production of goods and services such as transportation, communication, and housing; providing legal and administrative measures to control and regulate individuals' behavior; and adopting environment-friendly public policies. There is no doubt that all these external changes are necessary to achieve sustainability goals. However, it has been pointed out that external changes are insufficient without inner changes in people (Ulluwishewa 2014, 2016). The success of external changes in achieving sustainability ultimately depends on how individuals value sustainability, their sense of responsibility for the well-being of others and the natural environment around them, and the extent to which they are willing to sacrifice their own comforts for sustainability of the humankind.

This is where values can play a crucial role in achieving sustainability goals. Values generally refer to beliefs and ideals shared by the members of a society about what is good or bad and desirable and undesirable. They are psychological representations of what we believe to be important in life (Rokeach 1973). Psychologists, sociologists, and anthropologists (Schwartz 2006; Williams et al. 2000; Kluckhohn 1951) view values as the criteria people use to evaluate actions, people, and events. They influence individuals' behavior and attitudes and serve as broad guidelines in making choices in all situations. Schwartz (2006), in his theory of basic human values, identified ten basic human values that he claimed to be common in all cultures: self-direction, stimulation, hedonism, achievement, power, security, conformity, tradition, benevolence, and universalism. According to their congruities and conflicts, he clustered power and achievement into one group he calls the self-enhancement values, and benevolence and universalism into another group he calls the self-transcendent values. While self-enhancement values emphasize the pursuit of self-interests, the self-transcendent values involve concern for the welfare and interests of others. Kasser (2009), based on an extensive cross-cultural research, has identified a cluster of three materialistic values: (1) financial success, which concerns the desire for money and possessions; (2) image, which concerns the desire to have an appealing appearance; and (3) status, which concerns the desire to be popular and admired by others.

The present study, based on the origin of values and their relevance to sustainability, identifies two groups of values:

1. Intrinsic Values: These are the values that originate from within us and therefore are shared by all human beings regardless of their cultural, racial, or other personal differences. For instance, we all value love: everyone likes to love and be loved. In the same way, as it will be discussed in detail later in this chapter, all human beings value peace, truth, right conduct, and nonviolence. The intrinsic values encompass the values identified by Schwartz (2006) as transcendent values. As it will be pointed out later in this study, intrinsic values make our relationships with fellow human beings and with the environment selfless, loving, nonexploitative, and sustainable.
2. Extrinsic Values: Extrinsic values refer to the values that individuals learn from or are imposed by external sources such as the society, culture, education, and media, etc. Such values vary from culture to culture and over time. As cultures transform, their values change. Historically as human societies transformed from hunting and gathering to agricultural and then industrial and post-industrial, their extrinsic values have also changed. The extrinsic values encompass the values identified by Schwartz (2006) as self-enhancement values. As it will be pointed out later in this study, extrinsic values make our relationships with fellow human beings and with the environment self-centered, unloving, exploitative, and unsustainable.

This chapter first discusses the definition of intrinsic and extrinsic values and how they differ in origin and characteristics. Recent discoveries in quantum physics and neuroscience provide evidence that intrinsic values stem from the "Golden Rule" and altruistic love hardwired in the human brain, and are therefore common to all human beings. The discussion highlights the potential capacity of intrinsic values, also known as core human values, to guide individuals towards sustainability. The role of extrinsic values, with special emphasis on materialistic values in modern industrialized societies, is also discussed. With the help of empirical evidence, this section outlines how materialistic values cause our relationships with others and with the environment to be self-centered, unloving, exploitative, and unsustainable, and highlights the importance of shifting from materialistic to human values to achieve sustainability. The final section of this chapter introduces the program of education in human values (EHV), and drawing on evidence from the value-based water education undertaken by the United Nations Human Settlements Programme (UN-Habitat), demonstrates the potential capacity of EHV for planting the seed of sustainability in young minds.

Intrinsic and Extrinsic Values

As already mentioned, intrinsic values originate from within, from our inner reality, from what we really are. What is our inner reality? Perhaps, the best way to discover our inner reality rationally is to look into ourselves through a powerful microscope. If we do so, we will discover the energy which fills each and every atom of our body. Thus, our ultimate reality is energy and it represents 99.999% of what we call "I." The same energy that fills the atoms of our body fills the universe too. In spite of its vastness, the Universe, a mass field of energy, functions as an inseparable and indivisible single field of energy, or as "a single undivided whole" as described by Bohn (1980), Bohn and Hiley (1993), a renowned theoretical physicist. Recent scientific discoveries reveal that the Universe, "the single undivided whole," is alive, aware, and conscious. The view that the Universe is conscious, which has so far been a philosophical concept, is now supported by prominent scientists. Elgin (2009), a bestselling author and speaker, in his book *The Living Universe* brings together extraordinary evidence from cosmology, biology, and physics to show that the universe is not dead but rather uniquely alive. Referring to the consciousness of electrons, he quotes Freeman Dyson, a theoretical physicist, as saying "matter in quantum mechanics is not an inert substance but an active agent, constantly making choices between alternative possibilities. It appears that mind, as manifested by the capacity to make choices, is to some extent inherent in every electron." He thinks it is reasonable to believe in the existence of a "mental component of the universe." Using scientific evidence he shows that consciousness is present in molecules consisting of no more than a few simple proteins. Phillip Cohen, one of the researchers who made the discovery, has stated that "we were surprised that such simple proteins can act as if they had a mind of their own." Lanza (2009), a prominent American scientist, presenting his theory of "Biocentrism," says that the universe is fine-tuned for life and life creates the universe, not the other way around. The conscious and living universe is widely called Universal Consciousness.

It is our inner reality and what we really are. Some call it Ultimate Reality, Higher Self, Spiritual Self, and Infinite Self. This is what is called God in most religions, according to Haisch, a German-born American astrophysicist. He states in his book *The God Theory* (2009) that consciousness is not a mere epiphenomenon of the brain; it is our connection to God, the source of all consciousness. Ultimately it is consciousness that creates matter and not vice versa.

The Universal Consciousness (God), while remaining as an indivisible and inseparable single entity at the quantum level, manifests itself at the material level as separate forms, e.g., rocks, soils, plants, animals, and human beings, creating the material world. Thus, conscious energy manifests itself as matter. Though we perceive ourselves as separate individuals at the material level, at the quantum level, we all are interconnected and remain as inseparable parts of the Universal Consciousness. In the words of Lanza (2017):

> Our individual separateness in space and time is, in a sense, illusory. We are all melted together, parts of an organism that transcends the walls of space and time. This is not, you understand, a fanciful metaphor. It is a reality.

Recent discoveries in neuroscience suggest that this inseparability, interconnectedness, or oneness is hardwired in our brain as an inner urge to connect with others and serve others. Just as neurons connect and communicate with each other, brains strive to connect with one another says Cozolino (2006), an American psychologist and social neuroscientist. He considers brain as a social organ. Lieberman (2013), an American social neuroscientist, points out that our need to connect with other people is wired in our brain, and it is even more fundamental than our need for food and shelter. He argues that if people are motivated only by self-interest, how can we explain why folks cooperate, ensuring that they will earn less? He believes that people are even more motivated by something beyond self-interest: the drive for social connection. In addition to being self-interested, we are also interested in the welfare of others.

Pfaff (2007), an American neuroscientist, based on recent discoveries in neuroscience, says the human brain is hardwired to act according to the Golden Rule – one should treat others as one would like others to treat oneself – which represents common-sense ethics, and is the ultimate, all-encompassing principle for moral behavior. He explains how specific neural circuits in our brain cause us to perceive our actions toward another as they were going to happen to us, prompting us to treat others as we wish to be treated ourselves. In his recent book, *Altruistic Brain: How We Are Naturally Good*, Pfaff (2015) demonstrates that human beings are hardwired to behave altruistically in the first instance, such that unprompted, spontaneous kindness is our default behavior; such behavior comes naturally, irrespective of religious or cultural determinants. This view is further supported by the discovery of what neuroscientists call "Mirror Neurons," which enable us to experience others' pain and be empathetic (Rizzolatt and Crighero 2004). The mirror neurons instantly project into the other person's shoes and enable us to experience the other's feelings. The mirror neurons represent a basic biological mechanism inherent in all individuals and it is the biological foundation of the golden rule.

What is this force which comes from within us to prompt us to connect with others, serve others, be empathetic towards others, and act for their well-being altruistically? This is "Unlimited Love" according to Post (2003), a Professor of Bioethics and Family Medicine and President of the Institute for Research on Unlimited Love in USA. In his *Unlimited Love: Altruism, Compassion and Service* he defines love as:

> The essence of love is to affectively affirm as well as to unselfishly delight in the well-being of others, and to engage in acts of care and service on their behalf; unlimited love extends to all others without exception, in an enduring and constant way. Widely considered the highest form of virtue, unlimited love often demands a creative presence underlying and integral to all of reality: participation in unlimited love constitutes the fullest experience of spirituality. Unlimited love may result in new relationships, and deep community may emerge around helping behaviour, but this is secondary. Even if connections and relations do not emerge, love endures. (p. vii)

This is the purest form of love which is unselfish, unconditional, and unlimited. It does not expect anything in return: love for the sake of love. It is different from what we call love in our ordinary life: love of a mother toward her child which is affection, love that exists between wife and husband which is infatuation, love that exists among friends and relations which is affection, and love toward material objects which is desire. It manifests itself in various forms such as acceptance, forgiveness, compassion, kindness, tolerance, generosity, sharing, empathy, and selfless service. The foundation of love is not our feelings or emotions towards others, but our inner interconnectedness or oneness with others at the quantum level. It is the very nature of humanness and is natural to us. From this perspective, the opposite word to love is not hatred but separateness, individuality, or self-centeredness. It is the basis of our intrinsic values. Intrinsic values are what we would value if our thoughts are guided by the hardwired oneness, altruism and the Golden Rule. The UN-Habitat, in its Value-Based Water Education (VBWE) project which will be discussed later in this chapter, has identified five such values on the basis of the teachings of Sathya Sai Baba (1926–2011), a spiritual teacher who lived in India: love, truth, right conduct, peace, and nonviolence (Unhabitat 2002, p. 3). Since all these values are shared by all human beings regardless of racial, cultural, and class differences, they are called here "Core Human Values."[1]

If our thoughts are guided by the oneness of the Universal Consciousness hardwired in our brain, we would know that the separateness we perceive in our ordinary life in the material world is an illusion. Hence, we would not highly value "I" and what

[1](Sathya Sai Baba was a highly revered spiritual leader and world teacher, whose life and message are inspiring millions of people throughout the world to turn God-ward and to lead more purposeful and moral lives. His timeless and universal teachings, along with the manner in which he leads his own life, are attracting seekers of Truth from all the religions of the world. Yet, he is not seeking to start a new religion. Nor does he wish to direct followers to any particular religion. Rather, he urges us to continue to follow the religion of our choice and/or upbringing. For further information, please visit https://www.sathyasai.org/intro/message.htm)

"I" needs for its survival and to experience pleasure. Instead, we would highly value what is intrinsic to us: the core human values. These core human values are interdependent. Truth constitutes (1) the absolute truth, the things that never change such as the Universal Consciousness (God) and its oneness, and (2) the relative truth, things we perceive to be true and therefore value. Love arises from the understanding that we all are interdependent, interconnected, and integral parts of the same whole. Therefore, where there is absolute truth, there is love. Where there is truth and love, there are right conduct and nonviolence, because when we act from the understanding that we are all one, our conduct brings well-being to all and therefore it is right and it will never be violent. On the other hand, if one value disappears, then all the values will disappear. For instance, where there is no love, one will become selfish and act only for one's own benefit. Such a person is likely to engage in wrong conduct and violence. The VBWE program further subdivides these five core human values into their practical applications as follows (Unhabitat 2002, p. 4):

Love: Caring, compassion, dedication, devotion, friendship, forgiveness, generosity, helping, consideration, kindness, patience, sharing, sincerity, sympathy, and tolerance

Truth: Curiosity, discrimination, equity, honesty, integrity, intuition, memory, quest for knowledge, reason, self-analysis, self-awareness, self-knowledge, spirit of inquiry, synthesis, truthfulness, understanding

Right Conduct: Cleanliness, courage, dependability, duty, endurance, ethics, gratitude, goal setting, good behavior, good manners, healthy living, helpfulness, initiative, leadership, obedience, patience, perseverance, proper use of time, protection, resourcefulness, respect, responsibility, sacrifice, self-confidence, self-sufficient, serving, simplicity, teamwork, will

Peace: Attention, calm, concentration, contentment, dignity, discipline, focus, happiness, humility, individualism, inner silence, optimism, satisfaction, self-acceptance, self-control, self-discipline, self-respect

Nonviolence: Appreciation, appreciation of other cultures and religions, brotherhood, citizenship, concern for all life, co-operation, equality, fellow feeling, loyalty, minimum natural awareness, respect for property, service, social justice, unity, universal love, unwillingness to hurt

However, in our ordinary life, our thoughts are not guided by the oneness hardwired in our brain. When we experience ourselves and the rest of the world through our senses, we perceive ourselves as entities separated from each other and from the environment. The brain's neuroplasticity – the ability of the brain to change itself in response to our interactions with the external world – allows the perceived separateness to be "soft-wired" (Merzenich 2013) in our brain. Therefore, our thoughts are guided, not by the hardwired oneness but by the soft-wired separateness, and therefore we perceive ourselves as entities separated from others and from the environment. This gives rise to "I"-centeredness or self-centeredness. For each of us, "I" is the most important thing and we value "I" most, and then the things in the material world that "I" needs for its survival and pleasure, e.g., money, material possessions, power, social status, recognition, fame, and popularity, etc. These materialistic values or the "self-enhancement values" as Schwartz (2006) called them, become internalized. How do materialistic values become internalized (or soft-wired) in our brain? Kasser (2009) suggest two pathways:

1. Social Modeling: Social modeling involves the extent to which individuals are exposed to people or messages in their society suggesting that money, power, achievement, image, and status are important aims to strive in life. The empirical evidence provided by the authors show how the people's level of materialism is determined by that of their parents, friends, and peers; how advertising internalizes materialistic values in people; and how the exposure to advertising in schools promotes strong materialistic concerns.
2. Insecurity: Based on empirical evidence they have documented, the authors suggest that people tend to orient towards materialistic aims when they experience threats to their survival, their safety, and their security. For instance, children are more likely to be materialistic when they grow up in insecure environments, e. g., in broken families, controlling parents, and in poverty. Some experiments have revealed that economic hardships and poor interpersonal relationships lead people to care more about materialistic aims. In situations that promote insecurity, people tend to become self-interested and more concerned about acquisition of material possessions.

The neuroplasticity of the brain allows us to change these soft-wired values if we want to do so. Research findings in neuroscience reveal we have the capacity to re-wire our brain and change it permanently and transform ourselves (Begley 2007; Arden 2010; Newberg and Waldman 2015). This is particularly true for children. Research and experiments undertaken with children have shown that at a younger age, before materialistic values are firmly wired in the brain, their behavior is mostly guided by hardwired intrinsic human values. Tomasello (2008), an American psychologist, says that children show altruism in their behavior. They do not get this from adults but it comes naturally. According to his research findings, children have an almost reflective desire to help, inform, and share and they do so without expectations or desire for reward. Warneken and Tomasello (2013) believe children are naturally altruistic. His studies reveal toddlers as young as 14 months show spontaneous helping tendencies, the precursor to altruism. This evidence suggests that if we want a value-shift in our society, a shift from extrinsic materialistic values to intrinsic human values, it is wise to do it with our younger members, the future decision-makers and leaders and the best ambassadors to bring about this value transformation. The EHV is a program designed to achieve this goal.

Values, Relationships, and Sustainability

Values determine one's perceptions, attitudes and behaviors. Extrinsic materialistic values lead to the perception that we are independent entities separated from others and the natural environment, and the belief that it is okay to exploit others and the earth's resources for our pleasure and well-being. Such a perception and belief leads to unloving, self-centered, and exploitative relationships in which we take more from and give little to the other party, and to anti-sustainable attitudes and behaviors. On the other hand, intrinsic human values lead to the perception that we are dependent on and connected to others and the natural environment, and to the belief that our well-being is dependent on the well-being of others and the natural environment. Such a perception and belief lead to loving, selfless, and nonexploitative relationships, and pro-sustainable attitudes and behaviors. This view is supported by empirical evidence provided by psychologists. A cross-cultural study undertaken by Schwartz (1992, 2006) has revealed that materialistic values are associated with caring less about values such as "protecting the environment," "attaining unity with nature," and having "a world of beauty." In samples of American adults, both Richins and Dawson (1992) and Brown and Kasser (2005) have found that materialistic values are negatively associated with how much people engage in ecologically friendly behaviors such as riding one's bike, reusing paper, buying second-hand, recycling, etc.

Similarly, Gatersleben et al. (2008); Kasser (2005), based on their sample studies in the USA and UK, have reported that adolescents with a stronger materialistic orientation are less likely to turn off lights in unused rooms and recycle and reuse papers. Some have provided evidence that shows the correlation between values and exploitation of natural resources. Brown and Kasser (2005) have examined the ecological footprints of 400 North American adults and found that those who cared more about materialistic values used significantly more of the Earth's resources in order to support their lifestyle choices around transportation, housing, and food. Furthermore, Kasser (2011) obtained measures of the ecological footprints and carbon emissions of 20 wealthy, capitalistic nations and correlated these with measures of how much the citizens in those nations cared about materialistic values. As predicted, the more materialistic the citizens of a nation, the more CO2 that nation emitted and the higher that nation's ecological footprint. Research undertaken by Sheldon and McGregor (2000), using a resource dilemma game, has revealed that materialistic individuals are more motivated by greed for profit and they are more likely to make ecologically destructive decisions. This evidence suggests that to the extent individuals are materialistic, they are more likely to have negative attitudes about the natural environment, are less likely to engage in environment-friendly behaviors, are more likely to make behavioral choices that contribute to environmental degradation, and are more likely to have self-centered, unloving, and exploitative relationships with the environment.

There is also evidence to substantiate the perceived correlation between values and human relationships. Cohen and Cohen (1996), and Schwartz (1996) reveal that a strong materialistic value orientation tends to conflict with the desire to help the world be a better place and to take care of others. Kasser and Ryan (1993) and McHoskey (1999) show that people strongly focused on materialistic values are also lower in social interests, pro-social behavior, and social productivity and are more likely to engage in anti-social acts. Kasser et al. (2003) have provided evidence to show that the love relationships and the friendships of those with a strong materialistic value orientation are relatively short and are characterized more by emotional extremes and conflict than by trust and happiness. Drawing on evidence from research studies undertaken by psychologists, they have revealed that compared to those with a low materialistic value orientation, people who are strongly focused on materialistic values are less empathetic, more often use their friends to get ahead in life, score higher in Machiavellianism, (Using clever but often dishonest methods that deceive people so that you can win power or control (Cambridge Dictionary).) and are more likely to compete than cooperate with their friends. Furthermore, Kasser (2002), in his book *High Price of Materialism*, points out how materialistic values undermine our interpersonal relationships. This evidence suggests that to the extent that individuals value materialistic goals, their relationships with fellow human beings are more likely to be unloving, self-centered, and exploitative.

The anti-sustainable attitudes and behaviors driven by extrinsic materialistic values create obstacles to sustainability that cannot be resolved by technological, infrastructural, and institutional means used in conventional development. Such obstacles can only be resolved by the transformation of values from extrinsic to intrinsic values. The empirical evidence provided above reveals that those who hold materialistic values consume more and have larger ecological footprints. Overconsumption of the Earth's resources by the rich in consumer societies is now recognized as an obstacle to sustainability. It constrains sustainability directly by reducing the capacity of the Earth's resources to sustain human society and indirectly by worsening poverty and inequality. Mass production of the developed industrialized countries need mass consumption without which it cannot sustain itself. The consumer society and its overconsumption emerged to meet this need. Now it is often call "affluenza." Graaf et al. (2001) in their book *Affluenza: All Consuming Epidemic* describe it as: "a painful, contagious, socially transmitted condition of overload, debt, anxiety, and waste resulting from dogged pursuit of more" (p. 2). As Hamilton and Denniss (2005) pointed out in their book *Affluenza: When Too Much Is Never Enough,* overconsumption constitutes three aspects: (1) people consume more than their income allows so that they become indebted; (2) they have to overwork because they feel they have to work longer and harder to meet ever-rising aspirations, imposing severe cost and strain on health and relationships; and (3) heavy consumption generates a lot of waste, causing heavy pressure on the environment. This lifestyle of the consumer society is simply unsustainable. For instance, Americans constitute 5% of the world's population but consume 24% of the world's energy. It has been estimated that if China was to increase its car ownership to the US level, it would need to pave over an area for parking lots and roads equivalent to more than half of its current rice-producing land. On average, one American consumes as much energy as 13 Chinese, 31 Indians, 128 Bangladeshis, 307 Tanzanians, and 370 Ethiopians. It is said that if rest of the world would consume at the same rate as the USA, four complete planets the size of the Earth would be required. Clearly, sustainability cannot be achieved while maintaining this high level of consumption. It is widely accepted that a shift from fossil fuel to renewable energy is necessary to achieve sustainability goals, especially to reverse climate change.

But, Trainer (2007), with the support of substantial empirical evidence, revealed that renewable energy cannot sustain the consumer society. Hamilton and Denniss (2005) identified overconsumption by the rich as a prime cause of poverty and inequality which constrains sustainability. It is generally accepted that the world's resources are sufficient to meet our needs but not to meet our greed. If the rich and greedy consume more, the others will not be able to meet their needs. Then, poverty occurs, threatening sustainability.

There is a deep rooted belief in materialistic societies that having more money and the things that money buys make us happier. Therefore, one's income and material possessions are highly valued, and economic growth as measured by gross domestic product (GDP) is considered as the ultimate goal of development. But, the GDP does not take into account the environmental and social cost of economic growth. The high priority given to economic growth undermines sustainability in numerous ways. Economic growth generates more employment and higher incomes which in turn increase demand for goods and services. The increasing demand further stimulates economic growth. Hence, our drive for economic growth is endless. It continuously puts pressure on the Earth's limited resources and generates wastes and pollutants, reducing nature's capacity to regenerate resources and threatening sustainability. Governments of almost all countries want to further economic growth and corporations want to increase profit. In pursuing economic growth, governments set short-term targets – to accomplish goals before the next election. Corporations seek short-term profits. Both fail to pay attention to long-term environmental consequences of their actions. Corporate greed is an outcome of the values held by the executives. According to Hersh Shefrin, a professor of finance in the USA, executives view their personal millions and corporate profits as a way to measure their success relative to that of their peers, rather than as something to be spent (Fox 2010). Greed occurs when the natural human impulse to collect and consume useful resources like food, material wealth, or fame overwhelms the constraints that maintain the social ties in a group. When a person acquires resources, neurochemicals are released in the brain that causes pleasure. Greed is simply the addiction to that pleasure. "When we gather resources, we feel good. And because we feel good, we want more" says Andrew Lo, an MIT professor (Fox 2010). The environmental consequences of corporate greed for short-term profit are well documented.

The breakdown of the family which is becoming increasingly common in modern societies poses a threat to sustainability. Research evidence shows that while the couple's high focus on materialism leads to marital dissatisfaction and breakdown of families, the breakdown of families make children more materialistic. A study undertaken by Dean et al. (2007) on materialism, perceived financial problems, and marital satisfaction has revealed that husbands' and wives' materialism is positively related with increased perception of financial problems which is in turn negatively associated with marital satisfaction. They have also found that materialism had negative association with marital quality, even when both spouses were equal in their materialistic values. According to their findings, when both spouses hold equally low materialistic values, their marital quality is likely to be better off than the couples in which one or both spouses hold high materialistic values. Marital dissatisfaction of the couples with high focus on materialism is most likely to lead to divorce. In a survey undertaken by Aric Rindfleish and his colleagues, 165 participants from nondivorced families were compared with 96 from divorced families and found that the latter were more likely to be materialistic (cited in Kasser et al. 2003). As pointed out by psychologists, children of divorced parents are likely to be more materialistic. Kasser et al. (2003) states that

> When families experience divorce, parents' ability to engage in optimal parental practices often diminishes, leading children to experience lessened warmth and nurturance. As a result, many children turn to materialistic pursuits as a way of trying to fill this gap and feel safer, secure, and connected to others. This strategy does not seem to be very effective. (p. 32)

Can sustainability be achieved only by external means without shifting from materialistic to core human values? Evidence shows that the obstacles to sustainability discussed above cannot be resolved by technological, infrastructural, and institutional means. As already seen, they are products of materialistic values and they can be resolved only by changing values. Furthermore, the values that people hold and base their attitudes and behaviors on are probably the most crucial factor for deciding whether they do or do not support sustainability. This is because, arguably, if people do not support sustainability, all technological, institutional, and infrastructural measures undertaken for sustainability are bound to fail. "Most advocates of sustainable development recognize the need for changes in human values, attitudes, and behaviours in order to achieve a sustainability transition that will meet human needs and reduce hunger and poverty while maintaining the life support systems of the planet" (Leiserowitz et al. 2006, pp. 413–444). Ikerd (2015) believes that we have created an unsustainable economy and society because we have accepted as facts only those things that were based on materialistic value system that are inherently in conflict with the values of sustainability. Dahl (2001), former director of the United Nations Environmental Programme, recognizes values as the missing ingredient in most approaches to sustainable development. He states that "Grand declarations and detailed action plans, even when approved by all the governments, do not go far if people are not motivated to implement them in their own lives" (p. 5). Kasser (2009) suggested that "if we are to promote ecological sustainability, we must not focus solely on technological shifts and 'buying green,' but instead must consider the kinds of values that people hold, for these values can either lead individuals and nations to act in ecologically-destructive or ecologically-sustainable ways" (p. 199).

Education in Human Values and Sustainability

This section introduces the education in human values (EHV) program and then, drawing on empirical evidence from the Value-Based Water Education (VBWE) program implemented by the UN-Habitat in six African countries, highlights the potential capacity of the EHV to plant the seed of sustainability in young minds. Today's formal education is oriented towards imparting the knowledge and skills necessary for wealth generation, and so many educational institutions produce individuals who are rich in worldly knowledge and skills but with poor human values. Yet, education can be an effective tool for guiding the younger generation towards human values and for passing the understanding of human values to parents as well. Young minds are not yet fully conditioned by the materialistic values dominant in modern society. If guided at a young age, they will be able to bring out the human values from within. Some philosophers and spiritual teachers have already taken initiatives to develop human value-based educational institutions. Sathya Sai Baba has founded an education in human values program which later came to be known as Sathya Sai Education in Human Values (SSEHV). This is a multicultural, multi-faith self-development program designed for children and young people all over the world. The aim of the SSEHV program is to help children to realize their innate goodness, to bring out the inherent values of love, truth, right conduct, peace, and nonviolence and help them to sustain it by regular practice. In this program, it is emphasized that human values cannot be taught but have to be brought out from within. Teachers and parents both play a critical role in bringing out these values. While parents are the primary character trainers, teachers are responsible for incorporating human values teaching into lessons and leading by example.

There are over 70 Sathya Sai schools in India with an enrolment exceeding 16,000 students and 700 teachers. Outside India, there are 41 schools in 26 countries. Apart from the schools, the SSEHV program is introduced as appropriate in the public sector schools in 69 countries. The schools are philanthropically funded private schools. Literature published by the schools' and parents' testimonials document their children putting human values into practice in daily life. "Parents comment frequently that their children are calmer, more considerate and compassionate. There are even reports of children setting an example for parents to discontinue negative habits" (Rousseau 2013). The Sathya Sai School of Ndola, Zambia, has been awarded the International Gold Star Award for Quality in terms of leadership, innovation, training, and excellence in education. This school was started in 1992. Right from the start, they took only students who did not fit into standard schools – they had been failed and been thrown out. Many were trouble makers. Nevertheless, in the second year, the school had 93% passing rate on the national exams, and for the next 7 years, they had a 100% passing rate (Satyasaiuk 2017). This school came to be known as "miracle school." The school has become very famous and is often mentioned in the press (cited in Satyasaiuk 2017):

> Today to say the Sathya Sai Baba private school is the best school would be an understatement. The school which is situated in Ndola, Pamodzi Township has overshadowed all schools in Zambia again. The school has been hailed by many as a success story. *Times of Zambia* 12-02-96
>
> Sathya Sai has all it takes to be called a miracle school. Former truants, dunces and those considered untouchables are shaped into disciplined and hardworking students. *Times of Zambia* 26-04-97

Similarly, it has been reported that the Sathya Sai School in Toronto, Canada, has also produced excellent performance in results (Rousseau 2013). Drawing on evidence from Sathya Sai schools in Canada, Australia, and Thailand, the author reveals the capacity of the EHV values program to produce a generation of youths with goodness, love, and compassion.

It is with the collaboration of The Institute of Sathya Sai Education in Zambia that the UN-Habitat initiated the VBWE program as a part of its Water for African Cities Programme in 2001. There was a growing recognition that "improvements in water management cannot be accomplished by technical and regulatory measures alone; these must be complemented with changes in behaviour and in attitudes to the use of water in society" (UNESCO 2012, p. 436). A valued-based water education program was necessary to generate changes in attitudes and behaviors. The human values approach for water education was recommended by an expert group meeting which comprised of international and regional experts on urban water management and education. They have observed that the national goal – provision of adequate cost-effective and good water supply to all – cannot be achieved only by external means, and introduction and implementation of the human values approach to water education through formal, nonformal, and informal channels of learning is a promising strategy to bring a positive and lasting change in attitudes and behavior towards water at all levels of society.

The VBWE program was launched in six African cities: Abidjan in Cote d'Ivoire, Accra in Ghana, Addis Ababa in Ethiopia, Dakar in Senegal, Lusaka in Zambia, and Nairobi in Kenya over a period of 18 months. The broad aim of the project was to facilitate changes in behavior and personal attitudes among water consumers and to promote better understanding of the environment in a water context. The project is:

> a strategic entry point to bringing about positive attitudinal changes among both water consumers and providers, and in the longer term, can help develop a new water-use ethic in society. Children and youth are the best ambassadors to bring these attitudinal changes. Water education in schools and communities can therefore play an important role in bringing about a new water-use ethic in cities. (Unhabitat 2002, p. 2)

This section, drawing evidence from the documents published by the UN-Habitat on VBWE project, demonstrates the significance of human values in achieving sustainability in urban water management. It shows external measures are insufficient to achieve sustainability without bringing out the intrinsic core human values from within, how EHV brings out these values, and the potential capacity of the EHV in planting the seed of sustainability in young minds.

All technological, infrastructural, and institutional measures necessary for collection, purification, storage, and distribution of water for urban dwellers in all six cities were well in place, but sustainable use and management of water was impaired by a set of issues that cannot be addressed solely by technological, infrastructural, and institutional means. Some of these issues identified were water wastage, pollution of water, illegal connections, vandalism, nonpayment and late payment of water bills, tampering of water meters, corruption, and poor sanitation and poor hygiene. The root cause of these issues lies within people, specifically in their values. Kanu (2002) Director, African Institute of Sathya Sai Education, who acted in close collaboration with the project management, stated that:

> It is people who use water. People waste and pollute water; industry owners contaminate water; the wealthy monopolise available water at the expense of the poor and the less powerful; ignorance and misconceptions of the value of water – on the part of the poor – lead to wastage and result in unnecessary hardships. At the same time, institutional policing of water usage, to promote its efficient exploitation, has, on the whole, been costly and ineffective. (p. 19)

Wastage of water has been identified as one of the very serious issues. Andre, D., Programme Manager, Water for African Cities Programme, stated that "while the urban poor struggle for water, more than half the water abstracted and treated at a high cost is wasted due to leakage and profligate use. There is a growing understanding that regulation of much of this wastage cannot be accomplished by technical and regulatory measures alone" (Unhabitat 2002, p. 2). Gravity of the wastage of water has been identified in the all six cities. Barraque (2011) has highlighted the wastage of water in South African cities. He has identified wasteful domestic water use through community standpipes being left running and household water leaks being left unattended. While Hailu (2002) has attributed misuse and wastage of water in Addis Ababa (Ethiopia) to profligate use, unnecessary leakages, and evaporation, Oto and Alaye (2002) have identified some irresponsible behavior of the water consumers in Accra (Ghana). They have stated that:

> In our cities and towns, the main source of water is pipe-borne, which is supplied at great cost. But unfortunately some people misuse this facility. This has resulted in the wastage of millions of gallons of water. For example, taps are left running when the water is most needed; burst pipelines are ignored for hours before action is taken to remedy the situation; water hoses are used to wash cars and in the process they are left on the ground to go to waste; gardens are flooded with treated water because the hoses are left to run freely on the ground. (p. 24)

Corruption is another obstacle to sustainability that cannot be resolved by external means without changing values. Plummer and Cross (2006) identified four forms of corruption in water and sanitation sector in African counties: (1) abuse of resources – theft and embezzlement from budgets and revenues, (2) corruption in procurement which results in overpayment and failure to enforce quality standards, (3) administrative corruption in payment systems, and (4) corruption at the point of service delivery. The ultimate victim of all these forms of corruption is the poor. As pointed out by IRIN (2013) wealthy or politically connected people use their position to unduly influence the location of a water source at the cost of the poor. Most low-income squatter settlements in cities are not connected to water distribution networks, and poor people living in such areas have to seek alternative means for water. Water kiosks run by private providers is the common alternative for the urban poor. The providers purchase water in bulk from the authorities and establish supply points where householders queue for water. Referring to the sales of water by water kiosks in squatter settlements in Nairobi, Plummer and Cross (2006) revealed that the price for water is fixed and competitive within the squatter settlement, although it is five times the price of utility water, and varies according to the season and availability. The unduly high price of water provided by water kiosks is partly attributed to the corruption of the Nairobi Water Utility officials in meter reading, billing, and collection.

> There appears little the providers can do to bring the bills back in line. And so they tip the officials to revise the bills. The irregularity of the bulk water supply to the provider kiosks provides the utility with leverage over the providers and incentive for them to grin and bear the extortive demands. The losers are the poor who pay a higher price for their water each time this 'surcharge' is levied. (p. 18)

While legal kiosk operators bribe officials to obtain a more reliable and longer daily bulk supply, illegal operators bribe officials to connect into the network or deliver bulk water that they then distribute in a competitive market. Some common corrupt practices in the operation and maintenance of water services are officials providing illegal connections, using utility water for resale in utility vehicles, or offering preferential treatment for repairs or new services.

These kinds of obstacles to sustainable use and management of water in cities can only be resolved by changing values. This is the aim of the water education program based on values or the VBWE. It is not simply about teaching chemistry, physics, and economics of water; types of water; and their sources, uses, treatment, and management. It is also about other intangible aspects of water such as people's perceptions of water and their attitudes, cultural beliefs and practices, their sense of duty and responsibility to each other and to the use of water. In short, it is about human values. As Johannessen (2001) described it, water education is "about understanding the interactions of people and nature and the mutual love and respect that has to be a guiding principle in such a system. That is also how we will transmit the importance of water use, striking a chord in the beings and hearts of people, not in the minds" (p. 35). Such an education would hopefully generate incentives to preserve water and share it with fellow human beings with love and care.

> Value-based water education is an innovative approach that not only seeks to impart information on water, sanitation and hygiene but also inspire and motivate learners to change their behaviour and adopt attitudes that promote wise and sustainable use of water. The value-based approach to water education seeks to bring out, emphasise and stress desirable human qualities [core human values], which therefore help us in making informed choices in water resource management.
>
> Andre D. (2002), Programme Manager, Water for African Cities programme, UN-Habitat (p. 3).

The human values approach to water education uses two main methods: direct method and integration method.

1. The Direct Method: This method is to help children learn the values in an illustrative and enjoyable way. This method gives them an opportunity to explore and discover for themselves what right and wrong mean, develop greater empathy and therefore more compassion for others, take greater responsibility for their actions, and discover how to be happy, confident, and responsible members of society. This method includes five components:

- Silent sitting/meditation/guided visualization
- Quotations, proverbs, poetry related to water
- Stories about water
- Songs (local/international) about water
- Group activities relating to water

2. Integration Method: This method is more suitable in dealing with academic subjects in the school curriculum. Every academic subject has inherent values. Lessons are planned for each subject in such a way that their inherent values are drawn out during the course of teaching. The teacher looks for teachable moments to bring out values in any particular subject being taught, including extracurricular activities, sports, and field trips. For instance, when mathematics is taught in water education, values such as caring, sharing, compassion, love, and consideration can be emphasized by wording a mathematical problem as follows: "Mr. X draws 20 buckets of water from the well daily. If 7 buckets are given to the sick old lady next door, how many buckets will remain?" Similarly, environmental science can be taught in such a way that it generates in children love, respect, and reverence to natural resources, especially water.

Both methods have their focus on the key issues which impede sustainability of urban water management such as water wastage, pollution of water, illegal connections, vandalism, nonpayment and late payment of water bills, tampering of water meters, corruption, and poor sanitation and poor hygiene. Lesson plans are aimed at applying human values to address these issues and generating pro-sustainable attitudes and behaviors in children. The examples illustrated in Table 1 demonstrate the potential capacity of the VBWE in resolving some of the issues. The report of the evaluation undertaken by the Swedish International Development Cooperation Agency (Sida) which funded the program has found the VBWE quite successful and endorsed a continuance into Phase 2 (Norman Clark 2004). Its success has generated interest in other countries and led the UN-Habitat to extend the program to Asia and South America too.

Conclusion

The Universal Consciousness (God), while remaining as an inseparable and indivisible single whole at the quantum level, manifests itself at the material level as a multitude of separate entities: inanimate objects, plants, animals, and human beings. Though the material world appears to us as an assemblage of many separate entities, deep at the quantum level, they all are integral parts of a single entity. The separate entities we perceive at the material level are subject to change over time and space and therefore transient, whereas the oneness at the quantum level is changeless and therefore eternal. Hence, what we experience in the material world is true only relatively, whereas the oneness at the quantum level is true absolutely. Recent discoveries in neuroscience suggest that the absolute truth, the oneness at the quantum level, is hardwired in the human brain as the Golden Rule, altruism and

Table 1 Value-based management approaches, underlying human values, and teaching techniques

Issues	Value-based dilemma	Value-based solution	Underlying values	Teaching techniques
How could water and sanitation be made accessible and affordable to the poor in cities?	Am I willing to share the cost of providing water to the poor in the slums? This may mean that I will have to pay a higher price for water than I pay today	Yes, I care for my poor neighbor. I am ready to pay a higher price for water when I am convinced this will help extend water supply to poor neighborhoods. I will afford it by cutting down my entertainment expenses	Love: caring for and sharing with others Right conduct: self-sacrifice, respect for others, service to others	Story telling Group activities Prayer/ quotation
How to deal with corruption in daily life which ultimately affects sustainability of services in cities?	Should I pay the high water bill every month or make a deal with the meter-reader, who offers to under-read it or tamper with it so that I can pay a flat rate which will be less costly to me?	Yes, I will pay the actual cost of water I consume. If I follow unscrupulous means, this will set a bad example for my children, whom I want to see growing up as responsible citizens	Truth: truthfulness Right conduct: honesty Peace: integrity and self-respect	Group discussion Role playing Group singing Prayer/ quotation
How to deal with profligate wastage of water in households?	Should I stop watering garden and washing my cars when water is scarce? I can afford the water bill and I want my garden to be green and my car to shine even if it may mean less water available to others	Yes, I should take every opportunity to conserve water, even if it means a little inconvenience to me and even if I can afford a higher water bill	Right conduct: proper utilization of resources Peace: self-discipline Nonviolence: consideration of others	Quotation Story telling Group discussion Role playing
How to promote the concept of water as a social and economic good?	We are told that water is a gift of god. Then why are we asked to pay for water? Water in the river and the wells, after all, belong to everybody and should be freely available to all	Yes, I have an obligation to pay for water I consume. Water is a limited resource, to be shared by many users. Each must pay according to his need and ability, to cover the cost	Right conduct: respect for others' needs Nonviolence: awareness of responsibility towards common goods; readiness to cooperate; fellow feeling	Group discussion Group singing Prayer/ quotation

Source: Victor Kanu (2002)

unlimited love, prompting us to value love, truth, right conduct, peace, and nonviolence. However, these core human values cannot ordinarily guide our thoughts and actions because:

- We experience ourselves and the external world through our senses, thus perceive ourselves as entities separated from others and the natural environment.
- The neuroplasticity of the brain allows the perceived separateness to be soft-wired in our brain, preventing our thoughts and actions to be guided by the hardwired core human values.
- The soft-wired separateness leads us to value "I" and the things in the material world that "I" needs for its survival and pleasure such as money, material possessions, power, social status, recognition, and fame.
- These materialistic values guide our thoughts and actions.

This study highlights the significance of integrating the core human values into formal education. Empirical evidence provided by psychologists demonstrates that materialistic values make our relationships with others and with the environment self-centered, unloving, and exploitative, and our behavior anti-sustainable. On the other hand, the core human values make our relationships selfless, loving, and nonexploitative, and our behavior pro-sustainable. Therefore, if sustainability is to be achieved, a fundamental shift of values is necessary: a shift from materialistic values to the core human values. The conventional strategies of achieving sustainability based on technological, infrastructural, and institutional changes should therefore be supplemented with strategies aimed at value changes. The EHV is a strategy designed to bring out the core human values in young people. The value-based water education program implemented by UN-Habitat demonstrates the capacity of the EHV to plant the seed of sustainability in young minds. Without due attention to the core human values, the current education system will continue to produce individuals who are rich in knowledge and skills in their specific fields, but poor in core human values.

Cross-References

▶ Empathy Driving Engaged Sustainability in Enterprises
▶ Selfishness, Greed, and Apathy
▶ Social Entrepreneurship
▶ Sustainable Higher Education Teaching Approaches
▶ The Spirit of Sustainability
▶ The Theology of Sustainability Practice
▶ To Eat or Not To Eat Meat

Acknowledgments The author wishes to thank Richard Wallis and Kim Penny for providing editorial assistance.

References

Andre, D. (2002). Value-based water education: Project overview. In *Human values in water education: Creating a new water-use ethics in African cities* (pp. 1–10). Nairobi: UN-Habitat

Arden, J. B. (2010). *Rewire your brain: Think your way to a better life*. Hoboken: Wiley.

Barraque, B. (2011). *Urban water conflicts*. Boca Raton: UNESCO Publishing/CRC Press.

Begley, S. (2007). *Train your mind change your brain: How new science reveals our extraordinary potential to transform ourselves*. New York: Ballantine Books.

Bohn, D. (1980). *Wholeness and the implicate order*. London: Routledge & Kegan Paul.

Bohn, D., & Hiley, B. J. (1993). *The undivided universe*. New York: Routledge.

Brown, K. W., & Kasser, T. (2005). Are psychological and ecological well-being compatible? The role of values, mindfulness and lifestyle. *Social Indicators Research, 74*, 349–368.

Brudtland Report. (1987). *Our common future: Report of the world commission on environment and development*. www.un-documents.net/our-common-future.pdf. Accessed 20 July 2017.

Clark, N. (2004). *Water education in African cities (Sida Evaluation 4/21)*: United Nations Human Settlement Programme. http://www.sida.se/contentassets/94c8423f1b3245208dd27a6aa6dad10c/water-education-in-african-cities—united-nations-human-settlements-program_2129.pdf. Accessed 22 Aug 2017.

Cohen, P., & Cohen, J. (1996). *Life values and adolescents mental health*. Mahwah: Lawrence Erlbaum Associates, Publishers.

Cozolino, L. (2006). *The neuroscience of human relationships: Attachment and the developing social brain*. New York: W.W. Norton & Company.

Dahl, A. (2001). *Values as the foundation for sustainable behaviour*. Paper presented at the 5th annual conference of the International Environment Forum, Hluboka nad Vltavou, Czech Republic, 19–21 Oct 2001.

Dean, L. R., Carroll, J. S., & Chongming, Y. (2007). Materialism, perceived financial problems, and marital satisfaction. *Family & Consumer Sciences, 35*(3), 260–281.

Elgin, D. (2009). *Living universe*. San Francisco: Berrett-Koehler.

Fox, S. (2010). What causes corporate greed? *Live Science*. https://www.livescience.com/6394-corporate-greed.html. Accessed 20 Aug 2017.

Gatersleben, B., Meadows, J., Abrahamse, W., & Jackson, T. (2008). *Materialistic and environmental values of young people*. Unpublished manuscript. University of Surrey.

Graaf, J., Wann, D., & Naylor, T. H. (2001). *Affluenza: All consuming epidemic*. San Francisco: Berrett-Koehler Publishers.

Hailu, D. (2002). Country perspective Ethiopia. In *Human values in water education: Creating a new water-use ethics in African cities* (pp. 20–23). Nairobi, Kenya: UN-Habitat

Haisch, B. (2009). *The god theory: Universes: Zero-point fields, and what's behind it all*. San Francisco: Red Wheel/Weiser.

Hamilton, C., & Denniss, R. (2005). *Affluenza: When too much is never enough*. Crows Nest: Allen & Unwin.

Ikerd, J. (2015). *Sustainability and human values*. http://web.missouri.edu/ikerdj/papers/TSU%20–%20Human%20values%20-%20Sustainability.htm. Accessed 20 July 2017.

IRIN. (2013). In Africa, corruption dirties water. http://www.irinnews.org/analysis/2013/03/14. Accessed 22 Aug 2017.

Johannessen, A. (2002). Human values in water education: Application in water classrooms. In *Human values in water education: Creating a new water-use ethics in African cities* (pp. 34–37). Nairobi: UN-Habitat

Kanu, V. (2002). Contribution of value-based water education to national educational goals and objectives in Africa: A regional perspective. In *Human values in water education: Creating a new water-use ethic in African cities* (pp. 15–19). Nairobi, Kenya: UN-Habitat.

Kasser, T. (2002). *The high price of materialism*. Cambridge, MA: MIT Press.

Kasser, T. (2005). Frugality, generosity, and materialism in children and adolescents. In K. A. Moore & L. H. Lippman (Eds.), *What do children need to flourish?: Conceptualizing*

and measuring indicators of positive development (pp. 357–373). New York: Springer Science.
Kasser, T. (2009). Values and ecological sustainability. In S. R. Kellert & J. G. Speth (Eds.), *The coming transformation: Values to sustain human and natural communities* (pp. 180–204). New Haven: Yale School of Forestry and Environment.
Kasser, T. (2011). Cultural values and the well-being of future generations: A cross-national study. *Journal of Cross-Cultural Psychology, 42*(2), 206–215.
Kasser, T., and Ryan, R. M. (1993). A dark side of the American dream: Correlates of financial success as a central life aspiration. *Journal of Personal and Social psychology, 65*, 410–422.
Kasser, T., Ryan, R. M., Couchman, C. E., & Sheldon, K. M. (2003). Materialistic values: Their causes and consequences. In T. Kasser & A. D. Kanner (Eds.), *Psychology and consumer culture: The struggle for a good life in a materialistic world* (pp. 11–29). Washington, DC: American Psychological Association.
Kluckhohn, C. K. (1951). Values and value orientations in the theory of action. In T. Parsons & E. A. Shils (Eds.), *Toward a general theory of action*. Cambridge, MA: Harvard University Press.
Lanza, R. (2009). *Biocentrism: How life and consciousness are the keys to understanding the true nature of the universe*. Dallas: Ben Bella Books.
Lanza, R. (2017). Are we part of a single living organism? http://www.robertlanzabiocentrism.com/are-we-part-of-a-single-living-organism/#eSetzJgBzFeT8fiX.99. Accessed 25 Aug 2017.
Leiserowitz, A., Kates, R., & Parris, T. (2006). Sustainability values, attitudes and behaviours: A review of multi-national and global trends. *Annual Review of Environment and Resources, 31*, 413–444.
Lieberman, M. D. (2013). *Social: Why our brains are wired to connect*. New York: Crown.
McHoskey, J. W. (1999). Machiavellianism, intrinsic versus extrinsic goals, and social interests: A self-determination theory analysis. *Motivation and Emotion, 23*, 267–283.
Merzenich, M. (2013). *Soft-wired: How the new neuroscience of brain plasticity can change your life*. San Francisco: Parnassus Publishing.
Newberg, A., & Waldman, M. R. (2015). *How enlightenment change your brain: The new science of transformation*. New York: Hay House.
Oto, E. C., & Alaye, F. K. (2002). Country perspective: Ghana. In *Human values in water education: Creating a new water-use ethics in African cities* (pp. 24–25). Nairobi, Kenya: UN-Habitat.
Pfaff, D. (2007). *The neuroscience of fair play: Why we (usually) follow the golden rule*. New York: Dana Press.
Pfaff, D. (2015). *The altruistic brain: How we are naturally good*. Oxford: Oxford University Press.
Plummer, J., & Cross, P. (2006). Tackling corruption in the water sector and sanitation sector in Africa. In J. E. Campos & S. Pradhan (Eds.), *The many faces of corruption: Tackling vulnerabilities in sector level*. Washington, DC: The World Bank.
Post, S. (2003). *Unlimited love: Altruism, compassion and service*. Philadelphia: Templeton Press.
Richins, M. L., & Dawson, S. (1992). A consumer values orientation for materialism and its measurement: Scale development and validation. *Journal of Consumer Research, 19*, 303–316.
Rizzolatt, G., & Crighero, L. (2004). The mirror-neuron system. *Annual Review of Neuroscience, 27*, 169–192.
Rokeach, M. (1973). *The nature of human values*. New York: Free Press.
Rousseau, B. (2013). *Your conscious classroom: The power of reflection*. Bloomington: Balboa Press.
Satyasaiuk. (2017). Sathya Sai Schools Zambia. http://www.sathyasaiehv.org.uk/Zambia.htm. Accessed 20 Aug 2017.
Schwartz, S. H. (1992). Universals in the content and structure of values: Theory and empirical tests in 20 countries. In M. Zanna (Ed.), *Advances in experimental social psychology* (Vol. 25, pp. 1–65). New York: Academic Press.
Schwartz, S. H. (1996). Value priorities and behaviour: Applying of theory of integrated value systems. In C. Seligman, J. M. Olson, & M. P. Zanna (Eds.), *The psychology of values: The Ontario symposium* (Vol. 8, pp. 1–24). Hillsdale: Erlbaum.

Schwartz, S. H. (2006). Basic human values: Theory, measurement, and applications. *Revue Française de Sociologie, 47*(4), 929.

Sheldon, K. M., & McGregor, H. M. (2000). Extrinsic value orientation and the tragedy of the commons. *Journal of Personality, 68*, 383–411.

Tomasello, M. (2008). For kids altruism comes naturally, psychologist says. *Stanford News*. http://news.stanford.edu/news/2008/november5/tanner-110508.html. Accessed 20 June 2017.

Trainer, T. (2007). *Renewable energy cannot sustain a consumer society*. Dordrecht: Springer.

Ulluwishewa, R. (2014). *Spirituality and sustainable development*. Hampshire: Palgrave Macmillan.

Ulluwishewa, R. (2016). Spirituality, sustainability and happiness: A quantum-neuroscientific perspective. In S. Dhiman & J. Marques (Eds.), *Spirituality and sustainability: New horizons and exemplary approaches*. Switzerland: Springer International.

UNESCO. (2012). *Managing water under uncertainty and risk, United Nations world water development report, 4*. Paris: UNESCO.

UN-Habitat. (2002). *Human values in water education: Creating a new water-use ethic in African cities*. Nairobi, Kenya: UN-Habitat.

Warneken, F., & Tomasello, M. (2013). Altruistic helping in human infants and young chimpanzees. *Science, 311*, 1302–1303. http://www.eva.mpg.de/psycho/staff/tomas/pdf/Warn_Science.pdf. Accessed 20 June 2017.

Williams, G. C., Cox, E. M., Hedberg, V. A., & Deci, E. L. (2000). Extrinsic life goals and health risk behaviours in adolescents. *Journal of Applied Social Psychology, 30*, 1756–1771.

Utilizing Gamification to Promote Sustainable Practices

Making Sustainability Fun and Rewarding

Kristen Schiele

Contents

Abstract

Gamification is a growing phenomenon, described as one of today's top disruptive trends in technology. Utilizing traditional game design principles to affect user behavior, gamification is commonly used in many industries, such as marketing, health, and education. This chapter explores the potential of using gamification to motivate individuals to engage in sustainable practices. Prior studies have shown how utilizing gameful design can be a powerful strategy to convert serious real-world problems into engaging and meaningful user experiences. By promoting peer-to-peer education and behavior change through social

K. Schiele (✉)
California State Polytechnic University, Pomona, CA, USA
e-mail: krschiele@cpp.edu

S. Dhiman, J. Marques (eds.), *Handbook of Engaged Sustainability*,
https://doi.org/10.1007/978-3-319-71312-0_16

interactions, gamification can provide a positive solution to implement social change. Analyzing real case examples, Fit for Green, PowerAgent, Greenify, and PowerHouse, this chapter demonstrates how game design can make sustainability both fun and rewarding. Examining companies currently using gamification to promote sustainable practices, findings show that by creating positive peer pressure on sustainability issues, organizations can apply gamification principles to promote meaningful action and motivate users to adopt behavioral changes outside of the game.

Keywords

Sustainability · Gamification · Sustainable power · Green energy · Technology · Climate change · Gamification elements

Introduction

Since 2008, gamification has become a growing, innovative phenomenon in consumer engagement (Zichermann and Cunningham 2011). The Wall Street Journal described gamification as "a fast-moving hard trend of using advanced simulations and skill-based learning systems that are self-diagnostic, interactive, game-like and competitive–all focused on giving the user an immersive experience" (Burris 2014). Similar reports project that gamification will become a 5.5 billion dollar market by 2018 (Bloomberg 2014). If these predictions are correct, gamification requires further investigation on how to properly implement in various contexts, including sustainability efforts.

Gamification uses traditional elements of game design in nongame contexts (Deterding et al. 2011) and is defined by its use of game-like elements and principles to engage users in real-world activities (McGonigal 2011). Essentially, this trend takes the essence of game design and applies those principles to real-world objectives, rather than for only entertainment purposes (Palmer et al. 2012). By creating an experience similar to games, organizations can utilize gamification to affect user behavior, enhance user activity, and increase social interactions among users (Huotari and Hamari 2017; Hamari 2013).

In recent years, the popularity of gamification has skyrocketed with growing numbers of gamified applications, as well as a rapidly increasing amount of academic research and articles in trade publications (Blohm and Leimeister 2013; Marchand and Henning-Thurau 2013). A large body of literature has focused primarily on the use of gamification and game theory in business (Herbig 1991; McAfee and McMillan 1996; Reeves and Read 2009; Terlutter and Capella 2013) and computer sciences (Deterding 2012, 2015; Huotari and Hamari 2012). Although gamification has become a very popular topic, little has been studied on the use of game design to promote sustainability.

Many traditional approaches to educate the public on sustainability issues and promote environmentally conscious behaviors have not always proven effective.

The majority of these sustainability campaigns have tried to foster behavioral changes through information intensive means and have had little impact (McKenzie-Mohr 2008). Gamification offers an innovative alternative to these conventional campaigns, with an approach that blends together knowledge and tools from the field of game design with behavioral psychology. Since gamification has been successfully applied to a number of applications, this chapter will discuss why this can also be used to enable users to engage in sustainable behaviors.

The following sections first review the history of gamification and how the process works. Next is a discussion of current applications of gamification including contexts, such as marketing, health, ideation, and education. The subsequent sections will discuss how gamification can promote sustainability and give examples of best practices from companies who have been successful in using this method. Finally, the conclusion will outline the major takeaways and recommendations for organizations that wish to utilize gamification to promote sustainable practices.

History of Gamification

Although touted as the hot new business trend, gamification is not a new concept, as many believe. As part of a much broader phenomenon, play has been cited as an essential component in the formation of societies and civilization (Huizinga 1949), and historical predecessors to today's idea of gamification go back to the eighteenth century (Fuchs et al. 2014). Many of the early games for business purposes focused more on loyalty and rewards, such as the introduction of S&H Green Stamps in 1896. In the 1970s, Charles Coonradt founded a consulting firm, The Game of Work, which introduced sports game mechanics into the workplace.

The more recent idea of gamification as a term originated in the digital media industry in 2008, but gamification entered widespread adoption in 2010 (Deterding et al. 2011). The majority of studies on gamification have demonstrated how its application produces positive effects and benefits to users. Gamification has been seen to improve user engagement and enhance positive behavior patterns, such as increasing user activity, social interaction, or quality and productivity of actions (Hamari 2013; Hamari and Lehdonvirta 2010).

Practitioners forecast that gamification technologies will continue to be incorporated in more organizations. Gabe Zichermann, CEO of Gamification Co., asserted, "many enterprises have just scratched the surface of its potential. Over the next year, gamification is likely to morph from a tactical concept to a strategic imperative" (Zichermann 2013).

How Gamification Works

Like traditional games, applications that use gamification create a goal-oriented, competitive experience that also has the potential to provide real-life results (Carignan and Lawler Kennedy 2013; McGonigal 2011). A study by Conaway and Garay (2014) examined both customer and manager perspectives of gamification use, specifically in service marketing. Their results support a design theory that outlines four key characteristics that appeal to users including progress paths, feedback and rewards, social connection, and attractiveness of the site.

Gamification can be a powerful strategy to influence groups of people, by utilizing motivational affordances and behavioral outcomes to provide "gameful" experiences (Huotari and Hamari 2012). By applying traditional game mechanics to nongame activities, gamification has the power to influence user's behavior, by keeping score of points earned through various activities on the application. For example, Huotari and Hamari (2012) attribute gamification for the success of many mobile applications, such as Foursquare. Foursquare utilizes a rule-based system providing the user with feedback and interaction mechanisms, which supports the users' overall value creation.

According to Hamari et al. (2014), gamification can be described in the sequence of three main parts: (1) the implemented motivational affordances, (2) the resulting psychological outcomes, and (3) the further behavioral outcomes.

Gamification aims to improve users' motivation toward given behaviors and to improve both the quantity and quality of the output of the activities related to these behaviors (Morschheuser et al. 2017). Operant conditioning is also used in these games, using positive sound and visual reinforcements to encourage users. Implementing certain motivational affordances and holding them constant while varying the nature of the underlying service may give further insight into how the context affects the outcomes of the gamification.

The gamification process applies both psychological and sociological factors to drive intense gameplay by users (Donato and Link 2013). Psychological outcomes include aspects such as motivation, attitude, and engagement in the learning process, as well as users' enjoyment in the overall experience. Prior studies have shown many examples of motivational affordances including points, leaderboards (Halan et al. 2010), achievements/badges, levels, story/theme, clear goals, feedback rewards, progress, and challenges (Jung et al. 2010). Playing against others and elements of competition are also important to users. Other elements such as high score, personal path, and goals motivate users by setting up a road map for personal journey (Halan et al. 2010).

Sociological factors also play a part in acceptance of gamification, so it's important to recognize that there are differences across populations and cultures. For example, some studies have found that the younger population tends to respond more favorably to gaming techniques than older demographics (Terlutter and Capella 2013). Younger generations, such as Millennials and Gen Z, are often called "digital natives" since they were born after the widespread adoption of digital technology. Gabe Zichermann, the author of *Game-Based Marketing* and the CEO of Gamification.co, affirms that the millennial generation is one of the most significant factors in the increased success of gamification practices since they are more game attuned than previous generations. So if an organization wishes to reach a younger target market, incorporating digital games into their communication strategies may be one of the best ways to gain their attention.

Gamification also incorporates principles from game mechanics, behavioral economics, and design thinking to create a positive user experience. One important element to include in game design is creating progress paths that begin with an easy task and then evolve to more complex challenges over time as the user progresses (Palmer et al. 2012). With a relatively easy initial reward, a beginner is encouraged to continue playing, and since the challenges become more difficult with each level, the advanced users continue to stay engaged in the game.

One of the persuasive powers of gamification stems from the fact that games take place in a simulated environment. As explained by Bang et al. (2006), "within the micro-world of a game users can safely explore cause-and-effect relations and uncover new behaviors." Gamification can be used to direct the completion of specific tasks and gives users an opportunity to rehearse targeted behaviors in the simulation before conducting them in the real world.

"Shamification" is another emerging motivational trend that utilizes a more negative approach. The exact opposite of gamification, shamification makes people feel bad about a particular habit or behavior, so that they will be motivated to make a change in order to avoid feelings of shame. Although shamification has the potential to be an effective motivator due to one's competitive nature, positive reinforcement has traditionally been found to be a healthier and more sustainable way to motivate people.

Applications of Gamification

The gamification process merges the strategic thinking of a business manager with the tools and creativity of a game designer (Palmer et al. 2012). An increasing number of companies have focused on adding a gamified layer to their core activity. Other applications have been created to assist more traditional companies in gamifying their existing services. Gamification has been most widely used in marketing to promote customer retention, customer loyalty programs, customer engagement, positive word of mouth, and positive website usage behavior (Leclerq et al. 2017).

Gamification has also been successfully used in several human resource contexts, such as to motivate including workers' productivity, training, and development of specific job tasks or skills (Vesa et al. 2017). Primarily used with sales personnel, many organizations are using gamification techniques internally to motivate employee performance and allow employees to earn vacations and rewards. Gamification has also become part of many business packages, as companies attempt to recruit and retain talent from the millennial "gamer" generation (Zichermann and Linder 2013).

Over the last few years, marketing and consultancy agencies have actively promoted gamification as a potential revenue source for businesses. Despite its increased use by corporations to manage brand communities and personnel, gamification has become more than just a marketing buzzword. Over the past few decades, gameful play has become more widely utilized in various applications, including health and education.

In the health industry, gamification or "exergaming" has been utilized to promote healthy living and physical fitness. From Foot Craz by Atari in the 1980s to the recent popular mobile application Pokémon GO, exergaming has upended the stereotype of gaming as a sedentary activity and instead uses technology to promote an active lifestyle. For example, applications such as Fitocracy and QUENTIQ use gamification to encourage their users to exercise by awarding points for activities they perform in their workouts and gain levels based on points collected. Users can also complete quests (e.g., sets of related activities) and gain achievement badges for fitness milestones (Jeffries 2011).

Other examples of gamification include wearable technologies that motivate fitness, such as FitBit, NIKE PLUS, and Garmin watches. These wearable technologies allow users to track their activity and progress by measuring data such as the number of steps walked, heart rate, quality of sleep, and other personal metrics involved in fitness. There is also a gamified part of these technologies where users can set goals, join challenges, and connect with friends in the online community.

Gamification of education and learning has also become common, especially with games to test recall and learning. Educators that use games, such as Minecraft Education Edition, to teach part of their curriculum have reported improvement student engagement, collaboration, creative exploration, and tangible learning outcomes. Gamified applications, such as Everfi and Kahoot, have also been proven to empower unique and creative learning experiences for students, especially in elementary education. Using gamification, students are able to learn through a combination of observation, trial and error, and play-based practice. For example, applications, such as CodeCombat and Codecademy, give users game-like elements to help teach children how to create their own codes. This learn-by-doing gives students a sense of accomplishment when they can demonstrate their knowledge.

Using Gamification to Promote Sustainability

Gamification should not only be a profitable resource for designers and business people but can also be a tool to change the world (McGonigal 2011). Gamification is a concept that describes a new age where users can collectively apply their problem-solving skills not only to solve puzzles within a game but also to approach real social and political issues in the world (Fuchs et al. 2014). Game designers have an opportunity to become the new social entrepreneurs, by creating rewards for users, outside of gameplay, to benefit society as a whole. From this perspective, gamification can become an important method for enabling positive social change, such as increased sustainable efforts.

Over the past decade, many countries have begun programs to encourage the development of renewable energy sources, and many international agreements have been made to control emissions. Although these are positive steps forward in both the government and corporate sectors, unfortunately personal energy consumption behavior remains relatively unchallenged (Bang et al. 2006). One of the main challenges is how to best communicate to consumers on how to use energy efficiently in relation to everyday activities.

The objective of sustainability communications should be to encourage members of a community to engage in sustainable behaviors. In the past, most promotions to foster sustainable behavior have consisted primarily of large-scale informational campaigns that utilize education and advertising to encourage behavior change. Although education and advertising can be effective in creating awareness of a problem, and it has the potential to change consumer attitudes, studies show that behavior change rarely occurs as a result of simply providing information (McKenzie-Mohr 2008).

One of the theoretical cornerstones of gamification, *The Gameful World* (Walz and Deterding 2015), explores how social and political shifts can be made using pervasive game-like practices. Since studies have shown the effectiveness of gamification in various industries, sustainability education has the opportunity to implement similar innovative approaches such as utilizing new interactive, online platforms, and game elements. By leveraging games, these programs can create platforms to share new ideas in sustainability, unite like-minded people around a common goal, and empower people to take real-world action. But what are the best strategies to use games to promote sustainable practices?

Engagement by gamification can depend on several factors, such as the motivations of users or the nature of the gamified system. Prior studies indicate that pervasive games for behavior change and learning may also be appropriate to approach related domains such as environmental conservation and lifestyle-induced health problems. Gamification can be employed to raise awareness of energy-related issues and change the use patterns, as well as create a positive way of living where satisfaction becomes the center of a responsible practice. Game thinking in its purest form should give people direction about what they should be doing in small, incremental positive ways (Zichermann and Linder 2013). By doing so, gamification could become the name of a play practice that truly helps human beings in fulfilling their own lives and those of others.

Organizations need to reflect on how gamification might work as a method to inspire individuals in their everyday lives. Games can work as a supportive regulator of behavior since it offers positive feedback (e.g., rewards, leaderboard) rather than negative penalties (e.g., fines, prison). Many companies who have utilized gamification to foster positive social behaviors rely on one of the following: (1) individual's internal sense of right and wrong, (2) messages centered on resource usage and conservation, or (3) data derived from sensor-based systems (DiSalvo et al. 2010). To date, only a few organizations have worked on fostering sustainability through gamification.

To understand how gamification can influence users to engage in sustainable behaviors, it's important to take a look at real case examples. This chapter highlights some of the best practices from a few of these organizations: Fit for Green, PowerAgent, Greenify, and PowerHouse.

Fit for Green

Fit for Green has redefined wellness by uniting fitness and sustainability. The company created a line of exercise equipment that helps individuals track and improve their fitness levels while generating sustainable power during their workout. Unlike other leading gym equipment, Fit for Green machines reclaim power from users' workouts and channel this energy back into the power grid. The company currently provides commercial cardio equipment for university fitness centers but plans to expand to other markets over the next few years.

Their development team has dedicated many years to introducing this disruptive technology to the gym equipment market. According to Fit for Green founder, John Spirko, "for years skeptics have criticized the idea of humans generating any amount of meaningful renewable energy...Fit for Green changes the game from the individual to the 400 million plus gym members in the world, the more than 800 million active Facebook Users, and the 9,000,000 + pieces of self propelled gym machines in operation today."

When a user exercises on a piece of Fit for Green equipment, they create approximately the same amount of energy as a solar panel would in that same time period. Empowered with this knowledge, a user feels better knowing that their behavior not only has a positive benefit to their personal health but also has a positive effect on the environment. Fit for Green offers a unique experience incorporating responsive technology with collaborative play. Their mobile application also gives users convenient and comparable results while providing real-time motivation.

PowerAgent

PowerAgent is a game designed to encourage teenagers and their families to reduce energy consumption. PowerAgent is a pervasive game for Java-enabled mobile phones that was designed to influence everyday activities and the use of electricity in the home. The basic idea of the game is that users must compete in teams to collectively decrease consumption of household electricity and, in the process, learn how to conserve energy.

PowerAgent is connected to the user's actual household electricity meter reading equipment via the cell network, and this setup makes it possible to incorporate the user's real consumption data in the game system. Through cognitive and behavior learning, the properties of PowerAgent utilize persuasive technology components to change the users' behavior.

Greenify

Greenify was created at the Games Research Lab of Columbia University to foster sustainability and learning through real-world actions. This mobile, real-world action game (RWAG) utilizes the power of community-based interaction, crowdsourcing, and game mechanics to foster sustainability and learning through real-life actions. The Greenify system was designed with game elements, which allow players to address climate change in their day-to-day lives. These game elements include the ability to earn points for completing and creating real-world missions; a leaderboard that displays top scores daily, weekly, and all time; a player profile with progression mechanics; and a page that provides recognition for top-rated content and deeds.

The Institute of Sustainable Communities defined healthy climate and environment, social well-being, and economic security as the three essential elements of a sustainable community. Greenify addresses these in the following ways: (1) healthy climate and environment through user missions and game mechanics, (2) social well-being using game dynamics to strengthen social bonds, and (3) economic security with virtual currency and local rewards (Lee et al. 2013).

The PowerHouse

The PowerHouse is a computer game designed to motivate teenagers to increase their interest in energy-related issues and promote efficient energy use in their homes (Bang et al. 2006). In the PowerHouse, users manage a simulated domestic environment, similar to the one featured in the popular game the Sims. The activities in PowerHouse require the use of electrical energy (i.e., washing clothes, cooking, watching television), and the objective of the game is to direct characters to engage in energy-efficient behaviors. When users engage in sustainable actions, they earn virtual money that can be used to buy different game artifacts.

The user must work to balance the available resources and aim toward a more sustainable lifestyle. On the screen, the meters in the lower pane display a specific character's mental and physical state. In the upper right corner are the money and power meters that show the accumulated points and how much energy is being consumed.

Best Practices of Sustainable Applications

This section further examines real case examples utilizing gamification and discusses best practices employed by these organizations.

Make Sustainability a Fun and Rewarding Experience

Classic gamification design uses motivational affordances (such as points, badges, rewards) to encourage users to monitor their performance and compares it with performance of other users. Thus, using these game principles provides a platform for motivating sustainable actions in everyday life. When combined with social engagement, gamification can be a great opportunity to encourage desired behaviors and habits through positive motivational psychology.

In addition to motivating sustainable action, principles of good game design and contemporary learning theories should be used to guide users in the mastery of complex information (Gee 2005). Action-based learning in the area of climate change education and instructional design illustrates that effective games increase knowledge while accelerating the learning process through the completion of authentic tasks (Gee 2011). Feedback during tasks is very important, because it gives users a signal of success with virtual rewards (Palmer et al. 2012). Gamification typically rewards participants immediately after task completion, but some also use delayed gratification. The delivery of rewards depends on the path the user takes, since some users desire to have a level of power, leadership, or responsibility as they progress.

Fit for Green created a feedback and rewards program they call "The Triple Impact Workout." This program creates cardio machine workouts that are both fun and rewarding, especially for people that care about the environment.

This triple impact workout consists of:

1. Impact on climate change – creating green energy as people exercise and feeding that energy back into the grid
2. Impact on natural resource preservation – generating funds for charities that protect them
3. Impact on cardiovascular fitness – as any cardiovascular machine would

Fit for Green also utilizes a point system in the form of pulsed light generation. These lights are used to symbolize the flow of electricity from the person back into the power grid, as they pulse from the machines screen and transition on to an LED lighting strip. The lights indicate both current workout intensity and cumulative workout accomplishment. In other words the harder a user exercises, the higher frequency of the creation of lights. The lights can also be accumulated, so the end of a workout with 100 lights is more productive than a workout with 50 lights.

Greenify was designed to motivate adult learners to take informed actions regarding climate change. Greenify uses a system that encourages users to share knowledge about climate change and practical ways one could lessen their personal contribution to the problem. A team of educational technology experts at Columbia designed this effective, educational strategy to provide information that is relevant to the users' values and create a persuasive message that appeals to the target audience.

The user interactions provided by online game platforms, such as Fit for Green and Greenify, create a sense of community when users come together for a common purpose. Combined with the ability of social media to connect individuals with similar values, online environments show promise for creating social groups with a shared interest in climate change. With Greenify, several design features were developed in order to create a culture and community that value climate change discussions, sharing knowledge, and taking positive actions.

The Greenify system allows users to complete real-world missions to earn "tree points," the system's virtual currency tied to local retail incentives. Virtual currencies like this tend to elicit motivation, competition, and playfulness from the users (Lee et al. 2013). Missions can fall into one of four categories including personal (e.g., choosing green product choices), energy (e.g., transportation choices), resources (e.g., usage of water and electricity), and communication (e.g., debating issues and sharing knowledge with others). Users complete missions by presenting their completed deeds to other users in the form of written descriptions and photographs on the platform. The Greenify system was also designed to empower players to make a difference: Players are able to create actions for others to complete, and they receive points when others complete these actions. This also gives users a sense of ownership, since they listed as featured author of actions and articles.

The founders of Greenify surveyed users to learn about their overall experience with the program. According to their research, the program was viewed as a fun experience for nearly all participants, and the most commonly used words used by participants to describe their experience were informative, interactive, fun, practical, social, and engaging. Greenify participants asserted that the crowdsourced, social interactive aspects of the program were motivating, and participation affected their behaviors beyond the scope of missions (Lee et al. 2013).

Operant conditioning is an instrumental component to most effective games. The PowerHouse applied different kinds of conditioning, such as sound and visual effects, to motivate users to stay at the computer and continue to play the game. The display meters indicate how much time that is left of the game, as well as the users' accumulated points (e.g., virtual money). The display meter also includes a "high-score list," so the users can compare their results with others and enhance their motivation to compete.

Creating Positive Peer Pressure on Sustainability Issues

Fit for Green uses gym leader boards to create positive peer pressure by creating friendly competition across colleges. The company recently created a multiplayer, 100 light challenge that allows users to cooperatively work together. Fit for Green's platform also has the capability to display cumulative light competitions between gyms and quarterly competitions across fitness centers.

Fits for Green encourages members to join forces with friends and family in order to make working out while creating energy even more fun. For some social groups, this may be all about the competition, and for others it may be more about cooperation around some renewable energy creation goal that they have set as a unit. In either case the peers become a new driver to help members get healthier and at the same time contribute to a united renewable energy cause.

Similarly, Greenify uses gamification elements to facilitate the creation and completion of user-generated missions, encouraging interaction between geographically proximate communities of peers. Three elements were identified as necessary components to achieve sustainable communities: a healthy climate and environment, social well-being, and economic security. Studies on climate change education recommend social, accessible action-oriented learning that is specifically designed to resonate with a target audience's values and worldview. The Greenify game fosters the creation of peer-generated user content, motivated informed action, and created positive pressure that is perceived as a fun and engaging experience for its users (Lee et al. 2013). Greenify also added a social game layer to accelerate positive community impact through sustainability initiatives.

By leveraging the normative and committing power of social groups, Greenify crafted a culture of positive peer pressure. Just as support groups have proven to be effective in changing behavior in negative contexts (e.g., substance abuse), groups are also a viable strategy for promoting sustainable mind-sets and behaviors. Greenify uses social groups to provide accountability and transparency for the users' actions with platform features such as the ability to see recent activity by other users in a news-feed format, a publicly viewable profile and status, and the ability to show appreciation and give positive feedback.

Mechanisms of group behavior compel users to comply with group norms and values. Creating social groups with users, who normalize desired behaviors, can be a powerful strategy for affecting changes in behavior. Greenify users have expressed that sharing knowledge, ideas, and deeds within a social network was a positive and motivating experience for them. Social interactions, such as commenting on others' missions or deeds, were perceived by most players to be valuable (Lee et al. 2013). Social connection serves a key element to the attraction to gamification. Many successful game platforms leverage a user's social network to create competition and provide support through instant access to their friends and social connections (Palmer et al. 2012).

Related to the above issues of users' need to fit in socially, the PowerHouse found that it was important to create archetypical characters in the game that the users could identify with. PowerHouse game designers found that the persuasive power of the message increased with the use of physically attractive characters that match the identification processes of the target group. So from the initial stages of game development, the designers involved the target users in the design process meetings, where they discussed the game scenario and the teenagers provided designers with a set of archetypical characters and an outline for the graphic design.

Using Gamification to Promote Meaningful Action

With gamification, positive changes in user behaviors emerge when gameful experiences provide motivational affordances that are implemented into the program (Ryan and Deci 2000). By promoting the creation of messages that are accessible and relevant to the community of users, gamification works to promote meaningful action.

The objective of Fit for Green is to increase "healthily planet awareness." As people exercise and generate power, the experience is not just about creating enough energy to light a bulb or charge an iPod. The overall goal of the experience is to give users the realization of what energy is and how difficult it actually is to create it. For example, after a demanding 1-h cardio workout, a user's screen shows that 150 watt-h generated. This allows user to then relate work they did to something real such as "I can light a 75 watt bulb for two hours." This also allows users to conceptualize and better understand the impact they have on the environment, such as each time they needlessly leave a 75-watt bulb on, the same burden is also put on the planet mostly by burning coal.

"Fit for Green Charities" is another part of the program where the electricity that users collectively save is donated to environmental charities. The "Calories to Charities" screen on the platform allows sponsors the opportunity to contribute by pledging a donation for each watt-hour generated on Fit for Green. This benefits sponsors since their logo becomes associated with Fit for Green and becomes part of the engaged sustainability initiative.

PowerAgent uses engaging computer games and mobile applications, with the goal to change energy consumption patterns in the home. The objective of PowerAgent is to transform the home environment and its devices into a learning arena for hands-on experience with electricity usage and to promote engagement through team competitions. A study by Gustafsson and Bång (2008) examined the effectiveness of PowerAgent by evaluating teenagers and families that were playing the game for 10 days in two cities in Sweden. Data collection consisted of home energy measurements before, during, and after gameplay in addition to interviews with participants after the game sessions. The results suggest that the PowerAgent game was highly efficient in motivating and engaging the players and their families to change their daily energy consumption patterns during the game sessions.

Greenify communicates to the user practical everyday steps that can make a difference. Although the majority of climate change education strategies focus primarily on increased understanding of the broad problem, this knowledge does not necessarily lead to changes in a users' behavior (Kellstedt et al. 2008). Recommendations for future efforts in climate change education suggest that action-oriented learning should be used instead. This allows individuals to connect their understanding of the larger problem with actions that they can directly take in their own life. The Greenify system provides this practical and actionable knowledge and appears to be a more promising approach to encourage sustainable changes to one's lifestyle through repetition. As prior behavior studies have demonstrated, when people have done something once, they are more likely to do it again (McKenzie-Mohr 2008).

Built into the Greenify system are several design features that promote informed action and empathy-driven behavior change (Kim et al. 2010). For example, the users have polar bear pets whose happiness and status correspond directly to that user's actions. Similarly, the section with personal stories in which people can share videos or written testimonials of how climate change affects their lives also increases empathy-driven behavior. Greenify users reported that their overall game experience was heightened due to these design features, and they had become more aware of how their personal actions impacted the environment.

By combining general knowledge with specific actions that people could take, Greenify offers users increased personal relevance and accessibility, creating a sense of meaningful accomplishment. By providing missions that consisted of small actions, users felt less overwhelmed when dealing with such a large and complicated issue as climate change (Lee et al. 2013). Based on survey results, findings show that Greenify users felt a more personal relevance to sustainability issues, were empowered to create content and actions for others, and had a heightened awareness of how climate change was connected to their personal choices.

Similarly, in the PowerHouse users can explore the cause-and-effect relationships of different everyday activities and receive instant feedback. For example, if someone is taking a long hot shower, the energy meter will show this immediately and his or her money meter will drop. This immediate feedback allows users to try out different behaviors and learn from their mistakes, which reinforces their learning to promote meaningful change.

Conclusion

Gamification is a growing phenomenon and has been described as one of today's top disruptive technology trends. After examining companies currently using gamification to promote sustainable practices, there is a lot of potential for applications that promote sustainability efforts. These case examples provided suggest that gamification principles are compatible with sustainability education efforts and that game technologies can enable positive peer-to-peer pressure and ultimately motivate behavior change. Gamification can be a powerful strategy that converts serious real-world problems into engaging and meaningful gameplay that promotes peer-to-peer education and behavior change through social interactions.

One of the most important elements of gamification is that a user is having fun while engaging with the application. But companies are really missing out on some of the elements that make real games compelling when they focus only on leaderboards, achievement levels, and badges. User loyalty develops through the game designer's proper use of design principles and understanding of motivational psychology needed to change user behavior in favor of the organizations' objectives. However, if gamification platforms lack novelty, creativity or unexpected twists, and difficulty choices for the user, they may prove to lose users' interest over time.

Gamification is a powerful tool for increasing engagement, yet it is important to remember and address potential limitations. Game design is difficult to implement well since the games should be complex and involve an understanding of motivational information system design. Interface and user experience must be attractive to users, so unfortunately for many small and medium enterprises, complex design requirements can pose a challenge for proper implementation (Palmer et al. 2012).

Another layer that complicates the scope of game design is that the goal of gamification is to affect behavior outside of the game itself. Some studies predict that a majority of gamification implementations will fail due to poor understanding of how to successfully design games (Morschheuser et al. 2017). Poor design can lead to decrease of users' intrinsic motivation (Thom et al. 2012), potential cheating (Carignan and Lawler Kennedy 2013; Makanawala et al. 2013), and short-term engagement (Farzan et al. 2008). So it's important to remember that misused gamification can actually de-motivate users.

Popular game designers have criticized gamification as excluding important elements like storytelling and experiences and using simple reward systems in place of true game mechanics. It is crucial to have more than just badges and achievements but create a journey for the users that includes well-designed puzzles or challenges. Game designs should challenge the user to master skills over time, which is critical to the learning process. Gamification will work when the games reward productive struggle, which results in learning that's engaging and motivating.

With gamification, users must benefit beyond receiving rewards and experience an emotional connection with the messages being presented in the game. One of the big challenges with gamification is to balance the trade-off between persuasive methods and communicating sustainability information and the overall gaming experience. Some studies have shown that too much praise and explicit information can naturally impair the gaming experience negatively (Bang et al. 2006).

This chapter outlined a few of the organizations that are positively transforming sustainability by their use of gamification. After investigating Fit for Green, PowerAgent, Greenify, and PowerHouse, the gamification approach proves to give users more autonomy in achieving their personal goals while also engaging in sustainable behaviors. These examples show how it is important to frame sustainability as a positive idea that leverages intrinsic motivation and self-determination (Grant 2012). Through widespread use of game designs that promote sustainability, communities that share values and worldviews can more effectively exchange ideas for sustainable living (Leiserowitz 2006; Owens 2000). These strategic directions possess the potential to motivate communities to become more engaged and better equipped to take positive action. By embracing the theme of changing perspectives, organizations can begin to properly utilize gamification to achieve a sustainable future.

Cross-References

- ► Collaboration for Regional Sustainable Circular Economy Innovation
- ► Education in Human Values
- ► Moving Forward with Social Responsibility
- ► Smart Cities
- ► The LOHAS Lifestyle and Marketplace Behavior
- ► The Spirit of Sustainability
- ► Transformative Solutions for Sustainable Well-Being

References

Bang, M., Torstensson, C., & Katzeff, C. (2006). The powerhouse: A persuasive computer game designed to raise awareness of domestic energy consumption. In *International conference on persuasive technology* (pp. 123–132). Berlin: Springer.

Bloomberg. (2014). Badgeville doubles annual revenue fueled by strong global 2000 demand. http://www.bloomberg.com/article/2013-07-11/aoxJ9N6jr4C4.html.

Blohm, I., & Leimeister, J. (2013). Gamification. *Business and Information Systems Engineering, 5* (4), 275–278.

Burris, D. (2014). The tech trends that will disrupt, create opportunities in 2014. *Wall Street Journal.* https://blogs.wsj.com/cio/2014/01/16/the-tech-trends-that-will-disrupt-create-opportunities-in-2014/.

Carignan, J., & Lawler Kennedy, S. (2013). Case study: Identifying gamification opportunities in sales applications. In *International conference of design, user experience, and usability.* Berlin: Springer.

Conaway, R., & Garay, M. (2014). Gamification and service marketing. *Springer Plus, 3*(1), 653.

DiSalvo, C., Sengers, P., & Brynjarsdóttir, H. (2010). Mapping the landscape of sustainable HCI. In *CHI 2010 Proceedings of the SIGCHI conference on human factors in computing systems.* ACM Press.

Deterding, S. (2012). Gamification: Designing for motivation. *Interactions, 19*(4), 14–17.

Deterding, S. (2015). The lens of intrinsic skill atoms: A method for gameful design. *Human Computer Interaction, 30*(3–4), 294–335.

Deterding, S., Dixon, D., Khaled, R., & Nacke, L. (2011). From game design elements to gamefulness: Defining gamification. In *Proceedings of the 15th international academic MindTrek conference: Envisioning future media environments* (pp. 9–15).

Donato P., & Link M. (2013). AMA marketing research: The gamification of marketing research.

Farzan, R., DiMicco, J. M., Millen, D. R., Dugan, C., Geyer, W., & Brownholtz, E. A. (2008). Results from deploying a participation incentive mechanism within the enterprise. In *Proceedings of the SIGCHI conference on human factors in computing systems* (pp. 563–572).

Fuchs, M., Fizek, S., & Ruffino, P. (2014). *Rethinking gamification.* Lüneburg: Meson Press.

Gee, J. (2005). Learning by design: Good video games as learning machines. *E-Learning and Digital Media, 2*(1), 5–16.

Gee, J. (2011). Reflections on empirical evidence on games and learning. In *Computer games and instruction* (pp. 223–232). Charlotte, NC: Information Age Publishing.

Grant, G. B. (2012). Transforming sustainability. *The Journal of Corporate Citizenship*, (46), 123.

Gustafsson, A., & Bång, M. (2008). Evaluation of a pervasive game for domestic energy engagement among teenagers. In *Proceedings of the 2008 international conference on advances in computer entertainment technology* (pp. 232–239). Yokohama: ACM.

Halan, S., Rossen, B., Cendan, J., & Lok, B. (2010). High score!-motivation strategies for user participation in virtual human development. In *International conference on intelligent virtual agents* (pp. 482–488). Berlin: Springer.

Hamari, J. (2013). Transforming homo economicus into homo ludens: A field experiment on gamification in a utilitarian peer-to-peer trading service. *Electronic Commerce Research and Applications, 12*(4), 236–245.

Hamari, J., & Lehdonvirta, V. (2010). Game design as marketing: How game mechanics create demand for virtual goods. *International Journal of Business Science and Applied Management, 5*(1), 14–29.

Hamari, J., Koivisto, J., & Sarsa, H. (2014). Does gamification work? A literature review of empirical studies on gamification. In *System sciences (HICSS), 47th Hawaii international conference* (pp. 3025–3034).

Herbig, P. (1991). Game theory in marketing: Applications, uses, and limits. *Journal of Marketing Management, 7*, 285–298.

Huizinga, J. (1949). *Homo ludens: A study of the play-element in our culture.* London: Routledge & Kegan Paul.

Huotari, K., & Hamari, J. (2012) Defining gamification: A service marketing perspective. In *Proceedings of the 16th international Academic MindTrek conference* (pp. 17–22). Tampere: ACM.

Huotari, K., & Hamari, J. (2017). A definition for gamification: Anchoring gamification in the service marketing literature. *Electronic Markets, 27*(1), 21–31.

Jeffries, A. (2011). The fitocrats: How two nerds turned an addiction to videogames into an addiction to fitness. *The New York Observer*. 16 Sept.

Jung, J. H., Schneider, C., & Valacich, J. (2010). Enhancing the motivational affordance of information systems: The effects of real-time performance feedback and goal setting in group collaboration environments. *Management Science, 56*(4), 724–742.

Kellstedt, P., Zahran, S., & Vedlitz, A. (2008). Personal efficacy, the information environment, and attitudes toward global warming and climate change in the United States. *Risk Analysis, 28*(1), 113–126.

Kim, T., Hong, H., & Magerko, B. (2010). Designing for persuasion: Toward eco-visualization for awareness. In T. Ploug, P. Hasle, & H. Oinas-Kukkonen (Eds.), *Persuasive technology: 5th international conference, PERSUASIVE 2010, Copenhagen, Denmark, 2010, proceedings* (pp. 106–116). Berlin: Springer.

Leclerq, T., Hammedi, W., & Poncin, I. (2017). Engagement process during value co-creation: Gamification in new product-development platform. *International Journal of Electronic Commerce.*

Lee, J., Ceyhan, P., Jordan-Cooley, W., & Sung, W. (2013a). GREENIFY: A real-world action game for climate change education. *Simulation & Gaming, 44*(2–3), 349–365.

Lee, J., Matamoros, E., Kern, R., Marks, J., de Luna, C., Jordan-Cooley, W. (2013b). Greenify: Fostering sustainable communities via gamification. In *CHI'13 extended abstracts on human factors in computing systems* (pp. 1497–1502).

Leiserowitz, A. (2006). Climate change risk perception and policy preferences: The role of affect, imagery, and value. *Climate Change, 77*, 45–72.

Makanawala, P., Godara, J., Goldwasser, E., & Le, H. (2013). *Applying gamification in customer service application to improve agents' efficiency and satisfaction. International conference of design, user experience, and usability.* Berlin: Springer.

Marchand, A., & Henning-Thurau, T. (2013). Value creation in the video game industry: Industry economics, consumer benefits, and research opportunities. *Journal of Interactive Marketing, 27*(3), 141–157.

McAfee, P., & McMillan, J. (1996). Competition and game theory. *Journal of Marketing Research, 33*, 263–267.

McGonigal, J. (2011). *Reality is broken: Why games make us better and how they can change the world.* Londres: Penguin Press.

McKenzie-Mohr, D. (2008). Fostering sustainable behavior: Beyond brochures. *International Journal of Sustainability Communication, 3*, 108–118.

Morschheuser, B., Werder, K., Hamari, J., & Abe, J. (2017). How to gamify? Development of a method for gamification. In *Proceedings of the 50th Annual Hawaii international conference on system sciences (HICSS)*, Hawaii, 4–7 Jan 2017.

Owens, S. (2000). Engaging the public: Information and deliberation in environmental policy. *Environment and Planning A, 327*, 1141–1148.

Palmer, D., Lunceford, S., & Patton, A. (2012). The engagement economy: How gamification is reshaping businesses. *Deloitte Review, 11*, 52–69.

Reeves, B., & Read, J. (2009). *Total engagement: using games and virtual worlds to change the way people work and businesses compete*. Boston: Harvard Business School Press.

Ryan, R., & Deci, E. (2000). Self-determination theory and the facilitation of intrinsic motivation, social development, and well-being. *American Psychologist, 55*(1), 68–78.

Terlutter, R., & Capella, M. (2013). The gamification of advertising: Analysis and research directions of in-game advertising, advergames, and advertising in social network games. *Journal of Advertising, 42*(2/3), 95–112.

Thom, J., Millen, D., & DiMicco, J. (2012). Removing gamification from an enterprise SNS. In *Proceedings of the ACM 2012 conference on computer supported cooperative work* (pp. 1067–1070).

Vesa, M., Hamari, J., Harviainen, J. T., & Warmelink, H. (2017). Computer games and organization studies. *Organization Studies, 38*(2), 273–284.

Walz, S., & Deterding, S. (2015). *The gameful world: Approaches, issues, applications*. Cambridge, MA: MIT Press.

Zichermann, G. (2013). Gamification: From buzzword to strategic imperative. *Wall Steet Journal*. http://deloitte.wsj.com/cio/2013/05/15/gamification-from-buzzword-to-strategic-imperative/tab/print/

Zichermann, G., & Cunningham, C. (2011). *Gamification by design: Implementing game mechanics in web and mobile apps*. Sebastopol: O'Reilly Media.

Zichermann, G., & Linder, J. (2013). *The gamification revolution: How leaders leverage game mechanics to crush the competition*. Boston: McGraw-Hill.

Sustainable Higher Education Teaching Approaches

Naomi T. Krogman and Apryl Bergstrom

Contents

N. T. Krogman (✉)
Department of Resource Economics and Environmental Sociology, Faculty of Graduate Studies and Research, University of Alberta, Edmonton, AB, Canada
e-mail: nkrogman@ualberta.ca

A. Bergstrom
Department of Resource Economics and Environmental Sociology, University of Alberta, Edmonton, AB, Canada
e-mail: apryl@ualberta.ca

S. Dhiman, J. Marques (eds.), *Handbook of Engaged Sustainability*,
https://doi.org/10.1007/978-3-319-71312-0_29

Abstract

The goals of public higher education are to generate knowledge and transfer knowledge and skills to students to prepare them for making the world a better place. Given current severe threats to human well-being from climate change impacts, biodiversity loss, and global trends of inequality, we will need a strong commitment to sustainability education to achieve that "better place." This chapter focuses on several key concepts and teaching approaches that can engage students in sustainability challenges and give them some of the necessary knowledge and tools to become thoughtful leaders and followers, problem-solvers, and active citizens. It discusses how key concepts such as intrinsic and extrinsic values help students understand the role of values in human decision-making about addressing bigger-than-self sustainability challenges (e.g., global poverty). The concepts of overconsumption, social commodity chain, metabolic rift, the commons, polycentricity, and resilience allow instructors to traverse disciplines and help students recognize the complex, interdependent nature of social-environmental problems and solutions. The chapter also describes teaching approaches that help students understand how people, problems, and ecological conditions are interconnected and encourage them to move from individual to collective approaches to sustainability. These approaches include place-based experiential learning, project- or problem-based learning, case study conflict studies, collaborative learning, social learning, and community service learning. To effectively engage higher education students in sustainability, educators must provide interdisciplinary and experiential learning experiences and put students in positions where they imagine themselves using innovation, experimentation, trial-and-error social learning, and adaptive management to become future problem-solvers and change agents.

Keywords

Sustainability education · Core interdisciplinary concepts in sustainability · Pedagogical approaches to teaching sustainability

Introduction

There is a convergence of voices that call for sustainability, and a common adage they offer is "Everything is connected." While it may seem trite, it is actually very profound – internalizing this notion of how connected we are quickly moves us beyond our individual interests to realize our interdependence. We learn how the welfare of one species, such as bees, is connected to the welfare of food production, for example, in the ways bees pollinate plants. We see how the conflicts in a country like Syria reverberate in debates and policy changes throughout the world about how countries accept and treat refugees and immigrants. A multitude of voices from sciences, humanities, arts, Indigenous peoples, spiritual movements, and religions are saying the same things – that we are all connected and that long-term thinking and planning about how we can protect interdependent social and ecological systems will sustain us all.

The big questions nations face will require a huge amount of cooperation among people and organizations who see themselves in a web of relations that are shaped by the past and relations that are highly influenced by our collective view of the future. This requires that the students in higher education engage with deep questions of where they want to see the world change for the better, and generally this means striving for sustainability. Such striving requires increasing understanding across all disciplines, in ways that are believable and personally meaningful. A student's ability to see themselves as part of a connected web of relations counteracts the Western focus on individuality and individual rights, to a worldview of responsibilities for the collective. This requires higher education students view the present as a set of consequences from decisions of a past that was heavily influenced by colonialism, industrialization, differential contributions to environmental damage and resource scarcity across regions of the world, and impacts from the exploitation of people and resources. This peopled and ecological history of the world has to be part of sustainability education for students to understand the relations of power and decisions that have led to where the world is today. By understanding the past, students can more readily recognize the significant social and political shifts that will be needed to support ecosystem health and human well-being long into the future and the central role of human agency in doing things differently.

Consistent with Escrigas (2016), we believe higher education institutions have an intergenerational responsibility to engage students with sustainability because of the power of such institutions to shape civilization. Higher education reaches and influences large numbers of current and future leaders. Worldwide, an estimated 150 million students attend approximately 20,000 higher education institutions each year (UNESCO 2014). Higher education institutions possess a unique academic freedom that enables them to generate and advance new ideas, comment on societal challenges, and experiment in sustainable living (Cortese 2003). University scholars and students can share and discuss their ideas and knowledge with society at large through their service and outreach. They also have the ability to engage in multi-decadal partnerships to identify, study, and solve sustainability issues (Hart et al. 2015). Given their ability to generate information, advance ideas, and influence large numbers of people, we agree with Escrigas (2016) that higher education institutions must go beyond just helping students develop the necessary skills to earn a livelihood. They are also responsible for helping students become critically engaged citizens who are open-minded, creative, and critical contributors to society (Escrigas 2016).

Worldwide, there is a growing acknowledgment that higher education has an important responsibility in addressing sustainability challenges. This is demonstrated by an increasing number of universities and colleges around the world that are signing higher education declarations and charters on sustainability, such as the 1990 Talloires Declaration, the 1994 COPERNICUS Charter, as well as dozens of others (International Association of Universities 2018). Additionally, a large number of academic articles on sustainability in higher education call for universities to adopt a more prominent role in both sustainability and sustainability education (Stephens and Graham 2010).

The United Nations (UN) also recognizes the important role and responsibility of educational institutions in addressing environmental and development issues. The UN recognized that sustainability education and learning are necessary to achieve a more sustainable future by designating the years 2005–2014 as the Decade of Education for Sustainable Development (UNESCO n.d.). The United Nations continues to urge educational institutions and all sectors of society to engage with sustainability in reaching 17 sustainable development goals (SDGs), for which the UN has generated agreement among 193 world leaders (United Nations Development Programme 2015). The fourth SDG is specifically focused on quality education (UNESCO Media Services 2017).

The growing number of organizations dedicated to promoting and fostering sustainability in higher education also attests to the important role of higher education institutions. One prominent professional organization is the Association for the Advancement of Sustainability in Higher Education (AASHE). AASHE provides development webinars and workshops and holds an annual conference for those involved in sustainability education.

While education is not a panacea to engage an entire country in sustainability challenges (or convince everyone that climate change is real, for example), it is an essential part of social change. Most of the leaders of the world, in government, industry, nongovernmental organizations, and charitable organizations, spend several years learning about the world by earning a bachelor's degree or higher. Further, there has become a dramatic upward shift in education required for many occupations. Fifty-three percent of adults aged 25–64 have had some postsecondary education in Canada (Charbonneau 2014). In the USA, 40% of Americans have finished an associate's (usually 2 years) degree or above, and an additional 22% have attended some college but did not graduate (Kelly 2015). The Organisation for Economic Co-operation and Development (35 more developed countries of the world) average for at least some postsecondary education is 32% (Charbonneau 2014). A key opportunity to learn about sustainability thus rests in student experiences through higher education.

Sustainability education has the potential to unleash the imagination by teaching students the connections between thought, practice, and the conditions of life that must be understood to create lasting improvements to sustainability challenges. Education is in fact the most powerful force in bringing new science, with a focus on the public good, to bear on sustainability challenges. Better linkages between educational institutions and industry, government, and nongovernmental and civil society organizations are needed to use precious time and resources in the most well-informed way to create change.

Our suggestions in this chapter stem from our combined scholarship and experience. The first author has served for 20 years as a professor of environmental sociology at a research-intensive university and 4 years leading university initiatives to further sustainability research, curriculum, and transdisciplinary connections to communities of interest. The second author is currently an instructor of environmental sociology, served for 3 years as a research assistant for an academic arm of sustainability in higher education, and her thesis research is on how introductory sustainability courses are being taught in Canada and the USA. The first author has also served on a watershed management board, a Chamber of Commerce committee on sustainability, and has been engaged in numerous community efforts to further sustainability in city planning. As an associate dean in the Faculty of Graduate Studies and Research, the first author increasingly witnesses a call from industry, government, the nonprofit sector, and charitable organizations for educated employees who have collaborative intelligence, can work across disciplines and sectors, and accept that the world is changing and the need to change along with it by preparing for the future.

These skills are central to engaging with sustainability. Consistent with the World Commission on Environment and Development 1987 Report (commonly referred to as the "Brundtland Report"), we define sustainability as "development that meets the needs of the present without compromising the ability of future generations to meet their own needs" (WCED 1987, p. 43). The sustainable development definition in this report calls on decision-makers to focus on "the essential needs of the world's poor, to which overriding priority should be given" (WCED 1987, p. 43). Too often issues of poverty and equity are left out of technology and resource-driven agendas of sustainability. We believe this oversight is counterproductive (from the costs of human desperation and suffering) and may inadvertently disengage students from sustainability if they do not see how addressing human needs, social change, and better governance is central to the world's sustainability challenges.

Sustainability as a concept is contested (Caradonna 2014). Many are familiar with the term in regard to the "three E's" (environment, economics, and equity) or the "triple bottom line" (Hacking and Guthrie 2008). Those who follow the Natural Step philosophy hold that sustainability requires that fundamental conditions must be met so that humans do not extract resources, destroy habitats, and pollute in ways that undermine their capacity to meet their needs (The Natural Step 2016). For others, sustainability means "thriving" or creating more abundance and enrichment of individual and collective life (Edwards 2010) or "flourishing," where the focus is on nurturing possibility for all humans and other life on earth (Ehrenfeld and Hoffman 2013). Some academics view sustainability as a discourse or an emerging set of ideas as greater connections in our social and ecological system are understood and politicized (Dryzek 2013). Confusion about the meaning of the term also comes from the loose use of the term "sustainability" outside the classroom, where students in higher education often question to what extent the term is used to gloss over the cradle-to-grave issues in products advertised as sustainable, and in the glib treatment of sustainable lifestyles (that might focus on wearing second-hand clothing, for example, or recycling rather than reducing the volume of materials flowing through one's life). Some critics of the term "sustainability" think it means very little because the definition is so broad.

Sustainability as a concept has increasingly included more elements of systems (environmental, social, economic), although introductory sustainability courses are often taught with a specific focus. Most sustainability curriculum in higher education is interdisciplinary or engages sustainability from various disciplines. For example, if a course focuses on sustainable transitions in cities, it would likely include instruction from planning documents on transportation, sociology documents on social welfare, and generalist documents on housing, air and water quality, and energy and waste systems in the city. Many courses are also taught by instructors who seek to make them transdisciplinary or to include the voices and lessons from practitioners and other key knowledge holders (e.g., resource managers, local leaders, or Elders in Aboriginal communities) who are at the forefront of solving specific sustainability challenges.

Sustainability education often uses forms of constructivist learning (Barth 2015). Constructivism acknowledges that people gain their knowledge about the world through perception, thought, and language (Carolan 2005). It recognizes that knowledge claims about the objects of knowledge could be incorrect and that it is therefore worthwhile to critically examine them (Carolan 2005, p. 396). Sustainability education uses a number of different constructivist learning approaches, including active, experiential, problem- and project-based, self-directed, and collaborative learning approaches (Barth 2015), in which learners engage in constructing knowledge through their experiences (Moore 2005). Some approaches, such as the transformative learning approach, also encourages learners to revise their assumptions and how they interpret their experiences through processes of self- and critical reflection (Cranton 1996).

These teaching and learning approaches try to encourage students to question taken-for-granted perspectives, values, beliefs, and assumptions that could be shaping the way they and others understand and react to the world around them. They encourage students to explore complex problems using diverse perspectives to understand that there are alternative ways to approach problems (Barth 2015; Burns 2011; Moore 2005; Tilbury 2011). Burns (2011) attests that it is very important to include different perspectives in sustainability education because they provide diverse ways of understanding sustainability issues. Burns contends that perspectives should also include questioning and critiquing the dominant paradigms and the underlying power relationships and cultural assumptions that contribute to unsustainable practices. By doing so, learners may come to recognize the important role that culture plays in our sustainability crises (Burns 2011).

In the following sections, we will discuss key concepts and teaching approaches to engage higher education students with sustainability. Many educators and scholars are experimenting and writing about hundreds of concepts and approaches that are being used in higher education classroom ideas (e.g., Sipos et al. 2008; Tilbury 2011; Wals 2009, 2012; Wright 2013). In this chapter, we will focus on the concepts and approaches that can be used across disciplines. The concepts we have chosen involve social, environmental, and economic systems, across geographical scales and time frames, necessarily involving many disciplines. Indeed, teachers of interdisciplinary programs (e.g., environmental studies, conservation sciences, planning, or global health) need to better coordinate the sequence and complementarity of courses and more deliberately follow key learning objectives to illuminate the connections across human well-being, human decision-making, the built environment, land, water, atmosphere, and life systems. As it currently stands, university program leaders often throw a bunch of courses together and leave the integration and action part up to students (Clark et al. 2011; Foote et al. 2006). We emphasize concepts that put humans at the center of systems of sustainability because it is important for students to see the essential role of human agency, the ability to organize, the influence of power and politics, and the social learning from trial and error that drives sustainability action and inaction.

Key Concepts to Engage Students with Sustainability

Values and Bigger-Than-Life Problems

For students to have a sense of direction in how some part of the world can be more sustainable, they must understand "why" (Sinek 2009) changes to the status quo are necessary. Undergirding answers to "why" are the values embedded in human action, as every organized effort people make is normative (Heifetz et al. 2009). Values are conceptions of what one finds desirable. They tend to come in clusters and are able to coexist with competing values (Crompton et al. 2010). A reading we have found immensely useful to mobilize students to reflect on what they want to see changed, and why, is a report called "Common Cause: The Case for Working with our Cultural Values" (Crompton et al. 2010). In this report, Crompton and others convincingly explain how the cluster of intrinsic or self-transcendent values are far more likely to guide our behavior to address "bigger-than-self" problems (such as climate change adaptation, human rights violations, biodiversity loss, and global poverty) than extrinsic values. Intrinsic values include a sense of community, affiliation to friends and family, self-development and understanding, and appreciation and protection of people and nature. In contrast, extrinsic values emphasize social status, physical pleasure, material wealth, achievement, and power over others. While people hold both of these clusters of values within themselves, Crompton et al. (2010) prompt the reader to think about how these values are reinforced in sustainability rationales, political debates, public policy, career choices, advertising, and education, for example. When students are asked to identify their own values and ferret out the values undergirding social practices in society, they awaken to the role they play in reinforcing these values in their careers as citizens and as consumers.

Students learn that Western societies often overemphasize extrinsic values at the expense of intrinsic values and thereby diminish our motivation to address bigger-than-self problems or sustainability challenges. Students like to report their observations of how media coverage reinforces extrinsic versus intrinsic values. For example, many of our students point to the disproportionate media coverage of the career successes of high-status individuals (in sport, entertainment, business, and politics) over leaders that address bigger-than-self problems such as poverty alleviation and food security. Emphasis on extrinsic values also reinforces individualism. For example, students often give examples of product advertisements that emphasize attractive, employed, middle- to upper-class people who distinguish themselves by their leisure activities and what they consume. Students are increasingly interested in how messages from advertising and popular media intensify a feeling of social comparison and competition. They connect how these messages of individualism can discourage their sense of worth (and cause anxiety) as a human being in a community.

When students see themselves as part of a community, they more readily discuss how the fate of our environment is shared with others and the role of collective action, based on intrinsic values, that can embolden commitment to change sustainability systems. To personalize this understanding, we also invite students to reflect on the messages that were reinforced by their parents and extended family about success, living the "good life," and one's obligations to society (pay taxes, vote, volunteer, "give back" in some form, etc.) to see how one's connection to sustainability is influenced by the norms we abide to in our primary groups. Those norms are based on values that are sometimes hidden and unarticulated, and students often struggle with naming their core values and how they are manifested in their daily lives. When students grapple with naming the "why" behind what they do and do not do, as tied to their values, they also become more aware of the context in which certain intrinsic and extrinsic values are reinforced and honored. By addressing intrinsic values, students often recognize gaps between their own values and behavior and begin to contemplate ways they can more deliberately address bigger-than-self problems within their lifetime and with others. Students start to see themselves as change agents within certain communities by the behaviors they promote, demonstrate, and articulate as important and with whom they spend their time.

Overconsumption

A sustainability issue students immediately identify with is overconsumption. In fact, students all too often assume that a switch to green consumerism significantly offsets the profound influences of consumption practices on the planet's ecosystems. Overconsumption as a "bigger-than-self" problem engages students in sustainability by inviting them to discover the differential environmental and social justice impacts of consumption trends in relation to geographic scale, time, and rates of collateral damage (as especially driven by more wealthy countries) to our global ecosystem. Students can explore how the material production, use, and waste processes tied to consumption drive most environmental problems by creating pollution streams (in air, water, land) and degrading landscapes. By studying overconsumption, students learn about the "ecological rucksack" or the resource requirements for a good from cradle to grave, and when they learn this, they see that while to live is to consume, the way people consume has huge implications for the health of the planet (Sachs et al. 1998). Over the past few years, popular overconsumption subjects among students in our classes have been the human costs of electronic waste, fast fashion and the glut of clothing, and food waste.

Metabolic Rift

The term metabolic rift, developed by John Bellamy Foster from Marxist writings, points to the lack of connection humans have had with nature, where the intensification of production at one site creates an irreparable rift between the production region and where the product is consumed (Foster 1999). The rift is the lack of recognition and awareness by those who use the products and the living system from which the products were sourced, including the net loss of renewability at the extractive/production site. For example, there is often a metabolic rift between town and country. If labor conditions are poor in one region (country) (e.g., many miners experience this in countries like South Africa and Mongolia), the workers are sacrificed, in a sense, to support the consumptive habits of people in another region (city). When students engage with the notion of metabolic rift, they can look at the growing number of large cities and the consumptive habits across classes of people in those cities and explore the supply chain implications. Students engage with sustainability when they see that most of the needed resources in cities (food, construction materials, and household items) are sourced from rural places. The places and people in rural areas are often hidden from the students' view, but the people who live near the source of the good are often left with high risks from destructive extractive processes, resource scarcity, precarious employment, and contamination from waste.

To demonstrate metabolic rift, the first author has used Edward Burtynsky's dramatic photos of industrial landscapes in her classes to show the human faces of those gathering raw materials (e.g., farmers or miners) or cleaning up or living in the mess left behind (e.g., garbage pickers, cleanup crews). By providing examples of the metabolic rift, students see that sustainability is at once local and extra-local. They also realize that the relationship between consumption and sustainability must be examined across space and time and that there are significant implications for who benefit and who lose along the commodity chain of the product. Consequently, students internalize this awareness even more so by doing a social commodity chain project. In this project, they choose an item, such as a banana or household cleaning product, and trace out the material place of origin and throughput processes required to put that product on the shelf. They also investigate the safety and welfare of the workers at each part of the chain. As they trace each step, from the source of extraction through material processing, manufacture, distribution, use, repair, maintenance, disposal of the item, and its absorption back into the earth, issues of environment, health, livelihoods, trade, and lifestyles become even more apparent. Similarly, students can also learn about sustainability by doing a standardized life cycle assessment to calculate the lifetime environmental impact of a product, service, or human activity, an analysis often done by engineers (Curran 2013).

The Commons and Polycentricity

The commons and polycentricity are two important concepts that help students understand the importance of governance and the challenges of natural resource management. The commons refers to shared resources that morally or legally belong to everyone (Bollier and Helfrich 2014), such as the oceans, the air we breathe, most of the oil and gas resources in Canada, and many other public goods. The late Elinor Ostrom won the Nobel Prize in Economics in 2009 for her cumulative work on problems of collective action for individuals using common-pool resources. Common-pool resources are resources that are used by a group of people, such as forests, fisheries, and the air, that provide diminished benefits to everyone if each individual pursues his own self-interest (Ostrom et al. 1994). The gift of Ostrom's work, and that of many other scholars in this area, is that students can learn about successful ways that common-pool resources have been, continue to be, or can be sustainably managed. This work is an antidote to the rhetoric that it is only through privatization that a country can protect its stake in common-pool resources. By understanding that there are institutional features that are repeatedly associated with the protection of various natural resources, as well as features that undermine their protection, students dive deep into the nature of governance, fairness for stakeholders who rely on the resources, and alternative efforts that could be tried to correct deteriorating trends for many resources critical to the health of large ecosystems, such as wetlands and fresh water systems.

Elinor Ostrom, Michael Polanyi, and many others also brought us the term polycentricity, which captures the challenges of managing resources in social systems where many decision-makers are operating under an overarching set of rules. For example, watershed systems are often managed by multiple decision-makers, including farmers, industrial owners, wildlife managers, municipal regulators, utility companies, acreage owners, recreational operators, and any other active decision-makers for the land and waterways in that watershed system. There may be overarching rules in regard to water quality and streamflow standards that must be maintained, wetland environmental regulations, stream corridor protection zones, etc., but those protections notwithstanding healthy watersheds generally also require a place-based level of commitment, communication, cooperation, and collaborative intelligence to manage a watershed for particular values (e.g., riparian biodiversity or wildlife protection). By grasping the concept of polycentricity, students begin to carefully unpack what kind of cooperative approaches work, and are needed, for the particular sustainability challenges on that land base. The polycentricity concept invites students to see the different kinds of decision-makers on that land base, the nature of the resource as a common property, and the informal and formal rules operating in that cultural and political system.

Governance is at the heart of sustainable resource management, and students can learn a great deal from case studies that show the interactions that drive resource management decisions. By addressing the commons and polycentricity, this also opens the door to a large area of scholarship that combines social learning with sustainable resource management, often referred to as "adaptive management." Adaptive management refers to the experiments and lessons learned by managers, scientists, and other stakeholders to create and maintain sustainable ecosystems (Walters 1986). Case studies on adaptive management illustrate the trials and tribulations of decision-making in governance systems and the long-term commitment to manage a watershed, forest, or fishery sustainably, across geographic scales. It is only by making these governance decisions (and nondecisions) transparent in real places with real people that students can appreciate the complexity and channelization of decisions (where social learning is often suppressed) that often characterize the management of forests, fisheries, protected areas, agricultural areas, watersheds, and so on.

Resilience

Resilience is a term used across many disciplines, especially in reference to social-ecological systems and psychology. In the former, resilience refers to the ability of a social-ecological system to absorb stressors and continue to carry out crucial functions, self-organize, learn, and adapt (Holling 1973; Gunderson and Holling 2002). Similarly, in the psychology literature, resilience refers to the ability of an individual to bounce back after failure, tragedy, trauma, threats, or other significant sources of stress. Resilient individuals tend to be able to regulate emotions, see failure and hardship as setbacks from which to learn, and tend toward optimism (American Psychological Association 2017). Resilience is an engaging concept for students in higher education because of its basis in processes over time that highlight organizational behaviors and individual thoughts and behaviors that demonstrate adaptation and invoke hope (Youssef and Luthans 2007). By studying resilience in social-ecological systems, students learn how local people have adjusted their livelihoods to live sustainably within watersheds, delta systems, Northern tundra, mangrove wetlands, etc. By studying psychological resilience, students also learn about how people use social capital to rebuild their lives and communities after natural and technological disasters (Aldrich 2012). Like the common property resource management approach, students are invited to see alternatives that people have used to address sustainability challenges in specific ecosystems and cultures and in response to specific stressors. This broadens students' notions of pathways to repair ecosystems and support human dignity and health.

Again, case studies in particular provide students with an arsenal of examples where social-ecological systems are wisely managed or improve over time. Students need these examples to see a more sustainable future in places across the world and to imagine phronesis. Phronesis is a term from Aristotle (1980, Bk 6) that refers to wise practical reasoning built upon experience and sharp judgment. As students learn about how people across the world use phronesis to cooperatively and politically address their own sustainability challenges, it sets students up to ask, "How can I support that problem-solving from what I do in my own career and life?" An appreciation for the phronesis that different land managers possess, for example, helps students to envision local capacity for sustainability challenges and guards against the assumption that educated Westerners have all the answers. We have found that students show remarkable enthusiasm for learning about collaborative or community-based forms of natural resource co-management, as well as culturally appropriate approaches to sustainability improvements across the world.

Students also engage with sustainability when they investigate who the knowledge holders are in a place and how that place-based knowledge is maintained or built over time. In our Canadian context, where a recent federal Truth and Reconciliation Commission has stated renewed commitment to improving Aboriginal and non-Aboriginal relationships, we see a heightened student interest in Aboriginal traditional knowledge among hunters, fishers, trappers, healers, and agriculturalists and how their practical reasoning can inform sustainability plans for land, water, wildlife species, and protected areas. Much of this material can be effectively taught by incorporating neo-colonial and post-colonialist literature (Wisker 2006) that identifies the legacies of colonialism and invites students to consider epistemology, i.e., different ways of knowing that are historically, culturally, and politically informed and conscriptive about what actions, and by whom, sustainability should entail. Returning to an earlier point, when students contextualize a sustainability challenge in a historical framework, they can more critically understand the epistemology that undergirded the way a land, sea, or resource was managed. They can also see how social relations were embedded, relations that gave certain people the power or the right to make decisions, use the resources, and reap the benefits, while others live with the consequences.

Each of these concepts can be taught in a variety of discipline-based or interdisciplinary courses, such as in sociology, political science, human geography, environmental studies, conservation sciences, economics, history, and literature. In the next section, we turn to some of the ways in which these concepts can come to life for students across these disciplines.

Key Teaching Approaches to Engage Students with Sustainability

The context of teaching in higher education has changed significantly in the last 20 years, with far more student demand for courses that are relatable to their lives and in teaching formats that are entertaining and varied. Instructors are far more likely to use social media in the classroom, to use class websites to store various forms of digitized materials, and to require students to demonstrate knowledge in varied ways, beyond test-taking and standard research papers.

Sustainability in particular is best taught in ways that make sustainability challenges real and approachable. Early environmental education tended to rely on transmissive teaching approaches (Sterling 2004), where instructors tried to transfer predetermined knowledge to students through presenting, lecturing, and using supporting materials like workbooks and instruction forms (Wals 2012). The underlying assumption of this approach was that presenting information about the environment would be sufficient to encourage personal and social change (Sterling 2004). In practice, however, transmissive-based models of teaching have a limited capacity to engage people meaningfully in sustainability challenges (Wals 2012).

There is widespread recognition that we need to use alternative teaching and learning approaches if we are to work toward sustainable outcomes (Tilbury 2011). Worldwide, scholars and practitioners of sustainability education have proposed a number of alternatives, including participatory, collaborative, problem-based, interdisciplinary (Wals 2012), transdisciplinary (Evans 2015), project-based (Wiek et al. 2014), place-based, experiential, transformative, and service learning approaches (Sipos et al. 2008). These different learning approaches share a "family resemblance" in that they recognize that learning is more than just knowledge-based and that it requires more than a single discipline or perspective (Wals 2012). Thus, many of these approaches complement one another; they focus on real issues that engage learners, and they emphasize the importance of quality interactions with others and with the learning environment (Wals 2012). Rather than trying to transmit knowledge and train people to behave in particular ways, they engage in processes of inquiry, participation, and social learning that challenge current frameworks and practices that are unsustainable (Tilbury 2011). Through the use of active or participatory learning methods, they also encourage students to actively participate in the learning process itself (Wals 2012).

Many researchers and practitioners worldwide view active, experiential, and participatory methods as central to sustainability education because they encourage students to ask critical reflective questions, clarify values, think systemically, and envision more positive futures (Tilbury 2011, p. 29). Participatory learning is also thought to empower students and help them build their capacity to address sustainability problems (Burns 2011). Moreover, active and participatory approaches can potentially increase student engagement in sustainability. For example, Mintz and Tal (2018) looked at environmental courses and found that courses that address sustainability topics and include active and participatory learning activities report increased student motivation to promote sustainability. In this section, we will discuss several key teaching approaches to engage students in sustainability.

Place-Based Experiential Learning

Engaging sustainability is more likely when students learn about real and nuanced sustainability challenges facing a particular city, watershed, municipality, etc. Place-based learning situates the curriculum within the context of the learner's own life, community, or region, taking advantage of the learner's interest in the local (Sipos et al. 2008). Experiential learning is "learning by doing" (Domask 2007), where students have and reflect on direct experiences and then form ideas that are applied to new experiences (Sipos et al. 2008). It can include things like internships, simulations, fieldwork, or service learning (Domask 2007). In place-based experiential learning, students can be asked to investigate what they would do to address a challenge, identified by those who live in that place. By listening to residents or authorities in a particular place, students more readily see the social nature of moving toward sustainability improvements (Domask 2007; Sipos et al. 2008). The instructor might ask the students: How is the problem defined? Who are the actors most influential in its definition and controversy? What has been the history in that place that led to this state? What policies exist that constrain or support changing the situation? What is available to understand public opinion and engagement on the issue? Who benefits from the current contributors to the problem? Who would most benefit from a sustainable solution? Who has power among the stakeholders of the problem to change the status quo? What would have to change to implement certain solutions? Students who work on a sustainability challenge in a particular place begin to understand which values that are threatened by the sustainability challenge, the commons at stake, the key stakeholders, the polycentricity of the governance system, and the options to create more social-ecological resiliency. Additionally, by learning about and interacting with places, students can strengthen their connections with those places and with the people who live in them (Gruenewald 2003).

Problem- and Project-Based Learning

Similar to the approach above, many educators are finding that students learn more and can more readily draw on knowledge from across the disciplines if they are posed with a problem for which there is no one right answer, although there are well-informed and carefully reasoned answers (Wiek et al. 2011, 2014). Problem- and project-based learning combines two learning approaches: problem-based learning and project-based learning (Wiek et al. 2014). Problem-based learning helps students investigate a real-world sustainability problem and try to develop a deep, critical understanding of that problem. In project-based learning, students work on projects to understand and develop solution options to problems. In both approaches, students are self-directed and often work in small teams (Wiek et al. 2014).

As an example, instructors might pose this problem to students: "Climate change induced extreme droughts are expected to rise dramatically in this century. Choose a region of the world and propose how this area might prepare for such droughts to reduce human suffering and ecological degradation, and protect its ability to provide water, food and housing to the region's inhabitants." This can be even more transformative when students visit the place they are studying, such as doing a semester abroad. When they do this, they can embed themselves in the culture to imagine addressing a problem, such as drought in rural Italy versus the Australian outback. Rural areas in particular invite students to see where phronesis resides, such as in Indigenous Elders or other local knowledge keepers, and they often see where the metabolic rift is apparent in resource-dependent communities. When students integrate their academic or other expert knowledge from academia with the traditional or practical knowledge of stakeholders outside of academia, this moves student into transdisciplinary learning, which is considered essential for addressing sustainability challenges and solutions (Remington-Doucette et al. 2013). These are inherently complex, interdisciplinary challenges, and solving them will require people to think holistically, integrate knowledge of human and natural systems, and collaborate across institutional and disciplinary boundaries (Remington-Doucette et al. 2013).

One form that problem- and project-based learning can take is a "campus as a living lab" program, which uses the university and/or community as a real-world context in which students can experiment to solve local sustainability problems (Rowe and Hiser 2016). For example, the University of British Columbia (UBC) developed a program called the Social Ecological Economic Development Studies (SEEDS), in which students, staff, and faculty collaborate to develop and implement a number of different on-campus sustainability projects (UBC n.d.). Students can work on projects in many different themes, such as biodiversity, climate, energy, food, land, waste, water, and community. For example, students in one class helped UBC Food Services select and recover edible food from several residence kitchens, restaurants, and retail outlets, creating a buffet meal out of recovered food. This project led to a partnership between UBC Food Services and a local food bank for the ongoing recovery of food (UBC n.d.). Solutions-oriented, real-world projects like these can be very impactful because they help to generate hope and agency in students (Evans 2015).

Conflict-Focused Case Study Learning

Students can begin to understand the enormity of the stakes for people involved in vexing conflicts over natural resources by comparing ways in which different groups have confronted high-conflict natural resource or environmental problems (see examples in Clark 2002; Clarke and Peterson 2015; Daniels and Walker 2001). Sustainability conflict case studies may be over forest management, public transit infrastructure, large industrial developments, or predator control, for example. Students can be invited to examine the values held and threatened for different stakeholders and understand the relationship of the conflict to consumption drivers linked to increased housing, roads, mineral extraction, forestry or agricultural production, luxury item production (e.g., diamonds), tourist developments, and economic development in general. Students can use the case studies to learn how groups can define the problem. They also learn about processes for engaging stakeholders, meeting management, negotiation approaches, practices to sustain positive community relations, and policy influences and changes as part of the solution. The case studies can also involve role-play activities, which will give students a better understanding of the different perspectives involved and help them empathize with others (Tilbury 2011). By deep analysis of these case studies, students can begin to imagine how they might play a role in working with others to find a resolution. Additionally, they will better appreciate the sustained commitment for trust-building and communication that may be required to find longer-term resolutions to environmental conflicts.

We observe that students are generally eager to learn about different approaches to public consultation, public participation, and collaborative approaches for local and extra-local stakeholders for a particular sustainability challenge. There are a vast number of case studies to draw on from the literature on how to implement sustainability-focused policies, address conflict over natural resources, collaboratively manage natural resources, co-manage, and effectively work with government and/or industry advisory groups. Students learn that democracy is messy and any democratic process is imperfect. Case studies can reveal the finesse required to lead a sustainability change, and illustrate to students that incorporating the values and priorities of appropriate groups in sustainability actions is often finessed, based on a large number of factors in that location. By learning about the trial-and-error efforts of governing bodies to reflect citizen priorities, students in higher education recognize that sustainability must be credible to those who are expected to comply or adopt new behaviors, and widespread acceptance and adoption can take considerable time.

Collaborative Learning

Collaborative learning encourages learners to construct knowledge together, with an emphasis on questioning and exchanging ideas (Cranton 1996). Collaborative learning assumes that knowledge is not something that is transferred from one person to the next (Moore 2005). Instead, in collaborative learning, groups or communities socially produce knowledge by talking together and reaching a consensus (Cranton 1996). There is an emphasis on process, in which people exchange experiences, information, ideas, and feelings. Members listen to each other, ask questions, try to understand the different perspectives that are expressed, and negotiate points of view. The aim is to arrive at a shared understanding that all group members find acceptable (Cranton 1996).

In courses that include collaborative learning, the educator is responsible for creating an open and supportive environment that is open to self-reflection. The educator establishes the context and may provide materials but is not considered the expert, instead adopting the role of a co-learner or participant (Moore 2005). Course activities that can foster collaborative learning might include things like role-playing activities, group projects, and case-based learning (Remington-Doucette and Musgrove 2015).

Collaborative group work in sustainability education is believed to foster a number of skills that are needed to address sustainability problems, such as planning, communication, organization, negotiation, delegation, and conflict resolution skills. It can also increase empathy and tolerance of differences (Remington-Doucette and Musgrove 2015). However, collaborative group work can also be challenging, since group processes can be difficult and students may find it frustrating to depend on other group members for their grade. There is only so much time in any given academic term to work through differences for class projects. Thus, faculty need to plan group projects that are well-supported and structured (Remington-Doucette and Musgrove 2015).

Social Learning

The UN Decade of Education for Sustainable Development recognized that social learning is an important part of aligning education systems and practices with sustainability (Tilbury 2011). Social learning can be understood as "a special kind of learning that contributes to realising the learning society that is essential in realising a more sustainable world" (Wals et al. 2009, p. 11). Social learning is believed to be important to help develop the foundations that are needed for reflexivity and change (Tilbury 2011). It brings together people from different backgrounds and leads to an ensemble of knowledge, experiences, and perspectives. Wals et al. (2009) contend that this ensemble will be necessary for creatively seeking answers for questions that have no ready-made solutions in an ever-changing society.

An important characteristic of social learning is that people learn from and with each other. As a result of this learning, they become collectively better equipped to withstand setbacks and deal with complexity, risks, and insecurity (Wals et al. 2009). Social learning assumes that people will learn more from each other if everyone does not think or act alike. Thus, to encourage people to accept and make use of different perspectives, it is about building social cohesion and trust (Wals et al. 2009). When there is cohesion, the articulation of different perspectives can help people see that there are different ways of approaching and solving problems. When it works well, social learning can therefore contribute to creative and innovation solutions. Social learning is also concerned with collective meaning-making and with creating ownership of the learning process and solutions that people discover (Wals et al. 2009). Finally, social learning involves indeterminacy, since it cannot be known precisely what students will learn ahead of time and learning goals may shift during the process (Wals 2012).

Social learning is not limited to formal educational settings but can be used in many different contexts. For example, governments can create municipal- or regional-level forums where local actors develop actions for sustainable regional development through cooperation and consultation. At the neighborhood or community level, social learning processes can assist with improving the liveability or sustainability of a neighborhood (Wals et al. 2009).

In the context of higher education, Owens et al. (2015) discuss how a field school offered through the University of Victoria in Canada can provide a social learning and transformative experience that combines both critical reflection and practical action. In the month-long travel study program, students learned how sustainability was understood, practiced, and struggled over in the Pacific coastal region of Canada. Owens and colleagues (2015) describe how the program was carefully designed to expose students to multiple perspectives and to provide diverse learning experiences; students engaged with multiple actors, including scholars, planners, activists, other students, and the instructors. Students were encouraged to participate in ongoing self- and group reflection, as well as individual and group debates. Instructors acted to facilitate experiences rather than dictate or transmit information. Thus, in addition to providing structure and guidance, the program also included community-based learning and self-directed learning (Owens et al. 2015). The authors noted that, at the end of the field school, students were able to ask more qualified or conditional questions about sustainability and understand sustainability as a contested social construction.

Community Service Learning

Community service learning (CSL) is a form of experiential education that aims to balance student learning with service to the community (Taylor et al. 2015). In CSL, students collaborate with members of the greater community toward mutually beneficial outcomes (Sipos et al. 2008). CSL integrates "intentional formal learning activities" with learning that occurs through community service (Taylor et al. 2015, p. 5). Service learning is a very heterogeneous practice: it appears in a number of forms, takes place at many different kinds of sites, and involves collaborations with a variety of community partners (e.g., schools or not-for-profit organizations). Additionally, it is often combined with other pedagogical approaches, such as place-based learning, project-based learning, or community-based research (Taylor et al. 2015).

In sustainability courses, the use of community service learning allows students to grapple with real-world issues in their communities. At the University of British Columbia (UBC), for example, students participated in two short credit courses in 2003 and 2004 to initiate and then further develop a pilot ecovillage at the UBC Farm (Sipos et al. 2008). The 2003 course focused on collaborative group work, participatory decision-making, and nonformal education, while the 2004 course focused on personal sustainability and building practical skills (Sipos et al. 2008).

The literature reports a number of student outcomes from community service learning: among other benefits, it includes learning about oneself in relation to others, greater commitment to community engagement, commitment to social justice, reflective and collaborative learning, and an increased awareness of how theory and practice are linked (Taylor et al. 2015). Additionally, CSL initiatives have the potential to foster civic responsibility and critical thinking, as long as they are carefully organized, have a clear purpose, are relevant to the professional future of the students, and include ongoing student reflections that are guided by faculty (Taylor et al. 2015).

These outcomes are important for sustainability. To work toward a more sustainable future, we will need to change and to learn ways to live and work more in balance with nature. To make change, Burns (2011) argues that we will need to engage with ourselves, with others, and with places. Civic responsibility means actively participating in public issues in a manner that is informed, constructive, and that focuses on the common good (Dresner and Blatner 2006). The literature on sustainability education has consistently stressed the importance of critical thinking to enable learners to autonomously and authentically contribute to sustainability (Tilbury 2011). Critical reflective thinking helps learners clarify values, deeply examine the root causes of unsustainability, and recognize the biases and hidden assumptions in the knowledge, opinions, and perspectives of themselves and others (Tilbury 2011). Both civic responsibility and critical thinking are important for addressing the complex and challenging environmental and humanitarian problems that we face. If we are to democratically discuss, debate, and propose solutions to these problems, we will arguably need citizens who are informed, concerned, and actively engaged and who can understand and assess different information and perspectives.

Conclusion

Higher education has the power to shape civilization. Along with this power comes the intergenerational responsibility to help students become informed, active, creative, and critically engaged citizens who are concerned about the collective good.

There are a multitude of concepts from across the disciplines that can be used to engage students with the fundamental relationships that characterize or undermine sustainability. This chapter offers a few of those concepts, chosen in particular for the potential to use them as anchor concepts across numerous disciplines. Students in more developed countries in particular need to learn how overconsumption is harming humans and the environment, as well as potential alternatives to provide for our needs while preserving ecological integrity and human well-being. The concepts of the commons, polycentricity, metabolic rift, resilience, and adaptive management point to the many ways that groups at different scales have learned from past experiences and sustainably managed resources.

The chapter also explored some of the key teaching approaches that can help student learn about and engage in sustainability, including place-based experiential learning, project- or problem-based learning, case study conflict studies, collaborative learning, social learning, and community service learning. Students will more likely engage with sustainability if they can work on real-world challenges, examine issues in a particular place, struggle with the need to try and understand the problem from multiple vantage points, and draw upon successes elsewhere in process and outcome. In the current era, with vast and deep differences across the public in North America (and also Poland, Germany, France, Turkey) about how to engage "difference," curiosity and humility are key to be able to listen to mutually agreed-upon compromises and solutions. Thus, educators need to instill in students curiosity and humility about the uncertainty in offered solutions and the moral reasons to at least try. They also need to foster the courage to propose improvements to problems, learn from both the successes and failures in experiments (such as in campus as a living lab projects), and engage in their own efforts at adaptive management.

At the center of it all are human beings, and for students to imagine sustainability, they must also imagine a world in which improved systems of mutual respect, cooperation, collaboration, and sharing of the world's wealth in resources are possible. Thus, a central element of sustainability education is in the recognition of the role of values and how these are communicated and tied to principles that guide ethical practices that promote a healthy ecosystem, human dignity, and sustainable livelihoods across all countries. In years to come, higher education must find ways to more effectively engage students in sustainability, as leaders and followers, as active citizens, and as problem-solvers.

Featured Case in Point

One engaged sustainability practice to encourage in higher education is to start with where you are and ask students, "What are the sustainability challenges at your work place, in your community, or at your university? Many universities and colleges have an active Office of Sustainability with various student-centered awareness, engagement, and transformative learning opportunities (Vaughter et al. 2015). Many higher education institutions also have a program along the lines of "campus as a living lab" for students to become engaged in research or programming to make the campus more sustainable. Many student groups are addressing sustainability challenges, such as better access to healthy food. At our own university, students manage community gardens, a farmer's market in the Students' Union, and a student food bank. Community service learning opportunities, now offered at many universities, allow students to spend 20 hours or more a term to work with an organization to learn about how they are engaged with an issue and mobilize to change the status quo. Instructors could incorporate research internships, volunteer hours, community service learning, or, at the very least, participant observation credit, for students to learn about how people are organizing at various levels and intensity to make a workplace, community, or university more sustainable. We must expose students to collective organizing alternatives to engage students with sustainability.

Reflection Questions

The following are a number of questions that can help students think more deeply and reflectively about key concepts discussed in this chapter, including the commons, the overconsumption, and the role that values play in bigger-than-self problems:

1. How is flood protection related to the commons? What forms of governance influence flood risk? What values are threatened by floods, and who is most at risk in your state or province? How would government's recognition of the importance of climate change mitigation and adaptation affect its approach to flood risk planning? Who would be consulted to develop plans to mitigate harm from floods in your state or province?
2. How do your concerns for sustainability tie to your intrinsic and extrinsic values? What bigger-than-self problems draw your attention, and what is being done at the local, state, national, and international levels to address that problem?
3. Where do you find hope in the growing recognition of the importance of sustainable consumption? What trends of overconsumption most worry you, and in what social practices or policies do you find hope that these trends lead to a lowered ecological footprint?

Exercises in Practice

1. Case study in conflict. To engage students in a sustainability issue that is relevant to their lives, students could be asked to attend public meetings and read policy briefs and op-ed pieces related to an environmental conflict in the town or city in which they are attending college or university. Students could analyze the rhetoric and dialogue to see where there are arguments to protect intrinsic versus extrinsic values and how the solutions proposed suggest certain fundamental values. Students could identify the stakeholders and how the issue is related to protecting some form of the commons, observe informal and formal rules regarding social

practices and governance for the environmental issue, and identify who benefits and loses from the current status quo. Students could argue for a solution or set of solutions and provide evidence from other cases as to why they would take certain positions.

2. Project on sustainable consumption products. Students could look at the various ways in which coffee, bananas, and sugar have an environmental rucksack and how various organizations are addressing social justice and environmental impacts associated with these products through fair trade organizations, organic farming, eco-labeling, cooperative ownership, and local, small-scale production. What are the advantages and disadvantages for these efforts to address workers' well-being, community impacts for the production and harvest of these products, processing and distribution of environmental impacts, and consumer awareness and satisfaction?

Engaged Sustainability Lessons

1. Higher education has a unique role to play in engaging students with sustainability and shaping society by directly addressing ways to protect people and the planet for the collective good.
2. We must identify our values and encourage our students to do so to understand why we believe and advocate for lifestyles, social practices, and policies that have a profound effect on the sustainability of the planet. By priming intrinsic values more than extrinsic values, we elevate students' commitment to address the bigger-than-self problems of the world.
3. Students in more developed countries in particular need to learn about the role of consumption in driving environmental and human harm and the alternative ways people are organizing around providing for our needs to preserve ecological integrity and human well-being. Students need to understand how our systems of provision in housing, transportation, food, water, and infrastructure significantly influence land degradation, air and water quality, and pollution streams.
4. Concepts such as the commons, polycentricity, metabolic rift, resilience, and adaptive management point to the many ways groups at different scales have sustainably managed resources and learned from past experiences to improve on their management of resources and care for each other.
5. Students are more likely to engage with sustainability if they examine an environmental issue in a particular place and have the opportunity to draw across various disciplines to do a form of participatory learning, to understand the role of conflict, negotiation, and community engagement when they look at potential solutions.

Chapter-End Reflection Questions

1. How do current political debates about climate change reflect intrinsic versus extrinsic values?

2. What teaching or life experience is most memorable to you for igniting your interest in sustainability? What does that experience tell you about how you learn about sustainability or would teach sustainability yourself?
3. Who do students need to learn from outside of academia to address conservation? How can academia provide opportunities for students to engage in collaborative learning?
4. How can trends in sustainable consumption address multiple needs, such as lowering ecological footprint while at the same time enhancing a sense of community and willingness to organize around a collective problem?
5. How do you address conflict in your own life? How do you respond to conflict over environmental issues you care about? How can you address important challenges of our time, such as the many climate change actions that are necessary, and learn to work through the conflict with others, as difficult as that might be?

Cross-References

▶ Education in Human Values
▶ Moving Forward with Social Responsibility
▶ Sustainable Decision-Making

References

Aldrich, D. P. (2012). *Building resilience: Social capital in post-disaster recovery*. Chicago: University of Chicago Press.

American Psychological Association. (2017). What is resilience? http://www.apa.org/helpcenter/road-resilience.aspx. Accessed 6 Aug 2017.

Aristotle. (1980). *The nicomachean ethics*. (D. Ross, Trans.). Oxford: Oxford University Press. (Original work published 340 BCE).

Barth, M. (2015). *Implementing sustainability in higher education: Learning in an age of transformation*. London: Routledge.

Bollier, D., & Helfrich, S. (Eds.). (2014). *The wealth of the commons: A world beyond mark and state*. Amherst: Levellers Press.

Burns, H. (2011). Teaching for transformation: (re)designing sustainability courses based on ecological principles. *Journal of Sustainability Education, 2*. http://creativecommons.org/licenses/by/3.0/us/. Accessed 26 Jan 2018.

Caradonna, J. L. (2014). *Sustainability: A history*. Oxford, UK: University Press.

Carolan, M. S. (2005). Society, Biology, and Ecology. *Organization & Environment 18*(4):393–421.

Charbonneau, L. (2014). A glance at Canada's postsecondary education standings. University Affairs. https://www.universityaffairs.ca/opinion/margin-notes/a-glance-at-canadas-postsecondary-education-standings/. Accessed 24 Jan 24 2018.

Clark, S. G., Rutherford, M. B., Auer, M. R., Cherney, D. N., Wallace, R. L., Mattson, D. J., Clark, D. A., Foote, L., Krogman, N., Wilshusen, P., & Steelman, T. (2011). College and university environmental programs as a policy problem (part 1): Integrating knowledge, education, and action for a better world? *Environmental Management, 47*(5), 701–715. https://doi.org/10.1007/s00267-011-9619-2.

Clark, T. W. (2002). *The policy process: A practical guide for natural resource professionals*. New Haven: Yale University Press.

Clarke, T., & Peterson, T. R. (2015). *Environmental conflict management*. Thousand Oaks: Sage.

Cortese, A. D. (2003). The critical role of higher education in creating a sustainable future. *Planning for Higher Education, 31*(3), 15–22.

Cranton, P. (1996). Types of group learning. *New Directions for Adult and Continuing Education, 71*, 25–32.

Crompton, T., Brewer, J., Chilton, P., & Kasser, T. (2010). *Common cause: The case for working with our cultural values*. Surrey: WWF-UK. http://assets.wwf.org.uk/downloads/common_cause_report.pdf. Accessed 7 Aug 2017.

Curran, M. A. (2013). Life cycle assessment: A review of the methodology and its application to sustainability. *Current Opinion in Chemical Engineering, 2*(3), 273–277. https://doi.org/10.1016/j.coche.2013.02.002.

Daniels, S. E., & Walker, G. B. (2001). *Working through environmental conflict: The collaborative learning approach*. Westport: Praeger Publishers.

Domask, J. J. (2007). Achieving goals in higher education: An experiential approach to sustainability studies. *International Journal of Sustainability in Higher Education, 8*(1), 53–68.

Dresner, M., & Blatner, J.S. (2006). Approaching civic responsibility using guided controversies about environmental issues. *College Teaching, 2*, 213–219.

Dryzek, J. (2013). *The politics of the earth: Environmental discourses*. Oxford, UK: Oxford University Press.

Edwards, A. R. (2010). *Thriving beyond sustainability: Pathways to a resilient society*. Gabriola Island: New Society Publishers.

Ehrenfeld, J. R., & Hoffman, A. J. (2013). *Flourishing: A frank conversation about sustainability*. New York: Greenleaf Publishing Limited.

Escrigas, C. (2016). A higher calling for higher education. Great Transition Initiative: Toward a Transformative Vision and Praxis. http://www.greattransition.org/publication/a-higher-calling-for-higher-education. Accessed 7 Aug 2017.

Evans, T. L. (2015). Transdisciplinary collaborations for sustainability education: Institutional and intragroup challenges and opportunities. *Policy Futures in Education, 13*(1), 70–96. https://doi.org/10.1177/1478210314566731.

Foote, L., Krogman, N., & Spence, J. (2006). Should academics advocate on environmental issues? *Society and Natural Resources, 22*(6), 579–589. https://doi.org/10.1080/08941920802653257.

Foster, J. B. (1999). Marx's theory of metabolic rift: Classical foundations for environmental sociology. *American Journal of Sociology, 105*(2), 366–405.

Gruenewald, D. A. (2003). Foundations of place: A multidisciplinary framework for place-conscious education. *American Educational Research Journal, 40*(3), 619–654.

Gunderson, L. H., & Holling, C. S. (Eds.). (2002). *Panarchy: Understanding transformations in systems of humans and nature*. Washington DC: Island Press.

Hacking, T., & Guthrie, P. (2008). A framework for clarifying the meaning of the triple bottom-line, integrated, and sustainability assessment. *Environmental Assessment Impact Review, 28*(2–3), 73–89.

Hart, D. D., Bell, K. P., Lindenfeld, L. A., Jain, S., Johnson, T. R., Ranco, D., & McGill, B. (2015). Strengthening the role of universities in addressing sustainability challenges: The Mitchell Center for Sustainability Solutions as an institutional experiment. *Ecology and Society, 20*(2). https://doi.org/10.5751/ES-07283-200204.

Heifetz, R., Grashow, A., & Linsky, M. (2009). *The practice of adaptive leadership: Tools and tactics for changing your organization and the world*. Boston: Harvard Business Press.

Holling, C. S. (1973). Resilience and stability of ecological systems. *Annual Review of Ecology and Systematics, 4*, 1–23.

International Association of Universities. (2018). Higher education and research for sustainable development: HESD references. http://www.iau-hesd.net/en/ressources.html. Accessed 20 Jan 2018.

Kelly, A. (2015, April 28). The neglected majority: what Americans without a college degree think about higher education, part 1. Forbes. https://www.forbes.com/sites/akelly/2015/04/28/the-neglected-majority-what-americans-without-a-college-degree-think-about-higher-education-part-1/#2e182d3c72cf. Accessed 24 Jan 2018.

Mintz, K., & Tal, T. (2018). The place of content and pedagogy in shaping sustainability learning outcomes in higher education. *Environmental Education Research, 24*(2), 207–229. https://doi.org/10.1080/13504622.2016.1204986.

Moore, J. (2005). Is higher education ready for transformative learning? A question explored in the study of sustainability. *Journal of Transformative Education, 3*(1), 76–91.

Ostrom, E., Gardner, R., & Walker, J. (1994). *Rules, games and common-pool resources*. Ann Arbor: University of Michigan Press.

Owens, C., Sotoudehnia, M., & Erickson-McGee, P. (2015). Reflections on teaching and learning for sustainability from the Cascadia Sustainability Field School. *Journal of Geography in Higher Education, 39*(3), 313–327. https://doi.org/10.1080/03098265.2015.1038701.

Remington-Doucette, S. M., Hiller Connell, K. Y., Armstrong, C. M., & Musgrove, S. L. (2013). Assessing sustainability education in a transdisciplinary undergraduate course focused on real-world problem solving: A case for disciplinary grounding. *International Journal of Sustainability in Higher Education, 14*(4), 404–433.

Remington-Doucette, S. M., & Musgrove, S. (2015). Variation in sustainability competency development according to age, gender, and disciplinary affiliation. *International Journal of Sustainability in Higher Education, 16*(4), 537–575.

Rowe, D., & Hiser, K. (2016). Higher education for sustainable development in the community and through partnerships. In M. Barth, G. Michelsen, M. Rieckmann, & I. Thomas (Eds.), *Routledge handbook of higher education for sustainable development* (pp. 315–330). London: Routledge. https://www.routledgehandbooks.com/doi/10.4324/9781315852249.ch21.

Sachs, W., Loske, R., & Linz, M. (1998). *Greening the north: A post-industrial blueprint for ecology and equity*. London: Zed Books.

Sinek, S. (2009). *Start with why: How great leaders inspire everyone to take action*. New York: Penguin Group.

Sipos, Y., Battisti, B., & Grimm, K. (2008). Achieving transformative sustainability learning: Engaging heads, hands and heart. *International Journal of Sustainability in Higher Education, 9*(1), 68–86.

Stephens, J. C., & Graham, A. C. (2010). Toward an empirical research agenda for sustainability in higher education: Exploring the transition management framework. *Journal of Cleaner Production, 18*(7), 611–618.

Sterling, S. (2004). An analysis of the development of sustainability education internationally: Evolution, interpretation and transformative potential. In J. Blewitt & C. Cullingford (Eds.), *The sustainability curriculum: The challenge for higher education* (pp. 43–62). London: Earthscan.

Taylor, A., Butterwick, S. J., Raykov, M., Glick, S., Peikazadi, N., & Mehrabi, S. (2015). *Community service-learning in Canadian higher education*. Vancouver: The University of British Columbia. https://open.library.ubc.ca/cIRcle/collections/facultyresearchandpublications/52383/items/1.0226035. Accessed 26 Jan 2018.

The Natural Step. (2016). A science-based definition of sustainability. http://www.thenaturalstep.org/the-system-conditions/. Accessed 14 Jan 2018.

Tilbury, D. (2011). *Education for sustainable development: An expert review on processes and learning*. Paris: UNESCO. http://unesdoc.unesco.org/images/0019/001914/191442e.pdf. Accessed 25 Feb 2017.

United Nations Educational, Scientific and Cultural Organization (UNESCO). (n.d.). UN decade of ESD. https://en.unesco.org/themes/education-sustainable-development/what-is-esd/un-decade-of-esd. Accessed 20 Jan 2018.

United Nations Educational, Scientific and Cultural Organization (UNESCO). (2014). *Shaping the future we want: UN decade of education for sustainable development (2005–2014) final report*. Paris: UNESCO. https://sustainabledevelopment.un.org/content/documents/1682Shaping%20the%20future%20we%20want.pdf. Accessed 5 Feb 2017.

United Nations Development Programme (UNDP). (2015, September 25). World leaders adopt Sustainable Development Goals. http://www.undp.org/content/undp/en/home/presscenter/pressreleases/2015/09/24/undp-welcomes-adoption-of-sustainable-development-goals-by-world-leaders.html. Accessed 20 Jan 2018.

United Nations Educational, Scientific and Cultural Organization (UNESCO) Media Services. (2017, February 13). UNESCO tracks global progress on education for sustainable development and global citizenship. http://www.unesco.org/new/en/media-services/single-view/news/unesco_tracks_global_progress_on_education_for_sustainable_d/. Accessed 16 Feb 2017.

University of British Columbia. (n.d.). SEEDS Sustainability Program. Accessed 21 Feb 2017. https://sustain.ubc.ca/courses-teaching/seeds.

Vaughter, P., Wright, T., & Herbert, Y. (2015). 50 shades of green: An examination of sustainability policy on Canadian campuses. *The Canadian Journal of Higher Education, 45*(4), 81–100.

Wals, A. E. J. (2009). *Learning for a sustainable world: Review of contexts and structures for education for sustainable development*. Paris: UNESCO. http://www.unesco.org/education/justpublished_desd2009.pdf. Accessed 24 Feb 2017.

Wals, A. E. J. (2012). *Shaping the education of tomorrow: 2012 full-length report on the UN decade of education for sustainable development*. Paris: UNESCO. http://www.desd.in/UNESCO%20report.pdf. Accessed 19 Jan 2018.

Wals, A. E. J., van der Hoeven, N., & Blanken, H. (2009). *The acoustics of social learning: Designing learning processes that contribute to a more sustainable world*. The Netherlands: Wageningen Academic Publishers. http://library.wur.nl/WebQuery/wurpubs/fulltext/108487. Accessed 27 Jan 2018.

Walters, C. J. (1986). *Adaptive management of renewable resources*. New York: Macmillan Publishing Company.

Wiek, A., Withycombe, L., & Redman, C. L. (2011). Key competencies in sustainability: A reference framework for academic program development. *Sustainability Science, 6*(2), 203–218.

Wiek, A., Xiong, A., Brundiers, K., & van der Leeuw, S. (2014). Integrating problem- and project-based learning into sustainability programs: A case study on the School of Sustainability at Arizona State University. *International Journal of Sustainability in Higher Education, 15*(4), 431–449.

Wisker, G. (2006). *Key concepts in postcolonial literature*. New York: Palgrave Macmillan.

World Commission on Environment and Development (WCED). (1987). *Our common future*. Oxford, UK: Oxford University Press.

Wright, T. (2013). Stepping up to the sustainability challenge. In L. F. Johnson (Ed.), *Higher education for sustainability: Cases, challenges and opportunities across the curriculum* (pp. 201–213). New York: Routledge.

Youssef, C. M., & Luthans, F. (2007). Positive organizational behavior in the workplace: The impact of hope, optimism, and resilience. *Journal of Management, 33*(5), 774–800.

Expanding Sustainable Business Education Beyond Business Schools

Christopher G. Beehner

Contents

C. G. Beehner (✉)
Center for Business, Legal and Entrepreneurship, Seminole State College of Florida, Heathrow, FL, USA
e-mail: beehnerc@seminolestate.edu; beehner@embarqmail.com

S. Dhiman, J. Marques (eds.), *Handbook of Engaged Sustainability*,
https://doi.org/10.1007/978-3-319-71312-0_51

Abstract

Numerous business schools include sustainability as an elective or required curriculum component. Moreover, an increasing number of businesses are implementing sustainability initiatives and programs. However, the success of this top-down approach to sustainable business may be limited by lack of sustainability awareness among entry-level workers and managers. Therefore, sustainable business education should occur at multiple academic levels, specifically community colleges, wherein students prepare for entry-level trade, management, and professional positions at businesses which have implemented or will implement sustainability programs. In this chapter, we will examine sustainable business education efforts at several community colleges. This "bottom-up" sustainable business education approach will be presented as complementary to the "top-down" approach of business schools. Readers will be invited to reflect on their sustainable business education awareness and experience, and engage colleagues at community colleges to embrace sustainable business education, furthering the development of a cosmic sustainability vision.

Keywords

Sustainable business · Business sustainability · Community college · Sustainable development · Sustainable entrepreneurship

Introduction

The 1990s witnessed an increased level of inclusion of environmental literacy, civic engagement, and social responsibility requirements in higher education curriculum (Bridges and Wilhelm 2008; Rowe 2002). This trend was followed by an increase in the number of business schools including sustainability as an elective or required curriculum component (Bridges and Wilhelm 2008). Moreover, an increasing number of businesses are implementing sustainability initiatives and programs (Calder and Clugston 2003). However, the success of this top-down approach to sustainable business may be limited by lack of sustainability awareness among entry-level workers and managers. In other words, a gap may exist between middle and senior management, and entry-level workers and managers in terms of the understanding and implementation of sustainable business initiatives. Therefore, sustainable business education should occur at multiple academic levels, specifically community colleges, wherein students prepare for entry-level trade, management, and professional positions at businesses which have implemented, or will implement, sustainability programs and initiatives. This holistic, integrated approach to sustainable business education is necessary to ensure engaged sustainability occurs at all levels of the business – personal, departmental, and organizational level. Achievement of multilevel business sustainability engagement requires fundamental changes at the level of a common person on the street, a task which community colleges are best capable of accomplishing.

In this chapter we will explore the background of the sustainable business education movement, and the history and current state of sustainable business education at higher education institutions. We will then examine the community college system, and the current community college sustainability and sustainable business education efforts, primarily within the United States. While in the United States, community colleges are also known as city colleges, junior colleges, technical colleges, and more recently state colleges, in this chapter, the group of institutions will be referred to collectively as community colleges. A discussion outlining how community colleges might address the sustainable business education gap, including approaches, methods, tools, and best practices, will be followed by conclusions and recommendations for the future of community college sustainable business education. In this chapter, the reader should acquire an understanding and appreciation of the role of community colleges in the education of students who are sustainably engaged within the business environment.

Background

Sustainability education plays an important role in the expansion, promotion, and consistency of sustainable actions in organizations and society (Baccarin et al. 2015). The focus on sustainability in higher education began with the Stockholm Conference on the Human Environment in 1972 (Calder and Clugston 2003), gaining momentum with the 1978 United Nations International Environmental Education Programme. In 2002, a report was prepared by UNESCO and the International Work Program on Education, Public Awareness and Sustainability of the UN Commission on Sustainable Development (CSD) for the World Summit on Sustainable Development (WSSD) in Johannesburg, South Africa. The higher education section of this report identified key lessons for successful implementation of sustainability policies and practices in universities, including:

- Senior management commitment to strategy
- Include all components of the triple-bottom line, not just the "green" component
- Comprehensive sustainability strategy throughout institution
- Culture change across institution
- Monitoring, evaluation, and reporting process to ensure effective and continuous implementation (United Nations Educational, Scientific and Cultural Organization 2002)

In 1990, the US Congress passed the National Environmental Education Act, empowering the US Environmental Protection Agency (EPA) with increasing public literacy about environmental issues. Potter (2009) suggested the initiative be expanded to include sectors beyond the field of education. While the act expired in 1996, its programs and mandates continued despite the absence of a formal act. However, the future of these programs and mandates may be uncertain based upon to the Trump administration stance on environmental regulations and issues, and the 2017 appointment of climate change skeptic Scott Pruitt as EPA Director.

Academics have supported the notion of a higher education role in the creation of a sustainable society (Corcoran and Wals 2004; Jabbour 2010). However, barriers to the creation, adoption, and diffusion of sustainability principles in universities exist, including but not limited to: the abstract nature of the sustainability concept; the lack of staff with sustainability expertise; limited financial resources for sustainability programs; and, the perception that sustainability is only an issue for environmentalists (Leal Filho 2000). Sustainability is becoming a more concrete concept as sustainability content becomes more prevalent in academic literature. The increase in sustainability degree offerings, in particular at the graduate level may result in a corresponding increase in sustainability trained faculty. The financial barrier is ever present in academia; however, a growing change in perception that sustainability is an issue for all of society, not just environmentalists may result in additional financial resource availability for sustainability initiatives and degree programs.

Justification for Sustainable Business Education

Because the primary activity of business schools is teaching (Jabbour 2010), business school curriculum should include environmental content within the context of the classical disciplines of business management. Moreover, the compatibility of environmental management objectives with business objectives can create strategies for "win–win" opportunities in which both environmental and business performance improves (Jabbour 2010). While business schools are viewed as proponents of modern capitalism, sustainability students should learn to critique the current capitalist paradigm, considering and proposing viable modifications or alternatives (Von Der Heidt and Lamberton 2011). Moreover, while sustainability has become a prominent issue on the global agenda, a few business schools have clearly determined the key competencies required in this area (Adomßent et al. 2014). Further, business schools often do not keep pace with current trends such as sustainability (Bates et al. 2009). Nonetheless, there has been substantial growth of interest in sustainability in business, management, and organizational studies in recent years (Cullen 2016).

The inclusion of sustainability in business curricula is essential for the development of managers and leaders possessing a critical world view (Cezarino 2016). Contemporary business managers must possess an understanding of the critical interrelationship between environmental responsibility and good organizational performance (Aligleri et al. 2009). In order to prepare business students for the contemporary business environment, higher education must demonstrate to students that they can be successful business managers while simultaneously considering environmental facets of managerial decision-making and activities (Hoffman 1999). Jabbour (2010) suggested business schools can contribute to sustainable education through the adoption of environmental management programs on campus, and the development of environmentally knowledgeable faculty, staff, and students. Further, it is relevant that business faculty educate students about achieving high levels of professional and business performance, while simultaneously implementing the necessary actions and strategies for social and environmental problem mitigation (Gonçalves-Dias et al. 2009). Contemporary businesses need managers and employees capable of recognizing sustainability as an opportunity for strategic growth and innovation (Lans et al. 2014). Literature from the period 1994 to 2013 supports the inclusion of sustainability content in business school curricula, revealing evidence of multiple educational approaches used by faculty in providing sustainable business education (Cullen 2016).

In order to facilitate the transition to sustainable business, employees will require sustainability-focused skills which could easily be provided through undergraduate or graduate business education (Von Der Heidt and Lamberton 2011). Hundreds of international higher education institutions have incorporated sustainability into their curriculum, research, and scholarship (Calder and Clugston 2003). While colleges and universities often find it politically difficult to add additional courses for degree requirements, some higher education institutions have overcome this obstacle by integrating sustainability content into existing liberal arts and specialty courses (Rowe 2002). The growing business interest in sustainability suggests a need for an increased focus on sustainable business. However, current training approaches do not appear sufficient for meeting the challenges associated with a shift to sustainability (Hatfield-Dodds et al. 2008). Despite extensive research on organizational social responsibility, corporate social responsibility education is increasing, but still remains limited.

Support for Sustainable Business Education

The notion of sustainable business education is finding increased support outside academia. In order to guide universities and other educational institutions in the areas of management education addressing environmental and social concerns, the United Nations (UN) in partnership with several business schools created the project Principles for Responsible Management Education (PRME). Officially launched in 2007 by UN Secretary-General Ban Ki-Moon, the PRME initiative currently includes more than 650 business schools and academic institutions from more than 85 countries, including more than one-third of the Financial Times' top 100 business schools (PRME Secretariat 2017). Of the more than 650 PRME signatories, 110 are located in the United States.

While there are a number of nonprofit and industry organizations devoted to sustainable business, sustainability, and sustainability education, the results of a recent Internet search did not identify any organizations specifically devoted to sustainable business education. However, AACSB International – the Association to Advance Collegiate Schools of Business (AACSB) – is actively engaged in incorporating sustainability in business management education. In addition, the Association for the Advancement of Sustainability in Higher Education (AASHE) is the first professional higher education association devoted to the promotion of sustainability in higher education. In addition to colleges and universities, several of which are business schools or provide business degrees, AASHE members include businesses and non-profits devoted to advancement of the higher education sustainability movement.

In addition to support from PRME, AACSB, and AASHE, the case for sustainable business education is finding increased support among business practitioners. Ninety-three percent of global CEOs surveyed believe sustainable development is important to their company's future success (Accenture 2010). Corporate sustainability is increasingly being seen as both the right thing and the smart thing to do (Soyka 2013). While the goal of a traditional capitalist entity has traditionally been making money, there is a growing concern about how that money is made (Soyka 2013). Regardless of whether corporate executives altruistically and personally embrace the sustainability concept, many recognize the increased consumer demand for sustainability, and the increasing number of competitors incorporating sustainability into their business models. Numerous US-based multinational corporations (MNCs) have incorporated sustainability into their mission and vision, including household names such as The Walt Disney Company, General Electric, Johnson & Johnson, and Starbucks.

Sustainable Entrepreneurship Education

One specific area of sustainable business education still in the infancy stages is sustainable entrepreneurship. This field is related to the broader field of social entrepreneurship in which entrepreneurs develop a business model, often structured as a nonprofit organization wherein cultural, environmental, and social issues are addressed. Sustainable entrepreneurs seek to acquire a competitive business advantage through the promotion of sustainable approaches to business and societal problems. Higher education plays a significant role in the development of the growing number sustainable entrepreneurs who are capable of recognizing opportunities for sustainable development facilitation. However, academic institutions typically focus on either sustainability or entrepreneurship, with sustainable education housed within the environmental science department, and entrepreneurship education housed within the business department (Lans et al. 2014). For more discussion on the topic of sustainable entrepreneurship education, please see the chapter in this handbook entitled ▶ "People, Planet, and Profit" by Gil Domemech and Berbegal-Mirabent.

Challenges of Sustainable Business Education

Teaching sustainable business in a traditional business program can prove challenging (Beehner 2017). Moore (2005) identified four institutional barriers to the incorporation of sustainability in university curriculum: the disciplinary environment; the competitive environment; misdirected evaluation criteria; and, unclear priorities, decision-making, and power. The disciplinary environment often inhibits interdisciplinary collaboration or students taking courses outside major discipline. Administration and faculty are tasked with incorporating a comprehensive business curriculum within a fixed number of credit or semester hours. The competitive environment consists of student competition for grades, faculty competition for publication and grants, departmental competition for students and funding, and university competition for prestige and power. Evaluation criteria is often misdirected where faculty are hired and promoted based on publication, student are evaluated by jobs and salaries, and a lack of clear evaluation procedures for university policies and plans. Sustainability does not currently offer a broad metric for faculty, student, or institutional evaluation. Finally numerous priorities and unclear decision-making criteria may exist along with confusion concerning where on the hierarchy of power curriculum decisions should be made. While administration and faculty might equally embrace sustainable business curriculum, there may be disagreement concerning from where the curriculum should originate.

Numerous higher education institutions have embarked upon the sustainable business education journey. While some institutions offer dedicated degrees and certificates in sustainable business, others provide at least one course as a core or elective component to an existing business degree. The risk of teaching sustainability as a standalone course is the perception that sustainability is a separate issue, disconnected from core business concepts and curriculum (Stubbs and Cocklin 2008). The solution to successful sustainable business education is the incorporation of sustainability into the core business curriculum, a task deemed to be challenging by some sustainable business education scholars (Stubbs and Cocklin 2008).

Brief History of Sustainable Business Education

Having established the background of and justification for sustainable business education, a brief historical review of the sustainable business education movement is in order. In addition to a summary of sustainable business education history, the following section includes examples of institutions that have offered or are offering sustainable business courses, certificates, and degrees. The institutions explored in this section will not include community colleges because those institutions will be considered in a subsequent section of this chapter.

Although earlier literature suggests business schools were not developing environmentally knowledgeable managers (Hoffman 1999), sustainable business programs and courses are increasingly being offered at many institutions of higher education (Lozano et al. 2015). While colleges and universities have successfully embedded sustainability curriculum within environmental degree programs (Lozano et al. 2015), the incorporation of sustainability content into business programs has posed a significant challenge (Von Der Heidt and Lamberton 2011). The number of sustainable business elective courses being offered has increased, however, most implementations were compartmentalized (Lozano et al. 2015), with a limited number of programs in which sustainable business has been incorporated into key, functional business courses such as accounting, finance, or strategic management.

Sustainable Business Education Movement

The term education for sustainable development (ESD), originating from the 1992 Earth Summit in Rio de Janeiro, is a more broadly defined term than environmental education, encompassing the issues of cultural diversity, global development, and environmental and social equity (Calder and Clugston 2003). One of the initiatives of ESD is to provide training to all sectors of the workforce so that all public and private employees have access to the necessary knowledge and skills to make sustainable work decisions (UNESCO 2012). While there have been many United Nations declarations concerning sustainability education, two best provide support for the sustainable business education movement. The Tibilisi Declaration, sponsored by UNESCO and the United Nations Environment Program (UNEP), suggested that environmental education should be provided to people of all academic aptitudes. The Thessaloniki Declaration all subject disciplines must address environmental and sustainability issues. The previously discussed initiatives emphasize the promotion of sustainability throughout academia, including partnerships between universities and governments, nongovernmental organizations (NGOs) and businesses, and the moral obligation of higher education to promote a sustainable future (Wright 2002).

The Higher Education for Sustainable Development (HESD) movement received significant support from students, scholars, and administrators in Holland and Canada in the 1990s, with a focus on the sustainability of higher education institutions, and the implementation of sustainability curriculum (Calder and Clugston 2003). Interest expanded throughout Europe and to a lesser extent, the United States. Recognizing that colleges and universities are obligated by society to impart knowledge and skills in order to prepare responsible, discerning citizens who will make a positive contribution to the world (Corcoran and Wals 2004), these institutions have a further obligation to provide the moral vision and technical knowledge necessary to ensure high quality of life for future generations (Calder and Clugston 2003). Sustainable development is the current framework through which higher education must serve the greater society (Calder and Clugston 2003).

Sustainable Business Education Institutions (US)

Early sustainability courses were not housed within business schools, but rather within the fields of engineering, environmental science, and public policy. Brown University, Brandeis University, and Harvard University were among the early universities that began redesigning existing courses, and designing new courses to educate students about sustainability and environmental stewardship. A small sampling of current nonbusiness sustainability degrees includes: Master in Design Studies in sustainable design at Boston Architectural College; a Master of Architecture in sustainability at California State Polytechnic University-Pomona; both a Master and Doctorate in sustainable construction at SUNY College of Environmental Science and Forestry; and, a doctoral degree in sustainable development at Columbia University.

A recent review of the Degree Prospects (2017) website identified 67 institutions listed as offering sustainable business certificates, and undergraduate and graduate degrees. A summary of institutions delivering sustainable business education (which is by no means comprehensive) is provided in the following paragraphs to provide a perspective of the historic and current state of the curricula.

Early pioneers of sustainable business education include the Kenan-Flagler Business School at the University of North Carolina and Green Mountain College in Vermont. In 2001, the Kenan-Flagler Business School founded the Sustainable Enterprise Initiative, offering a Master of Business Administration (MBA) with a concentration in Sustainable Enterprise Enrichment, and undergraduate electives in sustainable business. Green Mountain College in Vermont, delivering environmental, social, and economic sustainability-focused curriculum for more than 20 years, launched the first online Sustainable MBA program in 2006. The Sustainable MBA format is unique in that sustainable business principles are incorporated in all courses.

Other institutions offering sustainability-focused MBAs include: Antioch University New England in New Hampshire; Bard College in New York; Humboldt State University in California; Maharishi University of Management on Iowa; Marylhurst University in Oregon; San Francisco State University; and, Presidio Graduate School in California with both an MBA and a Master of Public Administration (MPA) in Sustainable Management. Duquesne University in Pennsylvania and University of Oregon both offer MBAs in Sustainable Business Practices.

Sustainable business-related graduate degrees include: American University in Washington, DC, with a Master's in Sustainability Management; Arizona State University with an Executive Master of Sustainability Leadership; Brandeis University in Massachusetts with a Sustainable Development MBA; Franklin Pierce University in New Hampshire with an MBA in Energy and Sustainability; Rochester Institute of technology with an MBA in Environmentally Sustainable Management; St. Louis University with an MS in Sustainability/MBA dual degree program; University of Colorado Denver with an MS in Management, Managing for Sustainability; and, University of South Florida with an MBA in Building Sustainable Enterprise.

While the majority of sustainable business education degrees are offered at the graduate level (Bates et al. 2009; Bridges and Wilhelm 2008), a growing number of higher education institutions have developed undergraduate degree programs in sustainable development. In 2003, Aquinas College in Grand Rapids, Michigan, became the first US higher education institution to offer an undergraduate course in sustainable business. A sampling of US colleges and universities offering undergraduate sustainable business degrees includes: Arizona State University with a Bachelor of Arts in Business, sustainability focus; Stony Brook University in New York with a Bachelor of Science in Business Management, Sustainable Business; University of Wisconsin Extension with a Bachelor of Science in Sustainable Management; and, University of Wisconsin Superior with a Bachelor of Science in Sustainable Management. SUNY Empire State College offers an undergraduate certificate in Business and Environmental Sustainability.

Sustainable entrepreneurship is a topic within the sustainable business field which is gaining popularity. Sustainable entrepreneurs identify and solve environmental and social problems, in turn creating shared value – social and business value. The University of Vermont offers a Sustainable Entrepreneurship MBA, and Colorado State University offers a Global Social & Sustainable Enterprise MBA. Sustainable entrepreneurship is related to the field of social entrepreneurship in which entrepreneurs solve a pressing social problem, not necessarily environmental or sustainability-related. Roosevelt University offers a Bachelor of Science in Business Administration in social entrepreneurship MBA, and Pepperdine University offers a Master of Arts in Social Entrepreneurship.

Sustainable Business Education Institutions (Europe, Asia, and Pacific)

There continues to be a stronger interest in incorporating sustainability curriculum in higher education institutions in Europe than in the United States (Lozano et al. 2015). Unlike the United States, higher education sustainability initiatives in many developed nations receive substantial government assistance and offer curriculum supporting the triple-bottom line dimensions of sustainable development (Calder and Clugston 2003). However, the majority of governments, including the United States, have not formally embraced sustainable development as foundational to economic development or education (Calder and Clugston 2003).

European business schools offering Masters and MBA programs in sustainable business or management include: ESLSCA Business School in Paris; Business School Lausanne and Sustainability Management School (SUMAS) in Switzerland; Utrecht University in the Netherland; Bologna Business School in Italy; and, Coventry University and the University of Bath in the United Kingdom. Other European colleges and universities offering sustainability and sustainable development degrees include University of St. Andrews in Scotland; Bangor University in Wales; University of Nottingham and Aston Business School in the United Kingdom; and Ipaq Business School in France.

While sustainable business has made significant progress in becoming a common subject in Asia and the Pacific region business schools, the topic is frequently not prioritized, with a lack of systematic approaches to the incorporation of sustainability into business curricula (Ryan et al. 2010). Monash University in Australia offers a major or minor in sustainability in the Bachelor of Commerce business degree, a postgraduate certificate in sustainability, and a doctorate through the Monash Sustainable Development Institute (MSDI). Also in Australia, Charles Sturt University offers an MBA with a concentration in Sustainability. University of Waikato in New Zealand offers a Master of Management Studies in Management and Sustainability. TERI University (The Energy and Resources Institute) is a pioneer in sustainable and green management in India, offering an MBA in Business Sustainability. In addition, Amity School of Natural Resources and Sustainable Development in India offers an MBA in Natural Resources and Sustainable Development.

The focus of this chapter to this point has been with the background and history of sustainable business education in colleges and universities. The following sections will introduce the US community college system, present the case for a community college role in the delivery of sustainable business education, and provide examples of community colleges currently providing such curricula.

The US Community College System

A community college is an educational institution that typically offers adult high school, vocational training, college credit certificates, and 2-year Associate in Arts (AA), Associate in Science (AS), and Associate in Applied Science (AAS) degrees. In the United States, community colleges are also known as city colleges, junior colleges, technical colleges, and more recently state colleges. The term community college became commonplace because this class of higher education institution typically attracts and accepts students from the local community. Many community colleges also offer a university pathway wherein students graduating with an AA degree can transfer to, and in some cases are automatically accepted as juniors at State Universities. Many community colleges partner with 4-year colleges and universities to provide a limited offering of baccalaureate courses and degrees on campus.

Most community college missions have basic commitments to:

- Serve all segments of society through an open-access admissions policy that offers equal and fair treatment to all students
- A comprehensive educational program
- Serve its community as a community-based institution of higher education
- Teaching
- Lifelong learning (AACC 2012)

Recently, community colleges in several states have begun offering Bachelor of Science (BS), and Bachelor of Applied Science (BAS) degrees, primarily in workforce majors. The number and type of baccalaureate degrees offered by community colleges appears to be limited to specific industries and vocations not served by the traditional state university systems. While not a comprehensive list, a brief Internet searched identified community colleges in the following states currently offering one or more baccalaureate degrees: California, Colorado, Florida, Georgia, Hawaii, Illinois, New Mexico, New York, Nevada, Texas, Vermont, Washington, and West Virginia.

Community College History

The genesis of the US community college system occurred in 1895 when William Rainey Harper, first president of the University of Chicago, conceived a concept of local colleges providing the first 2 years of academic education, prior to attending university (Tillery and Deegan 1985). In 1901, Joliet Community College became the first public 2-year community college in the United States, founded by Harper and J. Stanley Brown, the superintendent of Joliet Township High School (Sullivan 2005).

Community colleges began offering job-training in the 1930s during the Great Depression. In 1944, the US Congress passed the Serviceman's Readjustment Act of 1944, also known as the "GI Bill" to provide soldiers returning from World War II with a wide range of benefits, including education and housing. In the post-war era, the conversion of military manufacturing to consumer goods manufacturing required a skilled workforce. The community college system greatly benefited from the post-war boom in higher education resulting from industry transition from military to consumer manufacturing, and veteran utilization of the GI Bill.

US Community colleges greatly expanded during the 1960s with the number of campuses doubling during that decade, and steadily increasing in the following decades (AACC 2012). The community college concept has experienced several generational changes, beginning as secondary high schools, junior colleges, community colleges, comprehensive community colleges (Tillery and Deegan 1985), and the current generation described as the learning community college (O'Banion 1997). The characteristics of a learning college include: creating substantive change in individual learners; engaging learners as partners in the learning process, who are responsible for their learning choices; creating and offering multiple learning options; assisting learners to form and participate in collaborative activities; defining the role of learning facilitators according to learner needs; and, success measured by the occurrence of improved, expanded learning occurs (O'Banion 1997).

Community College Characteristics

Community colleges vary in size and footprint, from small rural colleges to large multicampus, urban colleges (AACC 2012). The majority of Americans live within a 1-hr drive of a community college campus or extension center. Community colleges are distinct educational institutions, each with a community-specific mission, generally connected to each other by their common characteristics (AACC 2012). There are several common characteristics of community colleges, including offering open admission, affordable tuition, and occupational training.

The majority of community colleges in the United States offer open admission (Geller 2001), based upon the unique role of providing equal academic opportunity to all Americans with high school diplomas or the equivalent (Carnegie Commission on Higher Education 1970). While standardized testing may be a requirement of admission, applicants with a high school diploma or equivalent are admitted, with developmental coursework recommended or required for lower scoring students.

The community college system is more responsive to local community and workforce needs than any other higher educational segment. Community colleges have a successful history of providing occupational training (Geller 2001) through work-based, experiential, and applied learning designed to prepare students for entry-level technical work (Sullivan 2005). The Carnegie Commission on Higher Education (1970) observed that community college occupational programs are continually increasing in variety and scope, based upon the size and complexity of the labor market. This observation continues to be valid today. While offering AA degrees enabling transfer to a university for baccalaureate completion, many of the AS and AAS degrees offered by community colleges prepare students for entry-level employment in technical and occupational fields such as automotive technology, construction, healthcare, interior design, and legal studies (paralegal).

The primary sources of community college revenue are: tuition and fees; federal, state, and local funding; and grants, gifts, endowments, and contracts with local stakeholders. The diversity of revenue sources necessitate community colleges maintain responsiveness to the demands of local industry, legislature, and student population. Community colleges are most commonly recognized for providing an affordable tuition option to traditional 4-year institutions. In addition, the occupational programs and certificates usually require less credits or hours compared to AA, AAS, and AS degrees. As a result of low attendance costs, community college students typically incur lower student loan debt than students at public and private 4-year institutions.

Community College Student Characteristics

In 2012, there were 1,132 community colleges in the USA, with a total enrollment of 12.8 million students, 7.7 million of which were enrolled in college credit programs, 4.6 million of which were enrolled part-time, and 3.1 million were enrolled full-time (AACC 2012). In the 116-year history of US community colleges, more than 100 million people have attended community colleges (AACC 2012). When compared to university students, community college students are typically older, with an average age of 29, work more hours, attend part-time, and often have families to support (Kane and Rouse 1999). Community colleges serve approximately one-half of the US undergraduate population, and the majority of African-American and Hispanic undergraduate students (AACC 2012). Community colleges are the most common postsecondary education pathway for low income, minority, and first-generation college students.

Community college students are often less connected to their colleges than university students because community colleges are not residential, with students frequently attend part-time, and spending little nonclassroom time on campus (Kane and Rouse 1999). However, as local residents, community college students usually have closer ties to the local community than students of traditional colleges and universities. Community college students consist of local residents and commuters who frequently work for, or upon graduation seek employment with, local businesses and organizations. As such, students are in a position to see, experience, and have an impact upon the actions of corporations conducting business within their communities.

Nondegree seeking community college students may enroll in one or two courses for personal enrichment or a hobby.

From humble beginnings, the US community college system has expanded to become a significant provider of adult high school, vocational training, college credit certificates, and 2-year degrees offering open admission, affordable tuition, and occupational training. Community college students are typically older, work more hours, attend part-time, and may have families to support. The community college system is the most common postsecondary education pathway for low income, minority, and first-generation college students. Having introduced the background, history, and characteristics of community colleges and community college students, the subject of the next section will be the posited community college role in providing sustainable business education.

Case for Sustainable Business Education at Community Colleges

This author posits that in order for engaged sustainability to occur within business organizations, the strategies and initiatives developed by college and university-educated middle and upper management require support of entry-level trade, management, and professional employees, frequently educated at the community college level. This proposed framework consists of sustainable business education delivery at multiple academic levels, to students who may become employees and managers at multiple levels within business organizations. While sustainable business education is increasingly occurring at 4-year and graduate colleges and universities, sustainable business education still only occurs on a limited basis within the community college system.

In this section, the case for sustainable business education at the community college level will be explored. The community college role will be presented as an integral component of a higher education system in which employees and managers at all levels acquire an understanding of the sustainable business. Two common characteristics of community colleges support the case for providing sustainable business education at community colleges: the community focus of community colleges; and, the delivery of entry-level occupational training.

The "Community" in Community Colleges

The community aspect of community colleges may present both advantages and disadvantages in terms of teaching business sustainability. Two advantages include possible student connection with the local community and the associated environmental issues and concerns, and the possibility that students attending for personal enrichment might enroll in a sustainable business course, thereby developing a passion for the topic. One disadvantage is that community college students may be less likely to embrace sustainability if the adverse effects of these environmental issues are not clearly present within their own communities.

Because students attending community colleges typically do so within their own community, these students may have an attachment to the local community, and may also be witness to the effects of environmental actions within their communities. For example, a student of a community college near an industrial area might be concerned about air pollution or have read about local EPA brownfield sites. In this case, having students who are members of the community would be an advantage because those students might be more likely to embrace business sustainability as a method of mitigating adverse environmental activities and conditions. In addition, a number of community college students attend for personal enrichment or hobby purposes; therefore, the possibility that a student might enroll in a sustainable business course out of curiosity, and develop an interest in the topic.

The community aspect of community colleges might also present a disadvantage in that community college students might be less likely to embrace sustainability if the adverse effects of these environmental issues are not clearly present within their own communities, if the nature of these environmental issues seems geographically distant, or if the sociopolitical demographic of the community is not supportive of environmental topics such as climate change. For example, a student attending a campus in a more affluent suburban might seem isolated and insulated from adverse environmental circumstances. Further, students in rural areas may feel further disconnected from the effects of environmental actions that are occurring elsewhere (the Arctic Ocean, Antarctica, or a vanishing South Pacific island).

While the collective academic community is frequently labeled as more liberal, progressive, or open-minded, and as such, may be more likely to embrace topics such as environmentalism and sustainability, community college administration, faculty, and students may be more likely to mimic the political and ideological views of the local community. For example, in predominantly conservative Republican regions, students may be less likely to take interest in, or embrace environmental sustainability based upon the belief of a significant segment of that demographic group that climate change does not exist, or is not influenced by human activity.

Entry-Level Occupational Training

The business community requires and will only support curriculum that produces students with the necessary skills to ensure businesses remain competitive (O'Banion 1997). The role of community colleges in providing entry-level occupational and vocational training has been established. Community colleges are best able to respond to business community needs by demonstrating how those needs are met within the context of the curriculum (Geller 2001). The case for sustainable business education within the community college system is supported by the fact that community colleges provide the skills necessary for employees to become and remain competitive in the workforce (Sullivan 2005). While business practitioners have questioned the relevance of topics such as sustainable business to the community college focus and mission (Beehner 2017), business interest in and demand for sustainable business education exists, and continues to increase. Moreover, the entry-level career focus of community colleges provides a unique opportunity for entry-level sustainable business education, with no other higher educational segment addressing this gap. Unfortunately, the community college system is frequently overlooked in environmental and sustainable education discussions (Potter 2009). There are long-term consequences for academia ignoring business community requests (Geller 2001). For example, when community colleges do not provide future employees with required business skills, that role will be filled by another provider (Geller 2001), such as the for-profit college sector, which is currently addressing other academic needs not being met by traditional colleges (Ruch 2001).

There are two important reasons for providing sustainable business education to entry-level business employees and supervisors. First, while entry-level employees and supervisors may not be in a position to influence corporate strategy, they are frequently the actors who ensure sustainable behavior, performance, and policies are implemented (Soyka 2013). While the individual influence of employees depends on factors such as culture, leadership style, and union influence, an organization benefits by listening to and empowering employees, and conversely, may suffer by ignoring them (Soyka 2013). Second, these employees are the company's representatives in the local community (Soyka 2013). Therefore, it is important for these employees to understand and embrace the concept of sustainability. Moreover, these employees may bring practical ideas to the workplace because they may live and work on the frontline of environmental and sustainability impacts and concerns.

The increasingly diverse US student body and workforce might enable and further sustainable business education for two reasons: students from diverse countries and cultures may have learned about or experienced adverse environmental impacts in their home countries; and, demographic diversity has been demonstrated to have a significant positive impact on environmental issue attitudes (Dunlap 2008). In addition, the Millennial Generation and Generation Z are the generations most willing to pay higher prices for products and services from businesses demonstrating a positive commitment to environmental and social issues and problems (Nielsen Company 2015).

Community Colleges with Sustainable Business and Sustainability Curriculum

As the field of sustainability began to develop during the 1980s, postsecondary education institutions began implementing sustainability policies and practices, and later, curriculum (Vaughter et al. 2013). However, community colleges have been slower to embrace sustainability in their policies, procedures, and curriculum (Feldbaum 2009). An increasing number of community colleges have sustainability courses and programs, including offerings within the fields of environmental science, engineering, energy, agriculture, and more recently business. A summary of several of the nonbusiness courses and programs will be briefly enumerated in order to establish the history and direction of sustainability education within the community college system. However, the primary focus of this section will be with community colleges offering sustainable business courses, certificates, and degrees.

Sustainability Education in Community Colleges

The majority of community college sustainability courses, certificates, and degrees are offered in nonbusiness fields. A brief summary of nonbusiness community college sustainability activity is introduced in this section in order to establish a background on the movement. Kapi'olani Community College in Hawaii offers sustainability courses within a hospitality degree, in addition to having established sustainability plans, and a Service and Sustainability pathways program. Cascadia College in Bothell, WA, offers a Bachelor of Applied Science in Sustainable Practices which includes an interdisciplinary offering of courses. Gulf Coast State College in Florida offers an AS in building construction technology with a specialization in sustainable design. Monroe Community College in New York offers a Science Technology Engineering and Mathematics (STEM) Sustainability certificate that includes economics courses, but not business courses.

Sustainable Agriculture or Farming certificates are offered by Johnson Community College in Kansas, Lorain County Community College in Ohio, Wayne Community College in North Carolina, and Mendocino College in California. Associate degree in Sustainable Agriculture or Farming is offered by Lorain County Community College in Ohio, Tompkins Cortland Community College in New York, and Wayne Community College in North Carolina. Holyoke Community College in Massachusetts offers an AS degree and a certificate in Sustainability Studies with specializations in agriculture and energy. Certificates in Sustainable or Renewable Energy are offered by Manchester Community College in Connecticut, and Dallas County Community College District in Texas. Three Rivers Community College in Connecticut offers Sustainable Landscape Ecology and Conservation Technician, and Sustainable Facilities Management certificates.

Sustainable Business Education in Community Colleges

While the American Association of Community Colleges (AACC) is a member of the Higher Education Associations Sustainability Consortium (HEASC) of Association for the Advancement of Sustainability in Higher Education (AASHE), only one member is currently listed as offering a sustainable business degree (AASHE 2017). St. Petersburg College in Florida has offered a Bachelor of Science degree in Sustainability Management for several years. Moreover, while 32 of the 38 sustainable and sustainability-related Associate's degrees offered by AASHE members are delivered by community colleges, no community college members are currently offering an Associate's degree in sustainable business.

However, an increasing number of community colleges are offering courses and college credit certificates in sustainable business, especially in Washington State. Whatcom Community College and Shoreline Community College both offer certificates in sustainable or sustainability business leadership. Bellevue College previously offered a number of certificate level programs, including Sustainability Coordinator, Sustainable Business Accounting, Sustainable Systems Best Practices, and Sustainable Business Best Practices, but currently only offers a nonbusiness sustainability concentration. North Seattle College offers a Green Real Estate Certificate. Edmonds Community College in Washington State currently offers one sustainable business course Sustainable Business Practices.

Beyond Washington State, City College of San Francisco has a number of certificate level programs including college credit certificates in Green and Sustainable Business, and Green and Sustainable Travel, and a noncredit certificate in

Green and Sustainable Small Business. Chemeketa Community College in Oregon offers a Sustainability in Management Certificate. Bunker Hill Community College in Massachusetts offers a course entitled Introduction to Sustainable Business. Chandler-Gilbert Community College in Arizona offers an academic certificate in sustainability in the biological sciences department with a business and entrepreneurship specialization.

Florida State College Jacksonville offers the course *Sustainable Business Strategies* as a component of entrepreneurship concentration and general business administration concentration within a BS in Business Administration degree. Mesa Community College offers an academic certificate in Sustainability which, while not containing business courses, is promoted as being ideal for business professionals who wish to become change agents for sustainability within their respective organizations, or for students who already have a sound business foundation. SUNY Broome Community College offers one course entitled *The Sustainable Business* examining how large and small businesses can gain competitive advantage through incorporating sustainability.

In the Fall 2014 Semester, Seminole State College of Florida began offering a junior-level course, *Sustainable Business Strategies* as an elective within the Business and Information Management baccalaureate degree. During the first semester, the course filled to within two students of the 30-student classroom capacity. The course was subsequently modified for online delivery, after which enrollment per semester doubled to 60 students. In the Fall 2016 semester, the course was incorporated into the required coursework for a Supply Chain Management specialization offered within that same degree. For several years prior to the introduction of this course, Seminole State offered a technical certificate in Sustainable Engineering. Upon receipt of a National Science Foundation (NSF) EMERGE grant in 2015, the certificate was modified to be more interdisciplinary in focus, in order to be attractive to students with nonengineering majors. Along with sustainability courses in public policy and environmental policy, a sophomore-level *Sustainable Business* course was developed as an elective for the newly modified *Sustainability* certificate. Having developed both courses for Seminole State, this author will discuss the curriculum and pedagogy of the courses in the "Discussion" section of this chapter.

While community college participation in sustainable business education has been limited, an increasing number of institutions have embraced the necessity for entry-level employee and management training on the subject. The geographic areas currently represented by courses, certificates, and degrees, while limited suggests a growing acceptance of the need for sustainable business education at a level lower than, but certainly not less relevant than offerings at traditional colleges and universities.

Discussion

The following discussion outlines how community colleges might further address the gap in entry-level sustainable business education, and includes approaches, methods, and tools, based upon the literature and author experience. Best practices, and the role and influence of the Millennial and Z generations, and cultural diversity will be examined. While there are currently a limited number of community college courses and programs from which to extract best practices, this section will examine the limited literature on sustainable business education exist along with the sustainable business course development experience of this author. The challenges experienced by this author in the course development process, and the methods used to overcome those challenges will be discussed. The intent is to stimulate discussion and thought concerning how to operationalize the delivery of sustainable business education within the community college system.

Approaches for Teaching Sustainable Business

The various perspectives on sustainability exist on a continuum, categorized into three broad paradigms (Gladwin et al. 1995): ecocentrism, ecological modernization, and neoclassical economics. Each of these paradigms offers a lens through which sustainability may be examined, and assumptions drawn (Stubbs and Cocklin 2008). Ecocentrism is a philosophical paradigm that is nature-centered, and not human-centered. Ecological modernization is an increasingly popular social science paradigm based upon the view environmentalism is beneficial to the economy. Neoclassical economics is based upon the laws of supply and demand, individual rationality, and profit and utility maximization. While traditional MBA students have been primarily exposed to the neoclassical economic paradigm, sustainability necessitates a view from multiple perspectives in order to develop critical thought and reflection (Stubbs and Cocklin 2008).

Research findings suggest active learning to be a successful sustainable business education tool, by placing greater emphasis on personal responsibility, and less emphasis on faculty influence (MacVaugh and Norton 2012). Active learning is a teaching method wherein students actively participate in the learning process, instead of passively receiving instruction. Methods of active learning include role-playing, case studies, and experiential learning. Quality management has been also successfully used as method of conjoining environmental sustainability into business school curriculum (Rusinko 2005). The quality management goal of the minimization of defects in the production of goods and services resonates well with the sustainability goal of natural resource waste minimization.

Sustainable business courses should include a critical theory approach in addition to the curriculum content. Despite limited research on the suitability of critical theory in business education, this theory has been proven successful in promotion of radical change agendas, such as ecojustice and sustainability (Kearins and Springett 2003). Using critical theory, students can explore the benefits and limitations of current business practices, and consider alternative practices. With a focus beyond traditional management control, learners can focus on the underlying influences of business, and how they impact our collective lives (Kearins and Springett 2003).

In addition, a transdisciplinary approach, collaborative and transformative learning, and participatory evaluation are recommended pathways to achieving organizational change regarding sustainable business education (Moore 2005). A transdisciplinary or interdisciplinary approach would include, among other components: systems thinking as a method of understanding and reflecting upon the interdependency of systems and the effects and feedback loops; foresighted thinking, in order to analyze, evaluate, and develop a vision of the future and the corresponding impact of business decisions on the long term, collective future; and, strategic management, consisting of the ability to collectively design and implement projects, interventions, and strategies for sustainable development (Lans et al. 2014). The focus of collaborative and transformative learning... Participatory evaluation...is the creation of a participative, cooperative environment, instead of the traditional transfer of knowledge from faculty to students. The students exchange ideas and insights with an emphasis on listening and understanding alternative perspectives. The focus of participatory evaluation is the creation of transparent evaluation of academic projects and programs by all stakeholders (faculty, administration, and students).

Methods and Tools for Teaching Sustainable Business

Students pursuing community college business degrees usually follow one of the two paths: transfer to a university for bachelor degree completion; or, obtain entry-level employment in business. Because of the workforce training mission of community colleges, students attending these institutions may have taken limited coursework in the field of environmental science, and may have little or no interest in the environmental field. Therefore, sustainable business curriculum should be presented in a practical, nontechnical manner. The addition of sustainability content in business curriculum should demonstrate more of a business case than a social or philosophical case for embracing sustainability. Because community college students are frequently permanent residents of the local community, sustainable business case studies and success stories should be drawn from the local community, or at least from industries existing in the local community.

Sustainable business must be taught using the language of business. Because business students are taught within a capitalist framework, it is understandable that skepticism might exist regarding the relevance and role of sustainability in a capitalist business organization. Sustainability should be introduced as both a model of responsible business, as well as a method of managing waste and inefficiency for profit maximization. Sustainability may lead to cost-reduction through the reduction and mitigation of waste and inefficiency, and profit-maximization through marketing the business as being a sustainable, responsible member of society.

In reviewing the previously discussed sustainable business degree programs, two sustainable business teaching models emerge. With the more prominent model, sustainable business is taught as a separate, standalone course, either as a required or elective component of a degree. In the less prominent model, sustainability permeates multiple courses. As stated in a previous section, the isolated course model suggests sustainability is a separate issue, disconnected from core business concepts. Therefore, in order for sustainable business education to be successful within the community college system, the sustainability message should be present through the certificate or degree curriculum, in addition to courses specific to the topic of sustainable business.

Because the topic of sustainability is frequently new and foreign to many business organizations, introduction and implementation of sustainability initiatives represent change. Resistance to change is common, frequently motivated by anxiety, miscommunication, or misunderstanding concerning the proposed change. Change management content could be incorporated into the curriculum in order to prepare students for managing change and becoming change agents. Genuine and significant sustainability programs often require change at all levels, frequently requiring a significant shift in organizational culture. While middle and upper-management might be capable and prepared to create change, entry-level employees and managers must also be prepared to introduce change upward from their level.

Curriculum Development and Delivery Best Practices and Challenges

This author successfully developed curriculum for both a sophomore and a junior level sustainable business course by making the content practical and politically neutral. The sophomore-level course was intended to be an elective within a Sustainability certificate. The junior-level course was initially intended to be an elective within a BS degree in Business and Information Management degree, and was subsequently incorporated as a required course in a supply chain management specialization within the BS degree. Sustainability was presented from two perspectives: the need for a business model to be sustainable in the long-term if the enterprise is to remain economically competitive (Soyka 2013), and the relevance of environmental sustainability to contemporary consumer demands. This author suggests academics may bridge the perceived gap between sustainability and profitability by focusing on the symbiotic relationship of the two concepts. The two forms of sustainability were labeled the "two greens" (dollars and ecological), and presented as symbiotic components of a contemporary business model. This model was patterned after the structure of a sustainable business course offered by McGill University in Montreal, in which students were exposed to sustainability through a merger of the concepts of ecoeffectiveness and stakeholder effectiveness, including real examples of profitable sustainable business. Because the supply chain provides numerous opportunities for organizations to achieve both types of sustainability, and was the prior career field of this author, many of the case studies and examples used in the two courses were drawn from the supply chain management field.

Enterprise thinking, life cycle thinking (LCT), and life cycle analysis (LCA) were course content components, both of which are viable business concepts that also mirror environmental thought. These topics are representative of a circular economy, which is a regenerative system in which resource use and waste are limited through recycling, refurbishment, remanufacturing, and repair. In the contemporary capitalist business environment, consideration of the circular economic model is a credible model for cost reduction and profit maximization. For further discussion, see the chapter in this handbook entitled ▶ "Teaching Circular Economy" by Kopnina.

The challenges faced by this author in the development and delivery of the two sustainable business courses included: stakeholder concern about the need for or relevance of a business course in sustainability; and, the identification of appropriate curriculum and methodology for course development and delivery. Stakeholder concern about the need for or relevance of a sustainable business course was based upon two factors: the career-specific focus of a community college; and, the reluctance of many local business representatives to accept and embrace the business role in issues such as climate change and environmental responsibility. The course development challenge was based upon the limited availability of undergraduate level texts and course material (Beehner 2017). The majority of publishers offering sustainable business texts did not offer instructor resources, necessitating that this author develop test banks and Power Point presentations for both courses.

Because one of the purposes of business is to create and maximize value, it would be appropriate to approach the topic of sustainable business from a value creation and maximization perspective (Soyka 2013). The sustainable business value proposition is satisfied by the perceived increase in the value of a product or service that performs as intended, while minimizing environmental impact (Beehner 2017). In addition to value creation and maximization, sustainable marketing is encouraged as a method of reaching the consumer base desiring sustainable products and services. Sustainable marketing is encouraged as a means of business differentiation and value maximization. Because many consumers are seeking green and sustainable products and services, students should understand methods of marketing the sustainable aspects of businesses.

Generational and Cultural Role and Influence

As with any movement, hope is placed on the next generation to respond favorably to the notion of sustainable business. With the average age of a community college student being 29 (Kane and Rouse 1999), and community colleges currently serving approximately one-half of US undergraduate students (AACC 2012), community colleges are clearly positioned to serve the younger generations of current and future employees. It was previously noted that the Millennial Generation and Generation Z are more willing to pay higher prices for products and services from businesses with a positive commitment to environmental and social impact (Nielsen Company 2015), so it would follow that those generations would be more receptive to sustainability content within business curriculum. For more discussion on the engagement of younger students on sustainability, please see the chapter in this handbook entitled ▶ "Education in Human Values" by Ulluwishewa. For more discussion on the engagement of younger students on the topic of business sustainability, please see the chapter in this handbook entitled ▶ "Business Youth for Engaged Sustainability to Achieve the United Nations 17 Sustainable Development Goals (SDGs)" by De Feis.

Some of the challenge to business sustainability within the United States may be cultural in origin. The US culture is more individualistic than many European and

Asian cultures which may explain why residents of these regions have been more receptive to the message of sustainability. The US culture of individualism dates back to the founding of the original colonies which were established by early settlers in an effort to be governed independently from the European institutions left behind. This culture was reinforced by the Declaration of Independence in 1776, the Manifest Destiny movement that followed, and various political and ideological influences including the current populist, protectionist movement which supported the election of Donald Trump as President in 2016. In order for more of the US population to embrace sustainability, a cultural shift, or transition from individualism to collectivism might be required. For more discussion on the transition from an individualist to a collectivist approach to sustainability, please see the chapter in this handbook entitled ▶ "Sustainable Higher Education Teaching Approaches" by Krogman.

In this section, several methodologies and approaches for successful sustainable business education have been suggested, including active learning, quality management, change management, a transdisciplinary or interdisciplinary approach, collaborative or transformational learning, and participatory evaluation. While traditional business students are primarily exposed to the neoclassical economic paradigm, sustainable business education necessitates a view from multiple lenses in order to develop critical thought and reflection. Moving beyond traditional management theory, sustainability courses should be modified to include a critical theory approach in addition to the curriculum content.

Sustainable business content should be taught using the language of business, recognizing that community college business students might have limited environmental science background. In emphasizing the community focus of community colleges, attention should be focused on the local (in addition to the global) impacts of sustainability. Engaging and empowering the younger generation of students is essential given the greater likelihood that the Millennial and Z generations will embrace sustainable business. The practical business benefits of the circular economy model should be emphasized as credible methods of cost reduction and profit maximization. Cultural challenges (individualism versus collectivism) should be acknowledged and accounted for in the teaching focus.

Recommendations and Conclusion

While addressing future environmental issues will require environmentally literate managers, many business students currently graduate with an undergraduate degree absent environmental content (Rowe 2002). While an increasing number of colleges and universities are offering undergraduate and graduate coursework, certificates, and degrees in sustainable business, the success of this top-down approach to sustainable business may be limited by lack of sustainability awareness among entry-level workers and managers. Therefore, sustainable business education should occur at multiple academic levels, specifically community colleges, wherein students prepare for entry-level trade, management, and professional positions at businesses which have implemented, or will implement, sustainability programs. Community colleges provide an affordable higher education option, have a successful history of providing occupational training (Geller 2001) to prepare students for entry-level technical work (Sullivan 2005), and are able to respond to business community needs by demonstrating how those needs are met within the context of the curriculum (Geller 2001). Unfortunately, the community college system is frequently overlooked in environmental and sustainable education discussions (Potter 2009), and has been lagging in sustainable business curriculum offerings (Feldbaum 2009).

Based upon the identified gap in sustainable business education, community college faculty and administrators should consider implementing sustainable business courses, certificates, and degrees, incorporating the approaches, methods, tools, and best practices outlined in the previous section. As more community colleges participate in the sustainable business education movement, additional approaches, methods, tools, and best practices may emerge. Business may benefit by having sustainability initiatives originating at the top of the organization better understood at lower levels. The greater society may benefit from the increased awareness of sustainability, and the increased role business may play in promoting sustainability.

Future Research

Several areas for future sustainable business education research have been identified. The following are most relevant to the better understanding the role and effectiveness of community college education. The community college system clearly plays a significant role in providing education to a broad population segment. However, while higher education studies of environmental attitudes, behavior or antecedents of environmental behavior are lacking, studies focusing on those areas in community college campuses are nearly nonexistent (Hutcherson 2013). The lack of consistent assessment of sustainable business programs is a concern, necessitating the collection of valid assessment data by educational researchers in order to encourage curriculum reform and define best practices (Venkataraman 2009).

Future research in the area of sustainable business education has been recommended in the following areas: learning outcome measurement; consideration of geographical, political, and cultural contexts; and, prioritization of sustainable organizational change strategies (Adomßent et al. 2014). Sustainable business education learning outcomes should be measured in order to determine whether students are developing the intended sustainable business competencies. Because existing HESD research has originated in the USA, Western Europe, and the developed nations of Asia, it is essential to understand sustainable business education viewpoints based upon the geographical, political, and cultural characteristics of the underrepresented nations and regions. Finally, organizational aspects such as unit interaction with the overall organization, and internal and external stakeholder management should be better understood in order to develop curriculum for sustainable organizational change.

In addition to research focusing on curriculum and institutions, further research is needed about sustainable business education students. Limited research exists about the learning experiences of sustainable business education recipients (Cullen 2016), suggesting the need for future research on this topic. Future empirical research is also needed concerning how and why business students engage with sustainability principles (Cullen 2016). It would be helpful to understand the outcomes of sustainable business education with students who embrace the topic as compared to students who do not embrace sustainability. An understanding of what teaching methods and curricula are most successful in changing the viewpoints of sustainability skeptics might improve the success of sustainable business education curricula.

At time of press, there is no existing research examining community college sustainable business education programs, students, or outcomes. Future research could examine the successfulness of community college sustainable business programs, and whether graduates of these programs enhance sustainability initiatives within businesses as entry-level employees and managers, as posited by the author of this chapter.

Conclusion

The community college system plays a key role as benefactors of sustainable business education to a constituency essential for the development of a sustainability paradigm within the greater business community. While traditional colleges and universities prepare middle and senior management for the implementation of sustainable business strategies, the success of those strategies may be limited without the buy-in and support of up-and-coming entry-level employees and managers. This chapter presented the background of sustainability education and sustainable business education, and examples of higher education institutions participating in the sustainable business education initiative. The history and function of the community college system was examined, including the case for community college participation in sustainable business education, and examples of US community colleges currently participating in sustainable business education. Approaches, methods, tools, and best practices for successful community college sustainable business education strategies were explored along with conclusions and recommendations for the future. Best practices, and the role and influence of the Millennial and Z generations, and cultural diversity were examined.

It is the hope of this author that the reader concludes this chapter with a better understanding of the crucial role of community colleges in the delivery of a holistic platform of sustainable business education. The intent of this chapter was the stimulation of discussion and thought concerning operationalization of sustainable business education delivery within the community college system. If the "cosmic vision" of sustainability is to extend from neighborhoods to communities, states, countries, and the globe, community colleges play a fundamental role in expanding "big-picture awareness" of sustainable business at the neighborhood and community level.

In this chapter, a case has been presented for increased community college participation in sustainable business education. However, the community college role in sustainable business education is currently limited, with few models and examples to analyze or mimic. Further, because a large percentage of community college students earn their AA degree and transfer to 4-year colleges and universities, many of the students participating in sustainable business education may likely advance beyond entry-level positions, further mitigating the effects of the proposed gap community colleges would address. Finally, the inclusion of sustainability content in business curricula does not guarantee students will implement those concepts in the classroom or the workplace (Thomas 2005).

Cross-References

- ▶ Business Youth for Engaged Sustainability to Achieve the United Nations 17 Sustainable Development Goals (SDGs)
- ▶ Education in Human Values
- ▶ People, Planet, and Profit
- ▶ Sustainable Higher Education Teaching Approaches
- ▶ Teaching Circular Economy

References

Accenture. (2010). A new era of sustainability: UN Global Compact-Accenture CEO study. http://www.unglobalcompact.org. Accessed 4 May 2017.

Adomßent, M., Fischer, D., Godemann, J., Herzig, C., Otte, I., Rieckmann, M., & Timm, J. (2014). Emerging areas in research on higher education for sustainable development – Management education, sustainable consumption and perspectives from Central and Eastern Europe. *Journal of Cleaner Production, 62*(1), 1–7. https://doi.org/10.1016/j.jclepro.2013.09.045.

Aligleri, L., Aligleri, L. A., & Kruglianskas, I. (2009). *Gestão socioambiental: Responsabilidade e sustentabilidade do negócio* (1st ed.). São Paulo: Atlas.

American Association of Community Colleges (AACC). (2012). http://www.aacc.nche.edu/AboutCC/Pages/default.aspx. Accessed 3 June 2017.

Association for Advancement of Sustainability in Higher Education (ASHE). (2017). Academic programs. https://hub.aashe.org/browse/types/academicprogram/. Accessed 12 May 2017.

Baccarin, G., Cezarino, L. O., Fernandes, V., Liboni, L., & Martinelli, D. P. (2015). *Aplicação do Soft System Methodology na gestão de resíduos perigosos em laboratórios*. Paris: Anais do XI Congresso Brasileiro de Sistemas.

Bates, C., Silverblatt, R., & Kleban, J. (2009). Creating a new green management course. *Business Review, 12*(1), 60–66.

Beehner, C. G. (2017). Teaching sustainability to a traditional business audience. In Conference proceedings of the international academy of business and public administration discipline, 14(1). Paper presented at IABPAD conference, Orlando, pp. 298–308.

Bridges, C. M., & Wilhelm, W. B. (2008). Going beyond green: The "why and how" of integrating sustainability into the marketing curriculum. *Journal of Marketing Education, 30*(1), 33–46. https://doi.org/10.1177/0273475307312196.

Calder, W., & Clugston, R. M. (2003). International efforts to promote higher education for sustainable development. *Planning for Higher Education, 31*(3), 30–44.

Carnegie Commission on Higher Education. (1970). *The open-door colleges: Policies for community colleges*. New York: McGraw-Hill Book Company.

Cezarino, L. (2016). Teachers' opinion about sustainability on management education. *Business Management Dynamics, 6*(1), 1–8.

Corcoran, P., & Wals, A. (Eds.). (2004). *Higher education and the challenge of sustainability: Problematics, promise, and practice*. Dordrecht: Kluwer Academic.

Cullen, J. G. (2016). Educating business students about sustainability: A bibliometric review of current trends and research needs. *Journal of Business Ethics, 145*(2), 429–439. https://doi.org/10.1007/s10551-015-2838-3.

Dunlap, R. E. (2008). The new environmental paradigm scale: From marginality to worldwide use. *Journal of Environmental Education, 40*(1), 3–18. https://doi.org/10.3200/JOEE.40.1.3-18.

Feldbaum, M. (2009). *Going green: The vital role of community colleges in building a sustainable future and green workforce*. Washington, DC: Academy for Educational Development.

Geller, H. A. (2001). *A brief history of community colleges and a personal view of some issues (open admissions, occupational training and leadership)*. Washington, DC: US Department of Education. http://files.eric.ed.gov/fulltext/ED459881.pdf. Accessed 12 May 2017.

Gladwin, T. N., Kennelly, J. J., & Krause, T.-S. (1995). Shifting paradigms for sustainable development: Implications for management theory and research. *Academy of Management Review, 20*(4), 874–907. https://doi.org/10.5465/AMR.1995.9512280024.

Gonçalves-Dias, S. L. F., Teodósio, A. D. S. D. S., Carvalho, S., & Silva, H. M. R. D. (2009). Environmental awareness: An exploratory study into the implications for teaching business administration. *Revista de Administração de Empresas Eletrônica, 8*(1), 1–22.

Hatfield-Dodds, S., Turner, G., Schandl, H., & Doss, T. (2008). *Growing the green collar economy: Skills and labour challenges in reducing our greenhouse emissions and national environmental footprint*. Canberra: Report to the Dusseldorp Skills Forum.

Hoffman, A. J. (1999). Environmental education in business school. *Environment, 4*, 41.

Hutcherson, J. D. (2013). Community college students' pro-environmental behaviors and their relationship to awareness of college sustainability strategy implementation. Doctoral dissertation, Western Carolina University.

Jabbour, C. J. C. (2010). Greening of business schools: A systemic view. *International Journal of Sustainability in Higher Education, 11*(1), 49–60. https://doi.org/10.1108/14676371011010048.

Kane, T. J., & Rouse, C. E. (1999). The community college: Educating students at the margin between college and work. *Journal of Economic Perspectives, 13*(1), 63–84. https://doi.org/10.1257/jep.13.1.63.

Kearins, K., & Springett, D. (2003). Educating for sustainability: Developing critical skills. *Journal of Management Education, 27*(2), 188–204. https://doi.org/10.1177/1052562903251411.

Lans, T., Blok, V., & Wesselink, R. (2014). Learning apart and together: Towards an integrated competence framework for sustainable entrepreneurship in higher education. *Journal of Cleaner Production, 62*, 37–47. https://doi.org/10.1016/j.jclepro.2013.03.036.

Leal Filho, W. (2000). Dealing with misconceptions on the concept of sustainability. *International Journal of Sustainability in Higher Education, 1*(1), 9–19. https://doi.org/10.1108/1467630010307066.

Lozano, R., Ceulemans, K., Alonso-Almeida, M., Huisingh, D., Lozano, F. J., Waas, T., et al. (2015). A review of commitment and implementation of sustainable development in higher education: Results from a worldwide survey. *Journal of Cleaner Production, 108*, 1–18. https://doi.org/10.1016/j.jclepro.2014.09.048.

MacVaugh, J., & Norton, M. (2012). Introducing sustainability into business education contexts using active learning. *International Journal of Sustainability in Higher Education, 13*(1), 72–87. https://doi.org/10.1108/14676371211190326.

Moore, J. (2005). Barriers and pathways to creating sustainability education programs: Policy, rhetoric and reality. *Environmental Education Research, 11*(5), 537–555. https://doi.org/10.1080/13504620500169692.

Nielsen Company. (2015). The sustainability imperative. New York. http://www.nielsen.com/us/en/insights/reports/2015/the-sustainability-imperative.html. Accessed 20 May 2017.

O'Banion, T. (1997). Back to the future. In T. O'Banion (Ed.), *A learning college for the 21st century* (pp. 41–62). Phoenix: The Oryx Press.

Potter, G. (2009). Environmental education for the 21st century: Where do we go now? *Journal of Environmental Education, 41*(1), 22–33. https://doi.org/10.1080/00958960903209975.
PRME Secretariat. (2017). *Signatories*. New York: UN Principles of Responsible Management Education (PRME) Secretariat. http://www.unprme.org/participants/. Accessed 20 May 2017.
Prospects, D. (2017). *Sustainable business degrees*. Washington, DC: Degree Prospects LLC.. http://www.sustainabilitydegrees.com/degrees/sustainable-business/. Accessed 20 May 2017.
Rowe, D. (2002). Environmental literacy and sustainability as core requirements: Success stories and models. In W. Leal Filho (Ed.), *Reprinted from: Teaching sustainability at universities*. New York: Peter Lang.
Ruch, R. (2001). *Higher Ed, Inc.: The rise of the for-profit university*. Baltimore: Johns Hopkins University Press.
Rusinko, C. A. (2005). Using quality management as a bridge in educating for sustainability in a business school. *International Journal of Sustainability in Higher Education, 6*(4), 340–350. https://doi.org/10.1108/14676370510623838.
Ryan, A., Tilbury, D., Blaze Corcoran, P., Abe, O., & Nomura, K. (2010). Sustainability in higher education in the Asia-Pacific: Developments, challenges, and prospects. *International Journal of Sustainability in Higher Education, 11*(2), 106–119. https://doi.org/10.1108/14676371011031838.
Soyka, P. A. (2013). *Creating a sustainable organization: Approaches for enhancing corporate value through sustainability*. Upper Saddle River: Pearson Education.
Stubbs, W., & Cocklin, C. (2008). Teaching sustainability to business students: Shifting mindsets. *International Journal of Sustainability in Higher Education, 9*(3), 206–211. https://doi.org/10.1108/14676370810885844.
Sullivan, L. G. (2005). *National profile of community colleges: Trends & statistics*. Washington, DC: American Association of Community Colleges.
Thomas, T. E. (2005). Are business students buying it? A theoretical framework for measuring attitudes toward the legitimacy of environmental sustainability. *Business Strategy and the Environment, 14*(3), 186–197. https://doi.org/10.1002/bse.446.
Tillery, D., & Deegan, W. (1985). The evolution of two-year colleges through four generations. In W. Deegan & D. Tillery (Eds.), *Renewing the American community college: Priorities and strategies for effective leadership* (pp. 3–33). San Francisco: Jossey-Bass Publishers.
United Nations Educational, Scientific and Cultural Organization (UNESCO). (2002). *Education for sustainability – From Rio to Johannesburg: Lessons learnt from a decade of commitment*. Paris: UNESCO. http://unesdoc.unesco.org/images/0012/001271/127100e.pdf. Accessed 27 May 2017.
United Nations Educational, Scientific and Cultural Organization (UNESCO). (2012). *Education for sustainable development sourcebook*. Paris: UNESCO. https://sustainabledevelopment.un.org/content/documents/926unesco9.pdf. Accessed 27 May 2017.
Vaughter, P., Wright, T., McKenzie, M., & Lidstone, L. (2013). Greening the ivory tower: A review of educational research on sustainability in post-secondary education. *Sustainability, 5*, 2252–2271. https://doi.org/10.3390/su5052252.
Venkataraman, B. (2009). Education for sustainable development. *Environment: Science and Policy for Sustainable Development, 51*(2), 8–10. https://doi.org/10.3200/ENVT.51.2.08-10.
Von Der Heidt, T., & Lamberton, G. (2011). Sustainability in the undergraduate and postgraduate business curriculum of a regional university: A critical perspective. *Journal of Management & Organization, 17*(5), 670. https://doi.org/10.1017/S1833367200001322.
Wright, T. S. (2002). Definitions and frameworks for environmental sustainability in higher education. *International Journal of Sustainability in Higher Education, 3*(3), 203–220. https://doi.org/10.1108/14676370210434679.

Business Youth for Engaged Sustainability to Achieve the United Nations 17 Sustainable Development Goals (SDGs)

George L. De Feis

Contents

G. L. De Feis (✉)
Department of Management, Healthcare Management, and Business Administration, Iona College, School of Business, New Rochelle, NY, USA
e-mail: gdefeis@iona.edu

S. Dhiman, J. Marques (eds.), *Handbook of Engaged Sustainability*,
https://doi.org/10.1007/978-3-319-71312-0_39

Abstract

Sustainable development is often defined as "meeting the needs of the present without compromising the ability of future generations to meet their own needs." Three main pillars of "sustainability" are "social, environmental, and economic," which are often referred to informally as "people, planet, and profits." Indeed, there is no larger initiative to face the planet than achieving "global sustainable development" amidst globalization issues and *no larger untapped force than Business Youth*. Many youth programs exist – Enactus, the Future Business Leaders of America, Junior Achievement, Operation Enterprise, United Nations Youth Unit – all striving toward the United Nations 17 Sustainable Development Goals (SDGs), which include quality education (No. 4), industry innovation and infrastructure (No. 9), and partnerships for the goals (No. 17). Perhaps there could be no greater facilitator in this realm than the creation of a nonprofit organization, called *Business Youth for Sustainable Development* (BY4SD). An international business youth organization will have the energy to get it done by assembling the youth at business schools, state organizations, regional organizations, national organizations, and international organizations, which could help "shift the paradigm" for all. The concept of "sustainable development" was reborn in 2012 at the "Rio + 20 Convention" in Rio de Janeiro, where it was "christened" 20 years before at the 1992 Earth Summit. The concept, though, was born to the world in the 1987 book, *Our Common Future* from the United Nations World Commission on Environment and Development (WCED). This chapter will present an overview of existing business youth organizations, their work toward achieving the United Nations SDGs, and propose the concept and the need for BY4SD.

Keywords

Civic engagement · Corporate social responsibility · Earth Summit · Haves · Have-nots · Globalization · Malthusian theory of population · Personal social responsibility · Service-learning · Sustainability · Sustainable development · Sustainable Development Goals (SDGs) · United Nations

Introduction

The concept of "sustainable development" has been talked about for many years, but the conception of the term was put forth in 1987 by the Brundtland Commission, formerly known as the World Commission of Environment and Development (WCED), which issued their magnum opus, *Our Common Future*. The term has come to mean, "Meeting the needs of the present without compromising the ability of future generations to meet their own needs." *Our Common Future* was this UN body's assessment – after much analysis, synthesis, and expert testimony from industrialists, scientists, NGO representatives, and the general public – of the dismal future within our commonality. In other words, we were on an unsustainable path.

This 1987 work leads to the 1992 Earth Summit, formerly known as the United Nations Conference on Environment and Development, in Rio de Janeiro. The result of the Rio meeting was "hugs and kisses" for all who attended from 172 countries, with 108 heads of state, and a final document: Agenda 21. Agenda 21 set forth our uniform marching orders to reverse the unsustainable path we were making with our current development practices.

The 1996 Kyoto Protocol on climate change followed, with less participation than the Earth Summit – the United States did not participate for fear that participating would lead to a "reduction" it its way of life. The protocol's result was agreement that mandatory targets on greenhouse gas emissions for the world's leading economies were set and accepted. Not the United States. There the much less hugging and kissing.

An Inconvenient Truth (2006), "Rio + 20," and *An Inconvenient Sequel* (2017) followed – all striving to accomplish the elusive goal of "sustainable development." This goal may have been set forth by our senior colleagues, and we may all agree what the goal is, but we cannot achieve the said goal without the concerted effort of all individuals and from all classes of people. The "haves," "have-nots," industrialized, nonindustrialized, senior folk, junior folk, man, woman, and we must involve "youth." In fact, the United Nations put forth its "17 Sustainable Development Goals (SDGs);" all of them need the input of "youth." My concept is the development of "business youth" programs, worldwide, to achieve the sustainable development needs for which we have been seeking.

Three-Legged Stool of Sustainable Development

The confluence of the three needed constituent parts of the sustainable development includes social, environment, and economic (2006). Some consider this as the "triple bottom line" (3Ps): people, planet, and profits of sustainability (social/people, environment/planet, economic/profit) (Fig. 1).

We must remember that all three are needed in equal portions to accomplish sustainability. Anything less is not lasting, for it is either bearable, equitable, or viable, but not sustainable.

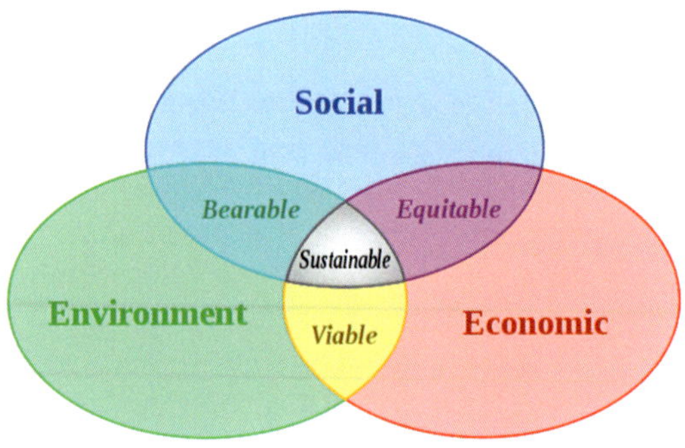

Fig. 1 Three-legged stool of sustainable development (http://www.uvm.edu/~jashman/CDAE195_ESCI375/What_is_Sustainable_Development.html)

Preceding Sustainable Development

The *environment* has been considered for years and years and years, from Thomas Robert Malthus, who wrote, *An Essay on the Principle of Population*, published in the 1800s to Teddy Roosevelt and John Muir, and our national parks have been considered greatly. Having just returned from Muir Woods (January 2018), one can imagine what the world would look like if we focused on sustainable development all of our lives, from infancy to youth to adulthood. Of course, we took a turn for the worst with *Silent Spring* (Rachel Carson 1962), Three Mile Island (1979), Love Canal (1980), Chernobyl (1986), Exxon Valdez (1989), and BP Oil Spill (2010). The world's environment is very considered and always current.

Environment Part

Population Growth (Thomas Robert Malthus)

The Malthusian theory of population shows that while our food production and resources grow at sort of an arithmetic rate, the population is growing exponentially. What will happen when the needs of the population outpace the availability of our food production and resources? Some say it has. Will conflict, violence, and wars become inevitable? A famous quote of Malthus from 1798:

> *The power of population is so superior to the power of the earth to produce subsistence for man that premature death must in some shape or other visit the human race. The vices of mankind are active and able ministers of depopulation. They are the precursors in the great army of destruction, and often finish the dreadful work themselves. But should they fail in this war of extermination, sickly seasons, epidemics, pestilence, and plague advance in terrific array, and sweep off their thousands and tens of thousands. Should success be still incomplete, gigantic inevitable famine stalks in the rear, and with one might blow levels the population with the food of the world.*

Not the kind of person you would invite to your party for levity and a good time.

Graph of Malthusian Theory of Population

But when the needs of the population exceed the available resources – see Figs. 2 and 3 – trouble may result.

Social Part

The *social* element came home to roost clearly with the advent of the Internet and advanced modern technology: Facebook, Twitter, and Instagram – noting Facebook's recognition when the discussion on the Arab Spring comes up. So

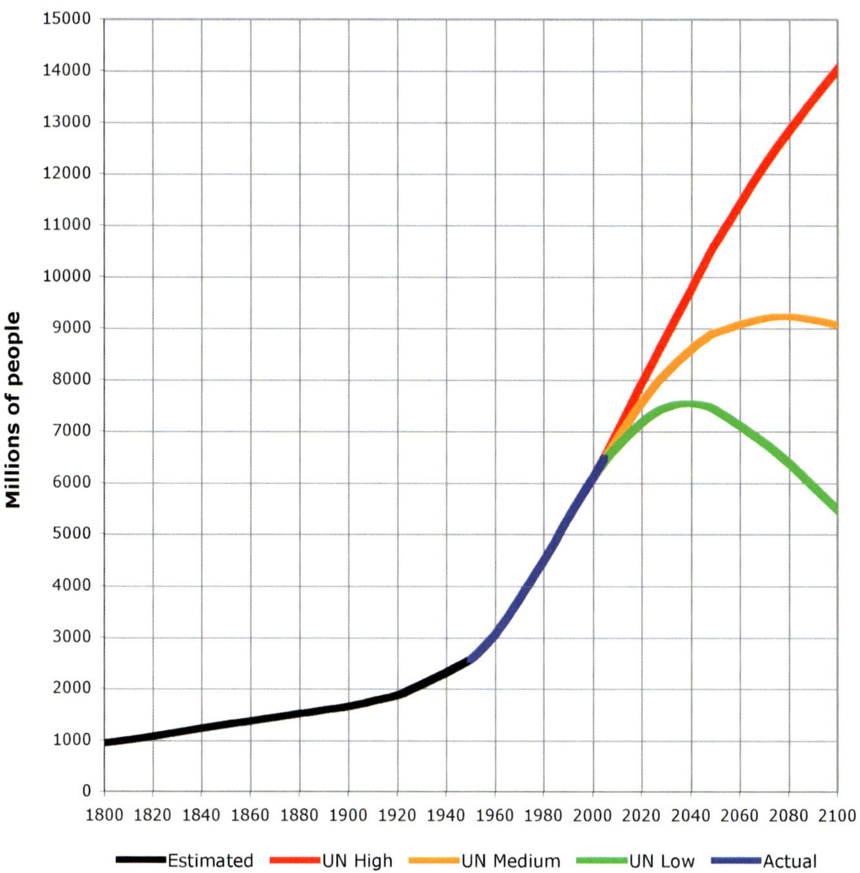

Fig. 2 World population from 1800 to 2100, based on UN 2004 projections (red, orange, green) and US Census Bureau historical estimates (black) (By Tga.D based on Aetheling's work – based on file:World-Population-1800–2100.png, but converted to SVG using original data from U.N. 2010 projections and US Census Bureau historical estimates, CC BY-SA 3.0, https://commons.wikimedia.org/w/index.php?curid=19813379)

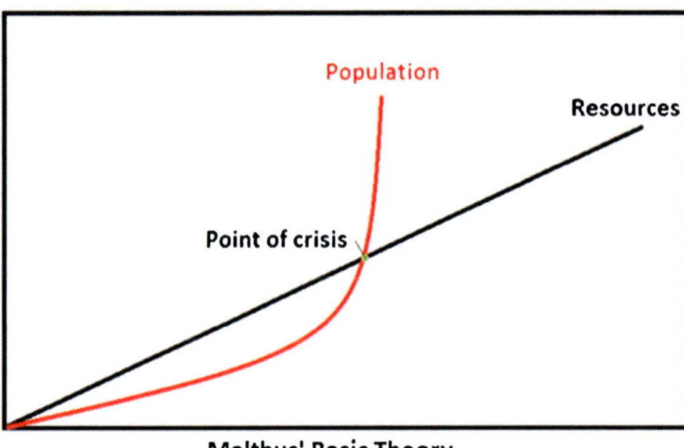

Fig. 3 Malthus' Basic Theory (https://netherhallgcserevision.wordpress.com/case-study-3-consumption-theories-malthus-boserup/)

with a "shrinking of the world" due to social media, ending of the Cold War, which caused many countries (and their people) bonding together in trade, EU, NAFTA, and ASEAN, hence the *social* element is indeed addressed.

The social part is still growing, and people are making significant headway in all sorts of realms, from the LGBTQ area to the Donald Trump Presidency to the 2018 "Me Too" movement.

Economic Part

But the *economic* component, which includes *business*, has not yet been as advanced as it should be. The fact is that someone has to pay, and it will take time to get there. Differences between haves and have-nots, developed and lesser developed, individualist and collectivist, and more remind us of differences which will take years to decipher. Thus, *business* and *business youth*, who have the time and longevity ahead, have a key role to play.

But how to get business youth involved???

Youth Programs for Achieving the United Nations 17 Sustainable Development Goals (SDGs) Amidst Globalization

There is no larger initiative for us to face than "Global Sustainable Development," and, indeed, no greater project than the creation of a nonprofit organization, called "Business Youth for Sustainable Development" (BY4SD). Nonprofits and social enterprises, which are not profit driven, are hence much more balanced in doing what is right and are at the heart of this project.

A youth organization will assemble the youth at business schools, first locally then statewide, then regionally, and then nationally, into a knowledgeable and then action-oriented cadre of cogs in the wheels grinding toward achieving sustainable development. The outcomes will be many, but here are a few:

- New business courses, highlighting "sustainability" to advance our schools and differentiate them from other business schools
- New articles and books co-written by students and faculty
- New financial support, attracted through grants, to advance "BY4SD" to become a "sustainable" part of the business community

The topic of sustainability interested the world's diverse interests in a multitude of standpoints – the haves and have-nots, collective interests and individual interests, long term vs. short term, etc. "BY4SD" will engage youth (students), who want to "DO," and engage in *service-learning* and make an impact in managing our collective efforts toward achieving sustainable development. The functions and concepts of the POLC of management are the same – planning, organizing, leading, and controlling – but the mode is very different. Getting individual youth and students involved and energized through service-learning will introduce the topic of *personal social responsibility* (PSR), in addition to CSR (corporate social responsibility), whose challenges will be addressed by individual. CSR is more well known and involves the corporation. PSR could rightly be a component of organizational behavior, which focuses on individuals within organizations. The corporate social responsibility and personal social responsibility components were based on the growing interest in Eugene Lang's vision of "service-learning and civic engagement," which is absolutely imperative for our growth in the globally interconnected world of the twenty-first century (https://www.newschool.edu/lang/our-founder/). This mantra will engage students, who want to "DO," which is service-learning, and make an impact in managing our collective efforts toward achieving sustainable development.

This concept will consider integrating and raising awareness of the issues around scarcity of resources, environmental costs (contingent liabilities, potential taxes on emissions), markets in environmental attributes (carbon credits, wetland rights, air pollution rights, etc.), employee morale (many employers believe that sustainable practices are important to the generation of new hires), green marketing (separating authentic brand equity in sustainability form claims based on little evidence), etc. To me, sustainability, like ethics, is best understood from a business perspective by integrating it into the basic disciplines.

Various pedagogical models have not been successful for this topic, so my opinion is based on my experience and observation about how sustainability is dealt with in a business context and how it would be most relevant to OUR students. United Nations Environment Program (UNEP); United Nations Development Program (UNDP); United Nations Department of Economic and Social Affairs, Division for Sustainable Development; Environmental Protection Agency (EPA); Environmental Advocates of New York; Environmental Defense League (NY); Nature Conservancy (NY); Citizens Campaign for the Environment (NY); Center for Clean Air Policy (NY); Earth Justice (NY); etc. are key influential players.

Globalization Considerations for Youth

First of all, we must discuss the issues that abound because of globalization, starting in earnest with the creation of the Internet and the end of the Cold War (De Feis, Grunewald, De Feis 2016) and the diminishment of communism. Surely, even China, which in the 1970s was referred to as "Red China," indicating extensive communism, is more or less "red, white, and blue China," now, with entry into the World Trade Organization in 2002, and the for-profit entrepreneurial success of Alibaba (http://www.alibabagroup.com/en/global/home), for instance. But how are the benefits of globalization shared by the industrialized versus the nonindustrialized world, leading to arguments for and against, posed emphatically by Bhagwati and Stiglitz. Needless to say, there are some adverse impacts of globalization on countries that do derive positive benefits. Still some countries fail to benefit, but youth will play a specific role.

Globalization is all around us, and it is here to stay, and it should be harnessed by the youth of the world, so we must work toward recognizing it as an influence in today's world. Also, globalization may be one of the moderating variables toward achieving sustainable development, which again is "development that meets the needs of the present without compromising the ability of future generations to meet their own needs" (World Commission on Environment and Development 1987). That is, the more globalization is embraced, the better are our chances for reaching the goals of sustainable development. If less globalization (and more protectionism) is supported, our youth will suffer, and indeed the worse are our chances (De Feis 2014, 2015).

With the fall of communism, culminating with the end of the Soviet Union, with the Soviet leaders voting communism out of existence in 1991; the Berlin wall being dismantled in 1989, further spurring the breakdown of the world's trade barriers; bonding together of countries (EU, NAFTA, ASEAN, etc.); and, now, social media, there have been resulting increases in international business, multinational business, and globalized pursuits. Globalization can be seen as "global competition characterized by networks that bind countries, institutions, and people in an interdependent global economy" (Deresky 2017). It is the youth today who will reap the bigger and more sustained benefits of the fall of communism.

But the question is: *will the benefits of globalization for the industrialized and nonindustrialized world aid in the ultimate quest to achieve "sustainable development?"* Long before the present day, the world was very simple and very static; now the world has become very complex and very dynamic. Our customers are no longer our neighbors for our local neighborhood, as they come from all over the world. Our suppliers, too, are from all over the world. If we think about the external environment – task (industry) environment, general (societal) environment, and natural environment – and the impacts that globalization has had on all aspects of it, we can see why it is the ubiquitous calling of our day.

Youth will see it through, as they will be around. When communism fell in the 1990s, and the world moved more toward a free-market realm, it is the older folks who yearned for the return of a "planned economy." Since it will take years for the vast benefits of a free-market mentality to take hold, the older folks do not have years to spare, so naturally they would prefer the "cradle to grave" guarantee, albeit there are the hardships of reduced freedoms.

The general (societal) environment has elements in it like technological issues, demographics, economic issues, and globalization – yet globalization impacts all of the other forces. Globalization and technological issues – we have the countries who are classified as "haves" and the countries that are "have-nots" – are brought together ever more forcefully with the Internet and the open communication of social media. Globalization and demographics – we have our changing population (less domestic and more global and international), changing desires (less homogeneity, more heterogeneity), and changing attitudes toward religions, cultures, and peoples. *Globalization and economic issues* – supply and demand issues, money supply issues, and currency exchange rates tie our countries together like no time before.

Trading blocks around the world are now common: the European Union, NAFTA (North American Free Trade Agreement), ASEAN (Association of Southeast Asian Nations), Mercosur (Southern Market), CARICOM (Caribbean Market), and others. Barriers are being broken down along the way, with the efforts of the WTO (World Trade Organization) and its predecessor GATT (General Agreement on Tariffs and Trade), the International Monetary Fund, the World Bank, and even the United Nations leading the way. Ever since the decline of communism and the rise of democratic, free-market reforms, globalization has become part of the way of life.

In the task (industry) or specific environment, we have our customers, who are now global; our suppliers, who are now global; and our competitors, who are now global. Is this a good thing? And, globalization, apart from economic activity refers to other aspects of life. The circulation and distribution of information mainly through the Internet and the facilitation of communications among people from different corners of the Earth (e.g., social media) are simple examples, which validate the globalization concept. This international communication network allows the transmission of political and cultural ideology, the spreading of fashion trends, and the dissemination of ideas worldwide (Bitzenis 2004). However, globalization and related issues such as outsourcing are hotly debated topics, as there are perceived costs and benefits (pros and cons) (Langenfeld and Nieberding 2005). So, globalization is here to stay, and the reduction of trade barriers will yield a more peaceful existence for everybody. Or will it?

Are the Benefits of Globalization Equally Shared?

How benefits are shared by the "haves" and the "have-nots" deserves some mention, as we find it is not entirely mutual, or perhaps I should say equal. As the movement toward technological progress and open markets increases, globalization offers the developed countries greater productive benefit. Globalization is a pervasive factor on industrialization in the developing world. As the technology progresses and more markets open, globalization offers huge productive benefits to developed and developing countries. However, its effects are fairly uneven, and it is driving a divide between the relatively few successful countries and the others (Sanjaya 2004).

Development policy has to address these growing structural gaps and to reverse (or relax) the stringent rules of the game that constrain the use of previously successful industrial policy. Such successful industrial policies have taken many different forms, and countries have to choose the combinations that suit the demands of current globalization. Great differences, however, between "have" and "have-not" countries, really between the countries of the northern and southern hemispheres, respectively, will lead to changes in politics, economics, and social issues (Le Veness and Fleckenstein 2003), and this may not be equitable.

Also, today we must add terrorism into the mix as it has imposed a new tax on international business. As a result, the costs of globalization are rising, while the benefits are declining (Weidenbaum 2002). Globalization is the trend toward the increasing interdependence of national economies, and the concept is being considered by individuals, corporations, countries at large, and regions of the world (European Union, NAFTA, and others), as they have all become impassioned critics of multinational corporations.

But this is the easier challenge, which business and society have begun to change. The second challenge is newer and more dangerous, as it arises from the spread of international terrorism combined with the properly strong response to it. With Al-Qaeda, ISIS, and homegrown rebels, the world may have become too globalized. This combination is burdening the cross-border flow of people, goods, services, and money and is also increasing the risk associated with overseas investments. Are we reaching a new and negative turning point? It is too soon to say.

Surely, we should not underestimate the inventiveness and resourcefulness of private enterprise and individual entrepreneurs in responding to new challenges. Nevertheless, a review of the history of globalization reveals that the trend of world commerce is not inevitably upward. Some may say that by some indicators, the planet was more globalized a century ago, relative to today. Measured by trade and investment flows, the world economy was more integrated in the late nineteenth century than it is today. Before passports were generally required for crossing borders, people were far freer to travel and migrate than is now the case. (Since 9-11, I have traveled to Brazil, China, India, and elsewhere, and there was such security and fear of terrorism that you might think there is a benefit to protectionism.) The extent of economic interdependence across national boundaries, that is, globalization, did not decline in the early twentieth century because of mass protests or a bad press. The causes were far more fundamental – World War I, the Great Depression of the 1930s, World War II, and 9-11.

At the present time, a major shift is underway in the external environment faced by global businesses. It is a movement away from reducing the role of government in business and toward greater public sector involvement in the private sector. The sense of openness in global markets may thus be a casualty of the terrorist attacks. Surely combating terrorism means reducing the openness of borders and restricting or at least slowing down business and financial activities that cross national boundaries. A company that viewed itself as a global citizen is now thinking of its organization as a citizen of a specific nation, albeit with global activities. As a consequence, a shrinking of the previous overwhelming position of American businesses in global marketplace may also be underway (Weidenbaum 2002).

The most powerful benefit of the global economy has not been economic at all. It is the ability of people to exchange the most strategic of all factors (i.e., new ideas). That process empowers individuals in ways never before possible. That is why it is essential to maintain an open global marketplace. Youth will benefit most, as they have the length of time ahead of them, and they are not jaded by past philosophies of failed governmental systems. The promise of a better way of life will be believed by the youth group, as opposed to "older folks." Yet, there are counterarguments "for globalization" and "against globalization."

The Case for Globalization

The many benefits of globalization include encouraging lower prices for goods and services, stimulating economic growth, raising the income of consumers, and helping to create jobs in all countries that participate in the global trading system (Hill 2017). All of these benefits will be enjoyed by youth in greater abundance than older folk. In his work, *In Defense of Globalization* (Bhagwati 2004), Bhagwati told us: "globalization is the most powerful force for social good in the world." Bhagwati could have focused the greater good on young people, than on older folks.

Furthermore, in his book, *Making Globalization Work* (Stiglitz 2006), the author puts forth radical new ways to deal with the severe indebtedness of developing nations and recommends a "new system of global reserves to overcome international financial instability that provides new proposals for addressing the current impasse in dealing with global warming – the most important threat to the world's environment." Arresting global warming is one of the benefits of achieving sustainable development (i.e., no fossil fuels). Again, Stiglitz could have focused his attention on the youth of the world. Stiglitz makes the case that treating developing countries more fairly is not only morally right, but it is also ultimately to the advantage of the developed world too.

There are many myths about the dangers of globalization (Weidenbaum 2001), and these include:

1. Globalization costs jobs.
2. Developed countries are hurt by imports.
3. Companies in developed countries are fleeing to low-cost areas overseas.
4. Companies in developed countries are polluting the environments of overseas areas.
5. Developed countries doing business overseas take advantage of local people (especially in poor countries).
6. Trade deficits hurt the economies of countries.
7. Trade agreements should be used to raise environmental and labor standards worldwide.
8. The manufacturing base is eroding due to unfair global competition.

Each of these myths will now be addressed as follows:

Globalization costs jobs – The facts are sound, in a globalized country (e.g., United States), for instance, outside of the 2008 recession, which really affected much more than just the United States, relative to other countries, employment is at historic highs, and unemployment is at a 30-year low. Look at the US economy and others around the world in 2018.

Developed countries are hurt by imports – This is the mercantilist approach discredited by Adam Smith 200 plus years ago. The fact is that all benefit from imports, as consumers get a huge array of goods and services at optimal prices.

Companies in developed countries are fleeing to low-cost areas overseas – The flow of money to buy and operate factories (and other businesses) is increasingly into high-cost, but stable, areas (e.g., United States).

Companies in developed countries are polluting the environments of overseas areas – United States-owned and operated factories in foreign countries are among the leaders in working conditions and better environmental standards than locally owned firms. Also, when "cap and trade" is considered, the total amount of polluted effluent is within the allowable standard.

Developed countries doing business overseas take advantage of local people (especially in poor countries) – Much has been written about the "sweatshops" in Southeast Asia, but there are so many developing countries that compete vigorously for US firms to be located in their country, so they follow similar operating standards as at home. However, when they do not and that is disclosed, they are called to justice.

Trade deficits hurt the economies of countries – Our trade deficit is at a record high, but that does not hurt our standard of living for the prosperity we enjoy is as great as most anyone we deal with.

Trade agreements should be used to raise environmental and labor standards worldwide – It would be counterproductive to impose costly social regulations on developing countries as a requirement to do business, and the fact is that most countries, including developing countries, oppose these regulations.

Manufacturing base is eroding due to unfair global competition – The fact is that total industrial production is much higher now than in 1992 (World Bank 2017).

The Case Against Globalization

There is also the anti-globalization mind-set, who sees job losses in industries under attack from foreign competitors, downward pressure on the wage rates of unskilled workers, environmental degradation (and thereby no sustainable development), and the cultural imperialism of global media and multinational enterprises (and the taking advantage of the culturally impoverished) (Hill 2017). In this book, *Globalization and Its Discontents* (Stiglitz 2002), Stiglitz states: "the way globalization has been managed, including the international trade agreements that have played such a large role in removing those barriers and the policies that have been imposed on the developing countries in the process of globalization, needs to radically be rethought."

Hence, there are certain negatives or consequences of globalization (Daly 2001), which are enumerated as follows:

1. Race to the bottom (standards-lowering competition)
2. Increased tolerance of mergers (leading to monopoly power)
3. Intense national specialization
4. Intellectual property right

Race to the bottom (standards-lowering competition) – Globalization does undercut the ability of nations to *internalize* environmental and social costs into prices, and instead economic integration, under free-market conditions, promotes *standards-lowering competition*. The externalization of environmental costs and social costs must be considered.

Increased tolerance of mergers (leading to monopoly power) – Encouraging the goal of "global competitive advantage" is used as an excuse for tolerance corporate mergers and acquisitions (leading to monopolies) in national markets. When recent US airline mergers are considered, in light of their international partners (alliances) with foreign carriers, oligopolistic tendencies could follow.

Intense national specialization – Free trade and free capital mobility increase pressures for specialization to gain or maintain a competitive advantage. As a result, globalization demands that workers accept a narrowing of the ways to earn a living.

Intellectual property rights – Shared knowledge increases the productivity of all labor, capital, and resources. Some may say, "What is wrong with that?" Also, international development aid should consist of "freely shared knowledge" and far less of foreign investment and loans, which countries default on. No one has ever defaulted on shared knowledge.

Adverse Impacts on Countries that Derive Positive Benefit from Globalization

Certainly, the United States and other countries have derived positive benefits from globalization, but that benefit has incurred some cost. You will never be able to convince a steel worker from Pittsburgh, who worked in a mill, that his job, which used to be located in Pittsburgh but, due to NAFTA, has now been outsourced to Mexico, that globalization is a good thing. The liberalization of trade, capital, and knowledge flows are having an adverse effect on employment.

Some argue that the pattern of increasing unemployment in most member countries of the European Community reflects a pervasive tendency toward delocalization of industries to low-wage countries and social dumping. "Social dumping" is a practice involving the export of a good from a country with weak or poorly enforced labor standards, where the exporter's costs are artificially lower than its competitors in countries with higher standards, hence representing an unfair advantage in international trade (European Foundation for the Improvement of Living and Working Conditions). The main finding in Europe is that the unemployment problem is rooted in the rigidness of the labor market itself. The increasing importance of international trade should provide an opportunity to reduce long-term labor market slack. To reap the potential benefits in this respect, European governments would need to reorient structural policies toward a better functioning of labor and product markets (Noord 1996). While we know the benefits of globalization are great, as we have technological advances and the spread of market-based economies, there are indeed risks.

There are new market opportunities with globalization, though competitive threats and diffusion of business models, associated with globalization, have existed for over the past decade (Jones 2002). Because of this, there are major forms of restructuring at the business level, and they include labor intensification, investment in new technologies, downsizing (also, reengineering), the formation of strategic alliances and networks, and a shift from international and multinational to global and transnational strategies. To be most effective, any type of restructuring must be clearly and explicitly aligned with a firm's business-level strategy in order to maximize the efficient and effective allocation of resources in pursuit of competitive advantage. A strategic use of restructuring which links such efforts to broader competitive strategy should result in more sustainable benefits.

Some Countries Fail to Benefit

In recent years, we have seen substantial advances in the use of benefit-cost analysis to analyze why come countries fail to benefit. The analysis is used for development projects in a wide variety of situations around the world. In his article, "Implementing Benefit-Cost Analysis," Jameson (1981) casts doubt on this easy assumption. It uses extensive empirical evidence taken from actual projects in one development institution, the US Agency for International Development (USAID), to indicate that the slippage between best and actual practice may be substantial.

In recent years, USAID's development assistance has focused primarily at the project level. There are a variety of arguments for such targeting, for example, efficiency in the use of funds or effectiveness of technical assistance. The great strides in theoretical treatments of benefit-cost analysis have not been translated into procedural and analytical changes in USAID. It is likely that other organizations are in a similar situation. The reasons for this situation are several: inappropriate use of the technique and misleading guidelines for its application, dominance of organizational interest over analytical requirement, inappropriate staffing and support for the analysis, and the use of the analysis as a tool in internal struggles.

United Nations: 17 Sustainable Development Goals – The Global Goals for Sustainable Development (http://www.un.org/sustainabledevelopment/sustainable-development-goals/)

In 2015, countries adopted the 2030 Agenda for Sustainable Development, which included the 17 Sustainable Development Goals (SDGs), which are:

Goal 1: No Poverty
Goal 2: Zero Hunger
Goal 3: Good Health and Well-being
Goal 4: Quality Education
Goal 5: Gender Equality
Goal 6: Clean Water and Sanitation
Goal 7: Affordable and Clean Energy
Goal 8: Decent Work and Economic Growth
Goal 9: Industry, Innovation, and Infrastructure
Goal 10: Reduced Inequality
Goal 11: Sustainable Cities and Communities
Goal 12: Responsible Consumption and Production
Goal 13: Climate Action
Goal 14: Life Below Water
Goal 15: Life on Land
Goal 16: Peace and Justice Strong Institution
Goal 17: Partnerships to Achieve the Goal

For each of these SDGs, there are international (and domestic) youth groups that address each SDG with gusto. Youth and youth groups are important to create a "change in life." Remember, driving in the 1970s and before? No seat belt was required. But in 1984, New York becomes the first state to require the wearing of seat belts. Within a short amount of time, with the regular encouragement of youth to promote safety for mom and dad in automobiles, all got used to fastening them. Remember the commercials to stop smoking? Youth were employed to encourage older folk to stop. Youth – in numbers – is powerful.

Goal 1: No Poverty – End Poverty in All Its Forms Everywhere

The target is to "eradicate extreme poverty for all people everywhere, currently measured as people living on less than $1.90 a day" (UN Division for Social Policy and Development Disability website). We need to take the lead from youth organizations like Enactus, which is "a community of student, academic and business leaders committed to using the power of entrepreneurial action to transform lives and shape a better more sustainable world" (ENACTUS website). Enactus is an acronym from "entrepreneurial, action, and us." For example, an Enactus team in Swaziland developed a plan for vegetable production that would meet the local needs and would generate profits.

Enactus (www.enactus.org) was founded 40 years ago and now has more than 72,000 student members, over 1700 college/university programs, in 36 countries. Enactus also has 550 corporate, organizational, and individual partners. The SIFE group (Students In Free Enterprise) merged with Enactus and now does activities all over the world.

Goal 2: Zero Hunger – End Hunger, Achieve Food Security and Improved Nutrition, and Promote Sustainable Agriculture

How we grow, what we grow, what we transport, and what we consume all need to be examined. The birth of "farmer's markets" introduces us to sustainability in agriculture. I am on the board of such a not-for-profit farmer's market – AirSoilWater (https://airsoilwaterorg.wordpress.com/), whose origination initiation involved youth at several schools in Pennsylvania. As we speak, soils, clean water, forests, and what we need to support our increasing population are becoming irreversibly degraded. We need to take the lead from the Future Business Leaders of America (http://www.fbla-pbl.org/), which was founded in Columbia University in New York City. FBLA, for instance, is the largest student business organization in the world with 230,000 members. These organizations do positive, creative initiatives for people all around to secure them sufficient and sustainable food in entrepreneurial ways.

Goal 3: Good Health and Well-Being – Ensure Healthy Lives and Promote Well-Being for All at All Ages

The United Nations Population Fund (UNFPA) (http://www.unfpa.org/), which is another youth program, has supported implementation plans geared to reduce adolescent pregnancy of the Caribbean, Ghana, Guyana, Myanmar, and elsewhere, working with UNICEF (www.unicef.org). UNFPA was created in 1969, as the UN General Assembly declared "parents have the exclusive right to determine freely and responsibly the number and spacing of their children." The UNICEF youth program was created over 70 years ago and now works in 190 countries and territories improving the lives of children and their families defending their children's right.

Goal 4: Quality Education – Ensure Inclusive and Equitable Quality Education and Promote Lifelong Learning Opportunities for All

The goal of *quality education* is a goal that all youth programs embrace – quality education – Enactus, the Future Business Leaders of America, Junior Achievement (https://www.juniorachievement.org/web/ja-usa/home), and United Nations Youth Unit, recognizing the indisputable positive correlation between education and development. Also, we see that the least developed countries in the world also hold the lowest educational achievement. It is as simple as that.

Goal 5: Gender Equality – Achieve Gender Equality and Empower All Women and Girls

The diversity statement of Junior Achievement says, "Junior Achievement is the recognized leader in 'empowering young people to own their economic success' through volunteer-led, experiential learning. We are dedicated to providing a positive, enriching learning experience free of bias. Junior Achievement welcomes K-12 students, volunteers and potential staff regardless of race, religion, age, gender, national origin, disability, sexual orientation or any other legally protected characteristic."

Junior Achievement boasts that it reached 4,845,904 students; 212,101 classes; 243,756 volunteers; and 21,955 schools in 2016–2017 with a diversity statement (above) that either sex would be proud to have. Thus, we should utilize these youth groups, who all have similar statements about equality.

Goal 6: Clean Water and Sanitation – Ensure Availability and Sustainable Management of Water and Sanitation for All

The Associate Member Forum (AMF) of the Metropolitan Section of the American Society of Civil Engineers (ASCE) (http://www.ascemetsection.org/branches/

younger-member-forum/history), which for 1 year (1989–1990) I served as the president, has clean water usage and sanitation provision. Such youth programs should be tapped for a much-needed volunteer work in this area.

Goal 7: Affordable and Clean Energy – Ensure Access to Affordable, Reliable, Sustainable, and Modern Energy for All

Kilowatts for Education (http://www.kw4ed.org/) has a mission statement as follows: "Offer a tremendous opportunity for educational institutions through the use of renewable energy projects to offset their power use with a sustainable and responsible resource while educating their students on the benefits of renewable energy." This student (youth) group would be wonderful to explore a program, which could even be sponsored by a grant-giving foundation.

Goal 8: Decent Work and Economic Growth – Promote Sustained, Inclusive, and Sustainable Economic Growth, Full and Productive Employment, and Decent Work for All

Karen Higgins, PhD, wrote, *"Economic growth and sustainability – Are they mutually exclusive: Striking a balance between unbounded economic growth and sustainability requires a new mindset"* (2013), which talked about present-day society and our dependence on water, oxygen, and other natural elements and the connection between the economy and earth. Obviously, the aspect of sustainability is important, and the long-term reversal of our non-sustainable ways, which requires the masses to retreat from the path we are on. When we talk of the masses, a mind-set, and aggressive pursuit with energy, it appears that youth are a natural.

Goal 9: Industry, Innovation, and Infrastructure

The American Management Association's *Operation Enterprise* (http://www.amanet.org/advantage/operation-enterprise.aspx), which is its youth program of which I served as executive director for a year and a half, has wonderful programs for youth interested in industry and entrepreneurship and through the website of the Small Business Administration (www.sba.gov) can accomplish much toward our failing infrastructure. President Trump highlighted failing infrastructure as one of America's priorities, which has been highlighted similarly by the American Society of Civil Engineers (ASCE).

Goal 10: Reduced Inequality

In the international community, the most vulnerable nations are the least developed with the least education, so their youth have the greatest time here to work toward a reversal. The reversal is not going to come from people in their 80s. While income inequality *between* countries may have been reduced, the income inequality within countries has risen. So has the resulting conflict between the haves and the have-nots. The three dimensions of sustainable development – economic, social, and environmental – need to be addressed by young leaders armed with the youth therein.

Goal 11: Sustainable Cities and Communities

The Future Business Leaders of America and *Junior Achievement*, with their network of over 470,000 volunteers serving more than 10 million students in over 100 countries, are two of the many national and international youth groups working toward sustainable cities and communities. As our country evolved, development grew from the coastal towns (Boston, New York, Charleston) to the internal cities or hubs, as rail and now air travel have expanded inward. At their best, cities have enabled people to advance socially and economically. Such youth programs prepare cities for the challenges they face, whether it be improving resource use, reducing pollution, and curtailing poverty.

Goal 12: Responsible Consumption and Production

The Enactus organization (www.enactus.org) engages in responsible consumption and production, which encourages sustainable resource usage, energy efficiency, minimalist infrastructure, and a better quality of life for all. Once again, Enactus is a worldwide youth organization. Enactus helps to achieve overall development plans; aims to reduce future economic, environmental, and social costs; and strives to instill economic competitiveness.

Sustainable consumption and production aim to "do more (and better) with less," increasing net welfare gains from economic activities by reducing resource use, degradation, and pollution within product life cycle while increasing quality of life.

Goal 13: Climate Action

The Climate Reality Project, whose well-known supporter, Al Gore, has a youth movement: https://www.climaterealityproject.org/blog/youth-leading-way-solving-crisis.

From their website: "Youth movements are emerging all over the world to combat climate change," including Australia, Nepal, Africa, and Canada, among others. Why? Because they are knowledgeable (from social media), with time on their side, and they are aggressive (youth!). Climate change is affecting every country on every continent, since we collectively have one climate. Climate change disrupts national economies, affects individuals' and groups' lives, and costs people, communities, and countries. Look around the world at what has happened recently.

Climate change knows no national borders.

Goal 14: Life Below Water

The National Oceanic and Atmospheric Administration (NOAA) in the Department of Commerce has its youth focus: http://www.noaa.gov/resource-collections/common-measures-definitions/youth-and-adults, focusing on lifelong learning to enhance their own knowledge, skills, and competencies from a personal, civic, social, and/or career-related perspective. Rainwater, drinking water, weather, climate, coastlines, much of our food, and even the oxygen in the air we breathe are all ultimately provided and regulated by the sea. Throughout our history, oceans and seas have been vital conduits for trade and transportation.

A very small amount of our water is fresh, clean water – maybe 2.5% – and most of that is inaccessible (e.g., polar ice caps), so only about 0.3% of the water on Earth is accessible. Careful management of this essential global resource is a key to a sustainable future.

Goal 15: Life on Land

The National Forest Foundation has a youth focus: https://www.nationalforests.org/get-involved/youth-engagement to encourage youth to become involved in forest sustainability. Forests cover 30 per cent of the Earth's surface, and, in addition to providing food security and shelter, forests are key to combating climate change, of which we just spoke, and protecting biodiversity and the homes of the indigenous population.

Goal 16: Peace and Justice Strong Institution

The promotion of peaceful and inclusive societies for sustainable development activities, the provision of access to justice for all, and building effective, accountable institutions at all levels could involve the American Bar Association (https://www.americanbar.org/aba.html), New York City Bar Association (http://www.nycbar.org/), for instance, and their youth groups.

Goal 17: Partnerships to Achieve the Goal

A successful sustainable development program will require partnerships between for-profit, not-for-profit, government, NGO, public, and private sectors. These inclusive partnerships will be built upon principles and values shared by all and shared goals that place people and the planet at the center. Entities are needed at the global, regional, national, and local level.

We need to mobilize, redirect, and unlock the transformative powers of energized people, particularly youth! Hence, we have a need for Business Youth for Sustainable Development (BY4SD).

2030 Agenda: A Plan of Action for People, Planet, and Prosperity (https://sustainabledevelopment.un.org/post2015/transformingourworld)

The SDGs just reviewed and their targets for the next 15 years in areas of critical importance for humanity and the planet become the 5Ps (people, planet, prosperity, peace, and partnership):

People – end poverty and hunger.

Planet – protect the planet from degradation through sustainability.

Prosperity – ensure that all human beings can enjoy prosperous and fulfilling lives.

Peace – foster peaceful, just, and inclusive societies (free from fear and violence).

Partnership – mobilize the means required to implement the agenda through a revitalized Global Partnership for Sustainable Development, with the participation of all countries, all stakeholders, and all people.

It is this last “P” (partnership), which comes directly from the United Nations Sustainable Development website, that calls for the establishment of the Business Youth for Sustainable Development (BY4SD) now.

Business Youth for Sustainable Development (BY4SD)

So *Business Youth for Sustainable Development* (BY4SD), combining the best efforts of the existing youth organizations, could be the answer. Let me, once again, take a look at some of these organizations to see more of their focus (agenda) and priorities.

Future Business Leaders of America (www.fbla-pbl.org/): Established 1940

The Future Business Leaders of America-Phi Beta Lambda is a not-for-profit, education association of students preparing for careers in business and business-related fields. The association has four divisions:

1. FBLA for high school students
2. FBLA middle level for junior high, middle, and intermediate school students
3. PBL for postsecondary students
4. Professional Alumni Division for business people, educators, and parents, who support the goals of the association

The FBLA-PBL mission is to bring business and education together in a positive working relationship through innovative leadership and career development programs. Business teachers/advisers and advisory councils (including school officials, business people, and community representatives) guide local chapters. State advisers and committee members coordinate chapter activities for the national organization.

FBLA-PBL sponsors conferences and seminars for members and advisers, which are designed to enhance experience initially developed on the local and state level. Among these programs are the Institute for Leaders and the National Fall Leadership Conference.

Junior Achievement (www.ja.org): Established 1919

In Junior Achievement, the growing number of volunteers, educators, parents, and contributors reaches out to 7 million students each year, in grades K–12. Junior Achievement has passionate people behind a movement that seeks to educate and inspire young people to value free enterprise, business, and economics to improve the quality of their lives.

Altogether, Junior Achievement reaches millions of students worldwide with their "age-appropriate" curricula. Junior Achievement programs begin at the elementary school level, teaching children how they can impact the world around them as individuals, workers, and consumers. Junior Achievement programs continue through the middle and high school grades, preparing students for future economic and workforce issues they'll face.

Operation Enterprise (American Management Association) (http://www.amanet.org/oe/): Established 1960s

Today, companies are constantly looking to do more with less in order to keep up with – and outpace – change and competition. Responsibilities may increase at a moment's notice and require different or greater skills. That is why training and education have never been more critical for advancing careers and achieving organizational success. Ongoing learning enables managers to continuously enhance their professional and personal development and increase their value to their organizations.

AMA/Operation Enterprise provides managers and their organizations worldwide with the knowledge, skills, and tools they need to improve business performance, adapt to a changing workplace, and prosper in a complex and competitive business world. AMA/Operation Enterprise serves as a forum for the exchange of the latest information, ideas, and insights on management practices and business trends. AMA/Operation Enterprise disseminates content and information to a worldwide audience through multiple distribution channels and its strategic partners.

Language, culture, and other barriers may separate global business communities. But every organization, regardless of its location, needs one thing: access to the best in management training. That is why the AMA/Operation Enterprise network now extends worldwide – reaching thousands of business professionals in the United States, Canada, Mexico, and Europe and in Japan, China, Southeast Asia, and the Middle East.

AMA/Operation Enterprise develops youth along three (3) paths: career guidance, career skills, and career development, with high school programs, college programs, and customized programs. Operation Enterprise has been a part of the American Management Association since 1963.

United Nations Youth Unit (www.un.org/youth or www.un.org/esa/socdev/unyin/): Established 1950s

The Youth Unit is a not-for-profit organization (and an NGO – nongovernmental organization), and it is the focal point within the United Nations system on matters relating to youth. It has been set up to:

- Enhance awareness of the global situation of youth and increase recognition of the rights and aspirations of youth.
- Promote national youth policies in cooperation with both governmental and nongovernmental youth organizations.
- Strengthen the participation of youth in decision-making processes at all levels.
- Encourage mutual respect and understanding and peace among youth.

The focal point within the United Nations system on matters relating to youth issues is the Program on Youth in the "Division for Social Policy and Development, United Nations Department of Economic and social Affairs." It has been set up to enhance awareness of the global situation of youth and increase recognition of the rights and aspirations of youth; promote national youth policies, national youth coordinating mechanisms, and national youth programs of action as integral parts of social and economic development, in cooperation with both governmental and nongovernmental organizations; and strengthen the participation of youth in decision-making processes at all levels in order to increase their impact on national development and international cooperation.

The United Nations has long recognized that the imagination, ideals, and energies of young women and men are vital for the continuing development of the societies in which they live. And, the member states of the United Nations acknowledged this in 1965 when they endorsed the "Declaration on the Promotion among Youth of the Ideals of Peace, Mutual Respect and Understanding between Peoples." Further than youth organizations, we should look at professional, social, and even sports organizations.

Professional, Social, and Sports Organizations

Many appropriate professional organizations exist which have a *youth* component – the American Society of Civil Engineers (established 1852), American Society of Mechanical Engineers, New York Academy of Sciences, Institute of Transportation Engineers, and Delta Sigma Pi (established 1907). Indeed, my first profession was that of a civil engineer.

Even social "club" or sports organizations, like the United States Chess Federation, United States Backgammon Federation, Boy Scouts of America, Girl Scouts of America, United States Tennis Association, and a slew of organized baseball, football, hockey, and soccer, focus on *youth*.

But, it is the professional (technical) societies that focus on the issue of sustainable development, so it is those societies and the *youth* organizations discussed earlier that must be aligned with, for example, the Academy of Management, and others, to get the right emphasis and management focus on the issue of sustainable development. As I am a New Yorker, I will highlight some activities from the metropolitan tri-state area.

Sample Environmental Issues, Relative to the Metropolitan New York Region, as Examples of Sustainability Needs

- Climate change – global warming, greenhouse gases, and ocean acidification
- Nuclear issues – *New York* (Indian Point Energy Center, Fitzpatrick, Ginna, Nine Mile Point), *New Jersey* (Hope Creek, Oyster Creek, Salem), and *Connecticut* (Millstone Nuclear Power Station)
- Water pollution (Hudson River, East River, Atlantic Ocean) – ocean polluting (dumping), acid rain, oil spills, thermal pollution, eutrophication/algal bloom, and mercury in fish
- Energy – fossil fuel depletion, conservation, renewable energy, and efficient energy use
- Overpopulation
- Air pollution – smog, indoor air quality, and particulate matter
- Land use – urban sprawl and habitat destruction
- Solid waste – E-waste, litter, medical waste, landfills, leachate, recycling, and incineration

Recommendations for a Unified Approach to Youth Organizations

1. Set objectives for aligning social and sports youth groups and management with professional (technical) organizations (really, their individual members).
2. Develop strategic alliances between these youth groups, management, and professional (technical) organizations.
3. Bring sustainable development attention to these youth and management groups – the professional (technical) organizations are already aware of sustainable development. For example, write articles, participate on the web, participate in conferences, hand out information, etc. – utilizing all available information technology.
4. Let these youth groups and management know that *they are the missing pegs*!
5. Convene meetings, regularly, between these youth groups and the others who are aware or who should be aware (i.e., everyone) of sustainable development.
6. Measure results to ascertain progress to sustainable development objectives.
7. First, we have to crawl, before we walk, before we run! Youth will be instrumental to the cause.

The United Nations' *Principles for Responsible Management Education* (http://www.unprme.org/) can certainly help in this regard.

Evidence shows that the industrialized world has the economic power to address the world's environmental problems, but the urgent desire and persistence will be brought to bear by the young people, who have a relatively "longer" standing in the world today (and tomorrow). These are young people from both the industrialized and the nonindustrialized world. Haves and have-nots together must uniformly embrace this agenda, which must be embraced into by the young. The young have the energy, long-standing (apolitical) commitment – if they are allowed to reason through by themselves – for it is for their children, their grandchildren, and their great grandchildren and so on. Albeit, we should know they will be influenced.

Some Closing Thoughts

Sustainable development is built on three pillars: economic growth, ecological balance, and social progress (Stigson 2000). All three pillars relate to globalization as well. Corporations face increasingly intense scrutiny to strive for sustainability, and to contend with this, they will have to enforce a set of globalized corporate values throughout their operations. Corporations must be able to demonstrate that sustainable development is good business, is good for globalization, and is good for the world's economy.

Globalization cuts across all of these three, and yet ecological growth and ecological balance receive the lion's share of attention. But, interest in social progress is again growing, helped with the ubiquitousness of social media. Corporations must show that with "globalized corporate values," the goals set forth in the sustainable development doctrine will be achieved. Balancing equity among the *needs of people*, *optimal resource utilization*, *the economy*, *and the environment* is at the heart of sustainable development. This "four-legged stool" needs to be managed well to achieve the objectives of each facet, albeit with compromise and consensus (De Feis 2006). By only considering the "needs of people," for instance, we may sacrifice tomorrow for today. By only considering "the environment," we may unnecessarily hinder today's pleasure while have unneeded excesses tomorrow. When thinking of tomorrow, we need to facilitate the understanding of and "light a fire" under those who will undertake action tomorrow: t*oday's globalized youth*.

Dissection of the Four-Legged Stool

Needs of People

The needs of people are what drive innovation today – without this we would still be doing math on a slide rule. But people are all different – some individualists, some democratic, some socialist, some collectivist, some free market driven, some government controlled economics, etc. The needs of 7.5 billion people in nearly 200 countries with all different levels of development and mind-sets cannot be overlooked.

Optimal Resource Utilization

When should I consume – now or later? Today or tomorrow? This year or next? Well, that depends – what will it be worth next year if I wait? Understand that the opportunity cost is important to our decision. What would you rather have a $10 cash today or $20 cash next year? If you consider "slack," how much slack have I got, which will indicate if I could wait or must I consume now.

Economy

How is the economy? Well, that is a relative question – compared to what? Compared to twenty years ago (1998), then the economy today is better. Compared to ten years ago (2008), today it is much, much better! Governments, businesses, and people must understand that there will either be *payment now* in the form of taxes on people (increase in tax revenues) and tax benefits to corporations (decrease in tax revenues) or *payment later* in the form of cleanups (De Feis 1994, 1991). There will be a payment.

Environment

An Inconvenient Truth (by former U.S. Vice President Al Gore) states that we cannot keep deferring our attention on the environment. We have problems, which must be addressed now. Just like the decision-makers, who were largely politicians, who decided not to strengthen or heighten the levies around New Orleans, Louisiana, as many knew could not withstand a category 4 or 5 hurricane, and instead decided the funds were needed for schools and housing and other "squeaky" wheels. And, so they argued, "If the levies had not faced a category 4 or 5 hurricane in the past 100 years, they could surely hold up until they are out of office …." But we now know what happened in 2005 when Hurricane Katrina came knocking.

Youth programs abound – United Nations Youth Unit, the Future Business Leaders of America, Junior Achievement, Operation Enterprise (youth program of the American Management Association), and others – and it is in these organizations that the sustainable development movement must take hold. Many of these groups embrace a management and business mind-set, with overtones of community service. The whole aspect of "service-learning" for youth is a good example of what could be. As such, they should be tapped to participate in the much-needed youth force of the sustainable development process.

Summary

Sustainable development is a quixotic quest! To develop in a sustainable way may be foolish for some – why can't I enjoy myself to the fullest extent *available*? Arguments will astound you, and indeed there are several legs to the "sustainable development" story, with each playing a different and important role. Therefore, communication, sharing information, through social media and the like is essential. Remember the Arab Spring? Credit was given to *Facebook*, among other entities! For first, the concepts of "I can't take it anymore!" spread to Tunisia, then to Egypt, then to Libya, and then to Iraq, and it continues it to spread.

Implementation of the vast challenges presents us with a great opportunity to manage of process – I argue that a key is to *involve youth early*. The role of government is certainly important to coordinate and facilitate the organization of such an immense network of youth. Note, though, that "governments must realize that they can either pay now (i.e., with tax benefits to companies, and therefore lost or decreased current tax revenues) or pay later, in the form of cleanups. Why not pay now?" (De Feis 1994).

Cross-References

- ▶ Collaboration for Regional Sustainable Circular Economy Innovation
- ▶ Education in Human Values
- ▶ Expanding Sustainable Business Education Beyond Business Schools
- ▶ Low-Carbon Economies (LCEs)
- ▶ People, Planet, and Profit
- ▶ Sustainable Decision-Making
- ▶ The Spirit of Sustainability

References

Bhagwati, J. N. (2004). *In defense of globalization*. Oxford: Oxford University Press.

Bitzenis, A. (2004). Is globalization consistent with the accumulation of FDI inflows in the Balkan countries? *European Business Review, 16*, 406–425. Bradford.

Daly, H. E. (2001). *Philosophy and public policy quarterly. 21*, 2–3 Spring/Summer, George Mason University.

De Feis, G. L., Grunewald, D., & De Feis, G. N. (2016). International trade theory of hyper-globalization and hyper-information flow conceived. *International Journal of Business & Applied Sciences, 5*(1), 22–28.

De Feis, G.L. (2015). Comparing the impact of globalization on the industrialized and non-industrialized world Vis-À-Vis achieving sustainable development. In *Proceedings of the XIV International Business and Economy Conference (IBEC)*, Bangkok, 5–8 Jan 2015.

De Feis, G.L. (2014). Sustainable 'Business' development today. In *Proceedings of the 9th Annual CIBER-HAGAN Summer Symposium at the Hagan School of Business at Iona College*, 21 June 2014.

De Feis, G.L. (2006). An insight to managing the four-legged stool of sustainable development: Involve youth organizations early.In *Proceedings of the Academy of Management/UN Global Compact/Case Western Reserve University: Business as an Agent of World Benefit: Management Knowledge Leading Positive Change Conference*, Cleveland, 23–25 Oct 2006 .

De Feis, G. L. (1994). Sustainable development issues: Industry, environment, regulations and competition. *Journal of Professional Issues in Engineering Education and Practice, 120*(2), 177–182.

De Feis, G. L. (1991). The double-edged sword of technological advancement. *Journal of Professional Issues in Engineering Education and Practice, 117*(3), 245–249.

Deresky, R. (2017). *International management: Managing across borders and cultures* (9th ed.). Upper Saddle River: Prentice Hall.

Higgins, K. L. (2013). Economic growth and sustainability – Are they mutually exclusive: Striking a balance between unbounded economic growth and sustainability requires a new mindset. Elsevier Blog Post.

Hill, C. W. L. (2017). *International business: Competing in the global marketplace* (11th ed.). New York: McGraw-Hill.

Jameson, K. (1981). Implementing benefit-cost analysis. *Journal of Management Studies, 18*(4), 411–422.

Jones, M. (2002). Globalization and organizational restructuring: A strategic perspective. *Thunderbird International Business Review, 44*(3), 325–351.

Lang, E. https://www.newschool.edu/lang/our-founder/.

Langenfeld, J., & Nieberding, J. (2005). Regionalisation for the case of FDI inflows in Bulgaria. *Business Economics, 40*(3), 41–51. Washington.

Le Veness, F. P., & Fleckenstein, M. (2003). North and south clash over the World Bank, International Monetary Fund, and the World Trade Organization. *Journal of Business Ethics, 47*(4), 365. Dordrecht.

Noord, P. J. (1996). Globalisation and the European disease. *The Economist, 144*(2), 195–222.

Sanjaya, L. (2004). Industrial success and failure in a globalizing world. *The International Journal of Technology Management & Sustainable Development, 3*(3), 189–213. Bristol.

Stiglitz, J. E. (2002). *Globalization and its discontents*. New York: W.W. Norton & Company.

Stiglitz, J. E. (2006). *Making globalization work*. New York: W.W. Norton & Company.

Stigson, B. (2000). Organisation for economic cooperation and development. The OECD observer. Paris: Summer Issue 221/222, pp. 36–7.

United Nations. https://www.un.org/development/desa/disabilities/envision2030-goal1.html.

United Nations. 17 Sustainable Development Goals – http://www.un.org/sustainabledevelopment/sustainable-development-goals/.

Weidenbaum, M. (2002). Whither globalization? *The International Economy, 16*(1), 50. Washington, DC.

Weidenbaum, M. (2001). Vital speeches of the day. Economic Club of Detroit.

World Bank. (2017). (The)/The International Bank for Reconstruction and Development. Washington, DC. www.worldbank.org.

World Commission on Environment and Development. (1987). *Our common future*. Oxford: Oxford University Press.

Part IV

Global Initiatives Toward Engaged Sustainability

Time Banks as Sustainable Alternatives for Refugee Social Integration in European Communities

Joachim Timlon and Mateus Possati Figueira

Contents

Abstract
In 2015, more than a million refugees crossed into Europe, which sparked an unprecedented displacement crisis as countries struggled to cope with the influx. Two years after, people are still seeking to escape the terror of war and extreme poverty on a sea path to the European Union. Integration is the next challenge.

J. Timlon (✉)
Ronneby, Sweden
e-mail: Joachim@Timlon.biz

M. P. Figueira
Arceburgo, MG, Brazil
e-mail: Mateus.Figueira@gmail.com

S. Dhiman, J. Marques (eds.), *Handbook of Engaged Sustainability*,
https://doi.org/10.1007/978-3-319-71312-0_26

While various governments sit on the inertia of the decision-making process, voluntary initiatives spike in many countries as creative ways to deal with what has been acclaimed as the *refugee crisis*. Community actions, such as in Scotland, provide cultural integration opportunities by introducing and matching refugees who are resettling to Europe with native people who have volunteered to help in adjusting to the local community; and in Sweden, where on large scale, the educational system is individualized to the specific needs of the refugees and facilitates the integration in the local labor market. In this pool of initiatives, time banking arises as a sustainable alternative for balancing social relations and improving social well-being while increasing the social capital. By redefining the concept of work and value of assets, it proofs itself as an ideal social dynamic that grounds on the equality, respect, and reciprocity values. It allows the refugee to work and be productive from the first moments in the country while absorbing the culture and social dynamics. The trade-offs are benefits that can be scaled up to the individuals involved, to the community, and to the country, to the level of empowering the *sustainable community development*.

Keywords
Refugee · Refugee crisis · Equality · Integration · Time bank · Time banking · Reciprocity · Social capital · Social empowerment · Sustainable community development

Introduction

Global Disruptive Trends

To be disruptive means to prevent something for continuing or operating in a normal way; a trend is a general development or change in a situation that affects many countries of the world. In this sense, the unprecedented displacement crisis, sparked by armed conflicts, resulting in the influx of refugees crossing into Europe, is a global disruptive trend. According to data published at the UNHCR Global Trends, at the beginning of 2014, Europe gave shelter to 1.771 million refugees plus 11,4 thousand in refugee-like situations. At the end of the same year, the number rose to 3,107 million, an increase with 74,3%, and the highest increase among other regions in the globe. The top host figures were reported from Germany, Hungary, Sweden, Austria, and Italy, followed by France, the Netherlands, Belgium, the UK, and Finland. In Italy, the catalogued population of refugees, refugee-like situations of people, and under statelessness mandate by the UNHCR were 93,715; 45,749; and 813, totalizing 140,277 people. In Sweden, the number reached 226,158 where 56,784 were asylum seekers and 27,167 considered stateless. More than half (53%) of the refugees worldwide came from just three countries: the Syrian Arab Republic (3.88 million), Afghanistan (2.59 million), and Somalia (1.11 million). The top six refugees' destinations hosted together an overall of 6.545 million refugees. Surprisingly none of the countries were from the European Union but instead are surrounding the conflict areas: Turkey, Pakistan, Lebanon, Islamic Republic of Iraq, Ethiopia, and Jordan. In 2015, Turkey became the largest refugee-hosting country worldwide for the first time, with 1.59 million refugees. At the end of the same year, the total number of forcibly displaced people worldwide went up to almost 65.3 million, which means that 1 out of 113 people on earth were displaced, the highest level since World War II (UNHCR Global Trends – Forced Displacement in 2015). Thus, the displacement of refugees is not solely a European crisis; it's a global crisis.

Integration Programs and Structural Barriers

Most societies in the EU acknowledge their responsibility to give shelter to refugees and provide *integration programs*, hence, also acknowledging its role in facilitating the integration process (see Fernandes (2015) for a discussion on (dis)empowering refugees and migrants through the participation in introduction programs in Sweden, Denmark, and Norway). However, there might be structural barriers or impediments that limits or hampers the integration of the refugees in the society. *Structural barriers* can be understood as aspects of the external environment that limits societal integration. According to Kenneth Waltz, a political scientist, structural impediments are formal and informal rules that regulate an entire system of interaction and the physical limitations of an environment. A set of formal and informal rules governing behavior in a particular sphere is commonly known as "the rules of the game" and can be found in integration policy fields (Waltz 2010). In Italy, four different governments opted for different approaches; some blamed the EU for its inefficiency to respond to Italy's calls for assistance, and others looked for buffer solutions while claiming that the crisis was not just a passing phenomenon but a structural fact. There is still a need for the introduction programs to acknowledge but also address the important role of structures and not only overemphasize the responsibility of the individual. This might reduce the potentially stigmatizing element of the programs and the framing of immigrants as a social problem that needs to "be fixed." Instead of enhancing and enforcing social control, "real" social change should be encouraged (see Pease 2002).

Social Empowerment

Empowerment is a multifaceted construct shared by many disciplines and arenas, such as studies of social movements and community development. Empowerment, by definition, is a social process, once it occurs in relationships among people. Empowerment is a process, similar to a path or journey, which develops when effort is given to it. Empowerment may vary according to the specific context and people involved (cf. Bailey 1992). Empowerment also occurs at various levels, such as individual, group, and community. An important implication is that the individual and community are fundamentally connected. Social empowerment can be defined as a social process that helps people gain control of their lives by fostering the capacity to act on issues they define as important and realize goals for use in the communities in which they live (cf. Page and Czuba 1999). A prerequisite for social empowerment is individual change, which makes a bridge to community connectedness and social change (Wilson 1996). Individual change enables the connectedness and for a community to complex issues that it is facing, such as the displacement crisis. If people have mutual respect and accept diverse perspectives, they can develop a shared vision and work toward creative and realistic solutions. Hence, inclusiveness of individuals and collective understanding of empowerment are essential for practices with empowerment as a goal, and the critical transition of individual change in combination with interconnectivity between individuals framed by a distinctive practice can have invaluable positive effect on refugee integration.

Sustainable Community Development and Building Community Capacity

A sustainable community goes beyond the mere focus on subsistence. According to Flint (2015), such a community emphasizes long-term benefits and the means to promote informed choices that moves the community beyond the status quo of just being livable. The livability of a community refers to the social and environmental quality perceived by its inhabitants, such as residents, employees, customers, and visitors. A sustainable community, on the other hand, promotes a conscious commitment, regarding how to make choices about dealing with different kinds of challenges, such as displacement of refugees. A long-term perspective in dealing with such matters is distinguishing a sustainable from a livable community.

Sustainable community development, then, is the action of continuously improving the social and environmental quality of living. It includes human interactions among the inhabitants that, for instance, promote physical and mental health but also environmental issues. This means that the limitations of earthly elements as well as their interconnectedness of human beings are recognized. Furthermore, equity is regarded as the foundation of a healthy community, in which all socioeconomic factors are grounded and ecological, economic, and sociocultural diversity within the community are contributing to its stability and resiliency. For the development to be conquered, Flint argues that these idealistic communities would have to continuously exercise the integration of three inseparable aspects of development – ecology, equity, and economy – establishing a balance between environmental protection, social well-being, and enrichment of human relationships and economic development that improves human welfare. Typically, evolving in a sustainable community is people's sense of well-being because there is a sense of belonging, a sense of place, a sense of self-worth, a sense of safety, a sense of connection with nature, and provision of goods and services that meet their needs within the ecological integrity of natural systems. A truly sustainable community provides for the welfare and health of the present community members as well as for future generations. Refugee integration touches upon all three pillars of sustainable community development. A distinctive feature, reflected in the case, pertains different aspects of equity, established in community in which the refugees are displaced. Here it becomes evident that people and their capacity to connect and develop together further develop a sustainable community. The result then, a kind of social capital, is created by the trust relationship that is developed among the citizens in one society (Kilpatrick et al. 2003). The more the citizens are willing to trust one another and the more connections and associative possibilities it is generated among one society, the bigger will be the social capital volume on this place, which is very well illustrated by the *Befriender's* practice further ahead in this chapter.

Time Banking

The concept of time banking was invented in 1980 by Edgar Cahn, a Distinguished Emeritus Professor of Law at the University of the District of Columbia School of Law. Time banking is a way of giving and receiving to build supportive networks and strong communities by addressing unmet societal needs. His ideas grew out of discontentment with the worldview on equity and opportunity regarding people in need of help and assistance as "objects of charity." This, he argued, would simply turn them into passive consumers of help. Instead he wanted to build in reciprocity in receiving help and use the dynamics of "pay it forward," an expression for describing the beneficiary of a good deed repaying it to others instead of to the original benefactor, which means that the recipient of help pays it back by paying it forward. A relation between two strangers has then changed to a relation that builds social networks and communities to tackle any kind of social problems.

Time bank systems offer an innovative way in which people can participate in society both as givers and receivers of services. They operate by using time as a form of currency. For every hour that some volunteers in giving services to others, they receive an hour of service in return; I earn a time credit by doing something for you. It doesn't matter what that "something" is. For example, an hour of gardening equals an hour of child care equals an hour of dentistry equals an hour of home repair equals an hour of teaching someone to play chess. This amount can be used immediately, be banked for future use or donated for use by other individuals. Running the system means to keep records of how much time individuals give or receive and also of what services people can offer or need and then matching up participants. Many groups have Internet sites that offer assistance to people who would like to set up new branches. All work is seen as equal because, regardless of its nature, it earns the same amount of hourly credits. In this way, time banking promotes equality and builds caring community economies through inclusive exchanges of time and talents with endless possibilities (Cahn 2000).

A study by Miller (2008) of the impact of time banks on the lives of older Japanese members shows how time banking groups can help senior members as well as society as a whole. The study shows that time banks could give people greater control of their lives and foster warmer community links. In this case, the benefits that older time bank members derive include formation of new friendship networks to replace those lost by retirement and the chance to use old skills and learn new ones. Furthermore, time banks can generate a new form of social capital that fosters traditional reciprocity, which in Japan is denoted as *ikigai* or sense of meaning in life. These groups can nurture alternate styles of human relationships in a complex society undergoing change. Such experimentstried at various levels of Japanese society are emerging as a significant force to bring about changes within the system itself.

How can this way of giving and receiving be used to address the displacement crisis of refugees crossing to many EU countries? How to prevent that those refugees in dire need of help and assistance merely are perceived as objects of charity work, turning them into passive consumers of help? How can reciprocity and the dynamics of pay it forward be built into integration programs? How would it strengthen social empowerment? These questions lead to the overarching one of how these inclusive exchanges of time and talents can benefit human relations and the society at large and be an alternative to the official programs for refugee integration. It would perhaps be presumptuous to think that these ideas can solve the unprecedented displacement crisis; the aim of this chapter is merely to contribute with real-life best practice stories that make sense, create solidarity and compassion, building more integrated local communities. We have found such practices in Scotland and Sweden.

Societal Integration of Refugees

During 2015, 1,321,050 people come to a EU country to seek asylum. One year later, in 2016, the situation was different. Today, the refugee crisis seems to have warded off. People are still seeking to escape the armed conflicts but now come via the sea road to the EU. Still integration challenges are expected. For instance, about 140,000 asylum seekers are waiting for decision by the Swedish Migration Agency if they are allowed to stay or not in the country.

A majority of them will most likely receive a positive decision and will have to establish themselves in the society. This means that their status as a refugee will change to a newly arrived migrant who has a residence permit and social security number, receives a daily allowance (about 30 Euros) from the government, and is enrolled into a Swedish municipality according to current legislations applied to refugee's integration.

Yet another country to which refugees are resettling is the UK. In different parts of the country, there are different initiatives to help with integration, which are usually conducted by the third sector or origins from community groups. There is a project being piloted by the Scottish refugee council, named "peer education" which gives help to connect local people with newly arrived refugees to support with the integration. This project is piloted in four different parts of Scotland and was highlighted on a public report from the parliament with what is happening with refugees in the UK.

However, the numbers of incoming refugees are tiny compared to Germany and Sweden, which gives it a completely different scale. The families are selected not from Syria directly but from refugee camps in Lebanon or Iraq. There is a careful screening process consisting of four or five interviews conducted and organized by representatives of the United Nations. Sometimes there are as much as three generations of one family. Having passed the selection process, they are flown to the UK and then resettle in different communities across the country. One such community is Edinburgh.

At The Welcoming center in this community, nearly 120 volunteers lead by seven full-time staff, along with four seasonal workers, deliver a diverse range of integration initiatives. The main activity area of the center is English language classes in order to help the refugees to overcome the language barrier and open up the gate to integration in the new country.

> It's the gateway to everything to job, to friendship, to hobbies, and so on. And that's why we have a huge demand for English language classes at the moment.

In the Edinburgh's community, there are structural barriers that impede integration initiatives and refugees to get a job in the regular labor market. Some major barriers are the local language, lack of understanding, validating qualifications and educational degrees that the refugees have acquired in the home countries, and even the name of the applicant. At the local center, Elaine Mowat, the director of this organization explains:

> I know that people often feel they have to start all over again here even have got a master's degree in their own country. It's not particularly recognized, understood or respected here even so that can be a barrier. And I guess discrimination as well. And sometimes it's difficult to know where that is happening. I know there's been lots of research showing that people submitting job applications with different names provoke a very different response. I mean the main challenge that we see for people trying to break into the job market is, it's not just getting jobs but is getting jobs that match their skills and experience and their qualifications.

The center is running 22 classes a week, and volunteers teach most of them. Distinctive about the program of language classes compared to other organizations is the "drop-in." This means that people don't book for a class, they come along if they're able to. In addition, the teachers do not know how many people they're going to get on each session which lacks on providing continuity between two classes. It is an interactive communicative style of teaching that each class stands alone. It works well for this kind of community as newly arrived often have quite a chaotic schedule, such as appointments in different places, and they may pick up temporary work or their schedules may change. Consequently, the center adapts the teaching for the newly arrived.

Another activity area for the center is employability support. Newly arrived looking for jobs can come and get help with their CVs. A difficulty is that many of them don't understand what a CV is needs for and to be like for the UK market. Yet another difficulty is when they reach the job interviews, how to prepare for them and how to conduct them. At the center, they get help to present themselves in the best possible way to local employers.

A third activity area is the welcoming friendship program or the so-called befriending program, which, in practice, is targeted at the Syrian refugee community once they seem to benefit the most from it. In this case, the center was contacted by the city of Edinburgh council and asked to provide services to the Syrians that are officially resettled in Edinburgh. The background was yet another structural barrier and seemingly big frustration for many refugees: how to connect with local Scottish people?

At the center newly arrived can meet people from other countries who come to the center and instantly become part of an open friendly multinational community that is a real support to the refugees. In fact, much support is given to the refugees by other refugees.

> A lot of the support they get here is not just from staff or from volunteers but it's from each other.

However, the next step of being part of the local Scottish community with local people is often much harder to make. To make this step happen, the center provides cultural integration opportunities through a welcoming friendship program by introducing and matching the refugees with local people who've approached the center and said they would like to volunteer and help refugees settle in Edinburgh. When accepted by the center, they become a kind of a social worker who works for the government. However, there is a difference between the council support worker and the welcoming "Befriender." This latter role at the welcoming friendship program is quite a new activity for the center.

> In Edinburgh, we were overwhelmed at the start of the year over 100 people were left desperate to be friends with refugees but we didn't have enough or we didn't really have capacity to support the program because we are quite careful about how we do it so that we will select, you will interview people and check that they seem appropriate we have training, so we can only take on a certain number of people. We have monthly peer support so that people who are befrienders can come to us and share their experiences and support each other. Because, well it is sensitive work and then there is the potential for things to start going wrong so as an organization we want to feel that we are supporting people back. But yes, the numbers grow.

The interest from the local community can be related with the LOHAS lifestyle, see chapter ▶ "The LOHAS Lifestyle and Marketplace Behavior," where individuals are increasingly concerned with social ethics, environmental health, and morality, incorporating practical actions to their daily lives and seeking a balance between individual, environment and society. Working with "befrienders" is, however, a preferred way, which is although quite different to how it originally was conceived. As a small organization, it is essential to be adaptable and to learn from doing.

> But that's the way that we like to work, because I think you can only really learn as you go along and adapt. And as a quite small organization we are able to be very adaptable, we can change things quickly if we find they are not working.

Through the "befriending program," some 40 matches have been made between local people and new families from Syria who have come through the official resettlement program as well as on their own accord.

> By far the main community we are working with is the Syrian community who've come through the resettlement program. We also work with some Syrians who've come here on their own, so not part of the official program.

A public report has recently been published by a parliamentary group who was looking to what was happening with the refugees in Britain. The report identified two tiers of refugees: the ones who've come through the official relocation schemes and another who've come on their own. The report further concluded that these refugees didn't get the same measure of support. Regardless, the center is open to all refugees.

> But we are open to everyone, we will work with all people and we have our Welcoming Friendship Program, we have matched Syrians who've come here under their own means as well as those through the Program.

However, the center works with a migrant population that comes in total from 77 countries, such as Spain and Italy where there is no displacement crisis due to armed conflicts. People from these countries are particularly in their 20s or 30s who are not finding work at home at the moment. Improving their English is one very concrete thing they can do. This means that in an English Language course class, there may be one or two Syrians in that class, but there are probably 25 people from different nationalities. In the past, the refugee population represented only a few percent of the migrants at the center, but that has changed over the last 2 years when the UK Syrian resettlement program started, which made a big difference for the center that now is starting to work much more with the refugee population. Instead of the government sponsoring newly arrived refugee's community groups, a provision/proposal can be put forward and an undertaking to financially support families to arrive in a local area. This new way of working comes from Canada.

Selecting the kind of program relies on a mix of criteria. Firstly, there must be a need to be fulfilled, such as the English language program. Secondly, the center is dependent on funding. For instance, in the last 3 years, the funding has been received from a climate fund, which is a Scottish government fund to help tackle carbon emissions and to help people live a more sustainable eco-friendly lifestyle. As a consequence, the center opened up a whole new stream of activities. One of the key programs became the home energy advice where an adviser goes to visit people in their homes and help them to understand how to use the home heating system and how to get a good energy supply. Although it may not be the most obvious activity in terms of integration, it actually seems to work well particularly for refugees coming to Scotland as preheated systems often are very different from those in warmer climates, such as in Syria. The newly arrived can waste a lot of money by spending poorly on home heating, but this advice helps them to save money. In addition, the home energy advice sessions are an opportunity for the refugees to raise all kinds of other issues. The home energy advisor is a kind of private counselor. It is a good way to help engage with people because the advisor enables the development of trusting relationship with people, opening up different opportunities to support them as well. However, this person is a different one from the one who is assigned to a family in the Befriender program.

Most recently the Faculty of Advocates in Edinburgh, Scotland's top lawyers, who are a powerful and influential body of people, contacted the center. They wanted to connect with refugees in Edinburgh with the help of the center. Their idea was to put on a Burns supper with a "haggis," which is a certain local traditional event that is celebrated on the 25th of January, the birthday of the national poet Robert Burns. The center invited the whole new Syrian community in Edinburgh, and the advocates themselves took on the role of volunteers at that event.

> We had some Syrian music. We had people sign 'Old Lang Syne', do you know? In Arabic, (laughs). There were many nice conversations between the local Scottish people and the Syrian community so that felt like a very special and nice event.

This was a unique event due to the mix of Syrian and Scottish culture. But even more interesting, the center has been able to follow up to help create some useful connections for some of the Syrians in Edinburgh.

> For example, that day of the Burns Super, one of the Syrian women started saying to me 'these are my colleagues', and I couldn't figure out what she's was saying and then I realized what she said. I hadn't realized that before. She used to study law in Syria and I think she only had time to do four months before she had to leave the country. But she identified hereself as a lawyer and here she was in the middle of the Scotland's most influential lawyers.

The event provided an opportunity for a Syrian woman, enabling her to meet and speak to different lawyers. Together with a representative from the center, she then went to visit the faculty on a separate occasion. This resulted in an invitation to her to take part in seminars with a possibility that she might start to study in the law library. However, she's not yet ready to pick up law studies because she's still learning English. Nevertheless, this young Syrian woman has now the chance to become a lawyer in Scotland thanks to the contacts, the encouragement, the kind of investment, and the generosity, creating an environment in which she can achieve her potential.

> I think it's always the unexpected things. I think so long as we as an organization are open to working with other organizations we can help connect and that's a big responsibility and a privilege but also such a joy to be able to do that.

> It's human conversations, how important it is to have personal contacts in the new country. We always say don't we, you know it's all about who you know. That's one of the major challenges for newcomers in the country. They don't know anyone. You don't have your auntie who knows your neighbour's cousin, who can put you in touch with... you know all those bigger social networks aren't in place for you.

Social networks are important, and for newly arrived who don't have them in place in the new country, it is even more important to develop new ones. So, one of the most useful things that newcomers can do is start to put into place those social networks that then can serve them on their personal and professional lives. Here the center plays an important role of providing the conditions for this to happen. Like the relational project environment that is suggested on the chapter ▶ "Relational Teams Turning the Cost of Waste into Sustainable Benefits," the social networks facilitated by the Befriender also comes as a response strategy to the social and environmental impacts that is often disregarded by the industry. With shared vision, values, and a common objective to generating human activity and bonds, both mitigate these practical outcomes, showing the industry the importance of shifting from a transactional to a relational team structure that is people-centered.

Another initiative to integrate the refugees where the center played an important role is Bikes for Refugees. It is an organization with an idea to collect bikes that are not being used by local people, fix them up, and then distribute them to refugees. Meaning that refugees instantly can be independent and empowered and enjoy discovering the city and have a completely free of charge way of getting about. This can lead to all sorts of other things, for instance, they can start to get connected with the cycling community in Edinburgh. One of the volunteers at the center, who's Syrian, got a bike from Bikes for Refugees and has now joined a local group, which is called the 20 Milers. Whereas before all his activities would mainly be with the Syrian community, now he's branched off and on; he's the only Syrian in this group of cyclists and it's all thanks to Bikes for Refugees being able to get him the bike to connect him into these communities.

A third initiative aligned with the two abovementioned is called "Code Your Future" which started in London and has now moved to Glasgow and Edinburgh A group of IT professionals wanted to support the refugee community, and created a 6-month course that would help refugees develop professional skills to take on jobs on programming. It's worked really well in London and they're just starting a course in Glasgow, which some people in Edinburgh are participating in. This initiative is not just about gaining the technical skills of programming, but it's about being connected to the community. As a matter of fact, the volunteers running the course all have good jobs in the IT world, and they're able to make introductions and get people to work placements. And suddenly people have got a new career.

Common for all these initiatives is that they are cost-effective, flexible, and friendly models that can adapt to the needs of the refugees. There is trust, reciprocity that facilities sharing among and contributing, a huge amount of professionalism, among people who make things happen, resulting in a huge amount of satisfaction for doing what they're doing. There is something really important about trust and reciprocity and how social capital works both ways, everyone gets something of it, the refugees who engage get the chance to build their networks and get to know different kinds of people, but also, the volunteers get a huge amount of interest and satisfaction from meeting people from all over the world. A common denominator for all the volunteers who come into the welcoming is that they have had an international component to their lives. Maybe they grew up abroad or worked abroad and came back to Scotland, but they just loved being surrounded by people from different countries.

Much of the work at the center is characterized as a learning process. This occurs as learning processes regarding challenging circumstances. For example, with the befriending, there is a peer support and development sessions where some of the volunteers have provided training on mental health for supporting people who've experienced torture. This means that the center is growing, not size-wise but in terms of knowledge- and understanding-wise from the more of this kind of work. This builds the capacity to support other people in the future and for the volunteers at building their capacity and their adaptability for different kinds of work.

> (We) are . . . truly, truly a learning organization that is always seeking to question assumptions and to learn where it can. I think it's really important to notice your capacity to learn and adapt as an organization particularly when you're working in challenging circumstances, when you are really busy, or you feel under threat because you might lose your funding. I think sustainability and development is an ideal that you've always got to be striving to work towards. And you've got to hold yourself account to make sure that happens because, you can easily miss opportunities along the way for that.

When is then a refugee integrated? A first step is to master new language to the level of getting a job to support oneself and the family. However, for some refugees, they might start to feel integrated when they look around and think "I've got friends here," "I'm starting to feel at home."

> With our befriending program, we were talking at a recent session "how do we know we're truly friends", you know because it's a very contrived relationship and it was interesting to come up with some examples. One was the Syrian I am working with, she was teasing me on WhatsApp the other day and I just thought oh my god we're friends because it's an indicator of a friendship. . . (and) when we first started thinking about our befriending program one of our participants said something. . . I think this is a lovely quote: 'friends make you feel at home'.

However, friendship is an individualist matter, in particular, if people have special interests.

In the case of the Syrian girl who met faculty members of advocates in Edinburgh at the Burns supper local event, it would be when she had the opportunity to visit them while reflecting and seeing her present as a natural continuation of her past with a feeling of belonging in that community. In other words, having the opportunity to choose to stay or not.

> I am thinking about the Syrian girl. She had the opportunity to visit their kind of the lawyers from city, and she was 'ok now I belong here, I saw my reflection of my future, I have a perspective. I don't see only my present and my past, but I see also a perspective of future here'. There might be something important around choices like when you first arrived particularly as a refugee. You've got no choice in your life anywhere. You've no choice to arrive here. But the minute when your life starts offering you choices I suspect, and that would be a really interesting thing to research, you know does that start to feel like. You're more integrated because you can either go this way or that way you can live there you can live there.

It's when people conquer freedom probably: "ok I'll be integrated now I can go anywhere."

Reflection Questions

Disruptive Global Trends

As a disruptive global trend, the unprecedented displacement crisis, sparked by armed conflicts, has resulted in the arrival of large number of refugees to many European countries. Shelter was given to more than one million refugees and in refugee-like situation. Many major European countries as well as smaller ones hosted these refugees who essentially came from Syria, Afghanistan, and Somalia. Notably, however, a country like Sweden with a relatively small population gave in 1-year shelter to 140,000 asylum seekers, which was the highest number per capita among the EU members. The numbers of incoming refugees in the UK are tiny compared to those in Germany and Sweden, which resulted in the displacement crisis causing a scale effect, that is, the displacement of the refugees is not in proportion to the size of a EU member country and its capacity to cope with the integration of them. Today refugees are still seeking to escape the armed conflicts but now they come via the sea road to EU.

The displacement of refugees is by many leaders as well as news people, perceived as a crisis. But perhaps in the drama there is an opportunity encapsulated in "immigration will help companies to grow and ensure long-term prosperity, which is good for economic growth." For instance, with Europe's lowest birthrate, rapidly aging population, and skilled worker shortage, Germany could lose its position as the powerhouse in Europe and one of the world's leading economies. Without significant immigration the working-age population will likely decrease from around 49 million in 2013 to almost 34 million in 2060 according to a government estimate. But it seems that the refugees will not plug the labor gap neither in Germany nor in Sweden. A major reason is the relatively low level of education among the refugees. Both Germany and Sweden have an advanced labor market with demands on well-educated workforce. Statistics show great difficulties for newly arrived to get a job on the regular labor market. In 2006, about one quarter of the labor force in Sweden was newly arrived. Ten years later, this group was for the first time the majority of unemployed. The most common reasons for the failing integration, like discrimination and lacking flexibility in the labor market, can only partly explain the exclusion of newly arrived in the society.

Integration Programs and Structural Barriers

Most EU member countries provide integration programs to facilitate the integration process. However, there can be structural barriers or impediments that limit or impede integration initiatives and refugees to get a job in the regular labor market, such as the local language, lack of understanding, accepting qualifications and educational degrees that the refugees have acquired in the home countries, and even the name of the applicant.

The case from The Welcoming in Edinburgh illustrates an engaged sustainability practice in the form of individualized language education for the refugees. Every week the center is running classes where volunteers teach English to the refugees. Distinctive about the language classes is the "drop-in," that is, the refugees don't book for a class but come along when they are available, and the "ad-hoc," that is, neither is there a preset amount of students nor continuity between two classes. Despite this very untraditional way of organizing education, it seems to be working quite well.

However, the displacement of refugees also results in an askew distribution of refugees among the European member countries. A relevant question is therefore if practices that work in one country that receives relatively few refugees also work in countries that receive relatively many refugees in relation to its population and integration capacity? In other words, would the language teaching practice in Scotland also work in, for instance, Sweden? This question introduces the notion of scale and individualization, normally a paradox, or, more specifically, *what kind of "rules of the game" favor scalable solutions that can be used for societal integration of the refugees adapted to their individual needs.*

To cope with the integration challenges, a majority of the political parties in Sweden have united around necessary efforts to be implemented in order to strengthen the establishment of newly arrived in the Swedish society. It has been suggested that rules and regulations in this respect should be simplified in order to increase the flexibility and thus facilitate the establishment in the regular labor market. A real need has been identified as finding better ways for low-educated newly arrived to receive professional education so that they can establish themselves in the regular labor market. The possibility to introduce educational duty for people who come to Sweden in the latter part of life and who have not achieved elementary school level education has been raised.

At present (2017) in Sweden, there are almost 300,000 newly arrived with, at the most, elementary school education who face difficulties to enter the labor market. This group of newly arrived poses the greatest integration challenge in Sweden. A relatively large number of the newly arrived has never attended a school or only a few years. They cannot read or write in their own mother tongue and have difficulties to assimilate information that is not concrete and specific, in particular, if it is in Swedish. A majority comes from countries where the everyday life, norms, and values are different from Sweden. Many situations are new and foreign. They may find it difficult to understand the current context, don't know what is going to happen and what is expected from them. Yet, newly arrived without school education and knowledge about the Swedish society are sent to different kinds of professional education and work life projects and programs that they cannot accommodate. The real challenge is, thus, *how to integrate newly arrived and in particular those with none or relatively low level of education*; a challenge that is increasing as digitalization and automation make even simplistic jobs more knowledge intense.

Folkuniversitetet, the public adult educational association in Sweden, offers about 38,000 places. Out of those who receive a study, only 16 percent finds a job after the study or continues to further studies; the rest remains inactive or in unemployment. In the public debate, discontent has been voiced over this situation, arguing the unsustainability in public efforts and official programs resulting in a majority of the students that end up inactive or unemployed. In addition, the queues to adult education at Folkuniversitetet are strained to the extent that the school law is violated in many municipalities, a pressure that is estimated to increase in the coming years.

In 2015, more than 70,000 children and juveniles in the age between 0 and 18 years applied for asylum in Sweden. The year after, the number was merely 4,000 students. A new kind of preparedness and knowledge has been required among the municipalities in order to receive and swiftly offer school placement to the newly arrived students. This could include a preparedness to introduce them to sustainability issues. Recent discoveries in quantum physics and neuroscience show that human values, which can play a significant role in achieving a sustainable society, are intrinsic to human beings and hardwired to our brains. How education can be an opportunity to plant a seed of sustainability in young minds is discussed in chapter ▶ "Education in Human Values" by Rohana Ulluwishewa.

The SSI has recently published a report about the language education programs in the Swedish high schools criticizing the use of standardized approaches for all students regardless of their schooling background. An eminent risk is therefore that students are losing knowledge they already have gained in their former home countries and that their knowledge progression is delayed. From the SSI report, it is evident that mapping is done by the schools of the students' knowledge but seldom encompasses other subject knowledge and abilities. Furthermore, the mappings are not shared to everybody working with newly arrived students and are not integrated and systematically used in the education. As a consequence, significant problems appear in the transfer of information about the students' knowledge levels if they change schools and/or municipality. For instance, in the program for language introduction in the high school, many students receive too little educational time. Full time can vary between 13 and 25 h. On several occasions, the syllabus is governed by the access to teachers and facilities rather than by the students' needs. Newly arrived students need a good overview and orientation about the offer, choices, opportunities, and ways forward in for them a completely new educational system. However, students' health and vocational guidance are not integrated into the education in sufficient degree.

Many newly arrived students have good subject knowledge from their previous home countries. It is reasonable to assume that they would benefit from stimulating varieties in the education that are equivalent to their level of knowledge. However, many schools face challenges with recruitments. The lack of accessible teachers and study supervisors in the students' mother tongue is today a critical factor that may make it difficult for the schools to provide the newly arrived students with individually adapted and varied education. If the Swedish schools cannot offer this, then there is a risk that their knowledge development progression will be delayed. Despite the difficulties some schools have gone through, systematic development of the education for newly arrived students within their traditional quality work found a well-functioning way of working and succeeded in accomplishing increasing variety and flexibility in the education.

This way of working involves how the mapping of newly arrived students' knowledge and how information about the students are being used. Newly arrived students meet many teachers, student health staff, and many others during their schooling. It is common that they change between classes, groups, and schools. Here it is of utmost importance that information about the individual students is passed on in order to avoid thresholds in the schooling and that training time is lost. Hence, much work has been done with integrating newly arrived students to the Swedish school system. However, there are shortcomings, which make it reasonable to assume that certain students today are studying on a too low level not equivalent with their level of knowledge, whereas others need more basic knowledge and support. As a consequence, their learning and knowledge development may be hampered and provide them worse conditions to continuous education and transfer to the professional life. There are good initiatives to provide individually adapted, varied, and stimulating education to newly arrived students, but the width, variation, and flexibility to fully meet their varying needs can be improved, and follow-up studies are needed. Initiatives to introduce sustainability models, such as circular economy (CE) and cradle to cradle (C2C), in business education, are discussed in chapter ▶ "Teaching Circular Economy" by Helen Kopnina).

Suggested solutions and enabling practices in the professional life and school system.

Many municipalities in Sweden have made huge efforts when it comes to the reception of newly arrived students. Most of them receive the students within the statutory time limits. Despite shortcomings (see above), the majority of the municipalities map the students' knowledge within 2 months. Some chief executives have argued that their organizations have been strengthened as a consequence of many newly arrived students, for instance, through competence development and new forms of cooperation. This is positive and important. However, in order for the newly arrived students to quickly receive good and equivalent education, the municipalities can do more. Firstly, the education is still not sufficiently developed to meet the newly arrived students' individual conditions and needs. Newly arrived students have very different school background, subject knowledge, language skills, and experiences. Yet many schools apply standardized collective solutions. For instance, standard decisions are common when it comes to the placement of students in preparatory classes. In addition, when it comes to the planning of the education, students who do language introduction in the high school are offered the same schedule and student pace regardless of previous knowledge.

A suggestion is therefore *a compressed and adapted elementary school education targeted at a specific group of newly arrived to provide the kind of competence enabling short-term educated people to qualify for a regular job.* More specifically, such an education could encompass lessons in Swedish combined with everyday life and social orientation in the mother tongue in addition to lessons in mathematics and, if needed, to improve the mother tongue. Normally such an education would take

3–6 years, which is costly. However, it would be a significant improvement compared to the current situation where it takes on average 8–9 years for newly arrived refugees to start working if they do it at all. It can be speculated that the majority of the people in this category still will not be able to complete such an education. Most likely it will be the younger and the most motivated who will do it. For the majority, the optimal solution might be a combination of a somewhat even shorter, compressed elementary education and less qualified job.

Also when it comes to unaccompanied juveniles with an inadequate school education, it would be a benefit if they could be offered an adapted education. Today they are integrated early in regular business operations, where a majority fails. A solution could be to give them some kind of study payment slightly above the (försörjningsstöd) government support and connected to the presence at the job. Especially for women without government support, who often remain at home, this could be an incentive to enter the professional life. To use current means, which today often result in fruitless activities for newly arrived refugees, for an alternative elementary school education for newly arrived analphabets and those with relatively low education and unaccompanied juveniles would mean huge savings of social costs. In addition, more would be integrated and have a chance on the regular labor market in Sweden. Yet another benefit with less qualified jobs would be that the newly arrived would have a job to go to and then become a positive role model for their children.

Exercises in Practice

Corporate and Citizens Social Investment: Scaling Up Community Initiatives

The Welcoming Association evidences that through the *Befriender* practice, reciprocity plays its role in society by conserving social norms, enhancing well-being, and increasing the exchanges within an environment. Those exchanges result in optimizing the utilization of idle resources, them being material or intellectual. Initiatives like those have always been a part of society, with the clear majority of them relying on financial support from government, corporations, faith institutions, or private funds. In the case of the time banking, reciprocity replaces the financial payback expected from those outcomes, and time is the currency mediating the relations between the individuals involved. Reflected on the learning acquired in this book chapter and exploring your maximum capacity of thinking outside the box, elaborate ideas and resolutions on how both private and public sectors can scale up and disseminate the practice of time banking within a certain society. Consider the two samples below to be inspired on some of the possibilities in both private and public sector.

Private Sector Initiative

NEFEJ co. is a private company from the luxury sector that decided to transform its business identity and renew its brand identity. Aiming a more sustainable organization, the company decided to transform its corporate social responsibility department into "private social investment on the community – PSIC." The head of this newly formed department started off by approving a series of adjustments on the corporate policies. One of them allowed its employees to convert up to 10% of monthly contractual hours into community projects managed by the organization. After some months of the policy, they decided to partner with a local time banking program that works with the integration of Syrian refugees on the same city as its headquarters. Distributing its more than 500 employees involved into teams, each department had the accountability of a project:

1. Skill mapping: HR departments mapped each of the time bank's account holder skills and capability to identify where they could be more productive.
2. Quality and training: These departments were put together to setup training courses that could fulfil the company's demands or lack of skills in high seasons.
3. Production KT (knowledge transfer): The production team was assigned with a group of refugees mapped by the HR department. This group of refugees possessed vast experience with a fabric that the company is starting to use, and they were designed to collaborate on trainings and support the initial months of the production.

Public Sector Initiative

The city of Polisburg is a state of the art city. Centered on the sustainability principles and balancing its economic, social, and equity pillars, it recognizes and rewards initiatives for a better quality of living of its citizens. Recently a group of citizens grouped to found the first time banking chapter of the municipality. The idea emerged when the city was appointed to receive a huge number of refugees along the year. The mayor was very concerned initially with the budgetary limitations and to the impact that those refugees would cause to the local community. Seeing no way out and embracing the challenge, he studied some failure and success stories and came up with a plan to support this integration process. His thoughts were that money would never be enough when dealing with a situation that goes beyond budget and reaches the social and individual well-being. He decided then to empower local citizens to participate on the decisions and for these citizens to have a saying on the integration process.

After recent studies, the city mayor found out that the public costs in a variety of departments dropped systematically. Some departments claim that there is a slowdown on the aim for public services and an improvement on the living conditions of citizens in that municipality. Some of the initiatives were:

1. Co-sharing of public infrastructure and local assets – The time banking community in that place could utilize public building to conduct time banking initiatives. The first project approved allowed them to conduct language courses and additional skills training on classrooms in public schools over the weekends.
2. Utilization of big data for mapping the skills of each region – Once the time banking system was implemented, the evaluators saw the tool as an opportunity

for the selection of the best professionals in the market. The public departments use this tool to map and assign skill demands in different areas of the city. This has helped the municipality to assign refugees to areas with the increasing aging population that is in need of caregivers or from where to relocate or assign skilled professionals in a moment of crisis such as mechanics and other handworkers for natural disasters, for example.

Time Is Not Always Money

Early morning Mrs. Pozzato patiently waits for the ring of the doorbell. She has been struggling with her boiler on the cold winter and she couldn't find any available plumber nor counted with the expenses of the services after the house acquisition. Luckily, she is part of the time banking community in her hometown, which gives her access to a list of skill masters offering a varied portfolio of services. There she spotted Mr. Lucco, a retired army sergeant that received full-mechanical training and enlisted to time bank to exchange a varied set of skill. He looked at the engine and soon after, eureka! "It seems your boiler's pressure was low, let's see if it works now!" Mrs. Pozzato, worried with her budget, asks for the cost of the service, which is promptly responded by Mr. Lucco: "One hour and a half." Mrs. Pozzato processes the payment from a cell phone app, evaluating Mr. Lucco's service in a range of aspects and recommending him to some friends. The app deducts from Mrs. Pozzato exactly one and a half hours she acquired by cooking on a nearby school. In fact, she is very active on the community and counts with a very positive social contribution factor, providing much more service than receiving. She enjoys participating of the time banking community because she has accesses to learning and sharing experiences she never would due to the budget limitations. In there, she finds it easy to meet and connecting with people she wouldn't naturally meet on her social cycle. Some time ago, she met Ms. Gonzalez, a Spanish teacher and actress that recently moved in to her town in Italy to accompany her husband that was transferred from work. Ms. Gonzalez encountered on the exchanges a possibility of connecting with the local community and makes a productive use of her time while she is not working. Through time banking, she managed to coach students on a play while learning how to cook local meals with Mrs. Pozzato and storytelling for kids in hospitals while joining a yoga class conducted by a skill master, and, through that, built a good reputation with the community. Her integration problems are being minimized since she is fully skilled in Italian language and currently count with some savings and her husband's support. The stories above are shared among citizens with common cultural backgrounds or with cultures that assimilated easily. Now imagine that this community receives a new member, his name is Mohammed and is father of three bright kids. His wife is on the last weeks of pregnancy and currently staying on the shelter they are living most of the time due to her locomotion limitations. Their family recently crossed into Italy from a long journey from Syria through the Mediterranean Sea and is experiencing all sorts of issues in this new homeland.

1. Imagine the following aspects of a human life: thought/feeling, choices, body/health, social context, and soul. Now identify the issues and barriers Mohammed and his family is encountering on each aspect to the integration into this new country.
2. How can Mohammed's family benefit from "opening an account" in the time bank?
3. What benefits can Mohammed's family bring to the community?
4. What are the main barriers that can be encountered by refugees on participating of the time banking scheme?
5. What would be the actions to be taken to overcome those obstacles?
6. On a political and global perspective, how can the country benefit from these exchanges?

Engaged Sustainability Lessons

This book chapter focused on how social initiatives can support government's actions to tackle the refugee crisis. Independently from the origins of the crisis and how the government structures have chosen to deal with this, local communities we forced to come up with fast solutions. Integration, which is a key aspect, raises as the first logical step for the increasing the quality of living of these citizens, enhancing the social well-being in that environment and, while doing that, bringing that whole community to a different way of living. But for that to happen, barriers from different levels of complexity must be overcome. Social barriers are the main causes of integration problems, and it attenuates its nuances on welcoming minorities within a formed social norm. For refugees the challenges are maximized, from communicating within that new environment to finding a job and start being productive, to be respected and has its differences acknowledged, accepted, and respected.

Those barriers can only be overcome when the individual is placed at the center of the priorities without diverging from the equality and reciprocity values. Due to this fact, more standardize initiatives as those usually provided by public services fail to fulfil those refugees and society needs of cost-efficient and community involving initiatives. The bigger the volume, the closer to standardization those initiatives get and, beyond that, the increase need of utilization of national funds to finance them, which has the potential to lead to contradictory public opinions, extremism, and inertia.

Through community involvement in practices such as the Befriender by the Welcoming Association, Code Your Future, Bikes for Refugees, and others, citizens can contribute to the construct of a more sustainable community. These communities transcend from the current social dynamic and invite us to reflect on this model where reciprocity and equality lay as a common ground of the value shared among citizens. On the Swedish initiatives to cope with the volume and the increasing impacts to its labor market and by consequence the economy, time banking aises on this pool of public and private actions as an alternative. Due to its potential to connect and share, it involves the whole community to act on a solution of a problem that impacts them directly. Its core values are essence for the sustainable solution of the refugee crisis. It redefines work by honoring equally the citizens' contributions, assets by shifting the focus to the individual, seeing everyone as potential assets, and recognizing that everyone has something to contribute. It promulgates reciprocity by implying that help works best as a two-way street. It respects by following the principle that everyone deserves to be heard and that conflicts must be avoided. It stimulates the social capital of the communities by grounding on the belief that together we can do much more than alone.

Citizens can participate by solely joining the reciprocity circle and adopt a positive behavior on employing its best skills and abilities. By doing that, refugees on that context get a chance to understand the social protocols of this new environment, get in contact with the local culture, and expand the limits of its social networks by transitioning from a virtual and, sometimes, isolated and segregate world to the reality of a community living. On the current society, these actions result on enhancing every participant psychological need to feel affiliated and appreciated. In a shared economy model, it has the potential to fulfil even the most basic human needs. Everyone is welcome to join a time bank, but thoughts grounded with economic, political, or religious values are not.

Chapter-End Reflection Questions

- Is it possible that this unprecedented displacement of refugees that has manifested in lackluster, mixed and at times outright inhuman responses to help people seeking refuge at European shores and borders, which has revealed deep underlying rifts within Europe, lack of solidary and commitment between member states, paradoxically, will make Europe a stronger and more cohesive union?
- How to go from shirking to sharing responsibility among European member states to handle the displacement of refugees? Can a country's capacity to host refugees, wealth, population, and unemployment rate be used as basic common sense non-complex criteria to give each EU member country their fair share?
- How to reframe immigrants as a social problem that needs to be fixed but rather as an opportunity?
- What are the practices in which refugees provide contributions that are valued by local communities and how can they be created, promulgated, and shared across nations (cross-national practices) in Europe?
- How can communities of practices develop and launch courses that help refugees to develop professional skills to take on jobs and at the same time being connected to and integrated into a society? What are cost-effective, flexible, and friendly means that can be adapted to the various needs of refugees?

Cross-References

▶ Relational Teams Turning the Cost of Waste Into Sustainable Benefits
▶ The LOHAS Lifestyle and Marketplace Behavior

References

Bailey, D. (1992). Using participatory research in community consortia development and evaluation: Lessons from the beginning of a story. *The American Sociologist, 23*, 71.

Cahn, E. S. (2000). *No more throw-away people: The co-production imperative*. Washington, DC: Essential Books.

Fernandes, A. G. (2015). (Dis) empowering new immigrants and refugees through their participation in introduction programs in Sweden, Denmark, and Norway. *Journal of Immigrant & Refugee Studies, 13*, 245.

Flint, R. W. (2015). *Practice of sustainable community development: A participatory framework for change*. New York: Springer.

Kilpatrick, S., Field, J., & Falk, J. (2003). Social capital: An analytic tool for exploring lifelong learning and community development. *British Educational Research Journal, 29*(3), 417.

Miller, E. J. (2008). *Both borrowers and lenders: time banks and the aged in Japan*. Ph.D. thesis Australian National University.

Page, N., & Czuba, C. E. (1999). Empowerment: What is it? *Journal of Extension, 37*, 24.

Pease, B. (2002). Rethinking empowerment: A postmodern reappraisal for emancipatory practice. *British Journal of Social Work, 32*, 135.

Waltz, K. (2010). *Theory of international politics*. Long Grove: Waveland Press.

Wilson, P. A. (1996). Empowerment: Community economic development from the inside out. *Urban Studies, 33*(4–5), 617–630.

Supermarket and Green Wave

An Urban Sociological Field Experience in Brazil

Josi Paz

Contents

Abstract

There is an ecological imperative established in the consumer society. Ecocriticism of the consumer society was raised when the environmental movement was born, but has become stronger lately as a repercussion of the anthropogenic hypothesis of global warming. As the greenhouse effect warms the planet, the material and symbolic effects of the global warming debate affect the issue of consumption. A green wave has taken over supermarkets in the West, where green products, eco-bags, and recyclable packaging, to name a few, have become more familiar. In this context, "human being" and "individual consumer" sound like equivalents, and the concept of "consumer society" becomes a metaphor for "humanity." This article points out the main lines of the sociological debate that emerges from this turning point, and approaches some theoretical perspectives on consumption. It also presents some excerpts from qualitative field experience developed in two supermarket chains in Brasília/Federal District, the capital of Brazil – one located in a high-income region and the other in a low-income region.

J. Paz (✉)
Consultant - Communication and International Cooperation, University of Brasília, Brasilia, DF, Brazil
e-mail: josi.ppaz@gmail.com

S. Dhiman, J. Marques (eds.), *Handbook of Engaged Sustainability*,
https://doi.org/10.1007/978-3-319-71312-0_11

Keywords
Consumption · Climate change · Environment · Supermarket · Brazil

Introduction

Consumption is a natural process; it arises from the profusion of objects and is presented to the individual as gifts from heaven. The consumption experience works as a miracle, although it is a human activity and not obedient to natural ecological laws. All this characterizes a type of fundamental mutation in the ecology of the human species, because even though objects do not constitute a flora and fauna, they give the impression of proliferating vegetation – a jungle where the new wild man of the modern era has a hard time re-encountering the reflexes of civilization.

This image was proposed by the French sociologist Jean Baudrillard to define the consumer society and is one of his most ironic metaphors on the semiotics of this society (Baudrillard 1995). His work, *The Consumer Society*, from which this text was paraphrased, was originally published in French in 1970 and became a reference for discussing the daily experience of consumption and understanding the historical unfolding of this cultural practice. In this sense, Baudrillard's ideas essentially contribute to a deeper analysis of the "green wave," a market trend that has hit supermarkets in the Western world as a repercussion of the global **climate change** debate.

Baudrillard's work is one of the main references in this article on green consumption in Brasilia, capital of **Brazil.** Two supermarket chains were visited, located in two unequal urban contexts, one of low-income consumers and the other of high-income consumers. Both supermarkets were exposed to the so-called green wave of consumption. This article discusses the wider repercussions of the environmental debate on consumer issues, especially the anthropogenic hypothesis of global climate change, which has become an ecological imperative in consumer society (Paz 2012).

About Consumption

Although Baudrillard refers to nature, he does not approach the environmental issue: his issue is the "consumer society." Still, he highlights the problems that emerge from a society full of things (Baudrillard 1995). The author refers to consumption as an order in the manipulation of signs, in which the sign "nature" allows consumers to escape from a wrecked empirical reality. In this perspective, the ecological appeal inscribed on green packaging, for instance, approaches nature as an escape in two possible ways: (1) by emphasizing the natural origin of the ingredients as an escape from industrialized products and (2) as an escape from a particular lifestyle. This is why the natural environment is also present in the green appeal of goods.

Consumption of the sign *nature* gives the possibility of "consuming" an absent natural environment or one that no longer exists.

The consumer society presented by Baudrillard is different from the perspective of the Polish philosopher Zygmunt Bauman. In *Consuming Life* (Bauman 2008), he prefers a "society of individuals that consume" instead of a "consumer society," alluding to the relational and dynamic nature of the interaction and the interdependence of individuals and objects, distinguishing this contemporary type of sociability. Bauman discusses the impact of the attribution of value to objects in human relations, which places consumption as a parameter of life itself. Bauman's perspective is different from that taken in the late 1970s by the French sociologist Pierre Bourdieu that led to research first published in French in 1979 in his book *Distinction: A Social Critique of the Judgment of Taste* (Bourdieu 2007). Baudrillard and Bourdieu have different theoretical and methodological perspectives and interests concerning this issue.

Bauman discusses the subject in a third line of approach. If "consumer society" is the object of reflection of Baudrillard, for Bourdieu the debate on consumption includes questioning its analytical framework, which combines qualitative and quantitative research on the formation of taste. The differential trait in Bourdieu's (2007) socio-analysis, however, is not only in the inflection of a theoretical–methodological slant on the theme of consumption, as Baudrillard (1995) does, but also in questioning the researcher and the instruments used to substantiate its worldview. Although he addresses the theme of taste formation tangled in the consumption debate, Bourdieu (2007) ultimately discusses the scientific methodology itself. Like Baudrillard, Bourdieu analyzes the power game – mainly in the cultural sphere, where acquisition practices interfere even in the meanings of works and arts: "Although it manifests itself as universal, the aesthetic disposition is rooted in the existence of particular conditions … it constitutes a dimension, the most rare distinctive and distinguished of a lifestyle" (Bourdieu 1983).

The three works, *The Consumer Society* (Baudrillard 1995), *Distinction* (Bourdieu 2007) and *Consuming Life* (Bauman 2008), have different readings on consumption that coincide with different contexts in the sociological debate on the subject, and therefore to different historical moments of this society. Baudrillard reacts to the advent of consumer society and its objects; Bourdieu reveals in what measure objects are valued in the consumer society; and Bauman, at a time when the profusion of objects and consumption as a way of life are crystallized in contemporary life, discusses consumption as a parameter of life itself. Consumption would be repeated, as well as being a parameter of the human condition. Bauman (2008) criticizes the importance given to the consumer as a parameter of contemporary subjectivity: our lives are now for consumption, he says.

On a timeline, consumer configuration as a sociological issue also goes back to Roland Barthes (1985) and Georg **Simmel** (1987, 1998), and certainly shows its influence in *The World of Goods: Towards an Anthropology of Consumption* by Mary Douglas and Baron Isherwood (2004). In *The Cultural Contradictions of Capitalism*, the American Daniel Bell (1996) pondered the symbolism and the material impact of one of the objects that would become the most iconic item in the consumer society: the credit card. In Latin America, the Argentinian Nestor **Garcia-Canclini** (1996) discussed consumption as a gesture of citizenship. However, before all these texts, *The Theory of the Leisure Class: An Economic Study of Institutions,* first published in 1899 by Thorstein Veblen (1974), founded the notion of **"conspicuous consumption"** that still explains the dynamics of the consumer society, in which the acquisition of certain objects aims to denote status and social differentiation. In a recent article, relying on Veblen's importance in understanding contemporary society, Currid-Halkett (2017) points out the extent to which the democratization of consumer goods has disrupted this process of demonstrating status through things; after all, regardless of differences and inequalities, both the rich and middle-class have modern TVs and purses, travel by plane, etc. For Currid-Halkett, because everyone can now buy designer bags and new cars, the rich have been led to use more tacit social position symbols. The consumption of environmentally friendly products is one of the forms of expression of this discrete consumption, or "**inconspicuous consumption**."

These man-made artifacts are used to distinguish the *status* of humans, to centralize the environmental critique of consumption. Climate has begun to be placed as a social issue, around the local consequences of human artifacts, remains, and remnants of evolution (Henson 2011). The idea of "global" warming is therefore only possible in a world that thinks globally about consumption, following the global paths of the economy and communications. David Harvey (1996) describes this historical process, which takes place as discourse and daily experience that shapes the notion of humanity in a global perspective. This is one of the impacts of the socio-historical configuration of globalization: an understanding of the human experience that passes through consumption. **Globalized goods** help make sense of a world much larger than the locality.

The Supermarket and the Planet

Green consumption refers to a positive outlook on consumption as an individual choice – amidst ambiguities and contradictions – with the possibility of changing the planet's future. In the supermarket, packages invite people to seal the Earth's destiny. Observing how individuals interact with objects helps in understanding the extent to which consumer culture (and therefore consumption and lifestyle) can work as a key to thinking about contemporaneity. As the Hungarian sociologist Agnes Heller (2004) says in her book *Everyday Life*, the assimilation of object manipulation is how people become adult in everyday life, and how an adult becomes able to live by himself in daily life, as social relations are assimilated by manipulating things. In this perspective, the daily manipulation of objects in a **consumer society** also implies socialization forms.

People learn how to be consumers, to understand visual codes, to recognize what is expected from their acts in different situations between individuals and objects in consumer society. Taking a product from the shelf, paying for an object, waiting in a queue, and the prerogative of not leaving the supermarket without exchanging goods for money are all acts that flow without being noticed. For Bourdieu (2007), this incorporated knowledge, the **capital knowledge** acquired or incorporated in the social trajectory, explains these gestures and the formation of lifestyle and taste. In *Distinction*, Bourdieu (2007) breaks with the notion of taste and of individuals' identification with a particular lifestyle as being something of their own. In his view, taste is the result of overlapping power relations grounded in the culture-transmitting institutions of capitalist society. The preference for certain products and the possibility of choosing environmentally friendly products, for instance, is about the existential question of whether to be or not be a conscious consumer: this is about income and recognition of the green value added to the product as being relevant and decisive regarding consumption.

There is a set of procedures that can be considered a socialization index related to the manipulation of objects, which contributes to create and legitimize some choices possibilities. In this sense, the green wave of consumption and its prescriptions (buy **organic**, recycle, reuse, use eco-bags instead of plastic bags, etc.) brings new consumption gestures that involve taste issues. By pushing a trolley through the supermarket aisles, consumers indicate the preferences and possibilities of consumption: the individual takes longer in front of some shelves, choosing items (removing one item from the shelf rather than others). Different strategies are implemented by the supermarket to retain consumers and to make them circulate, exploring their different buying potentials.

The level of interdependence of the global complex society in which we live in relation to the supermarket is not only an icon of the consumer society, but also its fundamental metaphor. In the supermarket (any supermarket) we are all consumers. The consumer can be found in this space, regardless of specific social contexts concerning individuals and the location of stores. The supermarket can be seen as a cultural and material expression of spatial overabundance in supermodernity:

> The second accelerated transformation specific to the contemporary world, and the second figure of excess characteristic of **supermodernity**, concerns space. We could start by saying – again somewhat paradoxically – that the excess of space is correlative with the shrinking of the planet: with the distancing from ourselves embodied in the feats of our astronauts and the endless circling of our satellites. (Augé 1994)

Auge emphasizes the difference between space and place, defining space as a more abstract notion, based on non-measurable magnitudes. Instead, Michel De Certeau (1998) defines place as stability, "the order [of whatever kind] in accord with which elements are distributed in relationship of coexistence" and space as "occurs as the effect produced by the operations that orient it." For Augé (1994), there is an opposition between place and **non-place**. For De Certeau (1998), space is not necessarily opposed to place. "The space could be to the place what the word becomes when it is spoken," he says, when defining space as "a practiced place." According to De Certeau's perspective, a supermarket can be understood as a place crossed by different spaces, but in Augé's perspective, a supermarket would be a non-place of supermodernity:

> Clearly the word "non-place" designates two complementary but distinct realities: spaces formed in relation to certain ends (transport, transit, commerce, leisure), and the relations that individuals have with these spaces. Although the two sets of relations overlap to a large extent, and in any case officially (individuals travel, make purchases, relax), they are still not confused with one another; for non-places mediate a whole mass of relations, with the self and with others, which are only indirectly connected with their purposes. (...) the non-places create solitary tension. (Augé 1994)

For Augé (1994), the individual of a non-place would be alone, resembling the others around, with identity suspended (a non-place characteristic) and remembered only on the basis of contractual relations. Personal identity is declined until the advent of a situation, which is limited to the supermodern experience. In the supermarket, for instance, the individual handles the **supermarket** trolley and moves with it through the supermarket without communicating to anyone; nobody asks the individual's name, where they come from, or what they are looking for. Furthermore, the individual does not need to ask any questions or interact with other individuals, who also take their own courses in the supermarket. The supermarket is a less prestigious non-place because of the higher frequency of female **consumers**, observed Augé (1994), but the supermarket is a place of general consumers, whatever their identity.

In the supermarket, the price of a product is given on a label or obtainable via a bar code reader, but it is assumed that individuals know that they must pay for the objects chosen; they know that they have to go to the cashier before leaving the supermarket. From Augé's perspective, until the individual goes through the cashier, all consumers are the same person, because the personal identity of the individual only becomes relevant at times, such as when using a credit card, cheque, or money, when then the minimum of interaction is necessary to identify the individual. According to Augé, the moment at the cashier's desk in a supermarket relates to the non-place's contract:

> Alone, but similar to others, the non-place user is (...) in a contractual relationship. The existence of this contract is remembered in some opportunities (...): the flight ticket he bought, the card that must be present at the cashier, or even pushing the trolley in supermarket aisles relates to a more or less strong contract. (Augé 1994)

The problem with Augé's view is that his emphasis on the idea of non-place as an architecture of modernity does not take into account the individual bonds that also constitute the dynamics of this space, especially in a supermarket. Although the author appropriately draws attention to the merely contractual relations of non-places, field experience in a supermarket allows one to observe that human relations of affection, recognition, and routine often affect those contracts. If the individual lives near the supermarket, the store may be visited with such frequency that supermarket employees become recognized and known. In a situation of doubt about a particular product, help might be given and, from there, a personal link could be formed.

In order for the supermarket chain to be recognized as belonging to some region in which stores are located (neighborhood, street), its facade, interiors, sections, thematic design, and even the commercial name used need to be linked in some way to the context surrounding the physical structure. Two examples from a sociological field study in Brasilia, capital of Brazil, are described to illustrate this.

Brasilia is the result of an architectonic and urbanism project created by **Oscar Niemeyer** and Lúcio Costa. Named a "Heritage of Humanity" by **Unesco**, the city's modernist architecture is peculiar: the central part of the city is an area called Pilot Plan, with residential areas called *Superquadras* or Superblocks, spread in north and south wings. From the air, the city resembles the shape of an airplane or a bird. Around the "airplane," there are other cities – administrative regions – also called satellite cities. The whole area is defined as the Federal District. In turn, the administrative region of Brazlândia is the furthest from the Pilot Plan and has mixed urban and rural characteristics. Its design resembles a small town, with a central shopping area, church, public square, residential neighborhoods, and rural areas. The supermarket visits described in this article occurred in two locations: Pilot Plan, Brasília, and Brazlândia, Federal District (Paz 2012) (Fig. 1).

The **green wave** of consumption referred to in this text refers to the marketing strategy that emphasizes aspects of composition, communication, and packaging associated with the commercialization of certain products, offered as better or with

Fig. 1 Map of Brasilia, including administrative regions; Federal District, Brazil

less impact on the natural environment. Such marketing suggests that consumers who purchase these products also have its qualities: one who buys green is green. In many cases, this offer appeals to an individual's care for their personal health, which evolves toward gestures of care for the health of the planet. Acts such as separating garbage and the use of returnable bags instead of plastic ones also are part of the green wave of consumption, once they integrate the status of good and desirable everyday practices.

The eco-sense consumer had been present for about three decades, since the emergence of the environmental movement (Tavolaro 2001; Dunlap et al. 2002; Garrard 2006), but there has been more evidence of marketing of these products and practices since the mid-2000s (Paz 2012; Portilho 2005) as a response to contemporary environmental problems. In the wake of antiglobalization movements and anticonsumerism (Klein 2002, 2014; Littler 2008), social distinction strategies also take place because the choice of more environmentally friendly products presents itself strongly as a matter of taste, implying style preferences and access possibilities as a function of income.

Although the onset of the green consumption wave is not very recent, its greater visibility is related to the repercussions of the hypothesis of the anthropogenic cause of climate change, as presented by the **United Nations** Climate Research Group, the Intergovernmental Panel on Climate Change Climate Change (**IPCC**). "Anthropogenic" comes from the Greek *anthropos,* meaning human, and *gennan,* meaning production or origin (Glacken 1967). It is in this context that notions and social practices such as "conscious consumption" and "environmentally responsible consumption" have established a direct link between the notions "consumer" and "human being," and between "consumer society" and "humanity." The IPCC does not carry out research: its conclusions are based on analysis of research carried out by several authors, in excellent research centers (peer review). Each group operates with a distinct thematic focus. According to the IPCC (2007), the increase in emissions of greenhouse gases into the **atmosphere**, taking as a parameter the historical milestone of the Industrial Revolution, is raising the average temperature of the planet.

"In the past, the concept of humanity was referring to an ideal image, distant, always peaceful and harmonious. Today, it refers to a reality rich in conflicts and tensions" stated the Polish sociologist Norbert Elias (1994), emphasizing the extent to which "retrospectively, humans often see only the apparently uniform progress of technique and not the struggles of elimination, which cost lives and those who are behind." "The affiliation to **humanity**," he says, "continued permanent and inescapable." However, "our ties with this universal unity" would be loosened so that "the sense of responsibility to humanity in danger is minimal."

In the debate on global warming, the sign humanity is inflected under the notion of a consuming humanity, which implies the development of a single gesture as a symbol of "responsibility" for the collective. It is up to each human being, understood as a consumer, to do their part and change their consumption habits; this represents a responsibility to humanity and to the planet. The limits of the consumer society would therefore be the limits of that affiliation to humanity: the mediation of consumption gives the possibility of holding their humanity. Changing the consumption habits of humanity, however, because of loosened ties, is not so easy.

For **Clifford Geertz** (1989), the understanding of themselves as a single species is shaped by culture: A human cannot be defined only by innate abilities (as did the Enlightenment), nor only by behavior (as does much of contemporary social science), but by the link between them. For Geertz, "when viewed as a set of symbolic devices for controlling **behavior**, extra-somatic sources of information, culture provides the link between what men are intrinsically capable of becoming and what they actually, one by one, in fact become." For the author, "to become human is to become individual, and we become individual under the direction of cultural patterns, systems of meanings created historically in terms of which we shape, order, aim, and direct our lives." Because the involved cultural patterns are not general but specific, global cultural practices such as consumption also need to be included in the individual perspective, although the routines of exchange of goods for money (plastic money) are similar around the world. At the same time, it is the scale of consumption and the environmental problems generated by consumption that are highlighted in current environmental criticism, due to the issue of global warming. It is in this way that the notions of consumer society and humanity, and consumer and individual, are approached. Recently, three global events have helped to make the climate–consumption debate even more visible and bring these notions closer: signing of the **Paris Agreement**, approval of the Sustainable Development Goals (**SDG**), and development of the **Anthropocene** theory.

When the Paris Agreement was signed in 2016, Ban-ki Moon, then Secretary of the United Nations, said in his speech, "the era of consumption without consequences is over." With these words, he reiterated the rhetoric of a civilization that, by devouring the natural resources of the planet in terms of consumption as a way of life, is causing extreme environmental consequences that are irreversible and of unprecedented scale. The same idea is present in the Sustainable Development Goals (SDG), approved in 2015. The Goals were presented as interconnected. Goal 12 directly addresses the sustainable production and consumption agenda, with the climate issue on the horizon. Goals 11 and 13 address the climate, with issues related to, among others, individual consumption. These two documents have boosted consumption issues to the scope of environmental climate governance and, as political acts, will have political consequences in the course of their mandates – such as the recent departure of the USA from the Paris Agreement.

Recent criticism of consumption as a way of life in a global warming context has also led to the proposal of a new geological epoch in the history of the world, the Anthropocene. The term "Anthropocene" was coined by biologist Eugene F. Stoermer and published by Paul **Crutzen**, Nobel Prize in Chemistry. Crutzen (2002) argues that, over the past three centuries, the effects of humans on the global environment have increased. The hypothesis of anthropogenic cause and the Anthropocene theory alludes to consumption as a human gesture that wears off, withdraws, and subtracts.

In Portuguese, as stated by the Brazilian consumer researcher Lívia Barbosa and the British Colin Campbell (2006), the word for consumption (consumo) derives from Latin consumere, "which means use everything, run out of and destroy," the same meaning of the English term "consummation." In Brazil, according to the authors, the meaning of the term "consumption" is closer to the first definition. The authors highlight, however, the ambiguity of the term, which has increased in recent years to the extent that "a new interest in the study of consumption, its meanings and consequences, considering both the exhaustion of the material goods of society and of the environment, as well as the addition, realization and creation of meaning." The notion of green consumption echoes this ambiguity inscribed in the word consumption.

"Conscious consumption" and "environmental responsibility," among others, are notions that suggest consumption as a gesture of preservation, saving, and replenishment. However, the debate on recent environmental issues denounces the environmental consequences of consumption as an urban lifestyle in which the effects of materiality unfold over generations. If, in the negative perspective, every consumer is responsible for global warming, in the positive perspective the gesture of the consumer is capable of saving the planet. Likewise, the identity of the consumer (the identity of the people) establishes a moral hierarchy among consumers, praising the conscious ones.

In ***The Consumer Society***, Jean Baudrillard (1995) takes a negative approach toward consumption. Aligned with Marxist criticism, the author's text incorporates the academic reflections of the time on a social phenomenon that arises after war – a consumer society. In this regard, as stated by Santos (2011), Baudrillard, like other French thinkers of the 1960s and 1970s, looks at the profusion of objects and the change of focus in the experience of the same. Baudrillard addresses the material dimension of consumption, but discusses this reality in his language. In Baudrillard's perspective, aided by design and publicity, objects circulate in the society of consumption as signs: "they form a code that subjects a whole society committed to consume and no longer to accumulate, as used to be." The modern consumer buys so that society can continue to produce (Baudrillard 1995, 2001). However, consumption does not appear in Baudrillard's view as a "passive relation of appropriation as opposed to a supposedly active production," as emphasized by Santos (2011). Baudrillard calls attention to the extent to which the logic of consumption goes beyond the objects themselves, "constituting itself as an idealistic practice and not as a material practice." Thus, the act of consumption always renews itself, because without necessity, without utility and without function, the continuum of consumption resides in its "cultural arbitrariness."

The green wave of consumption can be elucidated with the contribution of Baudrillard to the extent that it is understood as a consumption that, although presented by its purpose of caring for the natural environment, is more a dynamic within the system of meanings of the consumer society. These are the signs to which green consumption goes back, leading individuals to exchange their money for these products to the detriment of others. Some signs allude to green consumption and legitimize them, like the triangle that indicates **recyclable packages**, keywords such as **organic** and **biodegradable**, and others. Therefore, packaging plays a decisive role in the green wave of consumption: ultimately, it is the signs on the surface of the packaging that are consumed. The metaphor of the consumer society itself, its fundamental notion, gains rhetorical force in the wave of green consumption (Paz 2012). As Slade (2006) says, branding soon became closely associated with packaging as a strategy to create repetitive demand: "manufactuers of foodstuff could not screw a metal nameplate onto their products, but they could advertise their brand by enclosing those products in fancy packaging." At the same time, packaging is one of the most serious issues raised by the recent environmental debate because of the environmental impact of different materials, such as plastic, that take years, perhaps centuries, to decompose.

Therefore, the hypothesis of anthropogenic global warming, as proposed by the IPCC, allows consumption as a lifestyle to be discussed as the material and cultural reality of all human beings. But, as Baudrillard warns, the entire discourse regarding consumption seeks to transform the consumer into the Universal Man, in general incarnation, the ideal and definitive human species, and considers the consumer as the first of the "human liberation." The consumer, however, says Baudrillard, has nothing universal, but "emerges as a political and social being, a productive force and, as such, raises fundamental historical issues, such as ownership of the means of consumption (and no longer the means of production), economic responsibility (responsibility for the content of production), etc." (Baudrillard 1995). Consumption is not a universal value system, says the author; it is founded on the satisfaction of individual needs. Reflection on the green wave of consumption implies, therefore, reflection on the role played by consumption in forging identities, values, and contemporary practices. Understanding the centrality of consumption in the current environmental debate involves understanding the centrality of consumption in contemporary life and its role in the configuration of major issues.

Environmentalist Ambience

Marc Augé (1994) emphasized relations as merely contractual in non-places, but field experience in supermarkets in Brasília showed that, in consumer society, these contracts of the supermodernity described by the author can be broken in daily life. Examples of these social dynamics were observed in the supermarket chains visited in the Federal District: ***Pão de Açúcar*** (Sugar Loaf) in the Pilot Plan, central area and ***Pra Você*** (For You), a supermarket chain located in **Brazlândia**, far from the central area.

The name *Pão de Açúcar* alludes to a famous landmark in Rio de Janeiro, but the company began in São Paulo in 1929 as a candy store. The name was chosen as a tribute by its Portuguese founder to the landscape that he saw when he arrived in Brazil. The *Pão de Açúcar* supermarket has stores in several Brazilian cities and is the largest retailer in the country. In Brasília, in one of the stores we visited, located in the South Wing (Asa Sul) neighborhood, the walls show the modernist capital of Brazil (Rio de Janeiro was the previous capital of Brazil). Through this strategy of exposing images of Brasilia, the supermarket store in Asa Sul tries to create bonds with the residents of the region, who tend to be older and with greater income power. Many of the residents are bureaucrats, military people, and retired public servants, as identified by the interviewed manager (Paz 2012). Recently, the *Pão de Açucar* supermarket chain was sold to the multinational Casino group.

In Brasília, considering its 31 administrative areas, there are about 2.6 million inhabitants and the highest GPD in the country. However, because of the cost of living in the central area, although 47.7% of jobs are offered in the Pilot Plan, only 8% of these people work and live in the Pilot Plan.

Brazlândia, the most remote administrative city (about 59 km from the central area), has a population nearly of 54,000 inhabitants. Because of its natural springs, the city supplies 60% of all the water consumed by the Federal District. There are no *Pão de Açucar* supermarkets there. At the time of the field study, the supermarket chain *Pra Você* was the only one available. They had two stores: one in the central region of Brazlândia and the other in its farthest region, the São José district. Due to the rural aspect of Brazlândia, the products were arranged on shelves differently from how they were displayed in *Pão de Açucar*. Besides offers of individual items, common in all supermarket chains, *Pra Você* also offered products in "economic large packs." Thus, consumers living in the rural area could take more units of the same product and avoid returning often to the city.

The supermarket chain *Pra Você* does not belong to a large multinational group. There are five stores throughout Brasília, most of them located in administrative areas, outside the Pilot Plan. *Pão de Açucar* stores are mostly located in the Pilot Plan area (Figs. 2 and 3). According to each of the managers interviewed, the average amount spent by each individual consumer per visit to the supermarket was around R$ 30 at *Pão de Açucar* in South Wing neighborhood. This was half of the average amount identified in *Pra Você* supermarkets (Paz 2012).

In Brazil, the **average ticket price** is an indication of the country's emerging economy, with an affluent consumer society that saw millions of new consumers when 36 million people were lifted out of poverty. Average ticket prices, however, are impacted by the contraction and growth of the economy, as a result of financial and political crises.

The older Brasilia images hanging on the walls of *Pão de Açucar* in the South Wing neighborhood and the products offered in different packages sizes in the supermarket *Pra Você* give the supermarkets different ambiances. These two stores present themselves in similar ways (both as supermarket chains, focused on the retail

Fig. 2 Supermarket *Pão de Açúcar*, Pilot Plan, Brasilia, DF, Brazil

Fig. 3 Supermarket *Pra Você*, Brazlândia City, DF, Brazil

sector) and have the same market goals (the exchange of money for consumer items). The two stores are composed of a system of objects that result in an ambience. However, each ambience goes back to distinct local cultural characteristics – dynamics that escaped the criticism of Marc Augé (1994).

From this perspective, Jean Baudrillard's notion of **ambience** (1995) is a more suitable response to qualitative analysis of the cultural dynamics of the supermarket. This concept successfully contemplates both aspects of consumer society: what is transferable and repeated in the different supermarkets (such as the circulation of products with global origin, always the same brands and similar packages) and what happens in a singular way (such as the cultural reference to Brasília in the *Pão de Açúcar* supermarket). At supermarkets, personal ties are developed in a space that presents itself as impersonal, although impersonality gives the feeling of being always in the same space. To Augé (1994), people always circulate in the same space (the supermarket), wherever its location.

The consumption ambience is what gives meaning to this space that is presented as impersonal. For Jean Baudrillard (1995), it is the notion of ambience that defines the society of consumers and their experiences, like going to supermarkets. The ambience arises from a material–significant circuit that happens within the consumer society, in which is impossible to identify its starting point. Like Augé (1994), Baudrillard (1995) does not emphasize local cultural aspects in his analysis. Despite that, the notion of ambience as a system of signs can help to clarify the different meaning processes that emerge from different ambiences:

> All medium modern environments will thus block the level of a system of signs: the ambience, it does not follow more particularly treatment of each of the elements. Neither beauty nor ugliness. (. . .) In the current system is the level of constraints of abstraction and association that lies the success of a set. Baudrillard (1995)

For Baudrillard (1995), consumption is a **system of signs** in which the different objects in circulation are equivalent, despite any technological innovation and alleged utilities of products and brands, "no longer relating to a particular object in its specific utility, but to a set of objects." For the author, consumption loses its symbolic, poetic function of social practice in a consumer society because of the "eternal combination of 'ambience" (Baudrillard 1995). In line with the historical–materialism criticism, he points out that everyday restrictions and the limited horizons in terms of consumption, always related to signs and not too stuff, lead consumers to the "vertigo of reality" (Baudrillard 1995). The consumption sign works as a "refusal of the real, based on the avid and multiplied apprehension of its signs" (Baudrillard 1995).

The more enclosed the sign, the safer the individual feels, eliminating the emerging tensions of a society where the interdependence with consumption is increasing. In these terms, advertising is a pathway to happiness as myth to be consumed. It is not a coincidence in Baudrillard's analysis that the *Pão de Açúcar* slogan is "Place of happy people."

Following Baudrillard's notion of ambience, green consumption can also be understood as an ambience. It is realized through a particular dynamic in the system of objects, around the items offered under the label of conscious consumption. Outside the supermarket space, the planet is getting warmer; however, in the supermarket environment, there is no reference to this. The natural environment is represented in shelves and packages, but only through the sign of a nature that is far from being threatened.

Baudrillard (1995) mentions the residual effects of consumption; however, the author deconstructs the idea that the problem is in things that are more and more abundant. He mentions the abundance of objects, but also mentions planned obsolescence as a problem. The term "**planned obsolescence**"was coined in a positive approach by Bernard London (1932), a real-state broker who wrote a paper entitled "Ending the Depression through Planned Obsolescence." He proposed encouraging the consumption of new items, instead of using old ones, to overcome the 1929 crisis in the USA.

The critical sense of the expression only gained visibility with the North American cultural critic Vance Packard (2011). His book *Waste Makers* was first published in the 1960s in the golden era for advertising agencies, in the time of major exhibitions and big news, when products looked really different from each other. Packard criticized two aspects of the consumption euphoria in the American way of life: (1) the short product lifecycle as an internal strategy for production and commercialization; and (2) the consequent accumulation of discarded objects. In some aspects, the success of Packard's criticism at the time recalls the recent success of the Canadian journalist and economist Naomi Klein (2002, 2014).

In the bestseller *No Logo,* Klein (2002) deconstructs the image of successful global corporate brands such as Nike and Disney. She argues that people do not consume these brands; they live inside them, because they are announced not as goods, but as spirits. In *This Changes Everything,* Klein (2014) addresses global warming, defining it as a global debate that evokes a radical change in the capitalism system. The environmental crisis, she says, is capitalism as a systemic crisis, and she attributes an important role to the individual consumption gesture in the face of climate change reality, in objective terms.

For Baudrillard (1995), who has another view, the only objective perspective of consumption and consumers is their realization as myths – as "contemporary society's word about itself: it is the way our society is spoken." The only objective reality of consumption, in his view, is the very idea of consumption. Thus, not even the criticism of material consequences can escape from the logic of the system of objects that underlies the consumer society:

> Like every great myth worth its salt, the myth of 'Consumption' has its discourse and its anti-discourse. In others words, the elated discourse on affluence is everywhere shadowed by a morose, moralizing 'critical'counter-discourse on the ravages of consumer society and the tragic end to which it inevitably dooms society as a whole. That counter-discourse is to be heard everywhere. Not only is it found in intellectualist discourse, which is always ready to distance itself by its scorn for 'simple-minded values' and 'material satisfactions', but it is now present within 'mass culture' itself: advertising increasingly parodies itself, integrating counter-advertising into its promotional tecnique. France-Soir, Paris-Match, the radio, the TV, and ministerial speechers all contain as an obligatory refrain the lament on this 'consumer society', where values, ideals and ideologies are giving away to the pleasures of everyday life. (...)
>
> This endlessly repeated indictment is part of the game: it is the critical mirage, the anti-fable which rounds off the fable – the discourse of consumption and its critical undermining. Only the two sides taken together constitute the myth. We have, therefore, to allot to the 'critical' discourse and the moralizing protest their true responsibility for the elaboration of the myth. It is that discourse which locks us definitely into the mythic and prophetic teleology of the 'Civilization of the Object'. (Baudrillard 1995)

For the author therefore, the environmental consequences of consumption and the criticisms of consumption as a way of life only reiterate the consumption **myth**. In an interview on the same subject, the English researcher **Mike Featherstone** (Featherstone *apud* Paz 2010), a pioneer in the so-called sociology of consumption, said that the green appeal of consumption would also work (although this is not the only way) as an indulgence. Once the products are environmentally friendly and there is a green ambience in certain sections of the supermarket, the consumer could move forward in his gesture of consumption. The green light would work as a concession to advance, an allowance to continue, to buy without guilt – without fear of harming health or the **planet**. As long as it is "conscious," the consumer is allowed to be a consumer.

The environmental value and the supposed individual freedom to choose between environmental friendly goods and environmentally harmful goods is also part of the environmental ambience of a supermarket. Consumer green identity was criticized by **Raymond Williams** (1961). Writer, literary critic, and culture researcher, the famous Welsh author criticized the popularity of the term "consumer" in the 1960s, describing it as a way of representing the ordinary member of modern capitalist society. He criticized the minimization of humans in terms of markets and sales. In the economic fantasy, he warned, choices are made by corporations, but they are addressed as individual choices. This happens especially in advertising, to celebrate circulation of goods as "people's choices." He said that, in this controversial atmosphere, unfortunately, great decisions would be made.

Later, Williams (1989) also highlighted the nostalgia that was the effect of the pastoral appeal in the conception of the natural world facing the modern world and its machines: the gaze would turn to the past, so things and life would no longer be or could not be like before. Global climate change is irreversible. The possible effects of today's conscious consumption cannot have their positive consequences assured in the short or medium terms; even in the long-term perspective, any effect would depend on thousands of consumers' individual choices. Green consumption, therefore, has no connection to utility – at least in the economic sense of the term. The motivation to be part of the green consumption wave could be entirely cultural. This utility perspective would be created in a cultural order, in the terms in which Marshall Sahlins (2003) proposes. For him, culture is not built in connection with real activities and experiences or objective interests:

> All in capitalism conspire to hide the symbolic order of the system, especially those academic theories of praxis in which the world were conceived. Praxis of theory based on pragmatic interests and "objective" conditions is the secondary form of a cultural illusion. (Sahlins 2003)

The appeal of green consumption has gained more visibility in the current environmental debate, mainly around the repercussions of the anthropogenic hypothesis of climate change. Green is brought in response to the planet's **environmental crisis** – this is the marketing appeal – , but the imperative of green consumption can also be understood as a cultural expression of consumer society. Consumption as a lifestyle is also a cultural strategy to deal with the different aspects of contemporary life, which may glance on the ecological issue, but are far from being limited by it. The sociological field study in Brasilia suggested that the environmentalist consumption ambience concerns a diversity of cultural aspects.

There was a sector of organic products within the supermarket *Pão de Açucar* in Brasilia. One of the products marketed in this way was strawberries, sold in a transparent plastic package. Those strawberries were planted, harvested, and marketed by farmers from Brazlândia. The Strawberry Festival takes place annually in Brazlândia to celebrate one of the most important food practices of the region. In *Pão de Açucar* supermarket, strawberries from Brazlândia were labeled as an environmental friendly product. In the supermarket *Pra você*, the same item was promoted as the cheapest fruit: R$ 1.49 (US$ 0.45). In the *Pão de Açucar*, a section was decorated with greenish shelves exclusively for displaying organic products. There, the strawberries that came from Brazlândia were sold as organic strawberries for R$ 1.98 (US$ 0.59). The same item, sold with completely different ambiences and commercial approaches.

In the organic section of *Pão de Açucar*, a promotional poster hung from the ceiling, asking consumers, "Does living in harmony with nature make you happy?" A soft green color was everywhere on the shelves. Expressions such as "0% pesticides" and "100% healthy" appeared in large letters on the panels that separated organic products from regular products. The organic items were displayed under a large green umbrella. Under this umbrella were products that were friendly to the environment and customer's health, which implied that all other items in the supermarket were just the opposite. Among the texts splashed on the organic section in the *Pão de Açucar*, was the following: "Here you can find products without pesticides that respect the environment." They were saying that those organic products were good for individual health and for the environment at the same time. A good deal: do not hesitate and pay more in order to buy them.

In the *Pão de Açucar* supermarket, pale green colors were predominant in the organic products section, and they indicated where the more environmentally friendly products were, and the more expensive products: the greener the product, the higher the price – just like the strawberries from Brazlândia. The whole organic sector was colored with soft shades and pastel green tones. "The world of colors is seen as opposed to that of values. 'Chic' effaces appearance so that being might stand revealed. Black, white, and grey, the very negation of color, were the paradigm of dignity, control, and morality," says Baudrillard (2006). For him, the omission of contrasts and the return to the "natural" after the exaltation of red, blue, pink, and other colors in series of items (refrigerators, cars) reveals the modern ambiguity, as the "natural" is not real: it is an (impossible) calling to return to the state of nature." For the author, this is a color morality.

In the supermarket, the differences between a tomato produced with pesticides and a tomato organically grown are not so clear to the naked eye. Thus, the stamps, packaging, signs, texts to identify the products, labels assuring the origin of the products, and the association of green color with green products make this differentiation, increasing the value of organically grown tomatoes, without reminding the consumer that the other tomatoes are poisoned in some way.

Prices of organic products are higher. Consumers that go to *Pão de Açucar* tend to have higher incomes than consumers that shop at *Pra você*. The best products for the best consumers: this is the message that this distinction also suggests. The material perspective of consumption allows to observe how this cultural process works, associating people with some brands or kind of products, such as organics. To the English anthropologist Daniel Miller (2010), the material perspective of consumption must be taken into account in the **consumption sociology** because for him "objects make people":

> Things, not – note – individual things, but the whole system of things, with the internal order, make us the people we are. (. . .) But the lesson of material culture is that the more we fail to observe things, the more powerful and determinant they are upon us. This attributes the theory of material culture more importance to things than is expected. Miller (2010)

The supermarket was visited in this perspective too. Around the section of organic products, the supermarket chain *Pão de Açucar* carries out its environmental marketing. The corporate identity of the *Pão de Açucar* supermarket is directly linked to the rhetoric of social and environmental responsibility. This is the reason why they offer 750 different items, produced by over 100 farm producers (mainly from São Paulo) that sell their agriculture production directly to the supermarket.

The supermarket ambience is not restricted, however, to the environmental ambience. In both supermarket networks – *Pão de Açucar* and *Pra você* – there were different sections alluding to different consumption experiences, feelings, and memories. In both supermarkets, there was a whole section on the theme of black coffee, a very popular drink in Brazil. Also, in both networks, there was a demonstration of ready-ground coffee, offered to consumers as a prize – the smell invading the supermarket. This is also part of the supermarket ambience. In the *Pão de Açucar* supermarket bakery, bread was in the oven and the smell was in the air. Smells are not prevented; they are used to set up the ambience.However, although every supermarket has the same ambience, there are specific contexts that link every store to a specific consumer local reality.

In the supermarket *Pão de Açucar,* for instance, the wine section also stands out, occupying an entire shelf with bottles of many sizes and shapes, national and international brands. In the supermarket *Pra você* there was no organic sector and there was no wine sector. The two sectors built inside the *Pra você* store under the same ambience logic were the "barbecue corner" and the "**feijoada** area" – two very popular foods in Brazil.

In the meat section, the *Pão de Açucar* supermarket had a unique ambience, highlighting the program "Quality from the beginning." Meat with this label had a **QR code**: consumers could discover its meaning using their smartphones. They could access information about the origin of the meat. Animals, landscapes, and people involved in the meat production chain could be seen using the tech source. In the supermarket *Pra você,* the meats were displayed under the sign "Wednesday Swine," announcing the weekly sale, but without any advice about the origin of the meat. There was no green consumer appeal in the meat sold in the supermarket *Pra você.* The green wave of consumption only reached the meat sold in the supermarket *Pão de Açucar.*

Next to the meat section in the supermarket *Pra você*, in Brazlândia, there were some eco-bags offered to the consumer. They were produced by a local producer. Each eco-bag cost R$ 8 (US$ 2.30). There was a label on the bag giving the producer's address and phone number. Although it was produced in the midst of the wave of green consumption, the appeal of the *Pra você* supermarket eco-bag was for the transportation of meat, especially focused on the consumers who lived in the farthest zone in Brazlândia. In the *Pão de Açucar* supermarket, there were also returnable bags or eco-bags for sale; however, they were placed close to the cashiers. There was an exclusive cashier for those who used the returnable bags. The price of each bag in *Pão de Açucar* was R$ 2–3 (US$ 1). The bags showed some flower images; they were made of raffia and were not produced locally but imported from Vietnam (as written on the bag). The returnable bag or eco-bag is one of the most emblematic objects of the green consumption wave. It alludes to the importance given to packaging in the consumer society. The eco-bag is also a package.

On the one hand, **packaging** appears in the climate change debate because its decomposition affects the natural environment. Excessive packaging, unnecessary packaging, packages with the mere function of promoting corporate marketing are some of the criticisms. On the other hand, the green appeal of consumption needs some special packages to identify and promote an alleged better consumption to better consumers. From this perspective, packages can be understood not only as material objects that impact the natural environment, but also as a concept that defines the consumer society. In this sense, everything in the consumer society is wrapped in packages, whether material and/or symbolic.

Packages and Trajectory

During research at the *Pão de Açúcar* supermarket in the South Zone of the central district of Brasilia, a lady approached the **eco-bag** display, close to the cashier, and chose four units, for which she paid. The bags were empty, with no product inside. She was a university professor in the nursing area. When asked why she was buying so many empty returnable bags, she said, "Because they are beautiful. I use them for gifts. I put a flower inside and use the bag as packaging, as if it were a gift paper. Aren't they beautiful?" (Paz 2012). One of the bags was to present her daughter with a flower and the other bags would be in the car for when she wanted to give someone flowers. At the same store, one of the managers commented that there had been an attempt to sell the well-known and simpler "market bags," but they "would not sell." The price was equal to that of the raffia returnable bag, stamped with flowers and imported from Vietnam, but the market bags did not sell. This happened because people did not consider the market bags as beautiful as the others, he explained. The model imported from **Vietnam** was one of five models available in the supermarket. Some of the bags were produced domestically and the sale of one of these bags provided funds to the Brazilian NGO SOS Mata Atlântica Foundation; however, it was the model from Vietnam, with the visual appeal of the flower design, that sold the most.

These examples from a sociological field study in Brasilia describe another type of utility, not economic, for returnable or eco-bag bags: the value of the object depends on its aesthetics; the appeal is the natural environment. The British **ecocritic** researcher **Greg Garrard** (2006) describes this approach as deeply rooted in Western culture and deeply problematic for environmentalism, as it suggests the essential harmony of ideas of nature. This pastoral tradition, which is explored in the literature (especially English literature), is at the base of the modern environmental discourse and alludes both to refuge in the countryside and the idealization of rural life. Such exaggeration is constitutive of pastoral *tropos:*

> Classical pastoral therefore inclined to distort or mystify the social and environmental history at the same time providing a locus, legitimized by tradition, to the feelings of loss and alienation from nature that would be produced by the **Industrial Revolution**. (Garrard 2006)

The symbolic advertising game is a specific form of marketing discourse and the logic that underlies the consumer society in a broader sense. Its expression is evident in the rhetorical form of presentation – the "packaging" rhetoric of things and consumption of ideas. The lack of correspondence to a particular empirical reality does not prevent the advertising from undertaking poetic–persuasive journeys in which animals talk, cars fly, life concerns are restricted to buying things, and industrial product packaging suggests the "harvested" contents of trees.

The ambience of "things of consumption" is realized through this dynamic – the system of objects that circulate in the society of consumers, objects that include packaging. Packaging not only has the obvious functions of carrying, protecting, and identifying products, but also persuades consumers by suggesting flavors, textures, modes of use of the products, memories, and sensations, although not always related to the content.

Charting a timeline, Pedro Cavalcanti and Carmo Chagas (2006) in *The History of Packaging in Brazil* even say, "There have been those who pointed out nature itself as the first inventor of packaging, providing the pod to protect the beans and peas, straw to wrap the corn cob, and the shell to protect the egg and the walnut." Packaging, the authors propose, is to life in society just as prehistoric man is to the consumer.

Plastic is the biggest symbol of the packaging era, but has been the most attacked in terms of environmental criticism of consumption. In response to this criticism, the packaging industry conducted a number of surveys to assess support for the continued use of plastic bags, which have been banned in some Brazilian cities and around the world (e.g., in Paris). In Brasilia, to date, plastic bags are given away by supermarkets. In São Paulo, there was an attempt to ban the bags, but the reaction of the population in general and of the productive sector was not favorable.

Roland Barthes (1985), defined the advent of plastic like the idea of infinity and processing the first substance which if allowed would be mundane, between "Magic" substances, in an allusion to the rare gods' potions, which would be utensils and dreams of the housewife:

> And as an immediate consequence, the age-old function of nature is modified: it is no longer the Idea, the pure Substance to be regained or imitated: an artificial Matter, more bountiful than all the natural deposits, is about to replace her, and to determine the very invention of forms. A luxurious object is still of this earth, it still recalls, albeit in a precious mode, its mineral or animal origin, the natural theme of which it is but one actualization. Plastic is wholly swallowed up in the fact of being used: ultimately objects will be invented for the sole pleasure of using them. The heirachy of substances is abolished: a single one replaces them all: the whole world can be plasticized, and even life itself since, we are told, they are beginning to make plastic aortas. Barthes (1985)

Three points need commenting on Barthes (1985): (1) Barthes attributes to the material value luxury the quality of being "very grounded" in the sense of a precious stone. (2) Throughout the article, there is no mention of the scarcity element, finitude – a rarity would have a sense of the sublime, which refers to socio-historical conditions in which "rare" associated with the natural environment did not directly refer to the idea of scarcity of natural resources. (3) The author views utility as a value.

The value of plastic comes from its ability to respond and transform itself into diverse tools. However, his approach overlaps economic utility and utility in the cultural sense, a trend in the debate on consumption that is criticized by authors in different lines of sociological theory, such as **Appadurai** (1988), Kopytoff (1988),

Miller (2010), Braudel (1996), and Sahlins (2003). In view of the criticism that comes from material culture, the value of plastic is in its path that would engender different cultural biographies, not in the expectation of their pragmatic usefulness. The arrival of plastic in Brazil was full of enchantment, close to what Barthes (1985) attributed to Europeans. At the same time, the arrival of plastic was also seen as a negative symbol of so-called Yankee capitalism, according to Pedro Cavalcanti and Carmo Chagas (2006). The authors point out criticism that there was a mismatch between the national plastic industry and the international industry. There was resistance from dairy farmers to the arrival of plastic packaging, the "milk bags," which coincided with the opening of supermarkets in Brazil in the 1960s. With the arrival of plastic to package milk, manufacturers had to abolish glass containers and equipment for washing, bottling, and transporting in the milk trucks.

Dutch researcher Wiebe E. Bijker (1997) discusses the social construction of **bakelite,,** the first synthetic plastic, created by Leo Henricus Arthur Baekeland in 1907. Bijker states that even in the case of an "individual inventor" it is necessary to consider, for example, the relevance of the social group of designers and the complexity of meanings attributed by consumers to the new material. According to the author, it was not the practicality of plastic that determined its commercial success, but the perception of bakelite as a design object, a beautiful item that could be displayed in the most important rooms in the house and was not necessarily used for some task. The invention therefore was not confronted with strong environmental criticism at the time, as is the case today.

Currently, the environmental criticism of plastic bags emphasizes the long-term and very long-term effects of plastic in the wild. One of the most recent examples/symbols explored in this argument is the "plastic island" discovered off the Californian coast in the 1990s. It is an area in the sea of about 680,000 km^2, which is equal to the sum of the territories of the states of Minas Gerais, Rio de Janeiro, and Espirito Santo in Brazil, and reaches the Hawaiian coast, containing 100 million tons of debris. The oceanographer **Charles Moore** "discovered" the "sea trash"; since then, the foundation that bears the name of his boat, *Algalita*, began to be identified as the symbol of a world crusade against the use of plastic. Moore has become an international speaker on the subject:

> It is a floating mess of 3.5 million tons of waste, of which 90% are composed of plastic, bottle caps and toys to shoes, lighters, toothbrushes, networks, pacifiers, wrappers, packaging travel and shopping bags from all over the world. (Botsman and Rogers 2010)

The centrality of packaging in the debate that links climate and consumption is notorious, and includes issues such as selective collection, recycling, certification of origin of products, identification of ingredients, the use of returnable bags, and a ban on plastic shopping bags. Every day, millions of individuals make consumer choices by handling objects, suspending belief, or distrusting the information presented, but invariably establish silent dialogues with the things that pile up around them.

In places such as supermarkets, where goods are sold using a logistic strategy of self-service, the default color of product segments, the display, the arrangement of elements, and the performance of employees all appear as packaging. From this perspective, things and gestures are always "packed" in the consumer society, and advertising is a larger package, "system packaging," as suggested by Maria Arminda Arruda Nascimento (2004).

The appeal of green consumption gives packaging the role of organizing the consumer's gaze in the face of new offers of environmentally friendly items. The appeal to the natural, the lost paradise, and the suggestion that it can be rescued through consumption opens a very favorable trail for the poetic–persuasive journey of publicity. It is necessary to establish boundaries between environmental marketing and "greenwash" (laundering, deception, green lies). In Brazil, the National Council for Advertising Self-Regulation (Conar), which represents the advertising industry, acted against it, because the advertising itself is addressed regarding its dimensional reality.

In the environmental ambience of the *Pão de Açúcar,* therefore, in the green world that stands in its supermarket network, the color green has become a market position by identifying the supermarket as an environmentally friendly store. Assigning these terms of significance to the color green is part of the environmental consumer appeal, and releases the green sign to create an ambience or, as Baudrillard suggests (2006), a "systematic culture in terms of objects." Because of this freedom, which disconnects sign and meaning, there was a strategy of design in *Pão de Açúcar* that used the packaging of the products to favor the sensation of being, literally, in a green supermarket: the electronic bulbs were arranged along with Heineken beers. The packaging of the lamps was green in terms of its ecological appeal, which is the reduction of energy consumption through technological innovation, but the Heineken beer long-necked bottle, whose visual identity and packaging are green, does not have this appeal; the bottle is green so that it is permeable to light (Fig. 4).

Garbage, degradation, pollution, and moral decline versus purity, cleanliness, nature, and Eden are issues that refer to the impact of packaging on the natural environment; they are the *tropos* of the environmentalist rhetoric approach, especially in the media, which tends to confirm the presence of packaging in the natural environment as something evil. According Greg Garrard (2006), "the idea of the natural world, meaning nature in a state untainted by civilization, is the most powerful construct of nature available to the environmentalism of the **New World.**" However, he warns, this view has "pernicious consequences for our conceptions of nature and ourselves, because it suggests that nature is only authentic when we are entirely absenting from it." The price of purity of the natural environment would be the elimination of human history. In the way the packaging issue is addressed in the climate change debate, this is one of the aspects cited: Plastic packaging will survive the human odyssey on Earth.

According to market analysis surveys, which are frequently performed in Brazil, it is possible to identify the ways in which the green appeal of consumption is presented by packaging, mainly due to the problems observed in these presentations. In Brazil, spray cans and aerosols emphasize that the product "contains no CFCs," noting that it is harmless to the ozone layer. Presented in this way on the label, it seems that such absence is a choice of the brand, when, in fact, it is an international

Fig. 4 Environmentalist ambience: lamps and beer in the *Pão de Açúcar* supermarket

decision, in the ambit of environmental and legal governance, based on the country's own legislation. Indication of the absence of chlorofluorocarbon gas is a marketing appropriation of the ban on 14 September 2000, with the resolution of the National Environmental Council (CONAMA), 267 in view of the **Montreal Protocol** agreement on substances that deplete the **ozone** layer.

On the one hand, packaging represents the risk of consumption: look at the packaging, observe, distrust, look for signs that can authorize consumption, buy, use, and discard. There is always the risk of making the wrong decision. The Slovenian researcher **Renata Salecl** (2005) of Slavoj Žižek philosophical group, defined such a condition as the "tyranny of choice." The impasse in the face of seemingly mundane decisions such as buying products increases the alleged "variety" of possibilities in the society of consumers. The amount of goods and their velocity impact not only market dynamics but also emotions, the rhythm of life, and the body. Answers to existential questions are required at the same pace as a goods assembly line.

For Richard Sennet (2006), the permanent condition of choice leads to the heart of what he calls the "new economy": the product's promise (whether a political candidate or a thing) and the promise of democracy are at play. In his view, the greater the facility offered, the less involvement in making policy. Referring to the Wal-Mart supermarket as a metaphor that can be extended to an election campaign, the author states, "instead of considering the citizen just like an angry voter, we could see it as a policy consumer pressure to buy." Packaging and its persuasive message play with this individual vulnerability in the face of the things presented. When in doubt, the package itself becomes the purchasing differential, even making it the selling point.

For the Brazilian researcher Fátima Portilho (2005), two shifts have been decisive for the influx of an environmental critique of consumption in the 1990s, probably dating from the critique of industrial society. "Environmental issues begin to be redefined," she says, "to be identified, especially with the lifestyle and consumption patterns of the affluent societies." The two discursive displacements that led to this were (1) the critical population growth (mainly in the southern hemisphere), which shifted to the production model of affluent societies (especially in the northern hemisphere); and (2) the change of tone in the concern; the focus changed from environmental problems related to production to environmental problems related to consumption and lifestyles.

Outside the supermarket, however, the cultural trajectory of packaging continues. It is this trajectory, beyond the environmental ambience of things in the supermarket, that the global climate change–consumer debate emphasizes. Arjun Appadurai (2008) proposes "an approach of commodities as things in a certain situation, a situation that may characterize different types of things, in different parts of their social lives." In the supermarket perspective, the career of merchandise ends at purchase. From the perspective of collectors of recyclable packaging, the social trajectory of the package is just beginning. The Association of Collectors and Recyclers in Brazlândia assign other values to packaging; design elements, labels, and systematic disposition of colors and forms are all irrelevant. From this perspective, it is not the form element that gives value.

In the midst of **rubbish**, the packaging design presented at the grocery store is barely recognized. Far from the environmentalist ambience, the packaging does not resist the process of discarding and depositing; if they undo, they are dismantled. This does not mean that things cease to be for goods that deviate from the supermarket, but which become goods due to other commercial contexts or other social arenas. In the value schemes, such things would not take a place, even as a commodity (Appadurai 2008).

In the **Collectors** and Recyclers Association of Brazlândia, what was called the supermarket packaging, label, and design are awarded other names. The financial figures are also different. The names of manufacturers, logos, and consumer tips do not characterize the goods in the context of the Association, for what becomes of value is the material of the package (plastic, aluminum, paper, glass). The dynamics of the supermarket and domestic consumption streamlines around what is on the surface of the package, but collectors and recyclers have another dynamic: the divestment of the packaging becomes important. The packaging *in nature,* one might say, detached from any promise of consumption, is held as a commodity.

For Appadurai (2008), it is necessary to recognize the commercial potential of all things, instead of seeking in vain for the magical distinction between goods and other things. For the author, it means "break a categorical way with the Marxist view of the commodity, dominated by the prospect of production, and focus *throughout* history, from production, through exchange – distribution, to consumption."

In the Association of Collectors and Recyclers in Brazlândia, selective collection takes place every day, from the early morning hours. Cooperative recyclers head to the Association to await the arrival of the garbage truck. There is control of the material that each "picks" and of what is sold for recycling. The truck arrives and the bucket dumps the garbage collected in the city. That moment is called "eviction" by the scavengers. As the garbage is dumped and accumulates at the place for separation of the items, the collectors begin to separate bags, open containers, and check what was **discarded** by the other end of the production chain. Their only link with these social groups is that they touch the same objects in their different social trajectories. What is not suitable for recycling is intended for landfill. While they work, they talk. Does anyone remember how it was before and how it is different now? Now, the collector's work is more visible: "Before, people passed through like that," he said, pressing his nose with his thumb and forefinger. The others laughed. "Now everyone wants to know what we do." (Paz 2012).

Amid the culled **garbage**, it is possible to identify some brands and packaging from the *Pão de Açúcar* supermarket and *Pra Você*, but with some difficulty. There is an empty plastic container of Ypê biodegradable detergent. A waste picker explained that, for them, that detergent package is called "plot" and the black plastic bag that contained the garbage with the empty detergent package is called "silk;" "black Silk" and "white silk" according to the plastic's color. The table shown in Fig. 5 shows the date, the type of material collected, the name, and the amount involved in the commercialization of waste (Paz 2012).

In the middle of the street, a **trash picker** scans the asphalt; an orange uniform identifies the picker as contracted by Brazlândia City Hall. Another trash collector

Fig. 5 Collectors and Recyclers Association of Brazlândia, DF, Brazil

approaches and shows the backpack found among the items separated by street dwellers for disposal. The trash **collector** said he would take the backpack to his daughter who was in school; after all, the backpack was in the trash. "When this collection started [selective collection], Brazlândia was cleaner, but it fell," one of them said. "I even saw, on the TV, talking about the selective collection in Brazlândia yesterday and I told my husband that I was wrong, because I had no more, only if I was in another city, because really in Brazlândia it was not no," she commented. Brazlândia was the first city outside the Plano Piloto to institute selective collection, which is still in force; however, the residents' disregard for the disposal of items made them think that collection was no longer being performed. According to their testimony, people continued to throw a lot of garbage on the streets (Paz 2012):

> Women 1: *[thinking]* – Plastic cup, bottle of water *[she shows the bottle that she brings, to replenish water],* soda, too many soda cans.
> Women 2: [interrupting] They play all, play all, do not care, no.
> Women 1: – How they are advertising this collection, I do not know if I got it right or wrong, the trash was to be separated, is not it? It is not. It's all mixed up.
> Women 2: – There where I live, everything is mixed.

The volume of packaging disposal is associated with the consumption of certain items, matching lifestyles and income, usually associated with social groups with greater purchasing power. Although working every day in the collection of items for recycling, gathering what was discarded by others, the garbage pickers do not have this practice:

> Women 2: – When I have [*trash to recycle*] I separate. But I hardly do, because I do not have dry [*dry waste, packaging*], hardly have. I almost do not separate it, but when I have, together, I separate, yes.

Conclusion

The garbage collector's comments synthesizes the terms by which green consumption is a marketing appeal that implies lifestyle and taste issues. This does not mean, however, that the ecological problem to which this consumption qualification alludes relates only to the **higher income social** groups. Brazil is responsible for about 3% of global greenhouse gas emissions and most of this comes from deforestation (UN Environment Brazil). At the same time, Brazil's affluent consumer society needs to take into account the ecological impact of their new consumer habits. Consumer chains that start in the forest and in the countryside – producing meat, food from agriculture, and consumer items – cause the majority of emissions that affect different social groups. The effect of global climate change affects everyone. In the context of the consumer society, the ecological debate is more focused on higher income social groups.

On the one hand, arguments for the green appeal of consumption are outside the packaging issues. On the other hand, it is on the surface that green consumption takes place as a promise, affecting individual choices. The package synthesizes the dynamics of presentation, appropriate for a consumer society, but the social trajectory of goods and packages follow different value regimes, as demonstrated. The relationships between people and things that take place after discarding have gained more visibility in the context of the current environmental crisis and need to be elucidated further.

There are arguments and scientific evidence that justify the centrality of consumption in the global warming anthropogenic hypothesis. It puts the notions of human being and consumer closer together, and also the notions of humanity and consumer society. The consumption parameter is, by definition, exclusive, even in the aspects that make all consumers equivalent, such as the experience of going to a supermarket.

The **ecological imperative** that is placed on the consumer society establishes a relevant debate on consumption as a way of life, and not only around those items that are displayed in the supermarket. Observations from sociological studies do not relate to any particular experience in the consumer society; they are related to a particular city, Brasilia, but suggest the relevance of following the issue of the centrality of consumption in the present environmental debate and in contemporary life by following consumption itself.

Cross-References

▶ Gourmet Products from Food Waste
▶ To Eat or Not To Eat Meat
▶ Transformative Solutions for Sustainable Well-Being

References

Appadurai, A. (1988). *The social life of things: commodities in a cultural perspective.* Cambridge: Cambridge University Press.

Appadurai, A. (2008). *A vida social das coisas: as mercadorias sob uma perspectiva cultural* (pp. 27–31). Niterói: Editora UFF.

Augé, M. (1994). *Não-lugares: uma introdução a antropologia da supermodernidade. Travessia do século (col)* (pp. 35–201). Campinas: Papirus.

Barbosa, L., & Campbell, C. (Eds.). (2006). *Cultura, consumo e identidade* (pp. 1–208). Rio de Janeiro: Editora FGV.

Barthes, R. (1985). *Mitologias*. Algés: Difel.

Baudrillard, J. (1995). *A sociedade de consumo. N. 54, Arte e Comunicação (Col)* (pp. 17–35). Lisboa: Edições 70.

Baudrillard, J. (2001). *O sistema dos objetos. N. 70, Debates (Col)*. São Paulo: Perspectiva.

Baudrillard, J. (2006). *O sistema dos objetos.* São Paulo: Perspectiva.

Bauman, Z. (2008). Vida para consumo: a transformação das pessoas em mercadorias. Rio de Janeiro: Zahar Editores.

Bell, D. (1996). *The cultural contradictions of capitalism*. New York: Basic Books.

Bijker, W. E. (1997). The social construction of Bakelite: Toward a theory of invention. In W. E. Bijker, P. Hughes Trevor, & G. D. Douglas (Eds.), *The social construction of technological*

systems: New directions in the sociology and history of technology (pp. 101–102). London: The MIT Press.
Botsman, R., & Rogers, R. (2010). *What's mine is yours* (pp. 3–4). New York: Harpen Collins.
Bourdieu, Pierre (1983). *Questões de sociologia.* Rio de Janeiro: Marco Zero.
Bourdieu, P. (2007). *A distinção: crítica social do julgamento* (pp. 120–125). Porto Alegre: Zouk.
Braudel, F. (1996). *Civilização material, economia e capitalismo: séculos XV-XVIII, volume 3.* O tempo do mundo. São Paulo: Martins Fontes.
Canclini, N.-G. (1996). *Consumidores e cidadãos*. Rio de Janeiro: Universidade Federal do Rio de Janiero.
Cavalcanti, & Chagas (2006). *História da embalagem no Brasil.* São Paulo: Associação Brasileira de Embalagem.
Crutzen, P. (2002). Geology of humankind. *Nature, 415*, 23. https://doi.org/10.1038/415023a.
Currid-Hakkett, E. (2017). Saving for Yale. From silver spoons to fresh vegetables: why have elites' signals of status totally shifted?. Quartz. https://qz.com/1000565/from-silver-spoons-to-fresh-vegetables-why-have-elites-signals-of-status-totally-shifted/
De Certeau, M. (1998). *A invenção do cotidiano: artes de fazer* (Vol. Vol. I, pp. 200–210). Petrópolis: Vozes.
Douglas, M., & Isherwood, B. (2004). *O mundo dos bens: para uma antropologia do consumo*. Rio de Janeiro: Universidade Federal do Rio de Janiero.
Dunlap, R. E., Buttel, F. H., Dickens, P., & Gijswijt, A. (Eds.). (2002). *Sociological theory and the environment: classical foundations, contemporary insights*. New York: Rowman & Littlefield.
Elias, N. (1994). *A sociedade dos indivíduos* (pp. 184–185). Rio de Janeiro: Zahar.
Featherstone *apud* Paz, J. (2010). Interview. *Thesis supervision*. PhD Visiting Capes Brazilian Foundation Fellowship. Nottingham Trent University, Nottingham, England, The United Kingdom
Garrard, G. (2006). *Ecocrítica* (pp. 54–103). Brasilia: Editora UnB.
Geertz, C. (1989). *A interpretação das culturas* (pp. 37–38). Rio de Janeiro: Editora Guanabara Koogan.
Glacken, C. J. (1967). *Traces on the Rhodian shore*. Berkeley: University of California Press.
Harvey, D. (1996). *Condição pós-moderna: uma pesquisa sobre as origens da mudança cultural* (6th ed.). São Paulo: Loyola.
Heller, A. (2004). *O cotidiano e a história*. São Paulo: Paz e Terra.
Henson, B. (2011). *The rough guide to climate change*. London: Penguin.
IPCC. (2007). *Climate change 2007: Synthesis report. Contribution of Working Groups I, II and III to the Fourth Assessment Report of the Intergovernmental Panel on Climate Change*. [Core Writing Team, Pachauri, R. K. , Reisinger, A. (Eds.)]. Geneva: IPCC. http://www.ipcc.ch/publications_and_data/ar4/syr/en/contents.html
Klein, N. (2002). *Sem logo: a tirania das marcas em um planeta vendido*. São Paulo: Record.
Klein, N. (2014). *This changes everything: Capitalism vs the climate*. New York: Simon and Schuster.
Kopytoff, I. (1988). The cultural biography of things: commoditization as process. In: Appadurai, Arjun (1988). *The social life of things: commodities in a cultural perspective.* Cambridge: Cambridge University Press.
Litter, J. (2008). *Radical consumption: Shopping for change in contemporary culture*. London: Open University Press.
London, Bernard (1932). *Ending the depression through planned obsolescence.* Hathi Trust Digital Library.
Maria Arminda Arruda Nascimento. (2004). *A embalagem do sistema: a publicidade o capitalismo brasileiro*. Florinópolis: Edusc.
Miller, D. (2010). *Stuff* (pp. 53–54). London: Polity.
Packard, V. (2011). *The waste makers*. New York: Ig Publishing.

Paz, J. (2012). O clima do consumo: a sociedade de consumidores no debate sobre mudança climática. Doctoral thesis. Instituto de Ciências Sociais, Universidade de Brasília. repositorio.unb.br/bitstream/10482/14037/3/2013_JosiPaz.pdf

Portilho, F. (2005). *Sustentabilidade ambiental, consumo e cidadania* (p. 39). São Paulo: Cortez.

Sahlins, M. (2003). *Cultura e razão prática* (p. 151). Rio de Janeiro: Zahar.

Salecl, R. (2005). *Sobre a felicidade*. São Paulo: Alameda.

Santos, T. C. (2011). A sociedade de consumo, os media e a comunicação nas obras iniciais de Jean Baudrillard. *Galáxia, 11*(21), 125–136.

Sennet, R. (2006). *A cultura do novo capitalismo* (p. 126). Rio de Janeiro: Record.

Simmel, G. (1987). A metrópole e a vida mental. In O. G. Velho (Ed.), *O fenômeno urbano*. Rio de Janeiro: Editora Guanabara.

Simmel, G. (1998). O dinheiro na cultura moderna. In J. Souza & B. Öelze (Eds.), *Simmel e a modernidade* (pp. 23–40). Brasília: Universidade de Brasília.

Slade, G. (2006). *Made to break: Techonology and obsolescence in America*. Cambridge, MA: Harvard University Press.

Tavolaro, S. B. F. (2001). *Movimento ambientalista e modernidade: sociabilidade, risco e moral*. São Paulo: Annablume/Fapesp.

Veblen, T. (1974). *A teoria da classe ociosa – um estudo econômico das instituições. Tranlated by Olívia Krähenbühl*. São Paulo: Atica. (Os pensadores).

Williams, R. (1961). *Culture and materialism. Selected essays*. London: Verso.

Williams, R. (1989). *O campo e a cidade: na história e na literatura. 2a reimp* (P. H. de Britto, Trans.). São Paulo: Companhia das Letras.

Social License to Operate (SLO)

Case Review of Enbridge and the Northern Gateway Pipeline

Michael O. Wood and Jason Thistlethwaite

Contents

Abstract

With such overwhelming support, *why was the Northern Gateway Pipeline (NGP) project ultimately rejected?* Through the lens of social license to operate (SLO), this chapter explains why the NGP was not approved despite overwhelming support from government and industry as well as the economic viability of the project. This chapter shows that an SLO was not attained because of the disproportionate risks to benefits that would have been borne by local and Indigenous communities along the pipeline route, as well as the realized and potential impacts to the natural environment. This chapter concludes with key takeaways for organizations seeking an SLO as well as avenues for future research.

M. O. Wood (✉) · J. Thistlethwaite
School of Environment, Enterprise and Development (SEED), University of Waterloo, Waterloo, ON, Canada
e-mail: mowood@uwaterloo.ca; j2thistl@uwaterloo.ca

S. Dhiman, J. Marques (eds.), *Handbook of Engaged Sustainability*,
https://doi.org/10.1007/978-3-319-71312-0_45

Keywords
Social license to operate · Northern Gateway Pipeline · Stakeholder engagement · Co-ownership

Introduction

On November 29, 2016, the Prime Minister of Canada, the Honorable Justin Trudeau, announced that the Northern Gateway Pipeline (NGP) would be rejected (McCarthy and Lewis 2016b). The announcement countered the project's political support at both the provincial and national levels of government, as well as evidence of economic viability. Given the political support and economic incentives to develop the pipeline, *why was the NGP project ultimately rejected?*

To answer this question, this chapter explores the concept of social license to operate (SLO) as a framework through which to understand why the project failed to be approved. The concept of SLO developed in the context of the extractive industries, such as mining (e.g., Gunningham et al. 2004) and has been recently applied in the context of pipeline projects (e.g., Jijelava and Vanclay 2017). In recent years, SLO has received increased attention from both academics and practitioners as an important concept to help explain when and why projects, particularly those in the extractive industries, are able to receive and sustain stakeholder support for their projects and those that are not (e.g., Hall et al. 2015; Moffat and Zhang 2014; Parsons et al. 2014; Prno 2013).

An SLO differs from the legal or political licenses in that the former tends to be granted by stakeholders of the project and not public authorities (Hall et al. 2015). Thus firms need to go beyond compliance with regulations to show stakeholders through their interactions with them, that they are committed to building value beyond that of the firm, through commitments to principles like sustainable development (e.g., Mineral Council of Australia 2015). Importantly an SLO can only be attained when firms' activities receive acceptance from both local communities and stakeholders more broadly (Wilburn and Wilburn 2011).

This increase in scholarly research and industry focus is perhaps due to the fact that there was a high degree of uncertainty as to what constitutes an SLO and how to maintain it (Slack 2008). For firms whose operations have the potential for substantial social or environmental impacts, such as those in the extractive industries, an SLO is an important consideration given the dynamic nature of the relationship between stakeholders and the firm (Brown and Fraser 2006). Recent research has shown its relevance across industries (e.g., Hall et al. 2015) as a means of understanding the levels and boundaries of an SLO that firms must navigate with stakeholders in order reduce the potential for such impacts. More specifically, by considering the boundaries of legitimacy, credibility, and trust (i.e., Thomson and Boutilier 2011) between levels (acceptance, approval, and co-ownership) of an SLO with stakeholders of the project, a clearer picture emerges as to why the NGP was not approved. SLO is revealed as a useful and practical tool for organizations to deploy in the negotiation, implementation, and operational phases of any project.

The following section reviews the literature on the SLO to frame this assessment. Background on the NGP follows this section and then the application of the SLO framework to unpack why the project was not approved. The chapter concludes with key takeaways from the NGP experience for firms seeking an SLO and presents avenues for future research.

The Social License to Operate

Thomson and Boutilier (2011) define SLO as "the ongoing acceptance and approval from local communities and other stakeholders" (p. 1779) that relate specifically to nontechnical issues associated with a project (Smits et al. 2017). The initial push for firms to focus on SLO was as a form of risk management, which grew as a means of building understanding between the firm and the communities in which they operate (Thomson and Boutilier 2011). An SLO provides firms with a way to assess the degree of fit between stakeholders' expectations of the firm's behavior and their actual behavior (Salzmann et al. 2006). As a consequence, the attitude and approach to stakeholders on the part of the firm must be flexible and dynamic so as to accommodate evolving social and cultural paradigms (Nelsen 2006).

The SLO concept emerged out of industry, where practitioners were trying to understand why they were not able to gain or sustain stakeholder support for projects, despite having illustrated the technical proficiency (e.g., safety, efficiency, value proposition, etc.) of the project (Gunningham et al. 2004; Nelsen 2006). SLO has been shown to be particularly useful within industries such as energy (e.g., Smits et al. 2017), construction (e.g., Melé and Armengou 2016), forestry (e.g., Edwards and Lacey 2014), pipelines (e.g., Jijelava and Vanclay 2017), and mining (e.g., Thomson and Boutilier 2011) and has been argued as broadly applicable across industries (e.g., Nelsen 2006). However, evidence reveals that the uniqueness of the industry, context, and stakeholders must be taken into account when seeking to gain or sustain an SLO (e.g., Hall et al. 2015).

Industry leaders quickly realized that it was important to move beyond focusing solely on the legal, regulatory, and technical merits of a project (to which they had grown accustomed) and seek ways to identify, understand, and respond to stakeholders' concerns along the life-span of the project, in order to ensure a project's viability. Academics soon weighed in on the concept of SLO, seeking to explain *why* projects either succeeded or failed in gaining and sustaining an SLO (e.g., Hall et al. 2015; Jijelava and Vanclay 2017; Moffat and Zhang 2014; Parsons et al. 2014; Smits et al. 2017).

SLO emerged as an industry response to corporate social responsibility (CSR; Gunningham et al. 2004), which focuses on optimizing economic, social, and environmental value simultaneously, and is the firm-level operationalization of sustainable development (Bansal 2005). More specifically, given the centrality of stakeholders – defined as those who can affect or are affected by an organization's activities (Freeman 2010) – stakeholder theory provided useful insights for understanding not only who to consider as stakeholders but also their relative importance to the firm (Mitchell et al. 1997). SLO in return has provided a useful way for explaining how to go about gaining and sustaining stakeholder buy-in for projects and some of the missteps firms should avoid (e.g., Thomson and Boutilier 2011).

According to Parsons and colleagues (2014), the common thread that connects the SLO to these other CSR-related concepts is that of *legitimacy* (Suchman 1995), which is conferred by stakeholders to the firm for a given activity or project, for a period of time. Legitimacy is defined as a "generalized perception or assumption that the actions of an entity are desirable, proper, or appropriate within some socially constructed system of norms, values, beliefs, and definitions" (Suchman 1995, p. 574). It is through the firm's ability to illustrate and maintain their legitimacy in the eyes of stakeholders that affords them the "right" to continue their project unimpeded. Importantly, the duration of such legitimacy may or may not align with the life-span of a given project, especially, when considering capitally intensive projects such as those in the extractive industries. SLO has also emerged as a tool that stakeholders can embrace for making their concerns known throughout the negotiation, implementation, or operational phases of a project.

At the project level, SLO tends to be rooted in the perceptions and opinions of, first, the local community, which includes those living and working near the project site (Graafland 2002), and, second, stakeholders more generally. Given that stakeholders' perceptions and opinions are dynamic and can change over the life of a project, so too can their willingness to grant an SLO. Thus, SLO must not only be gained but also sustained.

The importance of gaining a social license from stakeholders has increased in recent years. For example, Goldman Sachs (2008) found that it took almost twice as long to bring major oil and gas projects to their operational phase, in comparison to projects from the previous decade, leading to increased costs to the firm. The single largest factor contributing to such project development delays is the relationships between stakeholders and the firm (Ruggie 2010). Thus, the business case for considering how to gain and sustain an SLO becomes clear.

Gaining and Sustaining the Social License

An SLO tends to be granted on a case-by-case basis, meaning that a firm may gain a social license from stakeholders for one project but not another, due in part to the complexity associated with identifying stakeholders for a given project (Prno 2013). Determining which stakeholders are affected by a project can be particularly challenging because it is a matter of perception, on the part of the stakeholder. Traditionally, a stakeholder is identified spatially (Gould et al. 1996); those near the proposed project would be considered stakeholders. This has been the focus of the majority of research in understanding the importance of SLO (with a few exceptions, e.g., Lacey and Lamont 2014) because it is those stakeholders whose buy-in to a project (or lack thereof) is most evident on the ground. When a social license has not been granted, it tends to be local stakeholders who engage in protests or other actions (see Hanna et al. 2016), to delay, derail, or deny the project altogether (Gould et al. 1996). Proximity, however, is not the only criteria for determining a stakeholder.

Stakeholders may be spatially proximate (i.e., local), yet proximity does not constrain stakeholders' effect on a project or the effect of a project on stakeholders (Moffat and Zhang 2014). Consider a pipeline project as an example. From an environmental perspective, if the project were to go ahead, there would be an increase in carbon dioxide emissions not only from the development of the pipeline but also through increased oil and gas products being made available for consumption. As such, the pipeline project will contribute to climate change. Because of climate change, which affects regions across the globe through rising sea level or increases in the frequency and intensity of extreme weather, is global, anyone in affected regions would be considered a stakeholder for the project. Thus, a stakeholder group, perhaps not even physically proximate to the project, may indeed be in a legitimate position to issue (or withhold) an SLO for the project.

Once the various stakeholders have been identified, they may hold differing opinions as to whether an SLO should be granted, and thus consensus may be difficult. Some stakeholders may prioritize economic development by way of job creation (Freeman et al. 2007), while others may focus on social justice issues (Scott and Oelofse 2005), and still, others may emphasize the importance of environmental sustainability (Kassinis and Vafeas 2006). In addition, Lacey and Lamont (2014) argue that increasingly, projects are being scrutinized across broader scales (e.g., regional, national, global) in that it is not only imperative for the firm to acquire an SLO from the local community but also stakeholders further afield. In the mining context, evidence suggests that stakeholders consider the operational track record across multiple operations through the value chain (Moffat and Zhang 2014), which has important implications for not only the focal firm but the industry more broadly. Furthermore, Hall and colleagues' (2015) cross-industry analysis of SLO found that when there were conflicting opinions toward a project between local stakeholders and those further afield, it became less clear which stakeholders were more important to the firm in securing their SLO. Therefore, identifying which stakeholders to pay most attention to becomes an increasingly difficult proposition. As such, the complexities associated with trying to determine what stakeholder issues to address, whom to talk to, and how to prioritize their concerns to be granted an SLO become a challenge.

The following section explores the levels of a social license and the boundaries between them.

Levels and Boundaries of a Social License to Operate

Thomson and Boutilier (2011) identify social license as a continuum from acceptance, through approval, to co-ownership. The authors argue that a firm's location along the continuum is dependent on stakeholders' perceptions of the firm's legitimacy, credibility, and trust, as realized through their mutually created social capital. For the firm, it is imperative to identify meaningful ways of participating in partnerships with stakeholders (e.g., local communities and others), which then, in turn, grant an SLO to the firm. For stakeholders, they must draw upon the social capital in their networks to ensure they have the capacity and buy-in to grant an SLO should the firm be seen as legitimate, credible, and trustworthy. Figure 1 depicts the thresholds (legitimacy, credibility, and trust) between levels (acceptance, approval, co-ownership) of an SLO.

According to Thomson and Boutilier (2011), the first level of a social license is *acceptance*. Acceptance is achieved when the firm responsible for the project can establish itself as legitimate in the eyes of stakeholders. The legitimacy of the firm (and its project) is based on its ability to identify, understand, appreciate, and respect established norms within the community of stakeholders (Kemp and Vanclay 2013). Legitimacy is also conferred through the benefits that the community derives from the partnership with the firm. Benefits can be in the form of the company investing in job, training, and other social programs (e.g., João et al. 2011). Firms must first do their due diligence to identify formal norms (e.g., economic and legal) during the project planning phase, which is important for understanding the rules of the game. Engaging with stakeholder at the community level allows the firm to share details of the project as well as giving stakeholders the opportunity to ask questions and share concerns. Such engagement with the local community – often in the form of face-to-face meetings – allows the firm to identify informal norms (e.g., cultural, social, and political), which can be incorporated into the project. A standardized approach for gaining acceptance is not recommended given that the makeup of the local community and stakeholders can vary considerably by the project (Prno 2013), which makes it difficult to identify their needs and wants (Suchman 1995). Therefore, the core principle for consideration by the firm in engaging with stakeholders is respect, respect for the health of the local community, to determine their future, their safety and security, their land and property, and the natural environment (e.g., Kemp and

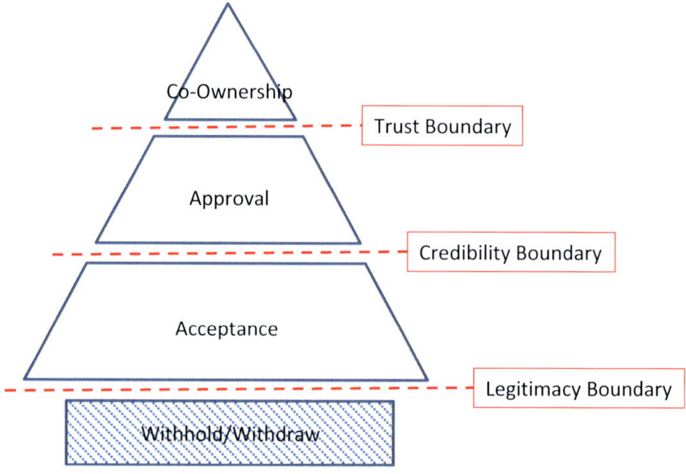

Fig. 1 Levels of social license (Adapted from Thomson and Boutilier 2011)

Vanclay 2013). Doing so builds legitimacy for the firm, increasing the likelihood of gaining acceptance for the project.

The second level of a social license is *approval* (Thomson and Boutilier 2011). Approval for a project is gained when the firm can establish itself as being credible. Credibility is established through open, honest, and transparent communication and engagement with stakeholders and a demonstration the firm's ability to consistently meet the commitments to the community. Importantly, credibility relates to the firm's social interactions as opposed to their technical credibility. Mistaking one for the other can result in failing to realize approval for the project. One type of engagement practice is the use of formal agreements, which can be a useful tool for defining and detailing the rules and responsibilities of all parties involved in the project. In doing so, commitments can be articulated so as to reduce the likelihood of there being a disconnect between commitments and expectations. Community engagement also comes in the form of face-to-face meetings. Such an approach enables the firm to build credibility with stakeholders as it gives the firm the repeated opportunity to showcase their competencies for executing the project (Dare et al. 2014) as well as the opportunity to engage in productive dialogue with stakeholder (Prno 2013). The engagement approach taken by the firm must match the expectations of project stakeholders. In some cases, a formal agreement may be preferable, whereas, in others, it is about building the personal connection through the repeated face-to-face meeting. Either approach will allow the firm to build their credibility with stakeholders and increase the likelihood of the project receiving approval.

The third and final level of social license is that of ***co-ownership*** (Thomson and Boutilier 2011). Co-ownership is realized when the firm has been able to develop their credibility to a point where they have trust with stakeholders. Trust is manifest through the continued interaction of shared experience between the firm and stakeholders, which is time-consuming to foster, thus challenging for firms to move beyond a transactional relationship with stakeholders to one that is transformational (Bowen et al. 2008). Importantly, trust acts to reduce the power distance between the firm and stakeholders, enabling each party to inculcate mental (and in some cases material) ownership of the project. Thomson and Boutilier (2011) identify two different types of trust: interactional and institutional. Interactional trust is the stakeholders' perception that the firm genuinely listens, responds, and respects their concerns surrounding a project. Institutional trust is the concern that mutual benefit is being realized for all stakeholders. Practically this means that the firm would be genuinely concerned that stakeholders realized the benefits committed to them, in whatever form those benefits may take for a given stakeholder group through the project. Conversely, stakeholders would be equally concerned that the firm's interests were being met. For both types of trust, Harvey (2014) suggests the firm needs to shift its efforts from project "outreach" to project "in-reach," which is working with stakeholders in developing projects as opposed to working independently and then telling stakeholders what they intend to do. In building trust, firms will be able to engage in a transformational relationship with stakeholder and build projects that are not only of mutual benefit but also instill a mind-set of co-ownership for the project.

The following section presents background information on the NGP, which is followed by an application of the SLO framework to understand why, despite significant support for the NGP, the project was ultimately rejected.

The Northern Gateway Pipeline

Canada is globally recognized as an energy exporter in large part due to the bitumen reserves in the Athabasca region of Western Canada. These reserves are also known as Alberta's oil sands, which represent 177 billion barrels in proven reserves (National Energy Board 2013a). The steep rise in oil prices since 2007 made bitumen production and export more lucrative than ever, spurring exponential production. The Canadian economy is closely tied to the oil and gas sector as Canada is the fifth largest oil producer and the fourth largest oil exporter (Natural Resources Canada 2014). Oil and gas products account for 20% of Canada's net exports, almost 25% of private sector investments, and 20% of capitalization on the Toronto Stock Exchange (Pineault and Hussey 2017). Record-breaking oil prices from 2011 to 2014 bolstered private investments in increased oil sand production capacity. In 2012, the oil and gas sector invested roughly $55 billion in new capital projects across Canada (Morgan 2013), which would increase production capacity up from 1.8 million barrels per day (bpd) to 5.2 million bpd when operational (Lemphers 2013). With production anticipated to rise over the next decade, restricted pipeline capacity (Fig. 2) could cost $1.3 trillion in foregone exports (Holden 2013). Hence pipeline proponents had a strong economic case for increasing pipleine capacity.

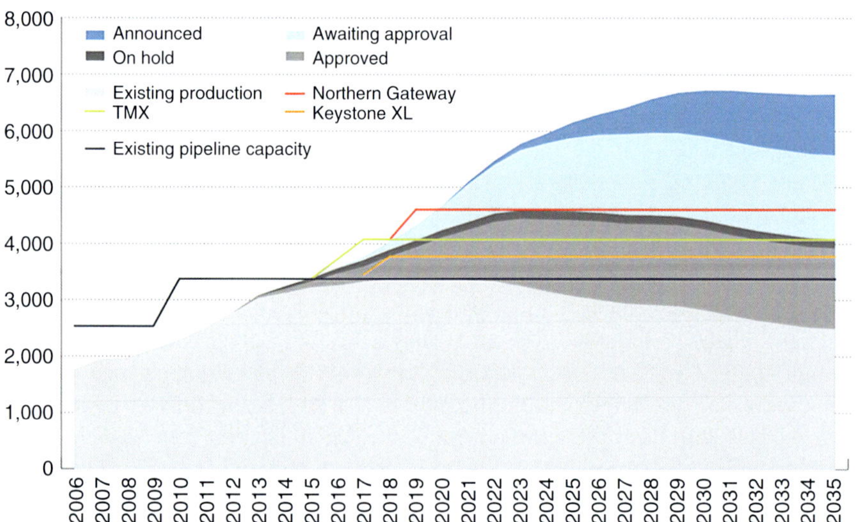

Fig. 2 Anticipated bitumen production relative to pipeline capacity (000 b/d) (Source: Holden 2013)

Pipelines that delivered Canadian crude oil for refining and export were predominantly directed south toward the United States, reducing the price of Canadian oil. As of 2013, the United States imported 71% of Canada's total crude production (Fig. 3), the majority of which was transported to the PADD II storage and distribution hub in Cushing, Oklahoma (Holden 2013). Stockpiles of oil inventory in the Midwestern United States, coupled with transportation bottlenecks in Canadian distribution channels, dampened Canadian crude demand and consequently depressed export prices relative to overseas benchmark rates (Hoberg 2013); Western Canada Select (WCS) prices were on average $19/barrel lower than the international West Texas Intermediate (WTI) price (Fig. 4) (National Energy Board 2013b). As a price taker on the North American market, Canadian oil producers faced a price discount that cost the industry billions in foregone revenues annually (Moore et al. 2011). As such, there was a strong incentive on the part of regulators to support Canadian oil producers in their efforts to seek alternative markets. To this end, oil producers, with the support of the Canadian government, needed to find a way to get Canadian oil to tidal water to be able to access Asian markets. After the United States, China and Japan represent the 2nd and 3rd most potential demand for Canadian crude (National Energy Board 2013a).

The Northern Gateway Pipeline (NGP) had the potential to open Canadian exports to Asian markets and secure the industry's profitability and growth into the future. The NGP was first proposed by Enbridge Inc. in 2004 to export Canadian oil from Alberta to the western coast of British Columbia for export. Enbridge is Canada's leading

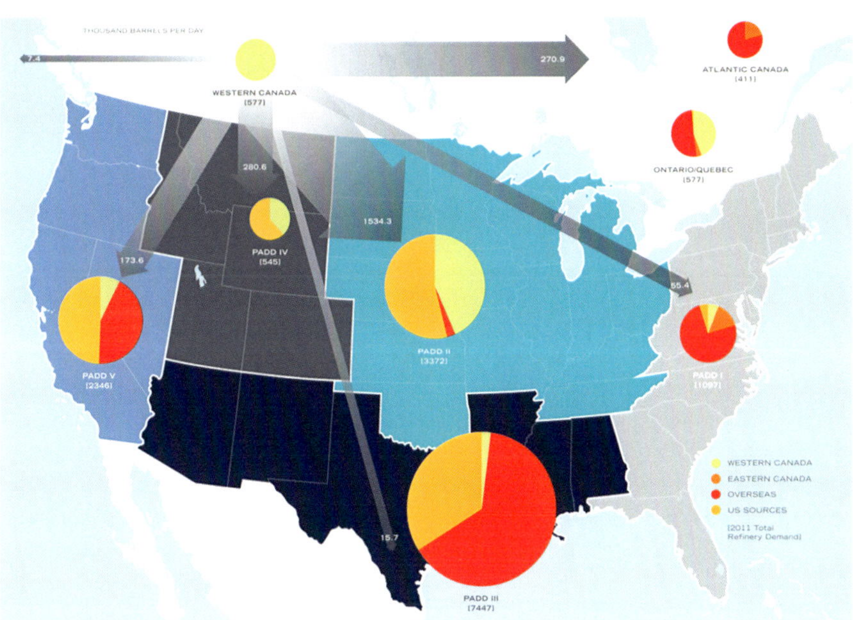

Fig. 3 Disposition of Canadian crude oil in 2011 (Source: Holden 2013)

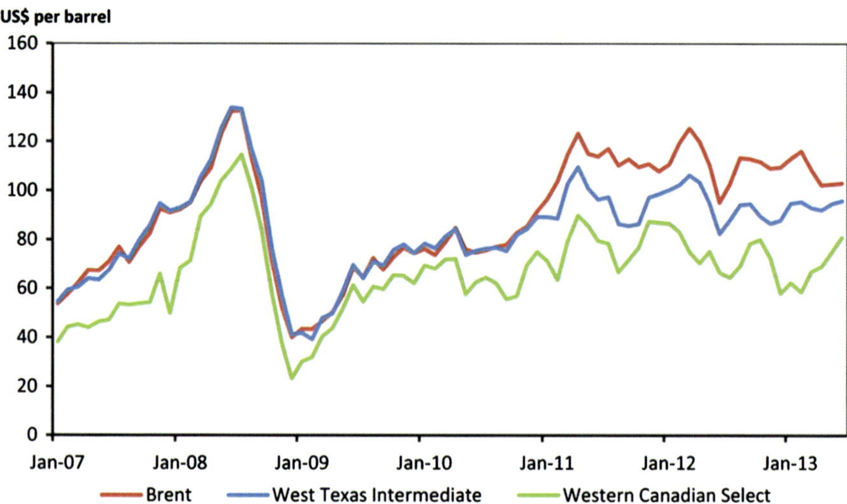

Fig. 4 Crude oil benchmark prices (US$/b) (Source: National Energy Board 2013b)

energy infrastructure provider and the project principal on the NGP. Enbridge currently operates more than 28,000 km of pipeline, delivering over 2.8 million barrels of crude oil from Alberta daily (Enbridge 2017). As proposed, the project involved two pipelines spanning approximately 1170 km from Bruderheim, Alberta, to Kitimat, British Columbia (National Energy Board 2013a). The project also included the construction of 2 tanker berths, 3 condensate storage tanks, and 16 oil storage tanks at the port of Kitimat, British Columbia. The final estimated cost for the project was $7.9 billion (National Energy Board 2013a). As early as 2005, Enbridge had signed an agreement with PetroChina to import half of the 400,000 bpd capacity (Jones 2008).

Economic Benefits

We explore the requisite boundary criteria (i.e., legitimacy, credibility, and trust) for the three levels of SLO (i.e., acceptance, approval, and co-ownership) to assess whether Enbridge was able to acquire an SLO for the NGP.

Despite having a very clear business and economic case for pursuing the development of the NGP, it proved to be one of the most divisive and controversial projects in Canada. On the one hand, proponents argued that the estimated economic benefits through new jobs and tax revenues, which would contribute $10 billion annually (National Energy Board 2013a) to the Canadian the gross domestic product (GDP), outweighed the risks of developing the pipeline. On the other hand, opponents argued that the risk and benefits were not equally distributed; the bulk of the risks would be borne by stakeholders along the pipeline route (e.g., communities and the natural environment), while the bulk of such benefits would accrue to Enbridge and Canadian oil producers.

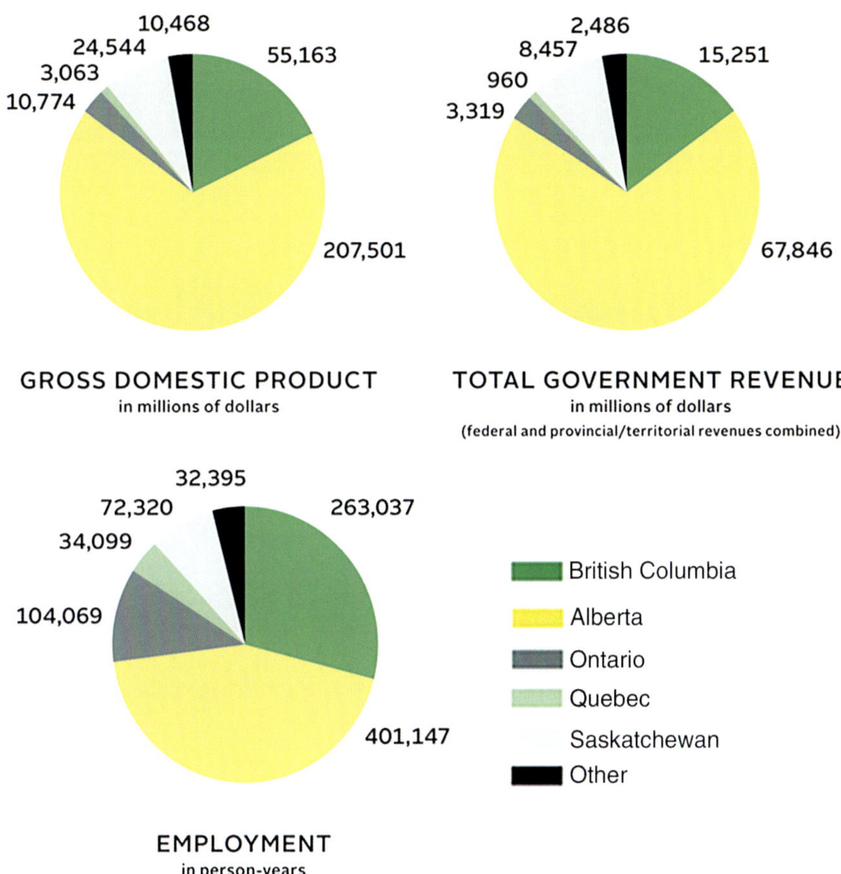

Fig. 5 Estimated economic benefits of the Northern Gateway Pipeline by province (Source: National Energy Board 2013a)

At the national level, the Canadian federal government has jurisdiction over issues relating to natural resources, when products like oil move across provincial or national boundaries. Northern Gateway claimed that the pipeline would generate 900,000 person-years of employment and more than 300 billion in GDP over the next 30 years, with 70 billion in labour income, 44 billion in federal revenues, and 54 billion to provincial governments (as shown in Fig. 5) (National Energy Board 2013a). However, environmental NGOs argued that the upstream environmental and social costs of oil sands developments were downplayed while the benefits were highly inflated (National Energy Board 2013a). Nevertheless, it was clearly in the federal government's interest to increase market access for Canadian oil.

At the provincial level, Alberta's economy and provincial government's revenue are closely tied to the oil and gas sector, which directly employs over 170,000 Albertans and accounts for 23.9% of the province's total gross domestic product

(KPMG 2013). Moreover, Alberta had long used its oil revenues to offer the "Alberta Advantage" – a flat 10% income tax and no provincial sales tax (Rubin 2015). But waning demand and discounted prices relative to international markets led to a $1.3 billion budgetary deficit in 2012 (Gerson 2013). The provincial government of British Columbia would have also benefited from a new pipeline, yet its compensation was not proportionate to the environmental risk incurred by the province. With 58% of the proposed pipeline to be laid in British Columbia, the envionrmental risk to salmon fisheries and remote watersheds in the event of a pipeline spill outweighed the financial benefits, with only 7.4% of the $81 billion in royalties going to the province (Government of British Columbia 2012). The government of British Columbia signaled its willingness to negotiate if the province were to secure an adequate share of the pipeline's economic benefits to compensate the province for the associated risks. The final decision as to whether the pipeline were to be built fell to the National Energy Board (NEB), the federal government's primary regulator of pipelines. Provincial legislative authority is superseded by the federal government when they deem a project to be in the national interest (Bankes 2012).

With the economic benefits of the NGP clear and Enbridge consistently ranked as one of the world's most sustainable corporations for their social and environmental commitments (Enbridge 2017), the project had considerable support. However, the environmental and social risks of the project challenged the legitimacy of Enbridge and the NGP on multiple fronts. Skepticism toward the pipeline was driven by concern that NGP did not consider the "big picture" implications, specifically environmental risks (Horter et al. 2009), such as oil spills, wildlife habitat, and climate change. Social risks, such those to human health, violations of Indigenous sovereignty, and the project's long-term economic viability called into question the Enbridge's legitimacy and threatened the acceptance of the project. Each dimension is discussed in detail below.

Environmental Risks

Oil Spills. The proposed NGP route would have crossed over hundreds of sensitive waterways and rugged mountains, before passing treacherous inner coastal and open waters (Swift et al. 2011). A study by Alberta's Energy and Utility Board on pipeline spills counted over 12,000 incidents of leaks and ruptures over a period of 15 years, 57% of which was caused by internal corrosion (Alberta Energy and Utility Board 2007), challenging the legitimacy of the industry. Enbridge's response to this reality was to commit to thickening pipeline walls through major tributaries; however these measures were considered inadequate given the varied terrains, unpredictable coastal weather, fragile ecosystems, and isolated pathways the pipeline would follow (Union of British Columbia Indian Chiefs 2012).

The proposed NGP route would have crossed more than 800 rivers and streams, upon which animals and humans depend. A spill along the route could have contaminated headwaters of three important watersheds – Mackenzie, Fraser, and

Skeena – which host critical marine migration routes and salmon habitats (Chia et al. 2015). A recent freshwater oil spill in Canada's Pine River exemplified this threat to marine life, where 1 million liters of petroleum resulted in a massive fish kill spanning an area over 20 km downstream from the spill site (Levy 2009), again calling into question the legitimacy of the industry. Downstream contamination not only threatens marine life but agricultural irrigation, revenues from freshwater fisheries, and British Columbia's booming recreational industry.

However, it was not only waterways that are at risk of oil spills. Permanent ground displacement, caused by landslides, faults, or liquefaction, represented one of the most significant geo-hazards associated with pipeline systems (Nyman et al. 2008). In its current state, "stress design requirements in the standard [Clause 4.2.4] do not include the effects of inertial earthquake loads, slope movements, fault movements, earthquake-induced earth movements, frost heave, or other loading sources" (Rathje 2011, p. 3).

The risk of spills carries past the pipeline route and into coastal waters, as supertankers would pass through Northern British Columbia's Douglas Channel and the ecologically rich Great Bear coastal temperate rainforest. The route to open waters was 185 km long, and less than a kilometer wide, through a treacherous pass of mountainous fjords and islands that often experienced hurricane force winds and dense fog (Honderich 2012). These routes were home to critical orca and humpback whale habitats, which were at risk of deadly collisions, toxic contaminants, and underwater noise impacts (Greenpeace 2012). In fear of the risk of oil spills, the proposed route – which was far narrower than where the Exxon Valdez ran aground – had never been navigated by supertankers due to an informal moratorium on all oil tanker traffic since 1972 (Honderich 2012).

Wildlife Habitat. The pipeline itself wass also believed to have adverse ecological and health-related effects. The pipeline's permanent "right of way," a narrow (25–30 m) slice of clear-cut forest, would have passed straight through the ranges of at least five of the most threatened herds of Woodland Caribou and eight grizzly bear populations in Canada (Cryderman 2013). Environmental NGOs lobbied to have the majority of these routes pass through previously disturbed lands, to minimize the ecological impact of the pipeline on native species.

Climate Change. Critics also opposed the NGP for its potential contribution to climate change. The proposed pipeline would have increased Canada's (well-to-wheel) greenhouse gas emissions by as much as 100 million tonnes of CO_2 equivalent ($MtCO_2eq$) per year, which roughly translates to 14% of Canada's 2008 emissions (Zickfeld 2011). Such an increase would have made it increasingly difficult, if not impossible, for Canada to meet its emission targets pledged under the United Nations Copenhagen and Paris Accords.

Social Risks

Human Health. Adverse impacts of hydrocarbons on human health could have also been significant, according to the Northern Gateway's Ecological and Human

Health Risk Assessment. Indigenous groups who rely on the lands were concerned about these results, citing concerns of degraded air and water quality as a result of increased industrial activity (National Energy Board 2013a). Concerns would have been amplified in the case of a spill – with disruptions to traditional diets, increased risks of chronic illnesses associated with processed foods, the potential for mental illnesses from psychological stresses, and diminished social well-being of affected communities (National Energy Board 2013a). Indigenous groups were most vocal in arguing that their lands and waterways – which have been a source of livelihood and rituals for thousands of years – were threatened by the pipeline (Gibson and Klinck 2005).

Indigenous Sovereignty. Indigenous groups opposed the NGP arguing that the pipeline's route through their territory was a violation of their sovereignty. Over 25% of the proposed pipeline and tanker corridor sat within 80 km of 69 Indigenous communities, tribal councils, and Metis organizations, who hold traditional titles to these lands (McCreary and Milligan 2014). Indigenous titles are acknowledged under the Canadian constitution, following a Supreme Court ruling in 1973 (Calder case) recognizing Aboriginal Peoples as distinct polities with distinct rights and claims (Godlewska and Webber 2007). A 2004 (Haida case) ruling enforced the government's duty to consult and accommodate Indigenous groups for developments that could negatively impact Aboriginal rights or titles (Newman 2009).

While the federal government called for consultations with Indigenous groups along the pipeline route (excluding downstream nations), Indigenous groups invoked international law (the United Nations Declaration on the Rights of Indigenous Peoples) to support their right to make free and informed choices about the development of their lands and resources (Boutilier 2017).The right to free, prior, and informed consent (FPIC) under the UN declaration is, however, a non-binding instrument to simply guide the behavior of states (McCreary and Milligan 2014). Indigenous groups thus requested that Indigenous rights be recognized on lands with title claims, through consultation and consent for all impacted Indigenous communities (Fine 2014). The NGP had gained acceptance from some Indigenous communities, but Enbridge faced its greatest challenges to its legitimacy among coastal Indigenous communities who feared a spill would destroy the marine environment (McCarthy and Lewis 2016a).

Long-Term Economic Viability. Increased reliance on fossil fuel operations comes at a cost to the larger Canadian economy. Higher oil prices result in inflationary pressures that influence domestic interest rates and business investments which lead to increased unemployment and declining production in non-oil-related industries like the export-reliant manufacturing sector (Allan 2012). Furthermore, foreign ownership of oil sands operations costs Canadians an even larger share of economic benefits, as profits and value-added refining capacities are exported offshore. For these reasons Canada's largest labor union – Unifor – opposed the NGP, arguing that pipelines not only export Canada's resources but also its profits and jobs (Unifor 2014). Thus, the legitimacy of claims to the economic benefits of the NGP for job creation were called into question.

Oil spills, destruction of wildlife habitat, climate change, risks to human health, Indigenous sovereignty, and the long-term viability of NGP resulted in the federal government's decision to not approve the pipeline in 2016, in part, due to inadequate engagement with stakeholders (McCarthy and Lewis 2016a). Despite the economic benefits of the NGP, and Enbridge's notoriety as being one of the world's most sustainable corporations (Enbridge 2017), they were unable to gain acceptance for the NGP due to the absence of legitimacy in mitigating the environmental and social risks of the project.

Discussion and Conclusion

This chapter explored the concept of social license to operate (SLO) as a framework to try to unpack why, despite the strong economic arguments in favor of and political support for developing the NGP, the project was ultimately rejected. Consistent with prior work that has applied SLO in the context of industries that experience a high degree of reputational risk (e.g., extractive industries), this chapter considered the boundaries of legitimacy, credibility, and trust (i.e., Thomson and Boutilier 2011) between levels of SLO (i.e., acceptance, approval, and co-ownership). Through SLO, it became clear that the NGP was not accepted by stakeholders, primarily because Enbridge was unable to gain legitimacy with local and Indigenous communities, environmental NGOs, and others, as to the sustainability of the project. It was revealed that the environmental and social risks associated with the project yielded an imbalanced distribution of risks and benefits across stakeholders, with the majority of the former being accrued by those proximate to the pipeline and the latter conferred to Enbridge and Canadian oil companies. The plethora of risks, both environmental and social, that were revealed through stakeholder engagement, proved to seed doubt among stakeholders as to the legitimacy of Enbridge mitigating the project-related risks; the risk of oil spills, threats to wildlife habitat, contibutions to climate change, potential impacts to human health, violations of Indigenous sovereignty, and the project's long-term economic viability. Although the federal government had much to gain through the approval of the NGP by way of tax revenue and royalties, they chose to reject the project because Enbridge was unable to counter challenges to their legitimacy by stakeholders that the risks (of NGP) outweighed the benefits, thereby failing to secure an SLO.

Through this assessment of Enbridge and the NGP project, it has become clear that from a theoretical perspective, there is great utility in the SLO framework for understanding decisions that seem counterintuitive (i.e., the federal government withheld approval of the NGP despite strong economic incentive to approve it). What the SLO framework does is illuminate the importance of stakeholder engagement in a very practical way. It provides a means for not only understanding the varying degrees of stakeholder "buy-in" to a project (i.e., acceptance, approval, co-ownership) but also what specific actions and boundaries (i.e., legitimacy, credibility, and trust) firms can take to enhance their level of engagement with stakeholders.

The SLO concept is a useful and important consideration for most firms, but particularly those in capitally intensive industries, such as the extractive industries or those engaged in large-scale infrastructure projects like pipelines. In gaining an SLO, firms can not only reduce the likelihood they'll experience project delays, disruptions, or damage to infrastructure associated with actions by local stakeholders such as protests (Goldman Sachs 2008; Ruggie 2010) but also reduce the potential for risks that manifest with stakeholders more broadly, including regulatory action (additional conditions for development or operation, fines, legal action), impacts to reputation, negative reactions of shareholders, or the rejection of future projects. Together these risks represent significant financial risk to the firm and therefore necessitate appropriate managerial attention and focus. For firms that are able to manage their social capital as operationalized through their SLOs, they can not only reduce said risks but also increase the likelihood that other projects will receive support.

The application of the SLO framework is a useful tool for firms across industries to manage and deploy their stakeholder engagement strategy. However, in applying the SLO framework, there may be a tendency to treat stakeholder groups as homogenous, which can be problematic. Evidence has shown that opinions and attitudes of stakeholders can vary by location (e.g., Prno 2013), through time (Dare et al. 2014) and within groups (e.g., Jijelava and Vanclay 2014). The findings of our inquiry were consistent in that some Indigenous communities accepted the development of the NGP, whereas those Indigenous communities on the Pacific coast did not (McCarthy and Lewis 2016a). For these reasons, it is important for the firm to build a high degree of familiarity with stakeholders so as to understand their concerns and answer their questions, which not only acts to fill an information gap but also illustrates the firm's respect for stakeholders, which is a central element for gaining an SLO. In doing so, the firm can foster a relationship with stakeholders, which not only reduces the potential for costly delays (or project denials, as in the case of the NGP) but also builds goodwill for gaining and sustaining an SLO over a project's life cycle, as well as building the firm's reputational capital as trustworthy partner.

The following section presents some key takeaways from Enbridge's NGP experience that could serve as a learning opportunity for other firms in their pursuit of an SLO.

Key Takeaways

Gaining and sustaining an SLO is particularly important for firms in industries with high risk to reputation. Enbridge was no exception. The following five key takeaways were gleaned from the NGP experience and should be avoided when seeking an SLO.

Unequal distribution of costs and benefits. One of the overarching takeaways from the NGP experience was that the distribution of costs and benefits were not equally distributed. The bulk of the economic benefits from the project would flow to Enbridge, its shareholders, and the Canadian oil industry more generally. The costs, on the other hand, would accrue with those who had the most to lose through the project, specifically, the natural environment, and the local and Indigenous communities along the pipeline route.

In championing such projects in the future, it is important for firms like Enbridge to seek ways making the distribution of costs and benefits more equitable. By seeking opportunities to create shared value (Porter and Kramer 2011), stakeholders, particularly those directly affected by the project, are more likely to grant an SLO given the firm has shown their commitment of thinking beyond their bottom line to include to needs and wants of stakeholders.

Overemphasis on economic value creation. The projected economic benefits of the NGP were no doubt significant, estimated at $10 billion annually (National Energy Board 2013a). However, a project cannot rely solely on its economic benefits to justify its development. Recognizing that societal and cultural paradigms change over time (Nelsen 2006), the central importance of economic benefits may be called into question. A recent study by Forbes found that more than 80% of respondents thought that business should incorporate social and environmental value into their business goals (Epstein-Reeves 2012), which is in stark contrast to the business-as-usual approach of profit maximization (Friedman 2007). As society's values shift, expectations as to what constitutes value creation may also change such that firms focusing solely on maximizing profits without considering social or environmental value may ultimately fail to secure an SLO. Incorporating elements of the triple bottom line (Elkington 1998) (i.e. the economic, social and environmental value creation) will build legitimacy with stakeholders and increase the likelihood of gaining an SLO.

One such action that Enbridge could have taken to balance the economic, social, and environmental values of the NGP would have been to give Indigenous groups a larger share of the equity in the pipeline or more authority in overseeing its construction. Although these actions would have increased costs, they might have improved the chances of securing an SLO.

Addressing complexity with a simple technical fix. One of the central environmental issues faced by pipeline companies is the threat of a pipeline leak. Pipeline safety has been called into question in recent years. Enbridge, for example, has experienced several leaks in recent history, but most notably was the leak that occurred in 2010 when 20,000 barrels of oil spilt into the Kalamazoo River, costing Enbridge more than $1.2 Billion USD to clean up the oil and hundreds of millions in fines (Hasemyer 2016). Given recent spills that have manifest from pipelines, it is understandable that the technical legitimacy of Enbridge, and the industry more generally, would be called into question by stakeholders when assurances that spills along the NGP would be mitigated by using a thicker pipe. Given the varied terrains, unpredictable coastal weather, fragile ecosystems, and isolated pathways, the pipeline follows, (Union of British Columbia Indian Chiefs 2012) a thicker pipe, may have indeed been the appropriate way to mitigate leaks. However, their solution failed to address the complexity of the natural environment in which the pipeline would be located and therefore was deemed illegitimate by stakeholders. Therefore, firms should strive to not only find the appropriate technical solution to the problem but not mistake their technical credibility for social credibility in addressing stakeholders concerns in their granting of an SLO.

Limited engagement with Indigenous communities in early project development. Large-scale infrastructure projects, and particularly those in the extractive industries, will inevitably come close to, intersect, or require the crossing of Indigenous lands. Therefore, it is essential to engage with Indigenous communities as early as possible. In the case of the NGP, Enbridge approached Indigenous communities after the project was conceptualized and at which point they were hoping to secure approval from Indigenous communities. According to Bowen and colleagues (2008), Enbridge's approach would be considered a transactional approach to stakeholder engagement where communication was mainly in the form of information sessions, which were infrequent and one-sided (from the firm to stakeholders). This is reflected in their strong commitment to the project, such that they intended to proceed with or without Indigenous communities buy-in (Paterson 2011).

For firms that might find themselves in similar situations, that is, seeking to develop a similar scale project, in order to secure an SLO, engagement with Indigenous communities needs to happen during the conceptualization stage. Asking communities whether they want a project should be the starting point for engagement. If there is no interest in the project, there is very little likelihood the project will be granted an SLO. If there is interest, or at least an interest to see how the project could unfold, this is the opportunity for the firm to build social legitimacy with Indigenous communities. By focusing on delivering reliable and accurate information as promised will go along way in showing communities that they are respected by, and important to, the firm. From there, significant time will be required to meaningfully and honestly engage with communities to move from acceptance, through approval, to co-ownership of the project as the firm builds its legitimacy, credibility, and trust with Indigenous communities.

Future Research

There are two main avenues for future research that have emerged from the application of the SLO concept in explaining why the NGP was not approved despite overwhelming support. First, it is important to explore in greater detail the tensions that emerge in the distribution of costs and benefits for large-scale projects such as pipelines. When the bulk of the costs are borne by stakeholders most proximate to the project and benefits tend to flow to those further afield, what might be an appropriate ratio for the distribution of costs and benefits? In order to make such a determination, and consistent with the results from the empirical study by Hall and colleagues (2015), there is a need for developing metrics that could be applied on a case-by-case basis to measure the dynamic nature of gaining and sustaining an SLO. With greater ability to measure costs and benefits, particularly those that are non-financial (e.g., social and environmental), more equitable distribution of costs and benefits could be realized, thereby increasing the likelihood of firms gaining and sustaining an SLO and projects being developed.

The second opportunity for future research is in how to identify and incorporate the changing societal and cultural paradigms in project planning in order to gain and sustain an SLO. In the case of the NGP, Enbridge emphasized the project's economic benefits while overlooking the importance of sustainable development to stakeholders and as such an SLO was not granted. However, when the project was conceived during the mid-2000s, the emphasis was on economic development as sustainability was just beginning to gain traction in business. Therefore, being able to identify trends that are substantive as opposed to merely fads would enable firms who are proposing similar projects to get out in front of the trend and be proactive in anticipating stakeholders changing attitudes. Doing so will increase the likelihood of projects being approved as SLOs can be gained and sustained.

Cross-References

- ▶ Empathy Driving Engaged Sustainability in Enterprises
- ▶ Responsible Investing and Corporate Social Responsibility for Engaged Sustainability
- ▶ Responsible Investing and Environmental Economics

References

Alberta Energy and Utility Board. (2007). Pipeline performance in Alberta, 1990–2005. https://www.aer.ca/documents/reports/r2007-A.pdf. Accessed 22 May 2017.

Allan, R. (2012). *An economic assessment of Northern Gateway*. National Energy Board. http://www.robynallan.com/wp-content/uploads/2012/02/Economic-Assessment-of-Northern-Gateway-January-31-2012.pdf. Accessed 12 June 2017.

Bankes, N. (2012). *British Columbia and the Northern Gateway Pipeline*. University of Calgary – Faculty of Law. http://ablawg.ca/2012/07/25/british-columbia-and-the-northern-gateway-pipeline/. Accessed 6 May 2017.

Bansal, P. (2005). Evolving sustainably: A longitudinal study of corporate sustainable development. *Strategic Management Journal, 26*(3), 197–218. https://doi.org/10.1002/smj.441.

Boutilier, S. (2017). Free, prior, and informed consent and reconciliation in Canada: Proposals to implement articles 19 and 32 of the UN Declaration on the Rights of Indigenous Peoples. *Western Journal of Legal Studies*, *7*(1), 1–21.

Bowen, F., Newenham-Kahindi, A., & Herremans, I. (2008). Engaging the community: a systematic review. *The Research Network for Business Sustainability*. http://nbs.net/wp-content/uploads/NBS-Systematic-Review-Community-Engagement.pdf. Accessed 14 Mar 2017.

Brown, J., & Fraser, M. (2006). Approaches and perspectives in social and environmental accounting: An overview of the conceptual landscape. *Business Strategy and the Environment, 15*(2), 103–117. https://doi.org/10.1002/bse.452.

Chia, A., de Sousa Jensen, T., O'Neill, J., & Zhan, J. (2015). Case study: Northern Gateway pipeline. http://environment.geog.ubc.ca/case-study-northern-gateway-pipeline/.

Cryderman, K. (2013). Northern Gateway panel notes 'adverse effects' for caribou, grizzly bears. https://www.theglobeandmail.com/news/politics/northern-gateway-panel-fears-adverse-effects-for-caribou-grizzly-bears/article16080402/. Accessed 2 June 2017.

Dare, M., Schirmer, J., & Vanclay, F. (2014). Community engagement and social licence to operate. *Impact Assessment and Project Appraisal, 32*(3), 188–197. https://doi.org/10.1080/14615517.2014.927108.

Edwards, P., & Lacey, J. (2014). Can't climb the trees anymore: Social licence to operate, bioenergy and whole stump removal in Sweden. *Social Epistemology, 28*(3–4), 239–257. https://doi.org/10.1080/02691728.2014.922637.

Elkington, J. (1998). *Cannibals with forks: The triple bottom line of 21st century business*. Gabriola Island: New Society Publishers.

Enbridge. (2017). About us. http://www.enbridge.com/about-us. Accessed 22 Apr 2017.

Epstein-Reeves, J. (2012). The "Pain" of sustainability. Forbes. https://www.forbes.com/sites/csr/2012/01/18/the-pain-of-sustainability/. Accessed 3 Dec 2017.

Fine, S. (2014). Supreme Court expands land-title rights in unanimous ruling. The Globe and Mail. http://www.theglobeandmail.com/news/national/supreme-court-expands-aboriginal-title-rights-in-unanimous-ruling/article19347252. Accessed 27 June 2017.

Freeman, R. E. (2010). *Strategic management: A stakeholder approach*. Cambridge, UK: Cambridge University Press.

Freeman, R. E., Harrison, J. S., & Wicks, A. C. (2007). *Managing for stakeholders: Survival, reputation, and success*. New Haven, CT: Yale University Press.

Friedman, M. (2007). The social responsibility of business is to increase its profits. In: Zimmerli W. C., Holzinger M., Richter K. (eds) *Corporate ethics and corporate governance* (pp. 173–178). Berlin: Springer. https://doi.org/10.1007/978-3-540-70818-6_14.

Gerson, J. (2013). Alberta budget 2013 marked by billions in deficit spending, service cuts. *National Post*. http://nationalpost.com/news/politics/alberta-budget-2013/wcm/e462ddc5-73a5-4ee1-aa2f-4787da5689eb. Accessed 23 Apr 2017.

Gibson, G., & Klinck, J. (2005). Canada's resilient north: The impact of mining on aboriginal communities. *Pimatisiwin, 3*(1), 116–139.

Godlewska, C., & Webber, J. (2007). The Calder decision, aboriginal title, treaties, and the Nisga'a. In H. Foster, H. Raven, & J. Webber (Eds.), *Let right be done: Aboriginal title, the Calder case, and the future of Indigenous rights* (pp. 1–36). Vancouver, BC: UBC Press.

Goldman Sachs. (2008). *190 projects to change the world*. London: Goldman Sachs Group.

Gould, K. A., Gould, K., Schnaiberg, A., & Weinberg, A. S. (1996). *Local environmental struggles: Citizen activism in the treadmill of production*. Cambridge, UK: Cambridge University Press.

Government of British Columbia. (2012). *British Columbia outlines requirements for heavy oil pipeline consideration*. https://news.gov.bc.ca/stories/british-columbia-outlines-requirements-for-heavy-oil-pipeline-consideration. Accessed 11 Feb 2017.

Graafland, J. J. (2002). Profits and principles: Four perspectives. *Journal of Business Ethics, 35*(4), 293–305. https://doi.org/10.1023/A:1013805111691.

Greenpeace. (2012). *7 reasons to say no to the Enbridge tar sands pipeline*. Greenpeace. http://www.greenpeace.org/canada/Global/canada/report/2012/05/Greenpeace%20stop%20the%20pipeline%20fact%20sheet.pdf. Accessed 21 Mar 2017.

Gunningham, N., Kagan, R. A., & Thornton, D. (2004). Social license and environmental protection: Why businesses go beyond compliance. *Law & Social Inquiry, 29*(2), 307–341. https://doi.org/10.1111/j.1747-4469.2004.tb00338.x.

Hall, N., Lacey, J., Carr-Cornish, S., & Dowd, A.-M. (2015). Social licence to operate: Understanding how a concept has been translated into practice in energy industries. *Journal of Cleaner Production, 86*, 301–310. https://doi.org/10.1016/j.jclepro.2014.08.020.

Hanna, P., Vanclay, F., Langdon, E. J., & Arts, J. (2016). Conceptualizing social protest and the significance of protest actions to large projects. *The Extractive Industries and Society, 3*(1), 217–239. https://doi.org/10.1016/j.exis.2015.10.006.

Harvey, B. (2014). Social development will not deliver social licence to operate for the extractive sector. *The Extractive Industries and Society, 1*(1), 7–11. https://doi.org/10.1016/j.exis.2013.11.001.

Hasemyer, D. (2016). Enbridge Kalamazoo spill saga ends with $177 million settlement. *Inside Climate News*. https://insideclimatenews.org/news/20072016/enbridge-saga-end-department-justice-fine-epa-kalamazoo-river-michigan-dilbit-spill. Accessed 4 Dec 2017.

Hoberg, G. (2013). The battle over oil sands access to tidewater: A political risk analysis of pipeline alternatives. *Canadian Public Policy*. https://doi.org/10.3138/CPP.39.3.371.

Holden, M. (2013). *Pipe or perish: Saving and oil industry at risk*. Canada West Foundation. http://cwf.ca/wp-content/uploads/2015/10/CWF_PipeOrPerish_Report_FEB2013.pdf. Accessed 4 Mar 2017.

Honderich, J. (2012). Why Northern Gateway shouldn't go near Great Bear Rainforest | Toronto Star. *thestar.com*. https://www.thestar.com/news/canada/2012/10/07/why_northern_gateway_shouldnt_go_near_great_bear_rainforest.html. Accessed 12 June 2017.

Horter, W., Wagner, D., Smith, M., Moss, P., Hill, B., O'Connor, O., et al. (2009). Letter to PM demanding public inquiry. http://www.ceaa.gc.ca/050/documents/38188/38188E.pdf. Accessed 2 July 2017.

Jijelava, D., & Vanclay, F. (2014). Social licence to operate through a gender lens: The challenges of including women's interests in development assistance projects. *Impact Assessment and Project Appraisal, 32*(4), 283–293. https://doi.org/10.1080/14615517.2014.933505.

Jijelava, D., & Vanclay, F. (2017). Legitimacy, credibility and trust as the key components of a social licence to operate: An analysis of BP's projects in Georgia. *Journal of Cleaner Production, 140*, 1077–1086. https://doi.org/10.1016/j.jclepro.2016.10.070.

João, E., Vanclay, F., & den Broeder, L. (2011). Emphasising enhancement in all forms of impact assessment: Introduction to a special issue. *Impact Assessment and Project Appraisal, 29*(3), 170–180. https://doi.org/10.3152/146155111X12959673796326.

Jones, J. (2008). Enbridge rekindles oil sands pipeline plan. http://uk.reuters.com/article/enbridge-gateway-idUKN2148130320080221. Accessed 5 Feb 2017.

Kassinis, G., & Vafeas, N. (2006). Stakeholder pressures and environmental performance. *Academy of Management Journal, 49*(1), 145–159. https://doi.org/10.5465/AMJ.2006.20785799.

Kemp, D., & Vanclay, F. (2013). Human rights and impact assessment: Clarifying the connections in practice. *Impact Assessment and Project Appraisal, 31*(2), 86–96. https://doi.org/10.1080/14615517.2013.782978.

KPMG. (2013). *Economic impacts of Western Canada's oil industry*. Fédération des Chambres de Commerce du Québec. http://www.fccq.ca/pdf/general/FCCQ-Economic-Impacts-of-Western-Canada-s-Oil-Industry_nov-2013.pdf. Accessed 2 June 2017.

Lacey, J., & Lamont, J. (2014). Using social contract to inform social licence to operate: An application in the Australian coal seam gas industry. *Journal of Cleaner Production, 84*, 831–839. https://doi.org/10.1016/j.jclepro.2013.11.047.

Lemphers, N. (2013). *The climate implications of the proposed Keystone XL oil sands pipeline*. Pembina Institute. https://www.pembina.org/reports/kxl-climate-backgrounder-jan2013.pdf. Accessed 12 Mar 2017.

Levy, D. (2009). *Pipelines and salmon in Northern British Columbia*. Pembina Institute. http://pipeupagainstenbridge.ca/images/uploads/resources/pipelines-and-salmon-in-northern-bc-report.pdf. Accessed 2 June 2017.

McCarthy, S., & Lewis, J. (2016a). Court overturns Ottawa's approval of Northern Gateway pipeline. https://www.theglobeandmail.com/report-on-business/industry-news/energy-and-resources/federal-court-overturns-ottawas-approval-of-northern-gateway-pipeline/article30703563/. Accessed 22 June 2017.

McCarthy, S., & Lewis, J. (2016b). Ottawa's pipeline approvals give Alberta boost, but rile critics. *The Globe and Mail*. https://www.theglobeandmail.com/report-on-business/ottawa-approves-trans-mountain-pipeline-line-3/article33094301/. Accessed 25 Mar 2017.

McCreary, T. A., & Milligan, R. A. (2014). Pipelines, permits, and protests: Carrier Sekani encounters with the Enbridge Northern Gateway Project. *Cultural Geographies, 21*(1), 115–129. https://doi.org/10.1177/1474474013482807.

Melé, D., & Armengou, J. (2016). Moral legitimacy in controversial projects and its relationship with social license to operate: A case study. *Journal of Business Ethics, 136*(4), 729–742. https://doi.org/10.1007/s10551-015-2866-z.

Mineral Council of Australia (2015). Enduring value: The Australian minerals industry framework for sustainable development. http://www.minerals.org.au/leading_practice/enduring_value. Accessed 1 Dec 2017.

Mitchell, R. K., Agle, B. R., & Wood, D. J. (1997). Toward a theory of stakeholder identification and salience: Defining the principle of who and what really counts. *Academy of Management Review, 22*(4), 853–886. https://doi.org/10.5465/AMR.1997.9711022105.

Moffat, K., & Zhang, A. (2014). The paths to social licence to operate: An integrative model explaining community acceptance of mining. *Resources Policy, 39*, 61–70. https://doi.org/10.1016/j.resourpol.2013.11.003.

Moore, M. C., Flaim, S., Hackett, D., Grissom, S. W., Crisan, D., & Honarvar, A. (2011). Catching the brass ring: Oil market diversification potential for Canada. *SPP Research Papers*, *4*(16). https://ideas.repec.org/a/clh/resear/v4y2011i16.html.

Morgan, G. (2013). Oil and gas vitality is key to Canada's economic success. *The Globe and Mail.* http://www.theglobeandmail.com/report-on-business/industry-news/energy-and-resources/oil-and-gas-vitality-is-key-to-canadas-economic-success/article8780664/. Accessed 30 May 2017.

National Energy Board. (2013a). Report of the joint review panel for the Enbridge northern gateway project. http://publications.gc.ca/site/eng/456575/publication.html. Accessed 15 June 2017.

National Energy Board. (2013b). Canada's energy future 2013 – energy supply and demand projections to 2035: An energy market assessment. https://www.neb-one.gc.ca/nrg/ntgrtd/ftr/2013/2013nrgftr-eng.pdf. Accessed 28 June 2017.

Natural Resources Canada. (2014). Energy markets fact book 2014–2015. http://www.nrcan.gc.ca/sites/www.nrcan.gc.ca/files/energy/files/pdf/2014/14-0173EnergyMarketFacts_e.pdf. Accessed 3 Apr 2017.

Nelsen, J. L. (2006). Social license to operate. *International Journal of Mining, Reclamation and Environment, 20*(3), 161–162. https://doi.org/10.1080/17480930600804182.

Newman, D. G. (2009). *The Duty to consult: New relationships with Aboriginal Peoples*. Vancouver, BC: UBC Press.

Nyman, D. J., Lee, E. M., Audibert, J. M. E., & Quest Geo-Technics. (2008). Mitigating geohazards for international pipeline projects: Challenges and lessons learned (pp. 639–648). Presented at the 7th International Pipeline Conference, American Society of Mechanical Engineers.

Parsons, R., Lacey, J., & Moffat, K. (2014). Maintaining legitimacy of a contested practice: How the minerals industry understands its 'social licence to operate'. *Resources Policy, 41*, 83–90. https://doi.org/10.1016/j.resourpol.2014.04.002.

Paterson, J. (2011). First nations tell Enbridge: No pipelines without consent. *West Coast Environmental Law.* https://www.wcel.org/blog/first-nations-tell-enbridge-no-pipelines-without-consent. Accessed 12 May 2017.

Pineault, E., & Hussey, I. (2017). *Restructuring in Alberta's oil industry Internationals pull out, domestic majors double down*. Parkland Institute. http://www.parklandinstitute.ca/restructuring_in_albertas_oil_industry. Accessed 14 May 2017.

Porter, M. E., & Kramer, M. R. (2011). The big idea: Creating shared value. *Harvard Business Review, 89*(1–2), 62–77.

Prno, J. (2013). An analysis of factors leading to the establishment of a social licence to operate in the mining industry. *Resources Policy, 38*(4), 577–590. https://doi.org/10.1016/j.resourpol.2013.09.010.

Rathje, E. M. (2011). *Geohazards issues for the Enbridge Northern Gateway project*. National Energy Board. https://docs.neb-one.gc.ca/LL-ENG/llisapi.dll/Open/776581. Accessed 14 Feb 2017.

Rubin, J. (2015). The sooner Alberta weans itself from its resource addiction, the better. The Globe and Mail. https://www.theglobeandmail.com/report-on-business/industry-news/energy-and-resources/

the-sooner-alberta-weans-itself-from-its-resource-addiction-the-better/article23681511/. Accessed 2 July 2017.

Ruggie, J. (2010). *Business and human rights: Further steps towards the operationalisation of the "protect, respect and remedy" framework. (No. 14th Session)*. New York/Geneva: United Nations Human Rights Council (Office of the High Commissioner for Human Rights).

Salzmann, O., Ionescu-Somers, A., & Steger, U. (2006). *Corporate license to operate (LTO): Review of the literature and research options*. http://m.imd.ch/research/publications/upload/CSM_Salzmann_Ionescu_Somers_Steger_WP_2006_23.pdf. Accessed 2 Dec 2017.

Scott, D., & Oelofse, C. (2005). Social and environmental justice in South African cities: Including "Invisible Stakeholders" in environmental assessment procedures. *Journal of Environmental Planning and Management, 48*(3), 445–467. https://doi.org/10.1080/09640560500067582.

Slack, Keith. (2008). Corporate Social License and Community Practice. Carnegie Council for Ethics in International Affairs. https://www.carnegiecouncil.org/publications/archive/policy_innovations/commentary/000094.

Smits, C. C. A., van Leeuwen, J., & van Tatenhove, J. P. M. (2017). Oil and gas development in Greenland: A social license to operate, trust and legitimacy in environmental governance. *Resources Policy, 53*, 109–116. https://doi.org/10.1016/j.resourpol.2017.06.004.

Suchman, M. C. (1995). Managing legitimacy: Strategic and institutional approaches. *Academy of Management Review, 20*(3), 571–610. https://doi.org/10.5465/AMR.1995.9508080331.

Swift, A., Lemphers, N., & Casey-Lefkowitz, S. (2011). *Pipeline and tanker trouble: the impact to British Columbia's communities, rivers, and Pacific coastline from tar sands oil transport*. Edmonton, AB: Pembina Institute for Appropriate Development.

Thomson, I., & Boutilier, R. (2011). Social license to operate. In P. Darling (Ed.), *SME mining engineering handbook* (3rd ed., pp. 1779–1796). Littleton: Society for Mining Metallurgy and Exploration.

Unifor. (2014). *Jerry Dias reiterates Unifor opposition to Northern Gateway pipeline*. Unifor National. http://www.unifor.org/en/ian.boyko%40unifor.org. Accessed 22 May 2017.

Union of British Columbia Indian Chiefs. (2012). Grandiose announcement by B.C. Government on Enbridge pipeline completely unrealistic. http://www.ubcic.bc.ca/News_Releases/UBCICNews07231201.html. Accessed 30 May 2017.

Wilburn, Kathleen M and Ralph Wilburn. (2011). Acheiving Social License to Operate Using Stakeholder Theory. Journal of Business Ethics. 4(2):3–16.

Zickfeld, K. (2011). *Greenhouse gas emission and climate impacts of the Enbridge Northern Gateway pipeline*. Written evidence submitted to the JRP for Living Oceans Society, Raincoast Conservation Foundation, and ForestEthics. https://ceaa-acee.gc.ca/050/documents_staticpost/cearref_21799/83858/Greenhouse_Gas_Emission_Report.pdf. Accessed 28 May 2017.

Intercultural Business

A Culturally Sensitive Path to Achieve Sustainable Development in Indigenous Maya Communities

Francisco J. Rosado-May, Valeria B. Cuevas-Albarrán, Francisco J. Moo-Xix, Jorge Huchin Chan, and Judith Cavazos-Arroyo

Contents

Abstract

Under the premise that the lack of economic growth is preventing the achievement of sustainable development in poor indigenous communities, this paper

F. J. Rosado-May (✉)
Universidad Intercultural Maya de Quintana Roo, José Ma. Morelos, Quintana Roo, Mexico
e-mail: francisco.rosadomay@uimqroo.edu.mx

V. B. Cuevas-Albarrán · F. J. Moo-Xix
Centro Intercultural de Proyectos y Negocios, Universidad Intercultural Maya de Quintana Roo, José Ma. Morelos, Quintana Roo, Mexico
e-mail: valeria_betzabe_c@hotmail.com; francisco.mooxix@hotmail.com

J. Huchin Chan
Desarrollo Empresarial, Universidad Intercultural Maya de Quintana Roo, José Ma. Morelos, Quintana Roo, Mexico
e-mail: jore_jorge@hotmail.com

J. Cavazos-Arroyo
Business, Universidad Popular Autónoma del Estado de Puebla, Puebla, Puebla, Mexico
e-mail: judith.cavazos@upaep.mx

S. Dhiman, J. Marques (eds.), *Handbook of Engaged Sustainability*,
https://doi.org/10.1007/978-3-319-71312-0_32

analyzes, and builds upon, the cultural bases of successful and unsuccessful local businesses, and presents a conceptual framework that helps to understand the differences in the processes that indigenous Yucatec Maya and nonindigenous people follow in their business. Observation and induction, using Western terms, seem to be at the core of the cultural process of coexistence, a driving mechanism that explains the success of local indigenous businesses. When coexistence is replaced by competition, a dominant paradigm that guides nonindigenous commercial processes, the community businesses have greater probabilities to not succeed. Understanding cultural processes related to business could help to design public policy and programs that can provide the economic growth needed to achieve sustainable development in poor indigenous regions.

Keywords
Intercultural business · Sustainable development · Intercultural education · Yucatec Maya · UIMQRoo

Introduction

Soon after the Brundtland Report in Rio, 1987, Mexico fully embraced the concept of sustainable development. Although in 1983 Mexico officially created the Secretary for Urban Development and Ecology, it was not until 1988 that an environmental legislation was decreed (Ley General de Equilibrio Ecológico y la Protección al Ambiente, LGEEPA); the idea was to promote economic growth under the concept of sustainable development (Escobar Delgadillo 2007). Almost 30 years later, in 2016, with a population of 125.4 million people, Mexico's economy represents 1.67% of the world's GDP, and is considered the 15th economy (ProMexico 2017).

On the other hand, the indicators of poverty in Mexico show little or no progress for a large segment of the population. For example, the report presented online by CONEVAL (2017), the Mexican council that evaluates social development as a result of public policies, indicates that in 2010, 11.3% of the population in Mexico was in extreme poverty, and 46.1% in poverty; in 2014, the population in extreme poverty declined to 9.5% and poverty was present in 46.2% of the population. Specifically, for the state of Quintana Roo, in 2010, 34.6% of the population was in extreme poverty and 6.4% in poverty; in 2014, extreme poverty was present in 35.9% of the population and 7:0% was in poverty. The same trend is still present in recent years (CEFP 2015). Using the Human Development Index for Indigenous Population in Mexico, Schmelkes (2013) reports a value of 0.7057, as compared to 0.8304 for the nonindigenous population; the trend is similar regardless parameters of education, health, income, and jobs. Prior to the embracement of sustainable development, 1950–1990, poverty was a great challenge for development in Mexico (Lusting and Székely 1997).

According to Merchand Rojas (2011), Mexico has promoted economic growth as a strategy to achieve sustainable development, under the premise that both are not incompatible, and financial resources are needed for development. However, so far in Mexico there is still highly unequal economic growth (Rodil Marzábal and López Arévalo 2011), there are few acknowledged pockets of sustainable development (De la Rosa Leal 2014), and several praises to indigenous communities for the sustainable management of their natural resources (García Frapolli et al. 2008; Toledo et al. 2007).

In 2010, OECD presented several key policies to achieve sustainable development in Mexico. In the economy section, this report recommends to "continue strengthening competition legislation, trough Senate passage of the current reform initiative, and facilitate implementation of the proposed changes"; it also devotes a segment on competition as a driving concept for the economy (pp. 17–20), and for the inequality section recommends the "replacement of subsidies by cash transfers for the poorest," as well as "to promote rural development, strengthening collaboration among numerous actors involved" (p. 23). Achieving financial inclusion, fighting poverty in other words, has been the target of several international organizations, thus acknowledging the importance of economic growth for sustainable development. Klapper et al. (2016), on behalf of the Consultative Group to Assist the Poor (CGAP), and the United Nations Secretary General's Special Advocate for Inclusive Finance for Development (UNSGSA), present a set of priorities to articulate actions in order to achieve a vision for a sustainable world in the year 2030. Access to a bank account and digital transactions for poor people are the strategies presented for financial aid to work better than in the past. One of the examples Klapper et al. (2016) use to sustain their strategies is the work by Angelucci et al. (2015), which demonstrated for Mexico that access to microcredit increased the ability of microentreprenours to cope with risk.

In 2001, Deruyttere presented a report for the Inter-American Development Bank (IDB) showing a strong correlation between poverty and indigenous ethnicity. This was the bases to promote investment from IDB not only for reducing negative impacts of projects financed by them on indigenous territories but to include opportunities for social and economic advancement in indigenous populations. With this approach, they achieved good levels of success in areas of intercultural education, land tenure, and management of protected areas. Basically, the strategy was to articulate development with participation of indigenous stakeholders. Deruyttere (2001) also presents the concepts of intercultural economy and intercultural environmental management, which are the result of the interactions between the indigenous system with the conventional [Western] market-driven system (pp. 9, 10).

Based on the above literature reviewed, poverty is a limiting factor to both economic growth and sustainable development in Mexico; thus, social entrepreneurship and social responsibility are at risk. The examples of successful strategies to overcome poverty, mainly based on public participation in the process, are not having the desired impact. Assuming that policies and strategies for economic growth and sustainable development in indigenous communities have been based on Western thinking, not understanding the cultural differences, and thus not having the expected results, this contribution examines examples of businesses in indigenous Yucatec Maya areas, and identifies the driving concepts that explain the success in the resilient ones. When those concepts are substituted by Western thinking, the businesses have greater possibilities to fail.

Trade and Commerce in Pre-Hispanic Maya

Trade and commerce were not strange to many indigenous civilizations. The sophistication included routes, products, timing, exchange, and the like, as described in a book published by Long Towell and Attolini Lecón (2009) for the Chichimeca, Maya, and Nahua people, all inhabitants of Mexico. Hutson (2017) edited a book that provides details of commerce and trade in ancient Maya times based on experiences from the Chunchucmil community in northwest Yucatan. In the description of commerce among the Classic Maya, Hutson et al. (2017) confirmed the hypothesis established by Rathje in 1971, who challenged the paradigm held in the field of cultural ecology, that complex societies emerged only in areas with environmental heterogeneity. The civilization of the Classic Maya in the Yucatan emerged, in an apparently homogenous and resource-poor environment, because the cooperation of households, especially for resources located far away from their communities. To Hutson et al. (2017), Chunchucmil was a gateway community for much of the northern lowlands, trading different goods originated in faraway places like the Gulf coast; they claim that competition among quadrangles (community marketplace like) engendered cooperation between them and households by sharing prosperity. This view of commerce among the Maya does not follow the two-tiered model of Maya economies that has prevailed in past decades, meaning managerial models or political economy models, both representing a Western view of a non-Western society, as described by Brumfiel (1994).

Studying the commerce among the pre-Hispanic Maya, Attolini Lecón (2009) points out that once the Maya adaptation to their great biodiversity was the specialization of diverse products that were concentrated and distributed strategically. In this system, a sophisticated network of terrestrial, fluvial, and marine routes was developed; and the driving concept was the exchange of products, needed by their communities, rather than price or market.

The concept of community market place developed by the Maya, and other indigenous cultures, is still present all over Mexico. It is, for indigenous communities, the place for social and economic activities, for exchange of goods, where families could sell or exchange their products with other families, share news, ideas, or knowledge (King 2015; Silva Riquer and Escobar Ohmstede 2000). A great variety of products are available in a market place: fruits, vegetables, roots, plants, and transformation of them for food or other house needs. These products are harvested mostly in home gardens and milpas (Mariaca Méndez 2015), two agroecological systems that have provided most of the food for the Maya, and play a critical role for food sovereignty and sufficiency in the culture (Rosado-May 2012b).

Other evidences of trade and commerce in the Maya culture are in the religion realm and in the language. *Ek Chuah* (*Ek* means star, *Chuah* means black), was the God for the merchants, as well as for the travelers and the cacao (Attolini Lecón

2013). Cacao beans were once used as currency by the Maya (Vail 2009, p. 3). In present times, the words "buy" and "sell" exist in Maya, *maan* and *koonol,* and are widely used; according to Gómez Navarrete (2009), the Maya have words for trade (*koonol*), for the initiative to start a business (*chumbesik*), for exchange of goods and bartering (*k'ex, paklan, jeel*). On the other hand, it is very common to hear the word *meyaj*, meaning work, but in two contexts; one is *múuch' meyaj*, meaning community work, usually nonpaid job, and only *meyaj*, mostly meaning a paid job. Although there is no solid evidence on how long the words *maan* (to buy) and *koonol* (to sell) have been used by the Maya, whether they are of pre-Hispanic of post-Hispanic origin, it would be safe to assume that the words *maan* and *koonol* are adaptations of the language to the system of commerce brought by the Spaniards. The other words reflect cultural attitudes for trade and commerce in pre-Hispanic times.

Based on the research on pre-Hispanic trade and commerce, it is possible to argue that the Maya reached a sophisticated and efficient system that built an important economy (Storey and Widmer 2006) and use concepts such as market, initiative to commerce, trade, and bartering, among others, to name the processes that support their commerce and economy. Thus, there seem to be a contradiction between a long tradition of successful trade and commerce in pre-Hispanic Maya communities and poverty in present days for lack of successful business.

Since the words that describe those concepts are still in use, it is possible to say that the cultural bases that sustain the processes of what we now call business are still in effect and have not received sufficient attention. This assumption can explain the success or failure of local businesses and government programs aimed at increasing the economic growth as an important step to achieve sustainable development in indigenous communities.

Trade and Commerce in Modern Yucatec Maya Communities, Implications for Sustainable Development

Modern Mayas, even in poor communities, have a great diversity of goods that are used for trade and commerce. Most of it is not visible but carries quite a potential for increasing the economy of their communities. The main sources are natural, forest and lakes or cenotes, and managed ecosystems, home gardens and milpas (See Ford and Night 2010 for details about Maya milpa).

One of the first researches on the commercialization of traditional vegetables grown in home gardens was reported by Rosado-May (1985) for Maya Chontales in Tabasco, Mexico; in the dry season, with no irrigation, a total of 51 species were sold in local market. According to a review of the literature by Toledo et al. (2007), there are between 2,500 and 3,000 plant species in Maya forest of the Yucatan Peninsula, and around 900 of them are known to be used for different purposes; home gardens may be from 500 to 5,000 m2 and have between 50 to 387 species including plants and animals; the milpa system may have up to 50 plant species for human use, mostly for food. In addition, the Maya people produce honey and other products from bee keeping practices that include European and stingless bees, as well as hunting and fishing; around 24 animal species have been documented in their hunting practices. In communities with water bodies, 14 fish species have been recorded, mostly for food, and a few species of small turtles and crocodiles.

A study in a Maya community, Punta Laguna, Quintana Roo, indicates that typically 52.6% of the annual work of a farmer is devoted to produce goods for self-consumption; the remaining 47.4% is for creating products (most likely to sell), providing services (e.g., house construction), or being employed. The milpa gets about 25% of the annual work, whereas the home garden gets around 20%; both activities represent about 40% of the return flow (based on the economic value of the goods and services produced), whereas honey bee products, handcraft, services and temporal employment represent another 40% (Toledo et al. 2007; García Frapolli 2006).

The above information indicates a high level of understanding the natural and managed ecosystems by modern Yucatec Maya. It also implies risks to a system that can be considered sustainable, if the equilibrium in the appropriation of resources in both, the natural and management ecosystems, achieved over time, is disrupted. One possibility for disruption is by increasing commercial activities using their natural resources as bases for economic growth; however, those studies do not offer insights about trade and commerce and potential impacts, especially when the government programs created to promote business in indigenous areas do not consider local cultural elements. Thus, learning insights about the processes on trade and commerce in the economy of modern Maya families is critical, and it is one of the objectives of the Intercultural Business Program at the Intercultural Maya University of Quintana Roo (UIMQRoo in Spanish).

Based on the pedagogy expressed by Rosado-May and Cuevas Albarrán (2015), students and faculty have conducted research on the business processes in several Maya communities, to identify processes that validate the move from the individual to the collective approach important for achieving sustainability. The students enrolled in the Intercultural Business Program at UIMQRoo, in the years 2015 and 2016, come from 21 Maya communities. Of those, five communities representing a wide array of population, based on the 2010 census, were chosen to learn about how trade and commerce are carried out and how important they are for the economy in the life of families. The communities are José María Morelos, with 11,750 inhabitants; Tzucacab, with 9,967 inhabitants; X-Hazil Sur, with 1,422 inhabitants; Kankabchen, with 1,083 inhabitants; and Miguel Hidalgo, with 676 inhabitants; (Fig. 1). X-Hazil Sur and Miguel Hidalgo have a lagoon.

Using a participatory research technique (Bergold and Thomas 2012), the first action was a diagnosis, collecting quantitative and qualitative data on activities related to the economy of the community. This process was facilitated by the fact that the students were natives of the community surveyed. After more than a year of work, all the students and faculty gathered together and used a focal group technic (Morgan 1997) to create a general understanding of business processes for each of the Maya communities chosen.

The first objective was to determine the percentage of families/community that carry out activities related to trade or commerce, both formal and informal. Table 1

Fig. 1 Yucatec Maya area includes the states of Campeche, Yucatan, and Quintana Roo. The communities studied, represented with dots, from north to south are Tzucacab, José María Morelos, Kankabchen, Xhazil Sur, and Miguel Hidalgo (Map from Attolini Lecón 2009)

shows very interesting results: between 80% and 90% of families in those communities are actively involved in trade or commerce; not only it was a surprising high number, but between 35% and 55% of the families generate through commerce at least 40% of their income. Between 20% and 60% of families, depending on the communities, transform primary products obtained from natural or managed ecosystems in their communities, for selling. Particularly in handcrafting, from 5% to 15% of families bring external elements in important quantities to transform them into products that are sold after the transformation (e.g., the making of hammocks uses tread brought from larger cities).

Knowing that trade and commerce is widely spread in Yucatec Maya communities, the next objective was to determine the role of the natural and managed ecosystems in the economy of the families. In this case, the figures shown in Table 2 are the combination of all communities, showing the average percentage of families using, selling or transforming for sale the primary products, and the range of percentages for each case. The data indicates that both natural and managed ecosystems are still very important for modern Yucatec Maya communities. From the forest, wood is a resource that is used heavily for self-consumption, for selling or transforming and then selling (e.g., charcoal); not less than 50% of the families depend on the forest, either for meat, plants, or even raising their cattle. Home gardens and milpas are also very important, obtaining all sorts of products, including animals, plants and different parts of plants and there is a combination of timing for harvesting goods. The milpas are the main source of primary goods, such as grains, fruits, and vegetables, during the moths of May–September, whereas the home garden provides all year round. Between 2% (for honey) to 100% (for fruits) of families, use products from those managed ecosystems for self-consumption, whereas for selling the range is between 2% and 55% of the families. The range of families that transform the primary products and then sells them (e.g., corn grain into corn dough or food) goes from 0.1% (honey) to 48% (wood).

Knowing the large percentage of families conducting some activity related to commerce (Table 1) and having identified the sources of the primary products (Table 2), the next objective was to have an idea of the diversity of sources of income from transformed products traded or sold in the Yucatec Maya communities studied. Table 3 presents a list of goods resulting from transforming products from natural and managed ecosystems that are then used for trade or commerce. All the products were grouped in four categories: food, a total of 106 possibilities were recorded; household needs (e.g., firewood), a total of 11 possibilities were recorded; handcraft (e.g., clothes, hammocks), a total of 83 possibilities were

Table 1 Percentage of families with activities of trade or commerce per community. In parenthesis, under each community, the population according to the 2010 census

	Community				
Families with activities of trade or commerce	JMM (11,750)	Tzucacab (9,967)	X-Hazil Sur (1,422)	Kankabchen (1,083)	Miguel Hidalgo (676)
Total	90	80	85	90	90
T/C representing at least 40% of their income	35	55	45	30	50
T/C that transforms primary products	43	60	30	20	40
T/C that uses important amount and number of elements from outside their community	15	15	10	5	10

Table 2 Sources and use of products in five Maya communities (José María Morelos, Tzucacab, X-Hazil Sur, Kankabchen, and Miguel Hidalgo). The values indicate the average and the range of percentages, in parenthesis, of families in the five communities studied. The tendency is that the larger the figure, the smaller the community

Source	Parts	Use		
		Self	Sell/trade	Transform and Sell/ Trade
Forest	Hunting (meat, skin, teeth)	35 (30–40)	49 (15–80)	35 (20–80)
	Soil	41 (10–100)	7 (5–15)	0.7 (0.5–1)
	Wood	70 (30–95)	50 (30–85)	48 (10–90)[a]
	Vines	4 (2–7.5)	0.8 (0–2)	0.2 (0–0.5)
	Whole plant	8 (5–10)	6 (4–15)	3 (3–7)
Lagoon[b]	Fish	16 (0–50)	18 (0–60)	2 (0–10)
	Shell	6 (0–30)	0 (0–0)	0 (0–0)
	Turtles	3 (0–15)	0 (0–0)	0 (0–0)
Home garden	Animal	100 (100–100)	51 (30–85)	33 (5–80)
	Bee keeping	2 (1–3)	2 (1–3)	0.1 (0–0.5)
	Whole plants	65 (50–80)	82 (20–50)	14 (5–30)
	Leaves	85 (70–100)	34 (30–35)	19 (10–30)
	Roots	28 (10–50)	10 (5–15)	10 (5–20)
	Fruits	100 (100–100)	55 (40–80)	26 (15–50)
	Vegetables	50 (20–70)	30 (20–50)	10 (5–20)
	Seeds	8 (5–10)	5 (3–5)	5 (3–8)
Milpa	Animal	65 (30–90)	35 (20–60)	12 (5–20)
	Bee keeping	26 (20–40)	24 (20–35)	2 (1–5)
	Whole plants	6 (3–5)	3 (2–5)	2 (1–5)
	Leaves	34 (20–50)	21 (20–25)	16 (10–25)
	Roots	33 (20–40)	23 (10–30)	8 (5–15)
	Fruits	45 (40–50)	43 (30–50)	6 (5–10)
	Vegetables	53 (40–80)	41 (20–80)	7 (5–10)
	Seeds	63 (40–80)	54 (35–80)	25 (20–40)

[a]Fire wood, charcoal, and carpentry
[b]Only for X-Hazil Sur and Miguel Hidalgo

recorded; and services (e.g., use of medicinal plants for health care), a total of 90 possibilities. The information on Table 3 indicates that there is a great diversity of goods that are sold resulting from transforming primary sources obtained from natural or managed ecosystems, a total of 290, but could be more with more detailed studies or more communities studied. A transformed product has changed the original shape, look or presentation. For instance, a corn seed, original, can be transformed in tortillas, corn dough, drinks, or different types of food. In addition, people use products from somewhere else (e.g., threads, fabric, jars) to make hammocks, clothes, handcraft, or preserved food, among other goods.

Most of the selling or trading, either the original or transformed products, is informal; there are no records of them. Through participatory observation, it was

Table 3 Goods resulting from the transformation of products from natural and managed ecosystems, for trade and commerce identified in five Maya communities

Food	Household needs	Handcraft	Services
Traditional food ready to eat (at least 12 varieties)	Firewood	Clothes for adults, children, men, women	Palapa (hut) construction (at least 3 types)
Traditional beverages ready to drink (at least 9 varieties)	Vines (liana)	Shoes (for men and women, adults and children)	House furniture (at least 9 types)
Tortillas	Lumber (madera)	Hammocks (at least 10 varieties	Milpa management
Corn dough	Gourd for drinks (Jícara)	Earrings (at least 10 varieties)	Grafting (at least for 15 species)
Chili sauce (at least 20 types)	Gourd for tortillas (lec)	Necklaces (at least 10 varieties)	Weeding
Seasonings (at least 10 varieties)	Scourer	House ornaments (at least 10 varieties)	Health (use of at least 35 species of medicinal plants)
Annato paste (colorant) (at least 5 varieties)	Charcoal	Religious symbols (at least 10 varieties)	Massages (for pregnancy, muscle pain, stress)
Snack (at least 10 varieties)	Water container	Kitchen and bathroom accessories (at least 10 varieties	Delivering babies
Dried and salted meat (at least 8 varieties)	Broom	Wooden figures (at least 10 varieties)	Therapeutic cleansing (using at least 3 types)
Sweets (at least 20 varieties)	Baskets	Dream catchers (at least 5 varieties)	Chiropractic (for at least 8 different injuries)
Preserved food, fruits and vegetables (at least 10 varieties)	Kitchen utensils		House cleaning
			Baby sitting
			Translation Maya-Spanish-Maya
			Escort for shopping
			Masonry
			Praying (at least for 5 diverse ceremonies)
$\sum$ 106	$\sum$ 11	$\sum$ 83	$\sum$ 90

possible to determine that the transformed products, from raw meat, or fruits, or leaves, or roots, or any of the parts produced in the milpas or home gardens, or collected in the forest, are in the form of edible food, sauces, seasoning products, colorants, ingredients for food, handcraft, or services.

Table 4 Fluctuation of time, minimum-maximum number of hours, devoted per day by each family member to activities related to business. Values reflect all five communities

Business	Husband	Wife	Children
Food	0–4	1–5	0–1
Household needs (e.g., furniture)	0–3	0–2	0–2
Handcraft	0–3	1–4	0–2
Services	0–6	1–5	0–3

This information and the data from Punta Laguna, presented earlier, seem to support Toledo (2003), regarding the appropriation of nature, in which the household articulates several activities carried out by the farmer and his/her family, for both self-consumption as well as for trade and commerce to meet family's needs. This point of view demands decisions not only regarding a sustainable management of their eco- and agro-eco-systems, but also the use of time, the organization of activities and participation of family members, the adaptations to changes around the community, and participating in community endeavors, including religious, political, and social activities. All of them reflect cultural values which in turn rest on concepts that need to be identified and described. This is the context in which trading and commerce are happening in Maya communities.

It is important to mention that the use of family as the unit of study in a community resulted not only from literature like Toledo (2003), and Morrison et al. (2014), but also based on observations and opinions from the students participating in the research. Thus, the use and distribution of time among family members, in activities related to transforming primary products to trade or sell, became the next objective. On Table 3 four groups of transformed products were identified; Table 4 shows the range of time devoted by members of the family, husband, wife, and children for production of food, household needs, handcraft or providing services, all for transforming primary products from natural or managed ecosystems. In general, the wife dedicates more time to each of the four groups; the husband does not allocate as much time because of his activities in the milpa or the forest or community services. Because of schooling, the children also participate a few hours in each of the four groups. Generally speaking, there is a family consent on the distribution of time and activities. Apparently, families that reach some kind of equilibrium in their distribution of time and activities are more able to succeed in selling their products.

The above information indicates that the modern Maya do have an economical system as a framework for their activities and carry out activities related to trade and commerce. Whether that framework facilitates or inhibits their potential for commerce and business, to promote economic growth and sustainable development, needs to be examined considering cultural processes. This is a very important aspect to consider, usually overlooked; if the programs aiming at promoting economic growth to achieve sustainable development are implemented in areas where the culture is different than the one from which the program was originated, then the possibility for success is very limited. Cultural differences lead to epistemological differences, which can explain lack of success of centralized policies designed to promote sustainable development, as demonstrated by Haenn (1999) and García-Frapolli et al. (2009), in the field of natural resources conservation.

In other words, trade and commerce, considered business in modern world, do exist in Maya communities, but under different processes and can be important component to achieve sustainability in the area. This paper claims that the potential is there for economic growth and sustainable development, but the cultural processes need to be better understood. This statement seems to contradict the present conditions of poverty and lack of economic growth that prevent sustainable development in Maya communities. How do they work? How are they different from conventional businesses? How to explain the failures of business in Maya communities? How do successful businesses scale up? How are business innovation and social innovation incorporated in successful business? What can we learn from them and propose changes to government policies regarding sustainable development? This paper contributes to answer the above questions from a cultural perspective. The information presented in the previous section needs to be interpreted and complemented with local cultural bases. To do so, an intercultural approach was used to understand local cultural processes involved in trade and commerce in Yucatec Maya communities. The intercultural approach, developed at UIMQRoo, is described in the following section. Based on this approach, information about the processes behind modern trade and commerce in Maya communities is presented to identify the driving concepts of commerce in these communities, which are rooted in their culture, and can be the bases for economic growth to achieve sustainable development.

The Intercultural Model for Higher Education in Quintana Roo, Mexico, and the Intercultural Business Program

Mexico embraced the intercultural model for higher education based on the reform of Art. 2 of the Constitution, in 2001. The first intercultural university was created in 2004, in the state of Mexico; nowadays 11 states in Mexico, out of 32, have an intercultural public university. UIMQRoo was created in October 2006, but started academic activities in August 2007 with the goal to contribute significantly to the sustainable development of the Maya region. To do so, UIMQRoo developed a specific intercultural model for the Yucatec Maya based on the following working definition: "Intercultural education is the result of a process in which different systems of creating knowledge (local and western, e.g., scientific method) coexist in a safe environment, thus creating conditions for synergy and new knowledge that articulates the views and understanding of the cultures involved, to understand the same phenomena/issue" (Rosado-May 2015, p. 130). In 2012 UIMQRoo opened the Intercultural Business Program and offered a BA degree. One of the arguments presented to support the opening of this program was that the region had no encouraging signs of overcoming poverty, and another argument was that without proper economic growth there will be no opportunity to achieve sustainable development; thus, there should be an opportunity to develop a new approach and paradigms on how to conduct successful business considering the cultural foundations of the communities in the region. These arguments assume that the indigenous Maya people are able to run business successfully, fight poverty, and increase their possibilities to achieve sustainable development, only if there is good understanding on how the cultural system, that supported successful commerce in the past and still present nowadays, can be incorporated in the training of future business people, intercultural business people.

For the intercultural business program, some of the most important components of the academic model are described next (see Rosado-May and Cuevas Albarrán 2015, for details). The students are selected mostly based on their interest on commerce, business, not necessarily using the results of the national admission test. To do so, the students take, in addition to the national test, a vocational test mostly designed to identify their ability to successfully face challenges. Also, whenever needed, the opinion of faculty in the program and leaders from the community where the student comes from are considered for admission. Although most of the students at UIMQRoo are from Maya communities, because it is located in a heavily populated indigenous area, the university is also open to students from other indigenous cultures or nonindigenous people as well.

Once the student is enrolled, and starts school (August) he/she must decide who from his/her community will serve as voluntary tutor (in Maya is called *nool iknal*, an elder who provides advice and guidance to young people). The university also assigns him/her an academic tutor. Thus, the student enters a system called *iknal* in Maya, which the Maya have used for centuries to provide excellent conditions for young people to learn and construct knowledge (for more details on this system see Rosado-May 2012a, 2017). The first academic year the students take courses designed to help them overcome issues that can limit their academic performance, and to help them to boost their self-confidence in their own culture, language, and knowledge system. They also have conditions to create, with other students from other communities, a network of support and problem solving, which substitutes the one he/she had built in previous years.

In 4 years, the student is trained to obtain the same competencies that students from other universities have, in business programs, but in addition he/she strengthens his/her abilities, competences, and knowledge on his/her own language and culture, which provides an added value that graduates from other schools do not have.

To receive the BA diploma in intercultural business, in addition to successfully complete all courses, a student must present and defend before an academic committee a written final report which is built upon the following steps (Rosado-May and Cuevas Albarrán 2015): (1) At the end of the first school year, in May, every student enrolls in a summer program, with credits, that takes place in his/her own community (June–July). He/she goes back full time with the objective to carry out a diagnosis on issues related to the business major. Each year a diagnosis is led by the academic program and sets goals with the purpose of collecting specific information, updating existing ones, or evaluating the result of programs financed mostly by the government. The students apply diagnosis techniques previously learned in class.

Each student reports to a faculty assigned to a specific community, as well as to his/her *nool iknal*. (2) At the end of the second school year, every student goes back full time with his/her community, enrolled in the second summer program, with credits, with the objective of defining a project to carry out with the consent of the community. At the end of the summer the student must have decided, with the support of the community, a project that must be developed and presented to different agencies for finance. In this process, the student has the support of a faculty who could be, later, the head of his/her graduation committee. The topic of the project must be the result of the interest of the student, the interest of the community and the possibility to obtain financing. This is very challenging but very important for the student, he/she is getting trained in real aspects of life, applying his/her knowledge from the program and meeting the expectations of his/her community with university standards. Not easy at all, but critical for the success of the intercultural model. (3) During the 3rd and 4th year each student has the time to carry out the project, evaluate results and present a report to both, the community and the academic committee. The students have opportunities to start working on his/her project not only until the 3rd summer; they can start earlier. The class schedule, during school term (August–May), is designed to have Fridays free to work on the project, and maintain contact with their families and communities. The final report will be defended in a public event. If the student has completed successfully all courses load and defended successfully the final project, then a BA diploma is granted by the university.

The process described in the previous paragraph sets conditions for an intercultural process to happen. In the classroom, the students are exposed to knowledge and ideas from other cultural sources, different than the local one, and at the same time re-learn and strengthen his/her own. It is a process in which a lot of internal and external discussion happens. There is a re-connection, a re-encountering, a re-validation of the local knowledge; there is an inevitable comparison of the local knowledge with the external one, not necessarily to decide which one is best but paving the road to a new understanding of processes that had been hiding. For instance, recognizing the important cultural elements that explain attitudes and reactions to external knowledge and policies and decisions on how to overcome critical issues such as poverty, and thinking of alternatives that are not culturally biased represents a critical step in understanding complex social processes.

Having students working in their own community in a well-structured, but flexible process, trained on techniques like participatory action research, the intercultural business program has been able to obtain basic but critical information on the ways that trade and commerce, meaning business for the non-Maya, that have shed light on the cultural bases that support their system. This describes a complete process behind the exercises from which the data for Tables 1, 2, 3, and 4 were collected in five communities, José María Morelos, Tzucacab, X-Hazil sur, Kankabchen, and Miguel Hidalgo, that were described in the previous section.

The information described provides important insight about the context in which business, trade, and commerce is carried out in modern Maya communities. Important elements that can be considered culturally based identified thus far are the following: All family members are involved in all activities; diversification seems to be an important component of the strategy, both in terms of producing primary goods as well as their transformation for commerce; there is a well-designed schedule of activities per day and season.

To identify driving cultural concepts behind the commerce/business, interviews were conducted with people running business that are locally considered successful for at least 3 years, and people who have failed in business. Some business did not have government support and others did. The students interviewed the stakeholders in different moments, using an open format, understanding as much as possible of the whole process (e.g., the motivation/inspiration to start a business, reflections on the cultural conditions that helped to succeed or to fail, strategies for scaling up, strategies for innovating, and vision for the future) and emphasizing the cultural elements of the success or failure, including analysis of the language used in the description and evaluation of the process. The students then exchange their findings with other students from the same and different communities, and with faculty.

Case Studies

Five case studies were chosen that illustrate the cultural bases that explain successful entrepreneurship in Yucatec Maya communities. A brief description is provided below. Each case had to overcome several hurdles from both the government/institutional and the community settings. By making visible the positive elements of their success, culturally speaking, it is expected that public policy could be adjusted and thus create conditions for economic growth much needed to achieve sustainable development in indigenous areas.

(a) JFCM, and his family, owns a chili sauce business, in Tzucacab, Yucatan, that has been in the local market for over 5 years.

> Since I was a kid I liked to watch with lots of attention how my father and my mother used to sell things to other people. My mother was known for making a delicious chili sauce. Before the chili sauce business started, I had a food stand where I used my own version of the sauce. It did not take long for people to ask extra sauce to take home, then I decided to try selling it in small bottles. People liked it and gave me suggestions. In the beginning, we use plastic bags to give the sauce to people, then we use small jars. Although I based the recipe on the original from my mother, over the years I learned some tricks to make the sauce tastier. The sauce was a success, people who knew me started recommending the sauce to other people. With time, I tried using different containers to make the product more attractive and I talked to local business to have them selling my sauce. It seems to be working fine. My wife and I run the business, sometimes we get help from other people, 2 or 3, especially when some of us is ill or I am traveling, but we know those people, we trust them. We want it to be a family business, not to grow much, not because if we grow we need to hire accountant, deal with other retailers in other cities, but because that is how we feel our business should be, a family tradition. Knowing our customers, talking to them, and them trusting that we deliver a fine product is probably our secret to succeed. The original recipe has changed a little, but before we make a full production, we let our customers know and

have them taste the new product, if they approve it, then we make the move for changing the whole production, a new generation. Nowadays I write down every formula for my children; in the future, they will decide what to do with the business. In the meantime, I encourage them to keep it in such a way that is family controlled, earning the confidence of our customers and getting to know as many as possible of them. Yes, I have been tempted to bring my product to Merida and other big cities, I go slow, take one step at the time, learn from customers and improve my product. I also have started other businesses; the idea is to diversify and have them connected. Here, in our community, I am not the only one selling chili sauce, there are at least three or four more, some of them from other cities; but so far, I do not have any problem selling my product, and I do not expect any in the near future, there seem to be enough room for all of us, none of us feel threated by the others.

(b) FJMX graduated at UIMQRoo and earned a master's degree in Business in a conventional technological institute. He owns a couple of businesses in Tzucacab, Yucatan, in which he is applying an intercultural approach successfully.

Now that I understand more broadly the concept of business and learning about the role of culture in commerce, I realized that business runs in my family for at least two generations. My mother still sells food and my father has tried several business avenues. When I was a kid I was not sure what would I be in my adulthood. I used to help my family in different ways, mostly related to their business activities. I listened carefully how the conversation for each sale went and how my father and the customer agreed on a price for the product, only words no papers. Some of the goods were produced by my father and others he bought from somebody else. My father had other businesses, like selling food, he did not fully rely on his milpa. My mother and my brothers used to help in the milpa, in the house and in preparing the things that my father would sell. I tried different ways to make money once I grew up, one of the activities was selling chicken, but I wanted to have a more formal business. By observing the context, talking and listening to people about the activity of selling chicken, I realized that with patience I could have my own farm business selling chicken or turkey. I figured out how to introduce the product understanding the culture of my community. Today I run two business, with the help of close friends and relatives, producing turkey and watermelon. In both cases I combine my learnings from conventional education and local Maya knowledge. It is working well and looks strong for the next several years. It took me a lot of thinking to realize the difference between competition and coexistence. Competition is quite a strong driving concept in business that is hard not to relate to it; however, most successful businesses in Maya communities do not even think about competition. The word competition does not exist in the Maya language. People think about the well-being of others in their community when doing what we call business. In Maya, we use the word múuch' meyaj, which means working, collaborating, together, and applies to all activities in the community.

(c) Several years ago, in 2001, in X-Hazil Sur, a Maya community located in the municipality of Felipe Carrillo Puerto, Quintana Roo, the state government promoted economic growth through the implementation of greenhouses to produce some chili species, tomatoes, coriander, and cucumber. The community accepted the government proposal as well as technical and financial support. Following is an account by RWPC, a native of X-Hazil and student of the intercultural business program.

Upon acceptance of the government's proposal, the community organized three groups, one of 14 women, and the other two of six men each. Each group had three large greenhouse structures. After less than two years one of the men's group basically quit, the other one was reduced to two members and the women's group was reduced to four members. The main reason for quitting were the high demand of work, very little earnings in the beginning, lack of technical support by the government, and, I'd say, not enough patience. After 16 years of work the remaining four women who decided to move on and make the project work, have been very successful. Their production is considered one of the best in the region. In the beginning, they sold their products to a state company, but now they sell to international retail stores. They explain their success basically because they knew that new projects require time and hard work to succeed, to be patient, something they learned from their parents. They used ways they learned from childhood to discuss and solve all kind of problems and make sound and collective decisions, including dealing with government requirements, community conflicts, technical skills for managing crops in greenhouses, accounting, personnel management and not making feel other community members that they were smarter or not appreciating the effort by others in the community to produce the same species. Talking to each other in an open and direct way, and willing to understand and to present solid arguments, helped greatly the process of understanding and solving many issues related to organization and management, because they trust each other. None of the women feel to be the boss or employee. In planting and harvesting season, they incorporated family members or close friends, trying not to treat them as employees but family. They also had to learn how to sell under contract to their big customers, not only using words. When their production is not enough to cover the expected demand, from areas such as Cancun or Playa del Carmen, big tourist resort areas, they talked to other producers in the community and treated them as equal as possible, sharing the market of their products. They had a long-term vision for their work, but not emphasizing the accumulation of capital but to satisfy their needs and provide well-being to their families and other community members.

(d) In 2013, the federal government approached the community of Miguel Hidalgo, Quintana Roo, to offer financial and technical support to those women interested in bee keeping for producing honey, candies, and marmalades. Following is an account by GAM, a native of Miguel Hidalgo and student of the intercultural business program.

After listening to the government representatives, several women showed interest in working together; voluntarily 10 of them decided to be part of the organization. Only a few of them had some experience in bee keeping, but the government offered training, technical and financial support. They had to start from nothing, preparing the field, which was located way outside of the community; learning, and not getting income from the project for several months. Before a year went by, a few women quitted the project, and just after the first year, many more did, frustrated for having to work very hard with basically no income. There was a moment in which only three of the original 10 were still working on the project, my mother, my aunt and my mother's sister in law; that is when they thought of incorporating myself and a cousin, because the government agency demanded that the group should be at least with five women. When I joined the group, I had some years working on my BA in intercultural business. First I decided to learn, by listening, observing and participating for as much as I could during the days that I did not have to attend school, which is located about three hours away. At the university, I brought questions to understand business processes and I read about conventional business. I notice that there was something not very clear, the books would mention strategies, methods and concepts that were not present in the women's organization. For instance, there was no formal structure (boss-employees), the decision making was participatory, they were willing to take risks and decisions were made based on trust and instinct. They set priorities based on some kind of vision that they had and shared, first in making sure to learn about bee keeping, about handling the products, about transforming fruits into candies, and so on. This required a lot of patience, hard work, and commitment, all of which my mother and aunts had to develop when they were kids, working with their families. My cousin and I were also learning about those skills along with the technical part of bee keeping. Their strategies worked; today, a couple of years after I joined the group, we have three sites with 50 bee colonies each; in the beginning, they had 30 in only one site. They started selling the honey, almost door to door, looking for people to get to know them and trust their product, today they distribute honey and candies to not less than five communities. Business is doing well, people in my community acknowledge the hard work of my mother's group and respect them; the original women now wish to have another opportunity to develop their own business. With my studies at UIMQRoo, I have been contributing with ideas on how to expand, although slowly, geographically, and how to meet the criteria set by national organizations to register a mark related to our products, and the like. It seems to me that understanding different cultures we can figure out how best to successfully do business without imposing one upon the other.

(e) Almost 40 years ago, a Maya mother started a small business in Felipe Carrillo Puerto, Quintana Roo, using coconut as the main ingredient to make empanadas for dessert. More and more people liked it, so the business became very important for the economy of the family. Years later, her older son moved to Dziuché and started a similar business, which has been around for not less than 25 years. Following is an account by LMY, a native of Dziuché, Quintana Roo, and student of the intercultural business program.

When I had to decide a project for my graduation, I chose the coconut empanada business that a close friend's family had been doing for several years in my hometown. I was curious to know how the system worked and how to make it better. At first I assumed that the business had no structure, no logic, and could grow much more and have greater earnings in a short period of time. My academic adviser cautioned me about this thinking and guided me to learn more about my own Maya culture. The reasoning that resonated in me was that if the business has been around for so long there is got to be an explanation which not necessarily follows the same logic that conventional businesses have. It is not because the empanada business has a well-structured system, nor it is because the brand is trademarked, nor it is because they are the best competitors in the region, nor it is because they have great facilities and effective marketing trough ads in papers, TV or online. I started to understand that the business runs smoothly because the family is well integrated and articulated, there is no need for a boss, everyone knows what to do, when to do it and how to do it so that things flow naturally; everybody relies on the other family member to do the actions that he/she should do in proper ways. The product is only sold in small community stores, not big business. This is not because the product does not have all the needed requirements to be displayed in big stores but because selling the product in community stores allows a closer contact with customers and thus learning about how to improve the product. The business has grown over the years but in a way that is not highly visible, stores from 21 communities now sell the products, without a contract. Not bad. After several interviews and participation in the process, I am learning that many aspects of the business, like its organization, management, quality and efficiency, are present but adapted to the local culture. There is no sense of competition, but to belong to the community; there is no vertical decision making, but more horizontal; there is no employee-boss relationship, but family and friendship ties; there is a system of improving the quality of the product by keeping close contact with customers, there is no need for advertisement trough regular media but the trust and mouth to mouth opinions about the products and relying on the community reputation of the family that runs the business. The business in Carrillo Puerto did not do as well as the one in Dziuché, they tried to go big and had the support from the government. I think that because they tried to compete with similar business they rushed their organization and production system but they were not prepared to fully leave behind a lot of their cultural principles; so, they failed. Based on this case, and other that I have heard from my classmates from other communities, I can perceive that there is room for improvement local indigenous business, using modern, western, concepts. Indeed, but I am still learning how to apply them without disrupting the traditional business system that has proven to be highly resilient.

Developing a Conceptual Framework

The construction of conceptual frameworks to incorporate businesses into sustainable development is not new; for instance, Schaltegger et al. (2011 p. 17) show interrelations between generic business model pillars (value proposition, customer relationships, business infrastructure, and financial aspects) and core drivers of business cases for sustainability (costs and costs reduction, sales and profit margin, risk and risk reduction, reputation and brand value); all in the context of market competition (p. 6). Competition is a powerful driving concept for businesses and development, according to a classic author (Porter 1990), and it is the bases for public policies (OECD 2010). But competition is not important in indigenous business systems. The findings by Morrison et al. (2014), in a report prepared for the Australian Research Council and Indigenous Business Australia, aiming at determining the factors influencing the success of private and community-owned indigenous business, suggest a new model of indigenous business development that involves helping early start indigenous business improve a range of business practices through in-depth skill development, mentoring and network engagement. They identify that indigenous enterprises play a critical role in indigenous economic, social, and cultural development; culture, networks, women, and mentorship play a critical role in business, in the context of family and social values. Indigenous business owners in Australia consider multiple criteria when measuring the success of their business venture, including conventional measures such as profitability and growth of business, self-satisfaction elements such as life-style and recognition, and contributions to the community (p. 142).

Based on the empirical information, quantitative and qualitative, presented in previous sections, and literature reviewed, there are elements to propose an initial conceptual framework to understand the differences between a business run successfully in Yucatec Maya communities as compared to conventional ones. The premise behind this exercise is that by understanding how successful business in

Table 5 Differences between successful and not successful business in Maya communities

Parameter	Business	
	Successful	Not successful
Labor	Family or close friends	Hired formally
Motivation	Role in the community	Increase wealth
Role of observation	Very strong, comprehensive, holistic	Not relevant, rely on business plan
Scaling up, increasing capital	Slowly, only if there is no disruption in the production system and the vision	As soon as possible Accumulation of capital
Cultural aspects	Cooperation, coexistence	Competition, exclusion
Vision for the future	Still family, small, community oriented	Selling in other markets, going international if possible
Time and schedule management	Diversified and distributed in other important activities, some related and some not related to the business	As much time, as possible, for the business

indigenous areas work, policies and programs aiming at sustainable development would have better chances to succeed.

The first step toward the conceptual framework was to identify parameters that both successful and not successful indigenous business share, based on the interviews. Table 5 shows seven parameters: labor, motivation, role of observation (as described by Rogoff 2014; Rosado-May 2017), scaling up and increasing capital, cultural aspects based on community values, vision for the future, and time and schedule management. The contrast between successful and not successful indigenous business are very noticeable. The successful businesses have similar features like the ones described, for example, by Dana (2007), Morrison et al. (2014), Schaltegger et al. (2011); for instance, whereas family and community values play a critical role for successful businesses, employing people and competition are for the nonsuccessful ones.

The second step was to identify the stages of indigenous business similar to the conventional ones. Three stages were identified: cultural principles that drive the process, operationalization, and innovation. The operationalization is divided in two parts: one relates to the system implemented for production, and the other is related to sales and marketing. The cultural principles, the foundations and drivers of the processes of any decision making, are different. Whereas coexistence is more important for indigenous business, competition is for conventional business; whereas community values and well-being are in the core of cultural principles for indigenous business, individual and accumulation of capital are for conventional business. The production system relies on family and close friends, and thus there is no need for a formal structure, for the indigenous business; whereas for the conventional ones, there is a relationship between boss and employees relationship that requires a formal, pyramidal structure. Patience and agreement based on words are

Table 6 Conceptual framework to understand the different processes that lead to successful business in indigenous Yucatec Maya and nonindigenous context

Stage	Indigenous	Nonindigenous
Innovation	Context	Context
	Slow process	Continuous process
	Trial and accumulation of experience	Scientific research
Operationalization, sales, marketing	Knowing personally the customers	Not relevant
	Trust	Branding
	Community acknowledgment	Not relevant
	Incorporates direct opinion from customers	Market analysis
	Trial and accumulation of experience	Scientific research
Operationalization, production system	Family/close friends	Employees
	No formal structure	Formal structure
	Natural flow mostly horizontal	Vertical flow
	Community is the basis for mission/vision	Mission/vision defined based on business and competition
	Trial and accumulation of experience	Scientific research
Cultural principle	Coexistence	Competition
	Observation	Strategic planning
	Community values	Individual/organizational values
	Well-being	Accumulation of capital and profitability
	Trial and accumulation of experience	Scientific research
	Patience, long-term expectation for results	Short-term expectation for results
	Agreements based on words	Agreements based on documents

other cultural principles present in Maya traditions contrasting with short-term expectation for returning the investment and signed contracts present in conventional business. As for marketing and sales, the indigenous business relies on trust, confidence, and knowing almost personally their customers; the conventional business rests on market analysis and the reputation of their brand. Innovation is present in both business systems; for the indigenous one, innovation is important but a slower process, whereas the conventional business rests on a continuous and dynamic process to keep its competitiveness (Table 6).

This conceptual framework offers sound explanations to nonsuccessful indigenous business. Based on the accounts from business owners and studies carried out by members of the intercultural business program at UIMQRoo, those indigenous people interested in developing a business based on principles not related to their culture face greater probabilities to fail. For instance, the idea of accumulating capital in short period of time brings along the need for implementing a system based on loans, employees, formal structure, accounting, etc., all of which most likely the business-interested person would not be prepared for. The same explanation can apply to government programs that encourage business in indigenous communities without considering the cultural context, imposing cultural principles that are not related to the local people, or not being flexible enough to carry out adaptations as required by the local people.

Discussion

This contribution presents a context that supports the assumption that sustainable development in indigenous areas cannot be achieved due to poverty, despite investment from the state and federal government to encourage decreation of business. Another assumption confirmed is that the Yucatec Maya traditional food systems are great sources for businesses. One explanation is that the cultural principles of the community might be in contradiction with the pillars of the types of businesses supported by the government. Based on quantitative and qualitative findings, successful and not successful businesses, ran by indigenous people, have been identified along with the processes that explain their performance. The findings suggest that businesses that work based on cultural principles developed by the Maya have greater possibilities to succeed; whereas those that rest on principles from conventional business reduce greatly their possibilities to succeed. A conceptual framework is constructed upon those findings designed to understand the differences in the processes driven by cultural principles, not for demonstrating which one is better than the other.

Presenting a conceptual framework to help understand the similarities and differences between the Yucatec Maya ways of conducting business and the conventional ways should not be interpreted that one is better than the other. It rather provides an opportunity to understand processes that can help to make better and sound decisions. Observation and induction, using Western terms, seem to be at the core of the cultural process of coexistence, a driving mechanism that explains the success of local indigenous businesses. When coexistence is replaced by competition, a dominant paradigm that guides nonindigenous commercial processes, the community businesses have greater probabilities to not succeed.

There is no question about the importance of business in sustainable development. The book on business strategies for sustainable development, published by the International Institute for Sustainable Development (IISD) in 1992, has encouraged research and education along those lines. The role of culture in business has received attention as well; the works by Schaltegger et al. (2011) and Morrison et al. (2014) are examples of changes, adaptations, and inclusion of the cultural dimension in business. On the other hand, the role of culture in sustainable development has also been investigated (Dessein et al. 2015), including the economic dimension and acknowledging that cultures offer possibilities of business in the areas of products, services, tourism, and place branding (p. 43). Another influential book, by Trompenaars and Hampden-Turner (2012), attempts to articulate business with culture and provides solid arguments grounded on a premise that it is never possible to understand other cultures (p. 1). In those examples the cognitive aspect of indigenous cultures, to identify the cultural principles that explain the processes in successful businesses, is not addressed fully.

This work indicates that we can understand at least the most important cultural principles that reflect cognitive aspects that explain processes in business carried out in different cultures. So many years of programs and public policies failing, not reducing poverty and thus making it almost impossible to achieve sustainable development, especially in indigenous areas, should at least open room for new ideas and research to set new understanding of those issues.

The cognitive and cultural aspects of business processes related to indigenous people are not new in the literature. For instance, assuming entrepreneurship as a function of opportunity is an ethnocentric approach, for the Canadian sub-Artic people; Dana (2007) demonstrates that the causal variable behind enterprise is not an opportunity, but rather one's cultural perception of opportunity. But for the modern Yucatec Maya, this is perhaps the first detailed report.

One of the most common comments by a foreigner, with university education, when visiting any Maya community is that too many people try to do the same kind of business. It is true. For instance, anybody going to the market place in José María Morelos (Quintana Roo) will notice that there are between five to eight people selling the same kind of food (e.g., salbutes, panuchos, and tamales – local names), and they had been there for not less than 5 years, doing the same thing. The foreigner might also notice that there are not many people buying that food, and could mention that if any of those people selling the same food make a different food or try to sell something else, they would have better chances for successful business. The foreigner's understanding of the situation is influenced by his/her culture, just like the activities of the five-eight people selling the same food are reflecting their culture. If the foreigner would have time and observe for a longer period, than just a short visit, he/she might notice that all five to eight business are doing enough to cover their needs and that they are not trying to compete but to coexist; eventually the foreigner could understand the cultural processes that explain how those local businesses determine when the number of eight sellers be increased or reduced and under what process of decision making.

Publications on intercultural business focus mainly on communication; for instance, Trompenaars and Woolliams (2003) based their book on the idea that knowledges of cultures are not enough for successful business across cultures, there is a need of knowledge for cultures. On the other hand, there is the work by Möller and Svahn (2004) who acknowledge the influence of ethnic culture on knowledge sharing in different types of intercultural business nets; one of the frameworks that they presented is based on the acknowledgment of a culture-classification scheme that identifies the individualism-collectivism and the vertical-horizontal dimension.

The works by Trompenaars and Woolliams (2003), Möller and Svahn (2004), and others in the literature on intercultural business base their analysis on the same cognitive platform regarding the understanding of business. In other words, the idea is how to understand and communicate between different cultures so that the basic cultural drivers of conventional business can be successful, regardless the culture in which the business is developed. There is nothing wrong with that approach. But there is another approach to expand the understanding of intercultural businesses, especially when it comes to indigenous people: understanding and encouraging the local cultural cognitive processes, which function as drivers, offers greater possibilities for business to succeed in indigenous areas. Education plays an important role to achieve sustainable development, but the educational model is much more important as presented in this contribution.

Hopefully the thoughts expressed in this contribution will find their way to influence public policies and programs to overcome poverty in indigenous areas and, thus, get closer to achieving sustainability in the region.

Cross-References

- Expanding Sustainable Business Education Beyond Business Schools
- Moving Forward with Social Responsibility
- People, Planet, and Profit
- Social Entrepreneurship
- Sustainable Higher Education Teaching Approaches

Acknowledgments The development of the field of intercultural business for the Intercultural Maya University of Quintana Roo (UIMQRoo), and thus for the students, mostly Maya, to have methodologies and concepts that help their training and increase their possibilities to succeed in business, has been with the support of the following organizations and people. The Kellogg Foundation who has funded different activities through the project "Training and research for young local college students to design and implement entrepreneurial activities that are socially and culturally sensible, while also environmentally sound"; Deborah Schimberg and Verve Inc., who funded a program with UIMQRoo that helps students get their first funding to develop businesses related to using their natural resources; and Steve Bell with whom the first author has discussed important issues that large international corporations face when it comes to the point of understanding the local culture, and thus paving the road for increasing their possibilities for success. The authors are grateful to Hilario Poot Cahun, for his support in the translation of Maya words, and to the students, juniors and seniors 2017, and faculty of the Intercultural Business Program at UIMQRoo for contributing with data, information and discussing important key aspects of this paper.

References

Angelucci, M., Karlan, D., & Zinman, J. (2015). Microcredit impacts: Evidence from a randomized mcrocredit program placement experiment by Compartamos banco. *American Economic Journal: Applied Economics, 7*(1), 151–182.

Attolini Lecón, A. (2009). Intercambio y caminos en el mundo maya prehispánico. In J. Long Towell & A. Attolini Lecón (Eds.), *Caminos y mercados de México*. México: Universidad Nacional Autónoma de México e Instituto Nacional de Antropología e Historia. Serie Historia General.

Attolini Lecón, A. (2013). Los placeres del paladar. Los caminos de las mercaderías entre los Mayas prehispánicos. *Arqueología Mexicana, 122*, 48–53.

Bergold, J., & Thomas, S. (2012). Participatory research methods: A methodological approach in motion. *Forum: Qualitative Social Research, 13*(1), Art. 30. http://www.qualitative-research.net/index.php/fqs/article/view/1801/3334. Accessed 30 Apr 2017.

Brumfiel, E. M. (1994). Factional development and political competition in the new world: An introduction. In E. M. Brumfiel & J. W. Fox (Eds.), *Factional competition and political development in the new world*. New York: Cambridge University Press.

CEFP. (2015). La pobreza y el gasto social en México. In *Centro de Estudios de las Finanzas Públicas*. México: Cámara de Diputados, LXIII Legislatura. http://www.cefp.gob.mx/publicaciones/presentaciones/2015/precefp0042015.pdf. Accessed 27 April 2017.

CONEVAL. (2017). Consejo Nacional para la Evaluación de la Política de Desarrollo Social. México. http://www.coneval.org.mx/Medicion/MP/Paginas/AE_pobreza_2014.aspx and http://www.coneval.org.mx/Medicion/EDP/Paginas/Evolucion-de-las-dimensiones-de-la-pobreza-1990-2014-.aspx Accessed 10 Apr 2017.

Dana, L. P. (2007). A comparison of indigenous and non-indigenous enterprise in the Canadian sub-Artic. *International Journal of Business Performance Management, 9*(3), 278–286.

De la Rosa Leal, M. E. (Ed.). (2014). *Retos y oportunidades del desarrollo sustentable y la responsabilidad social*. Aguascalientes: Universidad de Sonora, Universidad Autónoma de Aguascalientes y Consorcio de Universidades Mexicanas.

Deruyttere, A. (2001). Pueblos indígenas, globalización y desarrollo con identidad: algunas reflexiones de estrategia. Informe de trabajo para el Banco Interamericano de Desarrollo (IDB in English). http://www.bvsde.paho.org/bvsacd/cd27/puin2.pdf Accessed 27 Mar 2017.

Dessein, J., Soimi, K., Fairclough, G., & Horlings, L. (Eds.) (2015). Culture in, for, and as sustainable development. Conclusion from the COST action IS1007. Investigating cultural sustainability. University of Jyväskylä, Finland, and the European Cooperation in Science in Technology. https://www.culturalsustainability.eu/conclusions.pdf. Accessed 28 Mar 2017.

Escobar Delgadillo, J. L. (2007). El desarrollo sustentable en México (1980–2007). *Revista Digital Universitaria, 9*(3), 3–13. http://www.dgespe.sep.gob.mx/public/rc/programas/material/el_desarrollo_sustentable_en_mexico.pdf. Accessed 28 Mar 2017.

Ford, A., & Night, R. (2010). The milpa cycle and the making of the Maya forest garden. *Research Reports in Belizean Archaeology, 7*, 183–190.

García Frapolli, E. (2006). Conservation from below: Socioecological systems in natural protected areas in the Yucatan peninsula, Mexico. Chapters 5 & 6. Tesis Doctoral, Instituto de Ciencia y Tecnología Ambiental, Universidad Autónoma de Barcelona, Bellaterra, España.

García Frapolli, E., Toledo, V. M., & Martínez-Alier, J. (2008). Apropiación de la naturaleza por una comunidad Maya Yucateca: un análisis económico-ecológico. *Revista Iberoamericana de Economía Ecológica, 7*, 27–42.

García-Frapolli, E., Ramos-Fernández, G., Galicia, E., & Serrano, A. (2009). The complex reality of biodiversity conservation trough natural protected area policy: Three cases from the Yucatan Peninsula, Mexico. *Land Use Policy, 26*, 715–722.

Gómez Navarrete, J. A. (2009). *Diccionario introductorio, español-maya, maya-español*. Chetumal: Universidad de Quintana Roo.

Haenn, N. (1999). The power of environmental knowledge: Ethnoecology and environmental conflicts in Mexican conservation. *Human Ecology, 27*(3), 477–491.

Hutson, S. R. (Ed.). (2017). *Ancient Maya commerce. Multidisciplinary research at Chunchucmil*. Boulder: University of Colorado Press.

Hutson, S. R., Dahlin, B. H., & Mazeau, D. (2017). Commerce and cooperation among the classic Maya. The Chunchucmil case. In S. R. Hutson (Ed.), *Ancient Maya commerce. Multidisciplinary research at Chunchucmil*. Boulder: University of Colorado Press.

IISD. (1992). Business strategy for sustainable development, leadership and accountability for the 90's. Winnipeg: International Institute for Sustainable Development, Deloitte & Touche, and the World Business Council for Sustainable Development. https://www.iisd.org/business/pdf/business_strategy.pdf. Accssed 5 Apr 2017.
King, E. M. (2015). *The ancient Maya marketplace: The archaeology of transient space* (2nd ed.). Phoenix: The University of Arizona Press.
Klapper, L., El-Zoghbi, M., & Hess, J. (2016). *Achieving the sustainable development goals. The role of financial inclusion*. Washington, DC: CGAP & UNSGSA. https://www.cgap.org/sites/default/files/Working-Paper-Achieving-Sustainable-Development-Goals-Apr-2016.pdf. Accessed 25 Mar 2017.
Long Towell, J., & Attolini Lecón, A. (Eds.). (2009). *Caminos y mercados de México*. México: Universidad Nacional Autónoma de México e Instituto Nacional de Antropología e Historia. Serie Historia General.
Lusting, N. C., & Székely, M. (1997). *México: evolución económica, pobreza y desigualdad*. Washington, DC: PNUD, BID, CEPAL. https://publications.iadb.org/handle/11319/5293. Accessed 25 Mar 2017.
Mariaca Méndez, R. (2015). La milpa maya yucateca en el siglo XVI: evidencias etnohistóricas y conjeturas. *Etnobiología, 13*(1), 1–25.
Merchand Rojas, M. A. (2011). El estilo de desarrollo que hace inviable el "desarrollo sustentable". *Paradigma Económico, 3*(2), 33–60.
Möller, K., & Svahn, S. (2004). Crossing East-West boundaries: Knowledge sharing in intercultural business networks. *Industrial Marketing Management, 33*(3), 219–228.
Morgan, D. L. (1997). *Focus groups as qualitative research* (2nd ed.). Thousand Oaks: SAGE.
Morrison, M., Collins, J., Basu, P. K., & Krivokapic-Skoko, B. (2014). Determining the factors influencing the success of private and community-owned indigenous business across remote, regional and urban Australia. Final report prepared for the Australian Research Council and Indigenous Business Australia, December. Charles Sturt University & University of Technology, Sydney. https://www.csu.edu.au/__data/assets/pdf_file/0004/1311484/Final-Report-to-IBA-and-ARC-Linkage-LP110100698-Morrison,-Collins,-Basu-and-Krivokapic-191214.pdf. Accessed 27 Apr 2017.
OECD. (2010). OECD Perspectives: Mexico key policies for sustainable development. https://www.oecd.org/mexico/45570125.pdf. Accessed 5 Apr 2017.
Porter, M. E. (1990). *The competitive advantage of nations*. New York: The Free Press.
ProMexico. (2017). Importancia comercial de México en el mundo. Available at https://www.gob.mx/cms/uploads/attachment/file/43882/MEX_Ficha_resumen.pdf. Consulted on 24 Mar 2017.
Rathje, W. L. (1971). The origin and development of lowland classic Maya civilization. *American Antiquity, 36*, 275–285.
Rodil Marzábal, O., & López Arévalo, J. A. (2011). Disparidades en el crecimiento económico de los estados de México en el contexto del Tratado de Libre Comercio de América del Norte. *Economía UNAM, 8*(24), 78–98.
Rogoff, B. (2014). Learning by observing and pitching in to family and community endeavors: An orientation. *Human Development, 57*(2–3), 69–81.
Rosado-May, F. J. (1985). Comercialización de las hortalizas regionales en Cárdenas, Tabasco, México. *Horticultura Mexicana, 1*(2), 93–100.
Rosado-May, F. J. (2012a). Una perspectiva intercultural al concepto de tutoría académica. El caso de la UIMQRoo. In I. Deance & V. Vázquez Valdés (Eds.), *Aulas Diversas. Experiencias sobre educación intercultural en América*. Quito: ABYA/YALA Universidad Politécnica Salesiana, Deance-Vázquez y Universidad Intercultural Maya de Quintana Roo.
Rosado-May, F. J. (2012b). Los huertos familiares, un Sistema indispensable para la soberanía y suficiencia alimentaria en el sureste de México. In R. Mariaca Méndez (Ed.), *El huerto familiar del sureste de México*. Villahermosa: Secretaría de Recursos Naturales y Protección Ambiental del Estado de Tabasco y El Colegio de la Frontera Sur.

Rosado-May, F. J. (2015). The intercultural origin of agroecology: Contributions from Mexico. In V. E. Méndez, C. M. Bacon, R. Cohen, & S. R. Gliessman (Eds.), *Agroecology: A transdisciplinary action-oriented approach* (Advances in agroecology series). Boca Raton: CRC Press/ Taylor & Francis.

Rosado-May, F. J. (2017). Los retos y oportunidades de guiar inteligencia con inteligencia. El modelo de educación superior intercultural en Quintana Roo. In F. González González, F. J. Rosado-May, & G. Dietz (Eds.), *La gestión de la educación superior intercultural en México. Retos y perspectivas de las universidades interculturales*. Iguala: Universidad Autónoma de Guerrero, El Colegio de Guerrero A.C., y Ediciones Trinchera.

Rosado-May, F. J., & Cuevas Albarrán, V. B. (2015). El programa educativo "Ingeniería en Desarrollo Empresarial" de la Universidad Intercultural Maya de Quintana Roo. ¿Qué justifica su creación? In E. E. Brito Estrella (Ed.), *Empresa, sostenibilidad y desarrollo regional*. José María Morelos: Universidad Intercultural Maya de Quintana Roo.

Schaltegger, S., Lüdeke-Freund, F., & Hansen, E. G. (2011). *Business cases for sustainability and the role of business model innovation. Developing a conceptual framework*. Lüneburg: Centre for Sustainability Management, University of Lueneburg.

Schmelkes, S. (2013). Educación y pueblos indígenas: problemas de medición. *Revista Internacional de Estadística y Geografía, enero-abril, 4*(1), 5.

Silva Riquer, J., & Escobar Ohmstede, A. (2000). *Mercados indígenas en México, Chile y Argentina, siglos XVIII y XIX*. Ciudad de México: Centro de Investigaciones y Estudios Superiores en Antropología Social.

Storey, R., & Widmer, R. J. (2006). The pre-columbian economy. In V. Bulmer-Thomas, J. Coastworth, & R. Cortes-Conde (Eds.), *The Cambridge economic history of Latin America: Vol 1, the colonial era and the short nineteenth century*. Cambridge: Cambridge University Press.

Toledo, V. M. (2003). Los pueblos indígenas, actores estratégicos para el Corredor Biológico Mesoamericano. *Biodiversitas, 43*, 8–15.

Toledo, V. N., Barrera-Bassolos, N., García Frapolli, E., & Alarcón Cahires, P. (2007). Manejo y uso de la biodiversidad entre los mayas yucatecos. *CONABIO, Biodiversitas, 70*, 10–15.

Trompenaars, F., & Hampden-Turner, C. (2012). *Riding the waves of culture. Understanding diversity in global business* (3rd ed.). London: Nicholas Brealey Publishing.

Trompenaars, F., & Woolliams, P. (2003). *Business across cultures*. Chichester: Capstone Publishing Ltd..

Vail, G. (2009). Cacao use in Yucatan among the prehispanic Maya. In L. E. Grivetti & S. Howard-Yana (Eds.), *Chocolate, history, culture and heritage*. Hoboken: Wiley.

Low-Carbon Economies (LCEs)

International Applications and Future Trends

Elizabeth Gingerich

Contents

Abstract

After decades of living without regard to the health of the biosphere as evidenced by unbridled consumption, energy waste, disproportionate wealth, and disparate access to resources, the world's population is now collectively facing an existential threat. Global nation-states and territories universally acknowledge that climate change, or global warming, as marked by massive environmental degradation, species extinction, increasing incidences of

E. Gingerich (✉)
Valparaiso University, Valparaiso, IN, USA
e-mail: jvbl@valpo.edu; Elizabeth.Gingerich@valpo.edu

S. Dhiman, J. Marques (eds.), *Handbook of Engaged Sustainability*,
https://doi.org/10.1007/978-3-319-71312-0_15

extreme weather events, death and displacement of whole societies, and rising population growth with accompanying energy demands, must be addressed with full awareness, comprehensive engagement, and immediacy. Several nations, territories, and regions are stepping up to the plate to effectively wage the battle. This is evidenced by policies and strategies propounded by both the private and public sectors – and complemented by rapid advancements in clean energy generation and storage technologies – to achieve a low-carbon economy (LCE), one which answers the energy needs of its constituents in a manner of producing the lowest greenhouse gas (GHG) emissions possible with the ultimate objective of achieving complete carbon neutrality.

Keywords
Low-carbon economies (LCEs) · Renewable energy (RE) · Paris Agreement · Cap and trade · Carbon tax · Climate change · Global warming · Carbon emissions · Greenhouse gases (GHG)

Introduction: Transitioning to LCEs

Addressing climate change requires a transformative agenda and cooperative efforts of both the private and public sectors. Ostensibly, more countries are curbing emissions through the implementation of carbon reduction policies such as carbon pricing, regulatory intervention, and targeted assistance to promote innovation in low-carbon, sustainable technologies. But GHG emissions (comprised of carbon dioxide (CO_2), methane, nitrous oxide, hydrofluorocarbon, and perfluorocarbon – with carbon representing the most pervasive component) have risen to dangerously, unprecedented levels resulting in severe impact which warrants immediate action. Fossil fuels (i.e., coal, oil, and natural gas) have fueled global economic development since the dawn of the Industrial Revolution, leading to poor economic and infrastructure choices. For too long, public policy mechanisms have continued to be geared toward fossil fuel use and carbon-intensive activities. The failure to align modern-day policy strategies with the current science has largely failed to take the deleterious consequences of GHG emissions into account. Such misalignments have neglected action upon the development of low-carbon transport and energy systems. Decarbonizing electricity is the key to promoting cleaner air and water, better health, tightened national security, and a more diversified energy supply. This is most effectively done through governmental core climate change mitigation commitments, intelligent utility regulations, incentivizing long-term finance tools, and a well-trained workforce. Support of renewable energy (RE) technologies and low-carbon business models and sustainability are key to successful low-carbon transitions. Input from private businesses and implementation of a diversified host of incentivizing strategies are needed to embark upon combatting this existential threat.

Years of Excessive Consumerism

The transition to LCEs has failed to evolve uniformly among emerging and developed economies. Unfortunately, such streamlined approaches are nonexistent, and in certain areas, developing countries have advanced further than wealthier nations, even when carbon-reducing technologies are more prevalent and consumption rates uncircumscribed in more developed nations. For instance, while the United States has a population of approximately 326 million (4.5% of the world's population), it consumes more than 20% the world's nonrenewable resources, making it one of the highest GHG emitters in the world. China moved beyond the United States in 2007 in emissions, but it has a population (1.39 billion) of roughly four times that of the United States (Biello, *Scientific American* 2014). Energy consumption is almost as prevalent in Canada, Russia, Australia, and the EU countries. This changing market was a by-product of the Industrial Revolution (1880–2010) which was demarcated by a series of transitions: wood to fossil fuels, carriages to gas-burning and more currently electric vehicles (EVs), kerosene lights to LED bulbs, and paper journals to online news. These changes undoubtedly generated expediency and comfort for its beneficiaries but generated vast amounts of CO_2 in the process. Energy use is still soaring in developed countries, with the United States, much of Europe, and China carrying the dubious distinction of being the largest consumers and polluters (Worland 2017). Unfortunately, emerging economies are bearing the brunt of the consumption rates of wealthier nations yet are moving toward a similar phase. Those with growing populations and rising urban growth are especially in line to emulate such levels of consumerism and fossil-fuel-based energy consumption. World energy consumption rates and the resulting CO_2 emissions generated therefrom are shown in Fig. 1. According to data provided by Olivier et al. as part of a 2016 energy report presented to the European Commission, such findings verify that while China has moved ahead of the United States as the largest CO_2-emitting country worldwide, the United States remains the leader in per capita emissions.

Years of production and consumption have resulted in the pollution of waterways, aquifers, soil, and air. How – and to what level – governments and private industries are responding is, at least in several instances, predicated upon the degree to which each jurisdiction is being threatened by climate change.

Urgency of Decisive Action and Increasing Climate Vulnerability

According to the World Data Center for Paleoclimatology and the National Oceanic and Atmospheric Administration (NOAA), CO_2 levels are greater now than at any time in the last 800,000 years. Every ecosystem is currently in decline. In the healthcare sector, asthma rates in children have increased by 28% and nearly 80% of all new cancers that stem from nonhereditary factors (*US Centers for Disease Control*). Extreme weather events have killed or displaced hundreds of thousands of people worldwide, causing billions of dollars in physical damages. Sea levels are

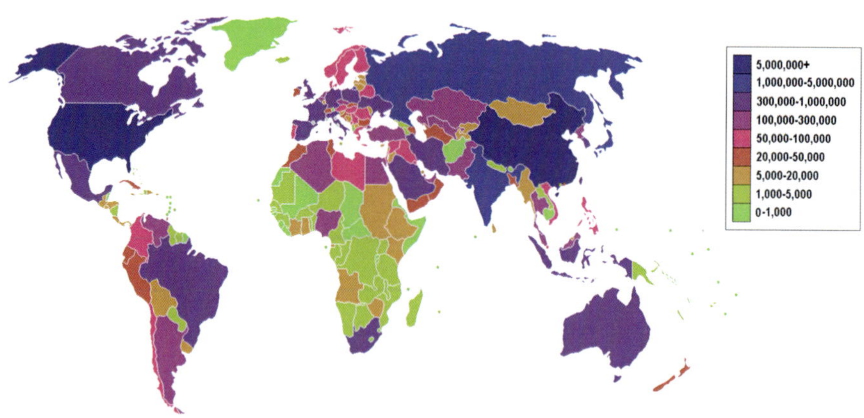

Fig. 1 Fossil-fuel-based energy consumption (Wikimedia 2006)

rising, and drought and diseases are proliferating (e.g., Zika virus originating in Brazil). Industrial sectors, including insurance, energy, transportation, healthcare, agriculture and forestry, construction, investment, and waste management, have all been severely impacted by the forces and consequences of climate change. And over 1.2 billion people – from a world population of 7.5 billion – completely lack access to electrical power (World Bank Group Report 2016). Specifically, according to the World Bank and other research institutions:

- The last decade was the world's hottest on record and 2016 was the hottest year yet.
- Fifty percent of all coral reefs have declined in the last 50 years.
- The world population of 7.5B is predicted to reach 9.7 billion by 2050.
- Factory (livestock) farming and landfill waste produce dangerous levels of methane gas – the most toxic greenhouse gas.
- Over the past 20 years, floods, storms, heat waves, droughts, fires, hurricanes, tornadoes, and record snows have resulted in more than 600,000 deaths, and 4.1 million people have been injured or left homeless.
- Every ecosystem in the world is currently in decline, and traces of mercury have been found in fish from every body of water on Earth (NOAA).
- Between 0.01% and 0.1% of all plant and animal species are facing extinction each year (World Wildlife Fund).
- Climate change costs the United States over $100 billion each year (CNBC Money).

Many governments and private businesses have steered their respective countries, provinces, and states to become low-carbon economies by focusing on the reduction of chiefly one of the major gases contributing to climate change: carbon dioxide (CO_2). Business operations and residences have been steered to minimizing their output of carbon emissions in a variety of ways to fulfill this commitment.

Low-Carbon Economies (LCEs) Defined

Also referred to as decarbonized, green, or low-fossil-fuel economies, low-carbon economies (LCEs) are premised upon the generation and use of energy which minimizes GHG emissions due to anthropogenic (human) activity – the primary cause of global climate change. One of the most often-used definitions of LCEs has been provided by the United Nations Environment Programme (UNEP) which states that a low-carbon economy is "an economy that results in improved human well-being and social equity, while significantly reducing environmental risks and ecological scarcities." It thus serves to not only reduce emissions to mitigate the destructive impact of climate change but also strives to create employment and investment opportunities, strengthen trading, improve infrastructure, and close the inequities of disparate wealth distribution.

Since most governments – particularly major oil producers – own 50–70% of global fossil-fuel-based resources (i.e., oil, gas, and coal), as well as assess and collect taxes and royalties on the portions they do not own or control, they are at a heightened risk of being abandoned in the new world economy. Therefore, these governments are in the best position to effectuate policy that could transition to cleaner environments and overall financial savings (Nelson et al. 2014a, b).

Many territories, countries, and regions throughout the world have designed and implemented low-emission development strategies (LEDS) which seek to reduce GHG levels while concomitantly achieving socioeconomic development goals. These governing bodies seek to achieve a carbon-neutral economy through mitigation strategies created to assign a cost per unit of production. These tools include **carbon taxing**, **emissions trading** or **cap and trade, auction markets, government subsidies/tax credits** in renewable energy, transactive energy projects, and green financing and investment (UN Division for Sustainable Development 2012). Many LCEs tout that implementing such low-carbon initiatives creates jobs, lessens rural poverty, encourages sustainable land use, furthers trade and industrial competitiveness, and safeguards energy security – all while buttressing ecosystem resilience. Low-carbon strategies often encompass technologies that produce goods and energy with low GHG emissions, mitigating carbon use from the energy, transport, manufacturing, and agricultural sectors. Fossil fuel systems are transitioning to low-carbon alternatives at an accelerated rate. Already by 2008, approximately 19% of all global energy consumption was generated by renewables, with grid-connected photovoltaics increasing the fastest (Renewables Progress Report 2010).

Collective Action: *The Paris Agreement*

Given the dire predictions of unmitigated climate change – i.e., the death and displacement of millions of people across the globe, abject poverty, famine and drought, and floods and rising ocean levels – both public and private stakeholders agreed to join together to craft climate action plans in an historic display of international cooperation at the 2015 Conference of Parties (COP 21). This level of collusion between signatory countries, intergovernmental agencies, and private actors was unprecedented in drafting the Paris Agreement (The Paris Agreement 2015). Bill Gates (Microsoft), Mark Zuckerberg (Facebook), Richard Branson (Virgin Group), Jeff Bezos (Amazon), and other business leaders comprising the *Breakthrough Energy Coalition* announced an initial pledge of 7 billion USD toward the advancement of clean energy, transportation, and agriculture. The initiative was launched in conjunction with *Mission Innovation* – a joint effort launched by 21 governments, including the United States, Britain, Australia, Germany, China, and Brazil, to double the amount of public money invested in clean energy innovation (The Guardian 2015). According to the United Nations Framework Convention on Climate Change (UNFCCC), the primary objectives of the Paris Agreement were to:

1. Hold the increase in the global average temperature to below 2 °C above preindustrial levels and to pursue efforts to limit the temperature increase to 1.5 °C above preindustrial levels.
2. Increase the ability to adapt to the adverse impact of climate change and support climate adaptation and low greenhouse gas emissions development.
3. Make financing consistent with a pathway toward low greenhouse gas emissions and climate-resilient innovation.

The Paris Agreement original signatories (excepting Syria and Nicaragua, the latter opting out for the express reason that the carbon control limitations had not gone far enough) left with a commitment to expedite efforts to move to lower-carbon economies. But in tandem with the United Nations' classification of nation-states according to particular socioeconomic indicators (i.e., developed, developing, and least developed countries (LDCs)), participating countries were not expected to design and deploy similar strategies at coordinated timetables. What *was* predicted was that wealthier nations would move more quickly to implement LEDS while emerging economies' attempts to transition would be more measured in comparison. Surprisingly, however, while many LDCs find themselves the greatest victims of climate change, many are leading the way in establishing low-carbon economies – largely due to the unchecked consumption practices of wealthier nations. Researchers and intergovernmental agencies have classified "more than 10% of countries as 'free riders,' ranking in the top fifth in terms of emissions and the bottom 20% in terms of vulnerability. These countries include the United States, much of Europe and Australia" (Worland 2017). Moreover, emerging economies face the multifaceted challenge of controlling greenhouse gas (GHG) emissions while reducing dependence on foreign sources of energy, growing their own economies, and solidifying national security interests. Sound governance and financial incentives are ostensibly imperative. Thus, the sustainable use of a nation's resources to satisfy a rising demand for energy without causing further environmental degradation is the optimal goal. Achieving this objective in abundance – i.e., producing an excess of clean energy for subsequent storage and transportation in the global market – is becoming more than a vision; it has transcended into a working reality for many governments, particularly those of LDCs, while leaving some of the wealthiest nations behind.

To fulfill the Paris Agreement commitments, many nation-states have implemented specific carbon-reducing strategies and policies, several of which are outlined below.

Low-Carbon Energy Strategies

How each country approaches control of its GHG emissions is generally distinguishable between more *developed* nations – most often demarcated by a democratic form of government, educated workforces, and a technologically advanced infrastructure – and emerging economies, often characterized by postcolonial struggles, poor infrastructure, underdeveloped utility grids, and disparate trading policies. However, with the signing of the Paris Agreement by nearly every country in the world, the collective commitment to lower carbon emissions and invigorate the clean energy sector (hydroelectric, wind, solar, geothermal, and biofuels) has established a parity of approaches of sorts. Many nations have engaged in emissions trading, subsidized renewable energy (RE) development, or imposed a carbon tax – or have adopted a combination thereof.

RE production has faced daunting obstacles in the past, most predominantly the cost of its manufacture and the ability to store any surplus. Within the last decade, however, two phenomena have occurred worldwide: (1) clean energy generation and storage technologies have rapidly advanced; and (2) the cost of constructing and installing renewable projects has declined sharply. With favorable topographical features and an abundance of particular mineral deposits available to use in the manufacture and storage of clean energy, LDCs are now emerging as formidable world leaders in this field. Although many of these nation-states (particularly located in Africa, and Central and South America) were former colonies of Western European powers, such nations have been participating in clean energy projects, often aligning with more advanced economies, including their former colonizers. In essence, they are leapfrogging over some of the deleterious aspects of the Industrial Revolution, including massive contamination and environmental degradation, to embrace modern LCEs. Yet all have one common, guiding principle: to reduce atmospheric carbon levels as effectively and expeditiously as possible.

Carbon Tax

There are several countries and regions that have adopted a carbon tax and by all accounts, successfully so. One of the leaders in this field is the province of *British Columbia, Canada* (BC) – the westernmost province of Canada – which is currently attempting to be one of the cleanest and healthiest territories in the world and is continuing to advocate a low-carbon future. In furtherance of this objective, its provincial government is reducing GHGs while fortifying its economy. BC has already supported the extensive planting of trees, construction of bike lanes and paths, and installation of rooftop gardens for both natural insulation and to encourage local food production. It has advocated lifestyle changes and stimulated new business strategies, including switching to cleaner forms of transportation, altering home utility usage, targeting fuel-efficient products and home heating/cooling systems, and promoting overall energy upgrades and retrofits. Many of these encouraged changes are further incentivized by financial savings on utility bills, rebates, and tax credits (British Columbia – Climate Change 2016).

British Columbia has been touted as a model for the rest of the world in promoting economic development while safeguarding natural capital (World Bank). With the exception of the transportation industry, BC boasts 100% clean renewable energy in its commercial and residential sectors. BC's renewables largely consist of hydroelectricity (97.7%), produced by that province's vast network of streams and rivers. While satisfying provincial energy needs through RE development, BC is also backing technological innovation to further reduce carbon emissions. This inexpensive and largely clean source of energy is supplemented by solar, wind, geothermal, and, most recently, marine shore power (Vancouver became the third location in the world to offer marine shore power for cruise ships in 2009). This service allows properly retrofitted vessels to plug into BC Hydro's land-based electrical power grid, thereby reducing diesel emissions.

In an effort to boost its economy while combatting climate change, BC launched a revenue-neutral carbon tax – now 9 years old – on July 1, 2008. The initial carbon tax added approximately 2.5 cents to a liter of gasoline based on a rate of CAN$10 per ton of carbon emissions. This increased to CAN$30 in 2012 and currently translates to a total of 7.4 cents per liter of gasoline and 8.2 cents per liter of diesel and home heating oil. This graduated tax has encouraged both homeowners and commercial entities to reduce emissions while incentivizing the adoption of cleaner energy use and development. As a revenue-neutral tax, all revenue generated by the tax is returned to the citizens of the province through credits and/or a proportionate reduction in other taxes (Porter 2016).

Covering approximately 70% of the provincial economy, BC's carbon tax has survived local elections and attempts to repeal it in both 2009 and 2013 and has stimulated the economy while reducing overall GHGs by between 5% and 15%. The tax does, however, exclude (1) exports (including that of carbon-rich coal and natural gas), (2) greenhouse growers (the BC government announced in 2013 that the commercial greenhouse sector would be given permanent carbon tax relief via a grant program to help offset carbon tax costs), (3) methane-producing landfills, and (4) agricultural operations. It also phases the tax in with respect to cement producers. It has been argued that while the carbon tax structure is revenue-neutral in design, the numerous exemptions may be inimical to the success of this tax lowering overall GHG emissions (Murray and Rivers 2015).

British Columbia presented its *Carbon Tax Plan and Report* as part of the larger *Balanced Budget Plan* in 2014 which outlined the tax reductions to individuals and businesses generated by carbon tax revenues. In pertinent portion, the report states: "Based on the revised forecast of revenue and tax reduction estimates, revenue neutrality has been met for 2013/14. In fact, the reduction in provincial revenue exceeds the $1,212 million in carbon tax revenue by $20 million" (British Columbia

– Ministry of Finance 2013/2014). For the present time, British Columbia has confirmed that it will keep this tax, after considering its economic impact on business and industry.

Emissions Trading (Cap and Trade)

Although 28 nations comprise the European Union, 31 (with the addition of Iceland, Norway, and Liechtenstein) actually participate in the *European Union Emissions Trading System* (EU ETS). Launched in 2005, the EU ETS is the world's first, and largest, GHG emissions trading mechanism that is premised on capping GHG emissions by covered power stations and factories pursuant to a government-determined threshold (European Union Emissions Trading System Fact Sheet 2016). Such restrictions and allowances delineate the amount of GHG that can be released from covered facilities. These entities then self-report to the government the actual amount of GHG emitted from that installation over a set period of time. Excess authorized emissions can then be sold, traded, or auctioned off internationally. This tool has been used to reduce the overall amount of GHG released into the atmosphere by infusing market-driven regulation and trading into business operations. Companies are authorized to receive or purchase emission allowances which can then be translated into a limited number of international credits (European Commission Fact Sheet 2016).

European nations initially established the Commission to combat the deleterious consequences of climate change gases and to promote the construction of sustainable projects. By 2008, nearly 50% of the EU's anthropogenic carbon emissions were reduced. Although emissions have not ceased altogether, they are effectively offset to a different geographical area. Thus, if a covered installation effectively reduces its emissions and comes under budget through credit allowances, any surpluses can be freely negotiated. While operators of covered installations must abide by the GHG limitations set by the government, they can continue to engage in business without pervasive governmental monitoring and serve as the beneficiaries of built-in incentives to develop clean energy alternatives using cost-effective measures and clean technologies. By allowing the purchase of international carbon credits, companies are able to incentivize other countries to adopt low-carbon schemes.

The emissions trading feature of the EU ETS constitutes a major component of the EU's climate action plan and currently covers approximately 11,000 factories and power stations throughout its participating nations. However, the system only operates during certain trading periods. The span of time covered by these trading periods has incrementally increased in length since January 2008. In comparison to restrictions in 2005 when the system was first launched, proposed caps for the current trading period to end in 2020 have successfully generated a 21% reduction in GHG. In July 2015, the European Commission presented a legislative proposal on a revision of the EU ETS for its next trading period (2021–2030), in coordination with the EU's 2030 climate and energy policy plan. The proposal aims to reduce EU ETS emissions by more than 43% compared to 2005 levels.

Emissions trading systems or schematics are recognized as one the most cost-effective tools for cutting GHG emissions in contrast to traditional regulatory schemes. Emissions trading harnesses market forces in the quest to develop the cheapest and most effective ways to reduce GHG emissions. One major advantage of this tool is that governmental intervention is minimal, restricting its participation to basically setting initial emission thresholds. Market forces take over from there and assess financial value to every ton of emissions saved. This has promoted long-term investment in clean renewable energy technologies and functions as a proven means to transition to a LCE (European Commission Fact Sheet 2016).

Transactive Energy Systems

LCE transitions may also involve the establishment transactive energy networks. These types of systems refer to the coordination and integration of energy generation and distribution – typically using various forms of renewable energy – premised upon consumer demand as opposed to a more traditional hierarchical grid structure. This structure allows for exchange of information, whereby all levels of energy generation and consumption are able to interact and operate, maintaining utility restraint while monitoring output flow. Removing carbon from the atmosphere and using more renewable energy is the bedrock of any strategy to transition toward more sustainable energy systems, often requiring the conversion of electrical grids and related infrastructure into more intelligent power systems. Large-scale smart grids must be designed to balance, coordinate, and integrate the fluctuating power generation of renewable sources. In this manner, transactive approaches are emerging as strong contenders for orchestrating a coordinated operation of multiple devices. Transactive energy systems use market-based consumption rates to guide operational decisions. In this exchange of value-based information captured in transactions between consumers and operators, different forms of renewable energy power and storage techniques are used to satisfy consumer demand and regulate distribution management.

One example of a transactive energy network is the *US Department of Energy Pacific Northwest Demonstration Project* – a 5-year government-funded project which was completed in 2015 (United States Department of Energy 2015). The project involved 2 universities, 11 utility companies, and several technology firms in an attempt to explore transactive energy concepts on a regional scale (more specifically, within the US states of Montana, Idaho, Washington, Oregon, and Wyoming). The dual project objectives were to test methods to reduce energy usage and introduce cleaner energy sources, integrating advanced storage technologies, automated power controls, and smart meters. Consumers were educated as to the best times of the day to use energy – all based upon communicated market demand (United States Department of Energy 2016).

Transactive signals were used to exchange information about predicted price and availability of power in real time with information updates given every 5 minutes. When peak power demands were registered, the transactive control mechanism was engineered to reduce power use. The project confirmed the viability of transactive control technologies, confirming their ability to improve energy efficiency and reliability, reduce energy costs, and encourage the deployment of clean renewables. As a result of this project, many participants have decided to continue smart grid programs.

Renewable Energy Storage Technologies

Wind, solar, hydro, geothermal, and biomass are the most common forms of non-fossil-fuel sources of renewable energy, but a steady production rate may be erratic at times due to various meteorological and topographical factors. Thus, RE storage is critical to embarking upon a transition to a LCE. While technological advancements in different methods of RE storage have been occurred, the use of the lithium-ion battery to perform this task has not wavered but rather has rapidly increased worldwide. Countries and regions with plentiful lithium deposits are expectedly on the forefront of propelling LCEs. Lithium is already used ubiquitously – from laptops to cell phones and electric vehicles (EV) to grid storage batteries. Demand for lithium is predicted to escalate precipitously within the next decade, largely as a consequence of unprecedented growth in the battery-driven transport and energy sectors (Lithium-ion Battery Market, MarketWatch 2016).

China is set on a course to lower its carbon emissions earlier than its Paris Agreement commitment date by simulating greater investment in clean energy technologies, forming alliances with other countries and among private sector actors, decommissioning over 100 coal plants within the last year, and witnessing a stark uptick of EV purchases (China and India Make Great Strides on Climate Change, New York Times 2017). China is the world's largest consumer of lithium (MarketWatch 2016) and is committed to cultivating low-carbon strategies. While the country houses some of the largest deposits of lithium, this type of lithium must be mined as is done with subterranean coal deposits – a costly venture considered to be both labor and capital intensive. Rather, the Chinese government has entered into agreements with other nations and foreign businesses to extricate lithium more cheaply through briny deposits and evaporation pools. As more than one-half of the world's lithium is currently being extracted from South America's "Lithium Triangle" – a South American region encompassing Argentina, Bolivia, and Chile – China has seized upon the opportunity to partner with area governmental powers and heavily invest in this region.

In 2014, the African country of Zimbabwe was reported to have produced the fifth most lithium at 1,000 metric tonnes in the world (The Top Lithium Producing Countries in the World, World Atlas 2017). As such, it occupies an opportune position to promote clean energy generation, storage, and world trade. Privately held corporation Bikita Minerals currently controls most of the country's lithium mining under the auspices of the Zimbabwean government. This company operates the Bikita mine, estimated to hold over 11 million tons of lithium. To date, Bikita has invested over 7 million USD to upgrade its facilities and is slated to double its production in 2018 (Bikita Minerals 2017).

Government Subsidies and Tax Credits

Global financial lenders, including Goldman Sachs, view the transition to low-carbon economies being propelled by a shift in transformative technology as much as it has been a response to environmental challenges. And in spite of the US exit from the Paris Agreement in late May of 2017, consecutive years of government sponsorship in the form of tax credits and direct subsidies have propelled the country's investment in clean, low-carbon energy sources, driving the costs associated with the generation of renewables dramatically downward. Currently, four technologies are dominating the low-carbon transition: electric vehicles, solar photovoltaic (PV), onshore wind farms, and LEDs. It has been estimated that between 2015 and 2020, solar PV and onshore wind alone will add more to global energy supply than US shale oil production did between 2010 and 2015 (The Low Carbon Economy 2015). By 2020, six in ten lightbulbs will be LEDs. By 2025, carmakers are anticipated to sell 25 million hybrid and electric vehicles, more than ten times than today. These technologies are estimated to save more than 5 gt of CO_2 emissions per annum by 2025 and to assist global emissions to peak earlier than expected (GS Market Report 2017).

The clean energy industry and its affiliated jobs are rapidly expanding in America, largely due to specific tax credits and progressive tax policies calculated to spur on this growth. Since 2005, federal tax credits for renewable energy (RE) development and deployment by both residential and commercial property owners – specifically in the forms of the Federal Production Tax Credit (PTC) and Investment Tax Credit (ITC). These credits, created in a joint effort to support lower carbon emissions by the Department of Energy (DOE) and the Internal Revenue Service (IRS), have represented the primary incentives for development and deployment of renewable energy technologies – especially wind- and solar-based power production. In December 2015, these credits were extended by 5 years as part of the *Consolidated Appropriations Act of 2016* – albeit in decreasing values. This extension through December 31, 2019, is anticipated to provide the market with sufficient surety to invest in long-term RE projects. Additionally, the US Advanced Energy Manufacturing Tax Credit (MTC) was formed to award tax credits to new or retrofitted manufacturing facilities that support clean energy development. MTC credits are based upon job creation, GHG reductions, and commercial viability (Goodward and Gonzalez 2010). The full impact of these federal credits has been claimed to have played an important role in the rapid growth of US renewables – by 2015, wind electricity generation had increased sevenfold, and photovoltaic (PV) electricity generation rose to 36 terawatts (TWh), stimulating "economic growth, energy security, job creation, energy price stability, and health and environmental co-benefits" (US – DOE 2016). Also, the National Resources Defense Council estimates that the federal tax credit extensions for wind and solar will add over 220,000 jobs and nearly $23 billion to the US economy in 2017 alone. Furthermore, economic forecasters believe that the clean energy industry will continue to accelerate in the United States as costs continue to decrease, carbon emissions lowered, and new jobs created due in large part due to these tax policies, although as the current administration espouses a reversal of the Clean Power Plan, this attempt to create an LCE may be in peril (Steinberger 2017).

Energy Auctions

Energy auctions typically refer to competitive bidding procurement processes for electricity from RE sources or even where renewable energy technologies are eligible and can be translated in either capacity (MW) or energy (MWh) measurements (IRENA 2015). In a renewable energy (RE) *forward auction*, buyers – typically governments – offer increasingly higher prices to attract energy providers. These bids are typically conveyed in a closed, confidential manner. Lucrative prices encourage RE development from both within and outside of the affected area. In an RE *reverse auction*, buyers and sellers exchange places; clean energy providers vie against one another, typically to gain a government or private energy contract (Maurer and Barroso 2011). Electricity auctions have increasingly become popular tools for emerging economies in RE development and procurement. A successful example is shown by Brazil – the largest auction user in Latin America.

While technically a developing country, Brazil maintains some of the world's largest hydroelectric power plants. Hydroelectricity production, which feeds off large reservoirs and controlled river systems, generates approximately two-thirds of Brazil's electricity despite continued water shortages due to one of the worst droughts in history. According to Brazil's Ministry of Mines and Energy, the country has pledged to increase its hydropower capacity by 27 gigawatts (GW) by 2024 irrespective of lingering weather patterns (Brazil Energy 2016). Brazil's hydro plants, coupled with wind power and bioelectricity derived from sugarcane bagasse, represent this country's main sources of RE. Consumers consist of both regulated populations and those businesses and individuals who are able to negotiate their own electricity suppliers (Brazil Energy Policy, Laws and Regulations Handbook 2015). The regulatory framework governing Brazil's quest for RE development has largely been the result of a combination of government- and private-owned electricity companies. These entities are monitored by a nearly independent regulatory agency which uses auctions to incentivize clean energy distribution companies to make efficient purchases on behalf of consumers (Maurer and Barroso 2011).

World Trade Organization (WTO) Support of LCEs

The commitments made by countries to reduce carbon levels by any of the tools aforementioned have received unequivocal support from the WTO. Following WWII, trading tariffs were high, most notably on manufactured goods, and fossil fuel dependence had become the hallmark of the postwar era. Within the next half century, wealthy nations, primarily driven by North America and Western Europe, liberalized trade through the establishment of the General Agreement on Tariffs and Trade (GATT) and its successor, the World Trade Organization (WTO) (Baldwin

2006). Regional trade agreements (RTAs) and free trade agreements (FTAs) provided the forum for trade liberalization, opened new markets, and governed international commerce, often incorporating certain environmental protections and incentives for RE project development. Underscoring these changes in trade is the WTO's declared support of the World Intellectual Property Organization's (WIPO) new division, WIPO Green – established in 2015 to guide and monitor how climate change-related technologies are developed and used worldwide. WIPO Green effectively complemented the WTO's agreement on Trade-Related Aspects of Intellectual Property Rights (TRIPS) – by providing a forum of innovation of ways to reduce GHG while encouraging sustainable economic development. WIPO Green functions as an interactive marketplace promoting both the creation and diffusion of green technologies by connecting technology with various service providers (WIPO Green 2017). Additionally, it is dedicated to the issuance and protection of patents worldwide – the majority representing clean energy generation and storage advancements. As of 2016, Germany, Japan, South Korea, and the United States led in the number of clean energy patents issued.

Leading Nations, Businesses, and Industrial Sectors in Carbon Reduction Strategies

The fate of the planet and long-term sustainability of its nation-states and businesses lies in the drastic reduction of GHGs – particularly carbon. Several countries and industries emanating from both developed and emerging economies have attracted attention by the level of progress made as well as the methods used to achieve this objective.

Sweden. Sweden claims title to the nation heralding the lowest amount of GHG emissions, ranking first in the world for carbon-neutral practices. Applying the "Polluter Pays Principle" – similar to the taxation structure of British Columbia – the Swedish government has two different taxes on fuels: one on petrol and diesel levied in the 1930s and the second on fossil heating fuels which began in the 1950s. Subsequently in 1991, the Swedish government introduced the CO_2 tax to curb use of fossil fuels at an initial rate of 27 € per ton of fossil carbon. To expedite the country's transition to an LCE, this carbon tax has been increased dramatically over the last decade to a level of 123 € per ton of carbon expended. And while the government has not fully earmarked the revenue generated by the CO_2 tax for specific purposes, a significant part of the national budget has been allocated to various projects including those which target lower emissions in the public transport sector and an increased use of biofuels in heating. Sweden has not sacrificed economics as the country continues to reduce emissions. From 1990 to 2013, its GDP increased by 61%, while its carbon emissions were reduced by 23%. However, since Sweden is a member of the EU and the EU Emissions Trading Scheme (EU ETS), its policies have had to comport with a community-wide economic instrument covering GHG emissions from energy-intensive industrial installations. But industries outside the EU ETS, already operating with relatively low-energy costs, have been taxed separately and have responded positively to the CO_2 tax by switching, *en masse*, to non-fossil-fuel heating sources of energy CO_2 (Akerfeldt and Hammar 2016).

Costa Rica. Costa Rica has already succeeded in leading the world in the protection of natural habitats and is on the cusp of becoming fossil-free nationwide. In 2015, energy generation plants and projects had already weaned off fossil fuels. Like Sweden, its government subsidization efforts have been substantially aided by a dramatic uptick in investment in both wind and solar, with massive geothermal and hydroelectric generation already established and growing. This Central American country has sustainably utilized its natural resources and topographical features to further its transition to an LCE. Harnessing the power of its heavy tropical rainfalls and pervasive river system, hydroelectric generation has, to date, gained the greatest success. According to the Costa Rican Electricity Institute (ICE), 99% of clean energy generation was achieved by 2015 (Costa Rica – AFP 2015). This small country, situate in a developing, often war-ravaged region comprised of six other Central American nation-states, has assumed an unforeseen position of leadership in the development of clean energy and LCE formation. Its government remains committed to energy independence policies which have accentuated the possibilities of excess capacity production and active participation in global trade.

Nicaragua. Costa Rica's northern neighbor, Nicaragua, is turning its overwhelming coastal winds and earthquake-prone land masses to its advantage by working with, once again, its natural resources. By mid-2015, Nicaragua was producing more than half of its electricity needs through renewable resources. Through early governmental commitments, this country had already invested the 5th highest percentage globally of its GDP in the development of clean renewables and stands to become nearly 90% fossil-fuel-free by 2020 (Climate Reality Project 2016). This success highlights the primary reason why this country is not a signatory nation to the Paris Agreement – the accord's voluntary commitment paradigm was simply not sufficiently stringent and binding.

Uruguay. Lacking entrenched interests in fossil fuel dominance which has undermined progress in clean energy development and trade, both the private and public sectors of this small South American country decided nearly a decade ago to invest – without any special subsidies – in wind and solar energy generation. This collaboration has weaned the nation nearly completely off all of its fossil fuel usage in both the residential and commercial sectors.

Morocco. One of the world's largest concentrated solar plants is in the final stages of construction and has been primarily funded by the EU. Located on the edge of the Sahara Desert in northern Africa, Morocco is taking full advantage of its intense sun exposure by building this RE installation. The plant consists of curved mirrors totaling approximate 16 million square feet – or the equivalent of 200 football fields (or soccer pitches) – and will have the ability to generate 580 MW to serve nearly 2 million people (Vast Moroccan Solar Plant 2016). Morocco represents a model for other African nations and for those countries with similar topographical and meteorological traits to move to a low-carbon-based economy.

Kenya. Kenya is a country currently plagued by devastating drought. It has historically been dependent upon foreign oil but nevertheless has managed to supply over one-half of its energy needs through geothermal power. Kenya, in fact, was the first African country to use geothermal power and still maintains the largest installed capacity of this form of renewable energy in Africa at 200 MW (Singh 2015). The Kenyan government has also recently added wind power as a supplementary power source as it is already home to Africa's largest onshore wind farm (Climate Reality Project 2016). In fact, winds are a dominant resource throughout the east central coast of Africa. Topping this country's energy mix is solar as Kenya represents a dominant world leader in the number of solar power systems installed per capita with more than 30,000 small solar panels sold in that country annually. Within the last 5 years, Kenya's nascent renewable energy sector has attracted significant foreign investment, primarily led by 20 British privately held companies. It is estimated that the renewable sources of energy could produce sufficient power for its own citizens with an excess to be channeled through foreign trade (Rubadiri 2012).

Denmark. Denmark, and other coastal countries including the Low Countries (Netherlands, Belgium, and Luxembourg), stand to lose the majority of its land mass with the melting of the polar ice caps and steadily rising sea levels (see Fig. 2).

Answering this urgent call to reduce its carbon emissions, Denmark's environmental and energy taxes contribute to a better understanding of the environmental costs of production, consumption, and disposal of goods and foodstuffs. This country's main energy initiatives have included the energy-efficiency labeling of construction materials, appliances, and entire buildings, the enactment of building codes (which directly focus on energy consumption), and the establishment and incentivization of electricity-saving financial trusts. Strategic planning for future RE grid investments has followed the current political energy agreement with adopted measures and hybridized policies toward the Danish government's long-term goal of full conversion to renewable energy by 2050. Denmark's planning procedures have been supported by a variety of different support mechanisms which include tax credits and exemptions, feed-in tariffs, and clean energy investment grants. The

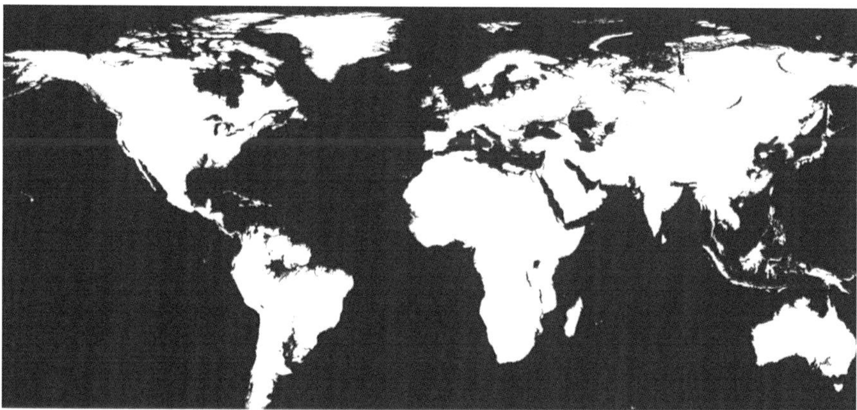

Fig. 2 Diminishing coastal shorelines depicted upon the melting of Earth's ice sheets (GoogleScholar 2017), paralleling *Business Insider* and *National Geographic* renderings

country's *Energy Agreement of 2012* and the *Climate Change Act of 2014* propose a transition to a fossil-free economy. In support thereof, the entire energy supply for all sectors – electricity, heating, manufacture, and transportation – is slated to be supplied by renewable energy by mid-century (Lilleholt 2015).

Tesla Motors, Inc. Born and raised in South Africa, Elon Musk emigrated to the United States where he was educated in engineering at Stanford and eventually earned his American citizenship in 2015. Elon Musk is a singular powerhouse in the areas of low-carbon and carbon- neutral manufacture and renewable energy products and technologies. With the announcement of the Tesla 3 in March of 2016, the activation of the Gigafactory in Fremont, Nevada, in July of 2016, the rollout of the Powerwall II and the solar-shingled roof in late 2016, and the acquisition of Solar City also in 2016, Musk's Tesla company is claiming dominant market shares in clean energy development in the United States and throughout the world. For instance, Tesla partnered with the State of California – the world's 6th largest economy – to lead clean energy initiatives and to reduce its fossil fuel reliance. One tangible example is the company's joint venture with Southern California Edison to construct the world's largest energy storage facility. The Mira Loma substation is comprised of nearly 400 Tesla Powerpack units (i.e., grid-connected, commercial-scale, lithium-ion batteries) which are used to offset a peak demand load. Tesla's commercial-grade storage battery is also being used in conjunction with 54,978 solar panels to help power the Hawaiian island of Kaua'i and has implemented in to a power grid used similarly to Mira Loma at the Vector utility company in New Zealand (Muoio 2017). Tesla's second Gigafactory in Buffalo, New York, is helping to make that state less dependent on fossil fuels, while negotiations continue in Europe with respect to the location of Tesla's third Gigafactory.

Siemens AG (Industrial Energy Development). As larger companies can rightfully be designated as truly global – operating throughout the world both with and without particular home countries – such entities have created spheres of low-carbon emission manufacturing. Albeit officially known as a German company headquartered in both Berlin and Munch, Siemens is a well-recognized world leader in low-carbon and renewable energy technologies. It has recently constructed the London Array off the coast of Great Britain – the largest offshore wind farm in the world, consisting of 175 wind turbines with the capacity to serve homes and factories. It has declared a primary objective to become climate neutral by 2030 (with CO_2 emissions scheduled to be cut by 50% as early as 2020) through energy-reducing measures to reduce its carbon footprint at its own production facilities (committed to invest €100 million to improve energy efficiency and increase automation), innovative technology investments (the company is the world leader in offshore wind turbine construction), energy-efficient drive systems for manufacturing, the substitution of low-emission vehicles for its global vehicle fleet, and working with nation-states to implement RE initiatives company (active in nearly every country in the world, focusing on the areas of electrification, plant automation, and process digitalization). According to the company's media releases, "Solutions from Siemens enabled customers to reduce their CO_2 emissions by 428 million metric tons – an amount equal to half of Germany's total carbon dioxide emissions" (Siemens Global 2016).

Conclusion: Future Trajectories

Leaders on the world stage are quickly emerging as pioneers of true change – engaging in serious attempts to remedy the long-term consequences of hundreds of years of carbon emissions. "Business as usual" is no longer sustainable, and those who acknowledge this must first educate those they lead about the consequences of unimpeded lifestyles and improper management of a nation's resources.

Both the passage of an agreement to reduce GHGs by 2030, officially entered into by174 countries on April 22, 2016, and the accelerated intensity and frequency of weather events with resulting loss of life and property, have incentivized world and business leaders to identify and acknowledge their own contributions to climate change and propound ways to mitigate its consequences. Those who have steered their respective countries, territories, and states to accelerate their transition to low-carbon economies have focused on CO_2 emission reduction through government actions including the levying of a carbon tax, tax credits and subsidies, emissions trading, and transactive energy projects. Many business operations and private individuals have adopted fossil-fuel-cutting measures in cooperation with presiding governmental strategy as well as on their own accord.

Continued emission of greenhouse gases will ostensibly cause further warming and long-lasting, devastating changes worldwide, increasing the likelihood of severe, pervasive, and irreversible impact for both people and ecosystems. The strategies that seek to balance social, economic, and environmental development goals while lowering long-term GHG emissions and increasing resilience to climate change impact will constitute the tools needed to effectively transition to low-carbon economies and hopefully avoid catastrophic climate change.

Cross-References

- ▶ Collaboration for Regional Sustainable Circular Economy Innovation
- ▶ Responsible Investing and Environmental Economics
- ▶ Social License to Operate (SLO)

References

Akerfeldt, S., & Hammar, S. (2016). Taxation in Sweden – Experiences of the past and future challenges Senior Adviser, Swedish Ministry of Finance, Tax and Customs Department. http://www.un.org/esa/ffd/wp-content/uploads/2016/12/13STM_Article_CO2-tax_AkerfeldtHam mar.pdf. Accessed 21 June 2017.

Baldwin, R. E. (2006). Multilateralising regionalism: Spaghetti bowls as building blocs on the path to global free trade. *The World Economy, 29*, 1451–1518.

Biello, D. (2014, November 12). Everything you need to know about the U.S. – China climate change agreement. *Scientific American*. https://www.scientificamerican.com/article/everything-you-need-to-know-about-the-u-s-china-climate-change-agreement/. Accessed 21 Apr 2017.

Brazil Energy (2016). *International Trade Administration*. https://www.trade.gov/top markets/pdf/Renewable_Energy_Brazil.pdf. Accessed 7 July 2017.

Bikita Minerals in $7mln Capex, to Double Output. (2017). *The Source*. http://source.co.zw/2017/03/bikita-minerals-7mln-capex-double-output/. Accessed 19 Mar 2017.

British Columbia – Climate Change. (2016). http://www2.gov.bc.ca/gov/content/environment/climate-change. Accessed 11 Feb 2017.

British Columbia – Ministry of Finance Carbon Tax. (2013/2014). http://www.fin.gov.bc.ca/tbs/tp/climate/carbon_tax.htm. Accessed 17 June 2017.

China and India Make Great Strides on Climate Change. (2017). *New York Times*. https://www.nytimes.com/2017/05/22/opinion/paris-agreement-climate-china-india.html. Accessed 13 June 2017.

Climate Reality Project. (2016). Follow the leader: How 11 countries are shifting to renewable energy. https://www.climaterealityproject.org/blog/follow-leader-how-11-countries-are-shifting-renewable-energy. Accessed 19 June 2017.

Costa Rica boasts 99% renewable energy in 2015. (2015). *Associated Free Press*. https://www.yahoo.com/news/costa-rica-boasts-99-renewable-energy-2015-210416028.html. Accessed 12 Mar 2017.

European Commission Fact Sheet. (2016). http://europa.eu/rapid/press-release_MEMO-16-2499_en.htm. Accessed 5 June 2017.

European Union Emissions Trading System Fact Sheet. (2016). https://ec.europa.eu/clima/sites/clima/files/factsheet_ets_en.pdf. Accessed 5 June 2017.

Goldman Sachs Market Report. (2017, June 5). The low carbon economy. http://www.goldmansachs.com/our-thinking/pages/report-the-low-carbon-economy.html. Accessed 18 June 2017.

Goodward, J., & Gonzalez, M. (2010). *Bottom line on renewable energy tax credits*. World Resources Institute.

IRENA, Renewable Energy Auctions. (2015). http://www.irena.org/menu/index. Accessed 15 May 2017.

International Trade Administration. (2016). Brazil energy. https://www.trade.gov/top markets/pdf/Renewable_Energy_Brazil.pdf. Accessed 7 July 2017.

Lilleholt, L. C. (2015). A Danish solution to a global challenge? UNEP: Minister for energy, utilities, and climate, Denmark. http://www.climateactionprogramme.org/climate-leader-papers/a_danish_solution_to_a_global_challenge. Accessed 9 Mar 2017.

Lithium-ion Battery Market is Projected to Reach US $77.42 bn in 2024. (2016). *MarketWatch*. http://www.marketwatch.com/story/lithium-ion-battery-market-is-projected-to-rea ch-us-7742-bn-in-2024-global-industry-analysis-size-share-growth-trends-and-forecast-2016—2024-tmr-2016-09-19. Accessed 19 Mar 2017.

MarketWatch. (2016, February 25). This emerging lithium producer could disrupt the entire industry. http://www.marketwatch.com/story/this-emerging-lithium-producer-could-disrupt-the-entire-industry-2016-02-25-920250. Accessed 12 Apr 2017.

Maurer, L. T. A., & Barroso, L. A. (2011). An overview of efficient practices. *World Bank*.

Muoio, D. (2017). Here are 15 things that Tesla's batteries are already powering. *Business Insider*. https://www.sciencealert.com/15-things-that-tesla-s-batteries-are-already-powe ring. Accessed 4 June 2017.

Murray, B. C., & Rivers, N. (2015). *British Columbia's revenue-neutral carbon tax: A review of the latest "grand experiment" in environmental policy*. Duke University Nicholas Institute.

Nelson, D., Zuckerman, J., Hervé-Mignucci, M., Goggins, A., & Szambelan, S. J. (2014a). Moving to a low carbon economy: The financial impact of the low-carbon transition. *Climate Policy Initiative*. http://www.eldis.org/document/A71406. Accessed 5 May 2017.

Nelson, D., Zuckerman, J., Hervé-Mignucci, M., Goggins, A., & Szambelan, S. J. (2014b). Moving to a low carbon economy: The impact of different policy pathways on fossil fuel asset values. *Climate Policy Initiative*. https://climatepolicyinitiative.org/wp-content/uploads/2014/10/Moving-to-a-Low-Carbon-Economy-The-Impacts-of-Policy-Pathways-on-Fossil-Fuel-Asset-Val ues.pdf. Accessed 5 May 2017.

New York Times. (2017, July 18). A brighter future for electric cars and the planet. https://www.nytimes.com/2017/07/18/opinion/a-brighter-future-for-electric-cars-and-the-planet.html. Accessed 21 July 2017.

Olivier, J. G. J., Janssens-Maenhout, G., Muntean, M., & Peters, J. A. H. W. (2016). *Trends in global CO_2 emissions*. The Hague: PBL Netherlands Environmental Assessment Agency. Ispra: European Commission, Joint Research Centre.

Porter, E. (2016). Does a carbon tax work? Ask British Columbia. *New York Times*. https://www.nytimes.com/2016/03/02/business/does-a-carbon-tax-work-ask-bri tish-columbia.html. Accessed 10 May 2017.

Renewables Progress Report. (2010). *Renewable energy policy network for the 21st century*. http://www.ren21.net/Portals/0/documents/activities/Topical%20Reports/REN21_10yr.pdf. Accessed 6 May 2017.

Rubadiri, V. (2012). Energy: Kenya's renewable energy sector attracts foreign interest. http://en.cchina.gov.cn/Detail.aspx?newsId=38323&TId=97. Accessed 12 Apr 2017.

Siemens Global. (2016). Siemens on target for 2030 climate pledge. https://www.siemens.com/customer-magazine/en/home/cities/siemens-on-target-for-2030-climate-pledge.html. Accessed 17 June 2017.

Singh, H. V. (2015). *International Centre for Trade and Sustainable Development* (ICTSD). http://www3.weforum.org/docs/WEF_GAC15_The_High_Low_Politics_Trade_WTO _Centrality_report_2015.pdf. Accessed 8 Apr 2017.

Steinberger, K. (2017). Renewable energy tax credits will power economic growth. National Resources Defense Council. https://www.nrdc.org/experts/kevin-steinberger/ renewable-energy-tax-credits-will-power-economic-growth. Accessed 9 June 2017.

The Guardian. (2015, November 30). Zuckerburg, Gates and other tech titans form clean energy investment coalition. https://www.theguardian.com/environment/2015/nov/30/bill-gates-breakthrough-energy-coalition-mark-zuckerberg-facebook-microsoft-amazon. Accessed 23 Feb 2017.

The Low Carbon Economy. (2015). *Goldman Sachs 2015–2025 market report*. http://www.goldmansachs.com/our-thinking/pages/new-energy-landscape-folder/report-the-low-carbon-economy/report.pdf. Accessed 17 Apr 2017.

The Paris Agreement. (2015). *The Guardian*. https://www.theguardian.com/environment/2015/nov/30/paris-climate-summit-in-numbers. Accessed 23 June 2017.

The Top Lithium Producing Countries in the World. (2017). *WorldAtlas*. http://www.worldatlas.com/articles/the-top-lithium-producing-countries-in-the-world.html. Accessed 19 June 2017.

UN Division for Sustainable Development. (2012). A guidebook to the green economy. https://sustainabledevelopment.un.org/content/documents/GE%20Guidebook.pdf. Accessed 15 February 2017.

United States Department of Energy. (2015). The pacific northwest demonstration project. http://www.pnnl.gov/news/release.aspx?id=4210. Accessed 11 June 2017.

United States Department of Energy. (2016). Leveraging federal renewable energy tax credits. https://energy.gov/sites/prod/files/2016/12/f34/Leveraging_Federal_Renewable_Energy_Tax_Credits_Final.pdf. Accessed 7 June 2017.

USA International Business Publications. (2015). *Brazil energy policy, laws and regulations handbook* (pp. 54–68). Washington, DC: International Business Publications.

Vast Moroccan Solar Power Plant is Hard Act for Africa to Follow. (2016). *Fortune*. http://fortune.com/2016/11/05/moroccan-solar-plant-africa/. Accessed 15 May 2017.

Wikimedia. (2006). World energy consumption map. https://commons.wikimedia.org/w/index.php?title=Special:Search&limit=20&offset=0&profile=default&search=energy+consumption+2006&searchToken=8uury1zkszlgcmzmr8cpuqw4d. Accessed 5 July 2017.

Worland, J. (2016, February 5). How climate change unfairly burdens poorer countries, *Time*. http://time.com/4209510/climate-change-poor-countries/. Accessed 3 May 2017.

World Bank Group Report. (2016). http://documents.worldbank.org/curated/en/896971468194972881/pdf/102725-PUB-Replacement-PUBLIC.pdf. Accessed 28 May 2017.

WorldAtlas. (2017). The top lithium producing countries in the world. http://www.worldatlas.com/articles/the-top-lithium-producing-countries-in-the-world.html. Accessed 19 June 2017.

WIPO Green: The Marketplace for Sustainable Technology. (2017). *World intellectual property organization*. https://www3.wipo.int/wipogreen/en/. Accessed 19 June 2017.

PGSTL